Modern Developments and Applications in Microbeam Analysis

Edited by
G. Love, W. A. P. Nicholson,
A. Armigliato

Mikrochimica Acta
Supplement 15

Springer-Verlag Wien GmbH

Dr. Glyn Love
Centre for Electron Optical Studies, University of Bath, United Kingdom

Dr. W. A. Patrick Nicholson
Department of Physics and Astronomy, University of Glasgow, United Kingdom

Dr. Aldo Armigliato
C.N.R. Istituto LAMEL, Bologna, Italy

Product Liability: The publisher can give no guarantee for information about drug dosage and application thereof contained in this book. In every individual case the respective user must check its accuracy by consulting other pharmaceutical literature. The use of registered names, trademarks, etc. in this publication does not imply, even in the absence of a specific statement, that such names are exempt from the relevant protective laws and regulations and therefore free for general use.

Typesetting: Thomson Press, New Delhi, India
Printing: Manz, A-1050 Wien
Graphic design: Ecke Bonk
Printed on acid-free and chlorine-free bleached paper
SPIN: 10674194

With 291 partly coloured Figures

Die Deutsche Bibliothek – CIP-Einheitsaufnahme

Modern developments and applications in microbeam analysis /
ed. by G. Love ... – Wien ; New York : Springer, 1998
(Mikrochimica acta : Supplementum ; 15)

ISSN 0026-3672
ISBN 978-3-211-83106-9 ISBN 978-3-7091-7506-4 (eBook)
DOI 10.1007/978-3-7091-7506-4

Preface

This supplement of *Mikrochimica Acta* contains selected papers from the Fifth Workshop of the European Microbeam Analysis Society (EMAS) on "Modern Developments and Applications in Microbeam Analysis" which took place from the 11th to 15th May 1997 in Torquay (UK).

EMAS was founded in 1986 by scientists from many European countries in order to stimulate research in microbeam analysis and into its development and application. The society now has over 350 members from more than 20 countries. An important EMAS activity is the organisation of biennial workshops which focus upon the current status and developing trends in microanalytical techniques. For this meeting EMAS chose to invite speakers on the following subjects: Standardless analysis, EPMA techniques for quantitative near-surface analysis and depth profiling, Matrix corrections in Auger electron and X-ray photon spectroscopy, X-ray analysis and imaging using low voltage beams, Scanning probe and near field microscopies, EPMA of frozen biological bulk samples, Environmental SEM and X-ray microanalysis of biological materials, Quantitative elemental mapping of X-ray radiographs by factorial correspondence, X-ray spectrum processing and multivariate analysis, Thin film analysis and chemical mapping in the analytical electron microscope, Wavelength dispersive X-ray spectroscopy, High resolution non-dispersive X-ray spectroscopy with state-of-the-art silicon detectors and Recent developments in instrumentation for X-ray analysis. These invited lectures were given by eminent scientists from Europe, the USA, and Australia

In addition to the introductory lectures there were poster sessions at which some 110 posters were on display. Short oral presentations of many of these posters were also given and a session was also introduced in which six younger scientists had an opportunity to give a more lengthy oral presentation. The prize for the best poster was awarded jointly to S. Richter from the Gemeinschaft für Elektronen Mikroskopie of the Rheinische Technische Hochschule in Aachen for "EPMA Sputter Depth Profiling: Part II: Theory and Evaluation" (by S. Richter, N. Lesch, and P. Karduck) and to G. Verlinden from the Department of Chemistry at the University of Antwerp for "Surface Analysis of Silver Halide Microcrystals by Imaging Time-of-flight SIMS" (by G. Verlinden, R. Gijbels, and I. Geuens) whilst the prize for the best oral presentation by a young scientist was awarded to N. Lesch also from the Gemeinschaft für Elektronen Mikroskopie of the Rheinische Technische Hochschule in Aachen for "EPMA Sputter Depth Profiling: Part I: Experiment" (by N. Lesch, S. Richter, and P. Karduck).

The papers presented in this volume have been divided into the following subjects: X-ray Spectrometry; Analytical Electron microscopy – Thin Films; Surface Studies – Coatings – Multilayers: Bulk Analysis – Materials Applications – Mineralogical Applications – Biological Applications including Environmental SEM; Fundamental Parameters – Quantitative Analysis; and other miscellaneous papers.

The issue contains the full texts of ten tutorial lectures and forty one papers originating from the posters presented at the workshop. All the contributions were subjected to peer review by two referees.

Proceedings of previous EMAS workshops have also been published, either in a regular issue of or a supplement to *Mikrochimica Acta.* We hope that this volume, as with others in the series, will form a key reference source for analysts using microbeam techniques.

February 1998 *G. Love, W. A. P. Nicholson, A. Armigliato*

European Microbeam Analysis Society

Members:

Jean-François Bresse
C.N.E.T. France Telecom, Département PAB / CDP / CMT
196, avenue H. Ravera, P.O. Box 107, F-92295 Bagneux, France
tel: +33-1-42.31.72.43
fax: +33-1-42.53.49.30
e-mail: jean-francois.bresse@bagneux.cnet.fr

Jacques Cazaux
Université de Reims Champagne-Ardenne, Lab. d'Analyse des Solides
Surfaces & Interfaces
Moulin de la Housse, P.O. Box 1039, F-51687 Reims Cedex 2, France
tel: +33-3-26.05.32.23
fax: +33-3-26.05.32.50
e-mail: lassi@univ-reims.fr

Erkki Heikinheimo
Helsinki University of Technology, Laboratory of Metallurgy
Vuorimiehentie 2K, FIN-02 150 Espoo, Finland
tel: +358-9-451.27.59
fax: +358-9-451.27.98
e-mail: eheikin@redhot.hut.fi

Francesc Salvat
Universitat de Barcelona, Facultat de Fisica (ECM)
Diagonal 647, E-08028 Barcelona, Spain
tel: +34-3-402.11.86
fax: +34-3-402.11.74
e-mail: cesc@smeagol.ecm.ub.es

Johann Wernisch
Technical University of Vienna, Institute of Applied and Technical Physics
Wiedner Hauptstrasse 8–10, A-1040 Vienna, Austria
tel: +43-1-58.801 ext. 5666
fax: +43-1-587.88.76
e-mail: wernisch@atpibm6000.tuwien.ac.at

Honorary members:

A. Boekestein, Wageningen (NL)
R. Castaing, Orsay (F)
P. Duncumb, Cambridge (UK)
A.W. Witmer, Eindhoven (NL)

Contents

Recent Developments in Instrumentation for X-Ray Microanalysis
P. J. Statham ... 1

High Resolution Non Dispersive X-Ray Spectroscopy with State of the Art Silicon Detectors
L. Strüder, C. Fiorini, E. Gatti, R. Hartmann, P. Holl, N. Krause, P. Lechner, A. Longoni, G. Lutz, J. Kemmer, N. Meidinger, M. Popp, H. Soltau, C. von Zanthier ... 11

Efficiency Calibration of a Si(Li) Detector by EPMA
K. Röhrbacher, M. Andrae, M. Völkerer, J. Wernisch ... 21

Wavelength-Dispersive X-Ray Spectrometry
S. J. B. Reed ... 29

X-Ray Spectrum Processing and Multivariate Analysis
J. M. Titchmarsh ... 37

Thin Film Analysis and Chemical Mapping in the Analytical Electron Microscope
D. B. Williams, M. Watanabe, D. T. Carpenter ... 49

On the Spatial Resolution in Analytical Electron Microscopy
A. Armigliato, D. J. Howard, R. Balboni, S. Frabboni, M. R. Caymax ... 59

Contamination in Analytical Electron Microscopy and in ALCHEMI
J. L. Lábár, M. Adamik, I. Dódony ... 65

Analytical Electron Microscopy of Diffusional Interfaces in an Al-22 at. % Zn Alloy
P. Zięba ... 73

Quantitative TEM-EDXS of Sol-Gel Derived PZT Ceramic Materials
G. Dražič, B. Malič, M. Kosec ... 77

Particulate Composites of TZP-Chromium Oxide and TZP-Chromium Carbide; Microbeam Investigations
M. Faryna, L. Lityńska, K. Haberko, Z. Pędzich, J. Babiarz ... 83

Cryo-Electron Spectroscopic Imaging, Electron Energy-Loss Spectroscopy and Energy-Dispersive X-Ray Analysis of Ag(Br, I) Nano- and Microcrystals
V. P. Oleshko, R. H. Gijbels, W. A. Jacob, A. J. Van Daele ... 87

Electron Energy-Loss Near-Edge Structure of Alumina Polymorphs
I. Levin, A. Berner, C. Scheu, H. Muellejans, D. G. Brandon ... 93

SPM Study of YBCO Films Prepared by Plasma Assisted Laser Ablation
Y. Fan, A. G. Fitzgerald, H. Xu ... 97

Surface Characterisation and Modification of YBCO Thin Films by STM
A. G. Fitzgerald, Y. Fan, H. Xu ... 101

Quantitative Near-Surface Microanalysis and Depth Profiling by EPMA
P. Karduck ... 109

EPMA Sputter Depth Profiling, Part I: Theory and Evaluation
S. Richter, N. Lesch, P. Karduck ... 125

EPMA Sputter Depth Profiling, Part II: Experiment
N. Lesch, S. Richter, P. Karduck ... 133

Quantitative Analysis of BN (C, O, Ar, H)-Coatings Using EPMA and SIMS
P. Willich, U. Wischmann ... 141

Quantitative EDS Analysis of $SiO_2/Al_2O_3/TiO_2$ Multilayer Films
D. G. Rickerby, N. Wächter, R. Reichelt ... 149

Surface Ionization of Thin Films on Substrates: Measurement and Simulation
X. Llovet, C. Merlet, F. Salvat ... 155

Comparison of Different Methods to Characterize Thin a-Si:H Films
H. G. Esser, P. Karduck, M. Rubel, N. Almqvist, L. Grobusch, J. von Seggern, F. Weschenfelder, P. Wienhold ... 163

EPMA Studies of the Growth of Thin Surface Coatings Produced by Evaporation
H. Hammer, S. Kulow, C. Schmidt, G. Schmidt, P. Zanel ... 171

Analysis of Thin Films with Slightly Rough Boundaries
I. Ohlídal, D. Franta, J. Hora, K. Navrátil, J. Weber, P. Janda ... 177

Effect of Chromium Substrate Pretreatment on Diamond Growth by the Chemical Vapour Deposition Method
O. Glozman, A. Berner, D. Shechtman, A. Hoffman ... 181

EPMA Determination of Arsenic Excess in Low Temperature Grown GaAs
V. V. Chaldyshev, V. V. Tretyakov ... 187

EPMA of Melted UO_2 Fuel Rods from the Phebus-FP Reactor Accident Experiment
P. D. W. Bottomley, F. Montigny, A. D. Stalios, C. T. Walker ... 191

Steels, Carbon Concentration, and Microhardness
P. Poelt, A. Fian ... 201

Determination of Chemical and Phase Composition of Fly-Ashes by Combined EPMA and XRD Methods
M. Żelechower, D. Smołka, M. Jabłońska, A. Dytkowicz ... 207

EPMA of the Composition of Opal-Based Nanostructured Materials
V. V. Tretyakov, S. G. Romanov, A. V. Fokin, V. I. Alperovich ... 211

NDIC and EMP Study of Plagioclase Mineral Zoning: An Example from Nea Kameni Lavas
A. P. Santo, R. M. Bomparola ... 219

Compositional X-Ray Maps of Metamorphic and Magmatic Minerals
A. Borghi, R. Cossio, F. Olmi, G. Vaggelli ... 227

Chemical Mapping of Weathering Stages in Laterites
W. Hachmann, G. Tietz ... 237

Electron Microprobe Determination of Minor and Trace Concentrations of Gold and Platinum Group Elements in Sulphides and Sulpharsenides: Problems, Solutions, and Applications
I. P. Laputina, V. A. Batyrev ... 247

Composition of 15–17th Century Archaeological Glass Vessels Excavated in Antwerp, Belgium
K. H. Janssens, I. Deraedt, O. Schalm, J. Veeckman ... 253

Potassium Migration in Silica Glass During Electron Beam Irradiation
K. Jurek, O. Gedeon, V. Hulínský ... 269

X-Ray Microanalysis of Frozen-Hydrated Biological Bulk Samples
A. T. Marshall ... 273

Environmental SEM and X-Ray Microanalysis of Biological Materials
D. C. Sigee ... 283

Effects of Electron-Beam/Gas Interactions on X-Ray Microanalysis in the Variable Pressure SEM
C. Mathieu 295

The Analytical Signal in EPMA and the Influence of the Electric Field Created by the Primary Beam
O. Gedeon, V. Hulínský, K. Jurek 301

Standardless Analysis
J. Wernisch, K. Röhrbacher 307

A New Technique for Standardless Analysis by EPMA-TWIX
M. Völkerer, M. Andrae, K. Röhrbacher, J. Wernisch 317

Stopping Power Factor for Standardless QEPMA
B. I. Losic, F. A. Tabak 321

On the Measurement of the Backscattering Coefficient for Low Energy Electrons
M. M. El-Gomati, A. M. D. Assa'd 325

Monte Carlo Simulations of Edge Artefacts in MULSAM Images
M. M. El-Gomati, A. M. D. Assa'd 333

Assessment of the Inelastic Scattering Model in Monte-Carlo Simulations
V. Starý 341

A Rapid Comparison of Matrix Corrections in AES and XPS by Means of Computer Programs
A. G. Fitzgerald 351

Fractals and $BaTiO_3$-Ceramic Microstructure Analysis
V. V. Mitić, L. M. Kocić, M. Miljković, I. Petković 365

Fragmentation of Sputtered Cluster Ions of Transition Metals: Distributions of Lifetimes and Internal Energies
A. D. Bekkerman, N. K. Dzhemilev, S. V. Verkhoturov, I. V. Veryovkin, A. Adriaens 371

Sputtering of Tantalum by Atomic and Molecular Gold Ions: Comparative Study of Yields and Kinetic Energy Distributions of Atomic and Cluster Ions
S. F. Belykh, U. K. Rasulev, A. V. Samartsev, S. V. Verkhoturov, I. V. Veryovkin 379

The Standards, Measurements and Testing Programme (SMT), the European Support to Standardisation, Measurements and Testing Projects
A. Boenke 387

Mikrochim. Acta [Suppl.] 15, 1–9 (1998)

Recent Developments in Instrumentation for X-Ray Microanalysis

Peter J. Statham

Oxford Instruments Microanalysis Group, Halifax Road, High Wycombe, Bucks HP12 3SE, U.K.

Abstract. This paper reviews the instrumentation advances for x-ray microanalysis 1991–1997, with particular emphasis on energy dispersive x-ray spectrometers. Most developments have been aimed at improved convenience and reliability while offering sensitivity well below 1 keV, particularly for semiconductor applications. Although EDX technology matured during the 1980's, previous methods of characterisation are now inadequate to reveal the variability in performance as a function of x-ray energy in this low energy region. Furthermore, at low beam voltages where K lines are not excited, computer processing of peak overlaps is the only way to obtain element intensities. In this situation, detector and electronic stability and reproducibility have to be substantially improved in order to achieve results anywhere near the limit of statistical precision.

Key words: microanalysis, x-ray, EDX.

Since this author's last review [1], developments in microanalysis instrumentation have been fuelled indirectly by consumer and business demand for digital electronics. Mass production of components for personal computers and telecommunications has brought down the prices of computer memory, cpu and digital signal processing chips. This in turn has enabled instrumentation manufacturers to provide more elaborate x-ray imaging modes and implement pulse processors using economic high speed ADCs with digital circuitry. Coincidentally, improvements in x-ray instrumentation have been driven by the needs of the very semiconductor manufacturers that are feeding the demand for information technology. In particular, x-ray microanalysis of submicron defect particles gives important clues for improving yield. Here, low SEM kV is desirable to reduce beam broadening and maximise the spectral contribution from contaminant particles rather than the substrate wafer. Operation in semiconductor fabrication lines has demanded alternatives to liquid nitrogen cooling for EDX systems but many options are unsuitable because vibration has to be eliminated to permit high resolution imaging. If low kV limits excitation to below 3 keV, many EDX spectrum peak overlaps have to be solved directly because well resolved K lines are no longer available to aid element identification. The improved resolution of WDX can be used to complement EDX by integrating displays using a common energy axis. However, current WDX designs are not available on all electron columns, particularly the high resolution FEGSEMs commonly used for semiconductor defect analysis. Effective EDX peak deconvolution can be achieved by computer processing provided calibration is reproducible and counting statistics are good but there are tradeoffs between rate and resolution. Given the substantial advantages of improved resolution when dealing with severe peak overlap, there is renewed interest in radical approaches to x-ray detection using superconducting technology which promises an order of magnitude improvement over conventional EDX. A recent publication [2] provides a comprehensive collection of papers summarising the state of the art which complements the current review.

X-Ray Imaging Improvements

X-ray mapping using fixed energy windows is rather limited. The operator has to choose which elements to map and after the experiment is finished, still may be concerned that some important elements have not been selected. The resultant images are invariably noisy so it is difficult to decide if an element is below the limit of detection or has been obscured by peak overlap. A useful extension has been to acquire a spectrum at each point and then do full least squares spectrum processing to resolve overlaps, calculate statistical precision and monitor for missing peaks [3]. Furthermore, with improvements in cpu speed, it has been possible to include x-ray corrections at each pixel for dwell times around 1 second per pixel [4].

Although "quantitative mapping" provides an effective compression of the data into useful information and can provide some diagnostic indication of missing elements, it still requires the operator to make a selection prior to the experiment and there is no opportunity to reprocess the data. Clearly it would be advantageous to store all the accumulated data. However, to acquire a 1000 channel, 16 bit, spectrum at each point on a 100×100 grid involves about 20 Mbyte of data which, though reasonable for some research experiments, is currently impractical for routine archiving on inexpensive diskettes or to transmit over the Internet. One approach to this problem, called "list mode" acquisition, has been used for many years in gamma camera imaging systems. Here, the x and y coordinates, energy and sometimes also time of arrival, of individual gamma rays are saved in a "list". At the end of the experiment, the "list" can be used like a tape recording to reconstruct single images of specific energies or time series on demand. Provided data rates are low and the images sparsely populated, data storage requirements are substantially less than that for a series of dedicated image arrays. A recent variant of this technique has been applied to EDX mapping [5]. The sample is exposed while the beam is scanning and a database of x-ray events, x,y and energy, is accumulated. At any time the user can choose a set of x-ray energy windows to construct elemental maps as with usual x-ray mapping. Alternatively, spectra can be reconstructed by specifying an area of interest in the field of view. While this offers flexibility to "replay" the mapping operation using different parameters, the data storage requirements can still be demanding. For example, with a modest data rate of 10 kcps, 9 million x-ray events would be recorded in 15 minutes and information on x, y and energy has to be saved for every one of these events. During mapping, the beam scans evenly over the field of view and if there is a small interesting feature, x-ray data will only be accumulated while the beam is on the feature. For example, in a 15 minute acquisition, the beam spends less than 0.1 seconds on a feature 1% of the field width; reconstructing a spectrum from such a small feature would therefore show the poor statistics associated with a 0.1 second acquisition time. To obtain a spectrum typical of 100 seconds acquisition time would require summing all the pixel data over a large area about 1/3 of the screen width across, and this would only be sensible for a large homogeneous feature of this size.

The concept of spectrum imaging, whether direct or by "list mode" techniques, is attractive because it seems to offer the possibility of acquiring data from a single exposure of the sample, followed by analysis at any time later. However, the dataset may not include sufficient data to do any sensible analysis on small features so it is unwise to leave interpretion of this data too late and simply rely on an archive. An alternative approach is to display all the data immediately in a way which guides the operator to areas where additional spectra would be useful. In this case, the operator can finish a live session on the sample with both a map and good data on all interesting features. This is the philosophy adopted in a true-colour imaging approach [6] which uses the complete x-ray spectrum generated at a point to reconstruct a colour response mimicking the eye's response to the optical spectrum. By encoding different x-ray wavelengths in different colours, a true colour picture is built up which reveals chemical compositional differences while maintaining a smooth transition of colour so that topography is still apparent. The colour signal is combined with a high resolution electron image signal, rather like the chrominance is combined with a luminance signal in broadcast TV. The final image exhibits the high resolution features revealed by the fine electron beam with a colour hue derived from the emitted x-rays. This true colour image is therefore intuitive and good results can be obtained in about 1 minute's acquisition time. Once the true colour image is obtained, it is easy for the operator to select the next course of action. For example, by selecting points of analysis for spectra, good statistical precision can be obtained on selected points. If no interesting features appear in the true colour image, the operator can move on to select a new field of view very quickly and not waste time accumulating map data which might never be used. Faced with a complex analytical problem the analyst thus has a useful cue for positioning the probe for subsequent more detailed analysis and a way of compressing topographical and compositional information into a single image for archiving.

Advances in Electronics

One of the key requirements for good x-ray mapping is a high data rate from the sample. The combination of time-variant pulse processing and low noise FETs has provided maximum acquisition rates in excess of 30 kcps while maintaining a resolution of 193 eV at Mn Kα [1]. Time variant processing requires digital circuitry, delay lines and analog switches to change filter characteristics during each event and recover the baseline afterwards. Although this approach is more effective than conventional analogue processing, there is still some overhead involved in baseline recovery. By using modern high speed ADCs, a digital "pipeline" can be implemented where the waveform is continuously available for digital processing. Since the signal is basically a step waveform, measurement of signal level before and after the arrival of the photon gives a measure of photon energy. The level after the event can be used as the baseline for the subsequent event and thus save the deadtime overhead. Thus, it has been possible to extend the output counting rate to over 50 kcps while maintaining resolution of 185 eV with a Si(Li) detector using digital processing [7].

A further benefit of the digital approach is improved linearity. X-ray events are superimposed on a ramp to form a staircase waveform [1]. If the ADC used is non-linear, the number corresponding to a given energy will vary depending on where the event appears on the staircase. Since there is no correlation of time of arrival of photons with the ramp, measurement of a particular energy will be converted into a spread of values. Although this degrades resolution of the final spectrum slightly, it does mean that the average conversion of energy to number is more linear than it would be with shaping plus digitisation. This has particular advantages when providing systems which are pre-loaded with peak profiles so that least squares fitting can be achieved "out of the box" without resort to standards.

New Detector Materials

A shorter shaping time or measurement time per event is required to achieve high data accumulation rate. This in turn degrades overall signal to noise ratio so the noise contribution to the spectrum will be higher and the resolution at MnKα worse. Achieving high acquisition rates has therefore traditionally been at great expense to resolution. HpGe detectors have now been produced where resolution at MnKα of better than 115 eV can be guaranteed. If high rates are required, shorter pulse measurement times can be chosen to give 10 kcps accumulation rate into the spectrum at a resolution of 133 eV, which is normally a "premium" specification for a Si(Li) detector only guaranteed at 1 kcps. Furthermore, it is possible to obtain over 50 kcps acquisition rate with a resolution of 165 eV [8].

HpGe detectors seem to offer the most potential for spectrometry at high energies $\gg$ 10 keV because of their improved stopping power for x-rays compared with Si(Li) and better detector resolution at high energies. Although high energy K lines are well separated, the peak to background is rather poorer than for the corresponding L lines excited in a TEM or STEM [9]. Furthermore, the Kα and Kβ escape peaks of germanium are a problem for all x-ray lines above about 10 keV. However, at energies below 10 keV, the only escape mechanism is for Ge L and since the L fluorescent yield is so small, and absorption of Ge Lα in Ge is high, low energy HpGe spectra are practically devoid of any escape artefacts. Thus HpGe is potentially a good material to use for an SEM detector, provided charge collection problems at low energies can be overcome.

Beyond germanium, there seem no immediate prospects for alternative EDX detector materials which offer improved resolution performance at 5.9 keV. However, noise can be reduced both by using structures which reduce capacitance and novel electronic processing techniques. For example, a silicon CCD using a multiple signal read technique has been reported to achieve 16 eV resolution on the Al L line at 72 eV and 120 eV fwhm at 6.9 keV [10]; unfortunately, this performance was obtained with liquid nitrogen cooling and count rate capability and efficiency at low energies do not appear to be adequate for SEM microanalysis. A room temperature detector is desirable because it avoids the need for cooling and the transducer can be exposed without fear of condensation. Developments in this area have been reviewed in a recent symposium [11]. Materials such as HgI_2 and CdZnTe are being developed and new structures such as the silicon drift detector pursued, but true room temperature operation rarely offers resolutions much better than 300 eV at MnKα.

WDX+EDX for Qualitative Analysis

The well established high resolution instrument for x-rays is the Bragg crystal spectrometer. So called wavelength dispersive x-ray (WDX) spectrometers have traditionally been used for microanalysis on electron microprobe instruments offering high beam current and light optics. EDX has been used in the past to complement WDX by providing rapid analysis for major element concentrations. Such systems were normally complex and involved high levels of automation and user programmability. Although modern EDX systems with automatic element identification seem to provide a push-button panacea for most routine applications, there are many pathological overlaps which cannot be resolved satisfactorily with fwhm resolutions of the order of 100 eV (e.g. S K + Mo L, Si K + W M, Al K + Br L, F K + Fe L, O K + V L etc). Clearly WDX can resolve such coincidences but has traditionally been operated quite separately from the EDX and systems reference wavelengths more often than energy. If operations are more closely coupled and WDX measurements are transformed to the same energy scale as used by EDX, qualitative identification can be made reliable. For example, a region of overlapping peaks can be identified in the EDX spectrum to initiate a WDX spectrometer scan over the same energy region. Figure 1 shows that the overlayed displays reveal true elemental content.

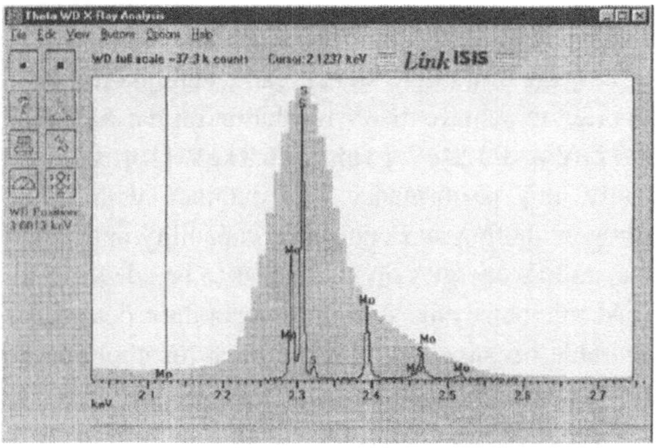

Fig 1. Screen display showing unresolved peak multiplet from EDX detector and WDS scan converted to same energy scale revealing Mo and S peaks

New Cooling Technologies

The advantages of push button spectrometry with EDX systems are offset by the inconvenience of dealing with liquid nitrogen so there has been a continual search for practical alternatives. Stirling engines can achieve very low temperatures and have been engineered to be extremely reliable, but are naturally prone to introduce vibration which is unacceptable for electron microscopy. Peltier cooling is at first sight the most tidy, electronic, vibration free solution. However, to achieve the cooling requirements for x-ray detection in a wide variety of microscopes requires a large heat sink and water cooling introduces its own potential for vibration. Furthermore, to date, the resolutions achieved with Peltier cooled detectors are slightly worse than those for liquid nitrogen cooling. A resolution of 145 eV fwhm at MnKα may sound only slightly worse than 133 eV but this implies a much larger electronic noise contribution so the resolution at low energies will be considerably worse and give dramatic differences in the ability to resolve an overlap such as C Kα/N Kα [12]. Joule Thomson coolers avoid the mechanical vibration associated with Stirling engines and can achieve lower temperatures than Peltier. Until recently, the problem has been that such coolers require a supply of high pressure gas which is used up very quickly so continuous operation is not possible. Recent progress in this technology has resulted in coolers with the capacity to hold a Si(Li) detector below 100 deg.K continuously using an electrical compressor. This recent generation of "non-LN2" detectors is capable of achieving resolution performance equivalent to that of a conventional LN2-cooled device. Although vastly reduced, vibration is still an issue requiring special design modifications to make the devices suitable for high resolution FEGSEMs operating at 200 kX.

EDX Spectrometry at Low Energies

There are potential advantages in using low beam voltages 1–5 kV, particularly in achieving high spatial resolution of the order of 0.1 μm [13, 14]. However, spectrometry of low energy x-rays by EDX systems requires careful consideration of many factors which have demanded further improvements in performance [12].

"Windowless" detectors clearly give maximum sensitivity but for SEM use in particular have largely been superseded by use of "Atmosphere-supporting Thin Windows" (ATWs) which seal the vacuum and reduce problems of ice and hydrocarbon build up on the detector surface. A strong support grid provides most of the force to support a pressure of at least one atmosphere and a very thin and fragile film is attached to the support to make the structure vacuum tight. Low energy x-rays have to penetrate the film and get through the holes in the grid to reach the detector. Since different films and support grids are in use, transmission at low energies can be very variable. For example, pure carbon has very poor transmission for N Kα while boron nitride absorbs C Kα strongly. Even the popular polymer film, which has overall good transmission, still has large "dips" in the transmission curve. The progress in ATWs has been recently reviewed [15].

Energetic electrons will penetrate an ATW and produce a high spurious background "hump" in the spectrum. Therefore, a strong magnetic field, produced by an "electron trap" or present in an "in-lens" design of SEM is essential to deflect electrons away from the crystal. Electron traps take up space and reduce the collection solid angle. Light will also penetrate such thin windows so a reflective coating of Aluminium is usually used to provide a light barrier. Obviously, a thicker layer is more effective but will also absorb more low energy x-rays. 40 nm of Al will block light from cathodo-luminescent samples while still giving good transmission for C and N but, in some detectors, thickness is reduced to 15 nm to improve transmission for Be and B. The penalty for this improved x-ray transmission is sensitivity to cathodoluminescence; light entering a detector generally causes severe peak broadening, peak shift and distortion [16].

Photons which penetrate the window reach the conductive electrode on the front surface of the crystal. The material, thickness and deposition method vary with detector recipe so although absorption in this "contact region" is small, it is difficult to predict. Textbooks also mention a "dead layer" of Si. However, a true dead layer would give rise to a predictable spurious Si peak due to fluorescence and this is not observed in all types of detector. The true "dead layer" is usually very thin compared to the "transition" layer of "incomplete charge collection", ICC, and it is this transition zone which typically accounts for the apparent absorption step which appears at 1.84 keV in a Si(Li) detector spectrum. ICC reduces signal output and thus affects peak shape, typically producing a small tail on the low energy side while broadening the peak and pushing the centroid towards lower energy. If x-ray absorption is high, x-rays are more often absorbed in the surface transition zone so peak distortion is worst at low energies. Performance may be influenced by any one of a plethora of manufacturing process details involving choice of materials, machining, diffusion, etching, surface chemistry and deposition. Although successful recipes are naturally kept secret, results of process breakthroughs are usually published to inspire competition. In a recent example [17], McCarthy looked at the "dispersion", or detector contribution to resolution, as a function of energy and showed how changes to the process can alter the peak shape and resolution for low energy peaks. With an original recipe, ICC distorted low energy peaks and affected linearity producing for example a −37.7 eV shift at C Kα. With a revised process, discrepancy in the C peak position was reduced to −11.8 eV and there were associated improvements in dispersion and thus resolution. A variety of detector structures and process recipes is in use today, all with different characteristics and optimised for different parts of the energy spectrum.

The resolution fwhm in eV of a peak at energy E keV is given by

$$\mathrm{fwhm}^2 = C.E + \mathrm{ICC}(E)^2 + \mathrm{fwhm}^2_{\mathrm{noise}} \qquad (1)$$

where the constant C is typically 2480 for a Si(Li) and 1750 for a HpGe detector. $\mathrm{fwhm}_{\mathrm{noise}}$ is the electronic noise contribution and ICC(E) represents the degradation of resolution at different energies due to ICC. In Fig. 2 the resolution is plotted as a function of energy

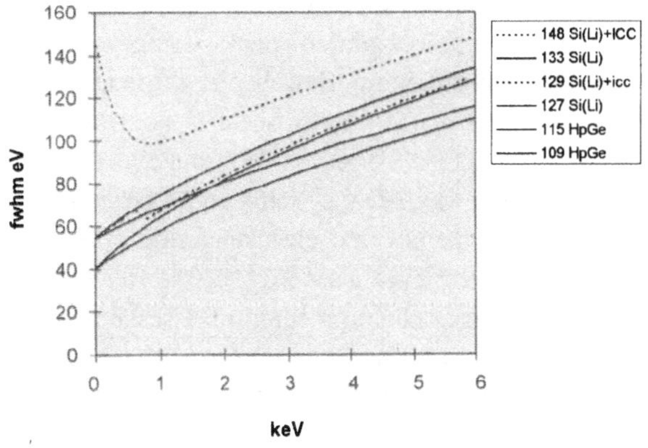

Fig 2. Variation of fwhm resolution with energy for some example EDX detectors

for a number of detectors. The one labelled "148 Si(Li)+ICC" is representative of a process with poor ICC, as reported in [17] and has noise of 85 eV, which is the sort of performance achievable on a Peltier cooled system, a larger area detector, or a detector with a poor FET for example. The one labelled "129 Si(Li)+icc" is a detector with the sort of reduced ICC as reported in [17] with good noise performance of 45 eV. The remaining detector examples are assumed to have perfect charge collection (ICC = 0) which is extremely difficult to achieve, and the "127 Si(Li)" and "109 HpGe" require electronic noise of only 40 eV which is also at the limit of what is currently attainable. On the right hand side of the graph in Fig. 2, the resolutions follow the order of the MnKα fwhm specification and remain in this order down to about 2 keV. Below this energy, the various influences of detector material, ICC and noise change the relative performance significantly. For example, below 1 keV, the "129 Si(Li)+icc" is no better than the "133 Si(Li)" and below 2 keV, the "127 Si(Li)" is better than the "115 HpGe". It is clearly impossible to characterise performance at low energies with a single parameter such as resolution at MnKα and it is important to assess the detector according to the application. For example, if Eq. (1) is used to predict the low energy performance of a 148 eV Si(Li) detector at the C Kα energy, assuming ICC = 0, the result, 89 eV, would suggest it could resolve the two peaks of C Kα and N Kα. However, because this detector exhibits some ICC, the resolution at C Kα would be 115 eV and it would not resolve these two peaks at all.

How to Validate Claimed Resolution

The full width at half maximum, fwhm, can be measured on an isolated peak using a simple geometric construction. However if the spectrum is mis-calibrated or the energy scale is non-linear, the fwhm may not be correct. Unfortunately, there are many reasons why EDX systems could be non-linear at very low energies and electronic side effects can distort the spectrum (e.g see Fig. 9.20 in Ref. [18]) so we need a more reliable method of assessing resolution. If we have overlapping peaks from two real x-ray lines, the peak-to-valley ratio will be independent of mis-calibration or non-linearity. If the real line separation is known, the overlap can be simulated on a spreadsheet by mixing Gaussians with the correct separation and different fwhms. The effective resolution fwhm is obtained when the peak-to-valley ratio of the simulation matches that of the observed spectrum. For example, experimental spectra of carbon and boron nitride specimens can be summed together to observe the valley between C Kα and N Kα peaks. There is still a chance that the system may drift or change characteristics between spectra so it is preferable if a single spectrum already has two close peaks. Figure 3 shows a spectrum from a carbon coated titanium standard acquired with a "Link GEM" HpGe detector. Note that the TiLl and TiLα peaks are just resolved. A simulation using Gaussians centred at 395 eV and 452 eV showed that the same P/V ratio is obtained when fwhm is 50.5 eV. Referring to Fig. 2 we see that this is indeed the sort of resolution expected for a high resolution HpGe detector with negligible ICC.

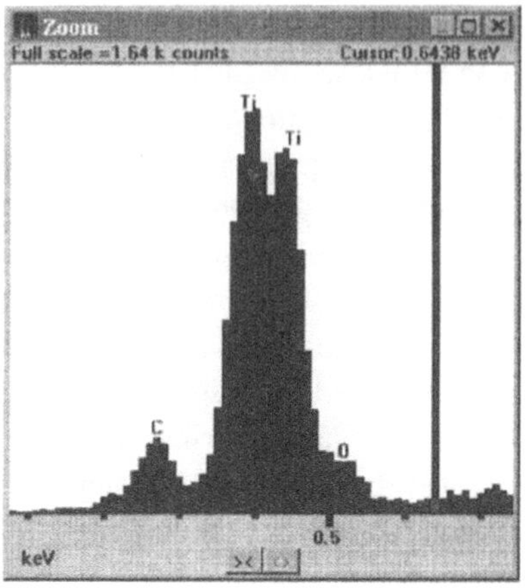

Fig 3. Spectrum from C-coated Ti specimen obtained with Link GEM detector

Spectrum Distortion Due to Electronic Effects

At low energies, there are special problems associated with separating and measuring the step signals in the presence of electronic noise. Typically, there is a "measurement" channel which uses long time constant filtering (or an equivalent weighted average in a digital system) to reduce the effects of electronic noise and one or more "pile-up inspection" channels which have to detect when pile-up occurs in the measurement channel. If we smooth a signal over a longer time interval, noise is reduced and we can lower the threshold for detection of low energy photons. However, the more smoothing we use, the

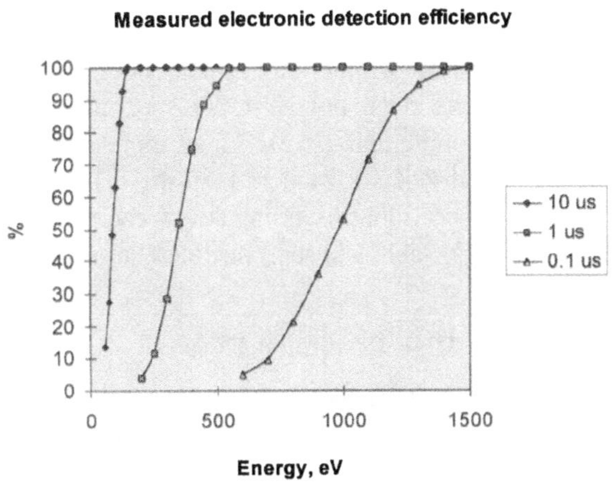

Fig 4. Pulse detection efficiency as a function of photon energy for 3 different channels with time resolutions of approximately 10 μs, 1 μs and 0.1 μs

worse the time resolution. Figure 4 demonstrates the tradeoff between time resolution and pulse detection efficiency for 3 different pile up inspection channels with different peaking times. The curves were obtained using an electronic pulser on a "Link Pentafet" Si(Li) detector and "XP2" pulse processor for a system with low noise. Although analogue shaping circuits were used in this example, even if a digital algorithm were used to detect pulses, the tradeoff would be similar.

The electronic noise as a function of pulse shaping time is likely to vary from system to system and between different manufacturing processes so although these curves represent current state of the art, it would be unwise to assume this performance is typical of every system. However, the general behaviour can be understood with reference to Fig. 4. The 0.1 μs channel will see any pulses >1.5 keV, the 1 μs channel will see any pulses >500 eV but only the 10 μs channel will have noise threshold low enough to see pulses below 500 eV. If two photons of energy E_1 and E_2 arrive within the time resolution of the inspector, a count appears in the sum peak. If t_{12} is the resolving time for E_1 followed by E_2, T_{LIVE} is the total livetime for acquisition and A_1 and A_2 are the areas of the respective peaks, then the area of the sum peak is given by [19]

$$\mathrm{Asum} = A_1 \cdot A_2(t_{12} + t_{21})/T_{LIVE} \qquad (2)$$

For a sum peak formed by coincidence of pulses from a single peak, the area is given by [19]

$$\mathrm{Asum} = A_1^2 \cdot t_{11}/T_{LIVE} \qquad (3)$$

In a test experiment using a magnesium oxide sample and a similar, but not identical, system used for Fig. 4, the following measurements were recorded: Mg = 315000 cts, O = 163000 cts, Mg+O sum peak = 2000 cts, T_{LIVE} = 66 secs. From Eq. (2) above, this gives $(t_{12}+t_{21}) = 2.6$ μs. The average time resolution is thus 1.3 μs which is what we might expect from Fig. 4 because the fastest channel that can recognise O Kα x-rays has about 1 μs time resolution. The following experimental data from a system using digital recognition has also been reported [20]: Mg = 137000, O = 145000, Mg+O sum peak = 2200 counts T_{LIVE} = 66 secs and this gives $(t_{12}+t_{21}) = 7.3$ μs. If we refer to the XP2 as an "analog" processor, then clearly, in this example, "analog" beats "digital". However such a conclusion would be a false simplification. In the above examples, the "analog" recognition unit happened to have a 1 μs channel but the "digital" recognition unit had an algorithm only capable of resolving pulses 3.8 μs apart. Obviously, the analog circuitry could be modified to change the time constants of each channel and the digital algorithms could be modified to improve the time resolution so this is purely a question of design optimisation. Whether "digital" or "analog", the effective pulse resolving time governs noise performance and thus detection efficiency of the channel along the lines described by Fig. 4. Furthermore, if the detector+FET noise varies from system to system, then the electronic detection efficiency will also vary, even if the same circuitry or algorithms are used.

The question remains: "what happens if a photon arrives which is not detected by any of the pile up inspector channels or algorithms?". A count appears in a sum peak when two photons arrive within the resolving time of the pile up inspector so that the measurement channel records the sum of the two energies. However, if an undetected photon arrives while a measurement is in progress, the result will be somewhere between the correct value for the first photon and the sum of the two photon energies. Thus, a pile-up continuum is produced above each major peak [1]. Pile up distortion is easy to check by experiment by acquiring a spectrum from the same sample at low and then high counting rate and looking for changes, particularly shift and distortion on the high energy side of peaks (see for example Fig. 11.23 in Ref. [16]).

Spectrum Processing and Statistics

With roughly 70 major element lines occuring below 3 keV, peak overlap is inevitable with EDX and spectrum processing is required to establish peak areas. Even with a perfect system and perfect software, statistics will limit the precision for resolving overlap and background corrections. Both resolution and count rate affect precision and detection limits and the ultimate limits of precision have been analysed for a number of pathological situations in [21]. As the peak to background gets smaller and the number of peaks in a multiplet increases, small improvements in resolution can have a substantial effect on statistical precision. The analysis in [21] assumes that peak shape and position are known exactly. It is therefore important to understand what accuracy and reproducibility are required for a spectrometer which has to be able to exploit the ultimate statistical precision available in a spectrum. The following worked example attempts to address this question. A typical spectrum showing an O Kα peak, 61 eV fwhm, area 4000 counts (background 50 counts at 10 eV/ch) was analysed as if it contained both oxygen and vanadium. The peaks involved were V Lα = 510 eV, O Kα = 523 eV, separation 13 eV. If we fit to V Lα and O Kα peaks, we should get a zero result for the vanadium and indeed we do if the model for the fitted O Kα peak is absolutely correct. If there is a small error between the model and true position of O Kα, a spurious amount of V Lα is obtained from the fit as shown in Table 1.

A statistical analysis of the same spectrum gave st. dev. in V Lα result to be 160 cts which corresponds to 4% of the O Kα area. This means, if we knew peak shapes accurately, we could consider sorting out overlaps involving a small V Lα peak and large O Kα peak. However, as Table 1 shows, an error of only 0.5 eV in peak position would produce a 4% error in V Lα area which is as large as the statistical limit in precision. Therefore, the spectrometer calibration has to be accurate to within 0.5 eV and count rate or temporal effects must not shift the spectrum after calibration by more than 0.5 eV. This simple example demonstrates that if we want to reap the benefits of sophisticated spectrum processing at low energies, we need extremely stable and reproducible equipment.

Table 1. Spurious result for V Lα obtained when fitting 2 peaks to a single O Kα peak with 4000 counts when using an incorrect model shifted from the true peak position

Shift, eV	Fitted area V Lα	% of O Kα
0	0	0%
1	295	7%
2	597	15%
3	905	23%
4	1217	30%

Prospects for High Resolution EDX

For some years, detectors operating below 4 deg K have been of interest to astronomers because excellent resolutions can be achieved either by calorimeters or superconducting tunnel junctions at these low temperatures. Thermal time constants have typically limited count rates to well below 100 cps but this is not usually a problem with weak x-ray sources in space. Recently, there has been renewed interest in optimising the parameters of these devices to make them suitable for SEM x-ray analysis. A useful symposium [11] reviews these developments. More recently, a working spectrometer has been mounted on a conventional SEM and successfully obtained spectra at about 100 cps showing resolution of 13 eV fwhm at 4.5 keV [22]. There are many technical difficulties to resolve, such as operation with a liquid helium cryostat which is larger than the electron column and achieving good light element sensitivity, but this is a promising development which offers parallel collection spectroscopy with resolution close to intrinsic line widths.

Conclusions

In recent years, EDX spectrometers have been increasingly used for low energy work yet conventional methods of characterisation are inappropriate. Detector efficiency and solid angle of collection are affected by the need to block light and electrons from reaching the detector and will be governed by detailed microscope geometry. Sensitivity does not roll off smoothly at low energies because of absorption steps within the detector window and the structure of windows and detectors is too complex and variable to characterise by simple generic absorption formulae. ICC and electronic noise govern the resolution at low energies and there is a big variation between different detector process recipes, structures and sizes; as a result, MnKα fwhm is a poor indicator of low energy performance. Resolution is particularly important at low energies to resolve K, L and M overlaps. 50 eV fwhm has been demonstrated at 0.5 keV with a state of the art HpGe EDX. Superconducting detector technology, though expensive, does show the potential of achieving 10 eV fwhm or less. ICC and electronic distortion at varying count rate make it more difficult to model peak shapes accurately at low energies and for extreme overlaps, commonplace below 1 keV, detector and electronics must reproduce peak position and width to well within 0.5 eV.

References

[1] P. J. Statham, *Inst. Phys. Conf. Ser. No. 119*: Section 10, Proc. EMAG 91, IOP Publishing Ltd, 1991, 425

[2] D. B. Williams, J. I. Goldstein, D. E. Newbury (eds.). X-Ray Spectrometry in Electron Beam Instruments, Plenum, NY, 1995.

[3] P. J. Statham, *Microbeam Analysis-1988*, San Francisco Press, CA, 1988 p., 429.

[4] Tech. Data Sheet, *Quantmap*, Oxford Instruments, UK, 1995.

[5] J. Friel, R. Mott, *Scanning* **1996**, *18*, 140.

[6] P. J. Statham, *Mikrochim. Acta* **1996**, [*Suppl.*] *13*, 573.

[7] R. Daniel, reporting ISIS-300/DXP50 performance at IUMAS-1, ACEM-14, 1996.

[8] R. Daniel, *Personal Communication*, Oxford Instruments.

[9] S. M. Zemyan, D. B. Williams, *Microbeam Analysis-1991*, San Francisco Press, CA, 1991, 134.

[10] R. P. Kraft, D. N. Burrows, G. P. Garmire, J. A. Nousek, J. R. Janesick, P. N. Vu, *Nucl. Instrum. Meth.* **1995**, *A 361*, 372.

[11] *Microbeam Analysis-1995*, VCH, NY, 1995, pp. 3–13.

[12] P. J. Statham, *Microbeam Analysis-1995*, VCH, NY, USA, 1995, p. 187.

[13] E. Boyes, *ICEM 13-Paris*, Les editions de physique, 1994, p. 51.

[14] C. E. Nockolds, *Microbeam Analysis* **1994**, *3*, 185.

[15] M. W. Lund, in: *X-Ray Spectrometry in Electron Beam Instruments*, Plenum , NY 1995, 21.

[16] D. E. Newbury, In: *X-Ray Spectrometry in Electron Beam Instruments*, Plenum NY, 1995, 167.

[17] J. J. McCarthy, In: *X-Ray Spectrometry in Electron Beam Instruments*, Plenum NY, 1995, p. 7.

[18] R. B. Mott, J. J. Friel, in: *X-Ray Spectrometry in Electron Beam Instruments*, Plenum, NY, 1995, 127.

[19] P. J. Statham, *X-Ray Spectrometry* **1977**, *6*, 94.

[20] S. A. Audet, J. J. Friel, T. P. Gagliardi, R. B. Mott, J. I. Patel, C. G. Waldman, *1994 IEEE Nuclear Science Symposium*, IEEE, NJ, 1994, p. 155.

[21] P. J. Statham, In: *X-Ray Spectrometry in Electron Beam Instruments*, Plenum, NY, 1995, p. 101.

[22] D. A. Wollman, G. C. Hilton, K. D. Irwin, J. M. Martinis, *Proc. Microscopy and Microanalysis 1996*, San Francisco Press, 1996, p. 488.

Mikrochim. Acta [Suppl.] 15, 11–19 (1998)

High Resolution Non Dispersive X-Ray Spectroscopy with State of the Art Silicon Detectors

L. Strüder[1,*], **C. Fiorini**[4], **E. Gatti**[4], **R. Hartmann**[1], **P. Holl**[3], **N. Krause**[1], **P. Lechner**[3], **A. Longoni**[4], **G. Lutz**[2], **J. Kemmer**[3], **N. Meidinger**[4], **M. Popp**[1], **H. Soltau**[3], and **C. von Zanthier**[3]

[1] MPI für extraterrestrische Physik, Halbleiterlabor, Paul-Gerhardt-Allee 42, D-81245 München, Federal Republic of Germany
[2] MPI für Physik, Halbleiterlabor, Paul-Gerhardt Allee 42, D-81245 München, Federal Republic of Germany
[3] KETEK GmbH, Paul-Gerhardt-Allee 42, D-81245 München, Federal Republic of Germany
[4] Politecnico di Milano, Piazza Leonardo da Vinci 32, 1–20122 Milan, Italy

Abstract. For the European X-ray Multi-Mirror mission (XMM) and the German X-ray satellite ABRIXAS fully depleted pn-CCDs have been fabricated, enabling high speed, low noise, position resolving X-ray spectroscopy. The detector was designed and fabricated with a homogeneously sensitive area of 36 cm^2. At 150 K it has a noise of 5 e$^-$ rms only, with a readout time for the total focal plane array of 4 ms. The maximum count rate for single photon counting is 100,000 cps under flat field conditions. Its position resolution is in the order of 100 μm. The quantum efficiency is higher than 90% from carbon K X-rays (272 eV) up to 10 keV. In addition, new cylindrical silicon drift detectors have been designed, fabricated and tested. They comprise an integrated on-chip amplifier system with continuous reset, on-chip voltage divider, electron accumulation layer stabilizer, large area, homogeneous radiation entrance window and a drain for surface generated leakage current. At count rates as high as 1,000,000 counts per sec and per cm^2, they still show excellent spectroscopic behaviour at room temperature operation.

These detector systems were developed for X-ray astronomy, synchrotron light sources and for X-ray fluorescence analysis applications. All systems are fabricated in planar technology having the detector and amplifiers monolithically integrated on high resistivity silicon.

Key words: silicon drift detector, x-ray charge coupled device, back side illuminated pn-CCD, room temperature x-ray spectroscopy, high count rate detectors, integrated JFETs

For the last few years, modern semiconductor detectors have been the prime choice for the measurement of nuclear radiation in various scientific fields. In satellite missions for astrophysics, silicon sensors are used as direct imaging and/or spectroscopic detectors for near infrared, visible and X-ray photons [1]. Equally in other fields of research, e.g. at synchrotron sources and in material characterization systems high "event throughput" detectors are required. Large quantities of silicon detectors are operational in high energy physics experiments as position sensitive particle detectors.

In contrast, until now, the industrial offer of semiconductor detectors for X-ray spectroscopy has been mainly restricted to the relatively old concepts of Si(Li) and HPGe sensors

This article will describe briefly the relevant concepts of modern high resolution silicon detectors recently being developed, fabricated and tested for photons in the energy band of 1 eV up to 20,000 eV. We include all sensors that are made in a planar fabrication process on high resistivity *n*-type FZ silicon. These detectors make available a sensitive volume of the whole wafer thickness of about 300 μm. The more sophisticated concepts allow electronic noise levels close to 1e$^-$ rms due to their small readout capacitance and the integration of the first amplification stage onto the detector chip. Several prototype

* To whom correspondence should be addressed

systems for industrial applications have already been manufactured and implemented in larger systems.

Two examples of detectors systems already operational, will be discussed:

1. The fully depleted X-ray pn-CCD for the X-ray Multi Mirror (XMM) satellite mission of the European Space Agency and for the German ABRIXAS satellite, both to be launched in 1999.
2. The silicon drift detector for X-ray fluorescence analysis in scanning electron microscopes and, recently, for applications in EXAFS experiments and in X-ray holography.

All detector systems described below were designed, fabricated and tested in the MPI-Halbleiterlabor, Munich.

Basic Concepts

In order to understand the basic functional principles of modern silicon radiation detectors four aspects must be mentioned:

(1) the charge generation in silicon, (2) the reverse biased diode as the basic radiation counter element, (3) the sideward depletion of a semiconductor leading to the silicon drift chamber principle, and (4) the amplification principles of on-chip-electronics to obtain noise figures close to $1e^-$ rms.

Charge Generation

The dominant interaction of photons up to 50 keV hitting silicon is the photoelectric process [2]. At higher energies Compton scattering and (above 1022 keV) pair production processes become more relevant. For energetic charged particles the Coulomb interaction dominates the energy loss.

For photon radiation with the energy E_{in} above 1.83 keV and the K shell ionization of the silicon atom is the most significant process. Due to the simultaneous conservation of energy and momentum the ionization is concentrated on the strongest bound electron. The electron generally leaves the atom with a kinetic energy of E_{in} – 1.83 keV. In the following processes, the ejected electron ionizes predominantly valence electrons of the neighboring atoms; core shell ionization is far less probable within the K-shell ionized Si-atom Auger and fluorescence processes are initiated, unless the Si atom is completely de-excited. The multitude of low energetic electrons and photons generated by these processes is responsible for the creation of electron-hole pairs. They free by ionization the valence band electrons from the covalent binding leaving behind the positively charged defect electron or hole.

Besides the generation of electron – hole pairs, a fraction of the energy E_{in} is transferred to the silicon lattice, resulting in an increase of the temperature. The average energy needed for the generation of an electron hole pair is thus 3.65 eV at room temperature and for energies E_{in} above 1 keV. Experimental data for the pair creation energy is given in [3].

The Reverse Biased Diode

In order to detect and analyze the radiation absorbed in the semiconductor material it is necessary that the charge, which is generated by ionization, can be collected without loss. In normal semiconductor material the high number of electrons and holes which are there due to the doping, make the measurement of an additional small number of charge carriers impossible. However, the depletion layers of rectifying contacts or pn-junctions are free of charge carriers and can be used for this purpose. By applying a reverse bias on the pn-junction, the width w of the depletion layer can be increased and the efficiency of the detector improves according to:

$$w = \sqrt{\frac{2\varepsilon_o\varepsilon_r(U - U_D)}{q} \cdot \frac{N_A + N_D}{N_A \cdot N_D}} \tag{1}$$

U is the bias voltage and U_D the diffusion or built-in voltage, typically in the order of 0.5 V for a standard p^+n junction, $\varepsilon_o\varepsilon_r$ the dielectric constants, N_A and N_D the concentration of acceptors and donors. As can be seen from the above equation for the abrupt pn-junction, the extension of the depletion layer into the n- and p-area of the semiconductor is inversely proportional to the doping concentration N_A and N_D, respectively. In practice, for semiconductor detectors highly asymmetric pn-junctions with a very shallow p^+ doping on a very weakly doped n-substrate are used. Under these conditions (and $U \gg U_D$) the depletion layer expands only into the high resistivity bulk material and the above equation simplifies to:

$$w = \sqrt{\frac{2\varepsilon_o\varepsilon_r U}{qN_D}} \tag{2}$$

E.g. for a 9 kΩ cm n-type silicon water about 30 V reverse voltage is needed to completely deplete a 280 μm thick wafer, resulting in a sensitive layer of the whole wafer thickness. If the Poisson equation, the basis of Eq. (1) is numerically solved in two dimensions, the electrical potential of a silicon drift chamber can be calculated as shown below.

The Silicon Drift Chamber Principle

In 1983 Gatti and Rehak proposed a new detector scheme based on sideways depletion [4]. The idea is that a large semiconductor wafer, of e.g. high resistivity n-type silicon, can be fully depleted from a small n^+ ohmic contact positively biased with respect to the p^+ contacts covering both surfaces of the silicon wafer.

In the standard configuration the depletion zones will expand from all rectifying junctions simultaneously as long as the ohmic access from the n^+ anode to the entire (non-depleted) bulk is not interrupted. At a given voltage the depletion zones propagating from the p^+ areas touch each other and the former conducting electron channel, symmetrically located in the middle of the substrate between the p^+ implants, will abruptly disappear. At this moment the depletion of the whole wafer is completed at a voltage which is four times lower than the voltage needed to deplete a simple diode of the same thickness (see also Fig. 2 and Eq. 2). Under this condition the electron potential energy in a section perpendicular to the wafer surface has a parabolic shape, with an electron potential minimum in the middle of the wafer.

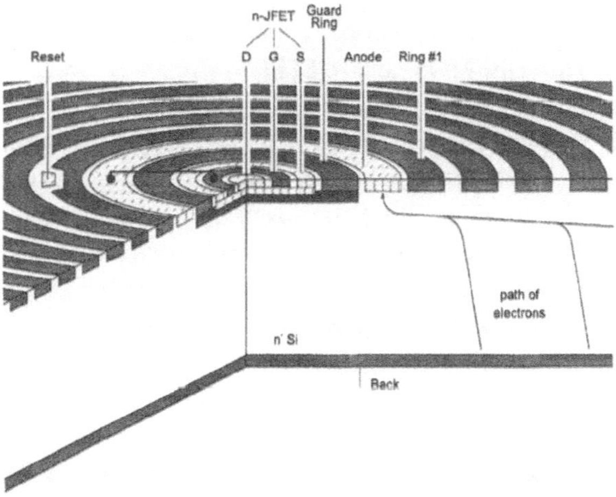

Fig. 1. Central area of a cylindrical silicon drift chamber with an integrated amplifier for spectroscopic applications. The enitre silicon wafer is sensitive to radiation

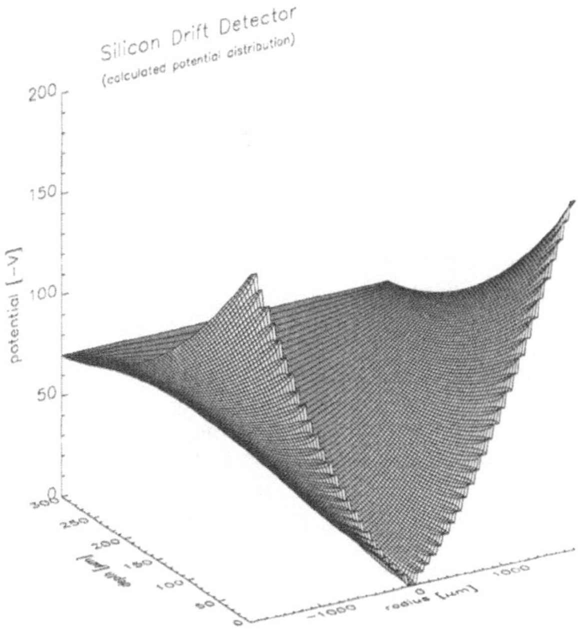

Fig. 2. Potential energy distribution in a silicon drift chamber. The simulation includes the whole detector shown in Fig. 1 including the electron collecting readout node

The silicon drift chamber is derived from the principle of sideward depletion by adding an electrical field, which forces the electrons to the n^+ readout anode. This is simply achieved by implantation of a parallel p^+ strip pattern at both sides of a semiconductor wafer and superimposing a voltage gradient at both strip systems. The direction of the voltage gradient is such that the n^+ readout anode has the highest positive potential, therefore collecting all the signal electrons accumulated in the potential minimum.

To simplify the fabrication technology, the strip system is replaced by a large area pn-junction at one side, which is used as a very homogeneous thin entrance window for the radiation [5], [6]. A further improvement of this type of drift chamber uses cylindrical drift electrodes which force the signal electrons to a very small anode in the centre of the device from where they are transferred to the gate of an integrated JFET (see Fig. 1).

The electrical potential of the cylindrical silicon drift chamber is shown in Fig. 2 in a two dimensional section perpendicular to the surface through the silicon wafer. It shows the potential energy for electrons of the device in Fig. 1 under test including all field strips and the central electron collecting anode. The equipotential of the homogeneously doped radiation entrance window can be seen on the back, the field strips with their decreasing (negative) potential on the front side. There is no field free region in the device and all electrons in the sensitive area are guided within less than 100 ns towards the readout node.

On-Chip Amplification

The traditional "external" front-end and shaping amplifier concepts as they are used in conventional Si(Li) and HPGe systems will not be treated. We restrict ourselves to amplification strategies realizing the first change of impedance or amplification directly on the detector chip. The advantages are: (1) by eliminating bond wires and pads the capacitance of the system can be much reduced resulting in a lower equivalent noise level and less pick-up, (2) problems of microphony are avoided and (3) more freedom in design is provided.

In our laboratory three concepts for on-chip amplification have been developed: (a) the n-channel single sided junction field effect transistor (SSJFET) [7], (b) the depleted PMOS or depleted p-type JFET detector-amplifier structures (DEPMOS or DEPFET) [8] and p-channel JFETs [9]. We have designed, fabricated and tested on-chip amplifiers of type (a) with an equivalent noise charge (ENC) of $2\,e^-$ rms only, at 150 K. We achieved the low noise level by avoiding all kinds of stray capacitances between the readout node and ground. Using the silicon drift chamber principle, which guarantees low detector capacitance, and the small on-chip FET's in the amplification stage, the total capacitive load C_{tot} of the detector – amplifier is as small as 100 fF. With

$$Q = C_{tot} \cdot U_{in} \tag{3}$$

we can easily see that a small capacitance results in a significant change of voltage on the collecting anode. Q is the total charge arriving at the readout capacitance and U_{in} is the voltage change upon arrival of the signal charges at the (total) input capacitance C_{tot}. One single electron arriving at the readout node would generate a voltage step of 1.6 μV, an already measurable quantity. State of the art read nodes succeed in reducing the total capacitance to 20 fF. The limits of further progress in noise reduction are mainly given by the fabrication process technology and material properties.

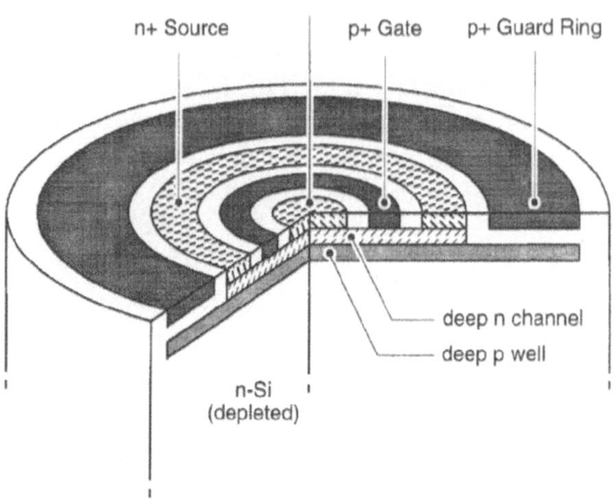

Fig. 3. Schematic cross section through a single sided junction field effect transistor

With the help of Fig. 3 the basics of the amplification process of a SSJFET can easily be understood. Let us assume that the electrons, generated by the ionizing radiation drift towards the readout anode. There they induce a negative voltage. This voltage is directly coupled to the p^+ gate of the n-channel transistor (source and drain are n^+ implants, the transistor channel is a deep n implant). The negative voltage on the p^+ gate reverse biases the junction, thus depleting the transistor channel, resulting in a current drop through the transistor. This change of current can be precisely detected.

The DEPMOS and DEPFET structures work a different way [10]. A potential minimum for electrons is generated in the silicon approximately 1 μm from the surface having the pixel structure. The p-channel of the MOS or JFET transistor is located directly above the electron potential minimum which is also called the "internal gate". If electrons arrive at the internal gate they influence positive charge – holes – in the transistor channel, giving rise to an increase in the transistor current. This change in current can be precisely detected.

At room temperature the noise figure of the SSJFET is about $8\,e^-$, for the DEPFET it is actually $15\,e^-$ rms. But the geometry of the DEPFET is not as optimized as the SSJFET design, yet.

Silicon Detector Systems

Two detector configurations will be briefly presented: (1) The fully depleted X-ray pn-CCD as built for the XMM satellite mission, and (2) the silicon drift chamber as built for spectroscopic investigations with synchrotron light, for fluorescent analysis, for particle counting and the parametrization of mixed radiation fields.

Fully Depleted Charge Coupled Devices (pn-CCDs)

For ESA's X-ray Multi Mirror mission, to be launched in the year 1999 we have developed a $6 \times 6\,cm^2$ large monolithic X-ray CCD [11] with high detection efficiency up to 15 keV, low noise level (ENC≈5 e^-) and an ultrafast readout time of 4 ms. A schematic cross section, already showing some of the advantages of the concept are displayed in Fig. 4. The concept allows for an optimum adaption of the pixel size to the X-ray optics, varying from 30 μm up to 300 μm pixel size. The energy response is higher than 90% at 10 keV because of the sensitive thickness of 280 μm. The low energy response is given by the very shallow implant of the p^+ back contact; the effective "dead" layer is smaller than 150 Å [12]. The good time resolution is given by the parallel readout. A high radiation hardness is built in by avoiding MOS structures and by the fast transfer of the charge in a depth of more than 10 μm. The spatially uniform detector quality over the entire field of view is realized by the monolithic fabrication of 12 individually operated $3 \times 1\,cm^2$ large pn-CCDs on a single wafer. As the telescope system resolution on XMM is in the order of 15 arcsec with a focal length of about 7.5 m, the extension of a point source in the X-ray sky is about 500 μm in the focal plane. That is why a pixel size of $150 \times 150\,\mu m^2$ for the CCDs was chosen.

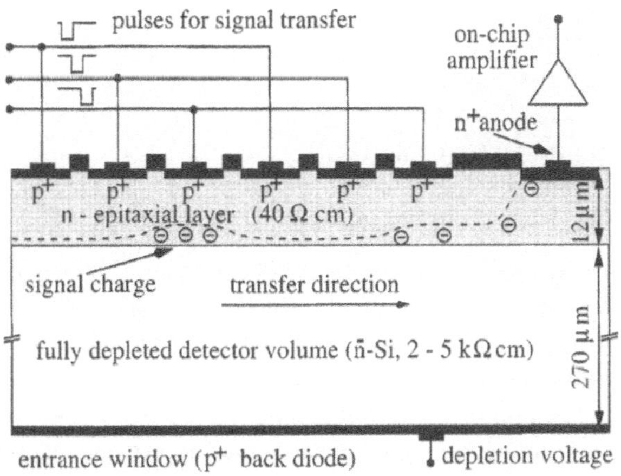

Fig. 4. Schematic cross section of a fully sensitive pn-CCD. The radiation enters through the planar thin p^+ back diode. The sensitive thickness is 280 μm

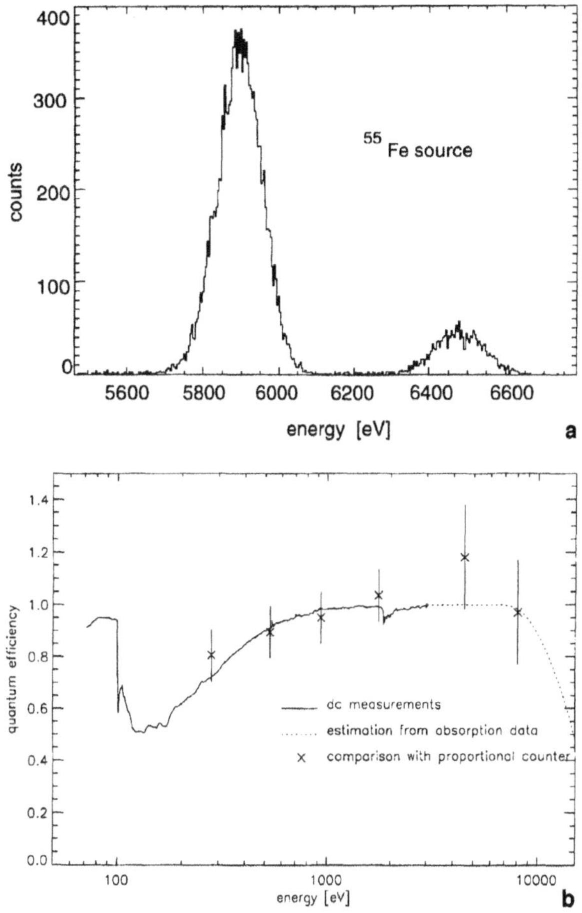

Fig. 5. a ^{55}Fe spectrum with a pn-CCD in a flat field measurement at a readout speed of 4 ms for a $3 \times 1\,cm^2$ large pn-CCD detector. The FWHM of the Mn K_α peak is 130 eV only. **b** measured quantum efficiency in a dc photocurrent measurement (solid line) and in a single photon counting mode (×signs)

The best values for the readout noise of the on-chip electronics is 3 e^- rms at 150 K, typical values scatter around 5 e^- rms. The charge transfer properties are as good as in standard CCDs, in the order of 1% charge loss from the last to the first pixel over a distance of 3 cm charge transfer. Fig. 5 a shows a ^{55}Fe spectrum of a pn-CCD in a flat field measurement resulting in a typical energy resolution of 130 eV (see Fig. 5) to 140 eV (FWHM) [13]. The impact of the material properties of silicon and related impurities and their consequences for the operation of scientific grade X-ray pn-CCDs, including the effects of radiation damage, is treated in detail in [14].

In a single photon counting mode the quantum efficiency was measured with respect to a calibrated position sensitive proportional counter. In addition, a dc photo current measurement at the BESSY synchrotron in Berlin with a calibrated monochromator and reference detector is shown. Fig. 5 shows the single photon counting data (×signs) and the dc photocurrent results, in steps of 5 eV from 60 eV up to 3 keV. The single photon data were taken with the C_K, O_K, Cu_L, Ti_K, $Cu_{K\alpha}$ and $Cu_{K\beta}$ from an X-ray tube (see EOBB Final Report, ESA and [13]). The quantum efficiency on the low energy side was further improved with respect to the measurements in Fig. 5 by increasing the drift field at the p^+ junction entrance window [15]. The read out electronics of the pn-CCD system is described in [16]. A charge sensing amplifier followed by a multicorrelated sampling stage, multiplexer and output amplifier (CAMEX64B JFET/CMOS chip) guide the pn-CCD pixel content as a voltage signal to a 10 MHz 12 bit flash ADC system. The whole system, i.e. CCD and CAMEX64B amplifier array dissipates a power of 0.7 W tolerable for the XMM satellite mission. A further increase of the readout speed can be made only at the expense of further increase of power, or a degradation of the noise performance.

We have tested the pn-CCD system in a single photon counting mode in flat field conditions at high count rates. If signal pile-up is to be avoided only one photon is allowed in one pixel and the maximum photon count rate is approximately 3.000 counts per cm^2 per second, leading to 10^5 counts per second (cps) for the full detector system. The experiment was performed with a mechanical chopper wheel, preventing photons hitting the CCD during readout. The open position on the chopper wheel added another 6 ms of "photon integration time" to the 4 ms readout time. Split events, i.e. events with electrons in more than one pixel, originating from a single photon, were reconstructed and added to one photon event. In total, 75% of all events are single pixel events, 23% are two pixel events and 2% are events with three and four pixels involved. One single X-ray hit never affects more than four pixels.

If more than one photon is allowed to be in one single pixel, only the full well capacity of the CCD pixel limits the maximum detectable photon flux. About 1,000 photons with an energy of 5 keV can be transferred per pixel, resulting in a count rate limit of more than 18^8 cps in the full detector. The charge handling capacity of the individual pixels was tested with the 5.5 MeV alpha particles form a radioactive ^{241}Am-source. More than 2×10^6 electrons can be properly transfered in every pixel.

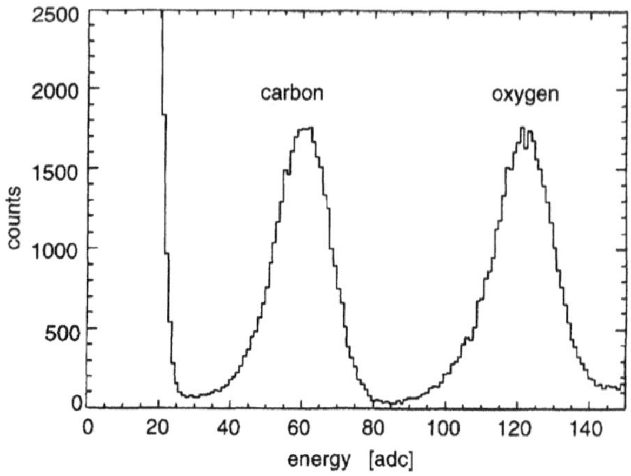

Fig. 6. Spectrum of carbon K (277 eV) and oxygen K (525 eV) recorded with a pn-CCD

Silicon Drift Chambers for Radiation Monitoring and Spectroscopy

Silicon drift detectors (SDD) have been used successfully in experiments in high energy physics for position measurements of minimum ionizing particles [17].

Large silicon drift detectors have been fabricated at the semiconductor laboratory of the Max-Planck-Institutes (MPI-HLL) with linear drift geometry up to $4.2 \times 3.6\,cm^2$ and a $55\,cm^2$ cylindrical geometry on 4 inch wafers. Both systems have been used for position measurements of minimum ionizing particles. We succeeded in a new development of fast SDCs as radiation monitors in our satellite test facility PANTER and for the synchrotron radiation research facility at Grenoble (ESRF), to include the on-chip amplifiers. The devices are used for spectroscopic investigations (EXAFS) with high count rates in the energy band between .5 keV and 9 keV, with an energy resolution of less than 150 eV FWHM over the required energy band at a temperature achievable with a single stage Peltier cooler.

The aim of the development was a compact silicon detector, operational with low noise at relatively warm temperatures (i.e. in the range of Peltier cooling systems), high count rate capability, radiation hardened and with a very thin entrance window [6], [8].

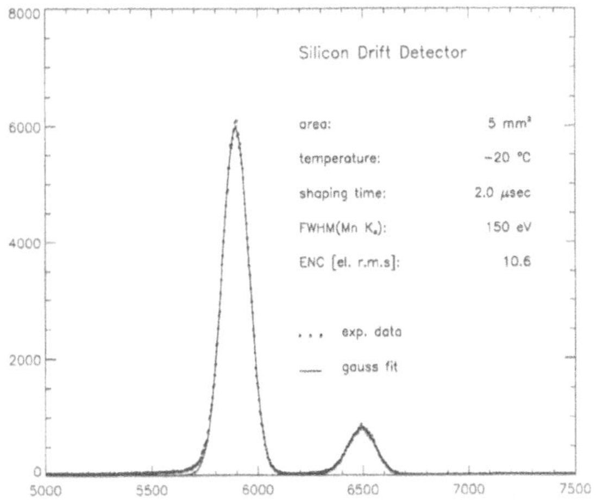

Fig. 7. ^{55}Fe spectrum recorded with a silicon drift chamber at 253 K (Peltier cooling) having a sensitive area of $5\,mm^2$ and a fully sensitive thickness of 280 μm. At a shaping time of 2 μs, the equivalent noise charge (ENC) was $10.6\,e^-$ rms. The FWHM at the Mn K_α line is 150 eV

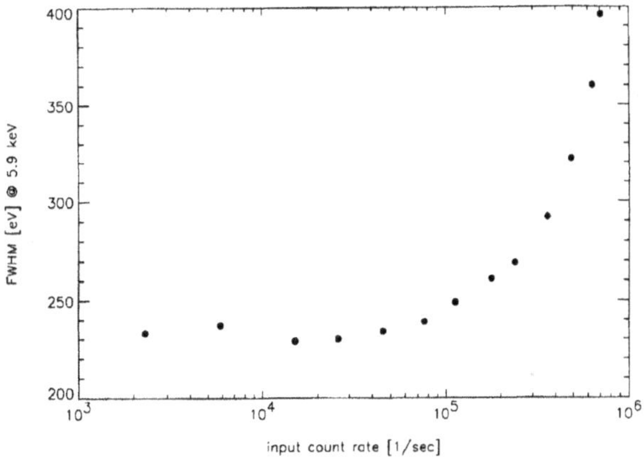

Fig. 8. Measurement at room temperature of the count rate dependent energy resolution with an ^{55}Fe source. The equivalent noise charge as a function of the incident photon rate is shown. At low count rates the FWHM is about 230 eV

The detectors are now used for all types of ionizing radiation, for charged particles, photons and neutrons if adequate converters are used.

We have designed, fabricated and tested the first detectors [19]. They are cylindrical SDCs with a central anode as shown in Fig. 1 and with an active area of $3.5\,mm^2$ and $5\,mm^2$.

The first experimental results of the devices show that at room temperature the intrinsic noise figures are in the order of $20\,e^-$, translating to a width of the ^{55}Fe Mn K_α line of typically 220 to 230 eV at a shaping time of 500 ns for the 3.5 and $5\,mm^2$ large detectors. The bulk leakage current at 300 K contributing to the system noise is less than 1 nA per cm^2. At 263 K the equivalent noise charge is reduced to about $12\,e^-$ rms, i.e. slightly less than 160 eV FWHM at the Mn K_α line. Fig. 7 shows the system performance with a Peltier cooled silicon drift detector at −20 °C with a FWHM of 150 eV at 5.9 keV for X-rays of Mn K_α from a ^{55}Fe source. At 220 K the noise decreases to $7.9\,e^-$ as shown experimentally, if optimum digital shaping is used [20]. The FWHM of an ^{55}Fe signal is then 137 eV, still being obtainable with thermoelectric cooling elements. All other detector relevant parameters are satisfactory: complete charge collection, spatially homogeneous response, reproducibility and long-term stability. Devices up to $1\,cm^2$ have been produced with one or more readout nodes. The largest

Fig. 9. Layout of the 39 cell silicon drift detector array with 39 integrated on-chip Amplifiers. The total active size of the system is $2\,cm^2$

device with 39 readout nodes is realized on a $2\,cm^2$ large detector chip to achieve the high count rate capability and the low noise level of the individual $5\,mm^2$ subsystems (see Fig. 9). Multiplexing VLSI amplifiers are actually being developed to keep the complexity for the user at a reasonable level. We have tested the decrease of the spectroscopic performance as a function of X-ray count rate with the SDD at room temperature and applying a signal shaping constant of 250 ns (see Fig. 8). 100,000 cps can be detected at room temperature without a significant increase in the equivalent noise charge, that means for example, without a broadening of the Mn K_α line of the ^{55}Fe spectrum. This translates to a maximum count rate of more than 2×10^6 cps per cm^2 without degradation of the spectroscopic behaviour. The increase of the FWHM from 10^5 cps (230 eV) to 400 eV at 8×10^5 cps arises from internal processes of the detector as well as from the signal processing electronics.

SDD devices have already been used as an X-ray fluorescence analysis detector in a scanning electron microscope (SEM) [21]. A detector module, comprising a SDD cooled by a single stage Peltier element, a ceramic substrate of $4\,cm^2$ with passive DC filters and on-chip thermometer was successfully operated in a SEM field test. A six-anode detector ($21\,mm^2$ of active area) has been operated in EXAFS measurements near room temperature at the ESRF beamline no. 6 [22]. The good energy resolution at short shaping times (e.g. 162 eV FWHM with an ^{55}Fe source with a 250 ns shaping at 150 K) makes the detector very attractive for those experiments, typically carried out at synchrotron radiation facilities, where a high rate capability is needed, together with good energy resolution [5].

The compact design, the absence of liquid nitrogen for cooling, the high count rate capabilities, the insensitivity to microphonic noise pick-up and the lower price make these systems very attractive for commercial applications.

Conclusions

Silicon detectors are able to perform radiation monitoring and spectroscopy at moderate ambient conditions with respect to contamination, low frequency accoustic noise and temperature. Because of the low capacitive readout anodes, noise levels below $5\,e^-$ rms at 150 K, and around $10\,e^-$ rms at 300 K are obtainable. The detector leakage current adds to this intrinsic on-chip amplifier noise a temperature dependent component, decreasing exponentially with decreasing temperature. Fully depleted pn-CCDs have shown an excellent performance for spatially resolved spectroscopy in a single photon counting mode. A multi purpose silicon drift detector (SDD) is operated for fast spectroscopic applications for synchrotron light users, for X-ray fluorescence applications, for dosimetry and as a low noise photo diode for optical and calorimetrical applications.

Acknowledgements. The author thanks the staff of the MPI Halbleiterlabor and the Max-Planck-Institute für Physik und extraterrestrische Physik for their continuous support. The perfect mounting and bonding of the devices by P. Solc and A. Kaltenberger is acknowledged. This research is supported by the German space agency DARA, the European Space Agency and the HCM program of the European Community.

References

[1] G. F. Bignami, G. E. Villa, G. Boella, G. Bonelli, P. Caraveo, L. Chiappetti, M. E. Quadrini, G. Di Cocco, M. Trifoglio, P. Ubertini, G. Peres, S. SCiortino, S. Serio, G. Vaiana, R. Rothenflug, L. Vigroux, L. Koch, Y. Rio, C. Pigot, J. Cretolle, A. Gabreil, B. Foing, J. L. Atteia, J. P. Roques, H. Bräuninger, W. Pietsch, P. Predehl, C. Reppin, L. Strüder, J. Trümper, G. Lutz, E. Kendyiorra, R. Staubert, A. D. Holland, R. E. Cole, A. Wells, K. Pounds, D. A. Lumb, J. Pye, M. J. L. Turner, C. V. Goodall, T. J. Ponman, G. K. Skinner, A. P. Willmore, *SPIE* **1990**, *1344*, 144.

[2] G. W. Fraser, A. F. Abbey, A. Holland, K. McCarthy, A. Owens, A. Wells, *NIMA* **1994**, *350*, 368.

[3] P. Lechner, L. Strüder, *NIMA* **1995**, *354*, 464.

[4] E. Gatti, P. Rehak, *NIMA* **1984**, *225*, 608.

[5] P. Lechner, S. Eckbauer, R. Hartmann, R. Richter, L. Strüder, S. Krisch, H. Soltau, D. Hauff, C. Fiorini, E. Gatti, A. Longoni, M. Sampietro, *NIMA* **1996**, 377, 346.

[6] J. Kemmer, G. Lutz, E. Belau, U. Prechtel, W. Welser, *NIMA* **1987**, *253*, 378.

[7] E. Pinotti, H. Bräuninger, N. Findeis, H. Gorke, D. Hauff, P. Holl, J. Kemmer, P. Lechner, G. Lutz, W. Kink, N. Meidinger, G. Metzner, P. Predehl, C. Reppin, L. Strüder, J. Trümper, C. v. Zanthier, *NIMA* **1993**, *326*, 85.

[8] J. Kemmer, G. Lutz, U. Prechtel, K. Schuster, M. Sterzik, L. Strüder, T. Ziemann, *NIMA* **1990**, *288*, 92.

[9] M. Sampietro, L. Fasoli, P. Rehak, L. Strüder, *IEEE Electron Device Letters* **1995**, *16(5)*, 142.

[10] P. Lechner, L. Andricek, N. Findeis, P. Klein, J. Kemmer, N. Meidinger, G. Lutz, K. Schuster, M. Sterzik, L. Strüder C. von Zanthier, R. Richter, *New DEPMOS Applications, NIMA* **1993**, *326(1,2)*, 284–290.

[11] L. Strüder, H. Bräuninger, M. Meier, P. Predehl, M. Sterzik, J. Trümper, P. Cattaneo, D. Hauff, G. Lutz, K. F. Schuster, A. Schwarz, E. Kendziorra, A. Staubert, E. Gatti, A. Longoni, M. Sampietro, V. Radeka, P. Rehak, S. Rescia, P. F. Manfredi, W. Buttler, P. Holl, J. Kemmer, U. Prechtel, T. Ziemann, *NIMA* **1990**, *288*, 227.

[12] R. Hartmann, D. Hauff, P. Lechner, R. Richter, L. Strüder, J. Kemmer, S. Krisch, F. Scholze, G. Ulm, *NIMA* **1996**, *377*, 191.

[13] H. Soltau, P. Holl, S. Krisch, C. v. Zanthier, D. Hauff, R. Richter, H. Bräuninger, R. Hartmann, G. Hartner, N. Krause, N. Meidinger, E. Pfeffermann, C. Reppin, G. Schwaab, L. Strüder, J. Trümper, E. Kendziorra, J. Krämer, *NIMA* **1996**, *377*, 340.

[14] N. Meidinger, L. Strüder, Heike Soltau, C. v. Zanthier, *IEEE - NS*, **1995**, *42(6)*, 2066.

[15] R. Hartmann, L. Strüder, J. Kemmer, P. Lechner, O. Fries, E. Lorenz, R. Mirzoyan, *NIMA*, **1997**, *387 (1,2)*, 241.

[16] L. Strüder, H. Bräuninger, M. Meier, P. Predehl, C. Reppin, M. Sterzik, J. Trümper, P. Cattaneo, D. Hauff, G. Lutz, K. Schuster, A. Schwarz, E. Kendziorra, R. Staubert, E. Gatti, A. Longoni, M. Sampietro, V. Radeka, P. Rehak, S. Rescia, P. F. Manfredi, W. Buttler, P. Holl, J. Kemmer, U. Prechtel, T. Ziemann, *NIMA* **1990**, *288*, 227.

[17] W. Chen, H. Kraner, Z. Li, P. Rehak, G. Bertuccio, E. Gatti, A. Longoni, M. Sampietro, P. Holl, J. Kemmer, U. Faschingbauer, F. Hess, C. Jacob, A. Wörner, J. P. Wurm, R. Baur, A. Drees, P. Fischer, P. Glässel, T. F. Günzel, D. Irmscher, L. Olsen, A. Pfeiffer, B. Schmidt, A. Schoen, H. J. Specht, Th. Ullrich, S. Tapprogge, N. Wermes, A. Breskin, R. Chechik, C. de los Heros, V. Steiner, I. Tserruya, *NIMA* **1993**, *326*, 273.

[18] J. Kemmer, G. Lutz, *NIMA* **1987**, 253, 365.

[19] R. Hartmann, D. Hauff, S. Krisch, P. Lechner, G. Lutz, R. H. Richter, H. Seitz, L. Fasoli, C. Fiorini, E. Gatti, A. Longoni, E. Pinotti, M. Sampietro, L. Strüder, G. Bertuccio, *IEDM Technical Digest*, 1994, pp. 535–539.

[20] M. Sampietro, A. Fazzi, P. Lechner, *Rev. Sci. Ins.* **1995**, *66(11)*, 5381.

[21] RÖNTEC GmbH, Berlin, Germany, *Product Information*, 1996.

[22] Ch. Gauthier, J. Goulon, E. Moguiline, A. Rogalev, P. Lechner, L. Strüder, C. Fiorini, A. Longoni, M. Sampietro, A. Walenta, H. Besch, H. Schenk, R. Pfitzner, U. Tafelmeier, K. Misiakos, S. Kavandias, D. Loukas, *A High Resolution, 6-Channel–Silicon Drift Detector Array with Integrated JFET's Designed for XAFS: First X-Ray Fluorescence Exitation Spectra Recorded at the ESRF, NIMA*, **1996**, *382*, 524–532.

Mikrochim. Acta [Suppl.] 15, 21–28 (1998)

Efficiency Calibration of a Si(Li) Detector by EPMA

Kurt Röhrbacher*, **Michael Andrae, Martin Völkerer,** and **Johann Wernisch**

Institute of Applied and Technical Physics, Technical University of Vienna, Wiedner Hauptstrasse 8-10, A-1040 Wien, Austria

Abstract. A $\varphi(\rho z)$-model, well known for its high accuracy, is employed for determining the detection efficiency of a Si(Li) detector by EPMA. Problems and advantages of this approach are pointed out and possible errors in the resulting thicknesses of the beryllium window, the silicon dead layer, the gold contact layer and the ice build-up on the detector surface are estimated. Furthermore the method allows the calculation of the cross section for L_3 subshell ionization to be made with high reliability.

Key words: detector efficiency, calibration, EPMA, standardless analysis.

Energy dispersive spectrometers offer the possibility of measuring a wide energy range from 1 to 40 keV simultaneously. On the other hand the analyst is confronted with the problem that the detector efficiency is not constant over the whole energy range. While analysis with standards compensates for this by the use of the so-called k-ratio, which is the ratio of sample to standard X-ray yields, an accurate knowledge of the instrumental function is essential for analysis without standards [1–5].

For this reason much effort has been expended on determining the detector efficiency to solve the problem of calibration in energy dispersive systems and many methods have already been presented:

- measurements of radioactive sources,
- methods involving secondary targets,
- comparison of calculated and measured background,
- proton or particle induced X-ray emission (PIXE),
- secondary fluorescence in the Au contact layer and the Si dead layer itself,
- Monte Carlo simulations,
- variation of the angle of incident radiation,
- electron induced K-X-rays in transmission electron microscopy TEM,
- synchrotron radiation.

A comprehensive compilation and a reference list of these methods can be found in [3, 6, 7, 8]. The absorption of the X-rays in the beryllium window, the gold contact layer, the silicon dead layer and in the contamination from ice build-up reduces the number of counts in comparison with the total amount of X-rays generated in the sample to

$$I = \varepsilon(E) \cdot I_0 = \frac{\Omega}{4\pi} \cdot \exp\left[-\left(\frac{\mu}{\rho}\right)_{\mathrm{Be}} \cdot \rho d_{\mathrm{Be}} - \left(\frac{\mu}{\rho}\right)_{\mathrm{ice}} \cdot \rho d_{\mathrm{ice}} - \left(\frac{\mu}{\rho}\right)_{\mathrm{Au}} \cdot \rho d_{\mathrm{Au}} - \left(\frac{\mu}{\rho}\right)_{\mathrm{Si}} \cdot \rho d_{\mathrm{Si}} \right] \cdot [1 - P_{\mathrm{esc}}(E)] \cdot I_0 \quad (1)$$

where Ω denotes the detector solid angle and the exponential functions allow for the attenuation of the emerging X-rays in Be, H_2O, Au and Si respectively [3, 9]. The second term describes the Si K X-ray escape probability according to [10]. The magnitude of an escape peak relative to the parent peak is for all energies less than 1.8% and of course zero for the

* To whom correspondence should be addressed

most interesting energy region below the excitation energy of the Si K shell [11, 12, 13, 14]. Because of its small size we did not include this effect in our investigations. Another term could be introduced into Eq. (1) to take into account of the fraction of the impinging high energy X-rays passing through the sensitive Si volume without being detected. Hence the detector efficiency decreases for energies typically higher than 15 KeV. This has not been implemented in our procedure, since investigations were made up to 10 keV only.

The measured K X-ray intensity from a pure element can be written as

$$I = n^{e^-} \cdot N_0/A \cdot \varepsilon(E) \cdot Q_k(E_0) \cdot \omega_k \cdot p_{k\alpha} \cdot (1+f_c) \cdot \int \varphi(\rho z) \cdot \exp(-\chi \cdot \rho z) \cdot d\rho z \quad (2)$$

where n^{e^-} represents the number of incident electrons, N_0 Avogadro's number, A the atomic mass, $Q_k(E_0)$ the ionization cross section of the K-shell by electrons with energy E_0, ω_K the fluorescence yield of the K-shell, $p_{K\alpha}$ the transition probability of the $K\alpha$ line relative to the sum of $K\alpha$ and $K\beta$ radiation from that element, f_c the fluorescence by the continuum, χ equal to $(\mu/\rho) \cdot \sin\psi^{-1}$, $\varphi(\rho z)$ the depth distribution function of the primary K-shell ionizations in the mass depth ρz and $\varepsilon(E)$ the unknown efficiency function [15]. The depth distribution function was formerly the least well known parameter in this equation, but there are nowadays many models available, which predict the emerging X-ray intensity with high accuracy [16–21]. One of these $\varphi(\rho z)$-models can thus be employed to determine the detector efficiency according to Eq. (1):

If eq. (1) for an element i is divided by the same equation for another element j, the prefactors are cancelled out, as occurs when calculating the k-ratio, and the ratio of two unknowns $\varepsilon(E_i)$ and $\varepsilon(E_j)$ remains. For example the ratio for Al $K\alpha$ to Zn $K\alpha$ intensities is given by

$$I_{\mathrm{Al}\,K\alpha}/I_{\mathrm{Zn}\,K\alpha} = \left[\varepsilon(\mathrm{Al}\,K\alpha) \cdot Q_K \cdot \omega_K \cdot p_{K\alpha}/A \cdot (1+f_c) \cdot \int \varphi(\rho z) \cdot \exp(-\chi \cdot \rho z) \cdot d\rho z\right]_{\mathrm{Al}} \Big/ \left[\varepsilon(\mathrm{Zn}\,K\alpha) \cdot Q_K \cdot \omega_K \cdot p_{K\alpha}/A \cdot (1+f_c) \cdot \int \varphi(\rho z) \cdot \exp(-\chi \cdot \rho z) \cdot d\rho z\right]_{\mathrm{Zn}} \quad (3)$$

If the detection efficiency at Zn $K\alpha = 8.639\,\mathrm{keV}$ is assumed to be 1.0, Eq. (3) can be rearranged and the function $\varepsilon(E)$ can be evaluated by

$$\varepsilon(\mathrm{Al}\,K\alpha) = \left[Q_K \cdot \omega_K \cdot p_{K\alpha}/A \cdot (1+f_c) \cdot \int \varphi(\rho z) \cdot \exp(-\chi \cdot \rho z) \cdot d\rho z\right]_{\mathrm{Zn}} \Big/ \left[Q_K \cdot \omega_K \cdot p_{K\alpha}/A \cdot (1+f_c) \cdot \int \varphi(\rho z) \cdot \exp(-\chi \cdot \rho z) \cdot d\rho z\right]_{\mathrm{Al}} \cdot I_{\mathrm{Al}\,K\alpha}/I_{\mathrm{Zn}}\,K\alpha \quad (4)$$

using Zn $K\alpha$ as the reference sample. In the case of an L line the net intensity

$$I_{L\alpha} = n^{e^-} \cdot N_0/A \cdot \varepsilon(E) \cdot Q_{L3}(E_0) \cdot (1+g_{ck}) \cdot \omega_{L3} \cdot p_{L\alpha/L3} \cdot (1+f_c) \cdot \int \varphi(\rho z) \cdot \exp(-\chi \cdot \rho z) \cdot d\rho z \quad (5)$$

has to be related to the reference intensity Zn $K\alpha$ in Eq. (3) to obtain the efficiency at the X-ray energy of the line concerned. Here ω_{L3} stands for the fluorescence yield and Q_{L3} for the ionization cross section of the L_3 subshell and

$$(1+g_{ck}) = 1 + 1/N_3 \cdot [N_1 \cdot (f_{12} \cdot f_{23} + f_{13}) + N_2 \cdot f_{23}] \quad (6)$$

is the enhancement factor induced by Coster Kronig transitions f_{ij} from subshell i to subshell j. The parameter $p_{L\alpha/L3}$ is the transition probability of the $L\alpha$ line relative to all radiative lines originating from the L_3 subshell. The primary vacancy distributions were set to $N_1 = N_2 = \frac{1}{2} \cdot N_3 = 0.25$ proportional to the number of electrons in the respective subshell.

Experimental

We used a JEOL JSM-T330A, equipped with a lithium drifted silicon detector from the company Link (now Oxford Instruments), model 5508. Its resolution is 140 eV at 5.9 keV. Neither the thickness of the Be window nor that of the gold contact layer have been specified by the manufacturer. All samples were analysed at 30 kV acceleration voltage and a probe current of $6 \cdot 10^{11}$ A with a live time of 400 seconds.

To eliminate any errors associated with elemental concentrations we used pure element standards for the determination of the thicknesses of ice, Be, Au and Si dead layer. We chose Al, Si, Sc, V, Cr, Mn, Fe, Ni, Cu, Zn, Ga and Ge for the K-lines and Zn, Ga, Ge, Y, Zr, Nb, Rh, Pd, Ag, In, Sb, Ta, Re, Pt, Au and Pb for the

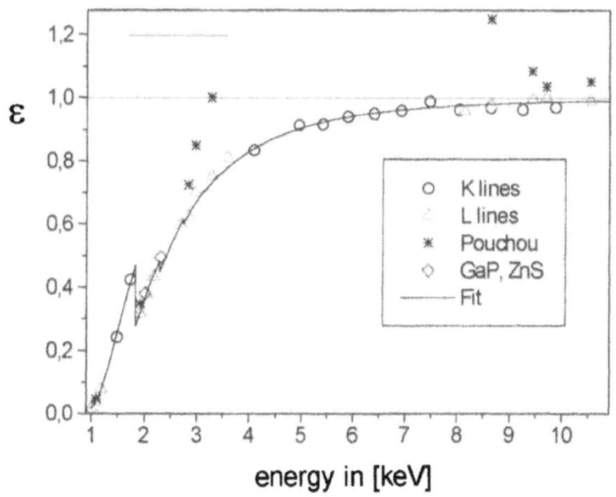

Fig. 1. The calculated efficiency data points according to Eq. (4): dots represents the K lines triangles the *L* lines, diamonds the efficiencies of *P* $K\alpha$ from GaP and S $K\alpha$ from ZnS and the resulting data fit is plotted as a solid line. Stars indicate the detector efficiencies, which were obtained by application of the absolute cross section of the L_3 subshell and its enhancement factor $(1 + gck)$ specified by Pouchou for several elements

L-lines to obtain efficiencies according to Eq. (4). in the region of interest from 1 to 10 keV. Additionally sulphur *K* intensity from ZnS and phosphorus *K* intensity from GaP were analysed to check the accuracy of the Si and Au layer thicknesses, which are affected by uncertainties in the fundamental *L* line parameters for the elements from Y to Mo. Q_K, ω_K and $(1 + f_c)$ were taken from Pouchou [15], ω_{Li} and f_{ij} for the calculation of relative $L\alpha$ intensities and the factor $(1 + g_{ck})$ from Chen et al. [22], $p_{K\alpha}$ and $p_{L\alpha/L3}$ from Scofield [23, 24] and mass absorption coefficients from Heinrich [25]. The value for $(1 + g_{ck})$ of indium was changed from 1.45 to 1.4, because otherwise the value from eq. (4) did not fit smoothly into the predicted efficiency shape. This indicates that the actual jump in the factor $(1 + g_{ck})$ between the atomic numbers $Z = 49$ and $Z = 50$ is probably not as abrupt as expected from theory.

For the depth distribution function we applied the model of Pouchou [21], because he has specified the only available *absolute* value for the ionization cross section Q_{L3} necessary for Eq. (4) [15]. Hence his modified values for the enhancement factor $(1 + g_{ck})$ were also adopted in our calculations. Using Pouchou's data we obtained detector efficiencies of more than 100% for indium and above, and application of the fitting procedure was impossible (see Fig. 1). Only reintroducing the theoretical values from Chen's Dirac-Hartree-Slater calculations [22] for the Coster Kronig transition rates led to a realistic dataset, precisely following the predicted shape of the efficiency function according to the standard efficiency model in Eq. (1). As a consequence the absolute cross section for L_3 subshell ionization of Pouchou had also to be modified from $q = 10.7\,\text{cm}^2\cdot\text{keV}^2$ to $q = 12.82\,\text{cm}^2\cdot\text{keV}^2$ (+3%) in

$$Q_{L3} = q \cdot 10^{-20} \cdot \ln(U)/U^{0.82}/E_{L3}^2 \tag{7}$$

The Fitting Procedure

The curve parameters d_{Si}, d_{Be}, d_{Au}, d_{ice} in Eq. (1) were varied until the minimum of

$$\chi^2 = \sum_i \left(\frac{\varepsilon(E) - \varepsilon_i}{\varepsilon_i}\right)^2 \tag{8}$$

was obtained. The weight function ε^{-2} in this sum guarantees that the resulting curve approximates the experimental data set especially in the low energy region, where the efficiency is very low and even small deviations can cause large errors when applied to standardless analysis. A multiplying factor for the detector function $\varepsilon(E)$ was introduced into Eq. (4) to evaluate the absolute reference efficiency at Zn $K\alpha$, which was found to fit best at 0.97. None of the problems with the simultaneous determination of the Si and Au thickness as described in [3, 4, 26] arose during our calculations, because each jump at Si K and Au M_5 could be identified and related clearly to the corresponding layer thickness. If the efficiency data points ε_b and ε_a directly before and after the respective edge energy from Eq. (4) are written as (Eq. 9a)

$$\varepsilon_b = k_b \cdot \exp\left(-\left(\frac{\mu}{\rho}\right)_b \cdot \rho d\right)$$
$$\varepsilon_a = k_a \cdot \exp\left(-\left(\frac{\mu}{\rho}\right)_a \cdot \rho d\right) \tag{9a}$$
$$d = \frac{\ln(\varepsilon_b) - \ln(\varepsilon_a)}{\mu_a - \mu_b} \tag{9b}$$

where absorption of the x-ray line in the other detector front layers has been considered to be constant in the small energy interval $(k_b = k_a)$, the corresponding layer thickness d can be directly derived (Eq. 9b).

However, if any unrealistic values are obtained during the fitting procedure, we recommend as a first step fitting the gold parameter d_{Au} to some data points above and below the critical energy of the M_5 edge and then, with the gold thickness fixed, fitting the silicon layer thickness d_{Si} according to the jump in the data points at the Si edge. As a second step the fitting procedure for all parameters can be applied to the whole data set using d_{Au} and d_{Si} as starting points.

Results

Problems and anticipated errors in the final thicknesses will now be discussed together with the estimated errors, which are also summarized in Table 1.

Table 1. The final thicknesses from the fitting procedure and estimated errors

	Si [nm]	Au [nm]	Be [μm]	Ice [μm]
Final thickness	723±10%	16.5±10%	11.4±30%	5.2±30%

Silicon Dead Layer

A decrease of the detector efficiency in the peripheral region of the detector has been reported in several publications [3, 27–29]. As a consequence the detector efficiency shows an increase in the energy region 6–20 keV, where a flat behavior is to be expected if a single uniform dead layer is assumed in the standard efficiency model [9]. Some authors tried to describe this effect by a radial dependent correction factor in the detector efficiency [27, 30]. Pajek et al. [3] have overcome this problem by introducing a peripheral 35 μm thick Si dead layer in addition to a uniform dead layer of 0.24 μm thickness: X-rays with a certain energy can pass the uniform dead layer when penetrating the detector area in the center, but get absorbed when they enter the detector aperture in the peripheral region (25% of the total detector area of typically 30 mm^2). With increasing energy more and more X-rays can pass even the surrounding peripheral dead zone, which means a higher detector efficiency, until the maximum of the efficiency is reached.

Hence the thickness of a Si dead layer, when determined in accordance to the standard efficiency model as in this work, never represents that of the real dead layer, but an average value for both the uniform and the peripheral zone [6, 31, 32]. Strictly speaking this region is not actually "dead", but is characterised by incomplete charge collection [33]. Nevertheless, we did not find any increase in our d_{Si} values with increasing energy above 6 keV anyway – probably because our investigations were carried out up to 10 keV X-ray energy only. This allowed any such effect to be neglected without a significant loss in accuracy.

The fitted efficiency curve approximates the measured data points excellently in the region of the Si K edge energy. But as a change from Al and Si K lines to Y, Zr and Nb L lines (see Fig. 1) occurs here the height of the jump and therefore the thickness of the silicon dead layer strongly depends on the accuracy of the corresponding fundamental parameters (ionization cross sections and Coster-Kronig values, fluorescence yields and relative line weights of Si $K\alpha$ and Y $L\alpha$ lines). On the other hand the additional evaluation of the phosphorus $K\alpha$ intensity from GaP, also lying in this energy region (Fig. 2), confirms the value of 723 nm for d_{Si}. Its relative error is estimated to be less than 10%, corresponding to uncertainties of K and L shell fluorescence yields [35]. The determination of the detector efficiency carried out by our group in 1990 by variation of the detection geometry [8] led to an average silicon dead layer thickness of 630 nm (with the assumption of a 8 μm Be window and a 15 nm gold layer at that time). This agrees well with the new layer thickness if the enlargement of the Si dead layer with time [3] is taken into account.

As can be seen in Fig. 1 the efficiency obtained by evaluation of Eq. (4) for Zn $K\alpha$ from ZnS is higher than the value predicted from the fit by 9%. The missing Au M_5 jump between P $K\alpha$ and S $K\alpha$ cannot be accounted for by the $\varphi(\rho z)$-model, because atomic number and absorption effects should be very similar for these two compounds and thus the net intensities predicted by the model can be regarded as being reliable. We are convinced that this deviation comes from the fact that a data fit for the fluorescence yield after K shell ionization has been employed in the fitting algorithm. As has been pointed out in a statistical consideration of experimental ω_K values by Langenberg and van Eck [34], these data deviate from the usually applied Wentzel parametrization depending on the electron configuration and conse-

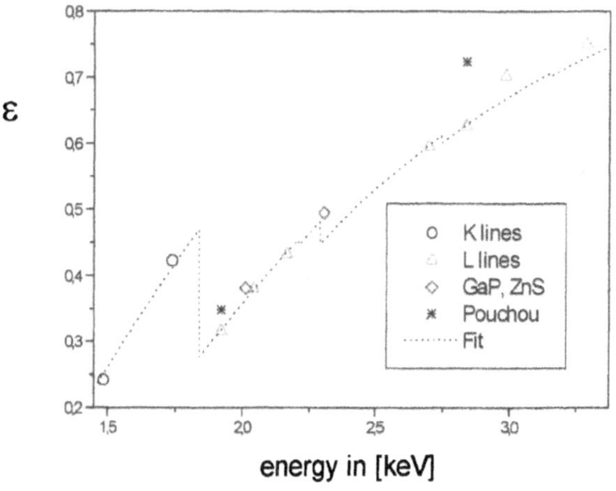

Fig. 2. As in Fig. 1, but for energies from 1.5 to 3.4 KeV. The jump in the efficiency data according to Eq. (4) at the Si K and Au M_5 edge energies can be seen clearly. The corresponding dead silicon and gold contact layer thicknesses were found to be 723 nm and 16.5 nm respectively and are implemented in the resulting fit curve

quently a data fit should not be used at all. In fact the experimental fluorescence yield ω_K of S cited by these authors is higher, by 10%, than the data here which would bring the respective efficiency point in excellent agreement with the fit.

Gold Contact Layer

Typical values for this layer to be found in the literature [1, 3, 4, 9, 11, 31, 36, 37, 38] range from 15 to 25 nm, or 0.29 to $0.48 \cdot 10^{-4}$ g/cm^2. In comparison with the ice contamination (up to $8 \cdot 10^{-4}$ g/cm^2) and the beryllium window (15 to $50 \cdot 10^{-4}$ g/cm^2) this gold mass thickness is rather small and several problems in determining its value accurately have been reported because it has little influence on the detection function [4, 26, 31, 36]. Although not high, the jump of the efficiency curve at the Au M_5 edge energy can be seen clearly in Fig. 2 and the corresponding gold thickness was determined to be 16.5 nm. Replacing the Coster-Kronig rates of Chen et al. [22] by those of McGuire [39] (which is also given in Bambynek's table [40]) and the mass attenuation coefficients of Heinrich [25] by those of Schmetterer [41] (which is a fitting procedure, developed at our institute, using Henke's [42] table for energies below 1 keV and McMaster's table [43] for energies above 1 keV) resulted in an error level of 10% relative, depending on the fundamental parameter set used.

Contamination on the Detector Surface

Oil from the diffusion pump as well as ice from condensation of residual vapours on the cooled silicon crystal can build contamination layers, which play an important role in the sensitivity of the detector especially for low energy X-rays. As has been shown in previous investigations [3, 44], the ice build-up increases tremendously within just a few months as the vaccum gets worse. While the ice layer grows up to 8 µm within one year, reducing the detector's sensitivity for Al $K\alpha$ X-rays by 50% relative, it can attain almost 40 µm after 3 years leading to a complete loss of any sensitivity for Al $K\alpha$. The consequences for standardless analysis as well as for long term experiments are obvious and make a frequent check of the detector sensitivity at low energies absolutely necessary. When a loss in sensitivity is observed the Si detector crystal has to be warmed up, repumped and the ice removed before it can be used again [44]. As proposed by Steele et al. [45] and Smith [46] a recontamination of the detector surface can be greatly inhibited by placing a miniature lamp with a total output of ~ 3W close to the detector window.

It has been shown by Lábár [2] that C and O yield similar absorption for all X-ray energies, which has allowed the omission of one of the two parameters (d_C and d_{ice}) in the fitting procedure. Since the main part of these contamination layers has been found to be H_2O [47], we included the attenuation of impinging photons in these contamination layers by a single, equivalent ice build-up layer in Eq. (1). The mass absorption coefficient of X-rays of any energy in hydrogen is almost zero, so we adopted

$$\left(\frac{\mu}{\rho}\right)_{\text{ice}} = \sum_{i=\text{H,O}} c_i \cdot \left(\frac{\mu}{\rho}\right)_i \approx 0.8881 \cdot \left(\frac{\mu}{\rho}\right)_O \qquad (10)$$

in our fitting procedure. The value for the ice layer of 5.2 µm is in very good agreement with the time dependence of the ice build-up given by the above mentioned authors, because our detector had been cleaned just a few months before the beginning of our investigation.

Nevertheless errors are expected to be high, since our Be window precludes the measurement of O K-intensities with an energy just below the critical oxygen K energy. Hence there is no jump in the efficiency data to define the thickness of the ice layer accurately (Eqs. 9a and 9b) unlike the case for the Si K and the Au M_5 edge energy (Fig. 2).

This problem does not occur in more modern "windowless" detectors, in which the Be window is substituted by materials of considerably reduced mass thickness which still withstand the full-atmosphere pressure differential. The measurement of ultra light elements like C or O is possible with those types of detectors, providing the researcher with the jump in the efficiency data at the oxygen edge energy required for accurate determination of the ice build-up. But other problems arise as these newer detectors are constructed of thin films of boron nitride, silicon nitride, diamond or a polymer deposited on a support grid of pure silicon or nickel having approximately 85% open space. The effective magnitude of this open space can only be roughly estimated and the measured net count rate thus lacks the accuracy required in Eq. (4). Further information about these new types of detectors can be found in Ref. [48].

Beryllium Window

This above mentioned uncertainty in ice thickness due to the absence of a jump in the efficiency data also applies to the thickness of the Be window itself. Both, the ice build-up and the Be window are determined during the fitting procedure by their attenuation influence in Eq. (4) only. Thus they depend on the accuracy of the other unknowns d_{Au} and d_{Si} in this equation and show a strong correlation between them. When we fixed the Be window thickness at 8 μm (−42%) during the fitting procedure this yielded small modifications for the Si dead layer (−2.5%) and for the Au conductive layer (−2.8%), but a large influence on the ice build-up (+17%) because of the strong correlation (all deviations are relative to the final thickness). In addition one must consider errors introduced by uncertain Coster Kronig transitions rates, different tables of mass absorption coefficients and – in the case of low energy X-rays – differing fluorescence yields for the L_3 subshell. Thus the values for the ice layer and the Be window thicknesses may still have errors of up to 30% when compared with the real thicknesses in the detector. Nevertheless standardless analysis is still possible with a high confidence level as will be explained in detail in the 'discussion' section.

The lowest deviations for the fitting function were obtained at 11.4 μm beryllium and 5.2 μm ice. This is in good agreement with the value of 8 μm for the beryllium window specified by many manufactures and also found in the literature [1, 2, 11, 44, 46, 49].

Discussion

An EPMA correction model utilizing Eq. (4) for the determination of the detector efficiency is a relative one because the measured Zn $K\alpha$ is used as a reference to cancel out the constant prefactors in Eq. (2) and (5). In order to evaluate the absolute height of the efficiency function $\varepsilon(E)$ at Zn $K\alpha$ we introduced a multiplying factor in Eq. (4), which was determined to be 0.97 during the fitting procedure. This is more realistic than the value of almost 1.0 at Zn $K\alpha$ predicted by the standard efficiency model [9], because each detector is at the most 99% efficient due to incomplete charge collection and a background shelf [11].

The new approach in Eq. (4) provides a set of thickness for all relevant layers in front of the active silicon zone in an energy dispersive detector. While d_{Si} and d_{Au} can be clearly identified by the jump in the efficiency data at the respective edge energy (Eq. 9a and 9b), d_{Be} and d_{ice} show a strong correlation between them due to the missing jumps. Errors for these two thicknesses can be up to 30%. Nevertheless, standardless analysis can be carried out with reliable precision.

With standardless analysis the experimental net intensity from the sample investigated must be related to the theoretical intensity in Eq. (2) for K lines and in Eq. (5) for L lines. If the detector efficiency in this theoretical intensity has been determined previously by the new approach, the whole data base of Coster Kronig rates, fluorescence yields, etc. is automatically included in the resulting $\varepsilon(E)$-function. Thus by application of

$$\varepsilon(\text{Sample } L\alpha) = 0.97 \cdot \left[Q_K \cdot \omega_K \cdot p_{K\alpha}/A \cdot (1+f_c) \cdot \int \varphi(\rho z) \cdot \exp(-\chi \cdot \rho z) \cdot d\rho z \right]_{\text{Ref } K\alpha} \Big/ \left[Q_{L3} \cdot (1+g_{ck}) \cdot \omega_{L3} \cdot p_{L\alpha/L3}/A \cdot (1+f_c) \cdot \int \varphi(\rho z) \cdot \exp(-\chi \cdot \rho z) \cdot d\rho z \right]_{\text{Sample } L\alpha} \cdot \left(\frac{I_{\text{Sample } L\alpha}}{I_{\text{Ref } K\alpha}} \right)_{\text{exp}} \tag{11}$$

for any L line in Eq. (5) all fundamental parameters of this L line cancel out and the theoretical net $L\alpha$ intensity is proportional to the fundamental parameters of the reference $K\alpha$ line. For this reason any errors from uncertainties in these parameters also cancel out and the accuracy of the analytical results is higher than that of the determined layer thicknesses themselves. This is a big advantage of the new approach, but one has also to be aware, that any change from one fundamental data set to another, e.g. from Bambynek's [40] to Krause's data [50], or the substitution of a table of mass absorption coefficients by any other [e.g. 25, 41, 51] can cause severe errors in the analytical results, especially in the low energy region, where uncertainties in the parameters are greatest and values often differ by more than 15 per cent from each other [35].

The method presented above is an easy way to determine the detector efficiency for people, who carry out their analyses by scanning electron microscopy. No special equipment except a SEM and no special know-how is necessay. Fluorescent samples as employed in some methods are expensive and anti-radiation precautions have to be taken. Furthermore there are only few radioactive sources with calibrated X-ray lines below 5 keV and self-absorption can be very high [3, 4, 52, 53]. A comparison of measured background spectra with calculated ones seems to be effective, but much longer counting times than in the case of characteristic radiation are necessary to achieve reliable counting statistics [54]. For characteristic radiation much higher count rates than for continuous radiation are obtained and thus much smaller statistical fluctuations can be expected. Net intensities can be calculated easily with high accuracy by several published correction models for the depth distribution function.

Conclusion

By application of the method presented, errors of approximately 10 per cent can be expected for the thicknesses of the dead silicon layer and the gold contact layer as a result of uncertainties in fluorescence yields and Coster Kronig transition rates. With a beryllium window neither Be K nor O K lines can be measured. This means that a jump in the efficiency data at the Be K and O K edge energy, by which thicknesses of the window and ice layer could be identified accurately according to Eq. (9b), cannot be measured and a strong correlation between those two layer thicknesses is the consequence. Hence, errors for these two thicknesses can be up to 30% relative. Nevertheless, as shown in the text, standardless analysis is still possible with higher reliability than for the layer thicknesses themselves. During our investigations a new value for the absolute ionization cross section of the L_3 subshell was obtained, which is expected to be highly reliable.

Acknowledgement. We acknowledge the support given by the Austrian „Fonds zur Förderung der wissenschaftlichen Forschung" (Project N° P10415-PHY).

References

[1] G. Hubricht, B. Knaf, G. Presser, J. Stähler, *Nucl. Instr. Meth.* **1977**, *144*, 359.
[2] J. Lábár, *Microbeam Analysis, Vol. 4*, 1995, pp. 65–83.
[3] M. Pajek, A. P. Kobzev, R. Sandrik, R. A. Ilkhamov, S. H. Khusmurodov, *Nucl. Instr. Meth. Phys. Res.* **1989**, *B42*, 346.
[4] J. Palinkas, B. Schlenk, *Nucl. Instr. Meth.* **1980**, *169*, 493.
[5] B. Rosner, D. Gur, L. Shabason, *Nucl. Instr. Meth.* **1975**, *131*, 81.
[6] J. L. Campbell, P. L. McGhee, *Nucl. Instr. Meth.* **1986**, *A 248*, 393.
[7] J. L. Campbell, L. A. McNelles, *Nucl. Instr. Meth.* **1975**, *125*, 205.
[8] J. Wernisch, A Schönthaler, H.-J. August, *Phys. Stat. Sol.* **1990**, *A 122*, 695.
[9] N. J. Zaluzek, in: *Introduction to Analytical Electron Microscopy* (J. J. Hren, J. I. Goldstein, D. C. Joy, eds.), Plenum P, New York, 1979, p. 121.
[10] S. B. J. Reed, N. G. Ware, *J. Phys. E5*, **1972**, 582.
[11] J. I. Goldstein, D. E. Newbury, P. Echlin, D. C. Joy, A. D. Romig, C. E. Lyman, C. Fiori, E. *Lifshin, 2nd Ed.*, Plenum, New York 1992, p. 310–319.
[12] U. Sennhauser, *Thesis*, Institut für Kernphysik, ETH Zürich, 1975.
[13] P. Sioshansi, A. S. Lodhi, *X-Ray Spectrom., Vol. 8, No. 2*, 1979 pp. 65–67.
[14] R. Woldseth, *X-Ray Energy Spectrometry*, Kevex Corp., Foster City, CA, 1973.
[15] J.-L. Pouchou, *Microchim. Acta,* **1994**, *114/115*, 33.
[16] H.-J. August, J. Wernisch, *X-ray Spectrom.*, **1991**, *20 (I, II)*, 131.
[17] G. F. Bastin, H. J. M. Heijligers, F. J. J. van Loo, *Scanning* **1986**, *8*, 45.
[18] D. A. Sewell, G. Love, V. D. Scott, J. Phys. **1985**, *D18*, 1245.
[19] C. Merlet, *Mikrochim. Acta* **1992**, *[Suppl.] 12*, 107.
[20] R. H. Packwood, J. D. Brown, *X-Ray Spectrom* **1981**, *10 (3)*, 138.
[21] J.-L. Pouchou, F. Pichoir, in: *Electron Probe Quantitation* (K. F. J. Heinrich, D. E. Newbury, eds.), Plenum, New York, 1991, pp. 31–75.
[22] M. H. Chen, B. Crasemann, H. Mark, *Phys. Review A* **1981**, 24, 177.
[23] J. H. Scofield, *Phys. Rev. A* **1974**, 9 (3), 1041.
[24] J. H. Scofield, *Phys. Rev. A* **1974**, 10 (5), 1507.
[25] K. F. J. Heinrich, *11th International Congress on X-Ray Optics and Microanalysis IXCOM 11*, August 1986, London, Canada, 1986, p. 67
[26] W. N. Lennard, D. Phillips, *Nucl. Instr. Meth.* **1979**, *166*, 521.
[27] D. D. Cohen, *Nucl. Instr. Meth.* **1980**, *178*, 481.
[28] J. L. Campbell, R. G. Leigh, W. J. Teesdale, *Nucl. Instr. Meth.* **1984**, *B 5*, 39.
[29] K. M. Barfoot, I. V. Mitchell, L. Avaldi, H. L. Eschbach, W. B. Gilboy, *Nucl. Instr. Meth.* **1984**, *B 5*, 534.
[30] M. L. Garg, J. Singh, H. R. Verma, P. N. Trehan, *X-Ray Spectrom* **1987**, *16*, 3.
[31] C. A. Baker, C. J. Balty, S. Sakamoto, *Nucl. Instr. Meth.* **1987**, *A 259*, 501.
[32] M. Geretschläger, *Nucl. Instr. Meth.* **1987**, *B 28*, 289.
[33] W. A. P. Nicholson P. F. Adam, A. J. Craven, J. D. Steele, *Analytical Electron Microscopy* (D. B. Williams, J. I. Goldstein, eds.), San Francisco Press, San Francisco, 1985, pp. 257–262.
[34] A. Langenberg, J. van Eck, J. Phys, B, *Atom. Molec. Phys.* **1979**, *12 (8)*, 1331.
[35] J. Wernisch, K. Röhrbacher, *Mikrochim. Acta [Suppl.]* **1998**, *15*, 307.
[36] C. R. Cothern, H. L. Manuel, R. J. Millette, *X-Ray Spectrom.* **1974**, *3*, 53.
[37] W. Maenhaut, H. Raemdonck, *Nucl. Instr. Meth.* **1984**, *B1*, 123.
[38] I. Orlic, W. J. M. Lenglet, R. D. Vis, *Nucl. Instr. Meth.* **1989**, *A 276*, 202.

[39] E. J. McGuire, *Phys. Rev. A*, **1971**, *3 (2)*, 587.
[40] W. Bambynek, B. Crasemann, R. W. Fink, H.-U. Freund, H. Mark, C. D. Swift, R. E. Price, P. Venugopala Rao, *Rev. of Mod. Phys.* **1972**, *44 (4)*, 793.
[41] L. Schmetterer, *Thesis*, Technical University Vienna, 1987.
[42] B. L. Henke, E. S. Ebisu, *Advances in X-ray Analysis*, *Vol. 17*, 1974 pp. 150–213.
[43] W. H. McMaster, N. K. Del Grande, J. H. Mallet, J. H. Hubbell, *Rep. UCRL-50174*, 1969 Sec. II.
[44] D. D. Cohen, *Nucl. Instr. Meth.* **1982**, *193*, 15.
[45] I. M. Steele, J. V. Smith, J. J. Pluth, T. N. Solberg, *Proceedings of the 10th Annual Conference of the Microbeam Analysis Society, Las Vegas, Nevada*, 1975, 37A–37D.
[46] D. G. W. Smith, *X-Ray Spectrom* **1981**, *10*, 78.
[47] R. G. Musket, *Nucl. Instr. Meth.* **1986**, *B 15*, 735.
[48] P. J. Statham, *Mikrochim. Acta [Suppl.]* **1998**, *15*, 1.
[49] J. Wernisch, A. Schönthaler, H.-J. August, *Microchim Acta* **1992**, *[Suppl.] 12*, 147.
[50] M. O. Krause, *J. Phys. Chem. Ref. Data*, **1979**, *8 (2)*, 307.
[51] W. M. Veigele, *At. Data Tables* **1973**, *5*, 51.
[52] S. J. Cipolla, M. J. Hewitt, *Nucl. Instr. Meth.* **1976**, *135*, 347.
[53] J. S. Hansen, J. C. McGeorge, D. Nix, W. D. Schmidt-Ott. I. Unis, R. W. Fink, *Nucl. Instr. Meth.* **1973**, *106*, 356.
[54] M. Procop, *Beitr. Elektronenmikroskop. Direktabb Analyse v Oberfl*, **1996**, *29*, 47.

Mikrochim. Acta [Suppl.] 15, 29–36 (1998)

Wavelength-Dispersive X-Ray Spectrometry

Stephen J. B. Reed

Department of Earth Sciences, University of Cambridge, Downing St., Cambridge, CB2 3EQ, England, U.K.

Abstract. The mode of operation of wavelength-dispersive (WD) spectrometers as used in EPMA is described. Factors governing the resolution and intensity obtained with different crystals are discussed, with a view to assessing the scope for improvement and helping the user to make suitable choices of operating conditions. High-order reflections and continuum background are also considered. Trace element detection limits as determined by peak intensity and peak-to-background ratio have been estimated. Computer modelling as an aid to selecting optimum conditions for WD analysis is discussed, with special reference to the 'Virtual WDS' program.

Key words: wavelength-dispersive spectrometry, Bragg reflection, trace elements, detection limits, virtual WDS.

The WD spectrometer acts as a monochromator which selects the wavelength (λ) satisfying Bragg's law: $n\lambda = 2d \sin\theta$, where n is the order of reflection, d the spacing of the reflecting planes in the crystal, and θ the glancing angle of incidence (Bragg angle). Crystals most commonly used are LiF (200 reflection), for which $d = 0.201$ nm, PET ($d = 0.437$ nm) and TAP ($d = 1.30$ nm). These have overlapping wavelength ranges covering 0.1 to 2.4 nm approximately, encompassing the K lines of elements of atomic number (Z) 9 to 35, L lines for Z up to 83, and all M lines. For the long wavelengths of the K lines of 'light' elements ($Z < 10$), synthetic layered structures such as lead stearate ($d = 5$ nm) and relatives with even larger d values can be used, but have been superseded by evaporated multilayers consisting of alternating layers of heavy and light atoms (e.g. W and Si) with various spacings [1].

For a perfect plane crystal, Bragg reflection occurs over a narrow angular range ($\sim 10^{-4}$ radians), as governed by the phase relationship between rays reflected from successive atomic layers, integrated over the depth penetrated by the incident X-rays. The crystal is curved so that reflection occurs over a relatively large area, giving a reasonable intensity from a point source. This condition is maintained for different Bragg angles when source, crystal, and detector all lie on a 'Rowland circle' of radius R, and the crystal planes are curved to radius 2 R. Usually the crystal moves along a linear track while its angle and the location of the detector (a proportional counter) are controlled by a suitable mechanism. The crystal-source distance is proportional to $\sin\theta$ and hence λ (Fig. 1).

The incidence angle varies somewhat over the face of the crystal for geometrical reasons. Variation caused by the displacement of the crystal surface from the Rowland circle in 'Johann' geometry can be corrected by grinding the surface to radius R ('Johansson' geometry). However, another form of 'aberration' is still present, namely the change in θ for rays inclined to the plane of the Rowland circle. This can be avoided in principle by double curvature of the crystal [2], but this is not yet available in practical form.

The width of the intensity profile obtained by scanning across a peak is determined by the natural line width (negligible except in the long-wavelength region), the width of the crystal reflection curve, and the range of angles for which Bragg reflection occurs somewhere on the crystal, of which this last factor is usually dominant. The resolution as determined by such purely geometrical considerations is a function

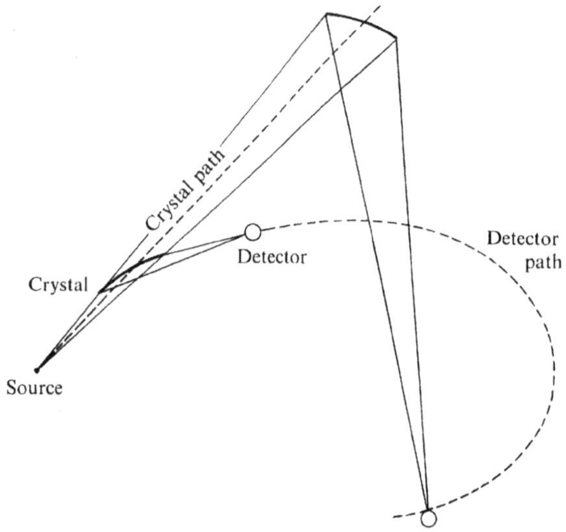

Fig. 1. Mode of operation of fully-focusing WD spectrometer. The crystal moves along a linear track while its angle changes so that $\sin\theta \propto$ source-crystal distance, where θ is the glancing angle of incidence (Bragg angle). This distance is thus proportional to the X-ray wavelength according to Bragg's law. The mechanism maintains equal angles of incidence and reflection, with detector-crystal and source-crystal distances equal, so that source, crystal and detector lie on a 'Rowland circle' of fixed radius

of the size of the crystal and its distance from the source: the larger and closer the crystal, the worse the resolution. The recorded intensity is a function of the solid angle subtended by the reflecting area of the crystal, which is greatest at low Bragg angles, when the crystal is closest, and is thus inversely related to resolution. It also depends on the efficiency of the detector.

Currently available electron microprobes are equipped with up to 5 WD spectrometers, each with 2 or more interchangeable crystals. For detecting short-wavelength X-rays a sealed proportional counter fitted with a beryllium entrance window and containing xenon (which absorbs X-rays strongly) is suitable. For long wavelengths a 'flow counter' with a thin window is used instead, argon being trickled continuously through the counter to compensate for leakage. (In some instruments, argon at high pressure, e.g. $\sim$ 3 bars, is used instead of xenon, though this absorbs short wavelengths less effectively.) Usually each spectrometer is fitted with one counter type and the crystals are matched to the appropriate wavelength range. (In the case of a single 'universal' spectrometer, as sometimes fitted to scanning electron microscopes, high detection efficiency for all wavelengths is obtained by using a flow counter with a rear exit window, mounted in front of a sealed counter.

The counter output pulse is proportional in height to the energy of the detected X-ray photon. With pulse-height analysis, X-ray photons falling within a prescribed energy 'window' are counted, whereas those lying outside are excluded.

Spectrometer functions which are readily susceptible to computer control include crystal-changing, moving to specified angles for peak and background measurements, and counting output pulses. Also, automatic routines for more complex operations such as peak-seeking may be utilised. Further, the computer can carry out the complete sequence of operations required for quantitative analysis and then calculate the results. Setting up such a procedure is facilitated by using stored 'default' parameters, but these are not necessarily ideal in a particular application. For example, the choice of crystal is influenced by whether or not high intensity is more important than high resolution. Also, default background positions may be unsuitable sometimes owing to the presence of interfering peaks. Obtaining reliable results, especially for complex samples, thus requires considerable expertise; embodying this in software is a possibility as yet largely unrealised. In the following sections various aspects of WD spectrometers relevant to choices available when devising analysis routines (whether by human operator or computer) are considered. The potential of computer modelling as an aid to optimisation is also discussed.

Resolution and Intensity

An understanding of the factors determining resolution and intensity is important for correct utilisation of WD spectrometers. From first-order geometrical considerations it can be shown [3] that the effective incidence angle at the edge of a crystal of width W (perpendicular to the Rowland circle) is equal to $\theta - \Delta\theta$, where $\Delta\theta = W^2/16R^2 \sin 2\theta$. For Johann geometry an additional angular 'error' $\Delta\theta'$ (of opposite sign) occurs owing to the departure of the surface from the Rowland circle, and at the ends of a crystal of length L, $\Delta\theta' = L^2/32R^2 \tan\theta$. The full width at half maximum (FWHM) of the profile of an X-ray line is approximately $(\Delta\theta + \Delta\theta')/2$, which typically ranges from 0.7 to 3 mrad for high and low Bragg angles respectively. (Note that $\Delta\theta$ and $\Delta\theta'$ are functions of W/R and L/R respectively and therefore

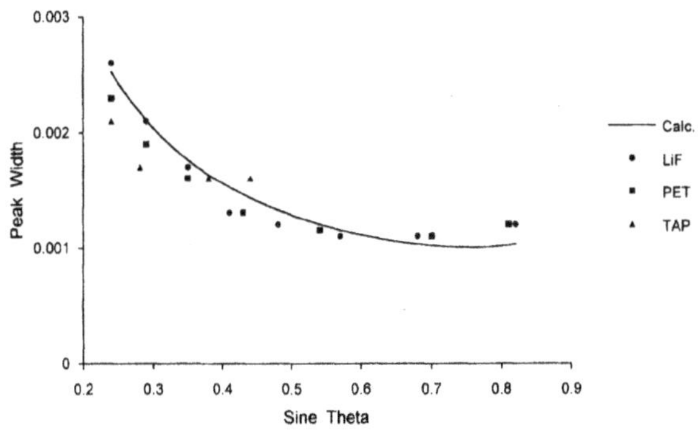

Fig. 2. Experimental FWHM peak widths obtained with a typical WD spectrometer (Johann geometry), showing reasonably close agreement with geometrical calculations (see text). (Note: data not given for wavelengths >1 nm, owing to significant natural line widths)

do not change when both Rowland circle radius and crystal dimensions are scaled by the same factor.) The second term does not apply to 'Johansson' geometry, in which the crystal surface lies on the Rowland circle, giving this configuration a theoretical advantage (especially at low θ). Experimental peak widths plotted in Fig. 2 agree quite well with the above expressions.

The efficiency of a WD spectrometer depends on: (1) the solid angle subtended by the crystal, (2) the fraction of the X-rays within this solid angle that are reflected, and (3) the efficiency of the detector. The solid angle is given to a sufficiently close approximation by $A/R^2 \sin\theta$, where A is the area of the crystal (typically $A \approx 400\,\text{mm}^2$ and $R \approx 150\,\text{mm}$, and the solid angle ranges from 20 to 80 msterad approximately). Crystal reflection efficiency is difficult to calculate from first principles but is found experimentally to be of the order of 10% (see below). The detection efficiency of a proportional counter is calculable as the product of the window absorption factor and the fraction of the incident X-rays absorbed within the gas. The efficiency of an argon-filled counter changes abruptly at the Ar K absorption edge. In the case of a xenon-filled counter there is a smaller step at the wavelength of the Xe L_3 edge, but the L_1 and L_2 edges are not significant.

Crystal Reflection Efficiency

As discussed above, the X-ray incidence angle varies over the face of the crystal. In the case of Johann geometry, the angular 'errors' cancel out along lines forming a cross. The angle between the arms varies with θ and reflection takes place over a zone representing a range of angles equal to the width of the reflection curve of the crystal (Fig. 3). The reflecting area in the case of Johansson geometry forms a strip parallel to the plane of the Rowland circle.

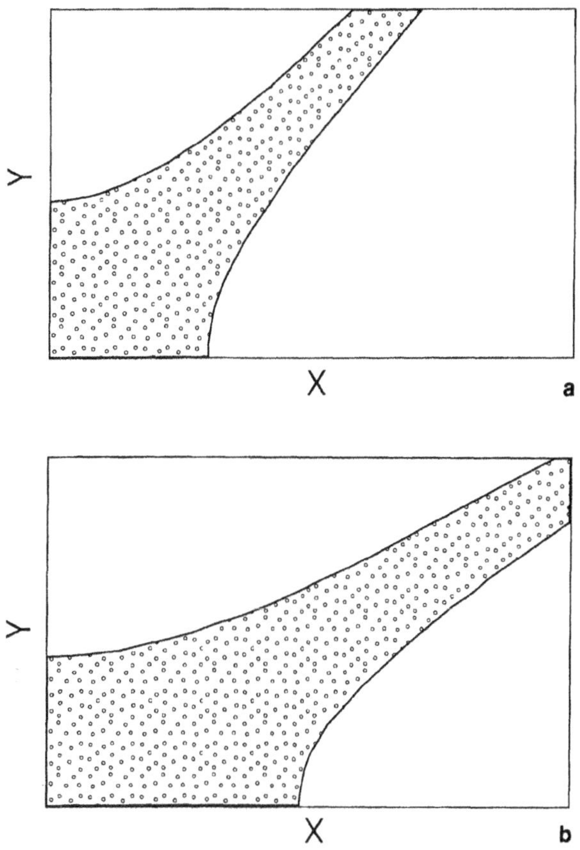

Fig. 3. Reflecting area of one quarter of Johann-type spectrometer crystal (based on idealised rectangular reflection curve); x axis = centre-line of crystal in plane of Rowland circle, y axis = centre-line perpendicular to Rowland circle; **a** $\theta = 15°$, **b** $\theta = 55°$

The reflected intensity is dependent on the product of the reflectivity and the width of the reflection curve, which determines the reflecting area and can be expressed as the 'integrated reflection', defined as the angular width of an idealised rectangle within which the reflectivity is 100%. The integrated reflection is enhanced if the crystal has 'mosaic' structure consisting of small slightly misoriented domains, since X-rays penetrate the crystal until they encounter a domain at the appropriate angle, so that reflection occurs over a wider range of θ than for a perfect single crystal.

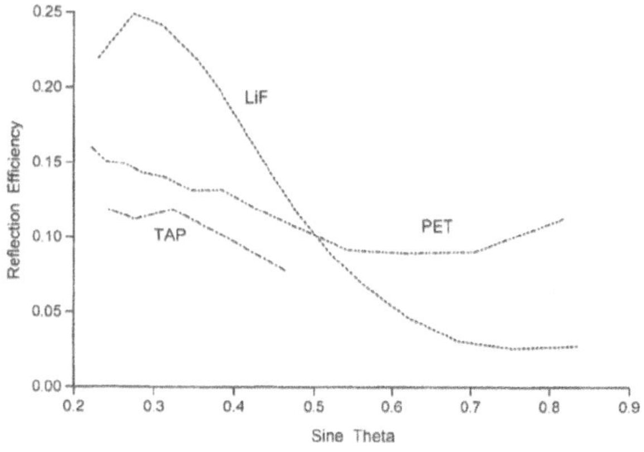

Fig. 4. Crystal reflection efficiency derived from comparison of WDS and EDS intensities

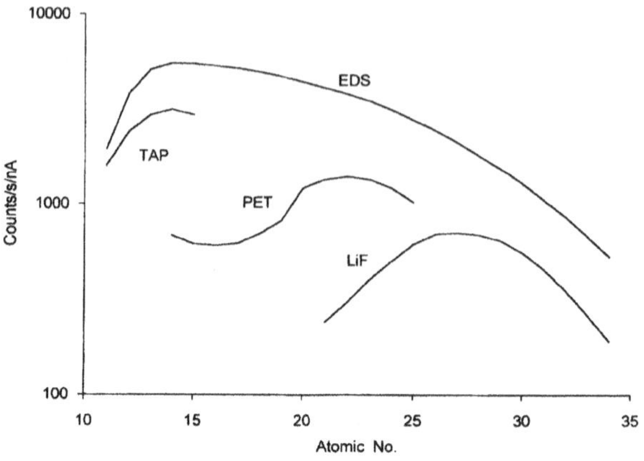

Fig. 5. Experimental $K\alpha$ line intensities as recorded by WDS with different crystals and by EDS with a detector of $10\,\mathrm{mm}^2$ area at 50 mm distance from the source (accelerating voltage −20 kV

Published crystal reflection data refer mainly to flat crystals and are thus not necessarily applicable to the present case. Experimental data recorded by an electron microprobe can, however, be used to estimate crystal reflectivity by comparing peak intensities measured using ED and WD spectrometers (see Fig. 4). A noteworthy feature is the marked change in the efficiency of LiF with $\sin\theta$ (and hence wavelength), compared to PET, which shows relatively little variation.

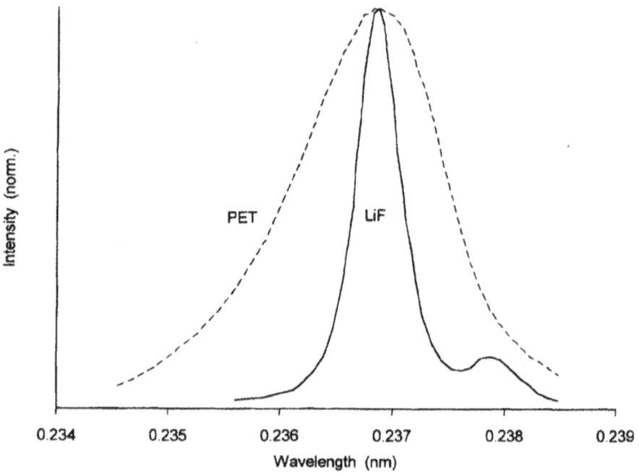

Fig. 6. Nd$L\alpha$ peak as recorded with LiF and PET crystals: the former gives better resolution but lower intensity than the latter (intensity scales are normalised in this plot)

Practical Spectrometer Performance

Of primary practical importance are the count-rates of peaks used for analysis, these being the product of generated intensity and spectrometer efficiency (Fig. 5). For a given accelerating voltage and beam current the count-rate varies strongly over the wavelength range of each crystal, the decrease in spectrometer efficiency with increasing Bragg angle being only partially offset by the opposite trend in the generated intensity (related to decreasing excitation energy).

In regions of overlapping wavelength coverage, a crystal with larger '*d*' gives more intensity than one with smaller '*d*' (for a given wavelength), the Bragg angle being lower. However, the resolution is relatively poor because not only is the angular peak width at a low Bragg angle greater but also a given angular interval represents a greater wavelength range, owing to the larger value of '*d*' in the Bragg equation. The peak width for PET is thus ~ 3.5 times greater than for LiF at the same wavelength (see Fig. 6), and the lower peak-to-background ratio which results is important in trace element analysis (see below). Also, resolution is a significant consideration when interfering lines are present.

Crystal Size

The reflected intensity can be increased by enlarging the crystal (with constant Rowland circle radius). In the case of Johann geometry the intensity is enhanced by extending the arms of the reflecting cross, which requires an increase mainly in crystal width at low Bragg angles and length at high angles (see Fig. 3), whereas for Johansson geometry length is of greatest importance at all angles. The usable width may be constrained by the length of the counter window; also

X-ray detection too near the ends of the counter results in undesirable distortion of the pulse-height distribution. Further, large size restricts the number of interchangeable crystals which can be accommodated. Enlargement is most beneficial for crystals with wide reflection curves and especially for multilayers.

Enlarging the crystal increases the range of incidence angles and thus worsens the resolution. Conversely resolution can be improved (with a sacrifice in intensity) by reducing the effective crystal area. This can be achieved by masking the crystal or placing a slit in front of the counter window, the effect of these being similar though not identical. When a slit is used, the best combination of resolution and intensity is obtained by shortening as well as narrowing the slit, in view of the geometrical factors previously described.

The above discussion refers to increasing the dimensions of the crystal *with constant Rowland circle radius*. Spectrometer characteristics remain constant if all dimensions are scaled either up or down by the same factor: there is thus no inherent advantage in either a large or small Rowland circle (with proportional crystal size). A large Rowland circle allows a lower minimum Bragg angle for a given closest crystal-source distance, giving a shorter wavelength limit. However, a large spectrometer is more costly. On the other hand, the adverse effects of crystal deformation are likely to be greater when the crystal is curved to a small radius.

High-Order Reflections

The relatively low intensity of high-order reflections rules out their use for analytical purposes but they are of interest (especially in trace element analysis) since they can cause interferences. The crystal reflection curve becomes narrower with increasing 'n', because the shorter the wavelength the more rapidly rays reflected from successive layers become out of phase with changing incidence angle. However, observed high-order peaks are not commensurately narrower than first-order peaks *at the same Bragg angle* (Fig. 7), their width being determined mainly by the geometrical factors already mentioned. The increased resolution observed for high-order reflections *for a given crystal* is due to the dispersion being proportional to 'n' (as deduced from the Bragg equation); also the geometrical broadening effect is less at high Bragg angles. For a different crystal with larger 'd' value, the dispersion is proportionately lower, compensating for the effect of increasing 'n': in practice the resolution thus does not increase continuously with 'n'.

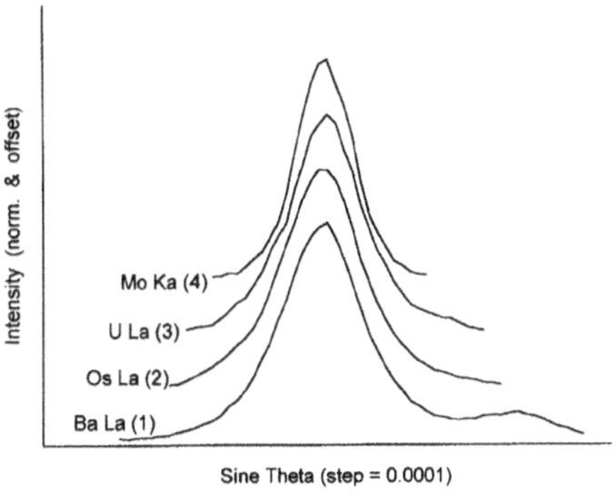

Fig. 7. Lines of different order (1 to 4) occurring at approximately the same Bragg angle (45°) with LiF crystal (Intensities normalized and plots offset for clarity)

Intensity shows a decreasing trend with increasing 'n' because of the decreasing reflecting area associated with the narrower reflection curve. It is also affected by the phase relations between X-rays scattered by atoms occupying different sites within the crystal structure, as described by the 'structure factor' (see standard texts).

Even though high-order peaks are relatively small, those originating from major elements can have significant effects on minor and trace elements. Such interferences can be minimised by pulse height analysis but are not necessarily totally suppressed. With an argon-filled counter this is particularly true for X-rays of energy above 3.203 keV (the Ar K absorption edge energy), since the 'escape peak' may lie partly within the p.h.a window set for the first order peak. (Escape peaks are relatively insignificant for xenon-filled counters.)

Continuum Background

The smooth *bremsstrahlung* or continuous X-ray spectrum (continuum) upon which characteristic emission lines are superimposed has an important effect on the detectability of trace elements. The Kramers expression commonly used to describe the shape of the continuum in EDS can be written in a form appropriate to WDS as follows: $I(\lambda) = \text{const.} (\lambda_o^{-1} - \lambda^{-1})/\lambda$, where $I(\lambda)$ is the intensity in photons/s per unit wavelength interval for wavelength λ, and λ_o is the wavelength of the Duane-Hunt limit (Fig. 8).

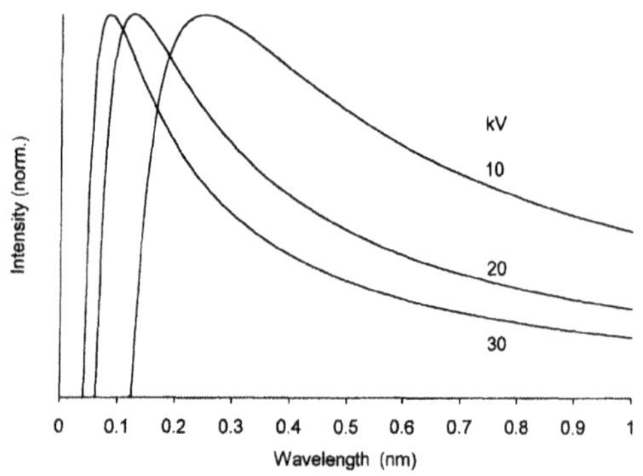

Fig. 8. Continuum intensity (normalised) as a function of wavelength for different accelerating voltages, calculated from Kramers' expression

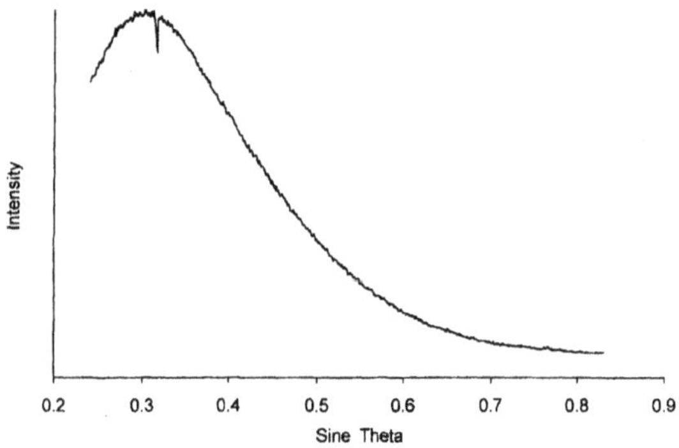

Fig. 9. Continuum recorded from pure Si at 20 kV using LiF crystal (note 'hole' caused by anomalous reflection)

The observed continuous spectrum is also influenced by the width of the band of wavelengths accepted by the spectrometer (i.e. the resolution) and the efficiency, which both reinforce the trend to decreasing intensity with increasing Bragg angle (the turn-down at short wavelengths is relevant only for LiF at low angles). The shape of the continuum in WDS can be determined experimentally using a sample with no lines in the wavelength range of interest (Fig. 9). To a fair approximation the continuum can be scaled in direct proportion to the mean atomic number for samples of different composition, though in some cases self-absorption effects are significant. With an argon-filled counter a step occurs at the wavelength of the Ar K Absorption edge.

Trace Element Analysis

The detection limit, defined as the concentration which can just be detected above background with an appropriate confidence level (e.g. 95%), typically ranges from tens to hundreds of ppm. Counting-statistical considerations dictate that the optimum strategy for trace elements is to use equal counting times for peak and background (the latter usually being divided equally between positions above and below the peak). The detection limit is proportional to $[(I_o/I_b)I_o t]^{-0.5}$, where I_o and I_b are the intensities measured at the peak and background positions

Table 1. Comparative peak intensities and peak-to background ratio for pure elements

Line	Crystal	Peak int. (Counts/s/nA)	Peak/bg.	$[I_o(I_o/I_b)] \times 10^{-3}$
a) Different crystals (measurements by author; accelerating voltage –20 kV)				
SiKα	PET	850	8700	7400
	TAP	2800	2550	7130
CrKα	LiF	340	680	230
	PET	1030	330	340
ScKα	LiF	145	730	110
	PET	1100	500	550
b) Effect of slit (data from C. T. Williams, Natural History Museum, London)				
FeKα	LiF (no slit)	430	310	140
	LiF (slit in)	230	540	120
c) Crystal size (data from Cameca; Fe –30 kV, Ti –20 kV)				
FeKα	LiF (normal)	1650	540	880
	LiF (large)	3400	480	1640
TiKα	PET (normal)	2650	440	1170
	PET (large)	6700	310	2070

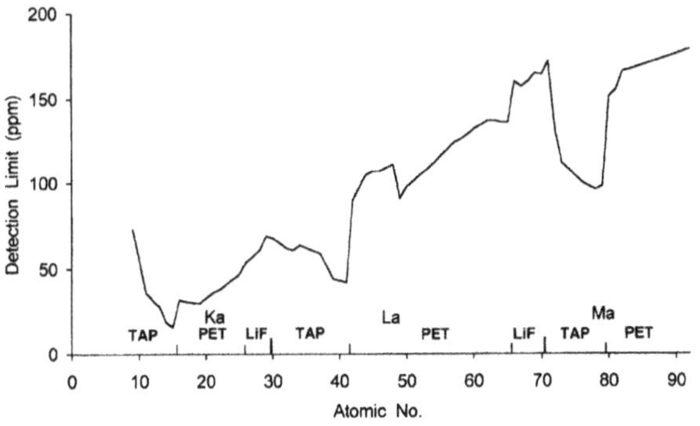

Fig. 10. Detection limit (ppm) plotted against atomic number of trace element in Fe for optimum choice of X-ray line and crystal (accelerating voltage –25 kV, beam current –100 nA, counting time –50 s each on peak and background). (Note: intensity data are derived from pure elements and practical detection limits will be somewhat higher when there is significant absorption in the specimen)

respectively: optimal conditions are therefore those for which the bracketed expression is maximised. It is always desirable to use the highest practicable accelerating voltage (thereby maximising I_o and I_o/I_b) as well as high beam current (maximising I_o) and long counting time (t). But, as already discussed, choices of spectrometer options generally entail a 'trade-off' between intensity (I_o) and resolution (hence I_o/I_b).

Table 1 contains examples of the application of the '*maximum peak intensity times peak-to-background ratio*' criterion and shows that detection limits can be lowered (at least in some cases) by appropriate choice of crystal where wavelength ranges overlap, also by choice of crystal size and slit, though where interferences are present high resolution may take priority. Optimum choices of emission line and crystal (neglecting interferences), and the variation of the detection limit with the atomic number of the analytical element, are shown in Fig. 10.

Conditions favourable for trace element analysis are not ideal for major elements. For example, a high accelerating voltage may result in excessive '*ZAF*' corrections; also peak count-rates may be unduly high, leading to excessive dead-time and distortion of the pulse-height distribution. One approach is to use separate routines for major and trace elements. Alternatively major element count-rates may be minimised by choice of crystal or by using a slit. Another possibility is to record major elements using an ED detector, keeping the count-rate within acceptable limits by means of a small aperture.

Standardless Analysis

Intensities measured on 'unknown' specimens can be converted to concentrations using pure-element count-rates derived either from experiment or by calculation, without the need to measure standards at the time of the analysis. Peak intensities are somewhat less reproducible in WD than ED analysis, where the concept is more familiar, but this approach is useful when a standard is not immediately available and some compromise with regard to accuracy is acceptable.

Computer Modelling

Selecting conditions for quantitative WD analysis is normally carried out by a combination of experience and 'trial and error'. Much time can be saved, and the reliability of the results improved, by computer modelling of spectrometer characteristics. The 'Virtual WDS' program [4] enables spectra to be synthesised on the computer screen, with provision for varying the relative concentrations of the elements concerned (Fig. 11). Interference at the peak position may indicate the desirability of using the β_1 rather than the α_1 line. Also, background measuring positions can be optimised to minimise interferences affecting either peak or background intensities (a much more effective approach than relying solely on wavelength tables). The presence of interferences may suggest measuring background only on one side.

When interferences cannot be avoided completely, the correction required may be estimated as a function of the amount of the interfering element present by

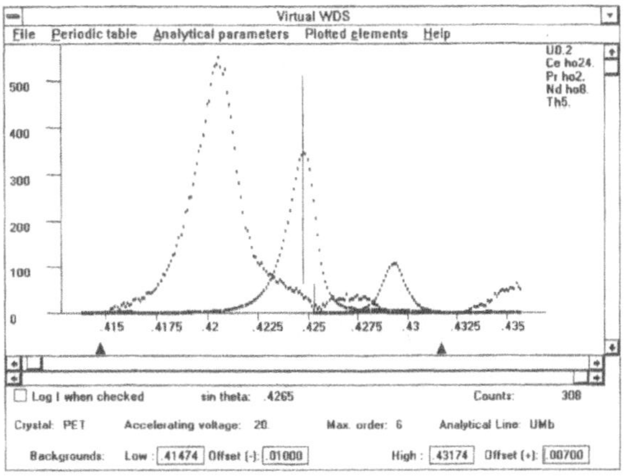

Fig. 11. Example of spectrum display using 'Virtual WDS' program: $U\ M\beta_1$ peak in monazite, with interfering spectra from other elements (on-screen display shows elemental spectra in different colours)

using 'overlap factors'. This approach is extremely useful in complex cases such as the determination of rare earths in minerals.

The choice of appropriate conditions can be facilitated by estimating the statistical precision of peak measurements from predicted intensities for given accelerating voltage, beam current, and counting time. By modelling the continuum, background intensities can be predicted, enabling trace element detection limits to be estimated and suitable conditions chose. Further, the size of the 'slope correction' required to correct for the continuum gradient in one-sided background measurements can be estimated (also the effect of curvature when measurements are made on both sides, though this is generally small).

Conclusions

WD spectrometers have been the mainstay of EPMA since Castaing's pioneering work in the early 1950 s and are still indispensable despite the introduction of ED spectrometers, the advantages of which (e.g. rapid parallel spectrum acquisition and high X-ray collection efficiency) are offset by disadvantages, including inferior spectral resolution and lower count-rate capability. The relatively high resolution of WDS is crucial for complex overlapping spectra, especially those of heavy elements, and the higher peak-to-background ratios obtained give lower elemental detection limits.

In this paper, the mode of operation of WD spectrometers is described and the factors governing their performance discussed, with emphasis on resolution and X-ray collection efficiency, which vary considerably (mainly as a function of Bragg angle). This information is useful to the EPMA analyst in making choices between different crystals, for example, and can be embodied in 'Virtual WDS' software which enables spectrometer characteristics to be simulated on a desktop computer with a view to devising analytical routines for specific purposes, obviating the need for a time-consuming 'trial and error' approach.

References

[1] R. Rybka, R. C. Wolf, in: *X-ray Spectrometry in Electron Beam Instruments*, (D. B. Williams, J. I. Goldstein, D. E. Newbury, eds.), Plenum, New York, 1995, p. 287.
[2] D. B. Wittry, S. Sun, *J. Appl. Phys.* **1990**, *67*, 1633.
[3] S. A. Ditsman, *Izvest. Akad. Nauk SSSR Ser. Fiz.* **1960**, *24*, 367.
[4] S. J. B. Reed, A. Buckley, *Mikrochim. Acta* **1996** [Suppl.] *13*, 479.

Mikrochim. Acta [Suppl.] 15, 37–47 (1998)

X-Ray Spectrum Processing and Multivariate Analysis

John M. Titchmarsh

Materials Research Institute, Sheffield Hallam University, City Campus, Pond Street, Sheffield S1 1WB, U.K.

Abstract. This paper reviews the extraction of quantitative compositional data from energy dispersive x-ray spectra acquired in the transmission electron microscope when a focused beam of high energy electrons is selectively positioned to excite fluorescence in a thin foil specimen. The influence of the parameters which control the generation of the characteristic and continuum contributions to the spectrum is outlined, together with the processing required to derive chemical composition. When high accuracy is required it may be necessary to consider the effects of self-absorption and secondary fluorescence, while unforeseen artefacts such as peak shape changes with time and coherent bremsstrahlung peaks can generate analytical inaccuracy. The potential of multivariate statistics is highlighted as a method for data analysis and information identification prior to the conventional processing routes, thereby reducing the influence of artefacts. The sensitivity of the energy dispersive technique is discussed in homogeneous and inhomogeneous specimens, together with spatial resolution and future prospects for improving the technique.

Key words: x-rays, energy dispersive analysis, TEM, multivariate analysis, microanalysis.

The measurement of chemical composition by energy dispersive x-ray analysis (EDX) both of bulk samples in the scanning electron microscope (SEM) and of thin foils in the transmission electron microscope (TEM) has been widely used for about two decades. However, EDX continues to present challenges to the microscopist as a consequence of the physical construction of the specimen/microscope/detector interface and the fundamental laws of electron scattering, x-ray production and detection. Highly energetic electrons lose energy through a combination of scattering, ionisation and fluorescence processes which are not confined only to the volume of the specimen selected for analysis. Some of the resulting artefacts can be minimised by good instrument design. Other problems such as surface oxide films, in-situ electron beam damage and ion-beam preparation artefacts are specific to particular materials and demand care and innovation in experimental technique. The most difficult challenges are probably those which are unexpected, such as channelling and coherent bremsstrahlung, which appear only occasionally and which may be impossible to overcome. Spectrum processing is merely the final stage by which inaccuracies may be introduced into the data. It is with some justification that a colleague once described thin-foil EDX analysis as 'a very good qualitative analysis technique'. In this paper the extraction of quantitative compositional information from EDX spectra is reviewed and some of the present limitations to accurate analysis discussed. The possible role of multivariate statistics as an analysis tool will be considered.

Production of Characteristic X-Rays

There are many parameters which influence the generation and collection of x-rays generated by electron beams with accelerating voltages in the range 100–400 keV, as used in most (S) TEMs, but the main factor is that the collected x-ray spectrum must contain sufficient characteristic intensity to determine the chemical composition of the feature(s) of interest with the required statistical significance. The production of characteristic x-rays is first considered.

The ionisation probability of a single atom of element, a, causing the ejection of an electron originally bound to the ith-shell, is measured by the cross-section, σ_i^a. An x-ray is generated during a subsequent electronic transition within the atom, with a very specific (characteristic) energy, as the atom minimises its energy. Not all electrons which are less tightly bound than the ejected electron can fill the vacant state. The so-called 'sum rules' specify in terms of the electron quantum numbers which transitions are allowed, and only a fraction, ω_i^a, the fluorescent yield, of a permitted transition results in characteristic x-ray generation because of the competing Auger electron production process.

The K and L-shell ionisation cross-sections [1, 2] are usually represented by the Bethe formula,

$$\sigma_i^a = 4\pi e^4 b_i \{\ln(c_i E_o / I_i^a) - \ln(1-\beta^2) - \beta^2\} / E_0 I_i^a \tag{1}$$

where e is the charge on the electron, b_i and c_i are constants, β is the ratio of the electron speed to that of light, E_o is the energy of the fast electron and I_i^a is the ionisation energy of the ejected electron. The last two relativistic terms are often ignored in quantitative EDX because their influence on cross-section ratios, in contrast to absolute values, is relatively small, even at TEM accelerating voltages. Various combinations of the Bethe parameters, b_i and c_i, have been used [1] but caution should be exercised when using such an expression over a large range of overvoltage ratios (E_o/I_i^a). While the form of the Bethe cross-section can be experimentally verified for variations in incident and ionisation energies [3], and values of b_i and c_i derived, absolute accuracy in experimental measurements of σ_i^a are typically only accurate to perhaps 10–20%.

The probability, p_i^a, of a fast electron causing ionisation in a foil a thickness, t, and atomic density, n^a, is the cross-sectional area per unit area of foil.

$$p_i^a = n^a t \sigma_i^a. \tag{2}$$

The total number, N_i^a, of ionisations in time τ by a beam current, i, is:

$$N_i^a = p_i^a (i/e)\tau = (i/e)\tau n^a t \sigma_i^a. \tag{3}$$

For a pure element of density ρ^a and atomic weight A^a, the atomic density, n^a, is:

$$n^a = \rho^a N_A / A^a \tag{4}$$

where N_A is Avogadro's Number. Of the fraction, ω_i^a, which generate a characteristic x-ray [4, 5], partitioning occurs between the possible lines within the relevant sub-shell, e.g. the K_α and K_β lines in the case of the K-shell, with the fraction a_i^a appearing in the K_α line [6–8].

Only x-rays which are emitted into the solid angle, Ω, subtended by the detector at the specimen can be detected, so the collected fraction is $\Omega/4\pi$. Fortunately, characteristic x-ray emission is isotropic so all intensities scale by the same factor. The detector efficiency factor, $e(E_K^a)$, at the x-ray energy, E_k^a, is determined by the fraction of x-rays that are subsequently absorbed within the active volume of the detector crystal. The final expression for the detected x-rays counts is [9]:

$$X_{io}^a = (i/e)\tau n^a t \sigma_i^a \omega_i^a a_i^a e(E_K^a)\Omega/4\pi \tag{5}$$

Analysis invariably involves measurements of binary or more complex specimens, so that ratios (X_i^a/X_j^b) are derived using Eq. (5), thereby eliminating some of the parameters:

$$\begin{aligned} X_{io}^a/X_{jo}^b &= n^a \sigma_i^a \omega_i^a a_i^a e(E_K^a) / n^b \sigma_j^b \omega_j^b a_j^b e(E_K^b) \\ &= c^a \sigma_i^a \omega_i^a a_i^a e(E_K^a) A^b / c^b \sigma_j^b \omega_j^b a_j^b e(E_K^b) A^a \qquad (6) \\ &= k_a c^a / k_b c^b \end{aligned}$$

where the atomic ratio, $n^a/n^b = c^a A^b / c^b A^a$. Hence:

$$\begin{aligned} c^a/c^b &= k_{ab} X_{io}^a / X_{jo}^b \\ &= k_b X_{io}^a / k_a X_{jo}^b, \qquad (7) \\ \text{where} \quad k_x &= \sigma_j^x \omega_j^x a_j^x e(E_j^x)/A^x \end{aligned}$$

for the characteristic line j.

The k-factors (Eq. 7) are constants of the microscope/detector system. Similar ratios can be used to define c^a/c^c, c^a/c^d, etc... for specimens containing several additional elements, c, d,... Ratios are usually referenced to a single element such as Si or Fe [8, 10, 11]. The simple relationship (Eq. 7) between the ratios of concentration and peak intensities is more commonly known as the Cliff-Lorimer equation [10].

Conventional Spectrum Processing

Measurements of elemental concentrations are usually derived from EDX spectra using a commercial spectrum processing software package. The word 'processing' is deliberately chosen here, rather than 'analysing' for reasons which are clarified later. Two different approaches are used for deriving a chemical composition from a spectrum using ratios such as those in Eqs. (6) and (7): (i) the 'standards' (S) method requires that peak count ratios in the spectrum are scaled by k-factors measured from standard samples [10] containing the elements of interest, ideally acquired under equivalent experimental conditions and close in time to that from the unknown spectrum and (ii) the fundamental parameters (FP), or 'standardless' method [12], whereby the peak intensities are scaled by theoretical k-factors generated from the values of σ_i^a, ω_i^a, a_i^a and A^a.

The S method implies the availability of standard thin-foil specimens which are homogeneous, stable, and do not contaminate or grow oxide films during storage and examination. Very few materials meet even some of these criteria in practice. In addition, the time taken to change samples and acquire spectra limits this approach, particularly in the high vacuum specimen environment of a dedicated field-emission gun (FEG)-STEM. The integrity of the FP method depends on the relative accuracies of the fundamental parameters, i.e. σ_i^a, ω_i^a, a_i^a which are limited for many lines, and are complicated by radiationless (Coster-Kronig) transitions [6]. A knowledge of the detector efficiency at each peak energy of interest must be available for the FP approach. This is an additional source of uncertainty in FP which is removed in the S method by being incorporated directly into the experimental ratios.

In practice, a compromise is often used which combines both approaches. k-ratios are measured for elements readily available in standard binary or multi-element specimens and ratios subsequently generated, as necessary. Interpolation between measured k-ratios permits estimates for elements which are unavailable from standards. Alternatively, ratios of pairs of k-ratios derived from FP can be used to scale a measured k-ratio in order to generate the best estimate for an element which cannot be obtained from standards. When the FP method alone is to be used then some comparisons with experimental measurements from standards are required to establish the integrity of the procedure.

For the S method, transportation of k-ratios from one microscope to another should only be contemplated for similar microscope/detector systems and for a common accelerating voltage. Even so, errors are likely because of the different efficiency factors of the two detectors, particularly for the low-energy x-ray region, below ~ 3 keV, and for the high-energy region above ~ 15 KeV. The accurate measurement of any k-factor is itself a major undertaking, requiring the preparation of an identified standard which is homogeneous on a scale significantly larger than the areas used for the measurements. Problems are likely to occur for the reasons mentioned earlier but, in addition, particular care is needed to avoid strong dynamic diffraction conditions in crystalline specimens, preparation artefacts, and poor specimen geometry. To investigate these it is often necessary to record measurements over a range of foil thicknesses and in several areas of the specimen to assess the contributions of inhomogeneities and artefacts [13].

Derivation of Peak Areas

For application of both the S and FP methods, or any combination, it is necessary to remove the background from under the characteristic peaks, deconvolute any peak overlaps and integrate the counts to determine X_{io}^a, X_{jo}^b, ... in Eq. (7). In addition to the specific energy of an x-ray, the resolution of a peak is a function of detector crystal type and the circuits which amplify and shape the electronic pulses which are generated by the conversion of the x-ray quanta into electron-hole pairs [14]. Resolution can be degraded by high input x-ray counting rates, and varying the pulse-processing circuit time constant, e.g. for enhancement of low energy x-ray measurements, will also affect the resolution. The exact zero and gain of the energy scale in the spectrum is subject to possible variation particularly over a period of several hours following the switching-on of power to the system.

One method of estimating the number of counts in a peak in a spectrum from an 'unknown' specimen is to scale a standard reference peak from a selected element to the equivalent peak in the 'unknown' spectrum, having first compensated for any differences in position and resolution, using a statistical test (i.e. χ^2) to indicate the optimum match. The reference spectrum should contain a large number of counts to minimise random noise. A variation of this approach is first to process both unknown and reference peaks by passing through a digital filter of appropriate shape [13], which effectively eliminates any continuum under the peaks in those regions of the spectrum where the continuum intensity is changing slowly as a function of energy. The optimised scaling of the filtered peaks can be performed to determine the background-subtracted counts in the 'unknown'. Peak overlaps can be deconvoluted by fitting combinations of two (or more) reference spectra to the unknown compound peak to derive an optimum. When a compound peak of low intensity is processed in this manner, however, considerable caution is needed because the fitting method is not very robust in some software packages. For example, a negative concentration of one element, even though this is physically unrealistic, when added to a positive concentration of another can produce a lower χ^2 value than more realistic positive combinations of both elements.

An alternative method of extracting the peak counts requires the generation and scaling of theoretical (Gaussian) peaks using the appropriate peak energy and detector resolution. An obvious problem arises when the experimental peak shapes are asymmetric as a consequence of detector artefacts such as incomplete charge collection [15, 16] or poor noise rejection in the electronic pulse gating. Such effects can be reduced in the S method by regular acquisition of reference spectra which will reflect long-term changes in detector response. Standard thin foil spectra are sometimes provided as part of the software package but these have almost certainly been recorded on a different microscope/detector system and are, therefore, of uncertain pedigree. While the use of these might be the only way to proceed when suitable reference materials are unavailable, they should only be used with caution and replaced by user-generated standard reference spectra as these become available.

The Continuum

In thin foil EDX analysis the continuum (or Bremsstrahlung) radiation is often considered to be an unwanted feature which detracts from the accuracy of analysis and, indeed, the sensitivity for trace element detection is limited by the accuracy with which the continuum can be subtracted (see below). However, the shape of the continuum should be monitored during TEM-EDX experiments, particularly when there are no characteristic peaks generated at low x-ray energies, because it can be a very useful indicator of the quality of the data.

Various theoretical expressions have been proposed for the partial differential cross-section, $d^2\sigma/dEd\theta$, for continuum x-ray production, and these have been systematically investigated by Chapman and colleagues [17–19]. It was concluded that the optimum form for TEM-EDX thin foil applications was that of a modified Bethe-Heitler theory (MBH). The contribution to the continuum from each element at any photon energy, E, is proportional to the concentration of the element and depends, additionally, on the energy of the fast electron, E_o, and the angle of elevation, θ, of the detector; the continuum is emitted anisotropically. The continuum exhibits a Z^2 atomic number dependence at any energy and, for any element, falls with increasing x-ray energy. It also decreases as θ increases, leading to a theoretical improvement in peak/background ratio. Unfortunately, the provision by some microscope manufacturers of EDX detectors mounted at a high θ has not always realised the theoretical improvement. Additional continuum x-rays can sometimes be collected from outside the electron-irradiated specimen because of the changed geometry, and the counting rates are generally lower because these detectors often subtend a smaller solid angle than more conventional geometries. Other advantages of the high θ geometry compared with the more conventional geometries are: collection of x-rays is possible over a much greater range of specimen tilt, and smaller corrections for self-absorption (see below).

The expression analagous to Eq. (5) for the number of continuuum x-rays generated in a photon energy window of width ΔE centred at an energy E, by element a, is:

$$B^a = (i/e)\tau\rho^a(N_A/A^a)t \\ (d^2\sigma^a/dEd\theta)\Delta E e(E)\Omega \quad (8)$$

where $e(E)$ is the detector efficiency at energy E and $(d^2\sigma^a/dEd\theta)$ is the continuum partial cross-section, differential in both energy and angle, for element a in the specific direction of the detector. These continuum x-rays are simultaneously generated with characteristic counts defined by Eq. 5. The continuum can be exploited by using it to assess the performance of the microscope/detector system. Theoretical ratios for the peak/background ratios of, e.g. the Cr and Ni K lines in Cr and NiO standards [20, 21], can be derived from Eqs.5 and 8 and compared with those determined experimentally. When the region of the continuum is chosen to be a single channel (either 10 or 20 eV in width) at the energy of the characteristic peak, the detector efficiency is eliminated from the ratio of X^a_{io}/B^a and self-absorption is the same for both characteristic and continuum signals, so that a knowledge only of the cross-sections and tabulated fundamental parameters (ω_i^a, a_i^a) is required. Typical values for pure elements such as Cr, Fe and Ni K lines are $\sim$ 2000 for a 20e V continuum window. X-rays generated in the condenser system of the microscope and multiply scattered electrons in more distant regions of the specimen can fluoresce the specimen holder/stage components [22, 23], leading to increased continuum detection and low peak/background ratios. The sources of extraneous radiation revealed by such an analysis should be identified and minimised.

The continuum and characteristic x-rays of low energy are strongly absorbed and it is this feature which can be exploited to monitor the quality of the EDX spectra during collection. If there were no absorption of x-rays the continuum intensity would continue to increase with reducing energy. In practice, the continuum reaches a maximum and then falls towards zero as the energy falls, due to absorption in the detector contact and dead layer, in any 'window' in front of the detector and in the specimen itself. When the continuum maximum from a detector with an ultra-thin window moves above 1–1.5Ke V then there may be problems with the specimen/detector geometry. For example, the specimen tilt may be very low, causing high self-absorption in the specimen which will affect the accuracy of analysis involving the use of peaks at low energies. Alternatively, there may be a partial cut-off in the line-of-sight between the detector and specimen by the cup of the specimen holder or by parts of the specimen if it is bent and broken. For a detector with a beryllium window the continuum maximum will always be above 1Ke V, but if the maximum moves above 2–2.5 Ke V then it is an indication that one of the above factors may be operating.

Self-Absorption

X-rays with energies of concern in EDX are absorbed primarily by the photoelectron effect. The x-ray transfers all its energy to a bound electron and ionisation occurs when the acquired energy exceeds the electron binding energy. This process underlies the operation of the EDX detector which is made sufficiently thick to ensure a high absorption efficiency. The self-absorption of x-rays by the thin foil specimen itself becomes more important as t increases. A self-absorption correction factor is usually applied to Eq. (5), such that the ratio of measured, X_i^a, to generated intensity, X_{io}^a, is:

$$X_i^a/X_{io}^a = \{1 - \exp(-(\mu/\rho)_{iabc...}^a \rho ft)\}/(\mu/\rho)_{iabc...}^a \rho ft \quad (9)$$

where $(\mu/\rho)_{iabc...}^a$ is the mass absorption coefficient (MAC) of x-ray i from element a by all the elements a, b, c .. in the sample, and which is a weighted sum of all the MACs, $(\mu/\rho)_{iy}^a$, for that x-ray in each of the separate elements, y, in the specimen [23].

$$\text{i.e.} \quad (\mu/\rho)_{i\,abc...}^a = c^a(\mu/\rho)_{ia}^a + c^b(\mu/\rho)_{ib}^a + c^c(\mu/\rho)_{ic}^a + \cdots \quad (10)$$

The factor, f, in Eq. (9) is determined from the specimen/detector geometry at the point of analysis. It is a function of (a) the local foil tilt angle, α, with respect to the horizontal, (b) the detector elevation, θ, from the horizontal and (c) the azimuthal angle, ϕ, between the planes containing the optic axis and (i) the line of steepest inclination on the foil at the point of analysis and (ii) the direction to the detector from the point of analysis. Of these, θ is likely to be known with some certainty, but α and ϕ can deviate markedly from the nominal values (derived by assuming that the plane of the area analysed is parallel to the plane of the specimen holder) when local foil bending occurs. It must also be remembered that the detector crystal subtends a large angle at the specimen, so that θ and ϕ correspond only to the central point of the detector. A 30 mm^2 detector with a collection solid angle of 0.2sterad is only $\sim$ 12 mm from the specimen and subtends an angular range of $\theta +/- 12°$ and $\phi +/- 12°$ at the specimen. Any absorption correction is, therefore, approximate.

The local foil density, ρ, is a function of the composition and crystal structure. The local foil thickness, t, can be measured by one of several methods, including EELS [24] and convergent-beam electron diffraction [25, 26], all of which are prone to error and are time-consuming in execution. Tables of the MACs are usually stored within the processing software. Alternatively, MACs can be calculated from parametric equations developed by optimum fitting to experimental absorption data [27]. As the x-ray energy falls, the uncertainty in both experimental measurements and parametric predictions increases so that errors of more than 10% should be assumed for energies below 500 eV.

The correction for self-absorption is performed by an iterative procedure. A first estimate of the concentrations, c^x, is derived either by the S or FP method, assuming no self-absorption. A value for the effective MAC, $(\mu/\rho)^{X}_{i\,abc...}$, for each element can then be calculated using Eq. (10), which is combined with the estimates of f, t and ρ in Eq. (9), to calculate an improved estimate for the x-ray intensity generated prior to self-absorption. This procedure is performed for all peaks of interest to generates an improved estimate of the composition, c^x. Further iterations are performed until the convergence is achieved. Corrections for self-absorption should be applied when measuring k-ratios from standards because it is the k-ratio at zero foil thickness which is specified in Eq. (7).

The approximate nature of the self-absorption correction should be already apparent. A further inaccuracy arises when the analysis point is adjacent to an interface between two phases of completely different composition, for example, analysis of a precipitate near a precipitate matrix boundary. It is likely that the self-absorption of x-rays generated in the precipitate will occur mainly within the matrix rather than the precipitate, particularly when α and θ are small, but software packages generally assume complete specimen homogeneity for the self-absorption correction. An extreme example of this failing occurs when FEG-(S)TEM probes, $\sim$1 nm in diameter, are used to measure segregation confined within a fraction of a nanometer at a boundary. The x-rays from the segregated atoms travel to the detector through matrix material of entirely different composition.

Several methods have been reported for overcoming the inadequacies of the correction factors generated using Eq. (9), particularly for eliminating the need to determine the detailed geometry of the specimen. These include: the estimation of the effective absorption path length by comparing the continuum intensity in a low-energy window with that predicted by the MBH method [28] and; fitting the variation of intensity of a low-energy characteristic peak as a function of specimen thickness, using a high-energy peak assumed to have minimal absorption to measure relative thickness, to derive the effective thickness for absorption [29].

Secondary Fluorescence

Self-absorption results in fluorescence within a volume which is much greater than that irradiated by the primary electron beam. Fluorescence corrections are most likely to be required when the characteristic x-rays of one element have an energy just above the ionisation threshhold of a second element. In an inhomogeneous sample, small characteristic peaks associated with one particular phase might appear when analysing another phase, while the ratio of characteristic peaks can be distorted in a homogeneous specimen. Corrections for the latter have been derived in the ideal case of a parallel-sided foil of infinite extent [30], which is rarely, if ever, encountered in practice. A more general consideration of real specimen geometries has recently been reported [31] which concluded that the ideal parallel-sided foil defines an upper limit to the fluorescence correction in [30]. Provided care is taken with the specimen geometry, in most applications the phenomenon can be, and is, safely ignored, because the likely corrections are small compared with other errors.

Coherent Bremsstrahlung

The MBH and other continuum theories predict that the intensity varies smoothly with x-ray energy over the range of interest. The experimental continuum may contain intensity steps at ionisation edges when there is significant self absorption, although such steps are rounded by the detector resolution function and are often concealed by characteristic peaks associated with the ionisation edges. In crystallographic specimens, the continuum can contain small coherent bremsstrahlung (CB) peaks [32, 33]. CB peaks occur when the incident electron beam direction is normal to a close-packed plane usually, but not always, close to a high-index zone axis. The rapid oscillations in acceleration/retardation which occur as the electron pass through successive atom planes cause the emission of CB x-rays in multiples of a fundamental energy which is a function of the E_o and the interplanar spacing. The actual measured CB energy is determined by a Doppler broadening which depends on the x-ray direction to the detector. Because the detector subtends a range of angles at the specimen, the fundamental CB energy is broadened into a peak, which is then further broadened by convolution with the detector resolution function. Hence, CB peaks are broader than characteristic peaks but, because they are small and noisy, they can sometimes be confused with the latter. Because CB is difficult to anticipate and the magnitude impossible to calculate, CB can cause significant processing problems both by overlapping small characteristic peaks and with continuum removal. The energy range over which the peaks occur is up to a few ke V for many materials, and they can interfere with the quantification of small characteristic peaks from segregating elements such as Si, P, S, Mo, Nb in many structural alloys [34].

Sensitivity

EDX sensitivity is quantified in terms of the minimum mass fraction (MMF) of one element which can be detected when homogeously distributed in a matrix of another element [35]. Sensitivity in TEM applications is also important in the case of inhomogeous specimens containing small particles and segregation. In the former applications the MMF criterion is established by measuring the statistical significance of a characteristic peak, P, from element, a, relative to the uncertainty in its measurement. The uncertainty depends on the background, B, in an energy range, ΔE, under the peak from the major components of the specimen. The significance criterion is usually chosen to be $P > h(B)^{0.5}$, where $h = 2$ or 3. For a simple binary alloy containing elements a small mass fraction, c^a, of element a in a matrix of element b (mass fraction, c^b, ~ 1), Eqs. (5) and (8) yield the expression:

$$\begin{aligned}(i/e)\ \tau c^a n^a t \sigma_i^a \omega_i^a a_i^a e(E_K^a)\Omega/4\pi \\ > h[(i/e)\tau n^b t(d^2\sigma^a/dEd\theta)\Delta E e(E_K^a)\Omega]^{0.5}\end{aligned} \quad (11)$$

i.e.

$$\begin{aligned}\mathrm{MMF} = c^a > [4\pi h/(\sigma_i^a\omega_i^a a_i^a)] \\ \times [d^2\sigma^a/dEd\theta\Delta E/\{(i/e)\tau n^b te(E_K^a)\Omega\}]^{0.5}\end{aligned} \quad (12)$$

assuming that n^a in the segregated region equals n^b in the matrix. MMF clearly depends on the inverse square root of parameters such as the probe current, analysis time, foil thickness, atomic density, detector efficiency which must be maximised to obtain the greatest sensitivity, subject to constraints imposed by detector count-rate saturation and self-absorption. With a thermal electron source, where the probe size is limited primarily by spherical aberration of the objective lens and $i \ \alpha \ d^{8/3}$ [36], the probe size can be increased to achieve the maximum possible x-ray counting rate. For many combinations of elements, a and b, MMF lies in the range $\sim 10^{-3}$ when analysis times of a few minutes are used.

The situation is more complex for inhomogeneous specimens because the ratio of characteristic to continuum counts is a function of the probe size, which is related to the probe current and position, through the distribution of the trace element. Additional practical constraints include the limited tolerance to probe/specimen drift, possible beam-induced damage and contamination. When the trace element is distributed as a fraction, S, in a uniform layer of interfacial segregation, one atom plane, δ, in thickness, then it can be shown [37] that for a gaussian probe current distribution, $i = (i_o/x_o(2\pi)^{0.5})$ $\exp(-x^2/2x_o^2)$:

$$\begin{aligned}S = \{4\pi h x_o/\delta\sigma_i^a\omega_i^a a_i^a\} \\ \times [(2\pi d^2\sigma^a/dEd\theta).\Delta E/\{(i/e)\tau n^b te(E_K^a)\Omega\}]^{0.5}\end{aligned} \quad (13)$$

The interdependence of current on probe size is complicated for a field emission electron gun [36], varying from a high power-law dependence at very small diameters, where the influence of objective lens spherical aberration is dominant, falling to a limiting 2/3 power law dependence at large diameters, when the aberrations of the gun and condenser system are limiting. This behaviour, together with the $x_o/i^{0.5}$ dependence predicted by Eq. (13), indicates that the probe size should be reduced to the smallest consistent with the aberrations of the objective lens for maximum sensitivity (minimum S).

Statistics and Multivariate Analysis

The processing package attempts to find an optimum fit between the experimental spectrum and a spectrum generated from standards or from first principles. A good package will also incorporate any variations in peak position and resolution into this process. However, when standards are used, account is unlikely to be taken of changes in peak shape due to detector response changes with time (ICC, count rate, etc) and changes in the relative self-absorption of lines within a multi-line peak which depend on specimen thickness. Counting statistics impose an ultimate limit in the accuracy of fitting. A most important point to appreciate is that the processing software forces a spectrum to fit a predetermined set of rules.

Experimental features which are not specifically incorporated into the software are simply ignored, or forced to fit the most applicable modelling routine. As the statistical limits of precision are approached the accuracy can be compromised by such shortcomings. A particular example of this would be when spectra contain CB.

When the EDX experiment is limited to the recording of a single spectrum there is no alternative but to process in the manner described in previous sections. The processing package can be regarded as a collection of physical information and experience which has been reduced and reassembled into a general form to assist the interpretation of a single spectrum. Other EDX experiments, however, require the comparison of compositions derived from two or more spectra, perhaps recorded over a long period of time from several specimens or over a range of crystal thicknesses and orientations, where small changes in sensitivity and accuracy are important. Examples of applications requiring the processing of series of spectra are the quantitative measurements of equilibrium segregation at interfaces and the deconvolution of diffusion profiles adjacent to point defect sinks [38, 39]. A series of spectra is recorded in such investigations, usually at points lying along a line perpendicular to the interfacial plane, to sample the full variation of composition. Spatial resolution requirements usually necessitate the use of a field emission electron source to produce an electron probe diameter of ~ 1 nm and x-ray generation and counting rates are relatively low because thin foils (< 50 nm) are ideally needed to minimise loss of spatial resolution through multiple scattering [40]. Total analysis time is limited by possible drift, contamination and beam damage. Each spectrum is determined by the product of the probe current distribution and the compositional distribution under the probe.

The variation in composition in such a profile sometimes results in a significant variation in foil thickness through differential chemical attack during electropolishing, and variable surface film composition and thickness. There might also be CB which abruptly changes at the grain boundary or phase interface. When such a spectral series is processed sequentially it can be very difficult to extract the segregation information of interest because the measured composition is a combination of contributions from different regions of the specimen and from the artefacts. Moreover, there is no way, a priori, of anticipating all the potential contributions and artefacts. The question then arises: is there an alternative approach which can compare more closely the variations within and between the spectra without the prior step of constraining each spectrum separately to conform to the set of processing rules? If so, can this benefit thin-foil EDX? One possible approach is to use chemometrics (multivariate statistical analysis) [41, 42].

Multivariate statistical analysis (MSA) provides a means of resolving a series of EDX spectra, S_1–S_n, into a set of component spectra, F_i, the number of possible components being no more than n. Each spectrum, S_i, can then be reconstructed from a linear combination of the components, F_j. Each component is related to an orthogonal (independent) source of information [43], which implies that in series of spectra where several phases and artefacts contribute in varying proportions to the spectra, it should be possible explicitly to separate the variations from the different phases present in any experimental spectrum and to follow the variation of an individual component throughout the series. A simplified algebraic representation of the n spectra, S_1–S_n, following MSA, is:

$$\begin{aligned} S_1 &= a_{11}F_1 + a_{12}F_2 + \cdots + a_{1n}F_n \\ S_2 &= a_{21}F_1 + a_{22}F_2 + \cdots + a_{2n}F_n \\ &\cdots \\ S_n &= a_{n1}F_1 + a_{n2}F_2 + \cdots + a_{nn}F_n \end{aligned} \tag{14}$$

MSA generates the components, F_i, together with the coefficients, a_{ij}, so that the variations of each component can be examined by plotting the spatial distributions of the coefficients.

There are few reports where MSA of EDX data has been attempted. Time dependent variations in spectra [44] and channelling effects [45] have been investigated, and recently analyses of interfacial segregation [46, 47]. For a segregated interface the probe diameter is typically much larger than the width of the segregated layer, and each spectrum contains contributions from the segregated layer and the two adjacent phases, in varying proportions depending on the exact probe position. MSA can extract a component explicitly related to the composition of the segregated layer [46]. The benefit of this analysis is that all the data in all the spectra are analysed simultaneously, without making any prior assumptions or constraining the spectra to conform to predetermined processing rules. Hence, unexpected phenomena are revealed which can be further investigated, real phenomena are separated from random noise, and spectra can be reconstructed from selected components, excluding both random noise and artefacts. Only when this analysis has been satisfactorily completed and the sources of the information within the whole data set understood is the conventional EDX processing performed. Segregation sensitivity appears to be improved significantly using MSA compared with conventional processing [47].

It is likely that the use of MSA will expand as the spatial and sensitivity limits for EDX continue to be extended. However, it is important that a thorough investigation is made to establish the extent to which all the information sources are independent and to recognise when the method might be inapplicable. The increasing use of FEG-TEMs for combined high-resolution imaging and microanalysis will cause limits to be pushed further towards the atomic scale, as interfaces and small particles are probed at ever improving resolution.

Spatial Resolution

The 'spatial resolution' of EDX measurements in thin foils refers to the lateral resolution perpendicular to the optic axis. When assessing resolution there is often an inherent assumption that the specimen composition is uniform in the beam direction. A fundamental resolution limit is imposed by the radial current distribution in the electron probe. The theoretical limit is determined by the wave optical point spread function (PSF) of the objective lens, which is a funciton of the lens aberrations, the convergence angle of the probe, α_p, and the focal setting of the lens [48]. The full-width half-maximum (FWHM) of the PSF can be ~ 0.2 nm in a modern microscope. However, the PSF relates to an infinitely small object source of electrons which, by definition, contains a current approaching zero. The PSF must be convoluted with the demagnified image of the electron source which forms the object for the probe-forming lens. When a thermal W or LaB_6 source is used, the effective source generated by the gun/condenser system has an approximate gaussian distribution with a diameter which, even after further demagnification by the objective lens, is still much wider than the PSF. The radial probe current distribution can, therefore, be regarded as gaussian with a diameter, d_o, defined as:

$$d_o^2 = 4i/\pi^2\beta\alpha_p^2 \qquad (15)$$

where β is the source brightness. Using the central limit theorem, the influence of spherical and chromatic aberrations can be included to estimate the actual probe diameter, d_p:

$$d_p^2 = d_o^2 + (0.5C_s\alpha_p^3)^2 + (C_c\alpha_p\Delta E/E_o)^2 \qquad (16)$$

where C_s and C_c are, respectively, the spherical and chromatic aberration coefficients and ΔE is the energy spread in the beam [36]. This is an approximation to the convolution of the PSF with an effective gaussian source in the specific case where the aberrations dominate and the influence of diffraction in the probe is minimal. Eqs. (15) and (16) imply that there is an optimum α_p which gives a minimum probe diameter when a specific probe current, i, is required to provide the necessary statistical significance for a chosen application. For thermal electron sources the minimum useful probe size is typically greater than $\sim$ 5–10 nm when a current of 0.1–1 nA is used and this can be achieved by demagnification of the effective source without degradation by the aberrations in the condenser lens system.

For a field emission source (FEG) with a brightness typically 10^3–10^4 greater than the thermal sources, probe diameters $\sim$ 1 nm containing $\sim$1 nA are possible. Diffraction contributions to the PSF can become significant when very small probes (< 1 nm) are generated and aberrations in the gun/condenser lens system can increase the source image incident on the probe-forming lens to a degree where the probe size (>5–10 nm) is limited by the gun/condenser rather than the objective. The latter can be reduced by limiting the angle of electron emission at the source, but this reduces probe current. FEG-STEMs are mainly used for EDX in a configuration where Eq. (15) still applies [49, 34] and smaller probe sizes only generated for electron energy-loss spectroscopy and imaging [50].

Future Prospects

EDX in the (S) TEM is now a mature technique, use of which persists in spite of the apparent shortcomings. The main reason for the continuing popularity is simply that there is no technique which can be readily interfaced to a TEM and which combines the equivalent (semi-)quantitative compositional information over such a wide range of the periodic table with equivalent sensitivity, accuracy, spatial resolution and rapid choice of analysis location. It is routine to select specific, individual features within a high resolution image and obtain almost instantaneous chemical information. Electron energy-loss spectroscopy ((P)EELS) is undoubtedly superior for low atomic number analysis [51] but has not replaced EDX for thin foil analysis to the extent predicted a decade ago by many people. This is probably due to the greater equipment costs, problems with multiple scattering and edge shape changes, difficulties with deconvoluting overlapping edges, a large range of cross-sections which are dependent on operational parameters, the need for thinner foils and lower contamination rates than for EDX, the huge dynamic range of the data and the difficulty in producing a fully automated processing package. However, the EDX method will continue to be threatened, so how can it be improved?

There is a fundamental energy resolution limit for the traditional Si and Ge EDX detectors, namely the stoichastic production of electron-hole pairs which limits sensitivity and, therefore, energy resolution [14]. This is particularly true for low-energy x-rays because the electronic noise in the amplifier circuits becomes significant compared with the pulse from a characteristic x-ray. A major improvement will only be achieved by radically new detectors with superior energy resolution. The bolometer [52, 53] is one such EDX system which is being actively developed at the present time. Instead of generating electron-hole pairs by photoelectron absorption in a semiconductor crystal, the x-ray energy is converted to a thermal pulse within a metal detector held in a calorimeter at a very low temperature. The temperature rise in the calorimeter is measured to determine the x-ray energy. Resolutions of < 20e V have been demonstrated. The major limitation of the bolometer system presently is a much smaller detector size compared with semiconductor detectors. When the detector solid angle can be increased to allow collection and processing at a rate of several hundred counts per second in the TEM then the order of magnitude improvement in energy resolution will increase the compositional sensitivity accordingly. The improved resolution might also allow investigations of low-energy peak shifts associated with valency changes, similar to those possible in bulk samples using wavelength dispersive systems in the electron microprobe analyser [54].

With respect to spatial resolution, a fundamental problem is the correction of spherical aberration in electromagnetic lenses. If attempts to build an aberration-free lens were to be successful then incorporation of that technology into both the condenser and objective lenses should enable the generation of probes with diameters on the atomic scale containing sufficient current to provide single atom sensitivity, particularly if combined with a bolometer detection system. However, as with previous advances, when the technology for these improvements is commercially available it is likely to increase substantially the price of the analytical microscope.

References

[1] C. J. Powell, *Rev. Mod. Phys. A* **1976**, *48*, 33.

[2] M. Inokuti, *Rev. Mod. Phys.* **1971**, *43*, 297.

[3] J. H. Paterson, J. N. Chapman, W. A. P. Nicholson, J. M. Titchmarsh, *J. Microscopy* **1989**, *154(1)*, 1.

[4] W. Bambenyk, B. Crasemann, R. W. Fink, H.-U. Freund, H. Mark, C. D. Swift, R. E. Price, R. Venugopola Rao, *Rev Mod. Phys.* **1972**, *44*, 716.

[5] A. Lagneberg, J. Van Eck, *J. Phys B* **1979** *12*, 1331.

[6] M. O. Krause, *J. Phys. Chem Ref Data* **1979**, 8.

[7] J. H. Schofield, *At. Data Nucl. Data Tables* **1974**, *14*, 122.

[8] T. P. Schreiber, A. M. Wims, *Ultramicroscopy* **1981**, *6*, 323.

[9] J. I. Goldstein, in: *Introduction to Analytical Electron Microscopy* (J. J. Hren, J. I. Goldstein, D. C. Joy, eds.), Plenum, New York, New York, 1979, Ch. 3.

[10] G. Cliff, G. W. Lorimer, *J. Microscopy*, **1975**, 103, 203.

[11] J. E. Wood, D. B. Williams, J. I. Goldstein, *J. Microsc.* **1984**, *133*, 255.

[12] W. A. P. Nicholson, *Mikrochim. Acta* **1994**, *114/115*, 53.

[13] D. B. Williams, *Practical Analytical Electron Microscopy in Materials Science*, PEI Mahwah, New Jersey 1984

[14] J. C. Russ, *Fundamentals of Energy-Dispersive X-Ray Analysis*, Butterworths, London, 1984.

[15] A. J. Craven, P. F. Adam, R. Howe, *Inst. Phys. Conf. Ser. No. 78* (G. J. Tatlock, ed.), IoP/Adam Hilger, Bristol 1985, 189.

[16] P. J. Statham, *J. Microsc.* **1981**, *123*, 1.

[17] J. N. Chapman, C. C. Gray, B. W. Robertson, W. A. P. Nicholson, *X-Ray Spectrom.* **1983**, *12*, 153.

[18] C. C. Gray, J. N. Chapman, W. A. P. Nicholson, B. W. Robertson, R. P. Ferrier, *X-Ray Spectrom.* **1983**, *12*, 163.

[19] B. W. Robertson, *Ph.D. Thesis*, University of Glasgow, 1979.

[20] S. M. Zemyan, D. B. Williams, *J. Microsc.* **1994**, *174*, 1.

[21] R. F. Egerton, S. C. Cheng, *Ultramicroscopy* **1994**, *55*, 43.

[22] L. F. Allard, D. F. Blake, in: *Microbeam Analysis* (K. F. J. Heinrich, ed.), San Francisco Press, San Francisco, 1982, p. 8.

[23] N. J. Zaluzec, in: *Introduction to Analytical Electron Microscopy* (J. J. Hren, J. L. Goldstein, and D. C. Joy, eds.) Plenum, New York, New York, 1979, p. 121.

[24] T. A. Malis, S. C. Cheng, R. F. Egerton, *J. Electron Microsc. Tech.* **1988**, *8*, 193.

[25] P. M. Kelly, A. Jostens, R. G. Blake, J. G. Napier, *Phys. Stat. Sol.* **1975**, *31*, 771.

[26] F. R. Castro-Fernandez, C. M. Sellars, J. A. Whiteman, *Phil. Mag. A.* **1985**, *52*, 289.

[27] K. F. J. Heinrich, *Proc. 11th ICXOM* (J. D. Brown, R. H. Packwood, eds.), Univ. Western Ontario, London, Canada, 1987, p. 67.

[28] A. J. Craven, P. F. Adam, in: *Electron Microscopy 1984, Proc. 8th European Congress on Electron Microscopy* (A. Csanady, P. Rohlich, D. Szabo, eds.), 1984, p. 371.
[29] W. A. P. Nicholson, K. M. Khan, in: *Microbeam Analysis-1991* (D. G. Howitt, ed.), San Francisco Press, San Francisco, 1991, p. 455.
[30] C. Nockolds, G. Cliff, G. W. Lorimer, *Micron* **1980**, *11*, 325.
[31] I. M. Anderson, C. B. Carter, J. Bentley, in: *Microbeam Analysis-1995, Proc 29th Annual Conf. of MAS, Breckenridge, Colorado* (E.Etz, ed.) VCH, New York, 1995, P. 285.
[32] J. H. C. Spence, J. M. Titchmarsh, N. Long, *Proc. 20th Annual Conf. MAS* (J. T. Armstrong, ed.), San Francisco Press, San Francisco, 1985, p. 349.
[33] K. S. Vecchio, D. B. Williams, *J. Microsc.* **1987**, *147*, 15.
[34] J. M. Titchmarsh, *Proc. 11th ICXOM* (J. D. Brown, R. H. Packwood eds.), Univ. Western Ontario, London, Canada, 1987, p. 337.
[35] D. C. Joy, D. M. Maher, in: *SEM/1977/I* (O. Johari,. ed.) SEM Inc, AMF O'Hare, Chicago, **1977**, 325.
[36] L. Reimer, *Transmission Electron Microscopy*, Springer, Berlin Heidelberg New York Tokyo, 1989.
[37] J. M. Titchmarsh, I. A. Vatter, *Proc. NATO Advanced Workshop: The Evaluation of Advanced Semiconductor Materials by Electron Microscopy* (D. Cherns. ed.) *NATO ASI Series B Physics* **1989**, *203*, 111.
[38] C. R. M. Grosvenor, P. E. Batson, D. A. Smith, C. Wong, *Phil. Mag. A* **1984**, *50*, 409.
[39] S. Dumbill, R. M. Boothby, T. M. Williams, *Mater. Sci. Tech.* **1991**, *7*, 385.
[40] J. I. Goldstein, J. L. Costley, G. W. Lorimer, S. J. B. Reed, *SEM/1977/1*, (O. Johari, ed.), AMF O'Hare, Chicago, 1977, p. 315.
[41] E. R. Malinowski, *Anal. Chem.* **1980**, *49*, 4.
[42] J. F. Hair, R. E. Anderson, R. L. Tatham, *Multivariate Data Analysis with Readings*, Macmillan, New York, 1987.
[43] P. Trebbia, N. Bonnet, *Ultramicroscopy*, **1990**, *34*, 165.
[44] O. Jbara, J. Cazaux, P. Trebbia, *J. Appl. Phys.* **1995**, *78*, 868.
[45] C. J. Rossouw, P. S. Turner, T. J. White, A. J. O'Connor, *Phil. Mag. Letters* **1989**, *60*, 225.
[46] J. M. Titchmarsh, S. Dumbill, I. A. Vatter in: *Microbeam Analysis-1995, Proc. 29th Annual Conf. of MAS, Breckenridge, Colorado*, (E. Etz, ed.) VCH, New York, 1995, p. 259.
[47] J. M. Titchmarsh, S. Dumbill, *J. Microscopy* **1996**, *184(3)*, 195.
[48] E. Munro, *Proc. 8th ICXOM, Boston*, (R. Ogilvie, D. Wittry, eds.) NBS, Washington, DC, 1977, p. 19.
[49] L. E. Thomas, *Ultramicroscopy* **1982**, *9*, 311.
[50] S. J. Pennycook, D. E. Jesson, *Ultramicroscopy* **1991**, *37*, 14.
[51] R. F. Egerton, *Electron Energy-Loss Spectroscopy in the Electron Microscope*, Plenum, New York, 1986.
[52] J. Martinis, in: *Microbeam Analysis-1995, Proc. 29th Annual Conf. of MAS, Breckenridge, Colorado* (E. Etz, ed.), VCH, New York, 1995, p. 3.
[53] E. Silver, M. LeGros, N. Madden, J. Beeman, in: *Microbeam Analysis-1995, Proc. 29th Annual Conf. of MAS, Breckenridge, Colorado*, (E. Etz, ed.), VCH, New York, *1995*, p. 5.
[54] J. T. Armstrong, in: *Microbeam Analysis-1995, Proc 29th Annual Conf. of MAS, Breckenridge, Colorado* (E. Etz, ed.), VCH New York, 1995, p. 189.

Mikrochim. Acta [Suppl.] 15, 49–57 (1998)

Thin Film Analysis and Chemical Mapping in the Analytical Electron Microscope

David B. Williams*, **Masashi Watanabe,** and **Derrick T. Carpenter**

Department of Materials Science and Engineering, Lehigh University, Bethlehem, PA 18015-3195, U.S.A.

Abstract. This paper reviews recent progress in the quantitative x-ray analysis of thin films for materials science and the development of quantitative x-ray mapping in the analytical electron microscope (AEM). Quantitative x-ray energy-dispersive spectrometry (XEDS) still relies basically on the Cliff-Lorimer ratio equation. There are two fundamental limitations to the routine application of this equation. The first problem is that the Cliff-Lorimer k factors are not standards, but vary from instrument to instrument. The second limitation is that the assumption of no significant x-ray absorption in thin specimens is not reasonable in many engineering materials. Methods to overcome both these limitations have been developed recently and are described.

The high spatial resolution due to the smaller analyzed volume in thin specimens means that quantitative x-ray imaging (mapping) in the AEM is generally not very accurate because of the low efficiency of x-ray generation and collection. These drawbacks can be offset by use of field-emission guns (FEG) and intermediate-voltage instruments to increase the beam current, ultra-high vacuum stages to reduce contamination, and careful stage design to maximize the x-ray collection angle and peak to background ratio. All of these design requirements are incorporated in a VG HB603 AEM at Lehigh and recent results are presented showing quantitative mapping of interfacial segregation with a spatial resolution <5 nm.

Key words: Cliff-Lorimer equation, absorption correction, zeta factor, x-ray mapping.

* To whom correspondence should be addressed

Quantitative thin-film microanalysis using x-ray energy-dispersive spectrometry (XEDS) in the analytical electron microscope (AEM) is performed by relating the intensities of the characteristic x-rays above background (I_A, I_B, etc.) to the compositions in wt. % (C_A, C_B, etc.) through the Cliff-Lorimer equation (Cliff and Lorimer [1]):

$$\frac{C_A}{C_B} = k_{AB}\frac{I_A}{I_B} \qquad (1)$$

where k_{AB} is the Cliff-Lorimer sensitivity factor, commonly called the k factor. This simple equation still remains the basis for quantification, more than twenty years after initial publication, despite the limitations of its inherent assumptions of no absorption or fluorescence and the difficulties in obtaining accurate, reproducible k factors.

Determination of k-Factors

Because determination of the k factor remains the crucial step in quantitative thin-film microanalysis it has been the subject of many previous investigations. There are three general methods for determining k factors. The first way, and that which gives the most accurate thin-film quantification is to measure k factors using standards, e.g., Cliff and Lorimer [1], Lorimer et al. [2], McGill and Hubbard [3], Wood et al. [4], Graham and Steeds [5], Wirmark and Norden [6] and Sheridan [7]. The use of standards produces k factors, with errors that can be as small as $\pm 1\%$ relative. However, the experimental approach requires the generation of a range of thin-film standards, many of which may not be ideal for the purpose since it is essential that the thin standards have known stoichiometry, are stable in the electron beam and unchanged by the method(s) of specimen preparation. This approach is not always possible because appropriate standards may not be available and, if they are, may take significant time to fabricate and to thin to electron transparency.

The second method is to calculate the k factors from first principles, e.g. Goldstein et al. [8], Schreiber and Wims [9], Wood et al. [4]. In contrast to the experimental approach, the calculation of k-factors is rapid and universally applicable. The expression for the theoretical k factor was first defined by Goldstein et al. (1977):

$$k_{AB} = \frac{(Q\omega a)_A A_B \varepsilon_A}{(Q\omega a)_B A_A \varepsilon_B} \qquad (2)$$

where the subscripts A and B indicate elements A and B, Q is the ionization cross section, ω is the fluorescence yield, a is the relative transition probability (i.e. the ratio of the intensities of $K_\alpha/[K_\alpha + K_\beta]$), A is the atomic weight and ε is the detector efficiency, which depends on the thicknesses (t) of absorbing layers in front of the detector (Be window, Au contact, Si dead layer, and ice or other contamination). Of the factors in the equation, ω, a, and A are well-known. Difficulties in calculating k_{AB} arise from uncertainty in the cross sections, Q [10] and the (generally unknown) thicknesses of the absorbing layers (required in the ε term) which can result in errors in calculated k factors as large as ± 15–20% relative, which may be too large for the quantitative application at hand.

The third method is to obtain values from tables of previous k-factor data (e.g. Williams and Carter [11]). However, since k-factor data strictly apply only in one laboratory on one instrument at one time, they should only be used for comparison purposes rather than for direct quantification.

Rapid and Accurate k-Factor Determination

Recently [12] a new method of k-factor determination has been proposed that overcomes the inherent dichotomy of the k-factor approach and allows for the determination of a range of k factors with errors similar to the experimental values, but with a speed approaching the calculation method. Several k factors are measured from a known standard then, from these, the best set of absorber thicknesses (contained in the ε term in Eq. (2)) and Q, are determined iteratively. Then a larger range of k factors (e.g. from Mg to Zn) can be fitted to the experimental data using the derived thicknesses. Basically, a very limited set of three experimental data points is used as the basis for creating a broad range of k factors. This approach works because of the availability of the National Institute of Standards and Technology (NIST) standard reference material (SRM 2063) [13]. This standard is an amorphous mineral glass film of fixed, known thickness and density and it contains accurately known amounts of Mg, Si, Ca, Fe and O permitting the generation of at least k_{MgFe}, k_{SiFe} and

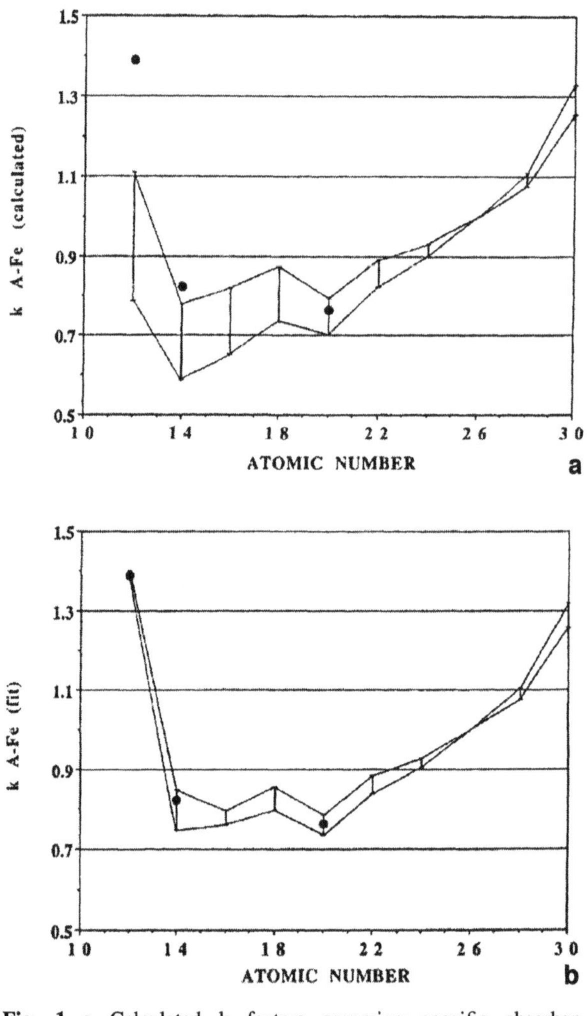

Fig. 1. a Calculated k factors assuming specific absorber thicknesses. The length of the lines indicates the range of the calculated values from eight different ionization cross section models. Comparison with experimental k factors (closed circles) from SRM 2063 shows the poor agreement particularly for the lighter elements. **b** Calculated k factors after fitting to the experimental values in (**a**). The range of the calculated k factors is significantly reduced after forcing the fit to the experimental k factors. (Courtesy S. M. Zemyan)

k_{CaFe} (and k_{OFe} if an ultra-thin window detector is available). Fig. 1 shows a comparison of the measured and calculated k factors before and after the iterative fitting to the best absorber thicknesses and the improvement in agreement between the experimental and theoretical values is clear. The errors in this approach are generally $< \pm 5\%$ and the time taken to gather the data is that required to obtain a series of spectra from SRM 2063 with reasonable statistics (i.e. several minutes maximum for each spectrum).

Absorption-Corrected k-Factor Determination

The second limitation of the Cliff-Lorimer approach is that it assumes no absorption or fluorescence of the characteristic x-rays. This assumption is unreasonable in many real engineering materials, particularly light-element systems such as Al and Mg-base alloys as well as intermetallics in alloy systems such as Ni-Al, Ti-Al and Nb-Al. This limitation has led to the development of a series of methods to determine k factors that are absorption corrected. The most common approach is that due to Horita and co-workers [14–16], Van Cappellen [17] and Nicholson and Kahn [18] who, over the last decade, have pioneered methods of determining absorption-corrected k factors through extrapolation methods. In this approach, the specimen must exhibit large differences in thickness and must generate at least two x-ray lines, one harder (higher energy) than the other. It is then assumed that the hard x-ray line is not absorbed while the soft line(s) are increasingly absorbed as the specimen thickness increases. The assumption of no absorption is reasonable for x-ray lines of energies $> \sim 5$ keV in specimens $< \sim 500$ nm thick. The thickness is then inferred from the intensity of the hard x-ray line.

Zeta Factor Approach

The culmination of this approach is the recent ζ-factor method which permits the determination of k factors without prior knowledge of the density or the thickness of the foil [19]. Concise features of the absorption correction using the zeta factor and the demonstrations of this approach are described as follows.

When x-ray absorption is significant in a thin specimen, the composition C is no longer simply proportional to the intensity, I. So, Eq. (1) needs to be modified as [8]:

$$\frac{C_A}{C_B} = k_{AB}\left(\frac{I_A}{I_B}\right)_m \left\{\frac{(\mu/\rho)^A_{sp}}{(\mu/\rho)^B_{sp}}\right\} \times \left[\frac{1-\exp\{-(\mu/\rho)^B_{sp}\rho t\,\mathrm{cosec}\,\theta\}}{1-\exp\{-(\mu/\rho)^A_{sp}\rho t\,\mathrm{cosec}\,\theta\}}\right], \tag{3}$$

where $(I_A/I_B)_m$ is the measured intensity ratio, $(\mu/\rho)^A_{sp}$ and $(\mu/\rho)^B_{sp}$ are the mass absorption coefficients of characteristic x-ray lines A and B, ρ and t are the density and the thickness of the specimen at the beam position respectively, and θ is the x-ray take-off angel. If the x-ray absorption is corrected directly using Eq. (3), unfortunately the values of ρ and t at the beam position must be determined. Generally accurate thickness determination makes the absorption correction and (hence quantification) difficult. Note that θ would also be uncertain unless the specimen shape is flat or parallel sided. Otherwise an azimuthal angle must be incorpolated into the geometrical term, or the geometric approach described in [18] can also be used.

For a thin specimen, where most incident electrons are not scattered and the energy loss of the incident electron is small enough, the characteristic x-ray intensity, I, generated from a given beam position can be assumed proportional to the mass thickness, ρt. So the mass thickness can be expressed using the generated intensity [19]:

$$\rho t = \zeta \frac{I}{C}, \tag{4}$$

where ζ is the factor connecting the characteristic x-ray intensity to I, and C is the weight fraction. In Eq. (4), the characteristic x-ray intensity is normalized by the weight fraction. Therefore, the ζ factor is independent of the specimen composition.

For the proportionality between ρt and (I/C) to be a reasonable approximation, the x-ray absorption should be negligible in the specimen. It is assumed that the absorption is negligible for the x-ray line from element B, but significant for the x-ray line from element A. For the x-rays from element B, Eq. (4) is rewritten as:

$$\rho t = \zeta_B \frac{(I_B)_m}{C_B}, \tag{5}$$

where $(I_B)_m$ is the x-ray intensity measured from element B. It is possible to eliminate ρt in Eq. (3) by substituting Eq. (5):

$$\frac{C_A}{C_B} = k_{AB}\left(\frac{I_A}{I_B}\right)_m \left\{\frac{(\mu/\rho)^A_{sp}}{(\mu/\rho)^B_{sp}}\right\} \times \left[\frac{1-\exp\{-(\mu/\rho)^B_{sp}(\mathrm{cosec}\,\theta)\,\zeta_B(I_B)_m/C_B\}}{1-\exp\{-(\mu/\rho)^A_{sp}(\mathrm{cosec}\,\theta)\,\zeta_B(I_B)_m/C_B\}}\right], \quad (6)$$

Therefore, an absorption correction can be conducted by using Eq. (6) without knowledge of ρt.

The determination of ζ is made together with the k factors, using a standard thin specimen. For the standard thin film with known composition, there are two unknown values in Eq. (6), namely k_{AB} and ζ_B. There two values can be determined by measuring the characteristic x-ray intensities at more than two different positions and by applying a nonlinear least-square fitting procedure. More details on the non-linear-fitting process of ζ- and k-factor determination are described in the original paper [19].

An example of this approach in the $NbAl_3$ intermetallic compound is shown in Fig. 2. Because the

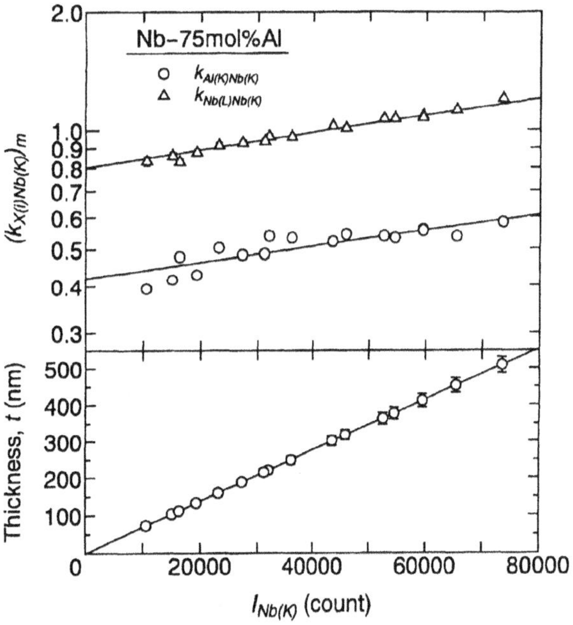

Fig. 2. Absorption-corrected k factors of $k_{Al(K)Nb(K)}$ and $k_{Nb(L)Nb(K)}$ obtained from a thin film of the $NbAl_3$ intermetallic compound. Absorption correction of the Al K and Nb L lines means that the k factors increase with increasing thickness to compensate for the decreased x-ray intensity. Extrapolation to zero counts by nonlinear least-squares fitting gives the absorption-free k factors necessary for quantification. The Nb K intensity increases linearly with thickness and zero thickness corresponds to zero counts

absorption of the Nb K line can be ignored in this system, the measured Nb K line intensity, $I_{Nb(K)}$, is proportional to thickness as shown in the lower diagram in Fig. 2. Thus the specimen thicknesses can be measured directly from the intensity of this x-ray line. Because the Al K and Nb L lines are absorbed significantly in the specimen, the measured k factors, $(k_{Al(K)Nb(K)})_m$ and $(k_{Nb(L)Nb(K)})_m$, increase with $I_{Nb(K)}$ as shown in the upper diagram in Fig. 2. However, direct extrapolation to zero thickness (at zero Nb K counts) by nonlinear least squares fitting is all that is required to determine the absorption-corrected k factors. For this measurement, the HB603 FEG-STEM was used at an accelerating voltage of 300 kV with a beam current of 0.5 nA.

Once the k and ζ factors are obtained, quantitative microanalysis of unknown specimens can be achieved through Eq. (6) with the combination of $C_A + C_B = 1$. In this composition determination, it is important to use the same beam current as that for the determination of the ζ and k factors. Additionally, an iterative calculation process is required for this quantification since Eq. (6) contains the composition-dependent terms such as C_A, C_B, $(\mu/\rho)^A_{sp}$ and $(\mu/\rho)^A_{sp}$. Furthermore, if the specimen density at the beam position is known, it is possible to determine the thickness through Eq. (4) instead of the mass thickness, provided θ is sufficiently well known.

Figure 3 shows a bright-field image, the concentration profiles for Nb and Al and the corresponding thickness variation in a Nb/Al multilayer specimen as a demonstration of the power of the zeta factor approach. A line scan was conducted using the same microscope conditions as in Fig. 2 and the trace of this line scan is indicated by the white line in the image. The concentration and the thickness were simultaneously calculated using the absorption-corrected $k_{Al(K)Nb(K)}$ and the $\zeta_{Nb(K)}$ determined from the $NbAl_3$ in Fig. 2. To obtain the thickness, the density at each measured point was calculated by the relationship $1/\rho = C_A/\rho_B + C_B/\rho_B$. The line scan reveals that the widths of both the interfaces were less than 4 nm and a sharp composition transition is exhibited regardless of the specimen thickness.

The main features of the ζ-factor method are as follows: it does not require knowledge of the thickness or the density for the absorption correction, it is possible to measure thickness if the density is known, and the determinations of composition and thickness are simultaneously achieved from a single measure-

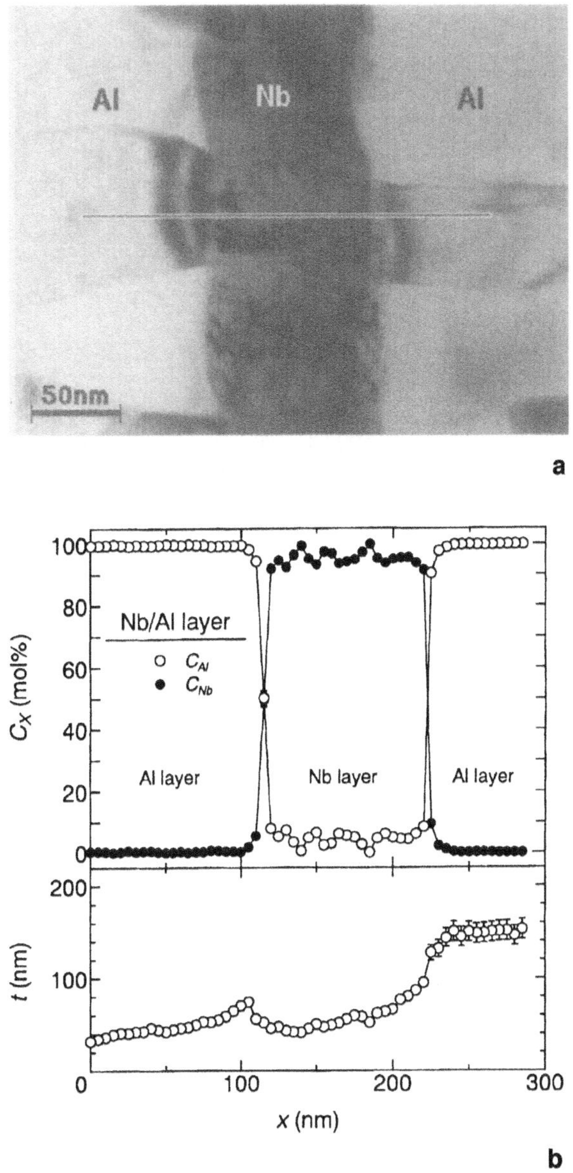

Fig. 3. **a** Bright-field image of a thin Nb/Al multilayer. The composition profile shows discrete changes at the interfaces. Application of the ζ factor to the data results in extraction of the thickness variation across the multilayer specimen shown in (**b**)

ment of x-ray intensities. The requirements for the ζ-factor method are that a standard thin foil with known composition should be used for both k- and ζ-factor determination and it is necessary to keep the total beam current the same between the determination of k and ζ factors from standards and the determination of composition and thickness in the specimens of unknown composition.

X-Ray Mapping

Thin-Foil Limitations

Generation of line profiles, such as in Fig. 3 is the standard way of presenting quantitative x-ray data in the AEM. Almost all analyses involving localized changes in concentration, such as segregation to interfaces, diffusion profiles etc., are performed by taking EDS spectra at a series of points across the region where the concentration change is occurring. While this approach is useful, it is highly selective and operator-biased. One way to overcome this limitation, to some degree, is to make a map of the x-ray intensity distribution as the electron beam is scanned over a larger area of the specimen, and convert this map into an image where the contrast is directly related to the amount of the element present. Such quantitative x-ray mapping has been a standard procedure in the electron probe microanalyzer (EPMA) for many years (see e.g. [20]). EPMA maps are limited by the poor spatial resolution of x-ray generation in bulk specimens, which typically is no better than 1 μm for beam energies in the range 20–30 keV. But it is precisely this poor spatial resolution that makes quantitative mapping feasible because the large x-ray generation volume in bulk specimens ensures that the x-ray count rate is substantial and thus sufficient counts can be gathered in a reasonable time to make quantification possible. Depending on the computer software, quantification can be performed 'on the fly' (i.e. as the data are gathered) or 'off line' after completion of the data gathering.

In thin-foil microanalysis, the price to pay for the improved spatial resolution of the AEM is a significantly reduced count rate which, hitherto, has meant that quantitative x-ray mapping is rarely attempted and most x-ray maps obtained in the AEM are simply dot maps. In such maps, the presence or absence of an x-ray is registered at the point where the beam is scanning. The density of dots gives some feel for the distribution of the element that is being mapped [21], but such maps are not quantitative and the bremsstrahlung intensity also contributes to the map and therefore, thickness variations can easily be misinterpreted as compositional variations. For good quantitative mapping, digital acquisition is desirable so that the background intensity contribution can be removed at each pixel in the image. It is generally assumed [22] that in excess of 10^6 counts are required in the map and if the instrument generates a probe current of at least 1 nA, quantitative mapping can be achieved in reasonable times. However, for comparable statistics, what takes less than an hour in the EPMA, would typically require many hours of mapping in a thin foil, during which time specimen drift may well have ruined the spatial resolution and beam damage may have changed the chemistry of the specimen. In the lower keV EPMA, beam damage is rarely a problem and drift is not usually a limiting factor when spatial resolutions of >1 μm are all that is required. So generally the quality of x-ray images from thin specimens does not approach that of bulk specimens in the EPMA.

Several of these limitations have been overcome in a relatively newly-designed AEM [23]. In the VG HB 603 AEM, the reduced x-ray count rate in thin specimens has been compensated in two ways. First, a 300 keV high-brightness field emission gun (FEG) electron source is used with a brightness of 10^{13} A $M^{-2}sr^{-1}$, capable of a beam current of 1 nA in a probe size of 1 nm (FWHM) and 100 nA in a probe size of 10 nm. Second the x-ray detection efficiency has been increased through use of two Oxford x-ray detectors; a windowless Si(Li) and an ultra-thin window Ge. These two detectors have large collection angles (0.3 sr and 0.2 sr respectively) compared to typical values of <0.15 sr on commercial AEMs. The two detectors can be used simultaneously if necessary. Also if the specimen is thick enough such that count rates result in increased deadtimes, use of a novel beam-blanking system ensures that output count rates up to ~10,000 counts per second are obtainable, approximately four times standard throughputs [23].

Experimental Maps

Both qualitative and quantitative x-ray mapping have recently been achieved on this instrument while maintaining the very high spatial resolution (<5 nm) that is the hallmark of FEG-AEM analysis. The particular problem to which mapping was first applied was that of grain boundary segregation. Typically, boundary segregants are limited to a monolayer or less (<0.5 nm), and detection of these thin layers can tax routine analysis with a stationary probe. Nevertheless, it has proven feasible to map the presence of S in Ni, which is a classic embrittling system. Sulfur impurities can be the cause of serious failures in Ni-base superalloys used in power-generation systems.

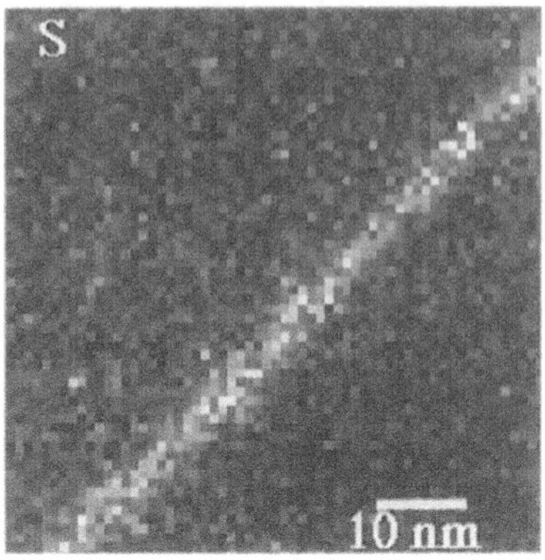

Fig. 4. Digital x-ray map showing the qualitative variation of S at a grain boundary in a Ni-S alloy. The increase in the S K signal along the boundary indicates that S is strongly segregated to the boundary but the absolute amount of S cannot be deduced from this map. (Courtesy V. J. Keast)

As shown in Fig. 4, the enhanced S K signal is clearly visible, located over a region <5 nm wide along the grain boundary, which is oriented parallel to the electron beam to optimize the signal from the segregant. The 64×64 pixel map was acquired on the VG HB 603 using the S K x-ray peak at 1 MX with a 200 ms dwell time per pixel (about 20% dead time), a probe size estimated to be < 1 nm, and a probe current of 0.5 nA. Typically each pixel contains 10 counts, with contributions from both the characteristic x-rays and the bremsstrahlung. The apparent increase in S K is not due to an increase in thickness since the electron image, obtained at the same time, revealed no significant thickness variations across the boundary.

To obtain quantitative maps requires that the bremsstrahlung intensity is subtracted from the characteristic peak intensity at every pixel and the resultant I_A, I_B etc. be modified by the k factor to produce C_A, C_B etc. This has been carried out for a similar investigation of the boundary segregation of Cu in Al-Cu alloys used in interconnects in integrated circuits. The high current densities carried by interconnects in service result in the phenomenon of electromigration, the preferential transport of mass, typically along grain boundaries, in the direction of current flow. As aluminum atoms are driven away from susceptible sites, voids form and failure of the interconnects results. The addition of Cu appears to stabilize the material and increase the time to failure because the Cu segregates to the boundaries and must be driven away before damage can occur. Fig. 5a is a low magnification bright-field image of an Al-4 wt.%

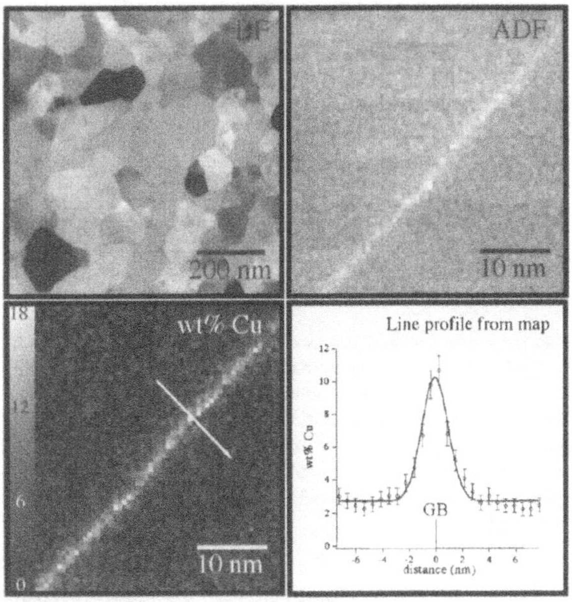

Fig. 5. **a** Top left: a low magnification electron image of the grain structure in a thin film of Al-4 wt. % Cu. **b** Top right: a high magnification annular dark-field image showing the increase in average atomic number at a single boundary. **c** Bottom left: the quantitative Cu map from the same boundary as in (**b**) including a look-up table to relate the intensity to the Cu content; **d** Bottom right, a Cu composition profile across the boundary extracted from the map in (**c**)

Cu film (ca. 100 nm thick) which had been aged to cause segregation of Cu to the grain boundaries. Background-subtracted Cu K and Al K x-ray maps were gathered from an edge-on grain boundary at 2 MX with a resolution of 64×64 pixels and a dwell time of 200 ms per pixel (about 20% dead time). An electron image was also collected from the annular dark-filed detector (ADF). The probe size was estimated to be < 1 nm, and probe current was 0.5 nA. The field of view is 40 nm wide, so each pixel is 0.625 nm wide. Calculated beam broadening through the 100 nm thickness results in an average exit-surface probe size of about 2.3 nm, so the scan oversampled by a factor of about 3.5. In the limit, to ensure oversampling as the probe diameter approaches the display pixel diagonals, the minimum probe size must be > pixel size. In the case of square display pixels (as in the Oxford Link eXL system), the minimum probe diameter should always be > square diagonal (i.e. 1.4 × display pixel diameter).

A large camera length was used to increase the acceptance angle of the ADF, reducing the effects of diffraction contrast and enhancing atomic number contrast in the image of the grain boundary (Fig. 5b). This image reveals a layer of higher atomic number located at the boundary, which appears bright against the matrix. A quantitative map of Cu concentration (in wt. %) was calculated from the raw Cu and Al intensity maps using a k factor measured from stoichiometric θ-phase (Al_2Cu) particles which were distributed throughout the material. This map (Fig. 5c) confirms that Cu has segregated to the boundary, with a local concentration of about 10 wt. % compared to the average matrix concentration of about 3 wt. %. The image is quite noisy due to the relatively small number of counts per pixel, but statistically meaningful information may be extracted by averaging over many pixels. For instance, by sampling 1600 pixels in the matrix, a matrix composition of 2.6 ± 0.1 wt. % Cu may be calculated. We expect the matrix concentration to be close to the equilibrium solubility of Cu in Al at 475 °C (the homogenizing temperature), which is 2.75 wt. % Cu. Line profiles may also be extracted from such a map, as illustrated in Fig. 5d. Each point in the profile is the average of 30 pixels in a line parallel to the grain boundary, giving it an effective width of about 20 nm. The line profile reveals the degree of local boundary enrichment, and shows that the spatial resolution of the composition variation is quite good, about 4 nm in this case (this would be expected to improve for thinner specimens). The error bars are based upon counting statistics and represent 95% confidence limits.

While mapping at high magnification is useful, it is only capable of showing a very small area of the specimen. Quantitative x-ray maps gathered at lower magnifications (ca. 200 kX) are very promising in that each image contains many different grain boundaries, each of which can be quantified after the map is gathered. This parallel acquisition of grain boundary data is essential in order to begin making generalizations about grain boundary segregation in an entire material. In addition to comparisons between different boundaries, it is also possible to quantify the homogeneity of single boundaries and measure the fraction of boundaries which contain segregant, neither of which is easily done using conventional line profiles. However, at lower magnifications, more pixels are required (for a given probe size) to fully sample the field of view, so the time required to create a map with the same statistics will rise accordingly. Maps were gathered from the Al-4 wt. % Cu specimen using several methods to increase the count rate, among them a larger probe size, beam blanking, and a shorter time constant for the pulse processor.

Using a larger virtual objective aperture, a larger probe (about 2 nm) with a higher usable current (about 3.5 nA) could be obtained, and fewer pixels were required to sample the area completely. Beam blanking was enabled in order to reduce dead time and increase the throughput of counts. However, the larger interaction volume resulted in a decreased detectability of Cu at the grain boundaries. That is, the small extra signal from Cu at the boundary was lost in the large background of Cu in the matrix, so the level of Cu at the boundaries was indistinguishable from that of the matrix. It was necessary to revert to the original <1 nm probe in order to maintain detectability, but by optimizing the extraction voltage and the focus voltage, a probe current of about 0.9 nA was attained. In addition, count throughput was increased by reducing the time constant of the pulse processor. This reduces the number of counts lost to pileup, and therefore also reduces dead time, but this is done with some loss of energy resolution. For the Al-Cu system, energy resolution is not important since the peaks are not subject to overlap, so the shortest time constant was used (Process Time $=1$ on the Oxford Link analyzer). Even with these improvements in the count rate, it was still necessary to count for quite a long time in order to gather reasonable statistics in each pixel.

X-ray maps of the Cu K and Al K peaks were gathered at 240 kX at a resolution of 256×256 pixels, with a dwell time of 40 ms per pixel (about 10% dead time), for a total frame time of about 1 hour. Maps were also gathered from background windows on either side of each peak, as well as an electron image from the bright-field detector. The field of view is 333 nm wide at this magnification, making each pixel 1.3 nm wide, so we have oversampled by a factor of about 1.8. The field-emission tip was cleaned before each map, and the total frame time was kept short, in order to ensure that the probe current remained reasonably constant during acquisition. In order to improve statistics, five separate acquisitions were performed from the same area and summed together off-line, giving a total dwell time of 200 ms per pixel. The pre- and post-peak background maps were averaged together and subtracted from their respective peak maps in order to remove the effects of the bremsstrahlung intensity before subsequent quantification.

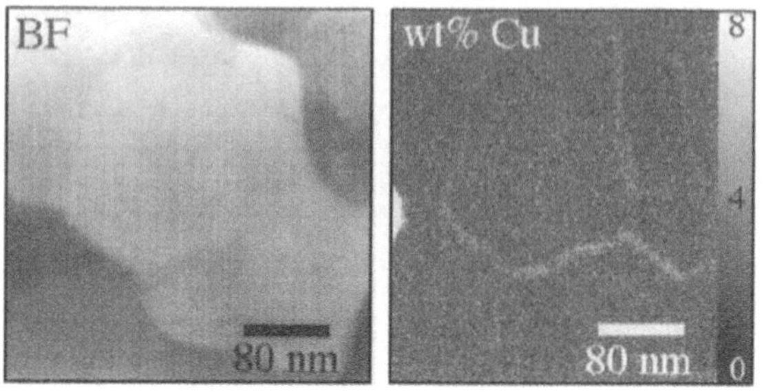

a b

Fig. 6. **a** A bright-field image of several grains in Al-4 wt. % Cu. A quantitative lower magnification map showing the Cu distribution across several grain boundaries. The lower intensity from the Cu-rich boundaries is due to the fact that the boundaries are generally not parallel to the electron beam. The white area at the edge of the image corresponds to a θ-phase particle, whose pixel values are off the scale (i.e. well above the maximum concentration)

The bright-field image (Fig. 6a) shows strong diffraction contrast, from which it can be seen that most of the boundaries in the region are not oriented in the ideal configuration parallel to the beam. In addition, all of the edges in the image appear to have periodic fringes parallel to the y direction. This effect is due to small amounts of drift (a few nanometers at most) in the y direction between maps. It can be avoided by shifting each image in the proper direction before adding them all together. However such image movement is not easily done manually. If drift correction were performed, some grain boundaries might appear to be slightly less inclined, but the effect is not a dominant one. Fig. 6b is the corresponding quantitative Cu concentration map, calculated using an experimentally determined k factor from nearby θ-phase particles. It shows several enriched boundaries of different apparent concentration, but since all are inclined with respect to the beam, explanations of these variations are non-trivial. One way to ascertain the actual reason for the variations in signal is to examine each boundary in the ideal configuration, but this would prove quite laborious and would likely cancel out any benefits of low magnification mapping. It is hoped that eventually it will be possible to correct for the effects of boundary inclination after the map has been gathered.

Conclusions

It is now possible to generate many accurate k factors in a rapid manner using the NIST SRM 2063. Quantification using the ζ factor eliminates the need to know the specimen thickness and density at the analysis point. Quantitative mapping with a spatial resolution $<\sim 5$ nm is achievable in modern intermediate-voltage FEG-AEMs.

Acknowledgments. DTC and DBW wish to acknowledge the support of NSF through the provision of a graduate research traineeship to DTC via grant #DMR 92-56322 as well as S. M. Zemyan and V. J. Keast for permission to use their unpublished data. MW wished to thank the Materials Research Center at Lehigh University for its support.

References

[1] G. Cliff, G. W. Lorimer, *J. Microsc.* **1975**, *103*, 203.
[2] G. W. Lorimer, S. A. Al-Salman, G. Cliff, in: *Developments in Electron Microscopy and Analysis* (D. L. Misell, ed.) Institute of Physics, Bristol, 1977, p. 369.
[3] R. H. McGill, F. H. Hubbard, in: *Quantitative Microanalysis with High Spatial Resolution* (G. W. Lorimer, M. H. Jacobs, P. Doig, eds.) The Metals Society, London, 1981, p. 30.
[4] J. E. Wood, D. B. Williams, J. I. Goldstein, *J. Microsc.* **1984**, *133*, 255.
[5] R. J. Graham, J. W. Steeds, *J. Microsc.* **1984**, *133*, 275.
[6] G. Wirmark, H. Norden, *J. Microsc.* **1987**, *148*, 167.
[7] P. H. Sheridan, *J. Electron Microsc. Tech.* **1989**, *11*, 41.
[8] J. I. Goldstein, J. L. Costley, G. W. Lorimer, S. J. B. Reed, *Scanning Electron Microscopy-1*, (O. Johari, ed.) IITRI, Chicago, IL 1977, p. 315.
[9] T. P. Schreiber, A. M. Wims, *Ultramicrosc.* **1981**, *6*, 323.
[10] D. B. Williams, D. E. Newbury, J. I. Goldstein, C. E. Fiori, *J. Microsc.* **1984**, *136*, 209.
[11] D. B. Williams, C. B. Carter, *Transmission Electron Microscopy: A Textbook for Materials Science*, Plenum, New York, 1996, p. 608.
[12] S. M. Zemyan, *Ph. D. Dissertation*, Lehigh University, 1995 p. 74; S. M. Zemyan, D. B. Williams, *J. Microsc.*, submitted for publication.
[13] S. D. Rasberry, *Certificate of Analysis for Standard Reference Material 2063*, National Bureau of Standards (now National Institute of Standards and Technology) Gaithersburg, MD 20899, USA, 1987.
[14] Z. Horita, T. Sano, M. Nemoto, *J. Electron Microsc*, **1986**, *35*, 324.
[15] Z. Horita, T. Sano, M. Nemoto, *Ultramicrosc.* **1987**, *21*, 271.
[16] Z. Horita, T. Sano, M. Nemoto, *Ultramicrosc.* **1991**, *35*, 27.
[17] E. Van Cappellen, *Microsc. Microanal. Microstruct.* **1990**, *1*, 1.
[18] W. A. P. Nicholson, K. Khan, in: *Microbeam Analysis-1990* (D. G. Howitt, ed.) San Francisco Press, CA 1990, p. 455.
[19] M. Watanabe, Z. Horita, M. Nemoto, *Ultramicrosc.* **1996**, *65*, 187.
[20] D. E. Newbury, R. B. Marinenko, R. L. Myklebust, D. S. Bright, in: *Images of Materials* (D. B. Williams, A. R. Pelton, R. Gronsky, eds.) Oxford University Press, New York, 1991, p. 290.
[21] C. E. Lyman, *Microscopy, the Key Research Tool*, Electron Microscopy Society of America, Woods Hole, MA, 1992, p. 1.
[22] J. I. Goldstein, D. E. Newbury, P. Echlin, D. C. Joy, A. D. Romig Jr., C. E. Lyman, C. E. Fiori, E. Lifshin, *Scanning Electron Microscopy and X-ray Microanalysis*, Plenum, New York, 1992, p. 525.
[23] C. E. Lyman, J. I. Goldstein, D. B. Williams, D. W. Ackland, S. Von Harrach, A. W. Nicholls, P. J. Statham, *J. Microsc.* **1994**, *176*, 85.

Mikrochim. Acta [Suppl.] 15, 59–64 (1998)

On the Spatial Resolution in Analytical Electron Microscopy

Aldo Armigliato[1,*], **David J. Howard**[2], **Roberto Balboni**[1], **Stefano Frabboni**[3], and **Matty R. Caymax**[2]

[1] CNR-Istituto LAMEL, Via P. Gobetti, 101, I-40129 Bologna, Italy
[2] IMEC vzw, Kapeldreef 75, B-3001 Leuven, Belgium
[3] INFM and Dipartimento di Fisica, Via Campi 213/A, I-41100 Modena, Italy

Abstract. The effect on spatial resolution of the spherical aberration of the objective prefield of a Philips CM30 (S)TEM has been investigated. For this purpose, the lattice mismatch of a 20 nm thick Si-Ge film in a cross-sectioned $Si/Si_{0.9}Ge_{0.1}/Si$ heterostructure has been determined by convergent beam electron diffraction (CBED), performed at 100 kV with a spot size of 20 nm. It has been found that, if the disk of minimum confusion is focused onto the specimen plane, the measurement is not affected by a change in the size of the condenser-2 aperture. Therefore, the mismatch information is contained within the Gaussian part of the electron probe. In strain profile determinations by CBED, this allows one to use higher beam intensities (larger C2-apertures) with improved signal/noise ratios and reduced sample drift due to shorter acquisition times.

Key words: analytical electron microscopy, spatial resolution, convergent beam electron diffraction, cross-sectioned samples, silicon-germanium alloys.

It is usually assumed that the spatial resolution in convergent beam electron diffraction (CBED) and in X-ray microanalysis of thin films for analytical electron microscopy (AEM) is given by the sum of the contributions due to the spot size and broadening of the beam.

The spot size is generally believed to be given by the image of the source in the Gaussian plane d_s and by the contribution of the spherical aberration d_{sa}, which is strongly dependent on the size of the second condenser lens (C2) aperture of the microscope. The term d_{sa} is very significant in the case of CBED experiments, where large C2 apertures (100, 200 µm) are employed. In this case the contributions to the spot size from diffraction aberration and astigmatism can be neglected, so the real size can be approximated as $d_{tot} = (d_s^2 + d_{sa}^2)^{1/2}$.

For instance, in a Philips CM30 AEM, equipped with a Twin objective lens, d_{sa} is 6.4, 51 and 406 nm for C2 apertures of 50, 100 and 200 µm, respectively. Therefore d_{tot}, and hence the spatial resolution in CBED experiments, should not be better than these values. In this paper, this will be investigated by measuring the local lattice mismatch through the shift of high order Laue zone (HOLZ) line patterns in different areas of cross-sectioned $Si/Si_{1-x}Ge_x/Si$ heterostructures at 100 kV using C2 apertures of 100 and 200 µm.

Experimental

$Si/Si_{1-x}Ge_x/Si$ heterostructures have been deposited by chemical vapour deposition (CVD) on ⟨001⟩ silicon wafers. Conditions for CVD include a wet etch of the wafer to remove the native oxide, an *in situ* high temperature bake in N_2, followed by CVD of a ≅ 1 µm Si buffer layer at 1050 °C, cooling of the wafer to 700 °C (in H_2), growth of a 20 nm thick $Si_{0.9}Ge_{0.1}$ film at 700 °C, and of a 110 nm thick Si-cap also at 700 °C. Si and Ge source gases were SiH_2Cl_2 and GeH_4. The fraction of Ge, x, was determined in thicker $Si_{1-x}Ge_x$ films, grown in identical conditions for longer times, by Rutherford backscattering spectrometry (RBS), which yielded a value $x = 0.105 \pm 0.005$. In addition, photoluminescence (PL) measurements of the heterostructure indicate that the $Si_{1-x}Ge_x$ layer is fully strained and without defects.

A Philips CM30 (S)TEM, equipped with a Gatan MSC 794 1-inch slow-scan CCD (SSCCD) camera and an EDAX 9900 energy dispersive spectrometer, has been employed for the CBED experiments and X-ray microanalysis of the heterostructures, respectively. The sample temperature has been kept at 100 °K, by

* To whom correspondence should be addressed

using a Gatan LN_2-cooled double-tilting holder, in order to both improve the HOLZ line visibility and to reduce contamination. The incident electron beam has been focused onto the sample, to obtain a disk of minimum confusion of about 20 nm (FWHM) in the specimen plane.

Results and Discussion

In Fig. 1 is shown a cross sectional TEM image of the Si-Ge heterostructure. From left to right are visible the surface, the 100 nm silicon cap, the 20 nm Si-Ge film and the top part of the 1 μm silicon buffer. The areas labelled A and B correspond to the regions selected for the CBED experiments. The Si-Ge film appears to be single crystalline and coherent with the Si buffer, and no dislocations were detected at the $Si_{0.9}Ge_{0.1}$/ buffer interface, confirming results from PL.

The sample is rotated about the [001] axis from the [110] direction, which is perpendicular to the plane of the TEM cross section, towards the [130] orientation. It has been proven that in the [130] orientation, at 100 kV, the HOLZ line pattern present in the central CBED disk allows accurate strain determination up to 20 at.% Ge in Si [1–3].

An electron beam of about 20 nm in diameter was then focused onto the heterostructure (point A in the figure) and two sets of CBED patterns were recorded by the SSCCD camera, with condenser 2 apertures of 100 and 200 μm, respectively.

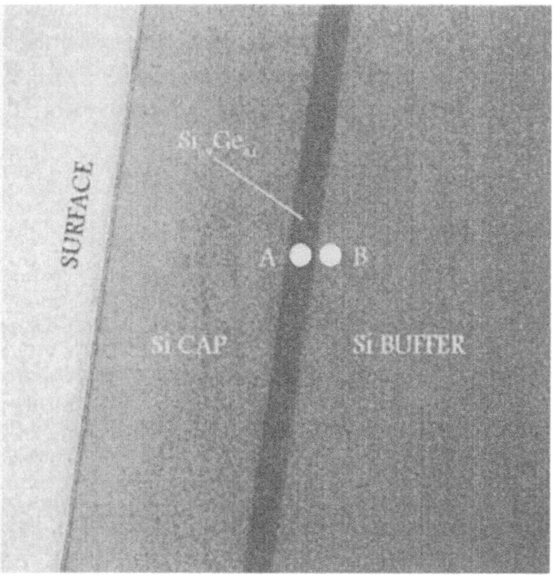

Fig. 1. Cross sectional TEM image of the 20 nm thick Si/ $Si_{1-x}Ge_x$/Si heterostructure ($x \cong 0.1$). *A* and *B* represent the points where the CBED patterns (and the X-ray spectra) have been taken. For the undeformed silicon, a region beyond the buffer has been chosen (outside the right edge of this figure)

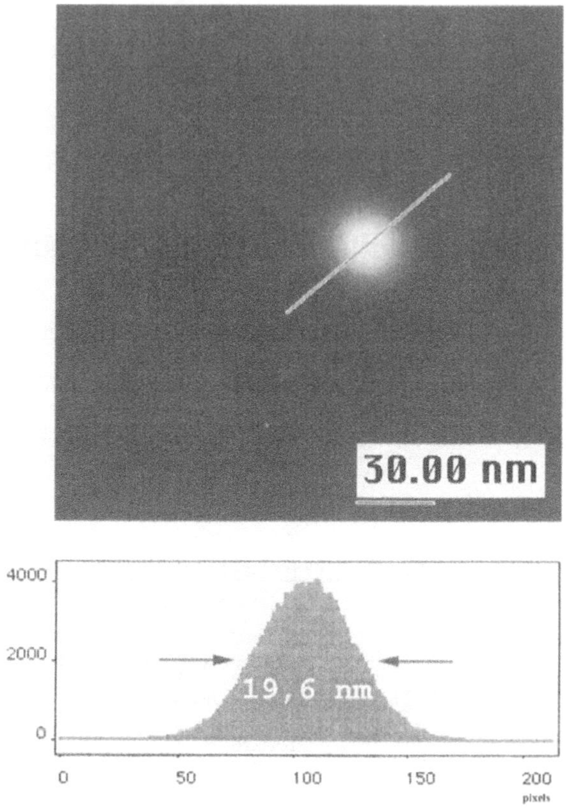

Fig. 2. Top: Image of the focused electron beam, in the case of a C2 aperture of 100 μm. Bottom: Beam profile along the line shown superimposed upon the image. The length of the line (51 nm) corresponds to the diameter of the aberrated halo, expected for the CM30 (S)TEM with this aperture (see text for details)

Fig. 2 shows the image of the focused spot in the case of the smaller aperture. The beam profile along the line superimposed onto the image is also shown. The full-width-half-maximum (FWHM) is 19.6 nm, whereas the full-width-tenth-maximum (FWTM) is 35.4 nm; the ratio is FWTM/FWHM = 1.82, which indicates that the spot is Gaussian [4, 5]. The profile has been drawn for a length of 50 nm across the spot, in order to include a region corresponding to the halo from spherical aberration of C2, as reported by Philips for the CM30/T (S)TEM [6]. This contribution, however, is not detectable in this profile. It is worth noting that all the images reported in this work have been taken in the so-called 'Gain Normalized Mode' of the Gatan Digital Micrograph software, i.e. the dark current of the CCD camera has been subtracted from each unprocessed image.

The CBED pattern obtained by locating this spot on the Si-Ge film (point A in Fig. 1) is shown in Fig. 3a. The arrows indicate the two pairs of (−1 1 13)/(2 0 −14) and (−1 1 −13)/(2 0 14) HOLZ lines, whose position is most sensitive to the lattice strain in the crystal. Their shift indicates a tensile strain, which is in fact expected for this heterostructure, due to the larger atomic radius of germanium with respect to silicon [7, 8]. The poor quality of the HOLZ lines in this pattern is due to the unfavourable ratio of the local thickness of the cross section t to the Si-Ge 'width' ($\lambda = 20\,\mathrm{nm}$). This results in a bending of the lattice planes in the illuminated volume and hence in a broadening of the corresponding HOLZ lines. According to Treacy and Gibson [9], this occurs when $\lambda/t < 0.6$, as in the present case. It is worth noting that this situation does not impair the purpose of this work, which is to demonstrate the achievable spatial resolution, rather than to measure accurately the strain. When $\lambda/t > 0.6$, it is possible to perform CBED measurements in areas far enough from the Si-Ge film edges to be sufficiently flat [9]; however, this criterion is impossible to fulfil in our sample, as it would imply a local thickness $t < \lambda/0.6 \cong 30\,\mathrm{nm}$, too small to give rise to detectable HOLZ lines.

The HOLZ pattern in Fig. 3b, taken in an area of the silicon buffer (point B in Fig. 1) quite close to the interface with the Si-Ge film, is markedly different from the previous one. Despite the poor quality of the lines, again due to the local distortion of the lattice planes, the two pairs of (−1 1 13)/(2 0 −14) and (−1 1 −13)/(2 0 14) HOLZ lines (see arrows) are now practically overlapping, giving the impression of only two visible HOLZ lines. This clearly proves that the spatial resolution in this CBED experiment is determined by the most intense part of the incident electron peak (Fig. 2), and the tails contribute only to the background in the central disk of the pattern. For comparison, the CBED pattern imaged in an undeformed region of the silicon substrate is shown in Fig. 3c. This CBED pattern is also required in the determination of the effective acceleration voltage [10]. The much narrower width of the HOLZ lines in Fig. 3c is due to the lack of local bending in the

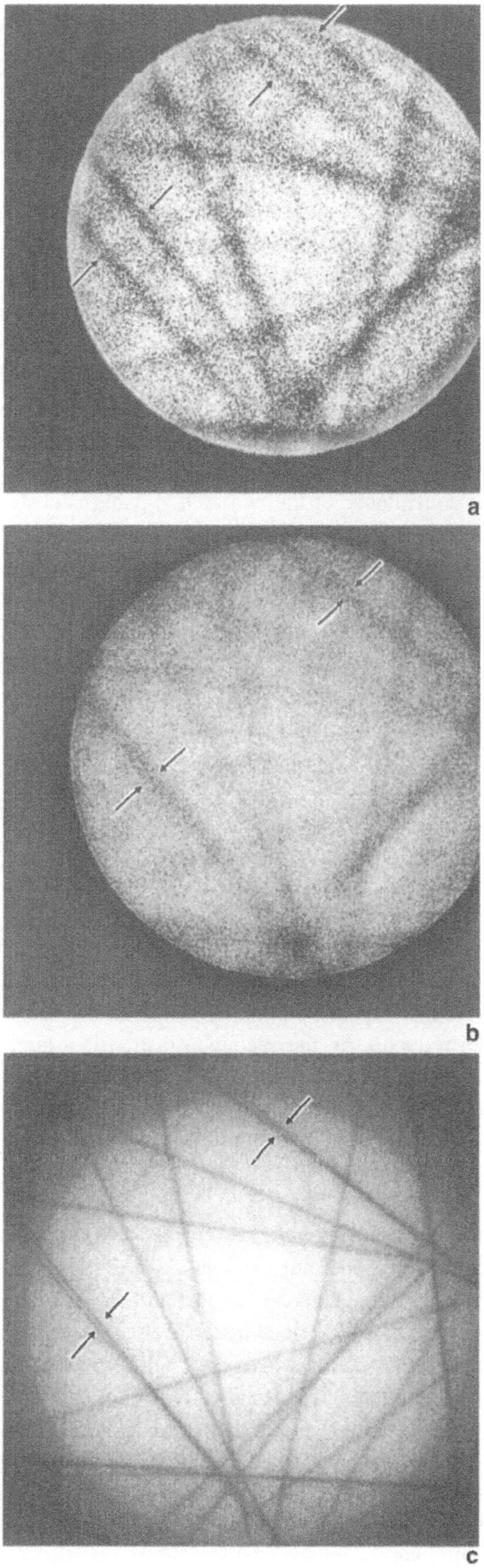

Fig. 3. [130] CBED patterns taken in **a** the region of the heterostructure (point A in Fig. 1), **b** the silicon buffer (point B) and **c** the silicon substrate. C2 aperture: 100 μm. The arrows mark the (−1 1 13)/(2 0 −14) and (−1 1 −13)/(2 0 14) HOLZ line pairs; they are widely separated in (a) and partially superimposed in (b) and (c)

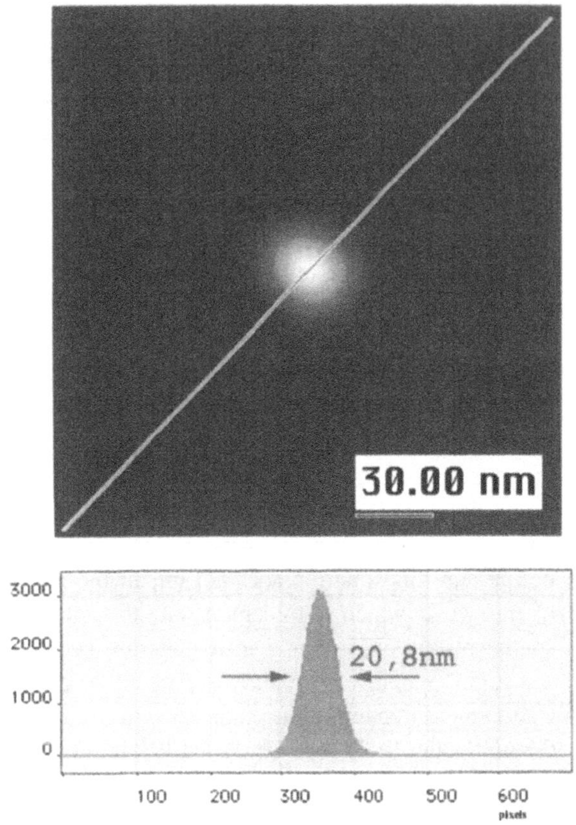

Fig. 4. Top: Image of the focused electron beam, in the case of a C2 aperture of 200 μm. Bottom: Beam profile along the line shown superimposed upon the image. The spot size is unaffected by the condenser aperture (compare with Fig. 2)

analyzed area; it is also evident that (−1 1 13)/(2 0 −14) and (−1 1 −13)/(2 0 14) HOLZ lines are quite close each other, as in Fig. 3b, which confirms that this latter pattern is generated by a silicon and not by a Si-Ge crystalline area.

As a further check that the spatial resolution in the local strain determination by CBED is mainly given by the Gaussian part of the beam rather than by the aberrated halo (if any), we have increased the size of the C2 aperture from 100 to 200 μm.

The image and the intensity profile of the focused beam is shown in Fig. 4. The FWHM of this spot is still 20.8 nm and the FWTM is 37.3 nm, again close to the theoretical ratio of 1.82. The profile reported has a width of 270 nm only, i.e. much smaller than the width of the expected halo (still invisible here), which should be of 406 nm for a 200 μm C2 aperture [6].

The CBED patterns obtained with the 200 μm aperture are reported in Figs. 5 a–c. Fig. 5a corresponds to the Si-Ge film, i.e. point A in Fig. 1. Due to the increased illumination, a higher diffracted intensity is available in the central disk which, in turn, results in an improved signal/noise ratio (compare Fig. 5a and 3a). What is important to note is that the HOLZ-line pattern is practically the same in the two cases (i.e. 100 and 200 μm C2 aperture): the two pairs of (−1 1 13)/(2 0 −14) and (−1 1 −13)/(2 0 14) HOLZ lines have the same separation, although this is not immediately apparent, due to the different magnification (camera length) chosen in Fig. 5, in order to keep the image of the C2 aperture within the 1-inch CCD camera. For the sake of clarity, the disk bounding the patterns in Fig. 3a–c (image of the 100 μm aperture) has been superimposed onto the pattern in Fig. 5c with the appropriate size.

The equivalence of the two HOLZ patterns clearly demonstrates that the spatial resolution is the same, regardless of the increased size of the C2 aperture. In orther words, the resolution is related to the peak intensity of the electron beam, rather than to the tails. This is further confirmed by the CBED pattern taken in Fig. 5b (point B in Fig. 1), which corresponds to the silicon buffer. The arrangement of the HOLZ lines is the same as in Fig. 3b, that is it is typical of a silicon lattice and not of a Si-Ge crystal.

For comparison, the CBED pattern taken in the same undistorted region of the substrate is shown in Fig. 5c; as expected, it is identical to the one in Fig. 3c, apart from the better signal/noise ratio.

There are two distinct advantages in using larger C2 apertures: (i) the beam intensity is higher, which allows reduced recording times of the patterns, with a parallel reduction in the specimen drift; (ii) a larger portion of the reciprocal lattice can be imaged, which increases the number of HOLZ lines available for comparison between the simulated and the experimental patterns, and hence permits a more accurate strain determination.

From the experimental patterns it is possible to deduce the tetragonal mismatch m_T, due to the Si-Ge/Si heterostructure. This can be accomplished according to a procedure described in detail elsewhere [1–3]. Briefly, a set of points where the HOLZ lines in the CBED pattern interact kinematically is chosen, then a number of distances among them are measured. The same distances are computed in patterns simulated for different values of the lattice parameters, according to an iterative procedure based on a χ^2-criterion [11]. From the values of the lattice constants (in the illuminated region) in the two directions – parallel to the film growth and perpendicular to the TEM sample – the bulk value of the distortion m_T can be deduced [3].

In our case, where the Si-Ge film has been coherently grown onto the buffer silicon, the bulk value obtained by the CBED method is $m_T = (6.2 \pm 0.4) \times 10^{-3}$. This is consistent with the Ge concentration of 10.5 ± 0.5 at.% (measured by RBS), as it results from the relation $m_T = 1.77\beta N x$, where $\beta = 7.53 \times 10^{-25}\,\mathrm{cm}^{-3}$ is the expansion coefficient, $N = 5 \times 10^{22}$ at/cm^3 is the silicon atomic density; in fact, $x = 15 m_T = 0.093 \pm 0.006$, i.e. C(Ge)$= 9.3 \pm 0.6$ at%.

The independence of the spatial resolution from the size of the C2 aperture is further confirmed by the X-ray microanalysis measurements. X-ray spectra have been collected at the same points A and B of the heterostructure in Fig. 1, again with the two different C2 apertures. The ratio between the GeKα and the SiKα intensities in the spectra have been evaluated; a value R = I(SiKα)/I(GeKα) = 7 has been found for both the 100 and the 200 µm C2 apertures. Moreover, no GeKα signal is detected when the beam is focused onto the Si buffer area (point B).

Although an accurate determination of the Ge concentration has not been performed, the above R value corresponds to a Ge concentration somewhat smaller than the expected one of 10 at.%. This can be attributed to the beam broadening effect, which increases the SiKα peak at the expense of the GeKα peak to a larger extent than the tails of the Gussian beam.

The reason why the aberrated halo from the prefield of the objective lens is practically invisible in the beam profiles in Figs. 2 and 4, unlike those shown by Cliff and Kenway [4], may be due to the fact that we have focused onto the specimen plane, rather than the Gaussian probe, the disk of minimum confusion (also Gaussian). This implies on one hand that, for a given excitation of the Cl lens, the corresponding minimum geometric spot size is not achieved, but, on the other hand that, for the large C2 apertures used in this work,

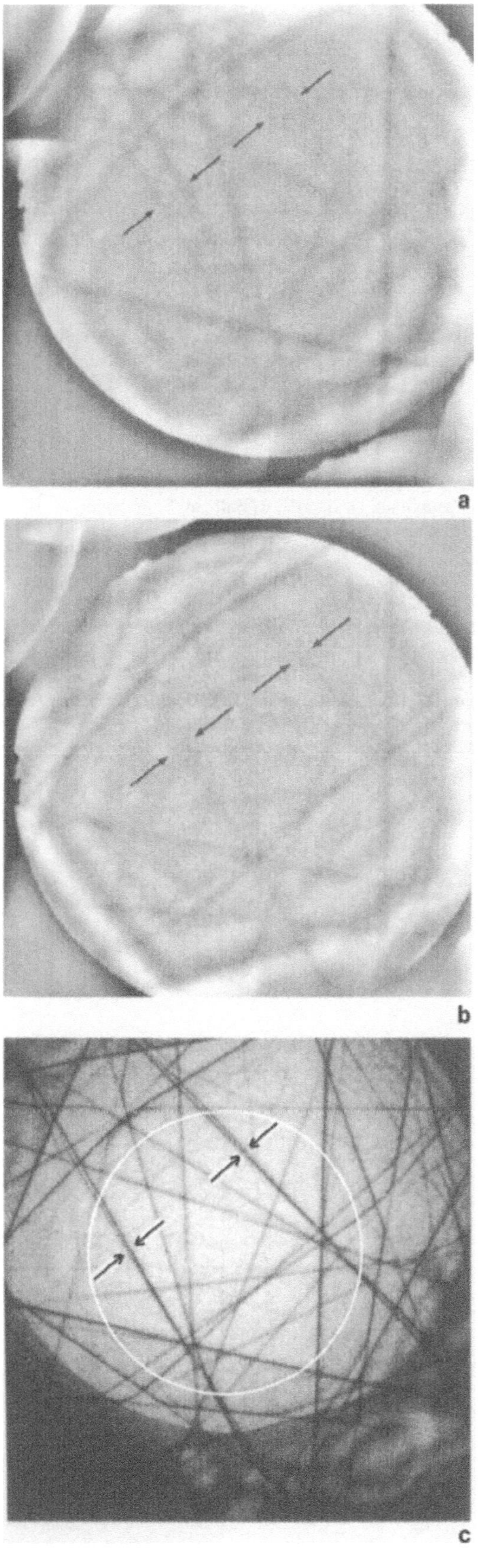

◄

Fig. 5. [130] CBED patterns taken in **a** the region of the heterostructure (point A in Fig. 1), **b** the silicon buffer (point B) and **c** the silicon substrate. C2 aperture: 200 µm. For the sake of clarity, the disk of Fig. 3a (image of the 100 µm aperture) has been superimposed onto the pattern in (c) with the appropriate size. The arrows mark the (−1 1 13)/(2 0 −14) and (−1 1 −13)/(2 0 14) HOLZ line pairs; note the strong similarity between those HOLZ patterns and the ones in Fig. 3

the overall beam diameter d_{tot} practically coincides with that of the disk of minimum confusion (i.e. smaller than $(d_s^2 + d_{sa}^2)^{\frac{1}{2}}$). This should be kept in mind when estimating the spatial resolution to be expected in a CBED experiment.

Conclusions

The spatial resolution attainable in the lattice strain measurements by CBED in an analytical electron microscope is limited practically by the FWHM size of the Gaussian electron probe rather than by the much larger aberrated halo, from the spherical aberration of the prefield of the objective lens. This has been clearly demonstrated by measuring the HOLZ-line shift in the central disk of CBED patterns taken inside and outside the Si-Ge film of a Si/ $Si_{1-x}Ge_x/Si$ heterostructure. This shift is unaffected by the use of different C2 apertures, and the resulting value of the bulk mismatch of the heterostructure is in good agreement with the Ge concentration (10 at.%).

The possibility of using larger apertures without impairing the spatial resolution offers the advantage of higher beam intensities, which improves the HOLZ line visibility and reduces the recording times of the CBED patterns.

Acknowledgements. The authors are indebted to P. Kenway for useful discussions and to S. Balboni for his helpful assistance.

References

[1] A. Armigliato, R. Balboni, S. Frabboni, in: *Materials Research Society Symp., Proc. Vol. 405* (S. M. Prokes, R. C. Cammarata, K. L. Wang, A. Christou, eds.), Materials Research Society, Pittsburgh, 1996, p. 467.

[2] R. Balboni, A. Armigliato, S. Frabboni, *Appl. Phys. Lett.* **1996**, *68*, 2831.

[3] R. Balboni, S. Frabboni, A. Armigliato, *Phil. Mag.* **1998**, *A77*, 67.

[4] G. Cliff, P. B. Kenway, *Microbeam Analysis-1982* (K. F. J. Heinrich, ed.) San Francisco Press, San Francisco, 1982, p. 107.

[5] D. B. Williams, C. B. Carter, *Transmission Electron Microscopy*, Plenum, New York, 1996, Ch. 33.

[6] *CM30 (S)TEM User's Handbook*, Philips Electron Optics, Eindhoven, NL, 1988.

[7] A. Armigliato, D. Govoni, R. Balboni, S. Frabboni, M. Berti, F. Romanato, A. V. Drigo, *Mikrochim. Acta* **1994**, *114/115*, 175.

[8] A. Armigliato, R. Balboni, F. Corticelli, S. Frabboni, F. Malvezzi, J. Vanhellemont, *Mat. Sci. Technol.* **1995**, *11*, 400.

[9] M. M. J. Treacy, J. M. Gibson, *J. Vac. Sci. Technol. B* **1986**, *4*, 1458.

[10] J. W. Steeds, in: *Introduction to Analytical Electron Microscopy* (J. J. Hren, J. I. Goldstein, D. C. Joys, eds.) Plenum, New York, 1979, p. 387.

[11] J. C. H. Spence, J. M. Zuo, *Electron Microdiffraction*, Plenum Press, New York, 1992.

Mikrochim. Acta [Suppl.] 15, 65–71 (1998)

Contamination in Analytical Electron Microscopy and in ALCHEMI

János L. Lábár[1,*], Miklós Adamik[1], and István Dódony[2]

[1] Research Institute for Technical Physics and Materials Science, P. O. Box 49, H-1525 Budapest, Hungary
[2] Eötvös University, Department of Mineralogy, Múzeum krt. 4/a, H-1088 Budapest, Hungary

Abstract. The amount of contamination was characterised quantitatively by the number of emitted carbon K_α X-ray photons. The magnitude of the channelling effect was plotted as a function of contamination. The Al/Si ratio varied by 80%, while the O/Si ratio only changed by 10%, ruling out x-ray absorption as the major cause of the observed variation. The randomised direction of bombarding electrons (caused by the contamination hill) was identified as the origin of the decreasing channelling effect as contamination increased. Dependence of contamination on the method of sample preparation and experimental condition were measured. Contamination affects a disc-shaped area of the sample with a diameter 10 times larger than that of the visible contamination hill. Flooding with a high dose of electrons was found to produce a 2 minute interval almost free from contamination for subsequent analysis. Reduced contamination facilitated the identification of a carbide inclusion in a steel matrix.

Key words: AEM, channelling, contamination, garnet, steel.

Introduction

ALCHEMI and Contamination

The usual assumption in electron probe X-ray microanalysis of thin samples in the analytical electron microscope (AEM) is that the unit cell of the single crystal is uniformly illuminated by the electron probe, i.e. that no information from the crystalline state needs to be taken into account. In that approximation, analyses are identical, irrespective of whether the analysis is carried out on a single crystal, polycrystal or amorphous material. In other words, the X-ray emission yield is treated as isotropic and independent of the incident beam direction. The observation that X-ray intensities do depend on the direction of the incident electron probe is called the channelling effect [1]. Such an effect can only be observed in single crystals oriented to a strong diffraction condition (i.e. close to a zone axes or to a two-beam condition). The origin of the effect is that dynamical elastic scattering concentrates electron intensity non-uniformly within the unit cell of the crystal. The subsequent inelastic scattering (ionisation) is proportional to the local electron intensity at the positions of the different atoms in the crystal. As the direction of the incident electrons is changed in relation to the crystallographic axes of the crystal, the local electron intensity is also changed since the excitation of the different Bloch-waves (the eigenstates of the fast electrons within the single crystal) varies accordingly. Variation of the electron intensity with orientation can be calculated by the Bloch wave method [2]. Total X-ray intensity follows the variation of this electron intensity and if the crystallographic positions of the elements in a multicomponent sample satisfy certain conditions, the ratio of the emitted characteristic X-ray lines also varies with the orientation (see Fig. 1). If this variation is disregarded, quantitative analysis will be in error. That is why strong diffracting conditions must be avoided in microanalysis. However, if the variation of the excitation is taken into account, not only can errors be eliminated but a new piece of information can also be

* To whom correspondence should be addressed

obtained from the experimental data. The method ALCHEMI (atomic locations by channelling enhanced microanalysis) uses this signal to determine the crystallographic position of a minor element within the single crystal [3]. X-ray intensities of major components of known crystallographic position serve as reference to probe the variation of electron intensity within the unit cell [3]. The intensity ratios are quantitatively compared in evaluating occupancy factors (i.e. that fraction of the examined minor elements to be found at certain crystallographic sites). In spite of the numerous publications applying either traditional [3] or statistical [4], [5] ALCHEMI, the effect of contamination on the channelling effect is either not examined or explicitly regarded as "not a problem" [4]. However contamination has two effects. On the one hand, the amorphous carbon layer in the contamination hill enhances absorption of the emitted X-rays on their way to the detector. (The hard radiation of Cr and Mn $K\alpha$ in Ref. [4] is insensitive to absorption in carbon.) On the other hand, scattering within the contamination hill randomises the directions of the originally collimated electron beam. Since the magnitude of the measured channelling effect depends on the collimation of the incident electron probe, the claimed insensitivity to contamination is in contrast to our expectations. Infact, contrary to the usual assumption, we found that the amorphous contamination hills do affect the magnitude of the channelling effect.

Variation and Reduction of Contamination

The analytical electron microscope (AEM) is able to characterise material at the 10 nm level (with an LaB_6 cathode). However, it has been known for decades that reduction of the probe size increases contamination [6]. Residual organic vapours in the vacuum, on the one hand, and contaminants residing on the sample, on the other hand, were long ago identified as the two main sources of contamination. Solutions were also suggested in the literature to reduce this contamination: namely, use of a stage which cools the sample to a temperature close to that of liquid nitrogen or increasing sample temperature above 200 °C [7]. A special means of "cleaning" is flooding the sample with high dose of electrons (provided the sample is not prone to beam damage). Maximum spot-size (current) and a large C2 aperture is applied prior to analysis in this latter case. The high current density over a large area results in depletion of a larger surface region from contaminant atoms around the point of interest, which is then analysed immediately after switching back to small spot size. This method proved to be efficient and to create about two-minute intervals in which the investigated area is almost free from contamination for subsequent examination [8].

Experimental

A CM-20 TEM equipped with an LaB_6 gun and double tilt (room temperature) analytical holder and a high purity germanium (HPGe) detector in a NORAN Voyager 3.2 energy dispersive (EDS) X-ray analyser was used for the experiments. The atmospheric thin window (ATW) and parameters of the detector facilitate analysis down to (and including) boron. Although analysis of boron is impractical for most of the samples, sensitivity for the other light elements (carbon, nitrogen and oxygen) is excellent; therefore the carbon count rate provided reliable data for quantifying contamination.

Contamination rate was examined on a pure Cu sample in order to measure quantitatively the area affected by contamination and the efficiency of flooding in reducing build-up of contamination. It was also of interest to establish how sample preparation influences the amount of contamination adhering to the surface of the sample when introduced into the microscope. Ion milling is a widely used and efficient method to produce TEM samples from a large variety of materials [9]. Two techniques, i.e. the preparation of cross sectional and plan-view specimens, were applied in the present experiments. In the first case, pieces of the sample are glued together into an appropriate holder and fixed into the miller after mechanical thinning. In the second case a disk of the sample material is subjected to the ion beam. The question arises if, in the case of cross sectional sample preparation, the remains of glue and/or solvents can influence the contamination rate during EDS analysis. In the present experiments pieces of a 25 μm thick copper sheet were prepared by both methods, since copper is known to be susceptible to electron beam induced contamination in the electron microscope [10]. The cross sectional preparation has beam carried out by gluing the sheet between two Si wafer pieces. For this purpose Araldite ATI, carbon (4:1) mixture has been used. The plan-view specimen served as a reference, since it was not treated with any substance which later could become a source of surface contamination. Condensation of residual vapours is assumed to be identical for the two samples. Experimental conditions were kept fixed and identical for the two sets of analyses. A liquid nitrogen cooled anti-contamination device was applied prior to and during the measurements. The beam was switched on after about 15 minutes following the insertion of the sample when $2*10^{-5}$Pa was reached. A fresh area of the sample was selected immediately after alignment of the microscope and spectrum collection started at once. New spectra were recorded after a controlled period of time and steps (and duration) of treatment (e.g. flooding, etc.) were documented. All the spectra were collected for 100 s live time. A nominal spot size of 27.5 nm produced a contamination spot of 100 nm in each case. The effect of re-sputtering during ion milling was also examined. Spectra collected at $2*10^{-5}$Pa are compared to those at $4*10^{-5}$Pa. In the first case, the plan-view sample was cleaner.

An austenitic steel sample was used to demonstrate the applicability of flooding to practical examinations. Different

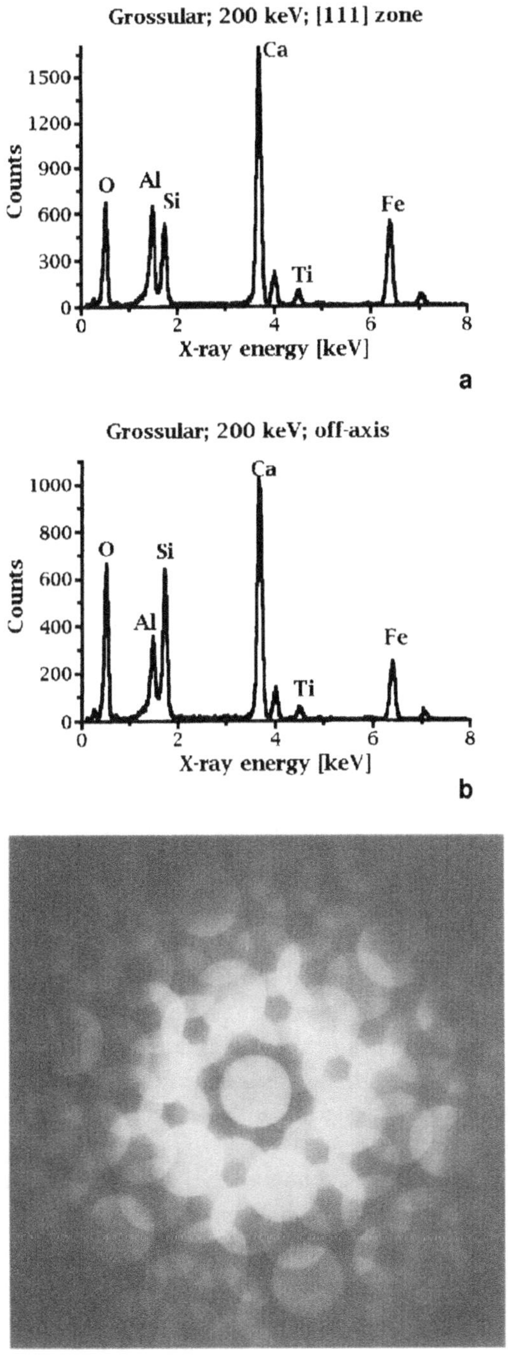

Fig. 1. Spectra of Grossular recorded at 200 keV at. **a** Exact [111] zone axis and **b** off-axis orientation. **c** CBED pattern of **a**, demonstrating the exact [111] orientation

inclusions were analysed in a search for a carbide phase. The channelling effect was studied on a single crystal of Grossular at 200 keV. Grossular has a garnet structure with the general formula $X_3Y_2Z_3O_{12}$, where X, Y and Z designate metallic components occupying dodecahedral, octahedral and tetrahedral positions in the crystal, respectively. X-sites are occupied by Ca, Y-sites by Al and Z-sites by Si, while Fe is incorporated as an additional element. The sample was pre-aligned to ensure that, after 20° tilt toward the detector, the orientation was close to the required [111] zone axis. Mechanical polishing and ion beam milling were the same as for the cross sectional Cu sample. No special precaution (except for the use of an anti-contamination device) was applied in this case so as to obtain conspicuous contamination and facilitate quantification of its effect. The exact orientation was set by observing the diffraction pattern using a convergent electron beam as shown in Fig. 1c. Beam convergence was 4 mrad, determined by a top-hat type thick C2 aperture used for EDS analysis.

Results and Discussion

Channelling

The aluminium/silicon intensity ratio was selected to characterise the magnitude of the channelling effect. It was plotted against the peak height of carbon K_α radiation which is a measure of contamination. Linear dependence can be observed in the given interval in Fig. 2a. The question arises as to whether the measured intensity really alters due to the change in the channelling effect or (as an alternative explanation) because absorption within a contamination layer of increasing thickness alters the ratio of the measured intensities. Mass absorption coefficients for Al K_α, Si K_α and O K_α radiation in carbon are 445, 704 and 12400 cm^2/g, respectively. Consequently, neither Al nor Si K_α radiation is sensitive to the absorption in carbon, in contrast to O K_α radiation which is severely absorbed. It follows that any given thickness of contamination would seriously decrease the measured O/Si ratio but only slightly decrease the Al/Si ratio. In our case however, the O/Si ratio decreased by contamination an order of magnitude less than the Al/Si ratio (compare Fig. 2b to Fig. 2a), confirming that the main cause for the change is not absorption. Examination of Fig. 2c further confirms this claim, since the same amount of contamination in an off-axis orientation does not change the Al/Si ratio significantly. In summary we can state that the change in the Al/Si ratio (i.e. in the channelling effect) is induced by a change in the distribution of directions of the incoming electron beam due to the scattering in the contamination hill.

One might wonder how the same hill can be a significant source of scattering for electrons and be insignificant for the absorption of X-rays simultaneously. On the one hand, the strength of interaction of electrons with matter is stronger by several orders of magnitude than that of X-rays. On the other hand,

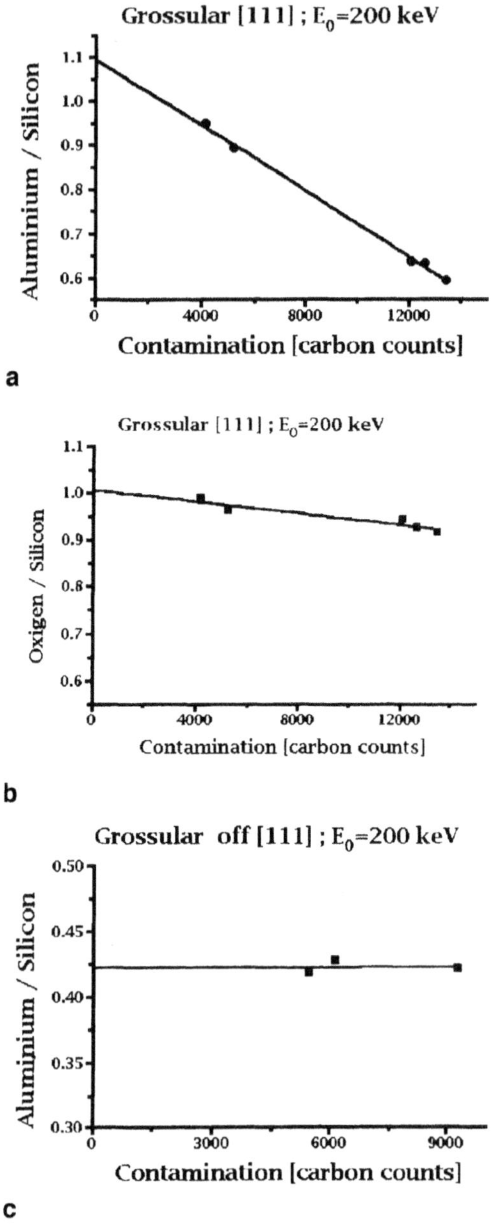

Fig. 2. Ratio of measured X-ray intensities as deduced from spectra of Grossular measured at 200 keV. **a** Exact [111] zone axis. Al/Si. **b** Exact [111] zone axis. O/Si. **c** Off-axis. Al/Si

electrons penetrate the full height of the contamination hill, while X-ray photons only need to cross a moderate thickness toward the detector which is situated at 14°. Fig. 3 illustrates this differences for the given geometry (20° tilt of the sample towards the detector resulting in 34° take-off-angle). Anyone who has seen a TEM photograph of a tilted contamination hill knows that it is extremely high and narrow with a tail at the bottom spreading in the direction of the

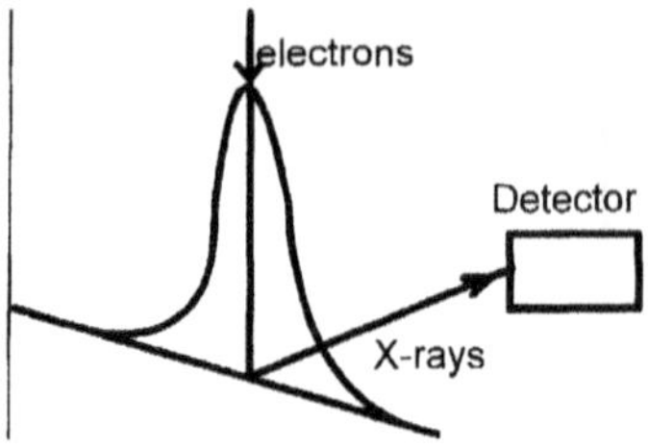

Fig. 3. Schematic geometry of a contamination hill for a tilted sample. Absorption path length for X-rays is also indicated together with the path of the incoming electrons

surface (which is not the direction of the detector in our case). It is expected that the intensity ratios measured at higher and higher contamination (i.e. higher and higher randomisation of directions) will approach the value measured at a random orientation (modified by differential absorption). The data in Fig. 2 are consistent with this expectation.

Altogether, the contamination hills have a major influence on the magnitude of the channelling effect and this is independent of (and adds to) the differential absorption of X-rays from individual components of the sample. Consequently, if the channelling effect is to be quantified, the effect of contamination must either be kept at minimum or must be taken into consideration during quantification. An effective method (flooding) is presented below for minimising contamination. An extrapolation method similar to that of Horita and co-workers can be used for estimating the contamination effect in an analogous way to that used for taking differential absorption into account [11].

As an alternative solution, the decrease of the channelling effect due to contamination can be used in a similar manner to an independent measurement carried out at a different channelling orientation. Details of such an approach are in preparation.

Reduction of Contamination

Fig. 4a, b demonstrate that contamination becomes very serious in a two minute interval even at a relatively good vacuum ($2*10^{-5}$Pa) in spite of the fact that the finest available spot size was not used. Contamination becomes even stronger if the minimum spot size is applied. Such a high artificial carbon peak would prevent us from establishing the carbon content of a sample.

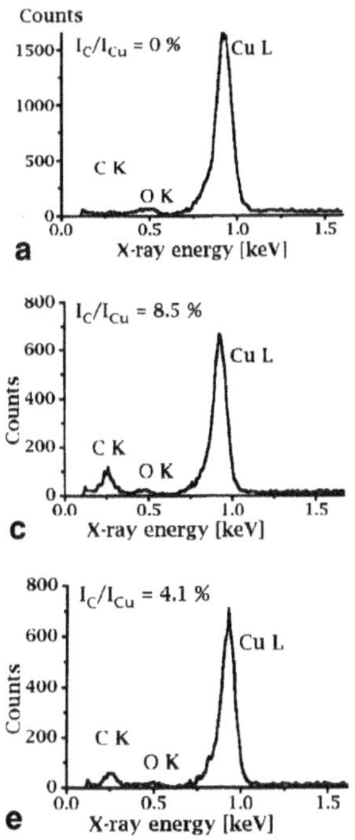

Fig. 4. Spectra from the plane-view Cu sample: **a** Immediately after insertion, $2*10^{-5}$Pa. **b** After 2 minutes measurement. **c** Fresh position $4*10^{-5}$Pa. **d** Right after positioning the beam 850 nm away from a contamination spot. **e** After flooding with electrons

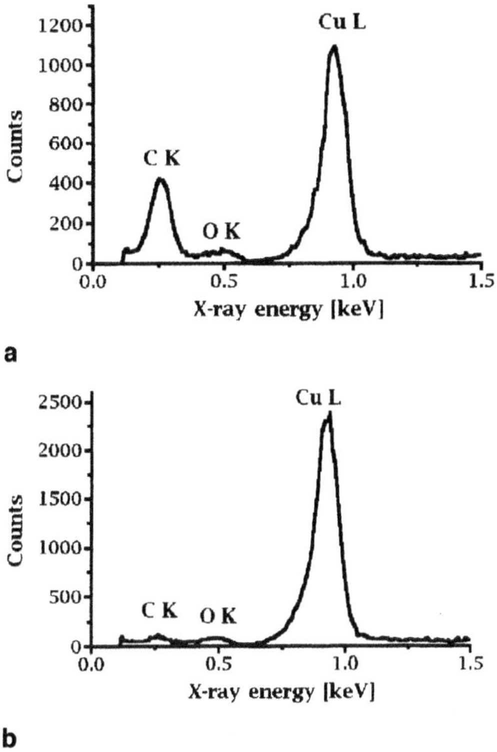

Fig. 5. Spectra form cross sectional Cu sample at $4*10^{-5}$Pa. **a** Fresh area close to the glue, **b** the same area after flooding with electrons

Contamination obviously increases in a worse vacuum condition ($4*10^{-5}$Pa) as can be seen in Fig. 4c. Although a contamination spot of 100 nm diameter was observed on the TEM screen, a much larger area is affected by contamination. Fig. 4d shows that any subsequent analyses, as far as 8–900 nm away from the previous one, are more seriously affected by contamination than an analysis on a fresh sample area. In other words, a diameter about ten times larger (i.e. an area of hundred times larger) than the diameter (area) of the visible contamination spot has accumulated carbon contamination as a consequence of the previous analysis.

Flooding with a high dose of electrons cannot remove the visible contamination spot (of cracked hydrocarbons). However, it effectively removes the accumulated (but not cracked) contamination from the large area. Not only can the level of cleanliness be restored to that existing prior to the first analysis, but an even higher level of cleanliness can be reached by this procedure (carbon signal in Fig. 4e is half of that in Fig. 4c).

Contamination is higher when a cross sectional sample is analysed close to the remains of the glue (Fig. 5a). However, the reduction of contamination by flooding is even more conspicuous for this more contaminated sample (compare Fig. 5a with Fig. 5b). It seems to be a general rule that the worse the vacuum and the more contaminated the surface of the sample, the larger the reduction of contamination due to flooding by electrons. (However, larger fractional reduction does not necessarily mean a cleaner surface at the end.)

The applicability of flooding to a practical problem was tested using an austenitic steel sample. A high contamination rate prevented examination of the carbon content in small inclusions when only the usual (liquid nitrogen cooled) anti-contamination device was used (see Fig. 6a). Flooding with electrons effectively reduced contamination enabling us to identify a carbide inclusion (Fig. 6b). Although a small carbon peak was also present in the spectrum (Fig. 6c) of the matrix (and this carbon is a sign of contamination because it is higher than that which can

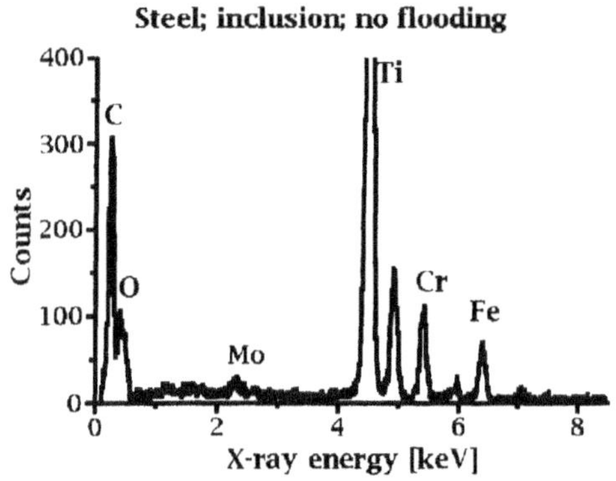

a

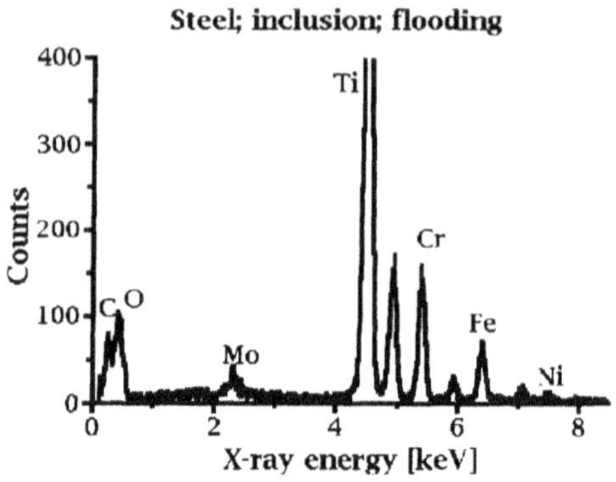

b

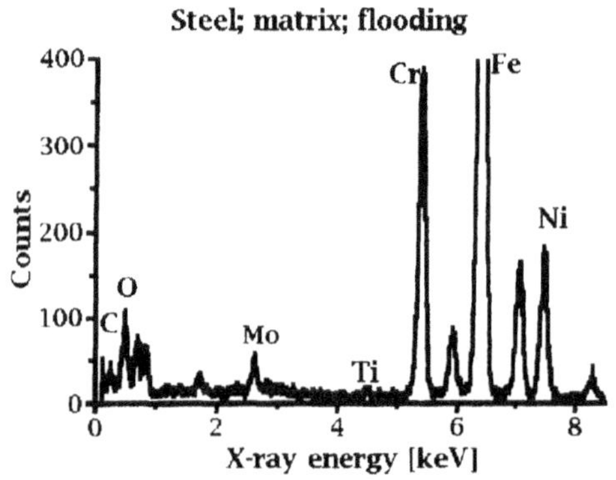

c

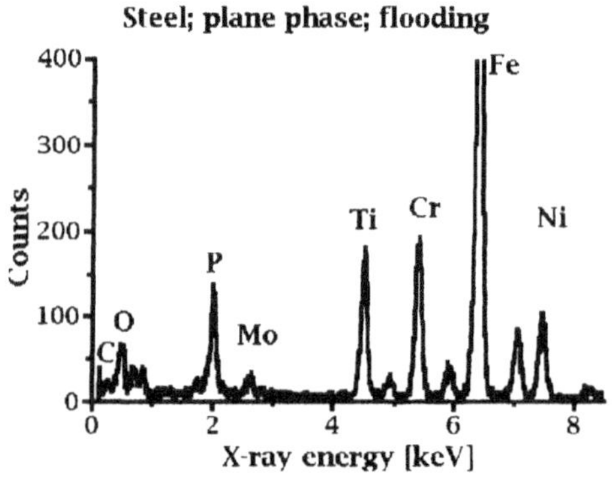

d

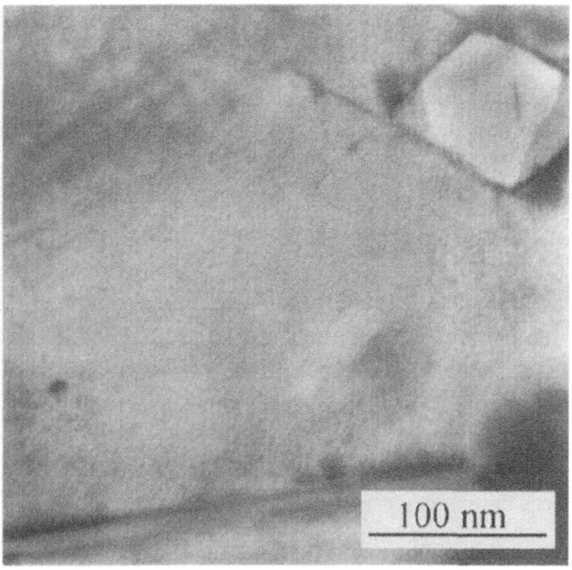

e

Fig. 6. Austenitic steel sample; 200 keV. **a** Inclusion without flooding. **b** Inclusion with flooding. **c** Matrix with flooding. **d** Plane shape phase with flooding. **e** TEM image of the analysed area

be expected from the true carbon content of the steel) this is significantly lower than observed in the inclusion. Similarly, the other (plane shaped) phase in Fig. 6e was analysed (Fig. 6d) and showed an insignificant carbon peak. The carbon peak in Fig. 6b and 6d can be regarded as spectral background from carbon contamination and the excess amount of carbon of Fig. 6c demonstrates that the inclusion is a carbide.

Conclusions

1. The usual treatment for cross sectioned samples may leave contaminant on the surface but this can be effectively removed by flooding with electrons. A significant fraction of this contamination can originate from re-sputtering during ion milling. Careful final milling with low energy ions can effectively reduce the amount of these resputtered contaminants.
2. The contamination rate is higher close to the glue.
3. A focused beam produces significant accumulation of contaminants in a region about 10 times larger than the diameter of the visible contamination spot; however this accumulated contamination can be removed by flooding.
4. Contamination spots themselves cannot be removed by flooding.
5. Effective reduction of contamination enables us to identify carbide inclusions in steel samples.
6. Reduction of the amorphous contamination hills on top of the sample enhances the channelling effect which is the basis of the method ALCHEMI (i.e. determination of the distribution of atomic species on crystallographic sub-lattices.)

Acknowledgement. Financial support of the National Research Fund (No. OTKA T015895) is acknowledged.

References

[1] P. B. Hirsch, A. Howie, M. J. Whelan *Phil. Mag.* **1962**, *7*, 2095.
[2] P. A. Stadelmann, *Ultramicroscopy* **1987**, *21*, 131.
[3] J. C. Spence, J. J. Tafto, *J. of Microscopy* **1983**, *130*, 147.
[4] C. J. Rossouw, P. S. Turner, T. J. White, A. J. O'Connor, *Phil. Mag. Lett.* **1989**, *60*, 225.
[5] J. Morgiel, *Proc. IX. Conference on Electron Microscopy of Solids, Krakow-Zakopane*, 1996, p. 133.
[6] J. J. Hren, J. I. Goldstein, D. C. Joy (eds.), *Introduction to Analytical Electron Microscopy*, Plènum, New York, 1979.
[7] L. Reimer, *Transmission Electron Microscopy*, Springer, Berlin Heidelberg New York. Tokyo, 1989, Ch.10.4.
[8] R. Voss, G. Lehmpfuhl, P. J. Smith, *Z. Naturforsch.* **1980**, *35a*, 973.
[9] Á. Barna, *MRS Conf. Ser.* **1991**, *254*, 3.
[10] G. F. Bastin, H. J. M. Heijligers, in: *Microbeam Analysis-1985* (A. D. Jr. Romig, and J. I. Goldstein, eds.), San Francisco Press, 1985, p. 1.
[11] Z. Horita, T. Sano, M. Nemoto, *Ultramicroscopy* **1987**, *21*, 271.

Mikrochim. Acta [Suppl.] 15, 73–76 (1998)

Analytical Electron Microscopy of Diffusional Interfaces in an Al-22 at.% Zn Alloy

Paweł Zięba

Alexander Krupkowski Institute of Metallurgy and Materials Science, Polish Academy of Sciences, 25 Reymonta Street, 30-059 Cracow, Poland

Abstract. High spatial resolution X-ray microanalysis of thin foils, in a transmission electron microscope, was applied to determine solute content profiles existing across the α lamellae formed by discontinuous precipitation (DP) in an Al-22 at. % Zn alloy. This technique, combined with the *in situ* observation of the kinetics of the process, allowed us to treat the DP reaction as a local phenomenon. The calculated diffusivity at a moving reaction front was found to be in good agreement with the data on diffusion at a stationary grain boundary and, in this way, the discrepancies resulting from the consideration of the DP as a global reaction were eliminated

Key words: analytical electron microscopy, EDX analysis, moving grain boundaries, discontinuous precipitation.

Some solid-solid phase transformations involve fast diffusion processes within a mobile reaction front (RF) such as a grain or interphase boundary. Discontinuous precipitation (DP) is the most widely studied case of such a transformation. The decomposition of the α° supersaturated solid solution, which takes place during DP, results in the formation of new solute-rich β and solute-depleted α phases, usually in the form of alternating lamellae. The investigations of the DP kinetics were focused on the estimation of diffusivity ($s\delta D_b$) at the moving RF based on the existing diffusion models [1–5]. Here, D_b is the diffusion coefficient, δ is the width of the grain boundary and s is the Gibbs segregation factor. Most of the results were obtained by assuming the DP reaction to be a global phenomenon averaged over a whole sample [6–10]. In such an approach, the thickness of the α lamella and the velocity of the RF were measured by means of quantitative metallography while X-ray diffraction was used to determine the average solute content within the α lamellae [6–10]. However, such a procedure applied to Cahn's model of DP [2] provided values of $s\delta D_b$ some 2–3 orders of magnitude lower than the diffusivity at a stationary grain boundary in the same material [6, 11, 12].

In the present paper, it will be shown that these discrepancies can be eliminated by the application of analytical electron microscopy to determine the residual solute concentration profiles across the α lamellae. An aluminium alloy containing 22.0 at.% of zinc was chosen for investigation due to reliable data on diffusivity at stationary grain boundaries as well as at the moving RF of discontinuous precipitates [8, 10, 13–16].

Experimental Procedure

An aluminium alloy containing 22.0 at.% of zinc was prepared from components of high purity (99.99%). Ingots were homogenised at 655 K for 60 hours and cold-rolled to 0.1 mm of thickness. Then, samples were heated in a salt bath at 645 K for 3 hours and quenched in ice water. Ageing was performed on a heating stage in a Philips 301 transmission electron microscope (TEM) with a simultaneous video recording.

The foils for transmission electron microscopy were prepared using a punching machine to produce discs of 3 mm in diameter. The final operation consisted of jet electropolishing using an electrolyte of 33% nitric acid in methanol at a temperature of −30 °C and a potential of 20 V. In order to minimize oxidation, the thin foils were either examined in the electron microscope immediately after preparation or kept in liquid nitrogen.

The energy-dispersive X-ray (EDX) analyses were carried out with a transmission electron microscope Philips CM 20 Twin operating at 200 kV in the nanoprobe mode with a LaB_6 source of electrons. The size of the incident electron beam was 5 nm in diameter at the "full width at half maximum" (FWHM) and it was moved manually across the α lamella in steps that ensured the

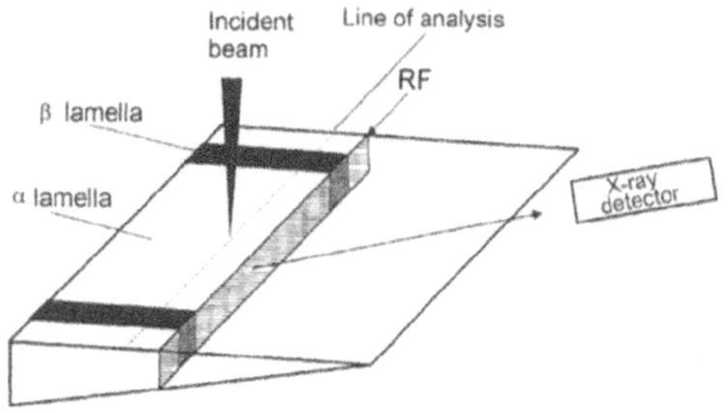

Fig. 1. Schematic view of the thin foil optimally located for TEM/EDX analysis across the α lamella formed due to the discontinuous precipitation

adjacent volumes investigated were spatially resolved. The specimen thickness in the region of interest was determined using a contamination spot method [17] and it was found to be between 100–170 nm. This method usually overestimates the foil thickness. Therefore, the results were randomly checked using convergent beam electron diffraction [18]. The resulting differences were within 20% relative error. During analysis the specimens were tilted approximately 20 degrees towards the energy dispersive detector which resulted in a 40° take-off angle. The intensity of characteristic X-ray peaks I_{Al} and I_{Zn} were obtained for X-ray acquisition time of 60 seconds using a Link eXl energy dispersive spectrometer and the ancillary electronics. This resulted in the accumulation of a few thousand counts for zinc and the associated relative error (3σ) due to X-ray counting was 3–4%. The atomic percentages of zinc at the points analysed were obtained from the observed peak integrals using the Cliff-Lorimer "ratio technique" [19]. In the present study the Cliff-Lorimer factor $k_{Al-Zn} = 1.26 \pm 0.1$ using a homogenised Al-22 at.% Zn standard.

The EDX analyses were performed only in those α lamellae which satisfied the condition shown in Fig. 1. It meant that the incident electron beam and X-ray detector were parallel to the β lamellae. The EDX detector was situated so that the profile was measured at the same foil thickness, and consequently, the X-ray path length through the specimen and the resulting absorption correction was minimised. The EDX analyses were carried out mostly in such regions of the cell where the β phase lamellae were removed by a selective etching.

The following procedure was applied during the experiment:

1. *In situ* ageing in the transmission electron microscope with video recording of the DP process in order to determine the growth rate of the discontinuous precipitates (v) and the thickness of the α lamellae (λ_α) for individual cells.
2. Cooling down and making EDX measurements of the solute concentration profiles across the α lamellae for which v and λ_α were previously evaluated.

In this way the discontinuous precipitation could be considered as a local phenomenon because the solute profiles were measured only for those α lamellae for which the reaction front velocity and the thickness had been previously estimated.

Results and Discussion

Fig. 2 presents a typical TEM micrograph of the Al-22 at.% Zn alloy aged at 450 K for 10 min and the corresponding result of EDX analysis. The experimental points were fitted to the solution of Cahn's [2] diffusion equation using a non-linear regression method:

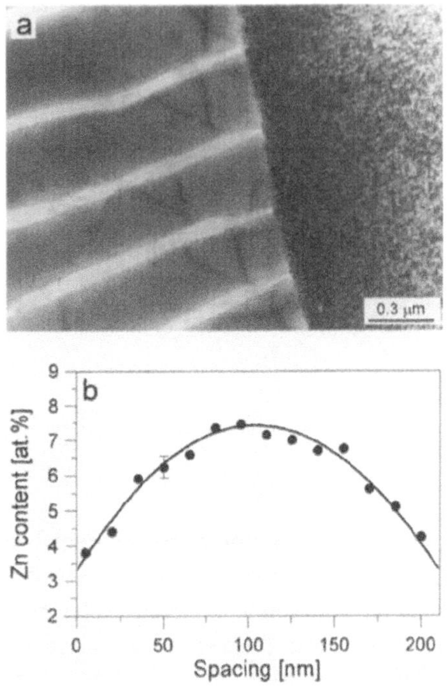

Fig. 2. a TEM micrograph of Al-22 at.% Zn alloy aged at 400 K for 10 min., **b** Zn concentration profile measured for DP process across the α lamella

$$x(y) = (x_e - x_o)\frac{\cosh\left((y-0.5)\sqrt{C}\right)}{\cosh(\sqrt{C}/2)} + x_o \qquad (1)$$

in which the dimensionless C parameter is expressed by:

$$C = \frac{\nu(\lambda_\alpha)^2}{s\delta D_b} \qquad (2)$$

Here x_o is the original solute content in the alloy, x_e is the solute content in the α lamella in contact with the β lamella, v is the growth rate of discontinuous precipitates, λ_α is the thickness of α lamella, $s\delta D_b$ is the diffusivity at the moving front of discontinuous precipitates and y is the normalised co-ordinate measured from the edge of the β lamella and running perpendicular to the α lamella.

The solid line in Fig. 2b represents the best fit of Eq. (1) to the experimental data, and provides the values of C parameters. Table 1 contains all details of the DP process obtained according to the procedure given in Section 2 for the samples aged at 400 K. The example shown Fig. 2b is relevant for lamella No. 1.

Table 1. Details of α lamellae analysed in Al-22at.% Zn alloy aged at 400 K for 10 min

Cells region	Lamella analysed	λ_α [nm]	C	x_e [at.%]	v [nm/s]	$s\delta D_b$ [m^3/s]
	1	210	2.14	3.28		$1.77 \cdot 10^{-21}$
	2	225	1.66	4.32		$2.62 \cdot 10^{-21}$
1	3	205	2.51	3.74	86	$1.44 \cdot 10^{-21}$
	4	195	1.95	3.51		$1.68 \cdot 10^{-21}$
	5	240	2.21	4.07		$2.24 \cdot 10^{-21}$
	1	308	2.74	4.02		$4.19 \cdot 10^{-21}$
	2	289	2.92	3.84		$3.46 \cdot 10^{-21}$
2	3	280	3.11	3.87	121	$3.05 \cdot 10^{-21}$
	4	295	3.01	3.69		$3.50 \cdot 10^{-21}$
	5	300	2.87	4.11		$3.79 \cdot 10^{-21}$
	1	154	1.84	3.54		$7.35 \cdot 10^{-22}$
	2	167	1.75	3.76		$9.08 \cdot 10^{-22}$
3	3	143	1.71	3.91	57	$6.82 \cdot 10^{-22}$
	4	148	1.68	3.62		$7.43 \cdot 10^{-22}$
	5	159	1.62	3.82		$8.90 \cdot 10^{-22}$

Knowing the most important kinetic parameters of DP, it is possible to determine the diffusivities at the moving RF. In order to accomplish this, Eq. (2) was used. The final result is presented in Table 1, and in Fig. 3 in the form of Arrhenius dependence. Figure 3 contains three sets of data: (i) values of tracer diffusivities at stationary grain boundaries for a wide range of aluminium-zinc solid solutions given by Hässner [13], (ii) values of the diffusivity at the RF of discontinuous precipitation obtained using analytical electron microscopy, (iii) values of the diffusivity at the RF of discontinuous precipitation obtained assuming the global approach to DP.

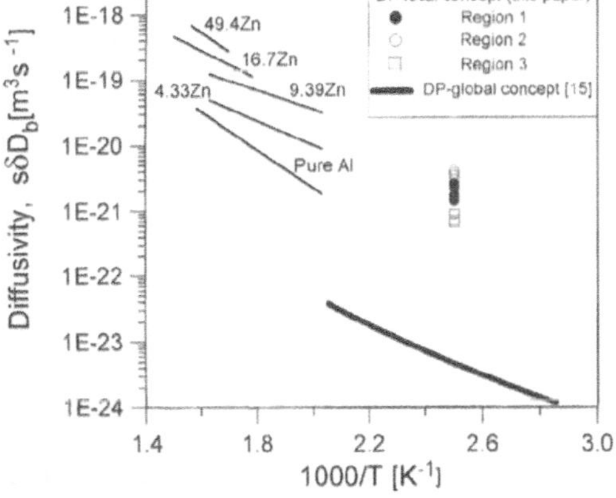

Fig. 3. Arrhenius graph for tracer diffusion at stationary grain boundaries [13] and data on diffusivity at moving reaction front of discontinuous precipitation

One can see that the discrepancies between the diffusivities at stationary and migrating grain boundaries disappear using the local, rather than the global treatment of the DP reaction. The discrepancy arises from the fact that the measurements of the RF growth rate do not allow us to distinguish the colonies which are still growing from those whose growth has already been completed. The measurements also take into account an initial growth distance which is accompanied by large variations of the growth rate of the RF, as well as α lamellae thickness and growth direction. Moreover, the values of average solute content (x_{av}) within the α lamellae cannot be precisely determined because: (i) it is almost impossible to perform X-ray phase analysis in the sample covered by 100% of DP products without any symptoms of discontinuous coarsening or competing continuous precipitation reaction, (ii) the measurements of x_{av} take into account not only the colonies which are still growing but also those whose growth has been completed because a driving force was used up or an impingement of various colonies occurred.

Conclusions

The measurements of diffusivity along the reaction front of discontinuous precipitation performed in the Al-22 at.% Zn alloy proved to be similar to the diffusivities at the stationary grain boundaries in a wide range of Al-Zn solid solutions. This result is attributed to the local treatment of the reaction associated with the application of analytical electron microscopy to determine the solute concentration profiles across the individual α lamellae. The profiles can thus be converted into the characteristic C parameters in Cahn's equation from which the values of diffusivities have been extracted with no discrepancies.

References

[1] D. Turnbull, *Acta Metall.* **1955**, *3*, 55.
[2] J. W. Cahn, *Acta Metall.* **1959**, *7*, 18.
[3] K. N. Tu, D. B. Turnbull, *Scripta Metall.* **1967**, *1*, 173.
[4] J. Petermann, E. Hornbogen, *Z. Metallkd.* **1968**, *59*, 814.
[5] M. Hillert, *Acta Metall.* **1982**, *30*, 1689.
[6] W. Gust, *DsC Thesis*, University of Stuttgart, 1980.
[7] T. H. Chuang, *Ph. D. Thesis*, University of Stuttgart, 1983.
[8] S. Abdou, *Ph. D. Thesis*, Suez Canal University, Port Said, 1987.
[9] C. F. Yang, G. Sarkar, R. A. Fournelle, *Acta Metall.* **1988**, *36*, 1511.
[10] G. R. Purdy, in: *Materials Science and Technology. A Comprehensive Treatment, Vol. 5-Phase Transformations in Materials*, (R. W. Cahn, P. Haasen, E. J. Kramer, eds.), VCH Vevlogsgesellschoft, 1995, p. 331.
[11] I. Kaur, W. Gust, *Defect and Diffusion Forum* **1989**, *66*, 765.
[12] A. Pawłowski, P. Zięba, *Arch. Metallurgy* **1990**, *35*, 401.
[13] A. Hässner, *Kristall und Technik* **1974**, *9*, 1371.
[14] V. Suresh, S. Gupta, *Z. Metallkd.* **1986**, *77*, 529.
[15] P. Zięba, A. Pawłowski, *Mater. Sci. Eng. A* **1994**, *187*, 57.
[16] P. Zięba, A. Pawłowski, *Arch. Metallurgy* **1988**, *33*, 627.
[17] G. W. Lorimer, G. Cliff, J. N. Clark, *Developments in Electron Microscopy and Analysis* (J. A. Venables, ed.) The Academic Press, London, 1976 p. 153.
[18] P. M. Kelly, A. Jostsons, R. G. Blake, J. G. Napier, *Phys. Stat. Sol. A* **1975**, *31*, 771.
[19] G. Cliff, G. W. Lorimer, *J. Microscopy* **1975**, *103*, 203.

Mikrochim. Acta [Suppl.] 15, 77–82 (1998)

Quantitative TEM-EDXS of Sol-Gel Derived PZT Ceramic Materials

Goran Dražič*, **Barbara Malič**, and **Marija Kosec**

"Jozef Stefan" Institute, Jamova 39, 1001 Ljubljana, Slovenia

Abstract. Qunatitative energy dispersive X-ray spectroscopy performed in a transmission electron microscope was used for characterisation of sol-gel derived lead zirconate-titanate bulk ceramics. Chemical composition fluctuations in submicron regions were examined. The influence of foil thickness measurement accuracy on results obtained using absorption corrections was estimated. The optimisation of the analytical parameters and procedures, and the correlation between the chemical composition, structure and processing parameters are described and discussed.

Key words: lead zirconate titanate, PZT, analytical electron microscopy, energy dispersive X-ray spectroscopy, AEM.

Lead zirconate-titanate ($Pb(Zr,Ti)O_3$, abbreviated as PZT) based ceramics are prospective materials for electronic and electro-optic devices [1]. The use of these materials in nonvolatile memories, piezoelectric SAW devices, microactuators, pyroelectric sensors and electro-optic modulators, requires miniaturisation which is achieved by using thin films of PZT on a substrate material [2]. For fabrication of PZT various methods are used, amongst which the sol-gel method appears to yield the best possibility of controlling the stoichiometry [3]. To ensure the required properties, dense, chemically homogeneous materials, consisting of the pure perovskite phase must be prepared [4].

The chemical composition of PZT specifically the ratio of Zr and Ti, which both and randomly occupy B sites in the perovskite structure, has a strong influence on its physical (electrical) properties. At room temperature two main perovskite phases are present, rhombohedral (R) and tetragonal (T). The phase composition is influenced by chemical composition; samples with higher amounts of Zr are rhombohedral and samples with higher amounts of Ti are tetragonal. In the vicinity of the so-called morphotropic phase boundary (MPB), where PZT materials have the approximate composition of $Pb(Zr_{0.52}Ti_{0.48})O_3$, both phases (R and T) could be found in the samples simultaneously. The width of the MPB strongly depends on the processing conditions [5].

A correlation between chemical composition, firing conditions and phase composition is obvious, although use of XRD methods, for phase composition study, failed to reveal if both phases coexisted in one grain or grains consisted of just one phase. There are some reports that both phases could coexist in one grain and that R and T phases form some kind of nanometer sized clusters [6]. The degree to which the phases are tetragonal or rhombohedral (their deviation from ideal cubic) is relatively small so that it is very difficult to see and measure differences in d-values and angles when carrying out electron diffraction experiments in the transmission electron microscope. Additionally, this material forms twins (ferroelectric domains), which further complicates the appearance of electron diffraction patterns. Having regard to the known relationship between the chemistry and phase composition, it is of interest to determine whether there are chemical composition fluctuations in the sample.

There are a limited number of analytical methods which can be used for the characterisation of the microstructure of fine grained PZT bulk ceramics or thin films because of the very small size of the phases present (<1 μm). Electron probe microanalysis could sometimes be used, but at relatively low accelerating voltages (below 8 keV) to avoid the electrons pene-

* To whom correspondence should be addressed

trating to surrounding grains or substrate material. Under such working conditions K lines of Ti are poorly excited and M lines of Pb and L lines of Zr (both series with energy around 2 keV) must be used for analysis. All of these limitations contribute to substantial errors in the final results.

To analyse chemical composition on a submicron scale the logical choice is to use analytical electron microscopy (transmission electron microscopy combined with energy dispersive X-ray spectroscopy, TEM-EDXS). The problem with quantitative TEM-EDXS is to achieve an adequate measurement reproducibility (precision) within a reasonable length of time. Performing analysis using a very small beam diameter means a low intensity signal and potential contamination and deterioration of the sample due to high electron exposure rates per area. To obtain results with a better than 5% relative standard deviation, EDXS spectra with peaks larger than 10^4 counts must be collected. Therefore, the thickness of the sample should be of the order of 100 nm to enable reasonable count rates to be attained and to overcome problems associated with the formation of an amorphous contamination layer during the ion erosion process. Using a foil thickness of this order means that results need to be corrected for absorption and in some cases for secondary fluorescence. The difficulty of measuring foil thickness at every point makes application of these corrections impractical for routine work.

In our previous studies [7] it was found that in a mixed-oxide, Nb doped PZT bulk ceramic, with a composition close to that of the morphotropic phase boundary (PZT(53/47)), the fluctuation in the Zr/Ti ratio depended strongly on firing conditions. The relative standard deviation of Zr/Ti (measured on various grains) was below 5% in samples fired at 1250 °C and close to 14% in samples fired at 1200 °C. This difference was ascribed to incomplete homogenisation at lower temperatures originating from the preparation technique (mixing oxide particles).

In the present paper the use of a quantitative TEM-EDXS analysis for the characterisation of sol-gel derived PZT bulk ceramics is described. Chemical composition fluctuations in submicron volumes within single grains and the chemical composition variations between different grains in PZT(50/50) and PZT(75/25) are examined. The influence of the accuracy of foil thickness determination on the results (corrected for x-ray absorption) is established. Optimisation of the analytical parameters and procedures, and correlation between the chemical composition, structure and processing parameters are described and discussed.

Experimental

Sol-gel derived PZT samples with compositions $Pb(Zr_{0.5}Ti_{0.5})O_3$ (PZT(50/50)) and $Pb(Zr_{0.75}Ti_{0.25})O_3$ (PZT(75/25)) were prepared from a solution of Ti and Zr n-butoxides and anhydrous Pb-acetate in n-butanol under a dry Ar atmosphere. The details of the preparation can be found elsewhere [8].

TEM samples were prepared by mechanical thinning, dimpling and ion milling using 3.8 keV argon ions. Samples were examined in a JEOL 2000FX transmission electron microscope (TEM), operated at 200 kV. The chemical composition was investigated using a Link AN-10000 EDXS system (Energy Dispersive X-ray Spectroscopy) with an ultra-thin window Si(Li) detector mounted perpendicular to the electron beam. The samples were tilted at an angle of 30.5° during EDXS spectrum collection. PZT(50/50), prepared by the mixed-oxide route and sintered at 1250 °C for 2 hours was used as a standard. The Cliff-Lorimer [9] method and absorption corrections [10, 11] were employed for quantitative analysis. Sample thickness, needed for absorption calculations, was determined using convergent beam electron diffraction and contamination spot methods. The estimated reproducibility of measurements (relative standard deviation) for the elements Pb, Zr and Ti was better or equal to 3%. The concentration of oxygen was calculated from the stoichiometry.

Results and Discussion

Microstructure

In Fig. 1 TEM micrographs of the sample PZT(50/50), prepared by the mixed-oxide route and fired at 1250 °C for 2 hours (Fig. 1a) and a sample with the same composition, prepared by sol-gel and fired at 1200 °C for 2 hours (Fig. 1b) are shown. In Fig. 1c a micrograph of the sample PZT(75/25) (sol-gel), fired at 1200 °C for 2 hours is displayed. The grain sizes in all cases were between 1 and 2 μm, and most of the ferroelectric domains (FE) were in the range of 50–200 nm. Besides perovskite no other phase was detected in the samples. Despite the addition of 2% excess PbO (to compensate for the PbO loss during the firing) in the case of the sol-gel derived PZT(75/25) sample, there was no evidence of an amorphous, PbO rich layer at the grain boundaries or at three-grain junctions. In Fig. 2 XRD spectra of samples prepared by the sol-gel method are shown. Samples with the composition PZT(50/50) consisted of the tetragonal phase only and in those with the composition PZT(75/25) only the rhombohedral phase was found. Mixed oxide PZT(50/50) ceramics had an identical XRD spectrum to the sol-gel sample with the same composition.

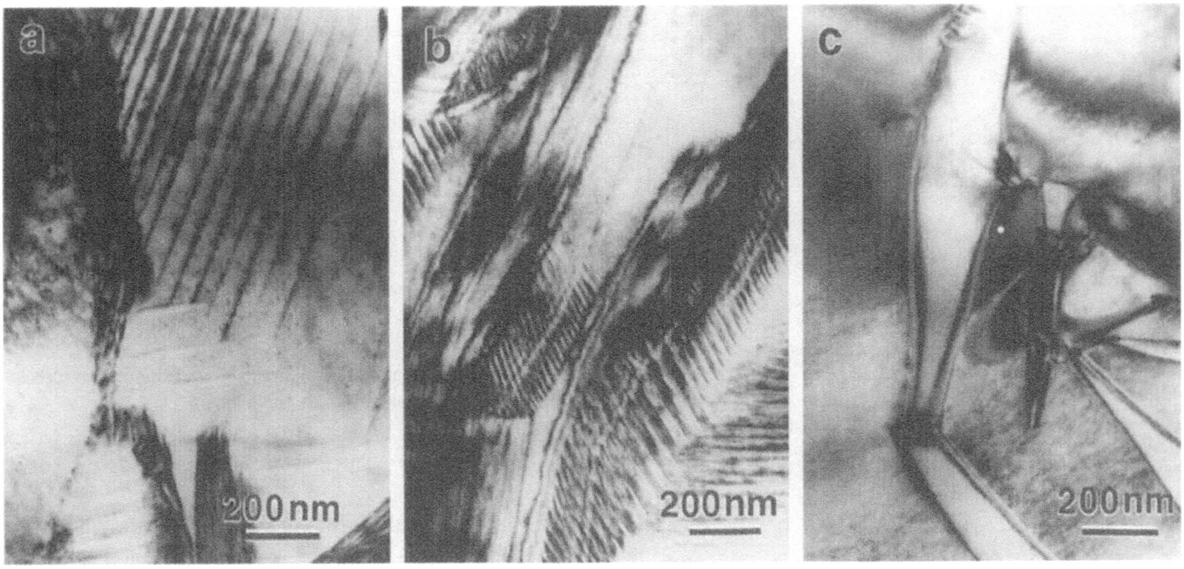

Fig. 1. TEM micrographs of: **a** PZT(50/50), prepared by mixed-oxide route and fired at 1250 °C for 2 hours, **b** sample with the same composition, prepared by sol-gel and fired at 1200 °C for 2 hours **c** PZT(75/25) (sol-gel), fired at 1200 °C for 2 hours

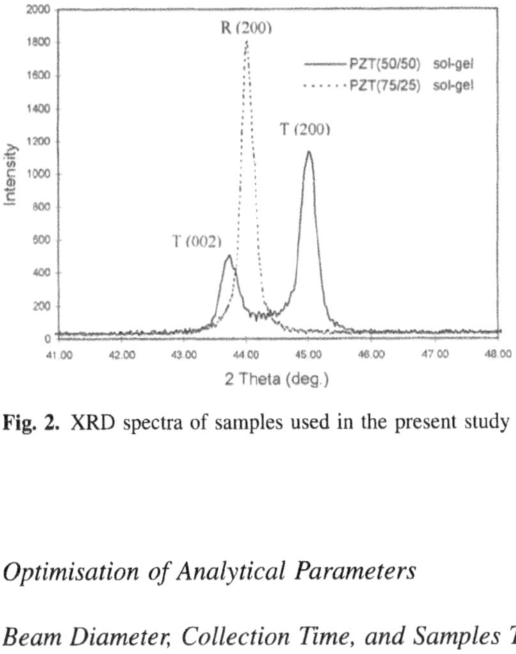

Fig. 2. XRD spectra of samples used in the present study

Optimisation of Analytical Parameters

Beam Diameter, Collection Time, and Samples Thickness. The shape of electron beam used was slightly elliptical, with diameters around 30 and 40 nm. According to an equation proposed by Reed [12], the beam diameter broadening for a foil thickness of 100 nm was estimated to be around 8 nm. When chemical fluctuations in one grain were measured, a matrix of points 300 nm apart were used. Experimentally it was found that during a period of 5 minutes the sample drift was less than 20 nm and the sample contamination, when using a liquid nitrogen trap, was relatively low. At higher collecting times, carbon build-up appeared and started to grow. To achieve a 1% uncertainty (using the 95% confidence limit) requires that at least 2×10^4 counts [13] are in the smallest peak (ZrKα in this case) and if the EDXS spectra collection time is limited to 300 seconds this, in turn, means that the foil thickness at the measuring point should be around 100 nm. At such a thickness the influence of the amorphous contamination layer, originating from the ion erosion process, is minimised. On the other hand, due to the relatively high mass absorption coecients of TiKα lines in Pb (1012) and in Zr (648) [10], and the high mass fraction of Pb (62%) and Zr (14%), absorption corrections must be employed if the errors are to be kept as low as possible.

Influence of thin foil thickness measurement errors on the results. We tried to estimate the influence of errors in foil thickness determination on the accuracy of the concentration measurements. We collected a spectrum from an 80 nm thickness in a PZT grain. The thickness was measured using the convergent beam electron diffraction technique (CBED), proposed by Kelly et al. [14]. Afterwards we made consecutive abosorption correction calculations using different values for foil thickness (in the range between 1 and 240 nm). In Fig. 3 a diagram is displayed of normalised concentrations (normalised to the stoichiometric PZT(50/50)) versus the foil thickness used in the calculations.

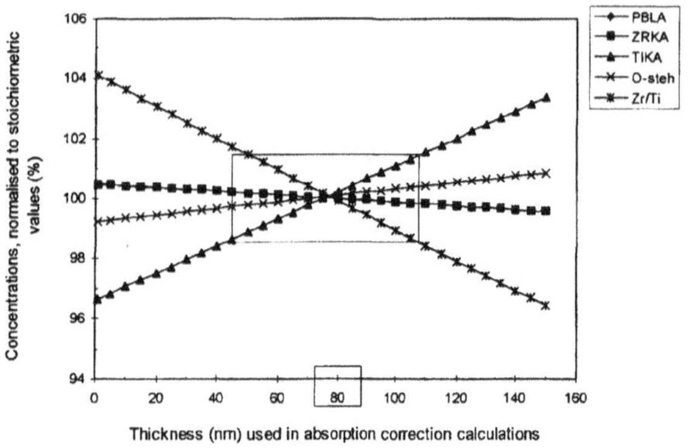

Fig. 3. The influence of foil thickness used in absorption correction calculations on the accuracy of calculated concentrations. Spectrum was collected on a mixed-oxide sample PZT(50/50), fired at 1250 °C for 2 hours, in a part of the sample where the foil was 80 nm thick

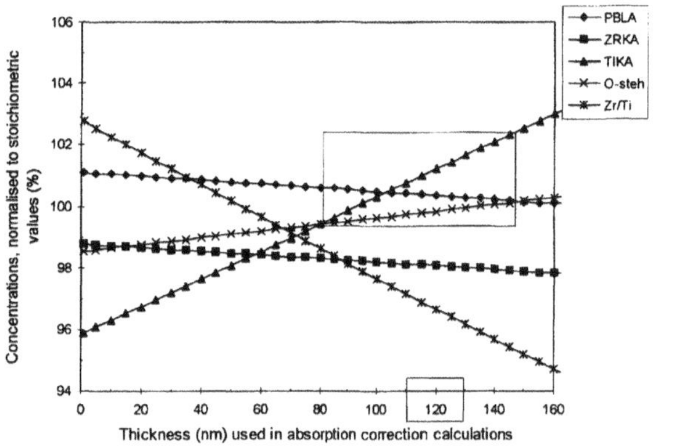

Fig. 4. The influence of foil thickness used in absorption correction calculations on the accuracy of calculated concentrations. Spectrum was collected on a sol-gel sample PZT(50/50), fired at 1200 °C for 2 hours, at a point where the foil was 120 nm thick and the composition was slightly Zr deficient

Underestimation of the Ti content in the range of 4% (relative) was observed if the absorption corrections were neglected. On the other hand the calculated results show errors are limited to ± 1.5% (box inside the graph) provided the thickness is in the range 80 ± 30 nm. From Fig. 3 we can conclude that the absorption correction should be used, but that the accuracy of the foil thickness measurements does not need to be very high (at least for foil thicknesses of this order of magnitude).

Similar results were found when we measured concentrations in an area where the foil thickness was 120 nm (Fig. 4). At this point lower concentrations of Zr and higher concentrations of Ti than expected in PZT(50/50) were found. Again, within the interval of

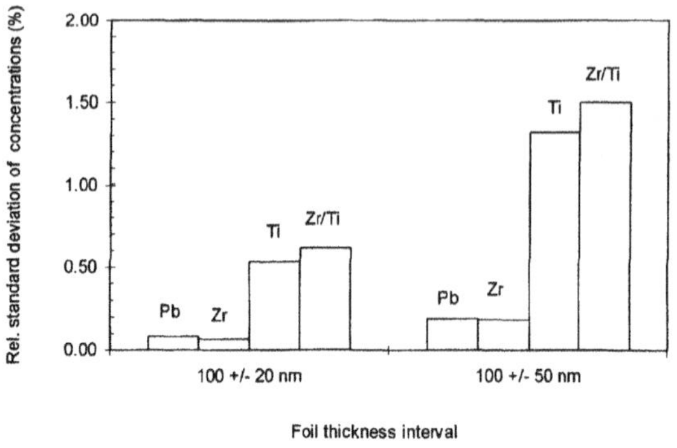

Fig. 5. Relative standard deviation of concentration measurements as a function of foil thickness variations

120 ± 35 nm (the thickness used in the absorption correction calculation) the errors in concentrations are in the range of ± 1.5%.

In the next experiment a spectrum was collected at a point where the thickness was 100 nm. In the calculation of the absorption correction the foil thickness was then varied randomly between 100 ± 20 nm in the first case and 100 ± 50 nm in the second case (50 values were generated for each case). From the results of the concentrations obtained we calculated relative standard deviations, which are shown in Fig. 5. Even in the second case the relative standard deviation of the concentrations did not exceed 1.5%.

These findings indicate that for PZT materials absorption corrections should be used to minimise the errors, but the accuracy of foil thickness measurement can be relatively low. This means that tedious measurement of foil thickness at every point using CBED or the contamination spot method is unnecessary if we use a semi-quantitative approach for thickness estimation. Based on experience (using the same instrumental conditions, such as accelerating voltage and condenser aperture) we found that in a sample region where CBED spots (circles), corresponding to d values around 0.15 nm, have 3 Kossel-Möllenstedt fringes in one half of the circle, the thickness was in the range of 100 ± 30 nm. In Fig. 6a CBED pattern from a PZT(50/50) sol-gel derived sample with a thickness of 120 nm is shown.

Based on calculations, we estimate that the errors due to the statistical nature of the method are around 1% and the errors due to foil thickness effects are in the range of 1.5%. We assumed that the total precision of the measurements should be equal or smaller than 3%.

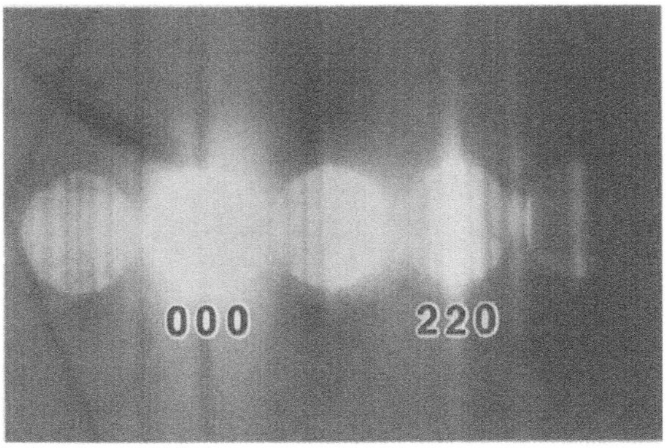

Fig. 6. CBED pattern from the 120 nm thick part of the PZT(50/50) (mixed oxide) sample.

Chemical Composition Fluctuations in PZT Samples

Using optimised analytical parameters and procedures described above, we measured between 27 and 40 grains in each sample (one point on each grain). In all samples we made also 9 to 17 measurements on chosen grains in the matrix where points were 300 nm apart. Measurements were made in areas where, according to the CBED pattern, the foil thickness was around 100 nm. All results were corrected for absorption using 100 nm as a foil thickness. In Fig. 7a–c the relative standard deviations of concentration measurements from different samples, measured in one grain and in several grains, are displayed.

In all three samples the relative standard deviation of measurements in one grain is lower than or equal to 3%. These values are very close to the estimated errors of the analytical method used in the present work. Based on these results we can state that the chemical composition fluctuations inside one grain are smaller than the analytical error. The relative standard deviation of measurements in different grains was between 4 and 5% for the mixed-oxide PZT(50/50) and sol-gel derived PZT(75/25) with 2% excess PbO, but about 8% in the sol-gel derived PZT(50/50) sample. If we can ascribe a standard deviation of approximately 5% to instrumental causes (variation of local surroundings during the collection of spectra), a standard deviation of around 8% indicates that the chemical composition of different grains are different. To exclude the influence of the TEM sample preparation (ion erosion) we repeated measurements on another sample and values in both were found to be quite similar.

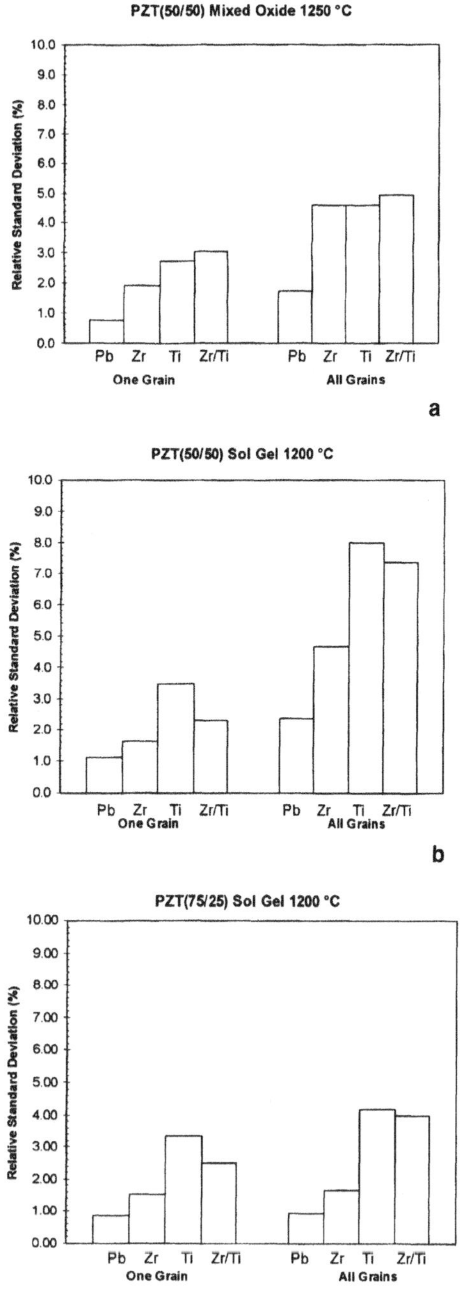

Fig. 7. Relative standard deviations of concentration measurements, measured in one grain and in various grains in mixed oxide prepared PZT(50/50), fired at 1250 °C for 2 hours (**a**), sol-gel derived PZT(50/50), fired at 1200 °C for 2 hours (**b**) and sol-gel derived PZT(75/25), fired at 1200 °C for 2 hours (**c**)

A tentative explanation for the variation between samples could be found in the different firing temperatures and PbO additions. Higher firing temperature and excess PbO promote the diffusion processes during the final stages of sintering and accelerate the chemical homogenisation.

For a better understanding of sintering process in this system, more samples with different chemical composition, fired at different temperatures and containing varying amounts of excess PbO need to be analysed using the analytical approaches described in the present work.

Conclusions

The degree of chemical homogeneity in PZT bulk ceramics, prepared by mixed oxide or sol-gel methods were measured using quantitative TEM-EDXS. EDXS spectra were collected at a matrix of points, 300 nm apart in one grain and from points in different grains (one point per grain) from which the chemical compositions were calculated. The possible errors due to counting statistics and uncertainty in foil thickness measurement were estimated to be equal or less than 3%. It was found that in all samples the chemical fluctuations inside one grain were smaller than the analytical errors. In the case of the sol-gel derived PZT(50/50) sample, fired at 1200 °C without additional PbO the relative standard deviations of chemical composition of different grains were close to 8% and were most probably the consequence of chemical inhomogeneity of the sample.

Acknowledgment. The financial support of the Ministry of Science and Technology of the Republic of Slovenia is gratefully acknowledged. The authors wish to thank Mrs. Medeja Gec for TEM sample preparation.

References

[1] B. Jaffe, W. R. Cook Jr., H. Jaffe, *Piezoelectric Ceramics*, Academic Press, London, 1986.
[2] R. Ramesh, T. Sands, V. G. Keramidas, D. K. Fork, *Mater. Sci. Eng. B, Solid-State Mater. Adv. Technol.* **1994**, *B22* (2–3), 283.
[3] M. Huffman, F. D. Gealy, L. Kammerdiner, P. Zurcher, G. J. Zhu, M. Al-Jassim, C. Echer, *Ferroelectrics* **1992**, *134* (1–4), 303.
[4] M. Kosec, Y. Huang, E. Sató, A. Bell, N. Setter, G. Dražič, S. Bernik, T. Beltram, in: Science and Technology of Electroceramic Thin Films, (O. Auciello, R. Waser, eds), Kluwer AP, Dordrecht 1995, pp. 179–186.
[5] T. Yamammoto, K. Okazaki, M. L. Dass, G. Thomas, *Ferroelecrics* **1988**, *81*, 331.
[6] P. G. Lucuta, *J. Am. Ceram. Soc.* **1989**, *72*, 933.
[7] G. Dražič, M. Kosec, *Ferroelectrics* **1997**, *201*, 23.
[8] B. Malič, M. Kosec, *J. Sol Gel Sci. Technol.* **1994**, *2*, 443.
[9] G. Cliff, G. W. Lorimer, *J. Microsc.* **1975**, *103*, 203.
[10] J. I. Goldstein, J. L. Costley, G. W. Lorimer, S. J. B. Reed, *SEM* 1977 (O. Johari, ed.), IITRI, Chicago, 1977, p. 315.
[11] K. F. J. Heinrich, *The Electron Microprobe* (T. D. McKinley, K. F. J. Heinrich, D. B. Wittry, eds.), Wiley, New York, 1964.
[12] S. J. B. Reed, *Ultramicroscopy* **1982**, *7*, 405.
[13] D. B. Williams, C. B. Carter, *Transmission Electron Microscopy – Part IV: Spectrometry, 1996*, Plenum, New York.
[14] Kelley, A. Jostsons, R. G. Blake, J. G. Napier, *Phys. Stat. Sol. (a)* **1975**, *31*, 771.

Mikrochim. Acta [Suppl.] 15, 83–86 (1998)

Particulate Composites of TZP-Chromium Oxide and TZP-Chromium Carbide; Microbeam Investigations

Marek Faryna[1,*], **Lidia Lityńska**[1], **Krzysztof Haberko**[2], **Zbigniew Pędzich**[2], and **Joanna Babiarz**[2]

[1] Institute of Metallurgy and Materials Science, Polish Academy of Sciences, ul. Reymonta 25, 30-059 Kraków, Poland
[2] Department of Special Ceramics, University of Mining and Metallurgy, Faculty of the Material Science and Ceramics, al. Mickiewicza 30, 30-059 Kraków, Poland

Abstract. Mechanical properties and microstructure of hot-pressed zirconia-based composites containing additions of chromium carbide or chromium oxide particles were investigated. The composites showed significantly higher elastic properties, hardness and fracture toughness compared with the pure matrix. The microstructure of the materials was investigated using scanning and transmission electron microscopy as well as energy-dispersive X-ray spectrometry.

Key words: tetragonal zirconia polycrystals, chromium carbide, chromium oxide, particulate composites, analytical electron microscopy.

The incorporation of hard carbide inclusions into polycrystalline tetragonal zirconia (TZP) seems to be an effective way of improving its mechanical properties and it has been shown [1, 2] that tungsten and silicon carbides increased hardness, fracture toughness and Young's modulus of the composite. However, due to the difference in coefficients of thermal expansion (CTEs) of the TZP matrix ($\alpha_{TZP} = 11.0 \times 10^{-6}\,K^{-1}$) compared with carbides such as α_{WC} ($5.2 \times 10^{-6}\,K^{-1}$) and $\alpha_{SiC}(4.7 \times 10^{-6}\,K^{-1})$, large tensile stresses are generated in the matrix [3, 4]. These stresses result in decreased toughness of the composite. In the paper, additions to TZP of Cr_7C_3 ($\alpha_{Cr7C3} = 9.4 \times 10^{-6}\,K^{-1}$) and Cr_2O_3 ($\alpha_{Cr2O3} = 7.5\,10^{-6}\,K^{-1}$) are considered since these should result in smaller stresses being developed during composite preparation.

* To whom correspondence should be addressed

Experimental

The chromium carbide (Cr_7C_3) and Oxide (Cr_2O_3) particles were incorporated into the $ZrO_2 + 2.9$ mol.% Y_2O_3 matrix (Y-TZP) by the coprecipitation technique. In each case 20 vol. % of the particles were introduced. In the case of carbide and chromia inclusions, attrition milling was applied in order to homogenise the mixture. The blended powders were hot pressed under 25 MPa at 1400 °C for 40 minutes and densities of the samples were determined by hydrostatic weighing. Their relative densities were estimated assuming the TZP density values of 6.1, 6.85, and 5.21 g/cm^3 for TZP, Cr_7C_3 and Cr_2O_3, respectively. The Vickers indentation technique was used for the hardness and fracture toughness measurements (the Palmqvist's crack model and the Niihara equation [5] were applied). Young's modulus was determined by ultrasonic measurements [6] and the phase composition of the sintered bodies was evaluated by x-ray diffraction (XRD) on polished surfaces of matrix and composite.

Cracks made by the Vickers indentation were studied using scanning electron microscopy (SEM), the samples being coated with a carbon film to improve conductivity and eliminate charging effects.

Specimens suitable for analytical transmission electron microscopy were prepared by cutting 3 mm diameter discs and polishing them to a thickness of 30 to 40 μm. These were then thinned by ion milling. The thin foils were examined in a Philips CM20 electron microscope operating at 200 keV. The local chemical composition was determined with high spatial resolution (~ 2 nm) for all elements with $Z > 6$ by use of an energy-dispersive x-ray spectrometer LINK eXL. Quantitative analyses of elemental concentrations were achieved using the ZAF correction programme on the LINK system.

Results and Discussion

Table 1 gives properties of the TZP matrix and the composites. The results showed that the hot pressing technique achieved high densification of the sample (>99% of the theoretical value).

Table 1. Properties of the TZP matrix and composites

	TZP	TZP+Cr_7C_3	TZP+Cr_2O_3
Density [% of theo.]	99.7 ± 0.1	99.7 ± 0.1	99.0 ± 0.1
Young's modulus [GPa]	208.9 ± 2.6	224.0 ± 2.6	224.5 ± 2.0
Vickers hardness [GPa]	14.1 ± 0.3	17.9 ± 0.7	19.6 ± 1.0
Fracture toughness [$MPam^{0.5}$]	4.6 ± 0.1	8.7 ± 1.3	6.2 ± 1.0

± denotes the 95% confidence interval

Both the carbide and oxide additions led to increased Young's modulus and hardness of the systems. The increase is more marked in the case of the oxide additive. Fracture toughness is also increased, the carbide being the more effective toughening agent.

SEM observations revealed crack deflection in the vicinity of carbide inclusions (Fig. 1) and transgranular cracking through chromia particles (Fig. 2).

Transmission electron microscopy indicated good bonding of the additions in both systems (Figs. 3 and 4). Fig. 3 shows chromia particles within zirconia grains and Fig. 5 illustrates a TEM micrograph with the corresponding indexed electron diffraction patterns of chromia and zirconia. The diffraction data indicate that the respective zone axes are [100] and $[\bar{1}\,2\,\bar{1}\,1]$. No such epitaxial arrangement was found in the Cr_7C_3/TZP system. EDX analyses revealed small amounts of Zr (up to 2.35 wt. %) in the chromia grains and Cr (up to 3.46 wt. %) in the zirconia grains. This is most probably due to a re-sputtering effect during the ion-thinning procedure (Fig. 6).

In both systems a certain part of the tetragonal zirconia grains transform to the monoclinic symmetry ($t \rightarrow m$), as it is indicated by TEM observations. It should be noted that no monoclinic phase is detected by the XRD measurements on the polished surfaces which suggests that the transformation results from

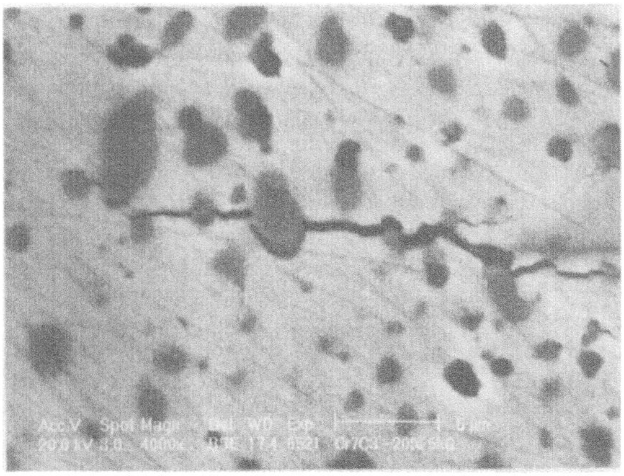

Fig. 1. SEM micrograph of the polished surface of the Cr_7C_3/TZP composite. Dark particles are the carbide grains

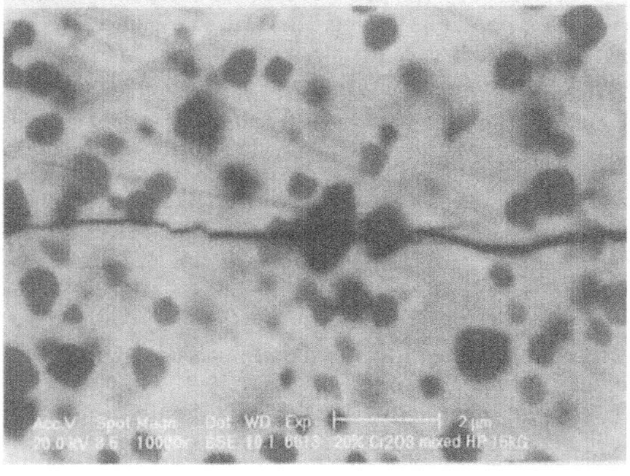

Fig. 2. SEM micrograph of the polished surface of the Cr_2O_3/TZP composite. Dark particles are the chromia grains

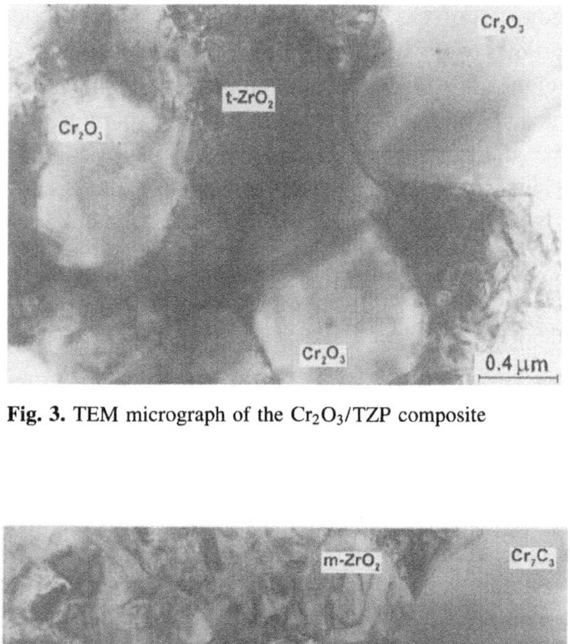

Fig. 3. TEM micrograph of the Cr_2O_3/TZP composite

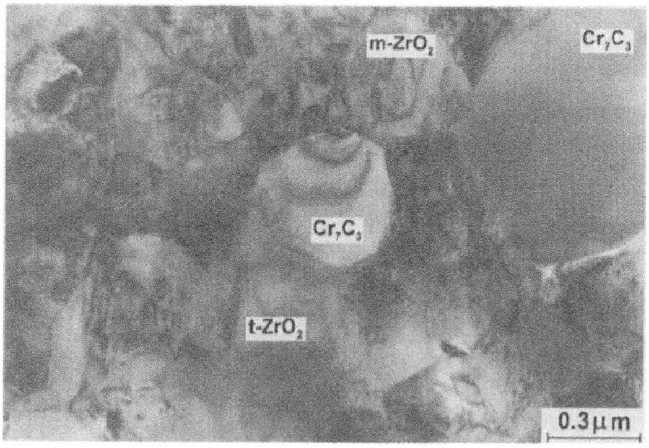

Fig. 4. TEM micrograph of the Cr_7C_3/TZP composite. Extinction contours in the Cr_7C_3 grain in the centre are caused by stresses related to the $t \rightarrow m$ transformation

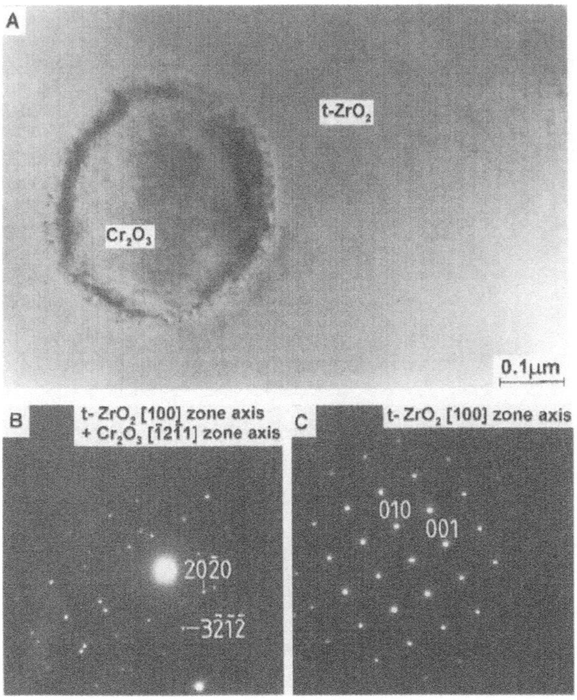

Fig. 5. **A** Chromia inclusion in the tetragonal zirconia grain, **B** SADP of Cr_2O_3 grain, **C** SADP of tetragonal ZrO_2

stresses introduced during the preparation of thin foils for the TEM studies.

At the Cr_2O_3-ZrO_2 interface cracks are observed in the vicinity of the transformed zirconia grains (Fig. 7). Such cracks do not appear in the case of non-transformed zirconia. So, it is concluded that they result from the stresses due to the $t \rightarrow m$ transformation. These stresses are manifested by strain contours present in the chromia and chromium carbide grains adjacent to the transformed zirconia particles (Fig. 4 and Fig. 8). No cracks are formed in the Cr_7C_3/TZP system even in the presence of the transformed zirconia grains. This seems to be evidence of the higher strength of the carbide-zirconia interface.

Conclusions

The hot-pressing technique results in dense composites of Cr_2O_3 / TZP and Cr_7C_3 / TZP systems with increased Young's modulus, fracture toughness and hardness compared with the TZP.

Both systems differ in the crack propagation mode; transgranular cracks through the chromia grains appeared in the Cr_2O_3 / TZP, whilst cracks surrounded the carbide grains in Cr_7C_3 / TZP. Good bonding between the addition particles and matrix is observed in both systems.

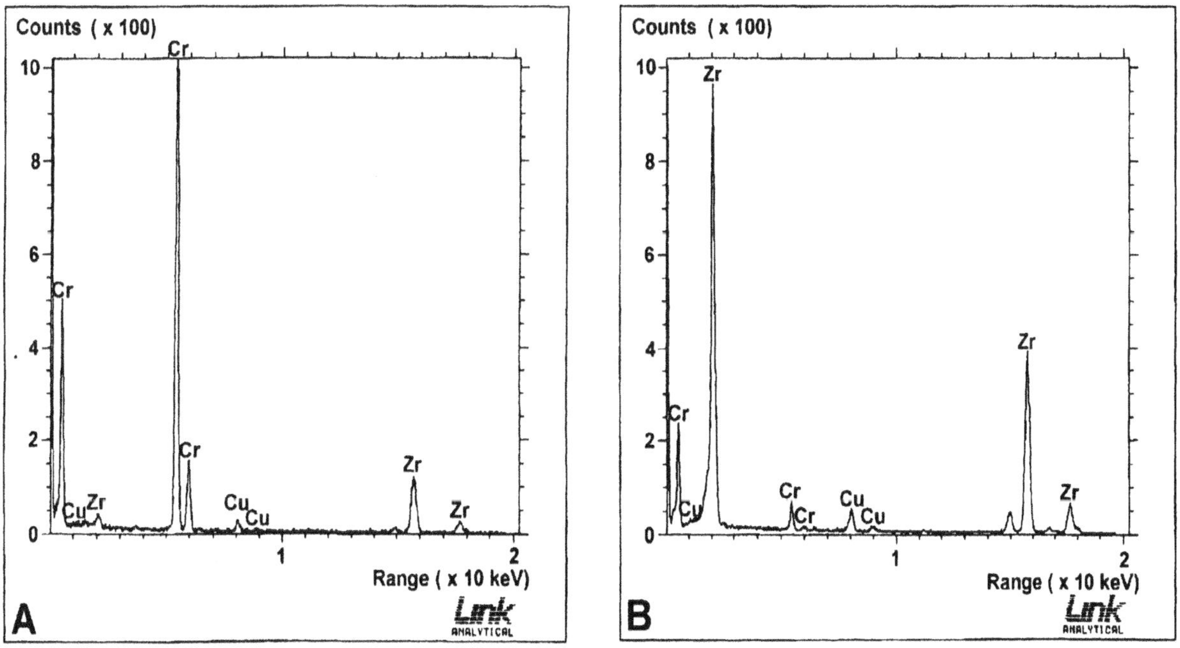

Fig. 6. EDX spectra taken from: **A** Chromia grain, **B** zirconia grain

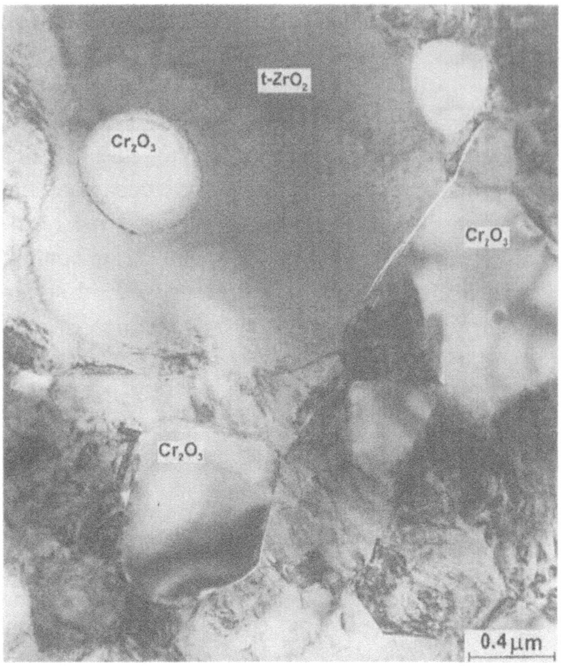

Fig. 7. TEM micrograph of tangential cracks in the Cr_2O_3/TZP composite

The $t \rightarrow m$ transformation is clearly visible in thin foils but not in the bulk material. This can be ascribed to the stresses introduced during the foil preparation for TEM observations. The phase transformation results in tangential cracks in the Cr_2O_3 / TZP system. However, no such cracks appeared in the Cr_7C_3 / TZP system, most probably because of the higher strength of the carbide-matrix interface.

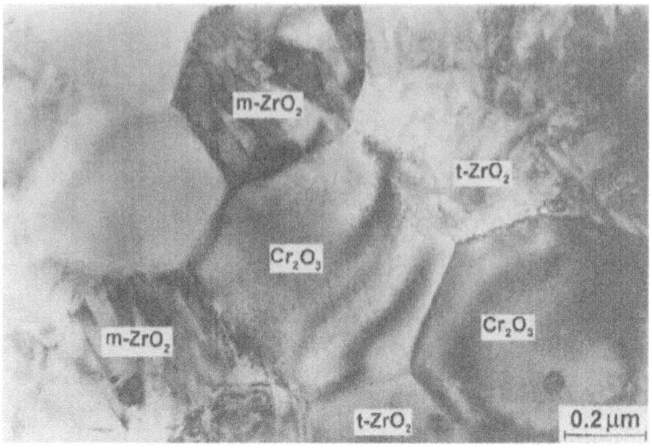

Fig. 8. TEM micrograph of strain contours in the Cr_2O_3/TZP composite

Acknowledgement. This work was supported by the Polish State Committee for Scientific Research under the grants no. 7 T08D 019 10 and 7 T08A 052 10.

References

[1] K. Haberko, Z. Pędzich, G. Róg, M. M. Bućko, M. Faryna, *Eur. J. Solid State Inorg. Chem.* **1995**, *32*, 593.

[2] Zh. Dingh, R. Oberacker, F. Thummler, *J. Europ. Ceram. Soc.* **1993**, *12*, 377.

[3] M. Taya, S. Hayashi, A. S. Kobayashi, H. S. Yoon, *J. Am. Ceram. Soc.* **1990**, *73 (5)*, 1382.

[4] Z. Pędzich, K. Haberko, M. Faryna, in: *Composite Science and Technology* (S. Adali, E. Verijenko eds.), University of Natal, Durban, South Africa, 1996, pp. 379–384.

[5] K. A. Niihara, *J. Mater. Sci. Lett.* **1983**, *2*, 221.

[6] J. Piekarczyk, H. W. Hennicke, R. Pampuch, *Ceramic Forum International/Berichte der Deutschen Keramischen Gesellschaft* **1982**, *59 (4)*, 227.

Mikrochim. Acta [Suppl.] 15, 87–92 (1998)

Cryo-Electron Spectroscopic Imaging, Electron Energy-Loss Spectroscopy and Energy-Dispersive X-Ray Analysis of Ag(Br, I) Nano- and Microcrystals

Vladimir P. Oleshko[1,3,*], **Renaat H. Gijbels**[1], **Willem A. Jacob**[2], and **Andre J. Van Daele**[2]

[1] Departments of Chemistry, University of Antwerp (UIA), B-2610 Antwerpen-Wilrijk, Belgium
[2] Department of Medicine, University of Antwerp (UIA), B-2610 Antwerpen-Wilrijk, Belgium
[3] Photochemistry Centre, N. N. Semenov Institute of Chemical Physics, Russian Academy of Sciences, 117421 Moscow, Russia

Abstract. The combination of cryo-electron spectroscopic imaging (ESI)/electron energy-loss spectroscopy (EELS) and cryo-energy-dispersive X-ray (EDX) analysis in the scanning transmission (STEM) mode was applied for the characterization of Ag(Br, I) nano- and microcrystals. Combined qualitative silver and halide distributions were obtained by ESI (a three-window method) and by EDX/STEM including a high-magnification mapping of nanocrystals. A low-loss fine structure in EEL spectra between 4 and 26 eV was attributed to excitons and plasmons possibly superimposed with interband transitions and many-electron effects. Contrast tuning under energy-filtering was used to image electron excitations in the silver halide particles. The real and imaginary parts, ε_1 and ε_2, of dielectric permittivity were determined by means of a Kramers-Kronig analysis. A proposed assignment of exciton peaks based on calculations of electronic band structure of silver halides is discussed.

Key words: silver halides, plasmons, excitons, AEM (ESI/EELS, STEM/EDX).

The silver halides exhibit many properties that have made them attractive for solid state research [1]. To a significant extent, this originates from the occurrence of a low energy indirect gap between the valence and conduction bands due to the influence of Ag d-states which indicates their close relationship to semiconductors. In addition to the unique character of optical absorption, the strong quadrupolar deformability of the Ag^+ is responsible for a soft crystal lattice which is significantly less rigid than one for cations of alkali halides. The Ag 4d-states and the halogen p-states in AgBr are very close to each other thus leading to considerable complexity in the valence-band structure. So, the upper valence band of AgX (X=Cl, Br, I) results from hybridisation of halogen p-states and Ag 4d-states [1]. AgX particles below 100 nm in size exhibit quantum effects of excitons which are largely studied by low-temperature luminescence and resonance Raman spectroscopies in order to test fundamental solid state concepts and applications related to the nonlinear optics [2, 3].

Due to its spatially-resolved nature, energy filtered imaging (electron spectroscopic imaging (ESI) and electron energy-loss spectroscopy (EELS)) [4, 5], may give an important new insight into the crystalline and electron structures of AgX [6–8]. The higher spectral resolution of EELS (0.1–2 eV) as compared to that of EDX (130–150 eV) offers an opportunity to probe electron excitations in the crystals induced by the electron beam. Low-loss excitations in Ag(Br, I) microcrystals arising from interband transitions and excitons, possibly superimposed with many-electron effects were reported recently [8].

However, EELS performance is limited by a lower signal to background ratio than that for EDX and by rather low intensities of major delayed silver and halide ionisation edges. Besides, limitations arising

* To whom correspondence should be addressed

from the insulating properties and damage of AgX under the electron irradiation hamper utilization, especially of stationary beam ESI/EELS techniques. Therefore, it is useful to evaluate the potential of a combination of cryo-ESI/EELS and EDX for studies of such labile species. The paper reports some recent applications of cryo-ESI/EELS and STEM/EDX techniques for the characterization of electronic properties and element distributions of Ag(Br, I) nano- and tabular microcrystals.

Experimental

The synthesis of Ag(Br, I) uniform nanocrystals of 20–120 nm in size containing about 5 mol.% of AgI and of composite multi-structured tabular microcrystals of 2–10 μm in size and of 100–120 nm in thickness containing 3–8 mol.% of AgI in the shells was performed by a computer-assisted double jet method [6]. The local crystal thickness was determined by a modified log-ratio technique with an accuracy of ±20% [8]. Procedures of specimen preparation for microscope studies were described elsewhere [7]. All preparation and handling were done under nonactinic light to avoid the formation of print-out silver. Studies were performed at an acceleration voltage of 80 kV using an analytical unit composed from a ZEISS CEM 902 computerized electron microscope with an integrated energy filter (ESI/EELS) and a JEOL JEM1200EX TEMSCAN (STEM/EDX) both connected to a Kontron IBAS-2000 image analysis system.

For ESI, images acquired with the exit energy-selecting slit of 10 eV were amplified by a DAGE SIT66X VIDICON TV-camera and transferred to the image processor. In order to obtain Ag distributions a three window technique was used [9]. The background contribution to the Ag $M_{4,5}$-edge for each individual pixel was removed. EELS spectra were recorded by a photomultiplier serial detector interfaced to a PC/AT for automatic data acquisition [10]. The collection semi-angle was varied from 6 to 35 mrad. The estimated energy resolution was 1.5–2.0 eV, based on the measured full-width at half-maximum (FWHM) of the zero-loss peak. The background fit according to the power-law function AE^{-r} was done to reveal the positions of edges followed by smoothing using Savitsky-Golay filters to remove noise from the spectrum profiles. The KRAKRO program [5] with single scattering distributions (SSD) applied as an input has been used to calculate the imaginary part $\mathrm{Im}(-1/\varepsilon)$ of the energy-loss function and the real part ε_1 and imaginary part ϵ_2 of the relative dielectric permittivity as well. Radiation damage of the grains during measurements was reduced using a top-entry cryo-stage at $t = -193\,°C$ supplied with a modified evacuated cooling trap.

The JEOL 1200EX electron microscope equipped with an EM-ASID10 scanning device and multidetector system and a NORAN TN5500 EDX-analyzer was applied to carry out X-ray mapping and spot analysis in the bright- or dark-field STEM mode. Elemental compositions of the crystals were estimated by a NORAN standardless Metallurgical Thin Films program based on the Cliff-Lorimer ratio technique with an accuracy of about 10%. The specimens were tilted 40° towards the Si(Li)-detector. To minimize crystal damage under electron bombardment, the temperature of the specimen was kept at $\leq -168\,°C$ by using a GATAN 636 double tilt cooling holder.

A KONTRON IBAS-2000 image analysis system was utilized to control acquisition parameters on-line and to perform automatic treatment of data. After input of the dwell time per pixel, frame loops, total number of pixels, a window of analysis, position of energy window, etc., the computer hardware took control over the scanning system (the JEM1200EX) and/or over the acquisition and processing of images (the ZEISS CEM 902). This allowed measurements to be made with higher signal-to-noise ratio and data to be stored and processed off line, thus decreasing the analysis time.

Results and Discussions

Figure 1 shows energy-filtered images of AgX particles and SSDs derived from EEL spectra by the use of a Fourier-log deconvolution technique [11]. Plural scattering effects can alter the spectra taking the form of one or more low-loss excitations if the specimen thickness is of order of the inelastic mean free path length $\lambda_{AgBr} = 72.9\,nm$ [8]. In order to minimize interference in the low-loss region of AgX and support, crystals deposited on holey carbon films were studied. Optimization of recording conditions allowed the detection of more intensive low-loss fine structure as compared with reported data [6, 8]. As is seen from the spectra in Fig. 1a, the electron intensity changes due to an intersection with low-loss features (4-17 eV), the bulk loss peak ($\sim 23\,eV$) which exhibits a shift to low losses as compared with the carbon support and a decrease of the background intensity and intersection with minor inner-shell excitations (55–70 eV). The spectra did not reveal significant differences between nanocrystals and a tabular grain except a 1–2 eV shift of a proposed exciton peak at 15 eV and volume losses at 22 eV to lower energy (expected shifts due to size confinement effects may also be of the same magnitude), which, however, lie within the accuracy of measurements. Changes of the electron intensity lead to contrast reversals when the selected energy is tuned over the range from zero to several hundreds eV. Therefore, tuning the energy loss may allow visualization of excitations concerned with certain energy losses. So, selection of proposed exciton losses at $E = 16 \pm 5\,eV$ resulted in a weak rim around the crystals because such losses can even extend to distances a few nanometers away from the particle, which corresponds satisfactorily to the expected exciton size of 2.5 nm for AgX (X=I, Br) [2]. At a first stage of a cascade of beam-solid interactions, electrons hit the crystal directly causing an inner ionisation with the generation of electron-hole pairs. Then, probably, fast electrons do not strike the particle directly but cause a polarization inside it. Images at $E = 75 \pm 5\,eV$ (beyond

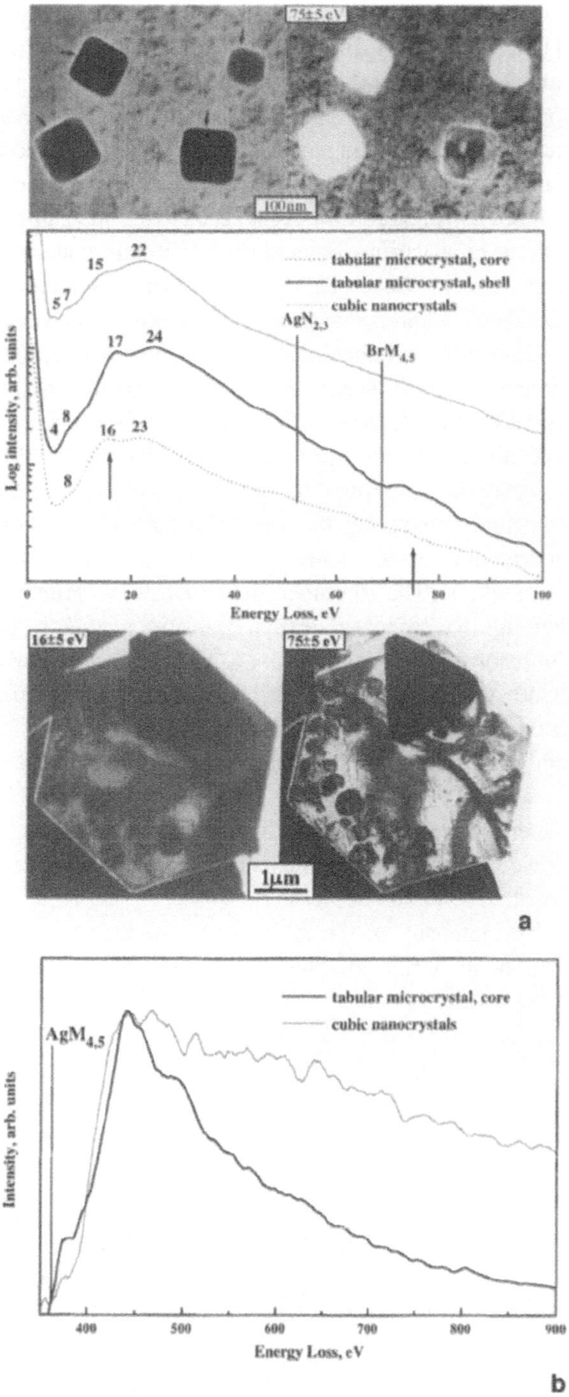

Fig. 1. Single scattering distributions (**a, b**) derived from energy-loss spectra and energy filtered images of Ag(Br, I) nanocrystals (top inset) and a composite tabular microcrystal (bottom inset) at 16 ± 5 and at 75 ± 5 eV shown by arrows in the spectrum a, $t = -177\,°C$: **a** the low-loss region; **b** the net Ag $M_{4,5}$-edge, spectra are normalized by intensity. The intensity is presented in arbitrary units. The numbers on the spectra indicate energies (in e V) of various features. Small arrows show rims around the nanocrystals at $E = 16$ e V assigned to excitons

the Br $M_{4,5}$-edge at 70 eV partially superimposed with the Ag $N_{2,3}$-edge at 62 eV) demonstrate an effect of constrast reversal (see top and bottom insets in Fig. 1a). Because of a decreasing background, this allowed visualization of excitations in the crystals caused by plasmons possibly superimposed with some contribution of core 4p (Ag^+) and 3d (Br^-) electrons. The excitation probability decreased exponentially with distance from the particle surface. In the range 40–100 eV fringes around particles caused by plasmons could be easily distinguished at distances of about 10–20 nm. The non-uniform contrast and dark areas observed inside the crystals may be referred to predominant excitations at the surfaces (particularly near the edges); this could decrease the probability of volume losses because of the sum rule [4]. Coupling of surface and volume losses resulted in some oscillation of their intensities with decreasing particle size. Observed excitations may be suitably described by the dielectric theory taking into account the band structure. The SSD intensity for low energy losses expressed as the differential energy-loss probability dp/dE is related to the imaginary part of the reciprocal of the complex relative permittivity $\varepsilon(\vec{q}, E)$ as a function of wave vector $\vec{q}$ and energy E, thus reflecting the local dielectric response of the media [11]:

$$\frac{dp}{dE} \propto \mathrm{Im}\left[-\frac{1}{\varepsilon(\vec{q}, E)}\right] \ln\left(1 + \frac{\beta^2}{\theta_E^2}\right), \qquad (1)$$

where β is the collecting semiangle and $\theta_E = (E/2T)(1 - (v^2/c^2))^{1/2}$ is the characteristic scattering angle; the term $T = (1/2)m_0 v^2 = (E_0(1 + E_0/2m_0c^2)/(1 + E_0/m_0c^2)^2)$: m_0 is the free electron mass: v is the velocity of incident electrons; c is the light velocity, and E_0 is the initial energy of the incident electrons.

Figure 2 shows the real and imaginary parts of dielectric permittivity, ε_1 and ε_2 and the energy-loss function calculated by a Kramers-Kronig analysis using the standard procedure [5]. Curves fall to 0 at $E \leq 3$ eV taking into account the position of the indirect band gap at 2.68 eV [1]. Here the relative permittivity is defined as $\varepsilon(\vec{q}, E) = \varepsilon_1(\vec{q}, E) + i\varepsilon_2(\vec{q}, E)$, where the real part describes the polarizability of the specimen (minima at 16–17 eV and above 23–24 eV), while the imaginary part $\varepsilon_2(\vec{q}, E)$ is related to absorption in the specimen (maxima at about 4, 8, 15–16 and at 21 eV). The energy-loss

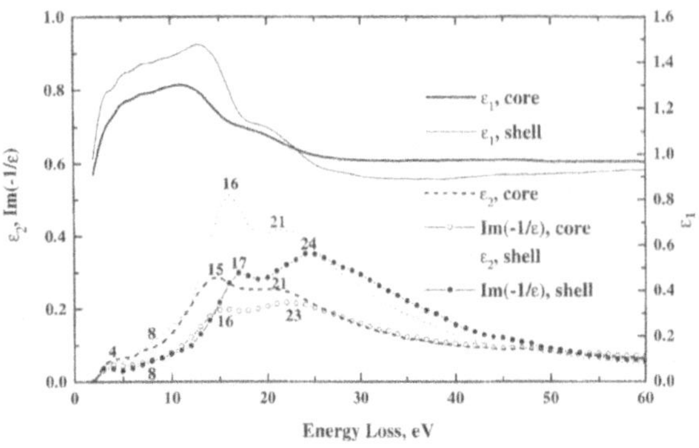

Fig. 2. Real (ε_1) and imaginary (ε_2) parts of dielectric permittivity and energy-loss function $\mathrm{Im}(-1/\varepsilon)$ derived by a Kramers-Kronig analysis of the EELS spectra of a Ag(Br, I) tabular microcrystal for the core and the shell region, respectively

function is expressed as [5]:

$$\mathrm{Im}\left[-\frac{1}{\varepsilon(\vec{q},E)}\right]=\frac{\varepsilon_2}{\varepsilon_1^2+\varepsilon_2^2}=\frac{\omega\Gamma\omega_p^2}{(\omega^2-\omega_p^2)^2+(\omega\Gamma)^2}, \tag{2}$$

where Γ is the damping constant, $\omega_p=\left[ne^2/(\varepsilon_0 m)\right]^{1/2}$ is the plasma resonance frequency, n is the electron density, e is the electron charge, ε_0 is the permittivity of vacuum, and $m=m_0$ is the electron mass. For low energy losses $E\le\hbar\omega_p=21.4\,\mathrm{eV}$ (the value calculated taking into account outer-shell 4d (Ag^+) and 4p (Br^-) electrons is somewhat less than experimental ones), the energy-loss function describes collective electron excitations against the ionic background caused by oscillations of bound electrons (interband transitions). When ε_1 has a local minimum (this points to the instability of the electron system against small external perturbations resulted in some excitations) and ε_2 is still small enough (this indicates small damping of oscillations due to absorption), the function $\mathrm{Im}[-(1/\varepsilon(\vec{q},E))]$ exhibits maxima corresponding to certain oscillations at 4, 8, 16–17 and 23–24 eV.

Although the energy of surface plasmons, E_s, is often described to be nearly $E_s=E_p/\sqrt{2}$, it can be argued that the peak at 15–16 eV in ε_2 can be considered as surface and/or defect plasmon, in view of its relatively high intensity as compared with the bulk loss peak at $E_p=21\,\mathrm{eV}$. Moreover, quasi free electrons usually taking part in these excitations are present in metals, but not in ionic insulating crystals such as AgX. Polarization waves formed by weakly bound electron-hole pairs of Mott-Wannier excitons are a prominent type of electronic excitation for AgBr [12]. Iodine in AgBr acts as an isoelectronic trap for holes at sufficiently low temperatures producing a positively charged centre which can subsequently be neutralised by binding an electron [13]. Thus, excitons bound on iodine can be formed. Interband transitions caused by mixing of the Ag 4d states with the halogen 4p states at various points in the Brillouin zone may also occur. Observed low-loss bands sometimes exhibited a composite arrangement, which, however, could not be revealed clearly because of overlapping of neighbouring peaks and limited resolution of the spectrometer. In line with studies of UV absorption and reflectivity on AgBr [12, 14], the band at 4 eV in ε_2 may be attributed to unresolved direct exciton transitions involving the spin-orbit split valence and conduction band states at Γ points, i.e., (Γ_8^-, $\Gamma_6^-\to\Gamma_6^+$). The shoulder at about 8 eV probably belongs to proposed exciton transitions at X point (unresolved $(X_6^-, X_6^-, X_7^-)\to X_6^+$) [14]. The intensive band at 15–16 eV may also be referred to exciton transitions particularly with $L_{4,5}^-$ and Γ_6^+ symmetries and/or to higher excited states at points different from Γ and L in the Brillouin zone as well, possibly superimposed with surface losses. Exact assignments, however, need additional experimental and theoretical work.

The intensity of the major $AgM_{4,5}$-edge shown in Fig.1b in terms of the dipole approximation is proportional to the differential ionization cross-section $d^2\sigma/d\Omega dE$. For small collecting angles Ω it can be written as [9]:

$$\frac{d^2\sigma}{d\Omega dE}\propto|\langle\Psi_f|\exp(i\vec{q}\,\vec{r})|\Psi_i\rangle|^2\rho(E). \tag{3}$$

Here $|\langle\Psi_f|\exp(i\vec{q}\,\vec{r})|\Psi_i\rangle|^2$ is the atomic transition matrix governing the overlap between the initial and final state wave functions Ψ_i and Ψ_f coupled by the dipole selection rule and $\rho(E)$ is the symmetry-projected density of final states. The matrix elements take into account that the 3d electron excitation involves a change in the orbital momentum $\Delta 1=\pm1$. The primarily atomic delayed edge shape, smoothly varying with energy-loss, is described by the first term, which is modulated by $\rho(E)$ representing here the result of probing of the preferentially f-like unoccupied states. So, the edge profile is affected by the energy distribution of unoccupied states above the Fermi level, the nearest-neighbor coordination of Ag ions and bonding characteristics. The Ag $M_{4,5}$-edge arises due to a centrifugal potential barrier making it difficult for $3d^{3/2}$ and $3d^{5/2}$ electrons with non-zero initial angular momentum ($l = 2$) to make a transition to a continuum state of low kinetic energy close to the threshold at 366.7 eV. Calculations of the edge profile by the Hartree-Slater method [15] indicate that atomic effects appear to dominate in the EEL spectrum rather than solid state factors. At low continuum energies and momentum transfer the edge intensities are small because only the 3d-4p transition contributes at the threshold. The intensity then rises steeply after the energy exceeds a certain value to enable the continuum wave functions to overlap with the 3d orbital. Optically forbidden transitions 3d-4s give rise to a maximum at nonzero momentum transfer near the threshold, but these are swamped by the 3d-4f transitions as energy loss increases. However, the spectra in Fig. 1b (the edge maximum for nanocrystals of 50 ± 4 nm in mean size is less pronounced as compared with one for a tabular grain due to the significant decrease of detected mass), exhibit additionally an electron loss near-edge structure (ELNES). It is governed by spin-orbital splitting between the $3d^{3/2}$-transition (M_4-edge at about 391 eV) and the lowest energy $3d^{5/2}$-transition (M_5-edge at 367 eV), which is not taken into account by estimations based on atomic models [15, 16]. The features are additionally broadened because of the finite lifetime of both the initial and final states and probably contain weak intensity modulations above 450 eV, which arise from the interference between the outgoing excited electron waves of a Ag^+ cation and the waves backscattered from the neighboring ions, an extended energy-loss fine structure (EXELFS) [4, 5].

Point X-ray spectra of the specimens contained bands of the Ag L-series (Ag $L\alpha_1$ at 2.99 ke V and Ag $L\beta_1$ at 3.15 ke V) and the Br L-series (Br $L\alpha$ at 1.49 ke V and Br $L\beta_3$ at 1.60 ke V). The weak I L-series (I $L\alpha$ at 3.92 ke V and I $L\beta$ at 4.22 ke V) was also detected although for nanocrystals this was possible only near the noise level for particles not less than 120 nm in size. Estimations of their elemental composition gave an iodide content of approximately 2–3% at. whereas the Ag and the Br contents were 48% at. and 48–49% at., respectively. Calculated absorption factors were 1.00 ± 0.07 for corresponding lines. At the same time it was impossible to detect iodide in the species by EELS because of the low concentrations. The major $BrL_{2,3}$-edge at 1550 eV was also very weak due to the low signal-to-noise ratio. Therefore

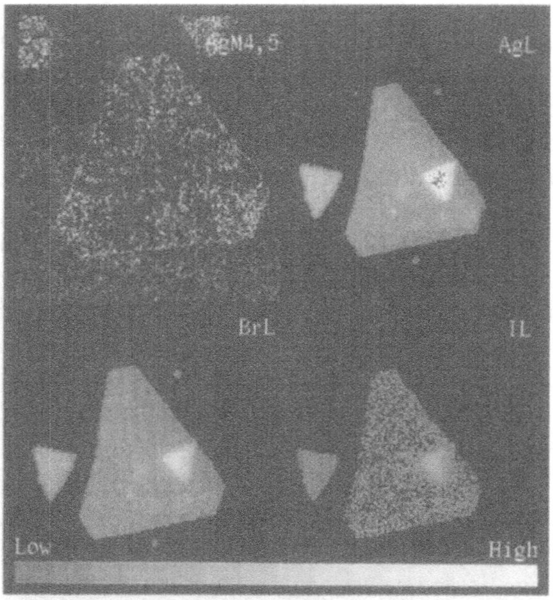

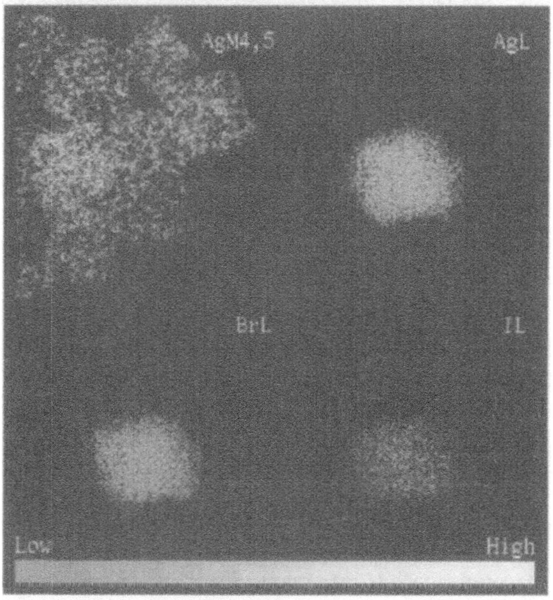

Fig. 3. Combined cryo-ESI/EDX maps of Ag(Br, I) particles: **a**-A tabular microcrystal; ESI, $AgM_{4,5}$-map, pre-edge 1 at 357 e V, pre-edge 2 at 367 e V and post-edge at 400 e V, acquisition time 1 min, X-ray AgL, BrL, and IL maps (the same population of microcrystals), 32 pixel repetition, dwell time 5 msec, acquisition time 190 min **b**-Nanocrystals, ESI, $AgM_{4,5}$-map, pre-edge 1 at 357 e V, pre-edge 2 at 367 e V and post-edge at 450 e V, acquisition time 1 min, X-ray AgL, BrL, and IL maps of a single nanocrystal (the same population of particles), 64 pixel repetition, dwell time 0.5 msec, acquisition time 40 min

combined elemental mapping by cryo-ESI/EDX (see Fig. 3) was applied to characterize Ag and halide distributions in the crystals of 100–120 nm in thickness which is more than the free mean path of inelastically scattered electrons. Earlier ESI data on Ag and I distributions by a two-window method have been reported only for 30 nm thick sections of converted Ag(Br, I) grains [17]. A qualitative Ag distribution in Fig. 3a exhibits intensity variations caused by bending a tabular crystal of a nearly constant thickness due to the local variations of the excitation errors [8]. In order to perform quantitative mapping, treatment of energy-filtered images should take into account dynamic diffraction and optimized conditions of specimen orientation. This effect is reduced for a population of nanocrystals that allowed $AgM_{4,5}$-distributions for particles of 50–80 nm in size to be obtained (see Fig. 3b). On the other hand, acquiring reasonable intensities by cryo-EDX/STEM especially for iodide distributions (for the tabular grain, it revealed a tendency for the migration of iodide ions from the shell to the core) required more than 3 hours as compared with 1 minute acquisition time for ESI. Measured intensities are also influenced by overlapping of particles. Fig. 3b presents high-magnification X-ray maps of a single 100 nm-sized nanocrystal collected at magnification $\times 500{,}000$ in the dark-field STEM mode. ESI and X-ray maps demonstrate a uniform silver and halide distribution in the nanoparticles. So, due to the complementary nature of information, combined utilization of cryo-ESI/EELS and cryo-STEM/EDX techniques including multielemental mapping may be of significant benefit for the purpose of comprehensive characterization of AgX micro- and nanocrystals of practical photographic emulsions.

Conclusions

The combination of cryo-ESI/EELS/STEM/EDX has been applied for characterizing electronic structure and element distributions in Ag(Br, I) nano- and tabular microcrystals. Qualitative Ag and halide distributions were obtained by ESI (a three-window method) and by EDX including a high magnification mapping of nanocrystals. The low-loss fine structure of EELS spectra between 4–26 eV was attributed to excitons and plasmons possibly superimposed with interband transitions. Contrast tuning under energy-filtering in the low-loss region was used to image electron excitations in the AgX particles. The real and imaginary parts ε_1 and ε_2 of the dielectric permittivity were determined by means of Kramers-Kronig analysis. The ELNES of the Ag $M_{4,5}$-edge governed by spin-orbital splitting of $3d^{3/2}$ and $3d^{5/2}$ states has been evaluated.

Acknowledgements. The authors are indebted to Prof. R. Egerton (University of Alberta, Canada) for the KRAKRO program for the Kramers-Kronig analysis of EELS spectra, to Prof. J. Cazaux (University of Reims, France) for helpful comments, to Prof. S. Wang (Institute of Photographic Chemistry, Academia Sinica, Beijing, China) and Dr. S. Brichkin (Institute of Chemical Physics, Russian Academy of Sciences, Chernogolovka, Russia) who provided some samples.

References

[1] J. F. Hamilton, *Adv. Phys.* **1988**, *37*, 359.
[2] A. P. Marchetti, R. P. Johansson, G. L. McLendon, *Phys. Rev.* **1993**, *B47*, 4268.
[3] M. I. Freedhoff, A. P. Marchetti, G. L. McLendon, *J. Luminescence* **1996**, *70*, 400.
[4] L. Reimer, in: *Energy-Filtering Transmission Electron Microscopy, Springer Series in Optical Sciences, Vol. 71* Springer-Berlin Heidelberg New York Tokyo, 1995, pp. 1, 347.
[5] R. F. Egerton, *Electron Energy-Loss Spectroscopy in the Electron Microscope, 2nd Ed.*, Plenum, New York, 1996, pp.1, 113, 131, 245, 407, 433.
[6] V. Oleshko, R. Gijbels, W. Jacob, F. Lakiere, A. Van Daele, E. Silaev, L. Kaplun, *Microsc. Microanal. Microstr.* **1995**, *6*, 79.
[7] V. Oleshko, R. Gijbels, W. Jacob, M. Alfimov, *Microbeam Analysis* **1995**, *4*, 1.
[8] V. Oleshko, R. Gijbels, W. Jacob, *J. Microsc.* **1996**, *183 (1)*, 27.
[9] F. Hofer, in: *Energy-Filtering Transmission Electron Microscopy, Springer Series in Optical Sciences, Vol. 71*, Springer, Berlin Heidelberg New York Tokyo, 1995, p. 225.
[10] J. Van Puymbroeck, P. Van Espen, W. Jacob, *Micron Microsc. Acta* **1989**, *20*, 155.
[11] R. F., Egerton, *Ultramicroscopy* **1989**, *28*, 215.
[12] N. J., Carrera, F. C. Brown, *Phys. Rev.* **1971**, *B4*, 3651.
[13] W., Czaja, A., Baldereschi, *J. Phys. C.* **1979**, *12*, 405.
[14] F., Bassani, R. S., Knox, W. B., Fowler, *Phys. Rev.* **1965**, *A4*, 1217.
[15] R. D. Leapman, P. Rez, and D. F. Mayers, *J. Chem. Phys.* **1980**, *72*, 1232.
[16] P. Rez, *Ultramicroscopy* **1989**, *28*, 16.
[17] J. L. Lavergne, C. Gimenez, G. Friour, J. M. Martin, in: *Proc. Intern. Congr. Phot. Sci. ICPS'94/IS7T's 47th Ann. Conf, Vol.1*, IS&T, Rochester, New York, 1994, p. 31.

Mikrochim. Acta [Suppl.] 15, 93–96 (1998)

Electron Energy-Loss Near-Edge Structure of Alumina Polymorphs

Igor Levin[1], **Alexander Berner**[1,*], **Christina Scheu**[1], **Harald Muellejans**[2], and **David G. Brandon**[1,3]

[1] Faculty of Materials Engineering, Technion 32000 Haifa, Israel
[2] Max-Planck-Institut für Metallforschung, Seestrasse 92, D-70174 Stuttgart, Federal Republic of Germany
[3] Wolfson Center for Interface Science, Technion 32000 Haifa, Israel

Abstract. In the present work the electron energy-loss near-edge structures (ELNES) of γ- and δ-Al_2O_3, both based on f.c.c. packing of oxygen anions, was investigated. The spectra were interpreted in terms of the relative occupancy of tetra- and octahedrally coordinated sites within the oxygen sublattice. This relative occupancy was found to be similar for both structures.

Key words: EELS, ELNES, alumina polymorphs

Alumina has a wealth of metastable crystalline phases in addition to the thermodynamically stable α-Al_2O_3 or corundum form [1]. These crystal structures may be divided into broad categories, based on either an f.c.c. or an h.c.p. arrangement of anions. Within each category, it is the distribution of cations that gives rise to the different polymorphs. Commonly observed polymorphs which are based on f.c.c. packing of the oxygen anions are γ (cubic spinel), δ (tetragonal/orthorhombic) and θ (monoclinic) alumina [1–3].

γ-Al_2O_3 possesses a spinel structure (space group $Fd\bar{3}m$) with 32 oxygen anions in a f.c.c. packing. The 64/3 Al cations (to satisfy stoichiometry) are distributed over the octahedral 16d (occupancy 1) and tetrahedral 8a (occupancy 2/3) interstitial sites in the oxygen sublattice [2]. The δ-Al_2O_3 structure is a superlattice of γ-Al_2O_3 with 64 fully ordered Al cations. Both tetragonal and orthorhombic unit cells have been reported for δ-Al_2O_3 [3, 4], but the cation positions in either of the proposed structures remain unknown. θ-Al_2O_3 has monoclinic symmetry described by the space group C2/m. The unit cell of θ-Al_2O_3 contains 8 ordered Al cations equally distributed between 4 octahedral and 4 tetrahedral sites in the oxygen sublattice [5]. The stable α-Al_2O_3 structure has trigonal symmetry with rhombohedral centering (space group $R\bar{3}c$). The oxygen anions are approximately hexagonally close packed and the Al cations are located in octahedrally coordinated interstitial sites within the oxygen sublattice.

Despite the amount of previous work published on the metastable aluminas, the exact atom arrangement of both γ- and δ-Al_2O_3 has eluded structural analysis. The major difference between these two structures is in the distribution of the cations over tetra- and octahedrally coordinated interstitial sites in the oxygen sublattice. The near-edge structure (ELNES) of Al $L_{2,3}$ in electron energy loss spectra (EELS), has been shown to depend on the local symmetry of the absorbing cation, and should provide information about the relative occupancy of differently coordinated sites [6, 7]. Such information is complementary to that derived from X-ray diffraction and high-resolution electron microscopy and should be helpful in developing atom models of the alumina structures.

Although the ELNES for γ-Al_2O_3 has been established in a number of studies [8–10], no spectra of other alumina polymorphs are reported in the literature. The present contribution presents experimental ELNES spectra for both γ-Al_2O_3 and δ-Al_2O_3 and attempts to determine whether these spectra can serve as a 'fingerprint' for either of the polymorphs.

Experimental

The ELNES of γ-Al_2O_3 was measured on self-supporting anodic alumina films. The thin films of amorphous alumina were prepared

* To whom correspondence should be addressed

by anodizing Al foil in a 0.01 M aqueous solution of ammonium tartrate at room temperature. The thickness of the film was controlled by choosing an appropriate anodizing voltage [11]. In all the reported experiments the applied voltage was 15 V, which gives a calculated film thickness slightly less than 20 nm. The purity of the Al foil was 99.999%. After anodizing the free-standing alumina films were released by etching away residual Al in 2 wt.% $HgCl_2$ solution. The films were retrieved on a TEM specimen support and dried in air. Films prepared by this route were placed in a pure alumina crucible and annealed at 8000 °C in air for 2 hours to form γ-Al_2O_3. The as-prepared films are of uniform thickness and transparent to the electron beam, so no additional ion-beam thinning is required.

The annealing of the anodic films at temperatures above 800 °C resulted in a mixture of metastable alumina phases (including γ- and δ-Al_2O_3) as has been reported elsewhere [12]. Two reciprocal lattice (rel) sections are usually required to distinguish between δ- and the other ordered metastable alumina structures. However, an absence of Kikuchi bands (due to the small foil thickness) together with the fine spatial scale and lack of distinct morphology for the metastable alumina phases makes controlled tilting of the anodic films extremely difficult. For this reason, plasma-sprayed alumina samples annealed at 1100 °C in air, were used to determine the ELNES of δ-Al_2O_3. TEM specimens of plasma-sprayed alumina were prepared by mechanical polishing followed by dimpling to a thickness of 30 µm. Further thinning was accomplished by ion-milling to perforation using a GATAN precision ion polishing system (PIPS) at 7°. These specimens also contained a mixture of metastable alumina structures, in this case with some α-Al_2O_3. The grain size was of the order of 0.5 µm, and the wedge-shaped specimen geometry permitted unambiguous diffraction identification of δ-Al_2O_3.

All TEM specimens were coated by a thin layer of carbon to improve electrical conductivity and both conventional and high-resolution electron microscopy were first applied to establish the phase composition. The ELNES of γ-, δ- and α-Al_2O_3 were measured in a conventional JEOL-2000FX TEM microscope equipped with a GATAN PEELS spectrometer (Model 766) and used in a diffraction mode. Conventional TEM, rather then dedicated STEM, was necessary to distinguish between the arrays of the weak superlattice reflections characteristic of the different alumina structures.

Both the O-K and the Al-$L_{2,3}$ edges, as well as the low loss part of the spectrum, were acquired from single grains of γ-, δ- and α-Al_2O_3. The spectra were recorded at 200 keV with a beam current of 2 µA, a spectrometer entrance aperture of 1 mm, and an energy dispersion of 0.1 eV/ch. The energy resolution as determined by the half-width of the zero-loss peak was about 1.3 eV. All spectra were corrected for dark current and for channel-to-channel gain variation of the detector. The spectra were deconvoluted to remove plural scattering effects and the background, extrapolated and substracted using the routines of the standard GATAN software package.

Results and Discussion

The annealed anodic alumina films were found to contain nano-sized grains of γ-Al_2O_3, as confirmed by selected area diffraction. No other alumina polymorphs were identified in this specimen. In plasma-sprayed alumina, the EELS spectra were acquired from single grains, confirmed by electron diffraction

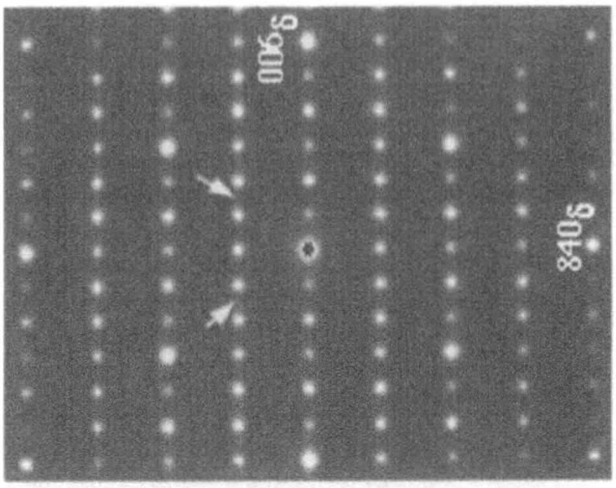

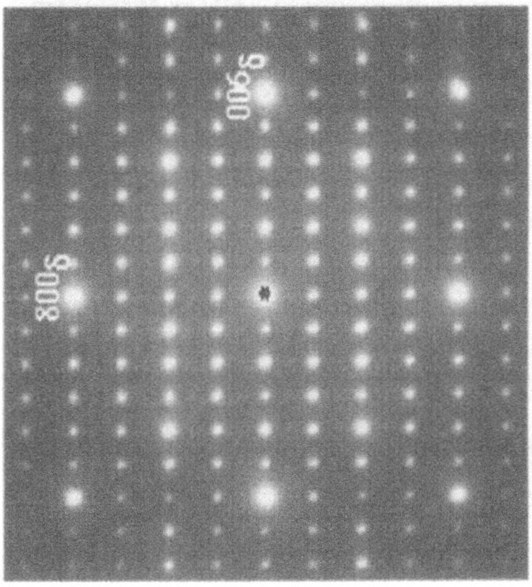

Fig. 1. Selected area diffraction patterns from the single grain of δ-Al_2O_3 in both $[100]_\delta$ and $[\bar{1}20]_\delta$ orientations (plasma-sprayed alumina). The arrows indicate the absence of all the former $\{111\}\gamma$ reflections in the $[\bar{1}20]_\delta \| \langle 110 \rangle_\gamma$ rel section of δ-Al_2O_3

to be orthorhombic δ-Al_2O_3 with lattice parameters $a_\delta = 2a_\gamma$, $b_\delta = a_\gamma$ and $c_\delta = 1.5a_\gamma$ (Fig. 1). The examination of two rel sections from the same crystal unambiguously identified the δ-Al_2O_3 phase. EELS spectra from single grains of α-Al_2O_3 present in the plasma-sprayed alumina sample were also recorded (Fig. 2b and 3b).

The ELNES of O-K and Al-$L_{2,3}$ edges for both γ- and δ-Al_2O_3 are presented in Figs. 2a and 3a, respectively. The ELNES for both edges measured in γ-Al_2O_3 on anodic alumina films are similar to those previously reported [13], while the ELNES of δ-Al_2O_3 is not appreciably different from that of γ-Al_2O_3.

The ELNES of O-K is primarily determined by the arrangement of the oxygen anions in the structure and the close similarity between the O-K ELNES for γ- and δ-Al_2O_3 reflects the same, approximately f.c.c.,

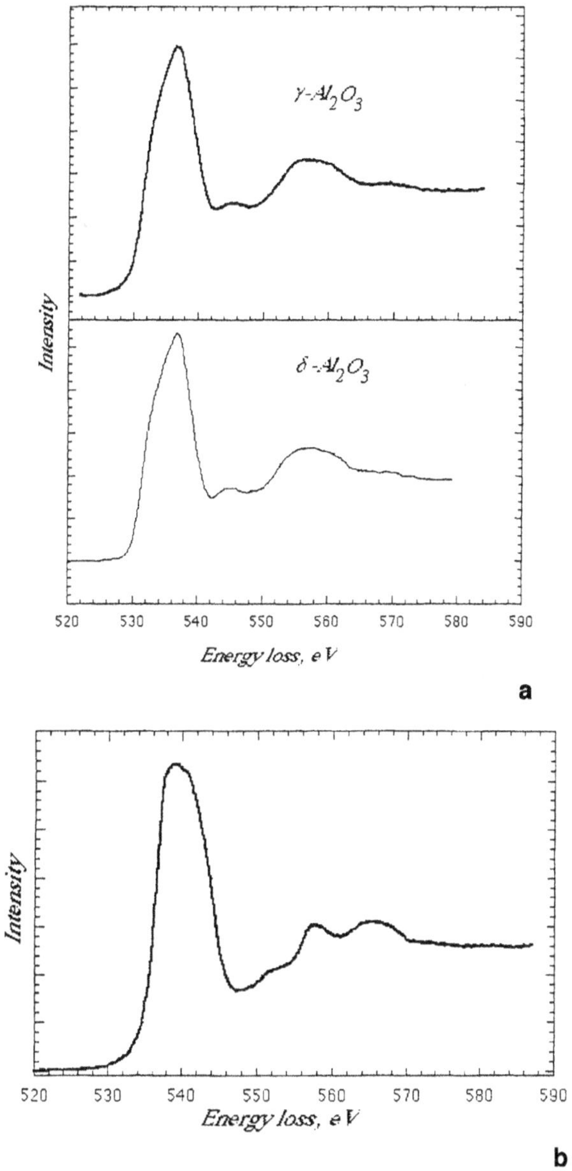

Fig. 2. ELNES of O-K edge for both γ- and δ-Al_2O_3 structures (a) and α-Al_2O_3 (b)

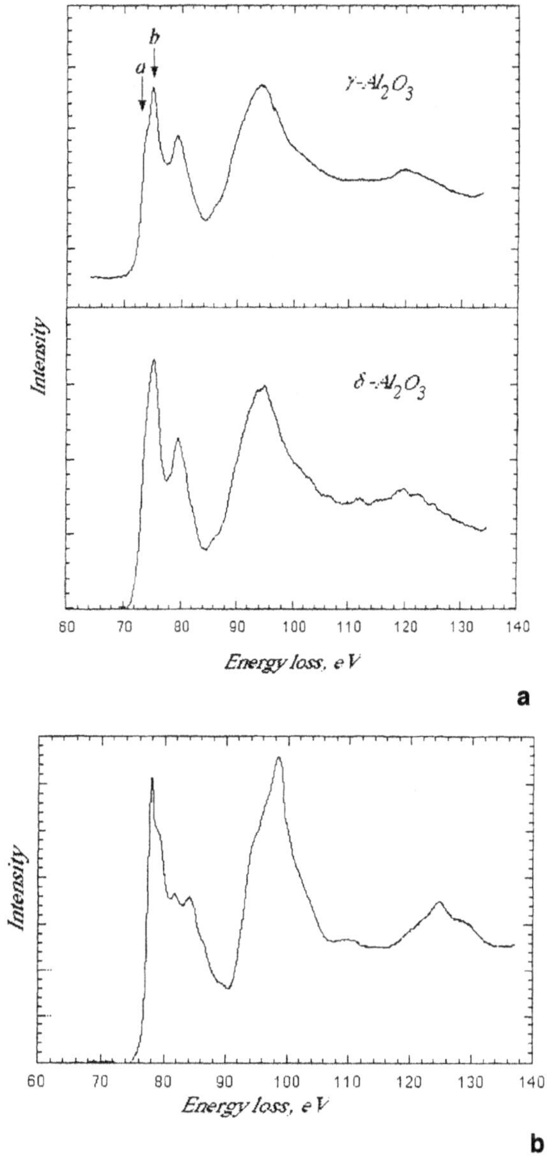

Fig. 3. ELNES of Al-$L_{2,3}$ edge for both γ- and δ-Al_2O_3 structure (a) and α-Al_2O_3 (b)

packing of the oxygen ions in these structures, as opposed to the hexagonal packing in corundum (Fig. 2b). The calculation of O-K ELNES reported by Brydson [13] for γ-Al_2O_3 predicts the presence of three distinct features located at 20, 25 and 30 eV above the threshold, but the observed experimental spectrum contains a single broad peak, centered at approximately 25 eV. It has been argued that the position of the central calculated peak should be strongly dependent on the distribution of the cations over the tetrahedrally coordinated interstitial sites, and a random distribution of cations over these sites in γ-Al_2O_3 could be responsible for the smearing of the three peaks into one broad feature. Following this argument, the ordering of cations in δ-Al_2O_3 would be expected to replace the broad peak observed at 25 eV above threshold by the three peaks predicted by calculations, although no such difference was observed in the present spectra from γ- and δ-Al_2O_3.

The main features of the Al-$L_{2,3}$ ELNES should depend on the coordination of the absorbing cation in the structure [11, 13]. The results obtained on the compounds with Al cations occupying sites with different coordination suggest that Al-$L_{2,3}$ ELNES may be considered as a linear superposition of ELNES from each type of cation position. Comparison of the present Al-$L_{2,3}$ ELNES of transition aluminas with those calculated for tetra-or octahedrally coordinated clusters suggests that the pre-edge shoulder in the experimental spectra should be attributed to the contribution of the tetrahedrally coordinated cations [13]. For example, Bruley et. al. [14] have shown that the intensity ratio of peaks 'a' and 'b' correlates with the relative occupancy of the tetrahedrally coordinated sites by Al in aluminate spinels, which belong to the same structural type as γ-Al_2O_3. Fitting of the Gaussian profiles to the features 'a' and 'b' in the present spectra has resulted in the same intensity ratio for both the γ- and δ-Al_2O_3 structures, indicating that the relative occupancy of the tetra- and octahedrally coordinated sites in δ-Al_2O_3 is close to that in γ-Al_2O_3, at least within the sensitivity of the ELNES.

Conclusions

In the present work the ELNES of Al-$L_{2,3}$ and O-K edges of both γ and δ-Al_2O_3 were presented. The ELNES of either of these edges for both phases are essentially identical and cannot be used to distinguish between these two structures. The similar relative intensities of the pre-edge shoulder in the spectra of γ- and δ-Al_2O_3 also suggest that the relative occupancy of the cation sites with octa- and tetrahedral coordination is similar in both structures.

Acknowledgements. The plasma-sprayed alumina for this study was kindly provided by Prof. K. Rajan (RPI, Troy). The study was supported by German-Israel Foundation for Scientific Research and Development under the contract # 040–551. We acknowledge financial support from the Volkswagen-Stiftung contract I/70082. I. L. is grateful to the Israel Ministry of Science for financial support. The authors would like to acknowledge the use of the facilities of the Wolfson Center for interface Science (Technion, Israel).

References

[1] K. Wefer, C. Misra, *Alcoa Technical Paper No. 19.* Alcoa Laboratories, 1987.
[2] B. C. Lippens, J. H. De Boer, *Acta Crystallogr.* **1964**, *17*, 1312.
[3] S. J. Wilson, *Proc. Br. Ceram. Soc.* **1979**, *28*, 281.
[4] V. Jayaram, C. G. Levi, *Acta Metall.* **1989**, *37* (2), 569.
[5] G. Yamaguchi, H. Yanagida, S. Ono, *Bull. Chem. Soc. Jpn.* **1964**, *37*, 752.
[6] R. Brydson, H. Sauer, W. Engel, J. M. Thomas, E. Zeitler, *J. Chem. Soc. Chem. Commun.* **1989**, *15*, 1010.
[7] P. L. Hansen, R. Brydson, D. W. Mc Comb, *Microsc. Microanal. Microstruct.* **1992**, *5*, 173.
[8] I. A. Brytov, Y. N. Romashenko, *Sovt. Phys. Solid State* **1978**, *20*, 384.
[9] M. Sternitzke, *J. Am. Ceram. Soc.* **1993**, *76*, 2289.
[10] J. C. Yang, E. Schumann, H. Mullejans, M. Ruhle , *J. Phys. D* **1996**, *29*, 1716.
[11] L. Young, *Anodic Alumina Films,* Academic Press, N Y 1991.
[12] I. Levin, L. A. Bendersky, D. G. Brandon, M. Ruhle, *Acta Mater.*, in print.
[13] Brydson R., *J. Phys. D* **1996**, *29*, 1699.
[14] J. Bruley, M.-W. Tseng, D. B. Williams, *Microsc. Microanal. Microstruct.* **1995**, *6*, 1.

Mikrochim. Acta [Suppl.] 15, 97–100 (1998)

SPM Study of YBCO Films Prepared by Plasma Assisted Laser Ablation

Yongchang Fan*, **Alexander G. Fitzgerald,** and **Hanchang Xu**

Carnegie Laboratory of Physics, Department of Applied Physics and Electronic & Mechanical Engineering, University of Dundee, Dundee DD1 4HN, U.K.

Abstract. High T_c and J_c $YBa_2Cu_3O_{7-x}$ (YBCO) superconducting thin films have been prepared by DC oxygen discharge plasma assisted excimer laser ablation on MgO (100), $SrTiO_3$ (100) and yttria-stabilized ZrO_2 (100) substrates. The main improvement obtained with this laser ablation scheme is that laser deposition can be carried out in the oxygen discharge environment. This is achieved by two ring-shaped electrodes which have been placed in the middle between the target and substrate. X-ray diffraction (XRD), atomic force microscopy (AFM) and scanning tunnelling microscopy (STM) techniques have been used to characterise the as-deposited films. A set of narrow (001) peaks in the x-ray diffraction patterns from these films indicate film growth with a preferential c-axis orientation. AFM image observations show that the number of the particulates observed on thin films prepared by this new laser ablation scheme is reduced by comparison with films prepared by the conventional laser ablation method. The majority of growth islands have merged and connected with each other and have formed much larger, denser and smoother flat plateaus. Similar to films fabricated by the conventional laser ablation scheme, STM surface characterisation reveals that the films prepared by oxygen plasma assisted laser ablation are also formed mainly by screw dislocation generated spiral growth.

Key words: oxygen discharge plasma, laser ablation, $YBa_2Cu_3\,O_{7-x}$ thin films, spiral growth, scanning probe microscopy.

Pulsed laser ablation, with its extraordinary ability to evaporate multi-component materials stoichiometrically, has developed to become one of the most widely used deposition techniques for the preparation of high T_c superconducting films [1–4]. With this deposition technique, high T_c and J_c c-axis oriented YBCO thin films have been reproducibly fabricated at relatively low substrate temperatures by many research groups around the world [5–10]. In these low temperature in-situ experimental schemes for epitaxial film growth, one of the key factors or common features is that certain kinds of reactive oxygen sources have been introduced into or produced in the deposition chamber. It is now recognised that reactive ionic or atomic oxygen species can facilitate the formation of the superconducting phase during the deposition process by incorporating enough oxygen into the films [10–13].

In the last few years, one of these improved laser deposition schemes, i.e. the so-called DC oxygen discharge plasma assisted laser ablation scheme [6] has been frequently used by the authors to prepare high quality YBCO superconducting films. It is found that this laser deposition scheme is particularly effective in ensuring film reproducibility, surface quality and the required larger critical current value. In this research work, AFM and STM techniques have been used to characterise these films. Large scan size AFM images show that the particulates often observed on the films prepared by conventional laser ablation have been reduced, and that the films are mainly composed of connected crystal plateaus seeded with some voids. Similar to the films prepared by the routine laser ablation, STM image observations reveal that the films prepared by oxygen discharge plasma assisted laser ablation on $SrTiO_3$, MgO and Y-ZrO_2 substrates are also formed mainly by the

* To whom correspondence should be addressed

well-known screw dislocation generated spiral growth [14–18].

Experimental

The oxygen discharge plasma assisted laser deposition experiment set up is shown in Fig. 1. The main improvement compared with the conventional laser deposition technique is that two ring-shaped electrodes are placed in the middle between the target and the substrate. During the deposition, positive and negative 160 V DC voltages are applied to these two electrodes with the electrode near to the substrate maintained at the positive potential. The DC discharge can effectively activate the oxygen by an ionising and exciting process.

In the experiments, an XeCl excimer laser (wavelength 308 nm, pulse duration 25 ns) was used to bombard the YBCO superconducting targets. The laser beam energy density on the target surface was 2 J/cm^2 and the laser operation frequency was 5 Hz. $SrTiO_3$ (100), MgO (100) and YSZ (100) single crystals were used as substrates and were maintained at 550 °C during the deposition. To maintain a stable oxygen discharge, the oxygen pressure in the deposition chamber was kept at 200 mTorr. After deposition, the oxygen pressure was further increased to 200 Torr. The DC voltage and substrate heating power supply were then switched off leaving the films to cool naturally to room temperature.

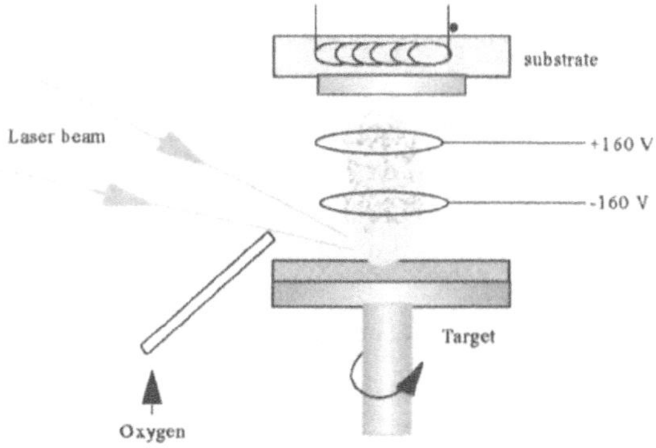

Fig. 1. Schematic diagram of oxygen discharge plasma assisted laser deposition

The prepared films were first examined by a Digital Instruments Dimension^{TM} 3000 atomic force microscope to obtain an overall view of topographical features. The films were then studied at high resolution with a Molecular Imaging scanning tunneling microscope operated by a Digital Instruments Nanoscope IIIa controller to enable observation of the detailed morphology of growth islands and deep voids which appeared at lower resolution in the AFM images. AFM images of the films were taken in the tapping mode with a standard etched silicon cantilever. STM observation was carried out in the constant current mode at room temperature under ambient conditions. All STM images of the film were obtained with mechanically cut Pt-Ir tips with a positive sample bias voltage of 800 mV and a tunneling current in the range of 50 to 200 pA.

Results and Discussion

General Properties

The as-deposited films prepared by this laser ablation deposition scheme were smooth and highly reflecting. Usually, the room temperature resistance which is measured between two arbitrary positions on the one centimetre square size films was less than one hundred ohms, much lower than that of films prepared by the conventional laser ablation deposition technique. This is a strong indication that the films had absorbed sufficient oxygen during the deposition process. The majority of films deposited on $SrTiO_3$, MgO and YSZ substrates have exhibited superconductivity above 85 K, and the measured critical current density J_c (at 77K and zero magnetic field) is in the range of 10^5 to 10^6 A/cm^2. From the x-ray diffraction patterns of the films, only a set of narrow (001) peaks can be identified and this is an indication of growth in a preferred orientation with the superconducting orthorhombic c-axis perpendicular to the substrate surface.

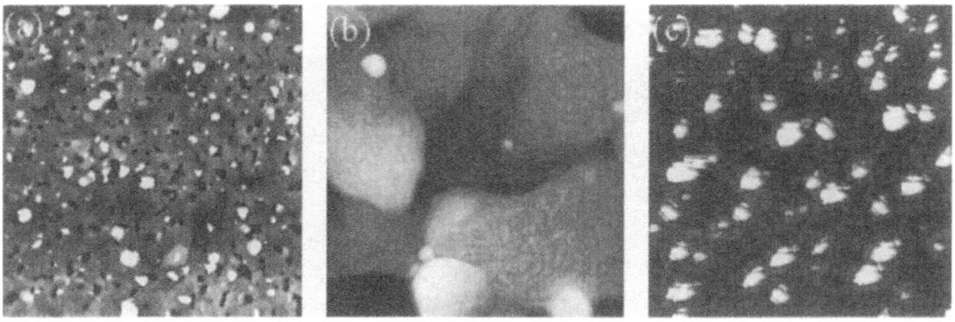

Fig. 2. AFM images of YBCO thin films on a YZS (100) substrates, film thickness 4500 Å. **a** and **b** prepared by plasma assisted laser deposition, **c** prepared by conventional laser deposition, scan size: **a** 10 μm, **b** 2.5 μm, **c** 10 μm

Morphologies

Figure 2a and Fig. 2b are typical AFM images of films prepared by the plasma assisted laser ablation on YSZ (100) susbtrates with a film thickness of about 4500 Å. An AFM image of the film prepared by conventional laser ablation on YSZ (100) substrates under the same deposition conditions but without the DC power supply to the two ring-shaped electrodes is also given in Fig. 2c. There are three obvious morphological changes which can be observed on the as-deposited films by comparison with films prepared without DC oxygen discharge. First, the density of particulates often observed on the routine laser ablation deposited films has been greatly reduced. Second, the majority of growth islands have merged and connected with each other to form much larger, denser and smoother flat plateaus. The third prominent feature is that some deep voids have appeared on the film surface. From these observations, it can be seen that the DC oxygen discharge process has the function of dissociating small particulates and facilitating nucleation of lateral island growth. Because preferential c-oriented crystal growth has been confirmed by x-ray diffraction analysis, the observed morphological features indicate that the DC discharge can speed up the growth rate in the a and b-axis directions.

The AFM image Fig. 2b does not have sufficient resolution to show the growth terraces in the YBCO film clearly. To obtain high resolution images the films have been observed using the STM. Fig. 3a to Fig. 3c show a set of different scan size STM images of the as-deposited films on an MgO substrate with a thickness of approximate 1200 Å. From these images we can see that the plasma assisted laser ablation deposited films have also undergone an island nucleation and growth process, and all of these islands observed in AFM images are in fact growth spirals as shown in these high resolution STM images. The terrace height of these growth spirals is the c-axis unit cell length. The deep voids observed on thicker films are originally developed from small open areas unseeded within the growth spirals.

With a prolonged deposition time, the film thickness increases by the addition of the deposited species on the ledges of the growth spiral as the spiral terrace continuously grows vertically. Meanwhile, the size of the spirals also increases laterally by accommodation

Fig. 3. STM images of YBCO thin film on MgO (100) substrate, film thickness approximately 1200 Å, scan size: **a** 2.0 μm, **b** 1.0 μm, **c** 500 nm

Fig. 4. STM images of the same YBCO thin film on $SrTiO_3$ (100) substrate, film thickness approximately 2500 Å., scan size: **a** 1.2 μm, **b** 1.3 μm, **c** 700 nm

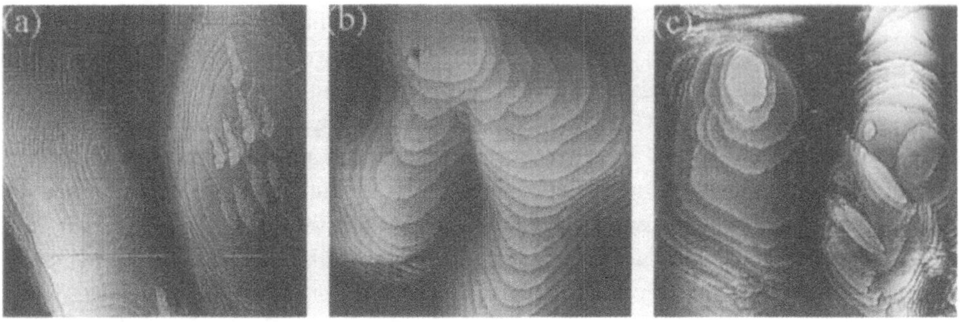

Fig. 5. STM images of growth spirals in two YBCO thin films on YSZ (100) substrates, film thickness in each case about 4500 Å, scan size: **a** 1.1 μm, **b** 900 nm, **c** 800 nm. **a** and **b** were obtained from the same film

of the deposited species on the ledges of the all the terrace layers. This growth tendency can be seen clearly in Fig. 4a to Fig. 4c. In these STM images, the film was deposited on the $SrTiO_3$ substrate with a thickness of about 2500 Å.

The detailed morphologies of the large growth spirals on some thicker films are shown in Fig. 5a to Fig. 5c. Here the films were deposited on YZS (100) substrates with a thickness of about 4500 Å. In Fig. 5a, the two separated growth spiral grains have dimensions of the order of one micrometer, the depth from the top surface layer to the bottom of the void is 59 nm, and the grains are formed by nearly 50 growth terrace layers. Fig. 5b and Fig. 5c show detailed features, for example, how growth spirals have merged and connected with each other and have grown into much larger crystal grains leaving any unconnected parts as deep voids.

Conclusions

In summary, YBCO superconducting films have been prepared by the oxygen discharge assisted laser ablation technique. This improved deposition method is particularly effective in obtaining film reproducibility, surface quality and the required larger critical current values. AFM and STM observations show that the films are also formed mainly of growth spirals generated by screw dislocations. The appearance of large crystal plateaus formed by the merging and connecting of the growth spirals suggests that the DC electrical field and the oxygen plasma discharge in the vicinity of the substrates promotes spiral growth laterally which results in a flat film surface.

Acknowledgements. One of the authors (Y. Fan) is grateful to the University of Dundee and the Commitee of Vice Chancellor and Principals of the Universities of United Kingdom for financial support for postgraduate research.

References

[1] D. Dijkkamp, T. Venkatensan, X. D. Wu, S. A. Shaheen, N. Jisrawi, Y. H. MinLee, W. L. Mcleean, M. Groft, *Appl. Phys. Lett.* **1987**, *51*, 619.
[2] A. Inam, M. S. Hedge, X. D. Wu, T. Ventatesan, P. England, P. F. Miceli, E. W. Chase, C. C. Chang, J. M. Trarascon, J. B. Wachtman, *Appl. Phys. Lett.* **1988**, *53*, 908.
[3] X. D. Wu, D. Dijkkamp, S. B. Ogale, A. Inam, E. W. Chase, P. F. Miceli, C. C. Chang, J. M. Tarascon, T. Venkatesan, *Appl. Phys. Lett.* **1987**, *51*, 861.
[4] J. Narayan, N. Biunno, R. Singh, O. W. Holland, O. Auciello, *Appl. Phys. Lett.* **1987**, *51*, 1845.
[5] X. D. Wu, A. Inam, T. Venkatesan, C. C. Chang, E. W. Chase, P. Barboux, J. M. Tarascon, B. Wilkens, *Appl. Phys. Lett.* **1988**, *52*, 754.
[6] S. Witanachchi, H. S. Kwok, S. W. Wang, D. T. Shaw, *Appl. Phys. Lett.* **1988**, *53*, 234.
[7] B. Roas, L. Schultz, G. Endres, *Appl. Phys. Lett.* **1988**, *53*, 1557.
[8] R. K. Singh, J. Narayan, A. K. Singh, J. Krishnaswamy, *Appl. Phys. Lett.* **1989**, *54*, 2171.
[9] E. Wiener-Avnear, G. L. Kenber, J. E. McFall, J. W. Spargo, A. G. Toth, *Appl. Phys. Lett.* **1990**, *56*, 1802.
[10] G. Koren, A. Gupta, R. J. Baseman, *Appl. Phys. Lett.* **1989**, *54*, 1920.
[11] J. P. Zheng, Q. Y. Ying, S. Witanachchi, Z. Q. Huang, D. T. Shaw, H. S. Kwok, *Appl. Phys. Lett.* **1989**, *54*, 954.
[12] C. Cirault, D. Damiani, J. Aubreton, A. Catherinot, *Appl. Phys. Lett.* **1989**, *54*, 2035.
[13] Q. Y. Ying, D. T. Shaw, H. S. Kwok, *Appl. Phys. Lett.* **1988**, *53*, 1762.
[14] M. Hawley, I. D. Raistrick, J. G. Beery, R. J. Houlton, *Science* **1991** *251*, 1587.
[15] D. G. Schlom, D. Anselmetti, J. G. Bednorz, Ch. Gerber, J. Mannhart, *J. Crystal Growth* **1994**, *137*, 259.
[16] J. Moreland, P. Rice, S. E. Russek, B. Jeanneret, A. Roshko, R. H. Ono, D. A. Rudman, *Appl. Phys. Lett.* **1991**, *59*, 3039.
[17] M. McElfresh, T. G. Miller, D. M. Schaefer, R. Reifenberger, R. E. Muenchausen, M. Hawley, S. R. Foltyn, X. D. Wu, *J. Appl. Phys.* **1992**, *71*, 5099.
[18] M. Yeadon, M. Aindow, F. WellhÖfer, J. S. Abell, *J. Crystal Growth* **1997**, *172*, 145.

Mikrochim. Acta [Suppl.] 15, 101–107 (1998)

Surface Characterisation and Modification of YBCO Thin Films by STM

Alexander G. Fitzgerald*, **Yongchang Fan**, and **Hanchang Xu**

Carnegie Laboratory of Physics, Department of Applied Physics and Electronic and Mechanical Engineering, University of Dundee, Dundee DD1 4HN, Scotland, U.K.

Abstract. The optimum conditions required to obtain a stable image and reliable morphological information for $YBa_2Cu_3O_{7-x}$ (YBCO) high T_c superconducting thin films are summarised. Some typical morphological images of YBCO thin films prepared by the conventional laser ablation technique are presented. By analysing the STM images, three growth modes i.e. spiral growth, layer-by-layer growth and step propagation growth have been identified. Scanning tunneling spectroscopy has been used to study the local area tunneling behaviour for different regions of the YBCO thin films. Distinct and representative I–V curves for spiral growth crystal grains, particulates and non-spiral growth islands have been obtained. Some initial experiments show nanometre scale modification and structure fabrication on the YBCO thin film surface obtained by scanning the tip.

Key words: $YBa_2Cu_3O_{7-x}$ thin films, surface characterisation, surface modification, scanning tunneling microscopy

Over the last few years, the scanning tunneling microscope (STM) has been widely used in the surface characterisation of high T_c superconducting (HTS) thin films [1–15]. The increasing interest and effort in using STM to study HTS thin films lies in its unprecedented lateral and vertical resolution for direct morphological observation even in ambient conditions; its unique ability to probe sub-unit cell electronic structures [16–20], and more importantly, its nanometre scale surface modification and nanometre scale microstructure fabrication capabilities [21–29].

In this paper, some of our recent research work on the STM surface characterisation of YBCO HTS thin films prepared by routine laser deposition technique are introduced. Typical morphological features of films on MgO (100), single-crystal (100) oriented yttria-stabilized zirconia (YSZ) and $SrTIO_3$ (100) substrates are presented. The scanning tunneling spectroscopy technique (STS) has been used to study the tunneling behaviour of the as-deposited films at room temperature in air. Different regions of the films such as growth spirals, particulates and some of the non-spiral growth islands have exhibited distinct tunneling characteristics. Examples of the use of the STM to modify the thin film surface and to fabricate nanometre scale structures are also demonstrated, and the potential applications of these capabilities of the STM in manufacturing nanometre dimension microelectronic devices are discussed.

Experimental

The YBCO HTS thin films were prepared by the conventional laser deposition technique [30] on MgO (100), $SrTiO_3$(100) and YSZ (100) substrates. During film deposition, the substrates were kept at 600 °C, and pure oxygen gas was introduced into the vacuum chamber which was maintained at 200 mTorr. The XeCl excimer laser which was used to ablate the YBCO HTS targets was operated at 5 Hz with a pulse energy density at the target surface of about 3 J/cm^2. After deposition, the oxygen pressure was further increased to 200 Torr, and then the substrate heating system was switched off letting the films cool to room temperature naturally.

The surface characterisation of the prepared YBCO HTS thin films was performed with a Molecular Imaging STM head operated by Digital Instruments NanoScope IIIa scanning probe microscope controller. The optimum imaging conditions for the YBCO HTS thin films have been determined by making STM image observations under a wide range of bias voltage and tunneling current combinations. Under the optimum imaging conditions i.e. bias voltage 650 to 800 mV and tunneling current

* To whom correspondence should be addressed

30 to 200 pA, the images were very stable during scanning in the same area over a long period. No image distortion or surface modification was observed. All of the images presented in this paper were captured with a positive sample bias voltage of 800 mV and a tunneling current of 150 pA using mechanically cut Pt/Ir tips in ambient conditions.

Scanning tunneling spectroscopy analysis of the films was also carried out at room temperature in air. When the STS 'plot I (v)' command was activated, the feedback loop ran to establish the tip-to-sample separation for the pre-set bias voltage and setpoint tunneling current (a constant tunneling current is then maintained by a feedback loop during scanning). The feedback loop was then disabled and the tip position was kept constant. At this moment, the tip-sample voltage was scanned and the change in the tunneling current as a function of tip-sample voltage was measured and plotted. The I–V curves allow the tunneling behaviour for different regions of the films to be compared and analysed.

When the bias voltage and tunneling current were set beyond the range of optimum imaging parameters i.e. by increasing the tunneling current meanwhile reducing or increasing the bias voltage substantially, the film surface can be modified by removing material away from the scanned area with the scanning tip. This capability was used to etch nanometre scale grooves on the top layer of the growth spirals.

Results and Discussion

Under the optimised deposition conditions, the majority of the prepared thin films have exhibited superconductivity with T_c above 81 K and critical current J_c in the range of 10^4 to 10^5 A/cm^2. X-ray diffraction analysis shows that these films have the orthorhombic crystal structure with the c-axis perpendicular to the substrate surface. The room temperature resistance measured between two arbitrary positions on the film surface is in the range of a few hundred to several thousand ohms. Although the conductivity of the films is not high, the films could be observed by scanning tunneling microscopy in ambient conditions.

Morphological Features

STM image observations indicate that the majority of high quality YBCO thin films ($T_c > 81$ K and $J_c > 10^5$ A/cm^2) on MgO (100) and $SrTiO_3$ (100) substrates exhibit a spiral growth mechanism with the films containing a high density of growth spirals generated from both left and right-hand screw dislocations (see Fig. 1a and Fig. 1b). These observed typical morphological features for films deposited by laser ablation are in agreement with observations reported by other researchers (3–5, 13–15]. For some films deposited under the same experiment conditions on YSZ (100) substrates, in addition to the well-known spiral growth, cake-layer like crystal grains have also been observed (see Fig. 2 a), and these grains are formed by a layer-by-layer growth mechanism. In some STM images of the films prepared on a few MgO substrates, high density directional terrace steps have appeared on the film surfaces (see Fig. 2b). These morphological features have also been observed by several researchers [3, 12]. It is believed that these features were caused by misoriented crystal faces on the substrate surface, and in this situation, the film growth occurs by step propagation [12] or what is described as ledge growth [3].

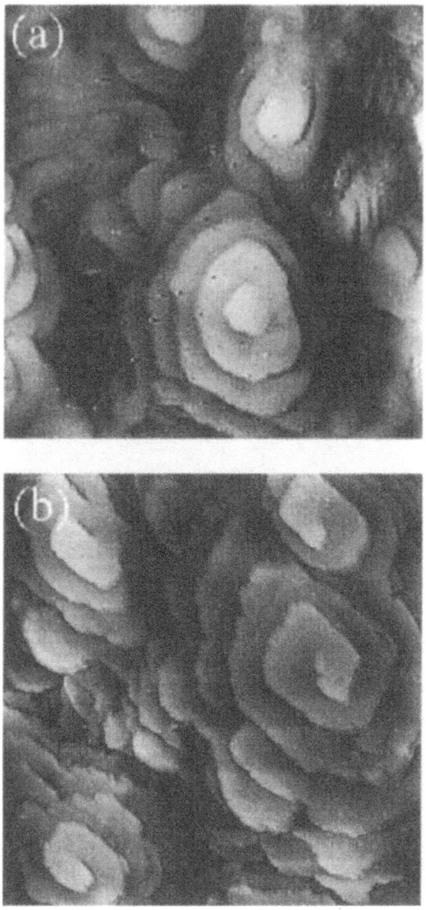

Fig. 1. STM images of the spiral growth morphologies of YBCO thin films on **a** MgO(100) and **b** $SrTiO_3$(100) substrates. Scan size: **a** 330 nm, **b** 300 nm

Cross-section analysis of the spiral growth grains show that the single terrace steps in the spiral growth grains correspond to the c-axis length in the unit cell. One of the typical spiral growth grains and the corresponding representative cross-section profile is

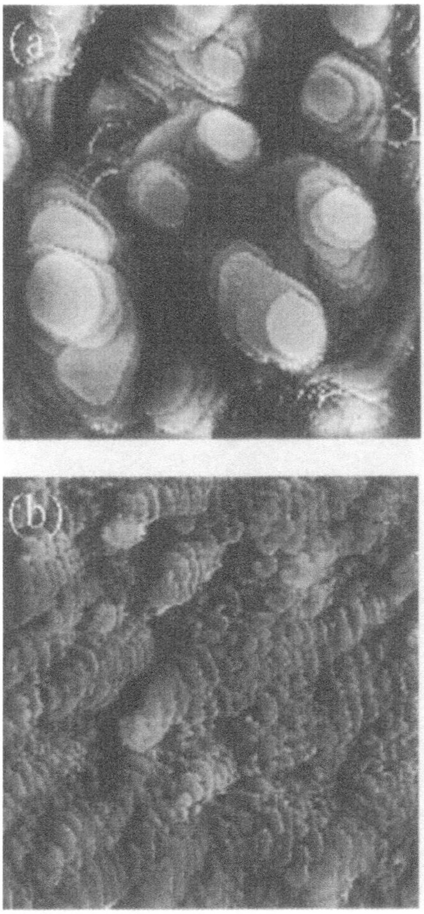

Fig. 2. STM images of **a** layer-by-layer growth morphologies of YBCO thin films on YSZ(100) substrate and **b** ledge growth morphologies of YBCO thin films on mis-oriented MgO substrates. Scan size: **a** 512 nm, **b** 312 nm

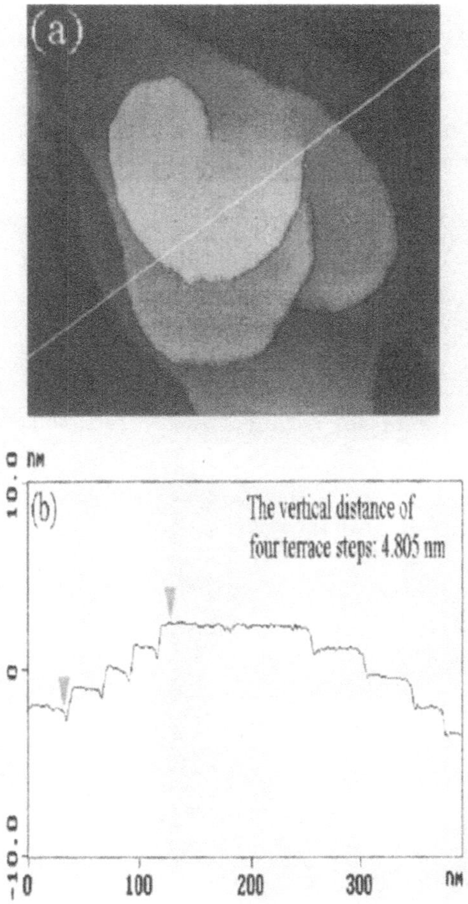

Fig. 3. STM images of a typical spiral growth grain and the correspondent cross section profile, **a** STM image, scan size 345 nm, **b** section profile

shown in Fig. 3a and Fig. 3b. In Fig. 3b, the measured total height of four well defined steps is 4.805 nm giving a single layer terrace height of exactly 1.2 nm, i.e. the c-axis unit cell length. From our experiments we find that in order to obtain reliable and accurate morphological information from STM images, the STM should first be engaged with a slightly higher bias voltage and a slightly smaller tunneling current than the optimum imaging parameters, and the area of interest is then located. Finally, the bias voltage and tunneling current are set to the optimum parameters required to capture the image. Following these procedures avoids the area of interest being modified by the scanning tip before the image is recorded. In these cases, the measured step height by cross-section analysis or bearing analysis from these STM images nearly always equals 1.2 nm.

One of the typical features of laser deposition is the large number of particulates on the deposited films. These particulates with diameters of a few hundred nanometers up to nearly one or two micrometres in size, have an adverse effect on the development of high T_c superconducting microelectronic devices. The details of the morphologies of these particulates on thin films have also been studied. From the high resolution STM images of these particulates as shown in Fig. 4a and Fig. 4b, some well defined crystal surfaces formed by cracking or cleavage can be clearly identified. These observations indicate that the removal of material by laser ablation has definitely undergone a violent explosive process. As seen in Fig. 4b, the existence of these particulates seriously hinders the epitaxial growth of the films in their vicinity resulting in breaking of the electrical

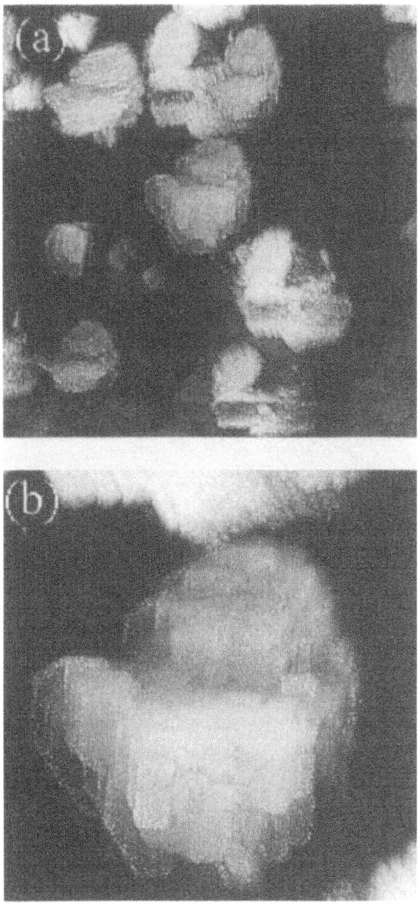

Fig. 4. STM image of particulates on laser deposited films. Scan size: a 2 μm, b 600 nm

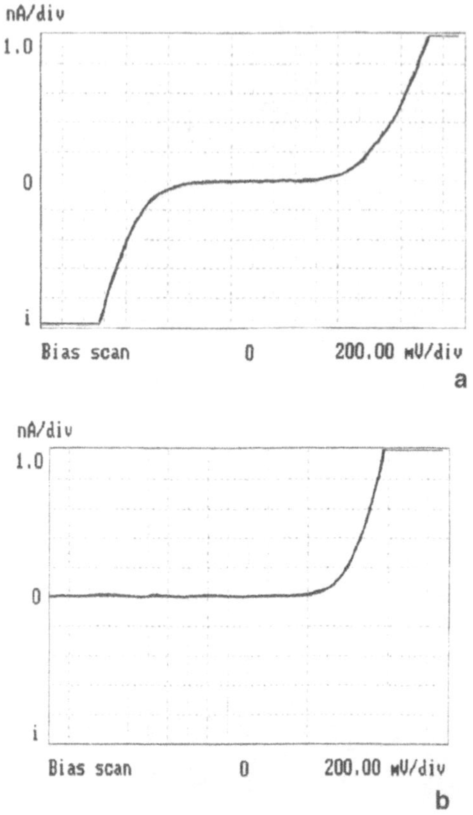

Fig. 5. Typical I-V curves for **a** spiral growth grains, **b** particulates

path and a lowering of the film current-carrying capability.

Tunneling Behaviour

The tunneling behaviour of the particulates and the surrounding film has been studied by scanning tunneling spectroscopy techniques. A typical I–V curve from a growth spiral grain is shown in Fig. 5a. from this curve, it can be seen that there is a threshold tip-sample voltage value for the tunneling current to transit through the potential barrier for a fixed tip-sample separation which is determined by the pre-set bias voltage and setpoint tunneling current. On setting the bias voltage and setpoint tunneling current to 800 mV and 200 pA, the corresponding threshold bias voltage was found to be approximately 550 to 600 mV when a scanning tunneling spectroscopy analysis of the films was undertaken following the procedure described in the experimental section. Another tunneling characteristic of the growth spirals reflected from this curve is that the tunneling current values are almost the same provided the dc bias voltage is the same. This is true whether the observed films are positively or negatively biased. Fig. 5b shows one of the typical I–V curves for the laser ablated particulates. Unlike the surrounding spiral growth grains, when the films are negatively biased, there is no detectable electron tunneling through the gap from the particulate surface to the tunneling tip.

Some growth islands which do not show any spiral growth features exhibit a similar tunneling behaviour to these particulates. At the moment, the mechanism which is responsible for this novel tunneling behaviour is not understood. Because the I–V characteristics for the particulates and the surrounding spiral growth films have been measured alternately with the same PtIr tip, the observed differences of the current vs voltage characteristics for the particulates and the films are definitely related to the local electron density in these regions rather than the properties of the tip. Hopefully, the observed distinct tunneling behaviour for the different regions of the films may be used to distinguish the superconducting phase from the non-superconducting phase in future superconducting device fabrication processes at the nanometre scale.

Surface Modification

When the STM imaging parameters are set beyond the optimum imaging conditions, the film surface can be modified by removing material away from the scanned area with the scanning tip. At a substantially lower bias voltage, i.e. less that 500 mV and with a fairly high tunneling current, e.g. larger than 200 pA, the surface modification seems to be caused by a mechanical miling process [24]. Usually, the scanning tip becomes blunt during the scanning process, possibly caused by mechanical contact with the surface. In most cases, some debris has been observed around the scanned area after surface modification.

At a higher bias voltage, i.e. larger than 1000 mV and with a tunneling current in the range of 200–2000 pA, material removal is believed to be caused by the so-called field induced evaporation mechanism [24, 29]. By using the field induced evaporation mode, the whole of a surface terrace layer, one unit cell in thickness has been gradually removed by scanning the tip over the same area consecutively as seen in Fig. 6a to d. In this process, the sample was positively biased at 1V and the tunneling current was kept at 250 pA. A nanometre scale groove has also been intentionally etched on the top of a growth spiral by setting the STM in line scan mode with a bias voltage of 1.5 V and a tunneling current 2 nA as shown in Fig. 7a. The width and depth of the etched groove were measured by section analysis (see Fig. 7b) and are 25 nm and

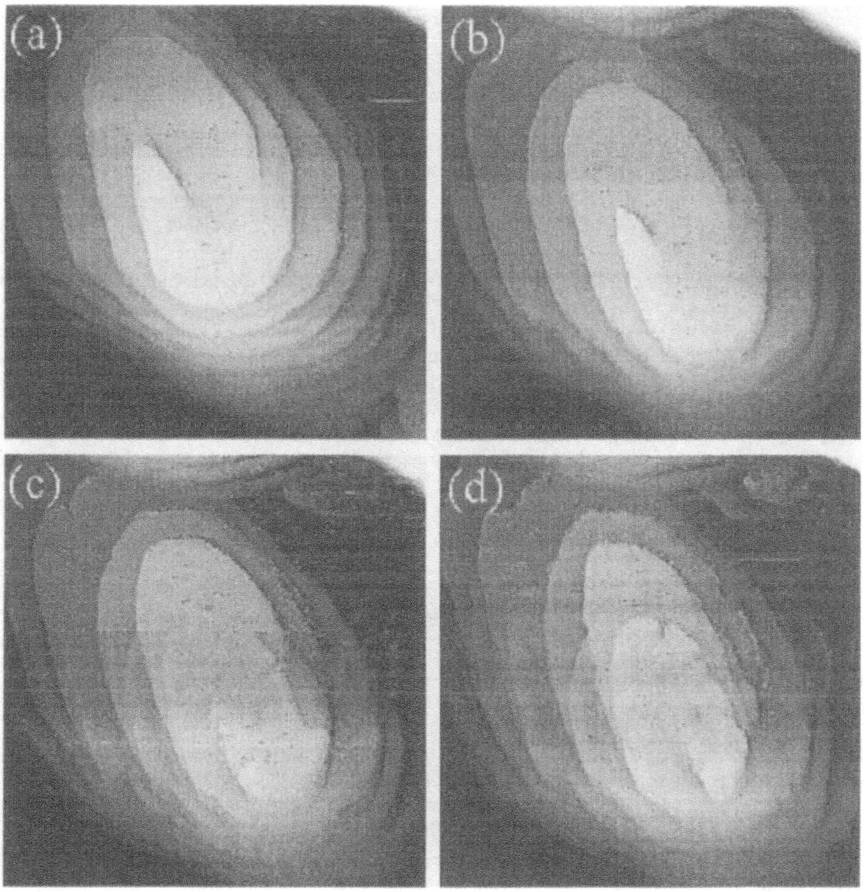

Fig. 6. STM images taken after scanning n times over the same area **a** $n = 2$, **b** $n = 4$, **c** $n = 6$, **d** $n = 8$, Scan size: 270 nm

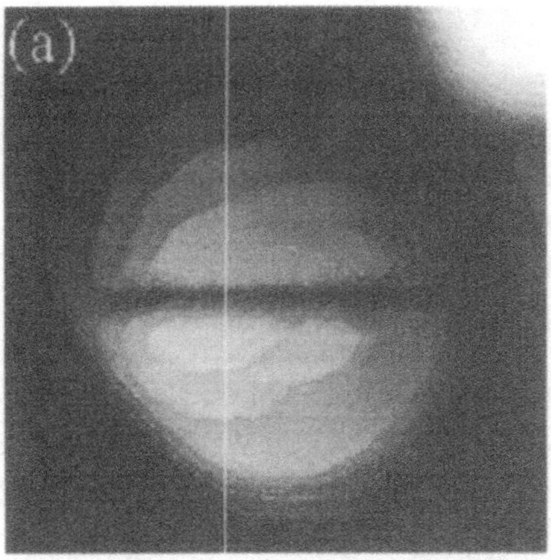

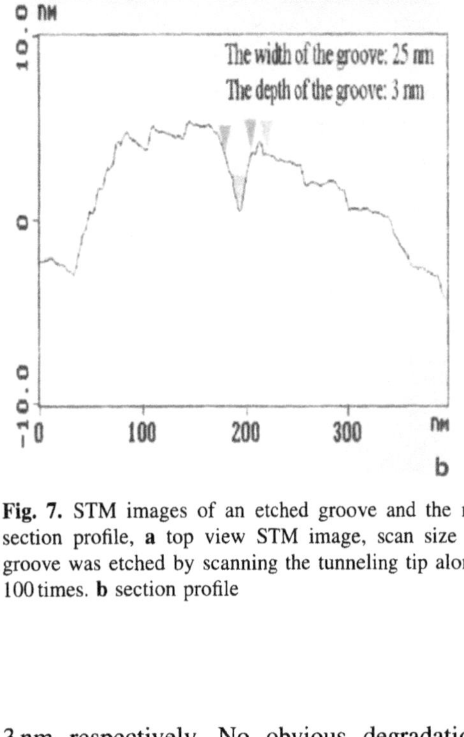

Fig. 7. STM images of an etched groove and the related cross section profile, **a** top view STM image, scan size 400 nm, the groove was etched by scanning the tunneling tip along a line for 100 times. **b** section profile

3 nm respectively. No obvious degradation of the tunneling tips was observed after the tips had been used to modify the surface or to etch a deep groove when the STM was set in this field induced evaporation mode.

Conclusions

STM image observations show that the most obvious features of laser ablation deposited c-axis oriented YBCO thin films are the granularity of the films and the spiral growth pattern of growth islands. Each grain grows by addition of atoms to spirally expanding steps on all of the exposed terrace layers of the grain, and the single growth terrace step height is equal to the c-axis length in the unit cell. STS analysis reveals that different regions on the films such as spiral growth grains, particulates and some non-spiral growth islands have a different tunneling behaviour, and these distinct tunneling characteristics may be used to identify components of the thin film at the nanometre level. Under certain experimental conditions, nanometre scale surface modification and nanometre structure fabrication can be realised in a controlled manner by line or raster scanning of the tunneling tip over the YBCO thin film surface. This surface modification capability of the STM will find important applications in the manufacture of nanometre dimension microelectronic devices.

Acknowledgement. One of the authors (Yongchang Fan) would like to express his thanks to the Committee of Vice Chancellor and Principals of Universities of United Kingdom and the University of Dundee for providing him with an ORS Award and a Postgraduate Research Award respectively.

References

[1] C. Gerber, D. Anselmetti, J. G. Bednorz, J. Mannhart, D. G. Schlom, *Nature* **1991**, *350*, 279.

[2] M. Hawley, I. D. Raistrick, J. G. Beery, R. J. Houlton, *Science* **1991**, *25*, 1587.

[3] J. Moreland, P. Rice, S. E. Russek, B. Jeanneret, A. Roshko, R. H. Ono, D. A. Rudman, *Appl. Phys. Lett.* **1991**, *59*, 3039.

[4] I. D. Raistrick, M. Hawley, J. G. Beery, F. H. Garzon, R. J. Houlton, *Appl. Phys. Lett.* **1991**, *59*, 3177.

[5] D. G. Schlom, D. Anselmetti, J. G. Bednorz, R. F. Broom, A. Catana, T. Frey, Ch. Gerber, H. J. Guntherodt, H. P. Lang, J. Mannhart, *Z. Phys. B* **1992**, *86*, 163.

[6] M. McElfresh, T. G. Miller, D. M. Schaefer, R. Reifenberger, *J. Appl. Phys.* **1992**, *71*, 5099.

[7] D. J. Lichtenwalner, O. Anciello, R. R. Woolcott, C. N. Soble, R. A. Poku, R. Chapman, S. H. Rou, J. Duarte, A. I. Kingon, *J. Vac. Sci. Technol. A* **1992**, *10*, 1537.

[8] J. R. Sheats, P. Merchant, *Appl. Phys. Lett.* **1993**, *62*, 99.

[9] L. Lou, M. E. Hawley, C. J. Maggiore, R. C. Dye, R. E. Muenchausen, L. Chen, B. Schmidt, A. E. Kaloyeros, *Appl. Phys. lett.* **1993**, *62*, 485.

[10] S. Zhu, D. H. Lowndes, B. C. Chankoumakos, S. J. Pennycook, X. Y. Zheng, R. J. Warmack, *Appl. Phys. Lett.* **1993**, *62*, 3363.

[11] R. E. Thomson, J. Moreland, N. Missert, D. A. Rudman, S. C. Sanders, B. F. Cole, *Appl. Phys. Lett.* **1993**, *63*, 614.

[12] D. G. Scholm, D. Anselmetti, J. G. Bednorz, Ch. Gerber, J. Mannhart, *J. Crystal Growth* **1994**, *137*, 259.

[13] M. Aindow, M. Yeadon, *Philos. Mag. Lett.* **1994**, *70*, 47.

[14] J. Summhammer, K. Kundzins, G. S. Hosseinali, R. M. Schalk, H. W. Weber, S. Proyer, E. Stangl, D. Bauerle, *Phys. C* **1995**, *242*, 127.

[15] M. Yeadon, M. Aindow, F. Wellhofer, J. S. Abell, *J. Crystal Growth* **1997**, *172*, 145.

[16] I. Maggio-Aprile, Ch. Renner, A. Erb, E. Walker, Ø. Fischer, *Phys. Rev. Lett.* **1995**, *75*, 2754.

[17] H. L. Edwards, D. J. Derro, A. L. Barr, J. T. Market, A. L. de Lozanne, *Phys. Rev. Lett.* **1995**, *75*, 1387.
[18] M. Tanaka, S. Takebayashi, M. Hashimoto, S. Kashiwaya, F. Hirayama, M. Koyanagi, *Jpn. J. Appl. Phys.* **1993**, *32*, 35.
[19] H. Edwards, D. J. Derro, A. L. Barr, J. T. Market, A. L. de Lozanne, J. Vac. *Sci. Technol. B* **1996**, *14*, 1217.
[20] F. Mielke, U. Memmert, A. A. Golubov, U. Hartmann, *J. Vac. Sci. Technol. B* **1996,** *14*, 1224.
[21] M. A. Harmer, C. R. Fincher, B. A. Parkinson, *J. Appl. Phys.* **1991**, *70*, 2760.
[22] J. A. Virtanen, P. Suketu, G. C. Huth, Z. H. Cho, *J. Appl. Phys.* **1991**, *70*, 3376.
[23] I. Heyvaert, E. Osquiguil, C. V. Haesendonck, Y. Bruynseraede, *Appl. Phys. Lett.* **1992**, *61*. 111.
[24] R. E. Thomson, J. Moreland, A. Roshko, *Nanotechnology* **1994**, *5*, 57.
[25] S. E. Russek, S. C. Sanders, A. Roshko, J. W. Ekin, *Appl. Phys. Lett.* **1994**, *64*, 3649.
[26] U. Geyer, G. V. Minnigero de, H. U. Krebs, *J. Appl. Phys.* **1994**, *76*, 7774.
[27] G. S. Shekhawat, R. P. Gupta, A. Agarwal, K. B. Garg, P. D. Vyas, *J. Appl. Phys.* **1995**, *78*, 127.
[28] G. Bertsche, W. Clauss, D. P. Kern, *Appl. Phys. Lett.* **1996**, *68*, 3632.
[29] H. J. Mamim, P. H. Guethner, D. Rugar, *Phys. Rev. Lett.* **1990**, *65*, 2418.
[30] D. Dijkkamp, T. Venkatesan, X. D. Wu, S. A. Shaheen, N. Jisrawi, Y. H. Min-Lee, W. L. Mclean, M. Groft, *Appl. Phys. Lett.* **1987**, *51*, 619.

Mikrochim. Acta [Suppl.] 15, 109–123 (1998)

Quantitative Near-Surface Microanalysis and Depth Profiling by EPMA

Peter Karduck

Central Facility for Electron Microscopy GFE, Aachen University of Technology, RWTH, D-52056 Aachen, Federal Republic of Germany

Abstract. This paper presents a review of developments in thin film analysis by EPMA during the last 15 years and reports on the capabilities and limitations of the technique when applied to the quantitative analysis of thin coatings, multilayers and oxide layers. During the development of thin film EPMA it was always intended to apply EPMA to the analysis of elemental depth distributions in the submicron range. It is pointed out that the technique of electron beam energy variation is clearly limited to cases where some pre-information on the structure of the material or the shape of the depth profiles is available. For real EPMA-depth profiling in the depth range of 1 μm to a few μm, mechanical bevelling of the sample and analysis with low, carefully chosen beam energies is recommended as a useful technique, which is demonstrated here by application to a multilayered hard coating.

For a more general application of EPMA depth profiling to the submicron range, a combination of X-ray microanalysis with ion beam sputtering has been introduced recently. This technique increases the sensitivity of EPMA to detect variations in element distributions with depth and combines its capability of accurately quantifying element coverage with a better depth resolution. It is shown that with EPMA sputter depth profiling quantitative depth profiles can be deduced from calibrated X-ray intensity measurements without needing a priori information on sputter rates. This development, with its ability to quantify both composition and depth scale, complements the well established methods of surface analysis. The paper reveiws first attempts in combining X-ray analysis with Auger depth profiling, introduces the basic principles and presents results from a quantitative depth profile analysis of an oxide layer on a technical Ti-Al-N hard coating; these results are compared with an SNMS analysis.

Key words: EPMA, thin film analysis, depth profiling, materials analysis

Electron probe X-ray microanalysis (EPMA) is a powerful tool for chemical characterisation of materials on a microscopic scale [1]. The most striking feature of EPMA is its superior ability to quantify composition reliably compared with other microbeam analysis techniques. The wavelength dispersive variant of X-ray spectroscopy also achieves good values of sensitivity and minimum detectable concentrations (20–200 ppm). However, for surface analysis EPMA has the drawback that its information volume has a spatial extent of some 100 nm to a few μm, depending on the material's composition and the electron beam parameters [1]. One reason for this is the rather large mean free paths of the characteristic X-rays measured. In Auger-electron spectroscopy the size of interaction volume of the electron beam with the material is identical but, since the escape depths of Auger electrons are in the range of a few atomic layers it is a real surface sensitive technique. This drawback of EPMA is the reason why most of its practical applications are in the field of analysis of microscopically homogeneous materials, i.e. analysis of phases, inclusions, precipitates with sizes > 1 μm or of element distributions with lateral resolutions ≥ 1 μm. Until recently less attention has been paid to the practical analysis of thin films or surfaces.

Nevertheless, only a few years after Castaing had published details of his pioneering invention, the electron beam X-ray microanalyser [2], some workers had already started to explore the applicability of EPMA to measure mass coverage (mass/area or area mass density) of thin metallic films on foreign substrates. Sweeney et al. [3] published a work in 1960 on electron probe measurements of metal films evaporated onto glass and showed a nearly linear dependence of X-ray intensity on the coverage of the metal film for thicknesses much smaller than the electron range. With suitable related standard films it was already possible at that time to determine film thickness experimentally. Cockett and Davis [4] additionally considered the radiation from the substrate for thickness determination and used theoretical calibration curves deduced from the experimental determination of $\Phi(\rho z)$ depth distributions for X-ray generation according to Castaing and Descamps [5]. They found the limits of detection for mass coverage to be $0.2\,\mu g/cm^2$ at that stage of instrumental development. Not only the thickness but also the composition of films were of interest, and a few years later the analysis of film composition was investigated [6]. Many other authors continued this work but until 1980 no real breakthrough occurred with respect to general applicability and straightforward calibration and quantitation.

At the beginning of the eighties, when some good analytical expressions for the $\Phi(\rho z)$ depth distribution were suggested and implemented into matrix correction models [7, 8], thin film analysis and quantitation were studied more extensively. Pouchou and Pichoir [16] were the first to propose a model fully capable of establishing the total mass coverage of a film on a foreign substrate and film composition, using bulk standards for the calibration of the X-ray intensities. Their model also included the correction of fluorescence effects by characteristic and continuum radiation [17]. A very interesting extension of the technique was the application of different electron beam energies to analyse more complex multilayer structures or even near-surface composition gradients (diffusion, depletion or implantation) [11, 15, 17, 18, 25]. The idea behind this was that with increasing energy the depth of maximum X-ray generation shifts to greater depths and so the emitted intensities carry information on the elemental depth distribution, even though only indirectly, i.e. convoluted, because of the rather large emission depth. Several other models were proposed subsequently based on the same general principle but using different appraoches to describe the generated X-ray depth distribution [9, 10, 11]. A summary of the capability and limitations of thin film- and multilayer EPMA, differentiated with respect to various boundary conditions, is given by Pouchou and Pichoir in [15].

This rapid development opened new approaches for the quick but reliable quantitation of thin film data, which directly met the demand of the growing activities in thin film technologies. The current interest nowadays is focused on extending the range of applicability of EPMA to depth profiling on a submicron scale, a subject which will be treated in detail later in this paper.

Nevertheless, when studying the literature during the preparation of this paper it became apparent that most of the papers on thin film EPMA deal with applications to known structures and with the more general aspects of assessing the reliability of the technique. Only a minority present real analytical solutions to unknown problems in surface science.

Therefore, one intention of the present work is to demonstrate the power of EPMA in thin film analysis using examples from current surface analysis. The second aim is to summarise the present state of the art in EPMA depth profiling and to suggest what might be feasible in the near future.

Application of Thin Film EPMA to "Real-Life Samples"

Analysis of Single Multi-Component Films on a Foreign Substrate: Coatings Produced by Laser Induced Chemical Vapour Deposition LCVD

An example of a thin layer analysis is given in Fig. 1. The optical micrograph in Fig. 1a depicts a copper layer deposited by laser induced chemical vapour deposition (LCVD) from metal-organic gas precursors [14]. This coating technique has received much attention in recent years because of its potential for replacing photolithographic metallisation processes. The aim of this technique is to deposit structured thin metallic layers directly in one step; in the present experiment a grid-like structure was grown on a silicon substrate. The experiment was run with an KrF-excimer laser [12]. The objective of the EPMA investigation was to determine the coverage of copper in the film, in the areas surrounding the grid not struck by the laser, and to analyse the role of impurities in the metal film originating from the metal-organic

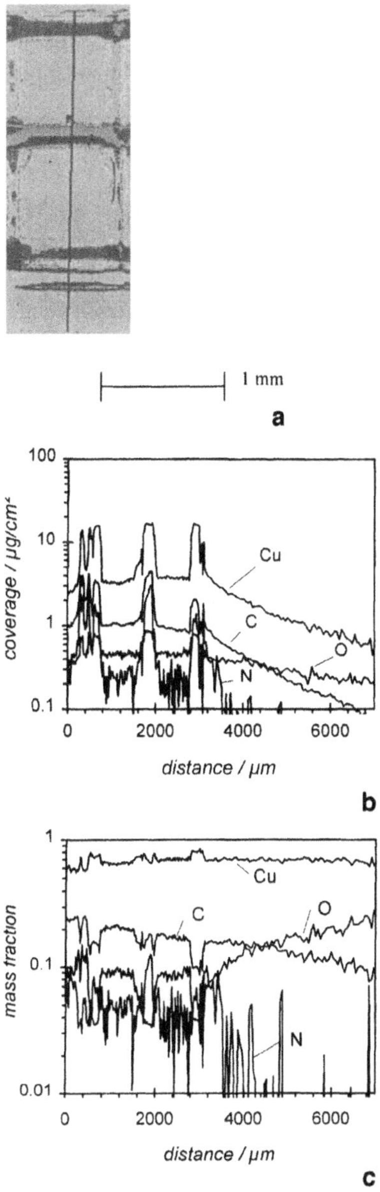

Fig. 1. Quantified EPMA-line scan **b** across the copper layer shown in the optical micrograph in **a**. The layer was deposited on silicon by a laser induced chemical vapour deposition. Its thickness amounts to around 30 nm at the maxima and 10 nm at minima. **c** Composition of the LCVD film shown in Fig. 1a. Calculation with the assumption of homogeneous composition with depth inside the film. $E_0 = 20\,\text{keV}$; $I_0 = 40\,\text{nA}$.

vapour phase. All these objectives could be achieved by one quantitative line-scan measurement along the line in Fig. 1a Resulting coverage of elements C, N, O and Cu are plotted in Fig. 1b. The data were evaluated from calibrated X-ray intensities by the procedure described by Ammann and Karduck [13].

As standards, pure bulk material (glassy carbon, Cu) or bulk compounds (Fe_4N, Fe_2O_3) were used. Particular data sets with various k-ratios (relation between measured intensities from sample and standard) were selected from line scan data and translated into mass coverage. From these data, calibration curves were determined for each element, leading to polynomial fits which were used to transfer the k-ratio line scan into quantitative line scans of partial mass coverage.

As a result of the laser induced photo dissociation, Cu was deposited in the ledges by a factor of 6 more than the area in-between, where pyrolytic deposition took place because of the heating of the substrate by the laser irradiation. Also impurities C, N and O are incorporated in the grid by factors 2–10 higher than in the surrounding area. With increasing distance from the grid the deposition rates of Cu and the impurities C, N and O decrease nearly exponentially because the temperature gradually falls with increasing distance from the area struck by the laser. Obviously, from their different slopes, the deposition rates of the impurities show different behaviour with temperature decrease and the data allow direct calculation of deposition rates. Furthermore, from the relative coverage of the elements present in the layer the mass- or atomic fractions, (i.e. the composition) of the film could be deduced giving more information about the Cu to impurity relations and the dependence on deposition rates and physical parameters of the experiment [14]. Fig. 1c shows the corresponding film composition. The values were calculated assuming that the film is homogeneous with depth, a fact which was proven by SIMS depth profiling.

Analysis of Double Films, Different Elements in Different Films: Coatings Produced by Laser Induced Chemical Vapour Deposition LCVD

A second example of a LCVD Cu-coating demonstrates the capability of EPMA to analyse multiple layered structures. If no element is present in more than one layer and the sequence of layers is known, the mass coverage can also be quantified from measurements with only one beam energy [15]. In the following example the Si substrate was partly precoated with approximately 1 nm Ti which was assumed to be present as TiO_x. This buffer layer should improve the growth of Cu-films by reducing

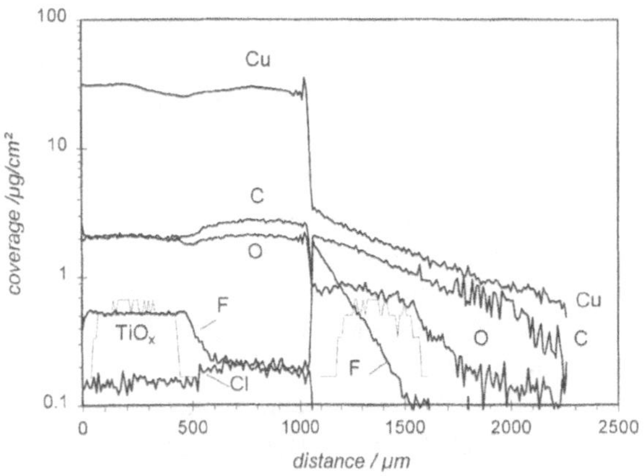

Fig. 2. Results of a line scan analysis at a double-layer system Cu (LCVD) on TiOx (ledges) on silicon substrate. $E_0 = 15\,\text{keV}$; $I_0 = 100\,\text{nA}$

the lattice mismatch and thereby improve the adhesion.

Figure 2 gives an example of a line scan across two ledges pre-coated with Ti (400 μm width) and an uncoated Si area (600 μm width) in between. One of the pre-coated ledges and the uncoated substrate were deposited with Cu by LCVD. The line scan crosses this area and the area outside the laser spot, which contains another Ti-ledge. The line scan data in this example were converted into mass coverage by applying Cameca software called *layerf*, which is based on the model of Pouchou and Pichoir [16]. This software was extended to convert line scan data (k-ratios) measured with a Camebax SX 50 microprobe analyser directly into coverage and layer compositions. Thus, high accuracy as reported by the authors is expected to be obtained by this extended software as well [16]. Figure 2 demonstrates that this measuring technique is able to detect the influence of the TiO_x-buffer layer on the deposition rates of Cu and impurity elements very sensitively. The thickness of TiO_x was determined to be 1.5 nm under the assumption that the total amount of Ti is present in an oxide state (TiO_2). The type of substrate – whether Ti-coated or not – only weakly influences the deposition rate of Cu and C, but inside the area struck by the laser the incorporation of F, C and Cl into the layer is strongly influenced by the nature of the substrate. Fluorine is deposited preferentially on the Ti-ledge whereas more chlorine is found on the silicon. On the area with the thermally induced deposition the coverage is influenced by the presence of pre-deposited TiO_x, but the exponential decrease of the deposition rates of Cu, C and F with increasing distance from the laser-irradiated area is not affected by the nature of the substrate at all.

In conclusion, these practical applications are striking examples of the capability of thin film EPMA with regard to sensitivity, accuracy and versatility. Also this second example demonstrates the importance of a reliable and quick quantitative analysis of elemental coverage with good lateral resolution to obtain a better insight into the processes underlying the laser induced chemical vapour deposition.

Same Elements in Film and Substrate: Oxidation Phenomena on Cold Rolled Steel Sheets

The analysis of oxidation phenomena on steel cannot be performed easily since, except oxygen, all possible elements of an oxide scale also are present in the steel itself. In the present example a failure analysis of cold rolled steel sheet is described. In a final production step a coil of steel sheet was annealed under H_2-atmosphere. After this treatment the sheet showed at the edges a few centimetres of brown tarnished surface, which was assumed to be produced during annealing by oxidation because of a non-oxygen-free H_2 atmosphere.

At first a measurement of elements known for their affinity to oxygen (Si, Mn) and oxygen itself was carried out along a line perpendicular to the edge of the sheet, across the brown coloured area ($\sim 3\,\text{cm}$) into the perfect surface ($E_0 = 9\,\text{keV}$; $I_0 = 200\,\text{nA}$). Before evaluating the X-ray measurements in terms of oxide thickness or composition, information about the composition of the steel substrate was required. Therefore, a small area of the sheet was polished to analyse the exact composition of the steel by conventional EPMA. The quantitation was performed by the method described above in the first example so that now, the composition of the substrate became a known quantity.

Thus, from the measured composition of the substrate the quantitation routine was able to calculate, in an iteration process, the portion of X-ray intensity excited in the substrate, and the coverage of the particular elements in the oxide scale could be deduced from the excess intensity emerging from the elements in the film.

The resulting line scan data are plotted in Fig. 3 as elemental coverage. Towards the edge of the sheet the

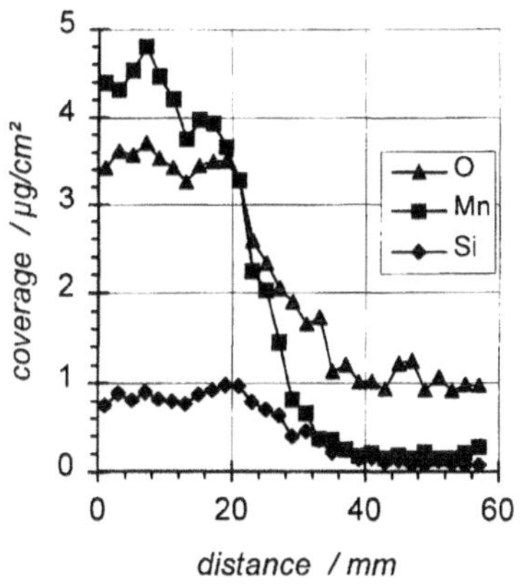

Fig. 3. Element coverages of an oxide layer on a partly oxidised steel after heat treatment in inert gas atmosphere measured by EPMA line scanning: left part of the profile: Brown tarnished area of the sheet due to oxidation. E_0: 9 keV; I_0: 200 nA. Standards: Fe_2O_3 (O), pure Mn and Si

coverage of O, Si and Mn increase significantly. It was therefore apparent that an oxide scale with high fractions of Si and Mn was causing the brown tarnished surface of the steel sheet. Furthermore, from the coverage data, the atomic composition of the scale could be determined assuming that the valences of Si and Mn are 4 and 2, respectively, and that the additional oxygen is bound to a certain fraction of Fe with valence 3.

The resulting composition is 7.6% Si, 21.0% Mn, 14.0% Fe and 57.4% O (atomic fractions). The scale had a thickness of 25–30 nm. Detection and calibration of the data is conventional and straightforward. Although the metallic elements are present in film and substrate it was possible to manage the problem with only one primary beam energy. However, this approach can only be successful as long as the sensitivity of the measurement is sufficient to distinguish the portion of X-ray intensity emitted from the oxide scale from the fraction emerging from the substrate. In the present case this holds for Si and Mn but not for Fe, which therefore had to be evaluated from stoichiometry considerations.

For this evaluation a software package is required, which handles information assumed about the structure as, for instance, the known composition of the substrate, the stoichiometry of single films or – as in the present case – the valency of elements in the oxide state and the calculation of an element by difference [15, 22].

Limits of Thin Film EPMA

Most of the factors limiting the power of thin film EPMA have already been discussed in the literature [15]. One important quantity, the limit of detection, has been of interest since the very beginning [4] and has been determined for the present state of instrument technology to be between 10^{14} and 10^{15} atoms/cm^2; i.e. less than an atomic monolayer if the film element is not present in the substrate [35–37, 42]. Of course this is a prediction based upon measurements of thicker films, and in the analysis of real-life samples surface roughness and adsorbed species on sample and standard sometimes can interfere with the validity of the results.

Generally it is assumed that single films on foreign substrates can be analysed as long as the film thickness is penetrated by the electron beam. It is advisable to measure as well the radiation from the elements in the substrate because film thickness also can be calculated by considering the absorption of substrate radiation in the films. One important condition for reliable thin film analysis has already been mentioned by Pouchou and Pichoir [15]. The radiation from elements of a layer or of the substrate must not be completely absorbed in upper layers. Indeed, strong absorption becomes a limiting factor when using soft X-ray lines. The following example will show, that even in case of rather thin layers the results of EPMA can become ambiguous: The mass absorption of C Kα in Si is 37000 cm^2/g, corresponding to a mean free path (mfp) of C Kα in pure Si of only 115 nm. This means that only 5% of C Kα photons reach the surface of a Si-rich matrix when generated in a depth of 220 nm (take-off angle is 40°). Because of the strong absorption the variation of the depth distribution of the emitted fraction of X-ray photons is very modest except for low electron energies.

The influence of such a restriction is shown in Fig. 4 for SiC-films with different thickness on pure titanium: The k-ratios of C Kα (Fig. 4a) behave completely differently from those of Si Kα (Fig. 4b), both with respect to a SiC-standard. The Si Kα k-ratios follow the well known dependency on beam energy and on thickness of film. There is a pronounced dependence on beam energy which would help to locate possible gradients in the Si depth distribution. The curves of C Kα are very sensitive to the film thickness, at least up to 200 nm, but the single

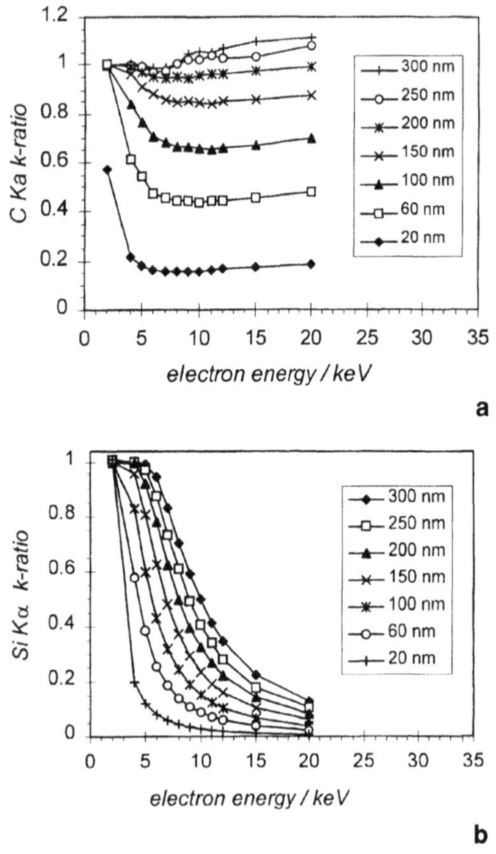

Fig. 4. Calibrated X-ray intensities versus primary electron energy for thin SiC-films with different thickness on a titanium substrate calculated with Monte Carlo simulation. **a** C Kα and **b** Si Kα

curves show no dependence on beam energy, except below 6 keV. At 200 nm there is almost no variation of the k-ratio with beam energy. For film thicknesses larger than 200 nm the k-ratio exceeds 1 at high beam energies which is a result of additional ionisation caused by primary electrons back scattered from the higher atomic number substrate (Ti) into the film. Thus, we can approximate the maximum film thickness to:

$$t_{\max} = 3 \cdot (mfp)_{\min} \cdot \sin\Theta \quad (1)$$

where $(\mathrm{mfp})_{\min}$ relates to the measured X-ray line with the smallest mfp and Θ is the take-off angle.

Additionally we learn from Fig. 4 that the k-ratios measured with relatively high beam energy only are a measure of the mass coverage of an element in a film but are not sensitive at all to any variation of the elements concentration in the film with depth.

Depth Profiling by EPMA

Variation of Primary Electron Energy

During the development of thin film EPMA it was intended to apply this technique also to the analysis of depth distributions in the submicron range. Since the total depth of information is large, compared, to the depth of concentration profiles, the desired information can be obtained only when there are significant variations in experimental parameters, which lead to changes in the X-ray intensities from which information about the depth profile is to be deconvoluted.

In principle there are two possible parameter variations:

- Variation of the incidence angle of the electron beam with respect to the sample surface and/or variation of the take-off angle.
- Variation of the energy of the primary electrons.

The latter is the most common technique [17–19]. Variation of the impact angle shifts the X-ray generation more towards the surface and thus enhances the sensitivity to detect elements in near surface zones [20]. The variation of take-off angle was also applied successfully to determine the centroid depth of distributions of ion-implanted analytes [21]. The most common way to determine depth distributions from X-ray data measured at various beam energies is to divide the depth of interest into several hypothetical sublayers and to determine the partial coverage $M_{i,j}$ of each element i in the layer j by an iterative procedure. From these values the composition of each sublayer j, $c_{i,j} = M_{i,j}/\Sigma M_{i,j}$, can be calculated. Pouchou and Pichoir described several iterative modes for different combinations of films or multilayers on substrates [15]. But, as pointed out [15] if one or more elements are present in several films and/or in the substrate the solution obtained by current iterative modes is not unique, i.e. more than one different variations of the histogram-type depth profile can lead to a minimum in the deviation between experimental k-ratios and those calculated by the particular model giving the solution (expressed by the root-mean-square value RMS).

A typical example of such a case is given in Fig. 5. The open symbols represent the k-ratio variation versus beam energy for a film of 100 μg/cm^2 pure zinc on pure copper (see Fig. 5b). In comparison, the curve

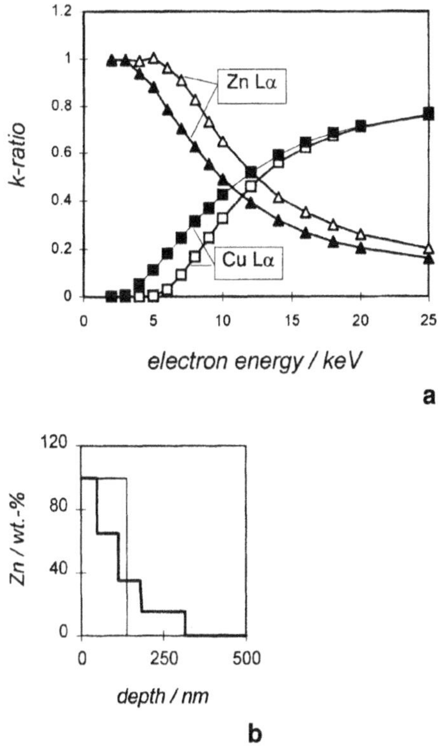

Fig. 5. k-ratio versus primary beam energy **a** for the two Zn-distributions in a Cu-matrix shown in **b** calculated by Monte Carlo simulation: open symbols represent the single layer of pure Zn; filled symbols for the stair-case like depth distribution of Zn in Cu

with the filled symbols is calculated by the same Monte Carlo simulation-model [13, 22] for a stair-case-like zinc profile in a copper matrix as depicted in Fig. 5b. For energies from 4 to 12 keV there is a significant difference between the two curves which in case of an unknown sample configuration could probably be utilised to determine variations of element concentrations with depth.

Figure 6 shows three possible results which are obtained by an optimisation procedure based on the same Monte Carlo code [13, 22] used for the calculation of the data shown in Fig. 5 and using the same set of k-ratios, but pretending it to be from a sample with unknown composition. As expected the different solutions increasingly differ from each other with greater depth, but the resulting RMS-values are still very small. In other words, there exists no pronounced minimum of the RMS-value for a single set of profile parameters, i.e. no unique solution, but there are several minima more or less equal in value leading to ambiguous solutions in the optimisation procedure. Nevertheless, independently of the first estimate, the upper layer was recognised in all cases as pure zinc and the total mass coverage of zinc was determined correctly.

We learn from this and other examples discussed in the literature that depth profile analysis of "real-life samples" by the method of energy variation requires much effort but – without any restriction of the number of unknowns or assumptions about the type of profile – does not always lead to unique results. Nevertheless, some examples of successful analyses are reported in the literature for ion implantation depth profiles [23–25].

Parameterisation of Depth Profiles

The above mentioned problem of ambiguous solutions by iterative procedures for depth profile determination from energy-variation measurements can be overcome in particular cases. One case is if the general shape of a depth profile is known and can be described analytically by a closed function with a limited set of parameters. This restriction leads to a reduction in

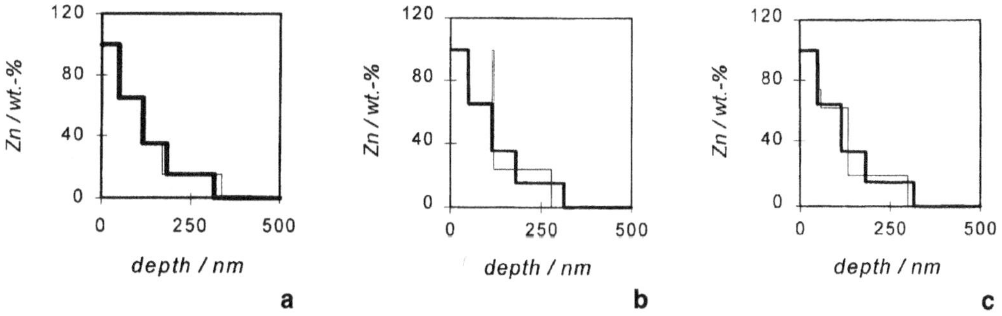

Fig. 6. Results of an iterative reconstruction of a staircase-like Zn depth profile in a Cu-matrix using the data shown in Fig. 5 (filled symbols). Broad line: Original profile for which the k-ratios were calculated. Thin line: Results after iteration from a first estimate: **a** 1 st estimate deviated arbitrarily from real profile by 10%, RMS = 0.2%; **b** 1 st estimate: Cu50Zn50 with the same Zn-coverage as in the original profile, RMS = 0.12%; **c** another run with same 1 st estimate as in **b**, RMS=0.1%

the number of unknowns to the number of parameters necessary to describe the profile. Depth information can be extracted from X-ray intensities measured at various energies of the electron beam corresponding to different depths of X-ray generation. This model can only be applied in cases when one element has a certain depth distribution and is a minor constituent in a matrix which, as a first approximation, can be regarded as having constant composition. This condition for instance holds for implantation profiles and near-surface diffusion profiles. Ammann [42], Ammann et al. [26] and Ammann and Karduck [27] applied this technique successfully to analyse depth profiles of implanted ions in semiconductor materials by assuming the shape of the ion implantation depth profiles to be Gaussian. A similar technique has been applied previously by other authors [21, 28, 29]. From calibrated X-ray intensities measured at various electron beam energies the dose density N_D, the projected range R_P and the standard deviation of R_P, ΔR_P, could be determined within $\pm 5\%$. The limits of detectability were determined to be below $10^{14}\,\mathrm{cm}^{-2}$ for elements with $Z > 10$ and about $10^{15}\,\mathrm{cm}^{-2}$ for light elements. Hence, the application of EPMA is restricted to dopant concentrations above $10^{18}\,\mathrm{cm}^{-3}$, but due to its accuracy EPMA can be a helpful tool for calibration of implanters or implanted reference materials.

Arbitrary depth profiles in the sub-micron range, which cannot be described easily by simple functions, cannot be resolved uniquely by the techniques described so far.

Depth Profile Measurements at Bevels

Analysis of Mechanically Prepared Bevels

Several approaches were proposed in the literature to improve the applicability of EPMA to the quantitative analysis of near-surface regions or concentration profiles near interfaces both with extensions around 1–2 µm.

Conventional line-scan measurement on polished cross sections is affected by artefacts caused by preparation and its resolution is limited by the size of the excitation volume to about 300 nm. Willich and Bethke demonstrated this by reducing the beam energy to < 5 keV and applying low beam currents. However, these parameters restrict the sensitivity as well as the limit of detection [30]. Under certain circumstances depth resolution of EPMA can be improved by bevelling the near surface region either by lapping or polishing at very small angles with respect to the surface ($\leq 1°$). EPMA-measurements can be carried out with reduced beam energy to decrease the total depth of X-ray emission. High beam currents can still be used since, because of the very shallow angle of the bevel, the depth increments corresponding to lateral distances on the bevel are rather small (≈ 15 nm/µm). Thus, the beam size is not the limiting factor for depth resolution. Baumgartl et al. applied a technique of surface polishing to obtain very shallow bevel-like transitions through the interface region of an aluminised steel sheet to detect AlN-diffusion barriers of 3–14 nm thickness between Al-coating and steel [31]. Willich and Bethke [30] demonstrated this technique by using a dimple grinding technique with a rotating steel ball and diamond suspension as polishing medium. Of course, one restriction of these techniques is that the depth profile cannot be recorded at one point perpendicular to the surface and therefore this bevel-depth analysis is only applicable to samples with laterally homogeneous structure and composition.

The technique can be improved by defined dimple grinding [32]. This can be demonstrated in an application of depth-profile analysis of a Ti-Al-N-O-multilayer with alternating N/O-ratio. Figure 7a gives a SEM-image of a cross fracture of this hard coating consisting of seven different sublayers on a hard metal substrate. The aim of the EPMA investigation was to acquire a reliable quantitative analysis of the composition of the single sublayers, particularly of the N/O-ratio, as a depth profile. The analysis was aimed to provide feedback for optimising coating parameters of the reactive PVD-process[1]. The preparation is schematically explained in Fig. 7b: a rotating steel ball with a smooth surface touches the surface of the sample with a diamond-oil suspension between. The position of the ball on the surface is fixed. The rotating ball polishes a dimple into the surface. The diameter D and depth d_d depend on the radius of the ball. With large radii compared to the depth of the dimple the angle between the surface of the dimple and the original sample surface is less than $1°$. The

[1] Sample was kindly supplied by CemeCon GmbH, Aachen, Germany.

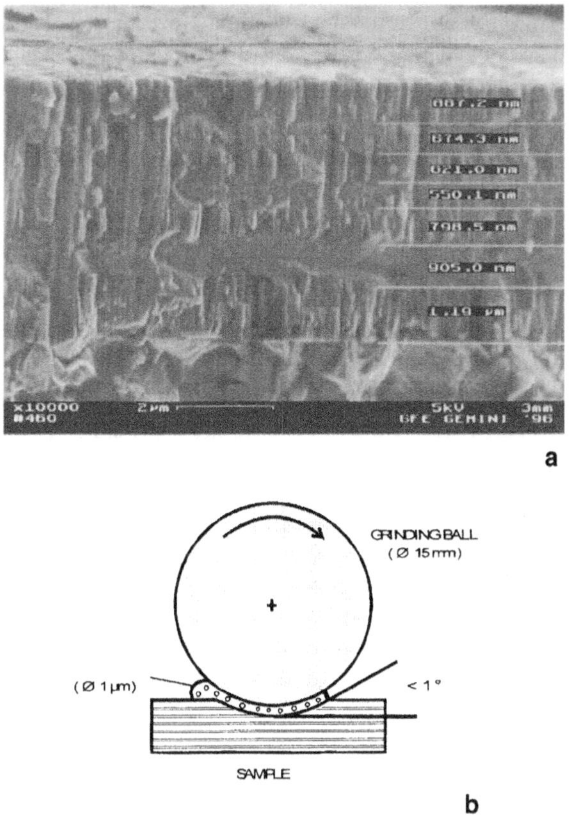

Fig. 7. a SEM-image from a cross fracture of a multilayer coating (Al, Ti) (N, O) with alternating N:O-ratio produced by reactive sputter ion plating. **b** Sketch to illustrate the dimple-grinding technique to produce very shallow and smooth bevels in thin layers

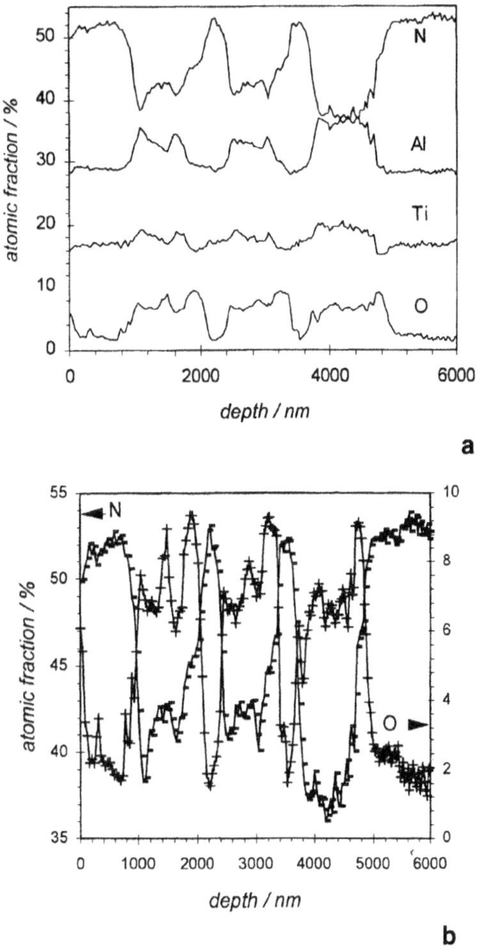

Fig. 8. Results of a quantitative EPMA measurement of a depth profile along the bevel of a dimple in the multilayer shown in Fig. 7a. $E_0 = 7\,\text{keV}$ (**a**). **b** Detail of (**a**) to demonstrate the high analytical sensitivity for light elements N, O.

depth d_d of the dimple at each point can be calculated easily by

$$d_d = \sqrt{R^2 - a^2} - \sqrt{R^2 - (x - a)^2} \qquad (2)$$

R: radius of the ball, a: projected radius of the dimple, x: distance from the rim of the dimple along a (point of measurement).

With this preparation technique care should be taken to prevent artefacts such as chipping where the wedges of sublayers become very thin [32]. Fig. 8a gives a quantitative depth profile through the system of layers shown in Fig. 7a with the depth coordinate rescaled according to Eq. (2). The corresponding measurement was carried out at 7 ke V and 40 nA in steps of 1 µm along the projected radius of a dimple. The N-profile, particularly, allows us to distinguish clearly between the seven sublayers with different N/O-ratio. The profiles give detailed information about the interrelation between changes in partial pressure of N_2 and O_2 in the atmosphere of the coating process and the Al/Ti-ratio or the ratio of metals to metalloids. The high sensitivity of the technique to even small changes of composition is demonstrated in Fig. 8b for the light elements N and O. The depth resolution can be estimated from the rather pronounced transitions in the single profiles assuming that interface regions are sharp compared to the depth of information of EPMA at 7 ke V. Values are between 150 and 200 nm. The real depth resolution is determined by the depth of information, i.e. the depth distribution of emitted X-ray signals. Once the composition of the material is known these $\Phi(\rho z)$-curves (absorption corrected) can be calculated by Monte Carlo simulation or analytical models [7, 8, 11, 13] and the depth resolution directly read from

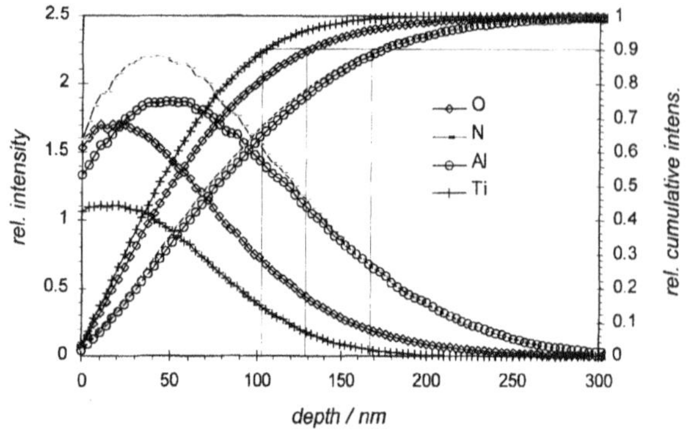

Fig. 9. Emitted fraction of $\Phi(\rho z)$ depth distributions of the Kα-lines of N, O, Al, and Ti in (Ti, Al)(N, O) at 7 keV beam energy and corresponding cumulative intensities as a function of depth. Vertical bars indicate the particular depth resolution for bevel depth profile analysis

them according to the procedure given in Fig. 9. For the present material the figure shows the $\Phi(\rho z)$-curves corrected for absorption and the corresponding cumulative depth distribution of total X-ray intensities produced up to a certain depth. If we define the total depth of information to be that in which 90% of the emitted intensity is generated, this depth can be regarded as the depth resolution of the bevel-line scan technique. It can be seen that these depths are different for each particular X-ray line according to the different shapes of the $\phi(\rho z)$-curves. In the present example that for Ti Kα with 100 nm is the lowest because of the low overvoltage ratio at 7 keV.

Thus, for depth profiling of complex layered structures in the thickness range of 1 to a few μm the technique presented offers an alternative to other depth profiling methods. One important advantage is that surface roughness or meshed structures in the layer system do not affect the results as with techniques like SNMS or GDOS. The depth resolution could further be improved by choosing even lower primary beam energies as will be seen in the next section.

EPMA Combined with Ion Beam Sputtering

In the previous section it was shown that line scan measurements with low electron energies across a mechanically produced bevel improve the depth resolution of EPMA to 100–200 nm and depth profiles of laterally homogeneous structures are feasible. The important advantage of EPMA in this context is its high reliability in quantification. Nevertheless, for the analysis of modern layered structures in semiconductor devices, in functional coatings on glass, in protective layers like hard coatings or anti-corrosion layers, a depth resolution is required down to 1 nm or less. Actually, this regime is the domain of real surface-sensitive techniques with information depths of one or a few atomic monolayers such as Auger electron spectroscopy, AES, X-ray induced photo electron spectroscopy, XPS, or secondary ion mass spectrometry, SIMS [33, 34]. But despite the rather large depth of information, wavelength dispersive EPMA is also capable of detecting less than an atomic monolayer of an element deposited on a foreign substrate material. Consistent data for the limit of detection in terms of coverage was reported by several authors [35–37]. Another important point with EPMA is the excellent quantitative knowledge about the distribution of signal generation inside the excitation volume and about X-ray absorption. Both are the keystones of current matrix correction procedures and of software packages for thin film analysis by EPMA. Thus, it seemed worth investigating in detail whether better resolved information about element depth distributions could be attained by combining low-voltage EPMA with a more sensitive type of bevelling, i.e. ion beam sputtering. Before reviewing the latest development in this rather new field called EPMA sputter depth profiling let us look back to an interesting technique proposed in the seventies as a supplement to Auger depth profiling to rescale "sputter time" into real depth units.

The principle idea of Kirschner and Etzkorn [38] was that for films on foreign substrates with thickness lower than the maximum depth of X-ray generation, the X-ray intensity emitted from the film was nearly proportional to the thickness of the film. Thus, with decreasing thickness of the film during a sputtering process the decrease of X-ray intensity was proportional to the sputter rate. In fact this proportionality had at that time already been known for more than 15 years, as seen above. But the novel idea was to combine X-ray spectroscopy with Auger depth profile analysis to make use of the dependence between X-ray intensity and sputter rate to obtain a true sputter depth instead of concentration versus sputter time profiles. Figure 10 is taken from the original paper of Kirschner and Etzkorn and gives a clear view about the technique the authors called KARMA („Kombinierte Auger/Röntgen Mikro-Analyse"): German for

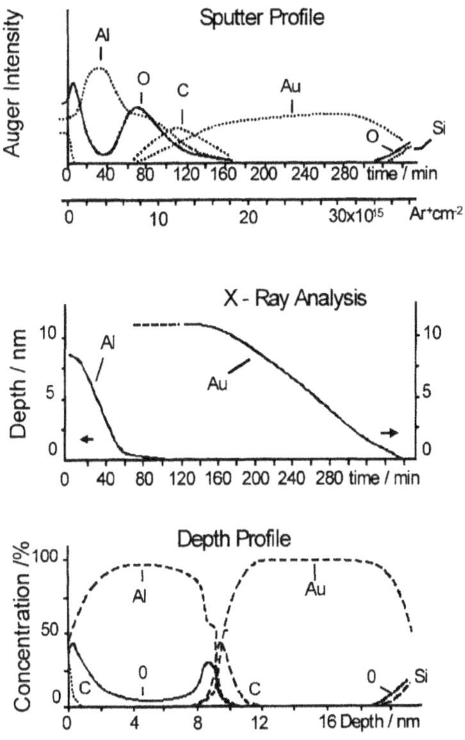

Fig. 10. Top: Sputter profile of a sandwich structure 10 nm Al/ 11 nm Au/Si-substrate. $E_0 = 10\,\text{keV}$. Middle: Film thickness as determined from a simultaneous X-ray measurement (Al and Au) as a function of sputter time. Bottom: Depth profile after quantitative evaluation and transformation of the "sputter-time scale" into a real depth scale using the X-ray calibration curve (middle) (according to Kirschner and Etzkorn [38])

"Combined Auger/X-ray Micro Analysis"). The upper figure shows the raw Auger depth profile through a double layer of pure Al on pure Au on a Si-substrate. It indicates significant oxide layers on top of the Al film and near the Al/Au interface as well as large carbon contamination on top of the Au film. Pure Al seems to be thinner than the Al-oxide part of the structure. The Au layer appears to be more than twice as thick as the Al film. The figure in the middle gives the course of the measured X-ray intensities of Al and Au with sputter time. The real depth was calibrated by using literature data [39]. The intensities were measured with a special UHV- compatible energy dispersive Si (Li)-detector [38]. The slope of the linear parts of these curves gives the relative sputter yields of the particular materials directly. Immediately it becomes obvious that the sputter yield of Al is very low in the oxidised zone compared to that in the pure Al. Furthermore, most of the Al coverage is pure and only a minor part is oxidised, in contrast to what is suggested in the upper figure. The lower diagram gives the final results after conversion of the time scale into a depth scale using the information from the X-ray measurement, and after quantification. It becomes clear that the aluminium oxide is rather thin at the interface as well as at the surface. This example clearly demonstrates the capability of EPMA to quantitatively analyse depth information about layered systems and how the quality of information from simple intensity versus sputter time profiles can be improved significantly, although there is no direct access to the depth profiles by the X-ray analysis.

Therefore, in the context of thin film X-ray analysis, it would be possible to transfer the X-ray data directly into quantitative data of coverage or thickness. Independently of the work of Kirschner and Etzkorn, von Richthofen et al. [40] studied the capability of low voltage EPMA combined with Auger depth profiling to analyse depth profiles in a very thin and complex Ti-Al-O-N-film ($\approx 25\,\text{nm}$) on a Si substrate. The aim of this investigation was to obtain information about the depth distribution inside the complex film only from EPMA measurements and to use AES as a complementary technique to prove the reliability of EPMA. X-ray analysis was carried out as a line scan across the wedge of a sputter crater, which was produced in an AES microprobe. Although the total film thickness was very small the calibrated X-ray intensities could be transferred into quantitative depth profiles, i.e. concentration versus mass coverage, since from the X-ray intensities the coverage can be deduced directly whereas calculation of thickness requires exact knowledge of the density of the structure. The so called reconstruction procedure is described in detail in [40]. The strength of this EPMA variant in combination with sputtering is its reliable depth calibration and quantitation. The accuracy of the composition data was estimated to be $\pm 10\%$ relative, i.e. less than in EPMA of bulk material or thin films. However, this is obvious because the information about depth profiles is not gathered directly from the absolute X-ray intensities, but indirectly from intensity differences acquired at different depths on the crater. Thus the ratio of the standard deviation of a calibrated intensity and the intensity difference $\sigma_i/(k_i\text{-}k_{i-1})$ is a measure of the accuracy of depth profile data instead of σ_i/k_i.

This technique of depth profile analysis was further applied by Karduck and von Richthofen to grown oxide scales on technical hard coatings Ti-Al-N [41]. Compared to the experiments described so far this task was more difficult since elements were present in the complex oxide layers as well as in the substrate. In the following, an analysis of a particular oxide scale on a Ti-Al-N hard coating will be described in detail[2]. A hard coating $Al_{0.6}Ti_{0.4}N$ was oxidised in air at 800 °C for 1 h to study its oxidation behaviour. The framework of this investigation was that wear resistant hard coatings on tools are more and more often applied without lubrication and thus degrade not only by abrasion but also by a combined process of oxidation and abrasion (tribo-chemical wear). The study of oxidation behaviour therefore is a key to optimising the composition and structure of the hard coating so as to give best resistance to this tribo-chemical wear. A crater was sputtered into the oxide scale by a controlled procedure inside an XPS instrument allowing the stepwise analysis of the residual elements on the surface. The procedure was stopped when the bottom of the crater was inside the original hard coating. The sample was transferred from this instrument into the microprobe analyser in a sealed container. Nevertheless, due to the high affinity of Al and Ti to oxygen a reaction of the bare bottom of the crater with the atmosphere could not completely be avoided.

The microprobe analysis was carried out in an Camebax SX 50 under the following conditions: Primary beam energy E_0: 2.5 and 5 keV; Beam current I_0: 150 nA; X-ray lines; N Kα, O Kα, Al Kα and Ti Lα; Standards: Fe_2O_3 (O), Fe_4N (N), Al, TiN (Ti); The well known overlap N Kα-Ti L1 was corrected by a CAMECA overlap correction procedure which is part of the SX 50 software. Stage line scans were measured from the original surface of the oxidised sample across the wedge of the sputter crater down to the bottom of the crater (Ti-Al-N-hard coating).

Figure 11a gives the calibrated X-ray data of N, O, Al and Ti as measured with a beam energy of 2.5 ke V across the edge of the crater. At this beam energy the maximum depth of X-ray emission is around 70 nm, but more importantly for the depth resolution, the depth of maximum X-ray generation – i.e. the maximum of the $\Phi(\rho z)$-distributions – is around

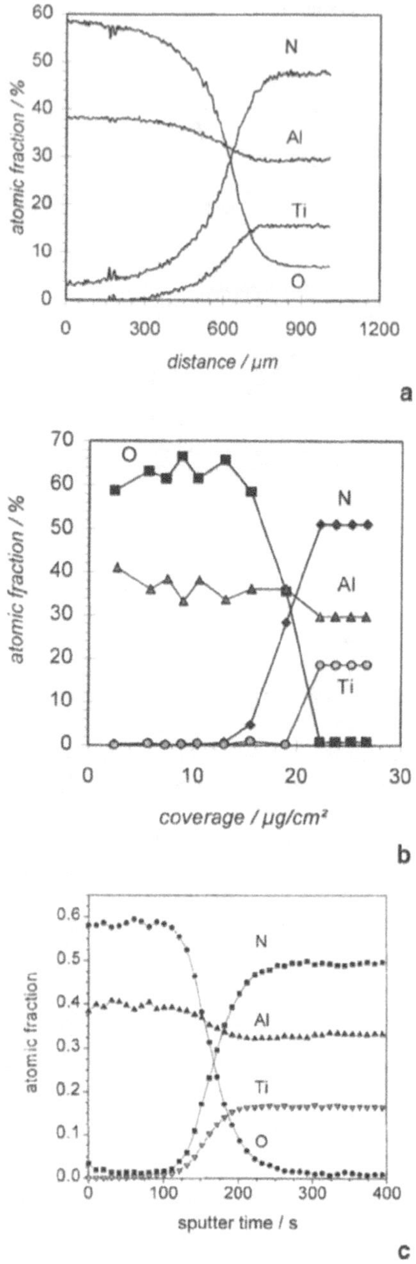

Fig. 11. **a** EPMA sputter crater line scan across the edge of a crater in an oxide layer on $Al_{0.6}Ti_{0.4}N$. $E_0 = 2.5$ keV, $I_0 = 150$ nA. Standards: Fe_4N (N), Fe_2O_3 (O), Al, TiN (Ti). Matrix correction PAP [16]. **b** Quantitative depth profile of the oxide layer on $Al_{0.6}Ti_{0.4}N$ determined by a reconstruction method from k-ratios measured with 5 keV and 150 nA. **c** SNMS-depth profile of the same oxide layer measured at the same sample as for **a**, **b**

12 nm [41]. Thus the data in Fig. 5a already give information about the oxide scale. In the surface it consists of Al_2O_3 which becomes obvious from the atomic fractions. No Ti- and only small N-signals are recorded at the surface. The increase in the N-signal

[2] These investigations form part of a research project at Lehrstruhl fur Theoretische Huttenkunde, RWTH Aachen, see [45]

and the corresponding decrease in O with depth indicates the transition from the oxide scale into the hard coating. The width of this transition is partly caused by the limited depth resolution, but also by the roughness of the original surface of the hard coating and consequently, the roughness of the interface between nitride and oxide scale.

The reconstruction technique [40, 41] was applied to a set of k-ratio data equivalent to the one shown in Fig. 11a, but measured at 5 keV. The reason for this is that the reconstruction method is based on the thin film analysis technique described above in which the electron range must be larger than the film thickness to be analysed. For the evaluation of depth profiles it follows that E_0 must be chosen in a way that the primary electrons penetrate one or even more fictitious sublayers. For the reconstruction the total line scan is subdivided into 8 portions representing 8 fictitious depth increments. The reconstruction begins with the conventional quantitation of the composition of the substrate by using the k-ratios measured at the bottom of the crater. The first set of k-ratios with slightly higher values for O Kα represent a superposition of X-ray intensities from the substrate plus the signals from the first oxide sublayer above the substrate. The corresponding intensity can be described as follows:

$$I_n \propto \underbrace{Q(E_0)\int_0^{t_n} c(z)\cdot\Phi(z)\cdot e^{-\chi_{t_n}z}dz)}_{\text{from top layer}}$$

$$+\underbrace{e^{-\chi_{t_n}t_n}\cdot Q(E_0)\int_{t_n}^{\infty} c(z)\cdot\Phi(z)\cdot e^{-\chi(z)(z-t_n)}dz}_{\underline{I_{t_n}^{\infty}}\text{ : from structure below top layer}} \tag{3}$$

n: number of fictitious sublayers; $c(z)$: concentration depth profile, χ: absorption coefficient; t_n: thickness of the uppermost sublayer at the particular point on the crater. The dependence of χ on the depth z in the second term considers that this integral could range over depth with varying composition.

The second term in this equation equals the contribution coming from the layers under the uppermost one and from the substrate. Since the composition of the substrate is known from the first step of evaluation the integral in this term can directly be calculated, and the unknown composition and total mass coverage of the first fictitious sublayer above the substrate (first term in (3)) is determined in an iteration process [41, 42]. This procedure successively is repeated for all fictitious sublayers up to the upper surface layer, resulting in the quantitative depth profile shown in Fig. 11b. Within the error range of ±5% the oxide layer consists of pure aluminium oxide. The mass thickness of the layer was determined to 20 μg/cm². This corresponds to a thickness of 50–60 nm, if we assume the density of the oxide to be around 3.5 g/cm³. In fact the actual density of the oxide could even be lower and accordingly the thickness higher. The nitrogen concentration, which in the line scan measurement with 2.5 ke V (Fig. 11a) was still between 3.5% and 4% at the surface, vanishes in Fig. 11b. The non-zero k-ratios of N kα are interpreted in the reconstruction as nitrogen enrichment in the transition zone between oxide and hard coating. To asses this interpretation a reference measurement was made by SNMS (Ar^+, 540 eV, 1.5 mA/cm², Ø4 mm). The resulting profile depicted in Fig. 11c clearly corroborates that an Al_2O_3-scale is grown on the Ti-Al-N-hard coating. The transition region is rather broad, but gives no indication of the existence of an intermediate zone with an enrichment of N. However, the SNMS-profile reveals a small amount of nitrogen (2–4 at.-%) throughout the oxide layer with a small maximum at the surface.

Thus, there is a small amount of N incorporated into the oxide scale, which was also detected by EPMA with 2.5 ke V in the crater line scan. The reconstruction technique misinterpreted this as an enrichment of N in the deeper part of the oxide scale. This is not surprising, since the roughness of the oxide scale is in the range of 100 nm, i.e. it does not possess the prerequisite flat surface for exact thin film EPMA. A general drawback of the reconstruction technique shows up, i.e. that errors made in the first steps of the evaluation are propagated through all further steps.

Nevertheless, the results are encouraging; EPMA is developing to become an important tool for quantitative depth profile analysis of structures on a nanometer to micrometer scale. Of course the methodology of reconstructing depth profiles from k-ratio data as well as the experimental techniques need further refinement and improvement. In this context, work is already being undertaken: Lesch et al. equipped a conventional microprobe analyser with an ion gun for in-situ sputter and analysis cycles [43]. Richter et al. discussed the theoretical background of EPMA sputter depth profiling and tried to apply algorithms from probability theory to extract depth profiles from k-ratio versus sputter time data [44].

Conclusions

For a long time, state-of-the-art EPMA has been developing from a versatile technique for quantitative microanalysis of bulk material into a sensitive and reliable tool for the analysis of thin films and multilayers on substrates. Of course, there are clear limits to its applicability and ability to provide a unique result. Depth profiling becomes feasible only, if – besides the possibility of beam-energy variation – additional parameters are introduced, for example "slicing" the sample from the surface in small steps, to make use of additional intensity variation connected to this process. This was carried out successfully for the depth range of a few microns by bevelling the near-surface region mechanically.

Recent developments combined EPMA-WDS with ion beam sputtering, which is a common procedure in Auger depth profiling. This was a step forward in exploiting the technical and physical capabilities of EPMA to analyse depth profiles on a sub-micron scale quantitatively. Of course, the applications described in this paper require good quality standards and software, but first of all experienced scientists to choose the right experimental conditions. The selection of measuring parameters has an important influence on the quality of the results and even on the feasibility of the techniques at all. The latter holds especially for the combination of EPMA with sputtering.

For further improving the capability and applicability of EPMA in surface analysis the following measures are proposed:

- Better vacuum conditions in the instruments: Contamination influences and diminishes the quality of results since with low beam energies the effects of contamination layers on electron stopping and absorption of soft X-rays become more severe.
- Implementation of ion guns in current microprobe analysers to allow in-situ sputtering for depth profile measurements.
- Application of appropriate numerical methods to deconvolute depth information form the X-ray-sputter profiles, also taking account of the statistical fluctuation of the measured intensities.

With respect to the rather high sensitivity – for example 1 nm of a buried metal layer can be resolved easily in an EPMA sputter profile – there is a great opportunity to realise the tasks mentioned above.

Finally it should be pointed out clearly that further development of EPMA towards surface analysis is not intended to compete with established surface sensitive techniques. It rather should become a complementary technique with its power focused on the field of quantitation, particularly direct quantitation of depth scales. A certain advantage could be that this new technique would open the possibility of depth profile analysis to institutions which so far do not have surface sensitive instruments at their disposal.

Acknowledgements. Thanks are due to Silvia Richter for her excellent support in evaluating thin film and depth profile analysis and to Martina Funken for carrying out the microprobe analysis. Furthermore I express thanks to the Fraunhofer Institut für Lasertechnik, Aachen, for giving permission to publish results on LCVD, and to the Lehrstuhl für Theoretische Hüttenkunde, RWTH Aachen for their permission to present results on oxidation of Ti-Al-N-hard coatings.

References

[1] S. J. B. Reed, *Electron Microprobe Analysis*, Cambridge University Press, Cambridge, 1993.
[2] R. Castaing, *Thesis*, University of Paris, 1951.
[3] W. E. Sweeney, Jr., R. E. Seebold, L. S. Birks, *J. Appl. Phys.* **1960**, *31*, 1061.
[4] G. H. Cockett, C. D. Davis, *Br. J. Appl. Phys.* **1963**, *14*, 813.
[5] R. Castaing, J. Descamps, *J. Phys. Radium* **1955** *16*, 304.
[6] B. Djuric, D. Cerovic, *X-Ray Optics and Microanalysis* (G. Möllenstedt, K. H. Gaukler, eds.), Springer, Berlin Heidelberg New York, 1969, pp. 99–103.
[7] R. H. Packwood, J. D. Brown, *X-ray Spectrom.* **1981**, *10*, 138.
[8] J. L. Pouchou, F. Pichoir, *La Recherche Aérospatiale*, n° 1984–93, ONERA.
[9] H. J. Hunger, W. Baumann, S. Schulze, *Cryst. Res. Technol.* **1985** *20*, 1427.
[10] R. A. Waldo, *Microbeam Analysis – 1988* (D. E. Newbury, ed.), San Francisco Press, San Francisco, pp. 310–314
[11] G. F. Bastin, J. M. Dijkstra, H. J. M. Heijligers, D. Klepper, *Microbeam Analysis* **1993**, *2*, 29.
[12] H. Frerichs, P. Karduck, D. A. Wesner, E. W. Kreutz, to be published.
[13] N. Ammann, P. Karduck, *Microbeam Analysis* **1990**, 150.
[14] H. Frerichs, *Doctoral Thesis*, RWTH Aachen, 1996.
[15] J. L. Pouchou, F. Pichoir *Scanning Microsc. [Suppl.]*, **1993**, *7*, 167.
[16] J. L. Pouchou, F. Pichoir, *La Rech. Aérospat.*, Publ. ONERA, 1984–5, pp. 47–65.
[17] J. L. Pouchou, F. Pichoir, *Scanning* **1990** *12*, 212.
[18] P. Willich, *Mikrochim. Acta* **1992**, [*Suppl.*] *12*, 1.

[19] J. L. Pouchou, F. Pichoir, in: *Electron Probe Quantitation* (K. F. J. Heinrich, D. E. Newbury, eds.), Plenum, New York, 1991, pp. 31–75.
[20] M. Wendt, *Fresenius J. Anal. Chem.* **1991**, *340*, 193.
[21] W. H. Gries, W. Koschig, *Surf. Interface Anal.* **1990**, *16*, 321.
[22] N. Ammann, *Diploma Work*, RWTH Aachen, 1989.
[23] J. M. Dijkstra, G. F. Bastin, H. J. M. Heijligers, N. Ammann, P. Karduck. *Beitr. Elektronenmikroskop. Direktabb. Oberfl.* **1992** *25*, 141.
[24] I. Berner and B. N. Yurchenko, in: *Proceedings 12th ICXOM* (S. Jasié nska L. J. Maksymowicz, eds.), Academy of Mining and Metallurgy, Cracow, 1989, pp. 92–95.
[25] A. Berner, G. Proaktor, *Mikrochim. Acta* **1994**, *114/115*, 195.
[26] N. Ammann, G. Gleitsmann, M. Heuken, K. Heime, P. Karduck, *Mikrochim. Acta* **1994**, *114/115*, 165.
[27] N. Ammann, P. Karduck, *Surf. Interface Anal.* **1994**, *22* 54.
[28] A. P. Alexejev, V. I. Zaporozchenko, *Fresenius J. Anal. Chem.* **1991**, *341*, 601.
[29] A. P. Alexejev, *Mikrochim. Acta* **1992**, [*Suppl.*] *12*, 229.
[30] P. Willich, R. Bethke, *Mikrochim. Acta* [*Suppl.*] *13*, 631.
[31] S. Baumgartl, L. Hachtel, G. K. Werners, *Prakt. Metallogr.* **1994**, *31*, 162.
[32] H. Bolm, *IAESTE-Trainee Work*, GFE, RWTH Aachen, 1996.
[33] D. Briggs, M. P. Seah, (eds.), *Practical Surface Analysis, 2nd Ed.*, Wiley Chichester, 1990.
[34] A. Benninghoven, F. G. Rüdenauer, H. W. Werner, *Secondary Ion Mass Spectrometry*, J. Wiley, New York, 1987.
[35] S. Baumgartl, P. L. Ryder, H. E. Bühler, *Z. Metallkd.* **1973**, *64*, 655.
[36] J. D. Brown., in: *Microbeam Analysis – 1988* (D. E. Newbury, ed.), San Francisco Press, San Francisco, pp. 271–272.
[37] N. Ammann, A. Lubig, P. Karduck, *Mikrochim. Acta* **1992**, [*Suppl.*] *12*, 213.
[38] J. Kirschner, H. W. Etzkorn, *Proc. 7th Intern. Conf. Solid Surfaces* (R. Dobrozemsky, F. Rüdenauer, F. P. Viehböck and A. Breth, eds.), Vienna, 1977, p. 2213.
[39] E. Haeussler, E. Saur, *Z. Naturf.* **1976**, *31a*, 1572.
[40] A. von Richthofen, M. Matsuo, P. Karduck, N. Ammann, *Mikrochim. Acta* **1994**, *114/115*, 511.
[41] P. Karduck, A. von Richthofen, *Microsc. Microanal. Microstruct.* **1995**, *6*, 421.
[42] N. Ammann, *Fortschr. -Ber. VDI Reihe 8, Nr. 399*, VDI Düsseldorf, 1994.
[43] N. Lesch, S. Richter, P. Karduck, *Mikrochim. Acta* [*Suppl.*] **1998**, *15*, 133.
[44] S. Richter, N. Lesch, P. Karduck, *Mikrochim. Acta* [*Suppl.*] **1998**, *15*, 125.
[45] M. Witthaut, R. Cremer, A. von Richthofen, D. Neuschütz, Fresenius' *J. of Anal. Chem.* **1998**, *361*, in press.

Mikrochim. Acta [Suppl.] 15, 125–131 (1998)

EPMA Sputter Depth Profiling, Part I: Theory and Evaluation

Silvia Richter*, **Norbert Lesch,** and **Peter Karduck**

Central Facility for Electron Microscopy, Aachen University of Technology, Ahornstrasse 55, D-52074 Aachen, Federal Republic of Germany

Abstract. Electron probe x-ray microanalysis is capable of the sensitive detection and the reliable quantification of small (partial)mass coverages on a submicron scale. To improve depth profiling so as to give reasonable depth resolution, the technique can be combined with a sputtering process. This paper presents theoretical aspects concerning the experiment of EPMA sputter depth profiling and the evaluation of the measured data. By means of a Monte-Carlo-Simulation program sputter-intensity profiles for multilayered systems can be calculated. It is shown that the calculated intensity-sputter profiles provide more information about elemental depth distributions than those obtained by the more common EPMA-method of electron beam energy variation.

For optimising the new technique, its sensitivity to the measuring parameters and the material's composition and structure is studied. As a result, at low electron beam energies the calculated profiles of intensity versus sputtered depth correspond approximately to the elemental-depth profiles. At relatively high energies the sensitivity is reduced but, due to the characteristic features in the intensity profiles such as kinks at sharp interfaces, there is sufficient information (although convoluted by the excitation volume) for structure determination.

For reconstructing elemental-depth distributions from intensity-sputter profiles an algorithm based on maximum entropy analysis is developed. The algorithm is tested using calculated x-ray data from a well-characterised multilayer structure and the data are overlaid by statistical noise. Then elemental-depth profiles can be determined quantitatively with regard to both composition and depth coordinate.

Key words: electron probe microanalysis, sputtering, surface analysis, quantitative depth profiling, maximum entropy analysis.

Electron probe x-ray microanalysis (EPMA) is a proven method in microbeam and thin film analysis. It renown for its high sensitivity and high accuracy in quantitative measurements of mass coverages. In the last 15 years enormous progress has been made in thin film and depth profile analysis at the submicron level [3–8]. One reason for the advance is the ability to establish accurately values for the depth distributions of generated x-ray intensities, $\Phi(\rho z)$, based on analytical or Monte-Carlo-Simulation models [1–6]. Depth profile analysis has been performed so far, either by variation of the incidence angle of the electron beam with respect to the sample surface [8] or, more commonly, by varying the energy of the incident electrons [9–11, 14]. However, the sensitivity of these non-destructive techniques turned out to be insufficient to resolve complex structures, particularly at larger depths [9].

Therefore, to improve the depth resolution we propose a combination of EPMA and sputtering of the sample surface: EPMA sputter depth profiling [16]. In first attempts [12, 13] it was demonstrated, that EPMA-SDP is capable of quantifying mass coverage of complex oxide scales on Ti-Al-N coatings by using EPMA crater edge profiling. But it turned out that the reconstruction algorithm used in thin film analysis is a time-consuming procedure and is restricted also in the number of depth profiles which can be described (limited to 9 films on a substrate) [1, 13]. Thus our

* To whom correspondence should be addressed

first intention was to obtain direct information about elemental-depth profiles from a better understanding of intensity-sputter profiles. Since in most cases the information depth is greater than the characteristic dimensions of the material's structure, an algorithm must be developed in order to reconstruct real depth profiles.

Principle of EPMA Sputter Depth Profiling

Until now depth profiling has been attempted in EPMA by varying the electron beam energy [9–11, 14]. In this technique the material is excited by electrons at various steps of primary electron beam energy and the corresponding variation of the k-ratios, i.e. the emitted characteristic x-ray intensities relative to those of reference materials, is measured. This variation from small to high beam energies causes an expansion of the ionisation depth distribution $\Phi(\rho z)$ and thus a shift of the mean information towards greater depths. To explain the limited sensitivity of this technique for detecting depth profiles the following fictitious example will be discussed: We assume a multilayer system zinc/copper/zinc, each film with a thickness of 50 nm, on a silicon substrate (see Fig. 1).

By using a Monte-Carlo-Program developed by Ammann and Karduck [1] for the simulation of electron interactions with complex, multi-component solids the emitted x-ray intensities for various incident electrons energies were calculated. Fig. 2 shows the calibrated Zn Lα- and Cu Lα-intensity profiles.

The maximum of the Cu Lα-intensity indicates that there is a buried layer of copper or copper-zinc alloy

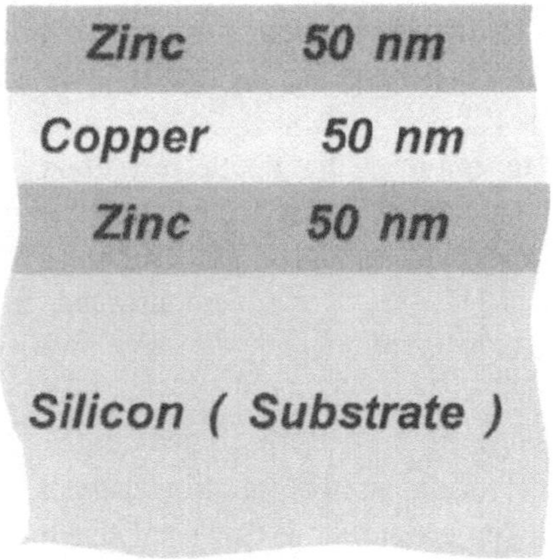

Fig. 1. Structure of a fictitious sample: A laterally homogeneous multilayer system Zinc/Copper/Zinc on a silicon substrate

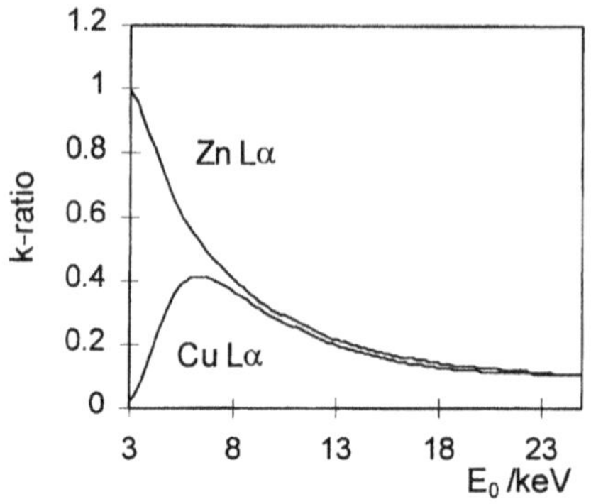

Fig. 2. Calibrated Zn Lα- and Cu Lα-intensities versus primary beam energy for the sample depicted in Fig. 1

at a certain depth. The monotonic decrease of the Zn Lα-profile only suggests a zinc coverage in the near-surface region, but the curve gives no obvious information on the existence of a second zinc layer.

Thus, the x-ray intensity-beam energy variation contains only sparse information about the structure of the specimen. It can be concluded that a Zn-profile with an arbitrary shape never could be resolved by this method uniquely. The problems of a unique and stable reconstruction of concentration-depth profiles of elements which are present in one or more layers have been discussed in the literature [14, 15]. In particular the detection of structures at larger depths needs rather high primary electron beam energies. However, at high beam energies the resolving power of the beam energy variation technique becomes weak.

Therefore in this energy range the intensity profiles are not sensitive enough to determine concentration gradients because electrons excite the whole structure [9]. Another limiting factor could be the strong absorption of soft x-ray lines, which reduces the depth of information also [15].

Therefore a new or additional variable parameter must be introduced to provide intensity variations which give more information about concentration-depth profiles. One possibility is to use the alternating procedure of measuring x-ray intensities at a rather low electron beam energy, E_0, and then to remove small parts of the material by sputtering with inert gas ions. The experimental arrangement is analogous to the principle of Auger sputter depth profiling [24]. This combination of EPMA with sputtering results in the new method: EPMA sputter depth profiling.

The experimental setup and further details of the procedure are described by Lesch et al. [16]. In this part of the paper the theoretical background of the new method has been worked out to support the experiment and evaluation of the measured data.

On the basis of the Monte Carlo Program mentioned above, a new routine was developed to calculate the x-ray intensities emitted from multilayer systems at various sputter depths. In this process the surface is sputtered by small depth elements dz' and the corresponding intensities $I(z')$ are calculated. For convenience in the following presentation the sputtered mass coverage $(\rho \cdot z')$ is interpreted as sputtered depth or analysis depth z'. The calculations were first carried out assuming that the ionisation depth distribution, $\Phi(z)$, generated by electron bombardment remains unchanged after each new sputtering step. This is a good approximation, if the elements examined do not differ much in their atomic number. For the sample of Zn /Cu /Zn on the silicon substrate the ionisation depth distribution of the initial (non-sputtered) state was chosen.

Actually, shifting the mean excitation towards the relatively light substrate Si requires a correction of $\Phi(z)$ with depth z. But for the subsequent theoretical study we have neglected this. Removing successively small amounts of the specimen with defined thickness dz' and calculating the emitted Zn Lα and Cu Lα intensities of the remaining structure, results in the intensity profiles represented in Fig. 3a for the electron energies 3 and 10 keV. At certain depths pronounced kinks appear in the curves, which are caused by the steep drop in the concentration at the interfaces. For a further interpretation we consider the $\Phi(z)$ depth distribution of the Lα lines at both beam energies. Fig. 3b shows these distributions of the Zn L_{III} ionisation for the present material. The range of $\Phi(z)$ depends on the overvoltage ratio, i.e. the ratio of the electron beam energy E_0 to the excitation energy of the characteristic line. The smaller this ratio the more the ionisation depth distribution $\Phi(z)$ is localised to the surface.

In addition to the concentration-depth profiles, this distribution is responsible for the variation of the intensity curves with respect to sputtered depth. Depending on the range of $\Phi(z)$ relative to the characteristic dimensions of the structure two different types of variation A and B can be distinguished (Fig. 3a). As an example of type A the Zn Lα intensity (3 keV) shows high sensitivity: at the beginning of the sputtering experiment only the Zn atoms of the upper Zn layer are excited. The intensity

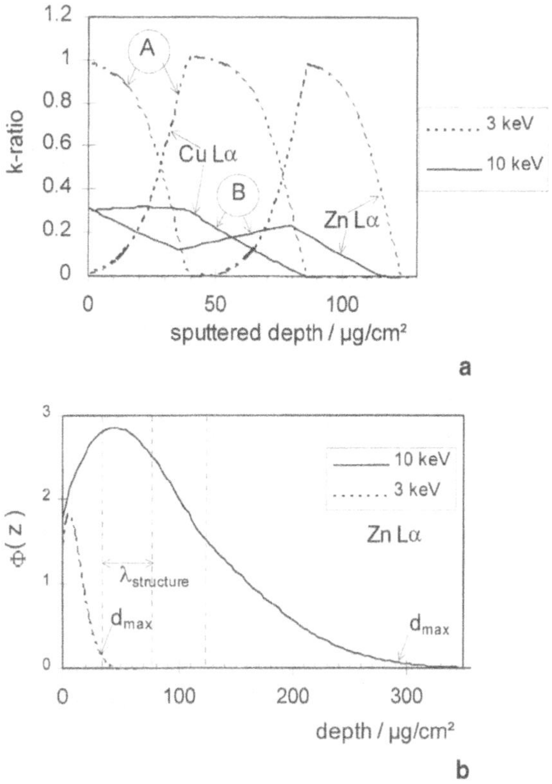

Fig. 3. **a** Calibrated Zn Lα-and Cu Lα-intensities versus sputtered depth at 3 and 10 keV beam energy for the sample depicted in Fig. 1. **b** $\Phi(z)$ depth distributions of Zn Lα-lines in the multilayer system Zn/Cu/Zn//Si at 3 and 10 keV beam energy

variation describes approximately the concentration-depth profile of zinc except for the smearing of the sharp interfaces due to the convolution with the ionisation depth distribution. The evaluated x-ray intensity sequence at 10 keV (type B in Fig. 3a) demonstrates smaller variations and a different shape caused by the larger depth of information (see Fig. 3b), but the kinks still reveal the position of the interfaces. Comparison with the intensity curves of the energy variation technique shows the significantly higher sensitivity of sputter-depth profiling for resolving structures with depth. At even higher electron beam energies the variations of intensity-sputter profiles would decrease still further. In addition when considering real experimental data these would be superimposed by statistical errors and hence the detection of elemental-depth profiles is limited. Therefore it is advisable to find optimal measuring conditions by adjusting the beam energy E_0 according to the structure of the sample.

Using the routine described above, intensity curves can be computed as a function of analysis depth z' for complex multilayer systems. If there are elements present with strong differences in atomic number and thus, different excitation energies, both types of curves (A and B) could appear in one experiment. As a result, this routine is able to evaluate optimal operating conditions prior to the measurement itself and therefore is suited to assess the performance of sputter depth profiling. Thus, by investing a small amount of time prior to an experiment, a first approximation to the expected intensity profiles can be calculated by roughly assuming a structure of the multilayer system. On the other hand, after a measurement the data can be compared with the theoretical values by a trial and error procedure to gain experience about the unknown sample.

Theoretical Considerations

The x-ray intensity of an element A is defined as a function of the corresponding concentration-depth profile $c(z)$ and of the sputtered depth z':

$$I_A(z') = \text{const} \cdot \int_{z'}^{\infty} dz \cdot \Phi_A(z - z') \cdot \exp(-\tilde{\chi}_A(z, z')) \cdot c_A(z) \tag{1}$$

The constant term takes into consideration solid angle, sensitivity of spectrometer and atomic constants. The explicit determination of this constant can be avoided, if we calibrate the net intensities emitted from the specimen relative to the corresponding one emitted from a standard sample; these ratios are called k-ratios. $\Phi(z - z')$ describes the depth distribution of the generated x-rays, but shifted relative to the depth profile $c(z)$ by the sputtered depth z': It depends on the incident electron energy E_0, the critical excitation energy E_{nl} and the structure as well as the composition of the material. The emitted portion of the generated x-rays is represented by $\exp(-\tilde{\chi}_A(z, z'))$ with:

$$\tilde{\chi}_A(z, z') = \frac{1}{\sin\Theta} \cdot \sum_{A'} \int_{z'}^{z} dz'' \cdot \mu_{\text{eff},A}(z'') \tag{2}$$

$\mu_{\text{eff},A}(z)$ is an effective absorption coefficient evaluated by a weighted average as follows:

$$\mu_{eff,A}(z) = \sum_{A'} \mu_{A,A'} \cdot C_{A'}(z) \tag{3}$$

where $\mu_{A,A'}$ is the mass absorption coefficient of the characteristic line of the element A in material A' and Θ is the detector take off angle. Since $\mu_{\text{eff},A}(z)$ depends on the concentration-depth profile $c_A(z)$, the relation between the emitted x-ray intensities and the profile $c_A(z)$ is non-linear. If the mass absorption coefficients of elements A' do not differ much Eq. (1) can be regarded as a convolution equation.

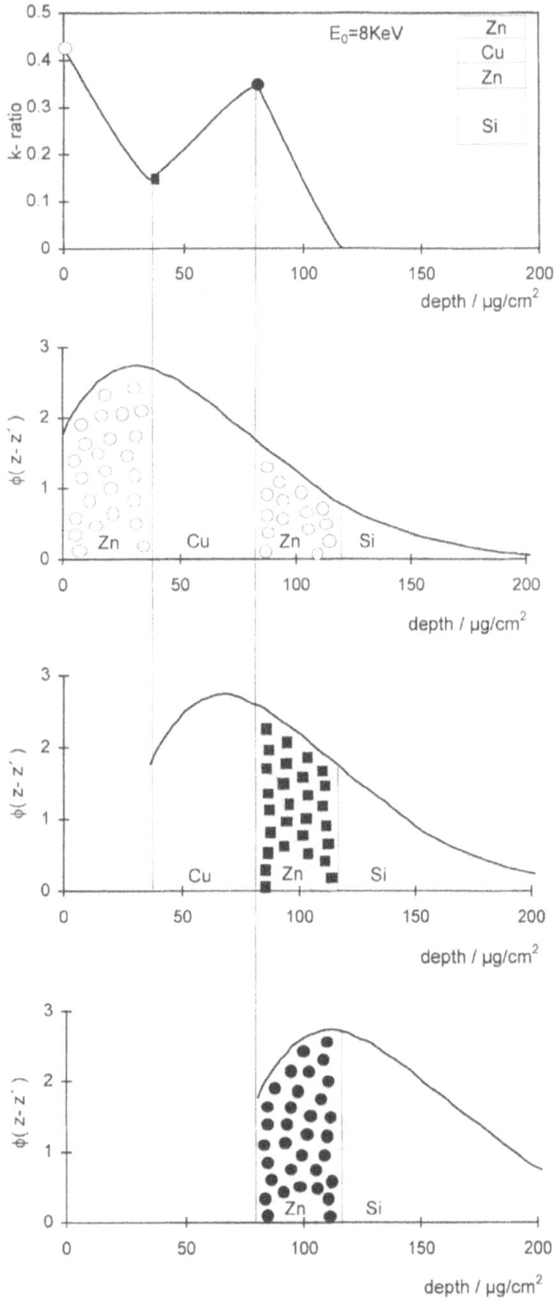

Fig. 4. Illustration of the convolution of $\Phi(z)$ with $c(z)$ at different sputtered depth z'. Symbols ○, ■ and ● represent the fraction of the Zn-distribution which is excited at the corresponding depth z'

The interrelation between the real concentration-depth profile $c(z)$ and the resulting intensity-sputter profile is explained in Fig. 4 at a beam energy of 8 keV.

There exists three pronounced characteristic intensity values marked in the diagram and for each symbol the corresponding $\Phi(z - z')$ is depicted at the particular depth z'.

The first point marked is the calibrated intensity measured on the initial material: Zn/Cu/Zn on a Si substrate, and both zinc layers are excited by the electrons. After removing the outermost zinc layer, generation of x-rays is reduced to that from the residual buried zinc layer, which is partly absorbed by the upper copper layer (2nd marked point). During sputtering of the Cu layer the absorption path gets shorter and the Zn intensity increases monotonically until the whole copper layer is removed (3rd marked point). A decrease then occurs caused by removal of the residual zinc layer.

Reconstruction of Depth Profiles

In the previous sections we discussed the behaviour of intensity-sputter profiles versus sputtered depth. However, in a real experiment, data are measured as a function of the sputter time or the length of the linescan along a crater edge. Since the sputter rate or crater gradient can vary strongly according to the structure of the unknown sample or to the local distribution of the sputtering ions, a distortion of the curves described above can appear. To simulate a sputtering experiment on the layered structure described above the calculated x-ray data were overlaid by Gaussian distributed statistical noise. The absolute amount of the standard deviation of the Gaussian distribution is one percent of the maximum intensity value. The result of this treatment is given in Fig. 5 for a 10 keV beam energy. We now will try to apply a reconstruction algorithm to these data with the aim of calculating a depth profile as close as possible to the real one. From the experiment we only know that the data were measured successively so that they corresponded with an increasing sputtered depth. As in a real experiment we pretend not to know anything about the real depth z' and therefore it is an additional unknown. This gives rise to a further problem, i.e. how to deduce the real depth scale from sputter-time or crater-length. Other problems are the statistical errors and the integral

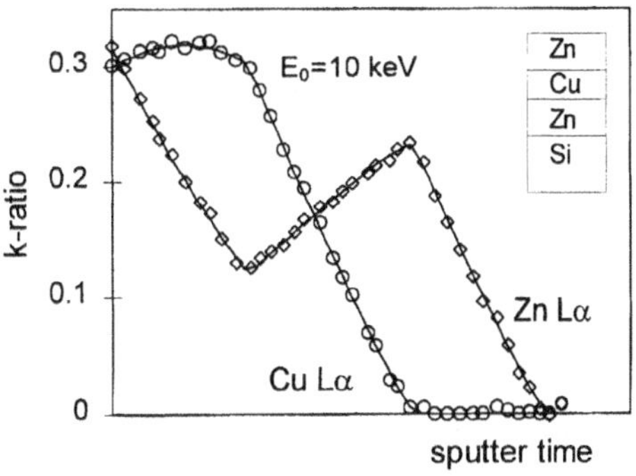

Fig. 5. Simulation of an EMPA-SDP experiment: Calculated Zn Lα- and Cu Lα- intensities versus sputter time at 10 keV beam energy with noise (data points) and without noise (solid line).

information about the structure as well as nonlinearity caused by the dependence of the absorption term on the concentration profile $c(z)$ (see Eq. 1). For the numerical treatment the multilayer system was subdivided into n_i fictitious sublayers, whose composition $c_{i,A}$ and thickness $\Delta z_{i,A}$ are described by the partial mass coverage $m_{i,A} = c_{i,A} \cdot \Delta z_{i,A}$ of each element A. The finer the thickness of the sublayer the smaller the variation in intensity from one sputter step to the next and thus, the larger the relative expected error of the reconstructed values. The sequential evaluation from one sublayer to the other (starting at the first and calculating the changes in chemical composition from one to the next) was found to succeed, but severe error propagation resulted, which had to be reduced further by appropriate measures. Therefore we chose a reconstruction procedure able to evaluate mass coverage of each sublayer in a quasi parallel procedure and to find a stable and unique solution. For this, maximum entropy analysis was selected. It is a fundamental tool in many branches of science and technology dealing with the reconstruction of distributions [17–22]. It is based on probability theory and tries to find a solution of the problem by introducing á priori information on the elemental-depth profile. Without detailed knowledge of the structure the best á priori estimate is the homogeneous distribution, i.e. the distribution with maximum entropy. This means it represents the most unbiased estimate. The Baysian probability theory takes into account both the á priori information as well as the errors $\underline{\sigma}$ in the 'measured' data $\underline{D}$ by maximising the equation:

$$\Lambda(\underline{m}) = \alpha \cdot S(\underline{m}) - L(\underline{m}) \tag{4}$$

$\underline{m}$ is a vector, the components of which consist of the partial mass coverage $m_{i,A}$ of an element A in the depth increments n_i.

Except for the factor 2, $L(\underline{m})$ is the well-known χ^2-term from the statistics.

$$L(\underline{m}) = \frac{1}{2} \cdot \sum_{i,A} \frac{\left(I_{i,A}(\underline{m}) - D_{i,A}\right)^2}{\sigma_{i,A}^2} \tag{5}$$

$S(\underline{m})$ represents the entropy of the structure,

$$S(\underline{m}) = \sum_{j,A} \left(m_{j,A} - m_{j,A}^0 - m_{j,A} \cdot \log \frac{m_{j,A}}{m_{j,A}^0} \right) \tag{6}$$

where $\underline{m}^0$ is the distribution with maximum entropy, the á priori estimate. α is a regularization parameter, the magnitude of which is determined in a way that the resulting depth profile represents the most probable values given the data, i.e. the most obvious solution [20].

For this maximum entropy method we used a software package from maximum entropy data consultants called Memsys5 [23].

This software package provides a means of solving ill-conditioned linear equation systems. After division into a discrete number of steps the integral equation can be regarded as a system of linear equations:

$$\underline{I} = \underline{\underline{R}}(\underline{m}) \cdot \underline{m} \tag{7}$$

The matrix i. a. $\underline{\underline{R}}$ consists of discrete values of the absorption corrected $\Phi(z)$ distribution. Because of the absorption term and the unknown sputtered depth, z', $\underline{\underline{R}}$ also depends on the structure of the sample itself. Therefore $\underline{\underline{R}}(\underline{m})$ must be updated after a few iterations by putting in the intermediate results. This process is repeated until convergence is obtained. At the start of the iteration we only need a rough estimate about the thickness of the whole range of the depth profile. Dividing this thickness by the number n_i of fictitious sublayers, which have equal composition, leads to a homogeneous distribution of partial mass coverage $\underline{m}^0$, which could be a possible initial choice for the calculation of the matrix $\underline{\underline{R}}$. For the present case the initial thickness estimate deviated 50% from the true value and the initial chemical composition was chosen to be the homogeneous distribution of Zn and Cu with 50 weight% each.

Fig. 6 shows the results of the reconstruction of the experimental data 'measured' in a simulated experiment.

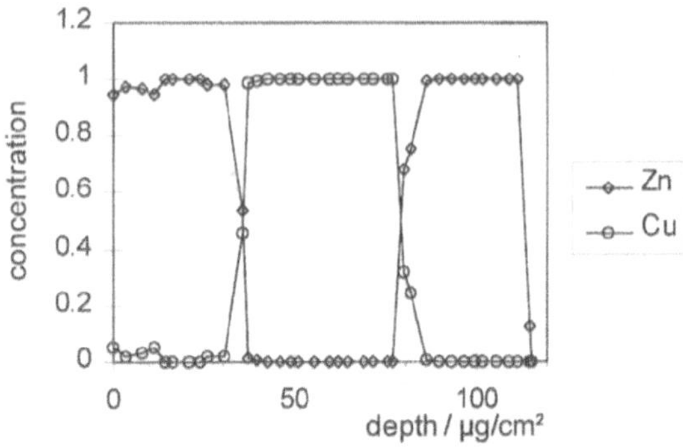

Fig. 6. Results of the reconstruction: Zn- and Cu-concentration versus depth

We can see that the assumed sample structure is retrieved in a very satisfactory way. The thickness and composition of each layer was reconstructed with an accuracy dependent on the statistical error. The less accurate values in the first layer can be explained as follows: Because of the weak absorption of Cu Lα in zinc the curve of Cu Lα is not very sensitive to the zinc mass coverage on top of the Cu-layer. As a result, the Cu Lα-intensity variations caused by the noise are interpreted as small *Cu*-coverages. The normalisation procedure, $\sum c_A = 1$, leads to the anticorrelated behaviour of the zinc concentration. Only a finite number of data could be used to calculate the intensity profiles, but this provides an insufficient description of the region around the discontinuities in the intensity profile. Therefore the depth resolution in our example is limited by the width of the sputter interval and in addition by the statistical uncertainty of the data. A measure for the depth resolution is the transition width, i.e. the depth region around the interface where the reconstructed concentration decreases from 84% to 16% of its initial value or vice versa. Thus, in our case, the transition width is about 5 nm assuming a mean density of 8.05 g/ml. However in a real experiment the depth resolution could be reduced further by sputter effects.

Finally to evaluate quantitative concentration-depth profiles only a knowledge of the sequence of successive measuring points is required, each of these corresponding to an increasing sputtered depth.

Conclusions

EPMA-SDP proves to be capable of resolving concentration-depth profiles and is superior to the energy variation method. On the basis of an already existing

Monte Carlo Program a new routine calculates sputter intensity profiles. This routine is able to evaluate optimal operating conditions and therefore to assess performance of EPMA-SDP. Its application gains first knowledge about the structure of the unknown sample by a 'trial and error' procedure (see Part II).

Using the new technique we have succeeded in reconstructing quantitative concentration-depth profiles, which necessitates conversion of a sputter-time scale or a crater-length scale into a depth scale.

The present results encourage us to apply the technique to the reconstruction of depth profiles in real-life samples. Here it will be necessary to improve the reconstruction algorithm by considering fluorescence effects and correcting for $\Phi(z)$ variations due to changes of atomic number with depth.

Acknowledgements. The present work was partly supported by the Deutsche Forschungsgemeinschaft (DFG).

References

[1] N. Ammann, P. Karduck, *Microbeam Analysis 1990* (J. R. Michael, P. Ingram, eds.), San Francisco Press, San Francisco, 1990, p. 150.
[2] P. Karduck, N. Ammann, *Microbeam Analysis 1990*, p. 21.
[3] J. L. Pouchou, F. Pichoir, *Recherche Aerospaciale* **1984**, *5*, 47.
[4] R. H. Packwood, J. D. Brown, *x-ray Spectrometry* **1981**, *10*, 138.
[5] G. F. Bastin, H. J. M. Heijligers, *Publ. of the Lab. for Physical Chemistry*, TH Eindhoven, 1984.
[6] G. F. Bastin, H. J. M. Heijligers, *Publ. of the Lab. for Physical Chemistry*, TH Eindhoven, 1984.
[7] N. Ammann, A. Lubig, P. Karduck, *Mikrochim, Acta* 1992 *[Suppl.] 12*, 213.
[8] H. Gries W. Koschig, *Surf. Interface Anal.* **1990**, *16* 321.
[9] N. Ammann, P. Karduck *Surf. Interface Anal.* **1994**, *22*, 54.
[10] P. Willich, R. Bethke, *Fresenius J. Anal. Chem.* **1995**, *353*, 389.
[11] P. Willich, R. Bethke, *Mikrochim. Acta.* **1996**, *[Suppl.]13*, 631.
[12] A. von Richthofen, M. Matzo, P. Karduck, N. Ammann, *Mikrochim. Acta* **1994**, *114/115*, 511.
[13] P. Karduck, A. von Richthofen, *Microsc. Microanal. Microstruct.* **1995**, *6* 421.
[14] J. L. Pouchou, F. Pichoir, *Scanning Microsc.* **1993**, *7*, 167.
[15] P. Karduck, *Mikrochim. Acta [Suppl.]* **1998** *15*, 109.
[16] N. Lesch, S. Richter, P. Karduck, *Mikrochim. Acta [Suppl.]* **1998**, *15*, 133.
[17] J. Skilling, *Proc. Maximum Entropy and Baysian Methods*, Kluwer Dordrecht, 1989.
[18] A. Mohammad-Djafari, G. Demoment, *Proc. Maximum Entropy and Baysian Methods*, Kluwer Academic Publishers, Dordrecht, 1989.
[19] W. S. Werner, G. C. Smith, A. K. Livesey, *Surf. Interface Anal.* **1994**, *21*, 38.
[20] S. F. Gull, *Proc. Maximum Entropy and Baysian Methods*, Kluwer, Dordrecht, 1989.
[21] W. van der Linden, *Appl. Phys.* **1995**, *A60*, 155.
[22] G. C. Smith, D. Part, O. Cochonneau, *J. Microsc* . **1994**, *178*, 48.
[23] J. Skilling, *Maximum Entropy Data Consultants Ltd.*, 33 North End, Meldreth, Royston, SG8 6NR, UK.
[24] D. Briggs, M. P. Seah, *Practical Surface Analysis, Vol. 1: Auger and x-ray Photoelectron Spectroscopy*, Wiley, Chichester, 1983.

Mikrochim. Acta [Suppl.] 15, 133–139 (1998)

EPMA Sputter Depth Profiling, Part II: Experiment

Norbert Lesch*, **Silvia Richter**, and **Peter Karduck**

Central Facility for Electron Microscopy, Aachen University of Technology, D-52056 Aachen, Federal Republic of Germany

Abstract. Electron probe x-ray microanalysis (EPMA) has been developed into a reliable technique for surface layer analysis but, due to the relatively large depth of information, it cannot yet be applied to resolve complex depth distributions, i.e. layers with concentration gradients or complex multilayer structures.

To improve the depth resolution of EPMA we combined an ion gun with a conventional microprobe analyzer, forming the technique of EPMA sputter depth profiling (EPMA-SDP). This arrangement allows the continuous removal of the surface and simultaneous measurement of emitted x-ray intensities as a function of sputtered depth. In this paper the experimental arrangement is described and first results are presented. The sample analysed is a Cu/NiCr/Cu/NiCr.../Si multilayer, a structure which could not be determined with a standard EPMA method such as beam energy variation. However with EPMA-SDP the mass coverage of each layer as well as its chemical composition was established directly from the measured data with an accuracy of about ±5% relative. The theoretical background of the new technique is described in part I of the paper (this volume).

Key words: quantitative depth profiling, EPMA, sputtering, multilayer structures.

In the past, EPMA combined with wavelength dispersive x-ray spectrometry (WDS) has been developed to provide an accurate tool for surface layer analysis. The detection limit of this technique can be as low as 10^{14} atoms per cm^2 as long as the particular elements of the layer are not present in the substrate [1, 2]. The technique enables the mass coverage and the chemical composition of the thin layers to be quantified from measured and calibrated x-ray intensities using special matrix correction techniques dedicated to this purpose (see for example [3–5]). All these methods are based on the assumption that the chemical composition of a single layer does not change with depth.

If this assumption is not valid, additional parameters must be introduced to obtain sufficient intensity changes to deduce trends in the chemical composition depth profiles. Usually, variation of electron beam energy or of angle of impact are applied [6, 14]. Variation of these parameters expands or contracts the excitation volume within the sample with respect to the concentration depth profile. Under certain circumstances the resulting intensity changes can be used to determine, approximately, the concentration depth profile. The maximum thickness of the structure that can be analysed has to be smaller than the maximum depth of ionisation and the depth resolution will depend on the amount of intensity variation that can be achieved. Such variations can be rather large at low electron beam energies, E_0, and can give information on near surface gradients in element concentrations, but with increasing E_0 the variation of x-ray intensity decreases and the resolving power of this technique for the concentration depth distribution at greater depths (> 300 nm) becomes weak [15]. Therefore these methods are applicable in daily analysis only to samples for which a parameterization of the profile shape can be assumed. Furthermore, use of the measured data does not always result in a unique solution so that ambiguous results [6, 13] may be obtained. In fact this is the reason why EPMA has

* To whom correspondence should be addressed

seldom been used for depth profiling of real life samples, i.e. samples on which a priori knowledge is minimal or absent altogether.

Thus, due to its large depth of information, EPMA cannot resolve complex depth distributions i.e. layers with concentration gradients or multilayer structures. Analysis of such structures is performed commonly by methods that are sensitive only to the very surface such as AES[1)], XPS[2)], SIMS[3)], or SNMS[4)] in combination with sputtering. But quantifying chemical composition and calibrating the depth scale with these methods is often inaccurate and troublesome. In particular the conversion of the sputter time scale into a true depth scale is subject to errors. This is due to the fact that the conversion is based on knowledge of material parameters, i.e. sputter yield, sensitivity factors, etc., which are known only for special standards and which differ strongly in samples with a complex composition [8].

Other authors have used energy dispersive EPMA to improve depth scale determination in Auger electron spectroscopy. They monitored the x-ray intensities of the elements to be analysed by AES during sputtering. These intensities were used to determine the sputter rates and to calibrate the depth scale [9, 10]. This technique allows the determination of the sputter rates only for homogeneous layers with known thicknesses. Thus, a common difficulty with all depth profiling techniques is the large uncertainty concerning the accurate determination of depth scale and of true composition.

Our approach is to combine the advantages of EPMA – i.e. high sensitivity and accurate determination of chemical composition and mass coverage of thin layers – with a sputtering technique. Basically there are two methods to achieve this. A crater can be sputtered into the sample and afterwards a line scan can be performed across the edge of the crater to obtain at least qualitative information on the concentration depth profile. First results obtained with such a combination have already been published by Karduck and von Richthofen [11]. The disadvantage of this, so called, crater edge line scan profiling is that lateral resolution is lost and only laterally homogeneous samples can be investigated. Furthermore, commercially available electron microprobe analysers are not equipped with an ion gun. Therefore contamination and oxidation of the freshly exposed surfaces can occur during transfer from the sputtering device to the microprobe. This causes additional errors in the measurement.

The second possibility is to alternately sputter the sample and perform x-ray microanalysis measurements in a procedure that is well established in Auger depth profiling. The advantage of this method is that the spatial resolution of EPMA, usually better than 1 μm, is maintained during the depth profiling. Therefore, laterally inhomogeneous samples could be used for analysis.

To carry out this kind of experiment we equipped an electron microanalyser with an ion gun. With this instrument we could either perform in situ EPMA sputter depth profiling or use the crater-line scan technique. Multilayered samples which are out of the scope of normal EPMA depth profiling methods were chosen to exploit the capabilities of this technique.

Experimental

Instrumentation

The general arrangement for EPMA-sputter depth profiling (EPMA-SDP) is shown in Fig. 1. The ion gun, the electron beam and the WDS spectrometers are arranged such that they share the same point of focus. With such a confocal arrangement x-ray intensities can be measured either during sputtering or with alternating sputtering and measuring cycles. The following equipment was used:

The microanalyser was a JEOL JXA 733 equipped with three WDS spectrometers, one with a multilayer crystal for the measurement of light elements. The x-ray take-off angle was 40°. This microprobe was chosen because its large specimen

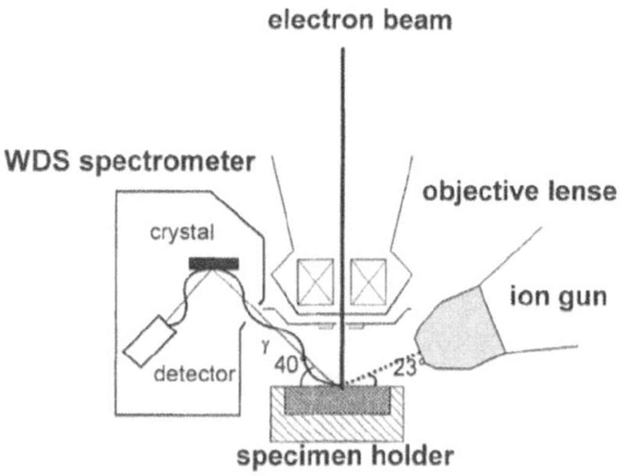

Fig. 1. General instrumental arrangement in a microprobe analyser for sputter depth profiling

[1] AES: Auger electron spectroscopy
[2] XPS: x-ray photoelectron spectroscopy
[3] SIMS: Secondary ion mass spectroscopy
[4] SNMS: Secondary neutrals mass spectroscopy

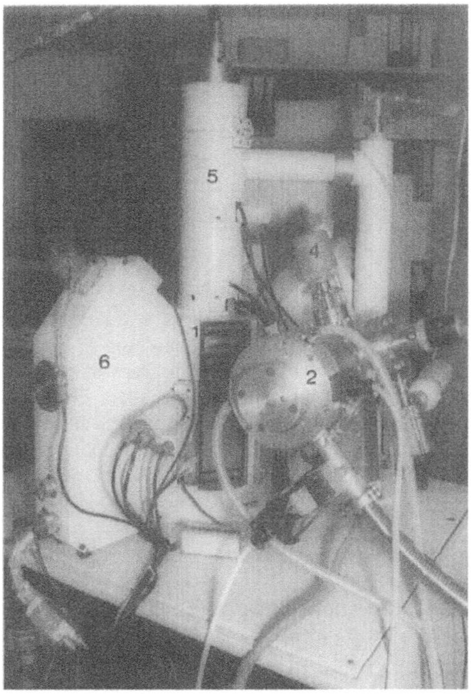

Fig. 2. Microanalyser with attached ion gun: *1* view through free spectrometer port, *2* turbomolecular pump, *3* ion gun, *4* vacuum gauge, *5* column, *6* WDS spectrometer

Fig. 3. View inside the specimen chamber: *1* objective lens, *2* tip of ion gun, *3* specimen holder, *4* electron deflector, *5* secondary electron detector, *6* slit for x-rays

chamber and its suitably sized spectrometer ports enabled the ion gun to be fitted to the instrument without extensive modification. Spectrometers, counting electronics and specimen stage were equipped with up to date computer control to allow automated data acquisition during the sputtering experiment. An ion gun of the type Riber CI50RB with a focus diameter of 250 μm FWHM and a working distance of 25 mm was available and was fitted to a free spectrometer port using an adapter specially designed for the purpose. For sputtering we used Ar^+ ions at energies from 0.5 keV to 5 keV. The ion gun could be operated either in scanning or stationary mode. The angle of impact was determined by the geometry of the specimen chamber and the ion gun, and was 67° with respect to the sample normal. At this angle, sample rotation should be used to minimise sputter induced surface roughening but this feature was not available on our instrument. The ion gun had to be equipped with a differential pumping system to attain a pressure in the ionisation chamber suitable for operation. Usually, when using an ion gun in UHV systems, the differential pumping system is used to prevent the pressure in the specimen chamber from rising when the sputtering gas is introduced into the ionisation chamber of the ion gun. In our case the pressure in the specimen chamber, typically 10^{-5} mbar, is higher than the operating pressure of the ionisation chamber, so the differential pumping system is used to decrease the pressure in the ionisation chamber to below 10^{-6} mbar. Fig. 2 shows the modified microprobe with the attached ion gun and the pumping stage. Fig. 3 gives a view into the specimen chamber. As can be seen, a part of the electron deflector, which prevents backscattered electrons from reaching the detector, had to be removed to give space for the ion gun. Because of the small working distance of this particular ion gun its tip is quite close to the objective lens of the microprobe. This causes the electrical fields of ion gun, objective lens and electron deflector to overlap which leads to a difficult adjustment procedure for the whole system. To prevent this an ion gun with a larger working distance would be preferable.

Influence of Contamination

The pressure inside the specimen chamber of a commercial microprobe is in the range of 10^{-5} to 10^{-7} mbar. Therefore contamination by hydrocarbons, which are cracked by the electron beam and deposited on the sample surface, is a serious problem in EPMA measurements, especially for carbon analysis when long measuring times have to be used. During EPMA-SDP some ten to a hundred sputtering and measurement cycles have to be performed, therefore special attention has to be paid to the effects of contamination. As mentioned above, EPMA-SDP can be performed in continuous or alternating mode.

When using the alternating mode the ion gun is switched off while X-ray intensities are measured and vice versa. But during x-ray measurement contamination is deposited at the surface of the sample, especially at the focus spot of the electron beam. When the ion gun is switched on again the contamination has to be sputtered off first before the actual sputtering of the sample can continue. Because of this effect the sputter rate at the point of beam impact is reduced compared with the rate on the surrounding area. This leads to errors when the electron beam is drifting during measurement. If the sputter rate is too low to remove contamination completely the x-ray intensities could be influenced by absorption in this contamination layer. These kinds of effects must be avoided if reliable measurements are to be obtained.

Table 1. Experimental conditions for EPMA-SDP measurement of a Cu/NiCr/.../Si multilayer

X-ray lines analysed	Cu La	Ni La	Cr Ka	Si Ka
Energy of line	0.928 keV	0.849 keV	5.414 keV	1.74 keV
Beam voltage	7 kV	7 kV	7 kV	7 kV
Overvoltage ratio	7.54	8.24	1.29	4.02
Max. depth of information	155 nm	165 nm	80 nm	170 nm
Standard	pure Cu	pure Ni	pure Cr	pure Si

When sputtering continuously, the ion gun is kept switched on during the whole measurement and x-ray intensities are recorded during sputtering. This prevents contamination at the surface almost completely. The drawback of this method is that the depth resolution is limited by the time, t, that is needed to acquire the x-ray intensities of all elements of interest. During this time a layer of thickness dz is sputtered and therefore the depth resolution is limited by dz. To obtain a good depth resolution, t should be as small as possible although this counteracts the requirement for good counting statistics, especially in EPMA-SDP [7].

The method of preference depends on the problem to be analysed. Generally, for materials tending to contaminate quickly, e.g. copper, continuous sputtering is the appropriate choice. On the other hand, if small concentrations are to be analysed and a good depth resolution is required, the alternating mode is the only way to achieve the required statistical confidence in the intensity values. If the sample is known to be laterally homogeneous, the crater edge line scan technique should be applied to prevent errors caused by contamination.

It is highly desirable for future generations of microprobes to be operated in the UHV pressure regime since this would dramatically reduce contamination problems, and be useful, not only for EPMA-SDP measurements, but also for light element analysis.

Measurement at a Cu/NiCr Multilayer

A multilayer structure, Cu/NiCr/Cu/NiCr/.../Si, of pure copper layers alternating with layers of a nickel-chromium alloy (nominal composition: 94.0 at % Ni, 6.0 at % Cr) was chosen to investigate the capabilities of EPMA-SDP. The sample consisted of ten layers with a nominal thickness of 100 nm each. The layers were deposited by physical vapour deposition, PVD, on a silicon wafer.

The EPMA-SDP measurement was performed using the instrument described above operating under the following conditions:

The settings for x-ray microanalysis are as given in Table 1. The electron-beam current was 100 nA and the beam was defocused to 20 µm. The primary energy of the Ar^+-ions was 2.5 keV. Scanning of the ion beam was disabled. Continuous sputtering was used to prevent contamination of the sample. The counting time was 1 s and, to reduce the time of measurement, no background correction was applied. The total acquisition time for all elements of interest amounted to 11 s. A break of 5 seconds was taken between two x-ray measurements, therefore a complete set of k-ratios was acquired every 16 seconds. During this time a layer of approximately 3 nm was removed limiting the depth resolution of each measurement to this value. Approximately 1.5 hours are required to complete the measurements.

To validate the EPMA-SDP measurements and to improve the data by correct background subtraction an additional crater edge line scan measurement was performed in a Camebax SX50 microanalyser. The length of the line scan was 800 µm and multilayer x-ray intensities were acquired in steps of 1 µm. The experimental setup for this measurement was as for EPMA-SDP except that now a background correction was applied, and use was made of a focused electron beam and a higher counting time of 10s.

Results and Discussion

Figure 4 shows the non-background corrected k-ratio data obtained by the EPMA-SDP measurement. The validity of these EPMA sputter depth profiles is confirmed by comparison with the crater-line scan data which are presented in Fig. 5. The depth scale in Fig. 5 is recalibrated by using the positions of the peaks and valleys in the individual curves as fixed points. These points correspond to interfaces as will be explained later. Taking the values for the thickness of each layer from a thin film evaluation (also described later) leads finally to the depth scale of Fig. 5. In the case of Ni- and Cu Lα there are only minor differences between the data shown in Fig. 4

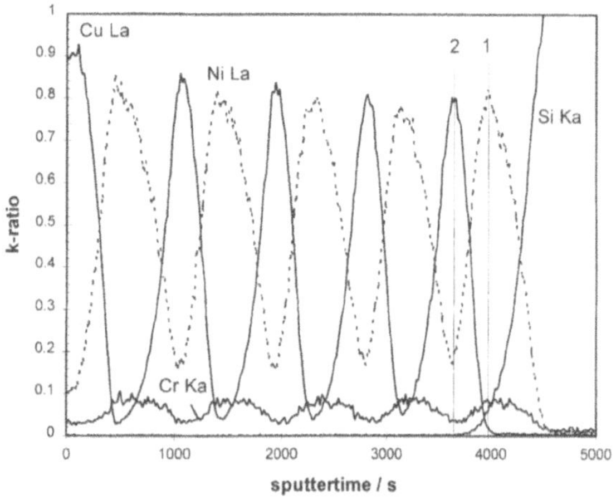

Fig. 4. Results of EPMA-SDP measurement, Cu/NiCr/.../Si multilayer, 7 keV

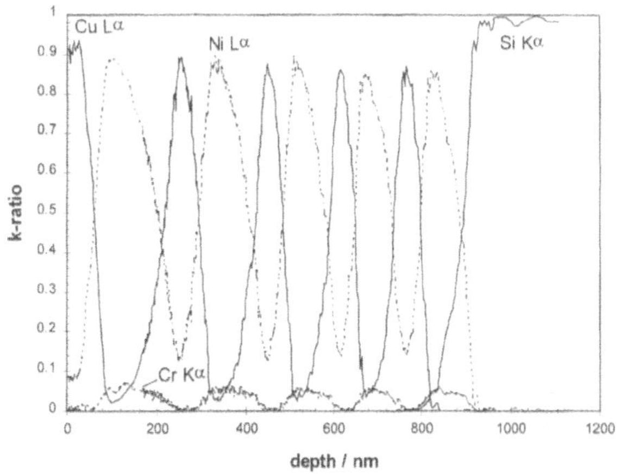

Fig. 5. Results of crater edge linescan measurement, Cu/NiCr/.../Si multilayer, 7 keV

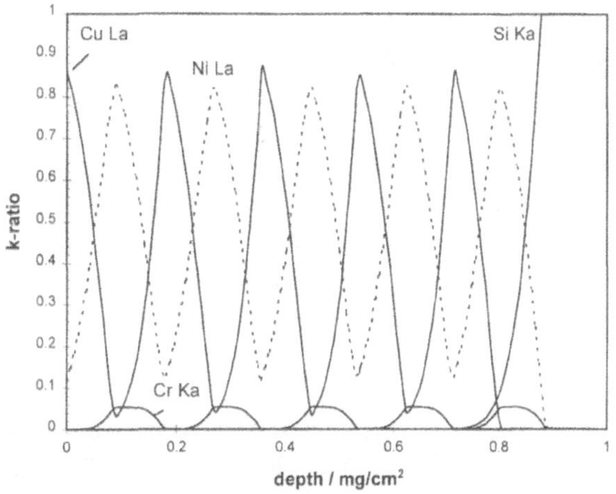

Fig. 6. Simulation of EPMA-SDP measurement, Cu/NiCr/.../Si multilayer, 7 keV

and Fig. 5 which can be explained by considering that the measurements have been carried out in different microprobes using different standard samples. The k-ratio data from the line scan measurement is believed to be more reliable because a background correction is applied and the counting statistics are better. This is shown particularly in the Cr Kα-curves by the decrease to zero in the region of the Cu-films and is to be expected because of the rather small ionisation depth for Cr Kα.

To understand the shape of the profiles we need to analyse the basic equation of EPMA-SDP:

$$I_A(z') = \text{const} \cdot \int_{z'}^{\infty} dz \cdot \Phi_A(z - z') \cdot \exp(-\chi^{\bar{A}}(z, z')) \cdot c_A(z) \qquad (1)$$

The intensity I_A that is measured after a layer of thickness z' is removed from the surface can be calculated by integrating over the ionisation depth distribution Φ_A times the chemical concentration c_A. X-ray absorption is accounted for by the exponential term. According to this equation EPMA sputter intensity profiles can be considered as the convolution of the chemical concentration depth profile with the absorption corrected ionisation depth distribution [7]. With Eq. (1) we are able to calculate the EPMA-sputter intensity profiles to be measured for a given sample composition at a given primary energy. For the Cu/NiCr.../Si multilayer this calculation has been performed by a modified Monte Carlo code [7, 12]. As long as the structure of the sample is known roughly, this code enables the EPMA-SDP measurement to be simulated before it is actually carried out and thus the best parameters for measurement may be determined. X-ray fluorescence effects have not yet been considered.

Figure 6 shows the results of the simulation for a 7 keV measurement. The profile for Cr Kα is flat in some regions and is rather similar to the real chemical composition profile of Cr whereas the profiles for Ni Lα and Cu Lα deviate strongly from real profiles. This difference in shape can be explained by studying the absorption corrected ionisation depth distributions for the elements (Fig. 7). These depth distributions represent the volume of information with depth which, in the experiment, is shifted step by step to larger depths in the sample. The shape of the absorption

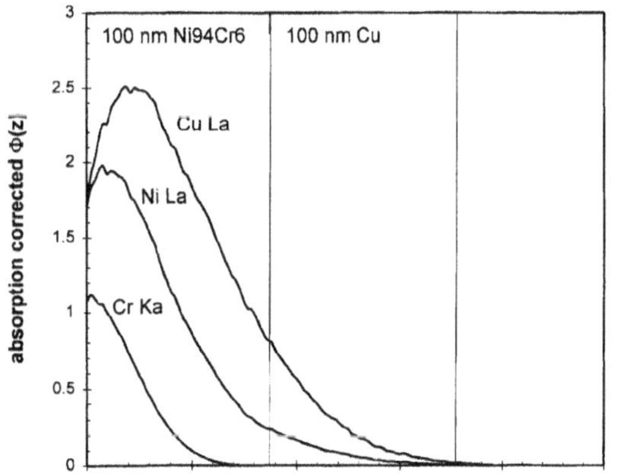

Fig. 7. Ionisation depth distribution calculated by MC-simulation (absorption corrected) for Cr-Kα, Ni-Lα and Cu-Lα; 7 keV; 100 nm Ni94Cr6 layer on Cu

corrected $\Phi(z)$ governs the shape of the resulting intensity depth profiles and the depth resolution of the technique. For example the maximum depth of information for Cr Kα in the present case is only 80 nm which is smaller than the thickness of a Ni-Cr or Cu layer. When the volume of information is shifted through the sample during sputtering the intensity of an x-ray line should be nearly constant as long as the whole volume of information is within a single layer. This is the case for Cr Kα. The maximum depths of information for Ni and Cu are larger than the thickness of a single layer. Thus flat regions do not occur in the profiles of these elements. Instead we find rather sharp peaks and valleys in the curves as predicted by Fig. 6. If we imagine the absorption corrected $\Phi(z)$ distribution of Cu and Ni Lα shifting through the layer structure, the curves can be explained as follows (see also [7]). Peaks and valleys occur when an upper layer is removed completely, i.e. when the sputtered depth z' has reached an interface. The branches in between indicate decreasing intensity when the particular layer with the emitting element is being removed and increasing intensity when the layer of interest is buried and the layer above it is being sputtered. For details see [7].

As can be seen, the EPMA-SDP and the crater edge measurements are in good agreement with theoretical predictions. The predicted peaks in the Cu and Ni curves as well as the flat parts of the Cr curve can be identified easily.

Calculating the compositional depth profile by solving Eq. (1) is a difficult task because of the statistical errors in the measured data and nonlinearity caused by the absorption term. First attempts were made by Richter et al. [7] but currently the technique is not generally applicable. Therefore, in the present case, we used methods of 'conventional' thin film EPMA to reconstruct the structure of the multilayer from the k-ratio data of both measurements presented. The following assumptions are necessary: (1) The sample is composed of individual layers with sharp interfaces in between. (2) The chemical composition of each layer does not vary with depth. (3) The positions of the peaks and valleys correspond to the interfaces.

As a first step we identify the intensities corresponding to the interface between the 1st and 2nd layers at the bottom of the structure (point 1 in Fig. 4). From the k-ratios for all elements at this point we can derive thickness and composition of the first layer on the Si substrate. Now, we identify the next interface (point 2 in Fig. 4) and calculate thickness and composition of the second layer using the k-ratios at point 2 and the knowledge about layer 1. We can continue this procedure until the composition and mass coverage of all layers are calculated.

The results of these reconstructions are shown in Table 2. The two measurements agree well in case of the Ni-Cr layers. The results for the thickness of the Cu layers obtained by the crater edge line scan are systematically higher by about 8 nm on average which indicates a systematic error in one of the two measurements. For the crater edge line scan the deviation in chemical composition from the nominal values is below 0.8%, and for the SDP measurement below 1.4%. The deviations from nominal thickness are below 16 nm, which, because of the agreement of the two independent measurements, can be attributed

Table 2. Comparison of results from EPMA-SDP and EPMA crater line scanning

Nominal values		EPMA-SDP		Crater edge linescan	
Thickness [nm]	Composition atomic%	Thickness [nm]	Composition atomic%	Thickness [nm]	Composition atomic%
100	Cu100	101.2	Cu98.8Cr1.2	110.0	Cu100
100	Ni94Cr6	102.2	Ni93.4Cr6.6	104.6	Ni93.5Cr6.5
100	Cu100	87.0	Cu98.7Cr1.3	96.0	Cu100
100	Ni94Cr6	95.1	Ni92.7Cr7.3	103.3	Ni94.8Cr5.2
100	Cu100	88.9	Cu99.9Cr0.1	95.3	Cu100
100	Ni94Cr6	95.9	Ni93.9Cr6.1	96.9	Ni94.3Cr5.7
100	Cu100	83.4	Cu99.9Cr0.1	94.1	Cu100
100	Ni94Cr6	89.2	Ni92.6Cr7.4	87.6	Ni94.3Cr5.7
100	Cu100	85.3	Cu99.9Cr0.1	93.3	Cu100
100	Ni94Cr6	107.0	Ni92.9Cr7.1	108.2	Ni94.4Cr5.6
$\Sigma = 1000$		$\Sigma = 935.6$		$\Sigma = 989.3$	

partly to deviations in the coating process. The EPMA-SDP measurement shows approximately 1% Cr in the copper layers. In Fig. 4 we find that the Cr signal is not zero at the maxima of the Cu signal. This is due to the missing background correction and does not occur in the line scan measurement, where a background correction has been applied to the data.

In principle the Ni Lα lines can be excited by Cu Lα radiation. Under the present measuring conditions this effect is expected to be pronounced when measuring on Cu as top layer because the Cu Lα radiation can excite Ni in the underlying Ni-Cr layer. Indeed the Ni Lα intensities of the valleys deviate slightly from the corresponding data in the simulated sputter intensity profile, where fluorescence effects were neglected. Since we also failed to take continuum fluorescence into account which, particularly for the standard, can contribute in the order of 2%, the effect of characteristic fluorescence seems to be partly compensated in the experimental k-ratios.

Conclusions

For the first time wavelength dispersive EPMA has been combined with in-situ sputtering with the aim of determining depth profiles from x-ray intensity measurements. The new technique of EPMA-SDP has been applied successfully to multilayer samples which normally are outside the scope of conventional EPMA techniques. It could be proven, that in the examples presented the measured intensity depth profiles were in good agreement with theoretical predictions obtained from Monte Carlo simulations. The measured data have been further evaluated by methods of EPMA thin film analysis. This at least provides information on composition and thickness of individuals layers. Nevertheless, this method does not reveal the full information content of the data. For further detailed evaluations special deconvolution techniques have to be developed in the future. Work in this field is already being done [7]. The addition of sputtering to EPMA has produced a new technique for quantitative depth profiling in the nm to μm regime. The outstanding feature of EPMA-SDP is the calibration of the depth scale directly from the measured data. This, together with the good quantification of chemical composition, makes EPMA-SDP a valuable supplement to other depth profiling techniques like AES or SIMS.

In future work the limits of the new technique require further assessment. Depth resolution and limits of detection will be determined using well defined samples which have already been analysed by other comparative techniques. Existing simulation codes must also be extended to include x-ray fluorescence effects.

Acknowledgements. The present work was supported by the Deutsche Forschungsgemeinschaft (DFG project KA843/2). The Cu/NiCr/.../Si multilayer sample was kindly supplied by the Institut für Festkörper- und Werkstofforschung, Dresden, Germany.

References

[1] S. Baumgartl, P. L. Ryder, H. E. Bühler, *Z. Metallk* **1973**, *64*, p. 655.
[2] J. D. Brown, in: *Microbeam Analysis* (D. E. Newbury, ed.), San Francisco Press, San Francisco, 1988, pp. 271–272.
[3] J. L. Pouchou, F. Pichoir, *La Recherche Aerospatiale* **1984**, 5, 47.
[4] R. A. Waldo, in: *Microbeam Analysis* (D. E. Newbury, ed.), San Francisco Press, San Francisco, 1988, pp. 310–314.
[5] G. F. Bastin, H. J. M. Heijligers, J. M. Dijkstra, in: *Microbeam Analysis* (J. R. Michael, P. Ingram, eds.), San Francisco Press, San Francisco, 1990, pp. 159–160.
[6] J. L. Pouchou, F. Pichoir, *Scanning Microsc.* **1993**, *7*, 167.
[7] S. Richter, N. Lesch, P. Karduck, *Mikrochim. Acta [Suppl.]* **1998**, *15*, 125.
[8] A. Benninghoven, F. G. Rüdenauer, H. W. Werner, *Secondary Ion Mass Spectrometry*, J. Wiley, New York, 1987, p. 824.
[9] J. Kirschner, H. W. Etzkorn, *Proc. 7th Intern. Conf. Solid Surfaces*, (R. Dobrozemsky, F. Rüdenauer, F. P. Viehböck, A. Breth, eds.), Vienna, 1977, pp. 2213.
[10] J. Fine, B. Navinsek, F. Davarya, T. D. Andreadis, *J. Vac. Sci. Technol.* **1982**, *20(3)*, 449.
[11] P. Karduck, A. von Richthofen, *Microsc. Mocroanal. Microstruct.* **1995**, *6*, 421.
[12] N. Ammann, P. Karduck, *Microbeam Analysis 1990* (J. R. Michael, P. Ingram, eds.), San Francisco Press, San Francisco, 1990, p. 150.
[13] P. Karduck, *Mikrochim. Acta [Suppl.]* **1998**, *15*, 109.
[14] H. Gries, W. Koschig, *Surf. Interface Anal.* **1990**, *16*, 321.
[15] N. Ammann, P. Karduck, *Surf. Interface Anal.* **1994**, *22*, 54.

Mikrochim. Acta [Suppl.] 15, 141–147 (1998)

Quantitative Analysis of BN (C, O, Ar, H)-Coatings Using EPMA and SIMS

Peter Willich* and **Ulrike Wischmann**

Fraunhofer Institut für Schicht- und Oberflächentechnik, D-38108 Braunschweig Federal Republic of Germany

Abstract. Electron probe microanalysis (EPMA) and secondary ion mass spectrometry (SIMS) are applied to the chemical characterization of boron nitride coatings with additional concentrations (1–20 at %) of H, C, O and Ar. Quantitative analysis of homogeneous films with a film thickness >0.3 μm is carried out by conventional "bulk" EPMA with an accuracy of about 5 % relative. A similar degree of accuracy is obtained for the simultaneous determination of film thickness (0.1–0.5 μm) and composition ("thin film" EPMA). SIMS based on monitoring MCs^+ molecular secondary ions is demonstrated as a useful scheme for quantitative in-depth analysis of matrix elements. In the case of BN (H, C, O, Ar) materials a single standard (e.g., defined by EPMA) is sufficient to establish a set of matrix-independent relative sensitivity factors. The accuracy of quantitative analysis by MCs^+-SIMS proves to be 10–20 % relative without being influenced by matrix effects arrising from variable (1–20 at %) oxygen concentrations.

Key words: EPMA, SIMS, BN coatings, ultralight elements, depth profiling.

Cubic boron nitride (cBN) shows a wide range of interesting properties with regard to hardness, chemical inertness and optical transparency. The most important industrial application is the deposition of cBN coatings on cutting tools for steel parts. Thin films with a high content of the cubic phase can be obtained by reactive r.f. sputtering using targets of hexagonal boron nitride (h-BN) or boron carbide (B_4C) [1, 2]. The use of different target materials and reactive sputtering in gas mixtures of N_2-Ar with impurities of O_2 and H_2 leads to complex compositions (B-N-C-O-Ar-H) of the coating materials. Accurate quantitative analysis is desirable to understand the influence of deposition parameters with respect to the formation of the cubic phase and the frequent problem of insufficient adhesion of the films on certain substrate materials (Si, steel).

Wavelength-dispersive electron probe microanalysis (EPMA) proved to be a reliable technique to study the composition of coatings containing ultralight elements [3]. The application to compositions of B-N-C-O may be regarded as an interesting example to test the accuracy of EPMA in a rather difficult analytical situation. This concerns quantitative analysis of sufficiently thick films (> 0.3 μm) in the mode of conventional "bulk" analysis and, in the case of thin films (< 0.3 μm) the simultaneous determination of film (mass) thickness and composition ("thin film" EPMA) [4, 5].

Although EPMA may be regarded as a useful technique to carry out accurate quantitative analysis, the chemical characterization of coatings, for which a variation of the composition as a function of depth can not be excluded, requires the application of a depth-profiling technique. Secondary Ion Mass Spectrometry (SIMS) may be a suitable choice with respect to sensitivity and is also able to detect hydrogen. But SIMS is well-known for severe problems in respect of quantification (matrix effects). However, several recent studies [6] have established that by employing Cs^+ primary ions and monitoring MCs^+ molecular secondary ions (here M stands for an element of the target material) a new route is opened leading towards more accurate quantitative analysis. A set of relative

* To whom correspondence should be addressed

sensitivity factors may be derived from SIMS-analysis of a coating material (=standard) defined by EPMA. The results of EPMA are also useful as reference data to estimate the accuracy of quantitative MCs^+-SIMS when applied to a wide range of "unknown" B-N-C-O-Ar coating materials.

Experimental

WDS-EPMA was carried out in a CAMECA SX50 microprobe using a multilayer monochromator of Ni/C (2d ~ 9.6 nm) for the x-rays of B and C and a multilayer monochromator of W/Si (2d ~ 6 nm) for the x-rays of N and O. The k-ratios of B and C were determined from integral recordings of the B-Kα and C-Kα peaks, emitted by standards (boron, vitreous carbon) and specimens. The elements N, O and Ar were calibrated by standards of Cr_2N, conductive $Y_3Fe_5O_{12}$ and Ar-Si (5 wt% Ar) [7].

SIMS-experiments were carried out in a CAMECA IMS 5f ion microscope. A Cs^+ primary ion beam of 5.5 keV impact energy was employed. The beam current was in the range of 50–100 nA and the beam was scanned across an area of 250×250 μm. Positive MCs^+ secondary ions were accepted from a circular area of about 60 μm in diameter. The mass resolution was in the order of 600 and the pressure in the target chamber amounted to $6–9\times10^{-8}$ Pa with the Cs-beam in operation.

Results and Discussion

EPMA: Quantitative Analysis of "Thick" Films

From a large collection of B-N-C-O coatings deposited on Si a set of specimens with a minimum film thickness of ~1 μm was selected. The samples, designated as Layer 1 to Layer 4 (Table 1) were investigated by SIMS to prove the homogeneity as a function of depth (Fig. 1). The set of test specimens was completed by binary compositions of B_4C, cBN (individual crystals with a size of ~250 μm) and hexagonal BN (hBN film prepared by CVD). EPMA was performed at electron energies of 5, 10 and 15 keV. The corresponding values for the depth of analysis (maximum depth of x-ray emission) are about 0.3, 0.7 and 1 μm. The model of Pouchou & Pichoir [8] was used to calculate the concentrations from experimental k-ratios.

The values for boron show a slight but systematic decrease of the concentration with increasing electron energy. The maximum effect, in the order of 5%

Table 1. EPMA of binary bulk materials and BN (C, O, Ar) coatings (film thickness ~1 μm) at different electron energies

Sample	E_0 [keV]	B [wt%]	C [wt%]	N [wt%]	O [wt%]	Ar [wt%]	Total [wt%]
B_4C(bulk)	5	78.8	21.4		0.6		100.8
B:78.3, C:21.7[a]	10	78.4	21.2		0.5		100.1
	15	77.7	21.4		0.5		99.3
cubic-BN (bulk)	5	44.1	0.2	56.4	0.5		101.2
B:43.5, N:56.5[a]	10	43.6	0.3	56.4	0.5		100.8
	15	42.8	0.2	56.1	0.6		99.7
hexagonal-BN	5	43.4	1.3[b]	56.9	1.8		103.4
CVD-coating	10	41.9	1.5[b]	56.7	1.5		101.6
B:43, N:56.5[a]	15	41.2	1.6[b]	57.2	1.7		101.7
Layer 1	5	46.2	18.6	21.9	1.3	14.3	102.3
	10	45.6	18.7	21.4	1.2	14.5	101.4
	15	44.6	18.7	21.4	1.2	14.2	100.1
Layer 2	5	37.2	21.8	12.9	20.2	6.1	98.2
	10	36.9	22.1	12.8	20.7	6.2	98.7
	15	36.8	21.7	13.1	20.8	6.3	98.7
Layer 3	5	34.7	17.4	36.9	3.9	8.2	101.1
	10	34.5	17.7	36.9	3.6	8.4	101.1
	15	33.8	17.4	36.2	3.6	8.4	99.4
Layer 4	5	31.4	14.6	34.3	13.3	4.8	98.4
	10	31.3	14.6	34.8	13.3	4.8	98.8
	15	30.4	14.8	34.5	12.9	4.6	97.2

[a] Nominal composition [wt%]. [b] Carbon coating on surface.

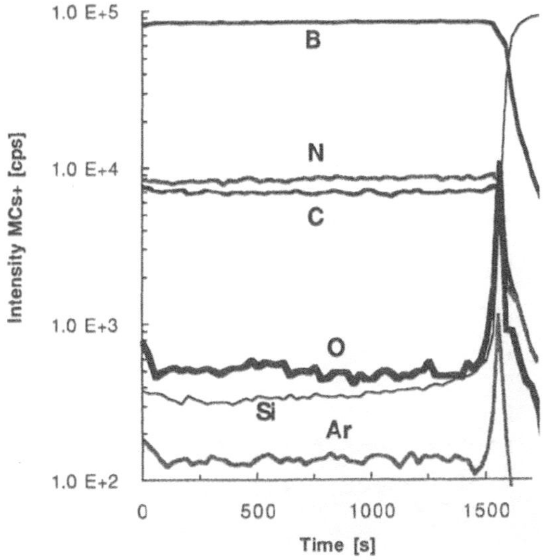

Fig. 1. SIMS depth profile of homogeneous BN (C, O, Ar) coating (Layer 3 of Table 1). Cs^+ primary ions and MCs^+ (M = B, C, N, O, Si, Ar) molecular secondary ions

relative, is visible for the sample of hBN. Although the surface was carbon-coated, the influence of electrical charging cannot be excluded. For all samples in Table 1 the total of all elements is in the range of 97–102 wt %. The concentrations of the binary materials are in good agreement with the corresponding nominal composition, deviations do not exceed 5 % relative. This degree of accuracy is in agreement with investigations of materials composed of several ultralight elements in combination with metals [3].

EPMA: Determination of Film Thickness and Composition

During the development of B-N-C-O coating materials a frequent requirement is to investigate very thin films, typically the film thickness is in the order of 0.05–0.3 µm. EPMA was successfully applied to the combined determination of film (mass) thickness and composition [4, 5]; however, the particular situation of ultralight elements has never been considered. Table 2 shows results for "thin films" (film thickness $\sim$0.2 µm) prepared from the "thick films" (film thickness $\sim$1 µm) of Table 1 by sputtering with the ion beam of the SIMS instrument. Experiments were carried out at $E_0 = 15$ keV and data reduction was based on the STRATA thin film software proposed by Pouchou & Pichoir [9]. The comparison of "bulk" analysis and "thin film" analysis shows, with the exception of oxygen, a reasonable agreement in the concentrations (Table 2). When assuming an average density of 2.5 g/cm^3 for the B-N-C-O materials, the geometrical film thicknesses derived from the mass thicknesses of EPMA agree with the data provided by a surface profile measuring system (Dektak) to within 10–15 %.

In general, the results of "thin film" analysis show, as compared with "bulk" analysis, higher values for the concentration of O, particularly visible for the samples Layer 1 and Layer 3 (Table 2) with absolute O-concentrations <5 wt %. This effect is more closely studied (Fig. 2) by the investigation of Layer 3 after "thinning" (SIMS) the original film thickness of

Table 2. EPMA of homogeneous BN (C, O, Ar) coatings. "Bulk analysis" (film thickness $\sim$1 µm) at $E_0 = 10$ keV and "thin film analysis" (film thickness $\sim$0.2 µm after sputtering) at $E_0 = 15$ keV

Sample	EPMA	Film thickness Dektak[a] [µm]	Film thickness EPMA[b] [µm]	B [wt %]	C [wt %]	N [wt %]	O [wt %]	Ar [wt %]
Layer 1	bulk	1.05		44.9	18.4	21.1	1.3	14.1
	thin film	0.19	0.21	43.8	18.7	22.1	2.9	14.5
Layer 2	bulk	0.98		37.4	22.4	12.9	20.9	6.3
	thin film	0.22	0.19	39.4	21.1	12.4	21.3	5.8
Layer 3	bulk	1.12		34.1	17.5	36.5	3.6	8.3
	thin film	0.21	0.21	33.1	17.3	35.5	6.2	7.9
Layer 4	bulk	1.05		31.7	14.8	35.2	13.5	4.9
	thin film	0.17	0.19	32.2	15.4	33.7	14.5	4.2

[a] Dektak surface profile measuring system.
[b] Calculated from mass thickness assuming density of 2.5 g/cm^3.

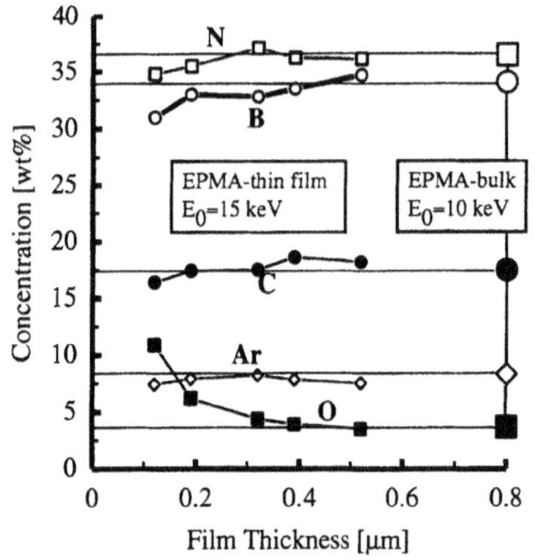

Fig. 2. "Thin film" EPMA ($E_0 = 15$ keV) of homogeneous BN (C, O, Ar) coating (Layer 3 of Table 1) at different levels (0.12–0.52 μm) of film thickness. "Bulk" EPMA at $E_0 = 10$ keV for film thickness of 0.8 μm

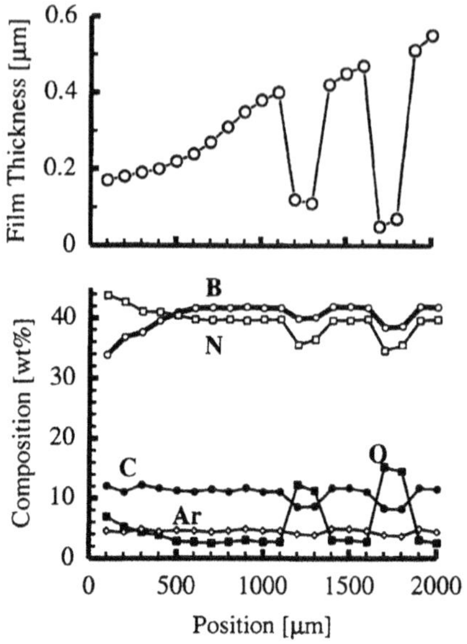

Fig. 3. Linescan in combination with "thin film" EPMA. Monitoring of local variations in respect of film thickness and composition

1.1 μm to levels of 0.52, 0.39, 0.21 and 0.12 μm as determined by mechanical surface profiling. The corresponding values derived from the mass thicknesses determined by EPMA are 0.5, 0.42, 0.19 and 0.1 μm. As it is shown in Fig. 2, the values of the O-concentration obtained by EPMA systematically increase from 3.5 wt% (film thickness ~0.5 μm) to 10.5 wt% (film thickness ~0.1 μm), although the SIMS depth profile (Fig. 1) gives a rather uniform O-distribution as a function of depth. On the other hand, the SIMS-profile of Layer 3 (Fig. 1) shows a local enhancement of O at the interface of the Si-substrate. In the case of EPMA applied to very thin films, the presence of Si-oxide at the interface obviously causes a significant contribution to the x-ray intensity of O, particularly disturbing in situations where the "intrinsic" O-concentration of the coating material is low. It can be concluded that "thin film" EPMA is rather sensitive to local effects, e.g., O-enhancement, at the substrate-film interface.

A useful practical application of "thin film" EPMA is the investigation of lateral homogeneity in respect of film thickness and composition (Fig. 3). The linescan across a coated area of 2 mm in diameter shows a continuous variation of film thickness and, at the left-handed periphery, a significantly enhanced concentration of N. In the region with a film thickness above 0.4 μm local delamination defects can be detected. Obviously, the delamination does not occur immediately at the substrate (Si) – film interface, a "residual film" (film thickness, of 0.05–0.1 μm) is still present. The O-concentration is clearly related to the local film thickness, due to the contribution of Si-oxide at the substrate-film interface.

SIMS: Quantitative In-Depth Analysis

Frequently, in the case of B-N-C-O coatings, the composition as a function of depth has to be studied to verify general homogeneity, to study in detail effects at the surface and the substrate-film interface, and to characterize multilayer structures. The characteristics of SIMS are its excellent sensitivity, good depth resolution (1–5 nm) and the ability to detect hydrogen. Although the positive aspects of quantification based on MCs^+-SIMS are generally recognized, the investigation of B-N-C-O coatings offers the unique opportunity to test the accuracy of quantitative analysis in situations where large variations of the matrix composition occur.

Table 3 summarizes the results of SIMS-quantification applied to the different B-N-C-O materials. The samples correspond with those of Table 1, the concentrations of EPMA are used as reference data. The sample of cBN was not studied by SIMS as handling small individual crystals proved to be difficult. Layer 3 was selected as a "standard" (Table 3) to establish a set of relative sensitivity factors

Table 3. Quantitative SIMS of homogeneous BN (C, O, Ar) materials. SIMS based on MCs^+ molecular secondary ions. Relative sensitivity factors (RSF) as defined by Eq. (1) (see text)

Sample		B [at %]	C [at %]	N [at %]	O [at %]	Ar [at %]	Si [at %]
Layer 3	EPMA	41.3	19.1	33.9	3.1	2.6	0.12
standard (RSF)	SIMS	(5.1)	(1)	(0.65)	(0.85)	(0.12)	(11.5)
B_4C	EPMA	80.1	19.5		0.5		
	SIMS	77.8	21.8	0.1	0.3		
hexagonal-BN	EPMA	47.6	1.5	49.6	1.3		
	SIMS	47.9	1.1	49.5	1.5		
Layer 1	EPMA	54.4	20.1	19.9	1.1	4.5	0.7
	SIMS	52.1	20.8	21.1	0.7	5.3	0.9
Layer 2	EPMA	44.9	24.1	12.2	16.9	1.9	1.1
	SIMS	46.2	21.9	11.8	17.4	2.7	0.8
Layer 4	EPMA	38.3	16.3	32.8	11.1	1.6	0.13
	SIMS	38.2	16.9	33.2	10.2	1.5	0.07

(RSF's) defined as

$$RSF_{M/C} = [(I_{MCs^+})/(I_{CCs^+})]/[c_M/c_C] \qquad (1)$$

where I_{MCs^+} is the intensity (MCs^+ molecular secondary ions) of the element M (M = B, N, O, Ar or Si) with the concentration c_M (at %). I_{CCs^+} is the intensity of the reference element (carbon) with the concentration c_C. Provided that the sensitivity factors can be regarded as matrix-independent, the use of a single and arbitrary standard (e.g., Layer 3 defined by EPMA) is sufficient. Once established the set of RSF's enables one to calculate the concentrations of "unknown" specimens according to Eq. (1) on the basis of experimental MCs^+/CCs^+ intensity ratios, with the additional assumption that the total of all elements is 100 at %.

Comparing the results of SIMS to the reference data of EPMA (Table 3) leads to the conclusion that the accuracy of SIMS quantification is at least 15 % relative. This is a rather positive result in view of the significantly different matrix compositions. Although the O-content of the samples ranges from 0.5–17 at %, no particular oxygen-related matrix effects are visible. For the samples Layer 1 and Layer 2 (Table 3) SIMS-analysis results in Ar-concentrations significantly higher as compared with the concentrations established by EPMA. An explanation may be the relatively high concentration of Si (about 1 at %) in these samples. This obviously causes the formation of $SiCCs^+$ molecular ions, unfortunately with the same mass number (173) as $ArCs^+$. For the sample Layer 4 the Si-concentration is low (about 0.1 at %), which enables an accurate determination of the Ar-concentration. The strong influence of mass interference is also visible in Fig. 1, where a sharp increase of the $ArCs^+$-intensity is visible at the interface to the Si-substrate. This artefact is caused by the formation $SiCCs^+$ molecular ions as a consequence of interface mixing by ion bombardment.

Figure 4 shows an example of a quantitative depth profile through a double layer system with a total film

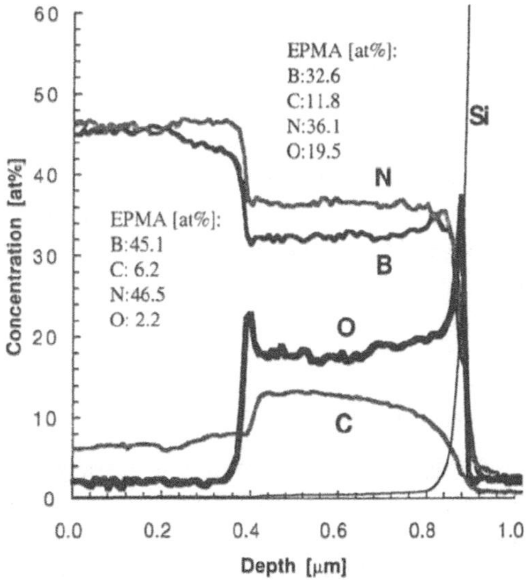

Fig. 4. Quantitative MCs^+-SIMS depth profile proving matrix-independance of relative sensitivity factors as given in Table 3

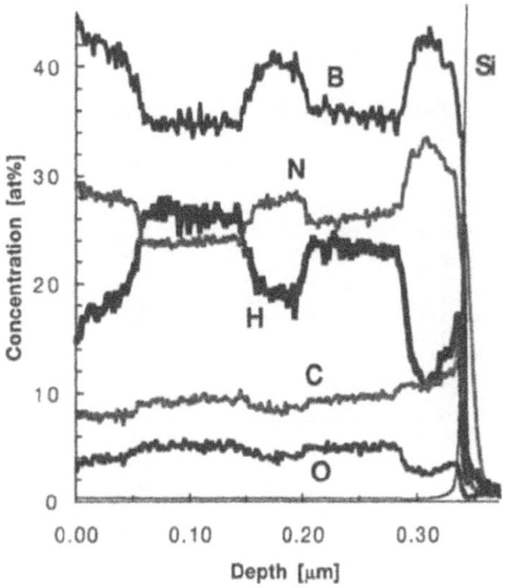

Fig. 5. Quantitative MCs^+-SIMS depth profile including hydrogen. Variation of H-content induced by temporary changes of H_2 partial pressure during reactive sputtering deposition process

thickness of 0.9 µm (determination of the crater depth using a surface profile measuring system). The depth scale represents a rough estimation, small differences between the sputtering rates of the adjacent materials are ignored. SIMS quantification was based on the RSF's as given in Table 3. The compositions of the surface layer and the buried layer were also determined by EPMA, in the case of the buried layer after removal (sputtering by SIMS) of the surface layer. Although the layers are not exactly homogeneous, a sufficient degree of accuracy for the quantification of SIMS based on MCs^+ secondary ions can be concluded. No significant matrix-effects can be detected as a consequence of different O-concentrations. Quantitative analysis of O, at least in the range up to 20 at%, causes no problems, although the matrix-dependance of the OCs^+ secondary ion yield is frequently discussed in the literature [6].

SIMS: Quantitative Analysis of Hydrogen

B-N-C-O coating materials are able to incorporate H from impurities in the reactive sputter gas mainly consisting of Ar-N_2. After it was realized that the hydrogen concentration influences certain properties of the coatings (formation of cBN, stability in respect of delamination), the determination of H was regarded as an important aspect of chemical characterization.

The procedure of MCs^+-SIMS can be applied to the quantitative determination of H, as was recently demonstrated by examples of amorphous-C:H and metal-C:H coating materials with H-concentrations in the range of 0.5 to 30 at% [10]. Elastic Recoil Detection (ERD) was used as a reference technique to prove the matrix-independence of the sensitivity factor for H relative to C. Typical values for $RSF_{H/C}$ are in the range of 0.65–0.8, which indicates a very similar secondary ion yield for CsH^+ in comparison with the secondary yields of CCs^+, CsN^+ and CsO^+ (Table 3).

The example in Fig. 5 represents a particular study to correlate the concentration of H in the coating material with the H-content of the sputter gas atmosphere and other parameters of the deposition process. The detection limit for H using monitoring of CsH^+ secondary ions is in the order of 0.05 at%, as estimated from the "background concentration" on the substrate of pure Si (Fig. 5). The detection limit for H by use of CsH^+ species is rather moderate when compared to the values (< 1 ppm) archievable by monitoring H^- secondary ions. However, the advantage of the detection scheme employing HCs^+ species is the accurate quantitative evaluation of high concentration levels.

Conclusions

The application of EPMA and SIMS enables a complete and quantitative chemical characterization of boron nitride coatings with additional concentrations (1–20 At%) of H, C, O and Ar. It is demonstrated that EPMA is a highly accurate technique to carry out quantitative analysis of ultralight elements even in the case of rather complex compositions. The minimum film thickness for applying "bulk" EPMA is about 0.3 µm, which corresponds to the maximum depth of x-ray emission for an electron energy of 5 keV. The technique of "thin film" EPMA (combined determination of film (mass) thickness and composition) is regarded as a suitable scheme to investigate coatings in the film thickness range of 0.05–0.5 µm. The determination of oxygen may be critical due to the influence of a possible oxide-layer at the substrate-film interface.

However, the application of EPMA is restricted to (in-depth) homogeneous coating materials, whereas

SIMS can be more generally employed to study the composition as a function of depth, including the determination of hydrogen. The use of Cs^+ primary ions in conjunction with MCs^+ secondary molecular ions opens a route towards rather straightforward quantification. In the case of boron nitride coatings, a single B-N-C-O-Ar reference material, the composition of which can be defined by EPMA, enables one to establish a set of matrix-independent relative sensitivity factors which can be applied to the determination of "unknown" compositions. The investigation of boron nitride materials confirms that quantitative in-depth analysis by SIMS can be carried out with an accuracy of 15 % relative or better. In general, EPMA proves to be a useful "quantitative" reference technique to support SIMS quantification procedures.

References

[1] A. Schütze, K. Bewilogua, H. Lüthje, S. Kouptsidis, S. Jäger, *Surf. Coat. Technol.* **1995**, *74–75*, 717.

[2] A. Schütze, K. Bewilogua, H. Lüthje, S. Kouptsidis, *Diamond Rel. Mater.* **1996**, *5*, 1130.

[3] P. Willich, *Microbeam Anal.* **1993**, *2*, 45.

[4] J. L. Pouchou, F. Pichoir, *Scanning* **1990**, *12*, 212.

[5] P. Willich, *Mikrochim. Acta [Suppl.]* **1992**, *12*, 1.

[6] See, e.g., A. Benninghoven et al. (eds.), *Secondary Ion Mass Spectrometry SIMS IX* Wiley, Chichester, 1994, pp. 377–425.

[7] P. Willich, D. Obertop, *X-ray Spectrom.* **1982**, *11*, 32.

[8] J. L. Pouchou, F. Pichoir, *La Recherche Aérospatiale* **1984**, *3*, 14.

[9] J. L. Pouchou, F. Pichoir, *Scanning Microsc. [Suppl.]* **1993**, *7*, 167.

[10] P. Willich, *Secondary Ion Mass Spectrometry SIMS X*, (A. Benninghoven et al., eds.), Wiley, Chichester, 1997, pp. 609–612.

Mikrochim. Acta [Suppl.] 15, 149–154 (1998)

Quantitative EDS Analysis of $SiO_2/Al_2O_3/TiO_2$ Multilayer Films

David G. Rickerby[1,*], **Norbert Wächter**[1], and **Rudolf Reichelt**[2]

[1] Institute for Advanced Materials, European Commission, Joint Research Centre, Via Enrico Fermi 1, I-21020 Ispra (Varese), Italy
[2] Institut für Medizinische Physik und Biophysik, Universität Münster, Robert-Koch-Strasse 31, D-48149 Münster, Federal Republic of Germany

Abstract. The application of the $\Phi(\rho Z)$ method to analysis of multilayer oxide films is described. Owing to the presence of oxygen in all the layers and the other elements in more than one layer, it is only possible to fit a unique solution to experimental x-ray intensities if the compositions of the individual layers are known or can be determined independently. By incrementally stepping the accelerating voltage, to vary the electron range and x-ray generation depth, information on composition and mass thickness may be obtained at successively greater depths within the specimen. Layer thicknesses can be determined to a high degree of accuracy by this technique, though with decreasing precision with increasing distance below the surface. The influence of the deposition of a thin conductive Pt-C coating on the emerging x-ray intensities and the thickness calculation is discussed.

Key words: multilayers, thin oxide films, EDS analysis, phi (rho-Z), Pt-C coatings.

Multilayer dielectric thin films are extensively used as antireflection coatings and interference filters. It has been demonstrated [1] that any sequence of alternating high and low refractive index layers of optical thickness smaller than the light wavelength is equivalent to a gradient index layer. This general principle can be applied to the design of efficient broadband antireflection coatings. The optical and mechanical performance of these coatings are dependent to a certain extent on the structure and composition of the component layers [2].

Examination of multilayer coatings by transmission electron microscopy entails preparation of replica specimens from cleavage fracture surfaces [2, 3] or cross-sectioned specimens by embedding, sectioning and ion beam thinning [4]. Both routes are time consuming and destructive. An alternative, non-destructive, technique is the use of wavelength dispersive or energy dispersive x-ray spectroscopy. By incrementally stepping the accelerating voltage, layer composition and thickness can be determined at successively greater depths in the coating, using a quantification procedure based on a physically accurate model of the depth distribution of x-ray emission [5, 6]. The precision of this technique for measurement of surface and near surface layers of a few tens of nanometers thickness is comparable to that attainable by sputter depth profiling using Auger electron spectroscopy [7] or x-ray photoelectron spectroscopy [8].

The present paper will discuss the application of EDS to the quantitative analysis of $SiO_2/Al_2O_3/TiO_2$ multilayer films. Insertion of intermediate Al_2O_3 layers 10–40 nm in thickness has been shown to increase abrasion resistance compared to SiO_2/TiO_2 stacks, while maintaining practically unaltered the spectral transmittance characteristics in the 400–900 nm wavelength range [9].

Difficulties may be encountered in x-ray microanalysis of non-conductors because of surface charging, which causes a high energy cut-off in the spectra, and distortion of the depth distribution function due to the internal electrical field generated by trapped charge within the specimen [10]. Deposition of a conductive carbon or metal coating will

* To whom correspondence should be addressed

generally be successful in avoiding surface charging, although it will not eliminate entirely the problem of internal charge build up. A thick carbon or metal layer may furthermore affect the emitted x-ray intensities in an unacceptable or unpredictable manner. The effectiveness of an ultra-thin Pt-C surface coating in preventing charging and its influence on x-ray intensities and thickness measurements in the oxide layers has therefore been investigated.

Experimental

Physical vapour deposited $SiO_2/Al_2O_3/TiO_2$ multilayer films on silicon substrates have been studied. Cross-sectional TEM specimens were prepared by low angle (<4°) ion beam milling using 3–5 keV argon ions [11]. These were observed at 120 kV in a Philips CM12 transmission electron microscope equipped with a Gatan 694 slow scan CCD camera.

X-ray microanalysis was performed with a JEOL 6400F field emission scanning electron microscope, operated at accelerating voltages of 2.5–18 kV, using a Link Pentafet Si(Li) windowless detector at a 35° take off angle. The specimens were analysed in both the as-received condition and after coating with a fine grain $\sim$ 1.5 nm evaporated Pt-C film [12]. The thickness of the coating was determined by a quartz crystal monitor and by EDS analysis of the Pt-C film deposited directly on a silicon substrate. Energy dispersive x-ray spectra were acquired in TV scanning mode at a magnification of 100× to reduce the effects of specimen contamination. Compensation for beam current variation was applied by electronically adjusting the detector live time in inverse proportion to the condenser aperture current, as described previously [13]. Pure element Al, Si, Ti, Pt and C standards were used, while the oxygen concentration was calculated by stoichiometry. The individual oxide layers were verified to be close to the ideal stoichiometric compositions by analysing single layers of several hundred nanometres thickness deposited on silicon, with reference to bulk samples of the pure oxides.

Experimental data were quantified by applying the STRATA thin layer analysis program, based on the XPP model developed by Pouchou and Pichoir [14, 15]. The depth distribution function $\Phi(\rho Z)$ of x-ray generation has the form:

$$\Phi(\rho Z) = A\exp(-a\rho Z) + [B\rho Z + \Phi(0) - A]\exp(-b\rho Z)$$

where $\Phi(0)$ is the surface ionisation. An iterative procedure is employed to minimise the deviation between the experimentally determined x-ray intensities and theoretically calculated values. For the analysis of a seven layer stack, using 20–30 experimental data, iteration times of up to several hours were required when the program was run on a 486 PC.

Results and Discussion

Uncoated Specimens

A cross-sectional TEM micrograph of a typical multilayer structure on silicon is illustrated in Fig. 1. This consists of seven discrete layers deposited in the sequence: $SiO_2/Al_2O_3/TiO_2/Al_2O_3/TiO_2/Al_2O_3/SiO_2$. The SiO_2 layers and Al_2O_3 layers appear featureless and amorphous, as confirmed by electron diffraction, while the TiO_2 layers are nanocrystalline and have a columnar growth structure. There is limited evidence of interdiffusion, in line with previous studies of vacuum evaporated TiO_2/SiO_2 layers by high resolution electron microscopy and electron energy loss spectroscopy [4]. The boundaries between the layers are well defined and the interfacial roughness is of the order of a few nanometres.

Fig. 1. Transmission electron micrograph of a seven component multilayer on a silicon substrate. The layer sequence is $SiO_2/Al_2O_3/TiO_2/Al_2O_3/TiO_2/Al_2O_3/SiO_2$

A pre-condition for analysis by the $\Phi(\rho Z)$ methods is that no element should have an unknown concentration in more than one layer. Due to the presence of oxygen in all of the layers, and silicon, aluminium and titanium in two or more layers, it is essential to know the concentrations beforehand in order to perform the iterative computation. It is not necessary that the layers are stoichiometric, only that their compositions are known or can be determined with reasonable accuracy. Small departures from stoichiometry, less than 10%, will lead to changes of <0.01 in the relative x-ray intensities (k-ratios) for the other elements at all accelerating voltages, corresponding to errors of just a few percent in the calculated layer thicknesses.

Experimental data from this multilayer are compared in Fig. 2 with theoretical curves fitted using the iterative procedure. The Si $K\alpha$ and Ti $K\alpha$ data show a gradual deviation at higher accelerating voltages with

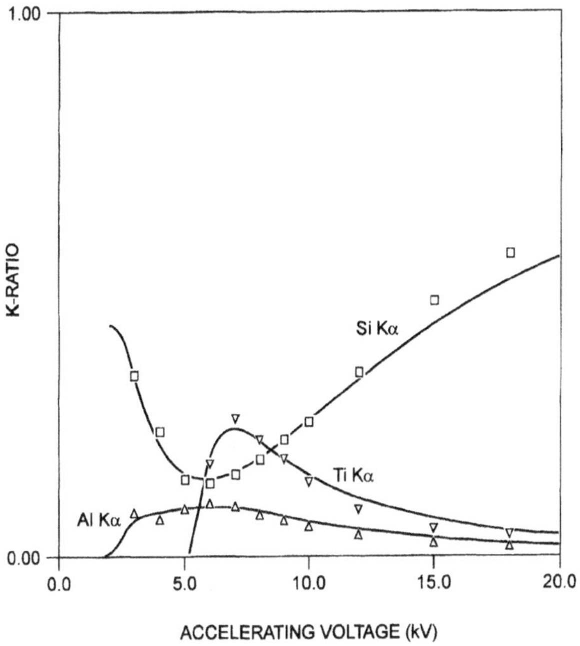

Fig. 2. Variation of the x-ray intensities with accelerating voltage for the uncoated multilayer

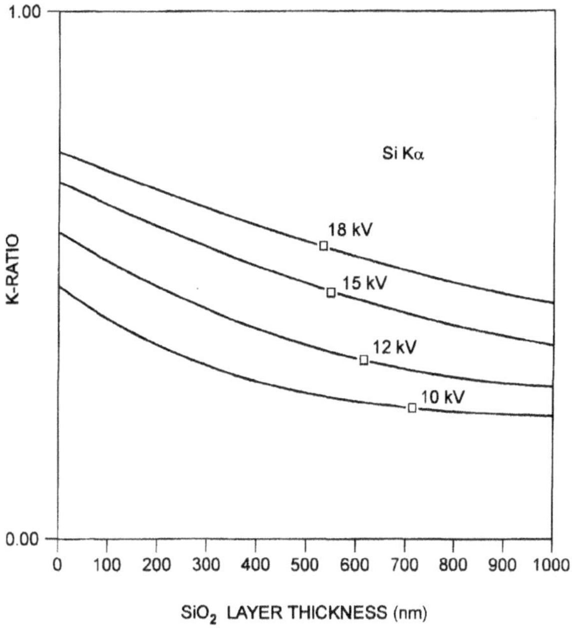

Fig. 3. Dependence of the Si Kα intensity on the thickness of the lower SiO_2 layer for accelerating voltages 10–18 kV

respect to the extrapolated behaviour of the experimental data obtained at 3–10 kV. The calculated thicknesses of the top six layers (Table 1), assuming that the densities of the vapour deposited material were 2.15 g cm^{-3} for SiO_2 [16], 3.96 g cm^{-3} for Al_2O_3 [17] and 3.76 g cm^{-3} for TiO_2 [18], are within a few nanometres of those estimated from the cross-sectional micrograph. The thickness measured for the bottom SiO_2 layer is approximately 12% too low if all the experimental data are included in the iteration and 9% too low if only those in the interval 3–10 kV are considered. At 10 kV the maximum depth of x-ray generation is equivalent to a mass thickness of $\sim 3 \cdot 10^2$ µg cm^{-2}, compared with a mass thickness of $\sim 2 \cdot 10^2$ µg cm^{-2} for the entire multilayer. The electron range extends approximately to the depth of first TiO_2 layer at 3 kV, the second Al_2O_3 layer at 4 kV, the top of the second SiO_2 layer at 5 kV and the SiO_2/Si interface at 8 kV.

When the Si Kα intensities for 10 kV and above are plotted as functions of layer thickness (Fig. 3), it is seen that the apparent thickness of the lower SiO_2 layer tends to decrease with increasing accelerating voltage. While this suggests a possible influence of specimen charging, the high energy cut-off in the continuum background was within approximately 0.1 keV of the nominal value at all accelerating voltages. The absence of visible surface charging does not however preclude that charge trapping takes place deeper within the specimen.

Table 1. Comparison of the calculated thicknesses (nm) of the individual layers with those estimated by TEM.

Layer	(a) TEM	(b) Uncoated specimen		(c) Coated with 1.5 nm Pt-C	
		3–10 kV	3–18 kV	3–10 kV	3–18 kV
SiO_2	65	66.5	65.1	63.3	62.3
Al_2O_3	15	13.8	13.0	13.2	12.4
TiO_2	100	106.4	98.7	101.3	96.0
Al_2O_3	15	11.9	12.6	10.1	11.2
TiO_2	8	5.6	7.3	6.7	8.0
Al_2O_3	35	36.2	34.9	35.4	34.4
SiO_2	680	618.6	595.3	641.9	604.7
Si substrate					

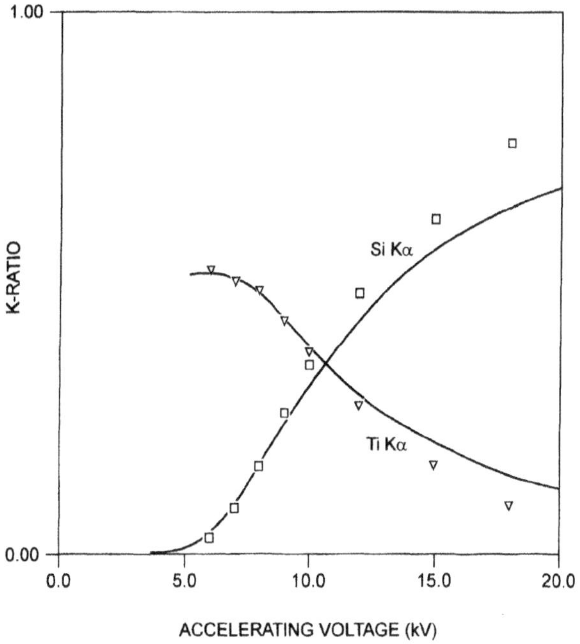

Fig. 4. Variation of the Ti Kα and Si Kα intensities with accelerating voltage for a TiO_2 film on a silicon substrate. The curves represent the theoretical behaviour for a 290 nm thick layer

Analyses of single layers of SiO_2 and Al_2O_3 deposited on silicon gave excellent agreement with the theoretical predictions over the entire voltage range and the high energy cut-off was within ~0.1 keV of the nominal value in both cases. Data from a ~300 nm TiO_2 layer on silicon deviated increasingly from the theoretical behaviour at accelerating voltages above 10 kV (Fig. 4). Only in the spectrum acquired at 18 kV was there a discernable high energy cut-off, at 17.3 keV, equivalent to a voltage drop of 0.7 keV at the surface. It is thus clear that significant modification of the emerging x-ray intensities can occur before surface charging becomes apparent. This is more pronounced for the single layer, which is a factor of three thicker than the ~100 nm TiO_2 layer in the multilayer, and is evidenced by the tendency of the Si Kα data to lie above the theoretical curve and a concomitant opposite deviation in the Ti Kα data above 10 kV in Fig. 4. The effect is partially masked by underestimation of the thickness of the SiO_2 layer in the calculation, due to interpretation of the increase in the Si Kα intensity as emission originating from the substrate.

Coated Specimens

In an insulating specimen the internal potential distribution is strongly dependent on whether the surface is coated with a conductive layer [19]. An effect of applying such a coating is to displace the inversion point of the electrical field down towards the maximum penetration depth of the primary electrons. The space charge distribution in a charged dielectric layer will vary according to whether the front or rear electrode is grounded [20]. Grounding the rear electrode will act to confine the space charge in the non-irradiated region, while grounding the front electrode will decrease the charge in this region. When the surface is grounded, depending on the degree of charging, the depth distribution of electron energy deposition may be restored to nearly that for an uncharged material [21]. Measurements of charge leakage in silica [22], with both the front and rear electrodes grounded have indicated that the current drawn from the top electrode decreases with decreasing sample thickness and that the internal charge is minimal when this is less than the electron range.

Figure 5 shows the results of EDS analysis of the Pt-C coating used in our work. Its composition was ~80% Pt, in good agreement with analysis of similar coatings by wavelength dispersive spectroscopy [12].

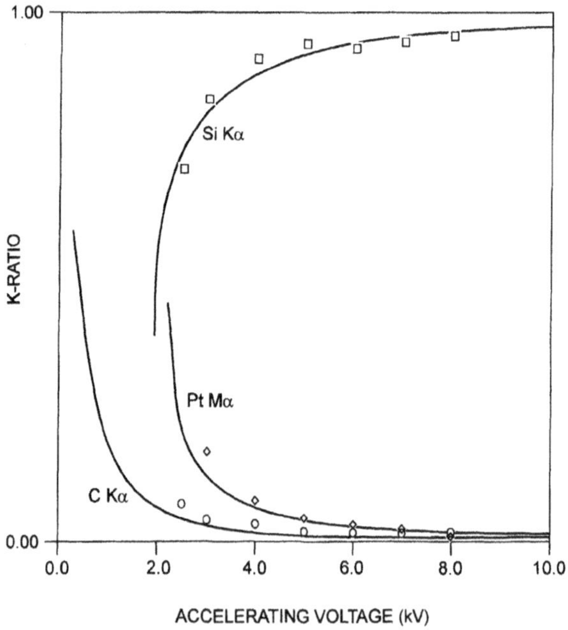

Fig. 5. Variation of the Pt Mα, C Kα and Si Kα intensities with accelerating voltage for a 1.5 nm Pt-C film deposited on silicon

The calculated thickness was ~ 1.7 nm, compared with 1.5 nm measured by the quartz crystal monitor. The C Kα data lie systematically above the theoretical curve due to the influence of specimen contamination. Attempts to allow for the contamination layer in the calculation were unsuccessful because the iterative procedure rapidly incorporated all of the available carbon into the Pt-C layer. An approximate correction for the influence of carbon contamination could be applied by subtracting the C Kα intensity for the uncoated substrate from that measured with the coating. This resulted in a shift of the C Kα data on to the theoretical curve with negligible change in the Si Kα and Pt Mα curves. The thickness of the contamination layer was calculated to be ~ 1 nm, which is a factor of ten smaller than in spot mode analysis [6, 13]. Hence, for consistency with measurements on uncoated specimens, the effect of this slight contamination was neglected.

X-ray intensities from the Pt-C coated multilayer are shown in Fig. 6; C Kα data have been omitted from this figure for clarity. For the purpose of calculation the thickness of the Pt-C layer was fixed, while those of the other layers were allowed to vary. The deviation in the Si Kα data above 10 kV is noticeably reduced with respect to the uncoated specimen but that in the Ti Kα data remains

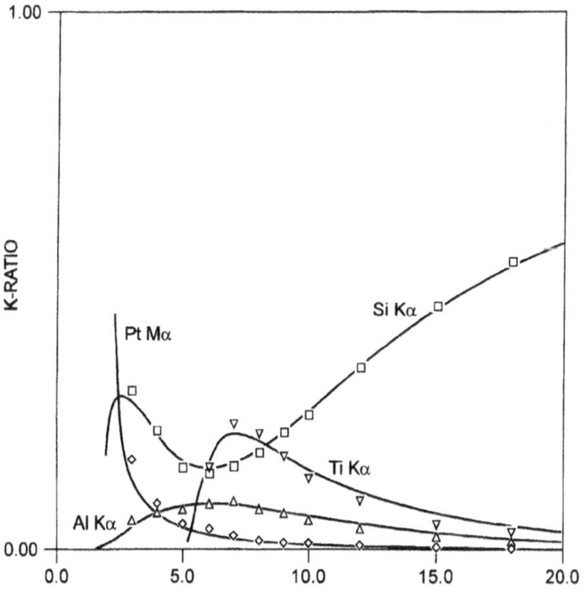

Fig. 6. Variation of x-ray intensities with accelerating voltage for the multilayer after coating with 1.5 nm Pt-C

practically the same. The calculated layer thicknesses, using the data obtained at accelerating voltages 3–10 kV and 3–18 kV, are reported in the last two columns of Table 1. Thicknesses of the top six layers differ relatively little from those for the uncoated specimen. The thickness of the lower SiO_2 layer, calculated with the data for 3–18 kV, is similar to that determined for the uncoated specimen but the thickness calculated with the data for 3–10 kV is significantly greater, reducing the error relative to the TEM measurement to $\sim 6\%$.

It has been reported that compositions of 360–500 nm SiO_2 and 320–400 nm Al_2O_3 films on silicon substrates could be accurately determined by wavelength dispersive spectroscopy without a carbon coating [23]. Accelerating voltages of 4–6 kV were used, so that radiation was excited in the film but not from the substrate. Analyses of uncoated SiO_2 films on InSb substrates at 8–20 kV [24], were found to be in close agreement with XPS data. However, analyses at 10 and 15 kV of 40–360 nm thick insulating Si_3N_4 films on titanium substrates, coated with 25 nm carbon or 4 nm gold, revealed an increase in the Si Kα intensity and a decrease in the Ti Kα intensity with increasing film thickness [25]. Monte Carlo simulations of the effect of a space charge distribution on electron trajectories and the x-ray depth distribution demonstrated that the maximum in the $\Phi(\rho Z)$ curve was displaced towards the surface as anticipated.

The present results confirm that SiO_2 and Al_2O_3 films on silicon are sufficiently conducting to allow analysis, whereas TiO_2 films apparently are not. The deviation in the Si Kα and Ti Kα intensities cannot be explained solely by distortion of the x-ray depth distribution. It is therefore suggested that the small high energy cut-off decreases x-ray emission from the TiO_2 layer but not from the underlying SiO_2 layer or silicon substrate because the effect of a reduction in overvoltage is less severe for Si Kα than Ti Kα and is compensated by an increased number of emerging x-rays photons due to a shift of the mean depth of generation in the direction of the surface.

One additional source of error in the measurement of thin layers should finally be mentioned: the formation of oxide layers on pure metal standards [26]. If it is assumed, for instance, that a 10 nm TiO_2 layer is present on the titanium standard, the increase in the effective k-ratio for Ti Kα would be $\sim 7\%$ at 6 kV but only $\sim 1\%$ at 10 kV. The resulting errors in the thickness calculations are relatively small, ~1 nm, because of the averaging effect of the iterative procedure.

Conclusions

Energy dispersive spectroscopy is a convenient non-destructive method for the analysis of multilayer coatings. For layers of thickness ~10 nm the attainable precision is comparable to that of conventional surface analysis techniques. The principal sources of error in the EDS measurements are departures from stoichiometry in the vacuum deposited films, distortion of the x-ray depth distribution function caused by the internal electrical field in non-conducting films, variations in density due to porosity, and the presence of oxide layers on metal standards.

Thin films of SiO_2 and Al_2O_3 on silicon are adequately conductive for analysis up to 18 kV without a coating. TiO_2 layers of similar thickness exhibit deviations in the measured x-ray intensities at accelerating voltages above 10 kV. Deposition of an ultra-thin Pt-C coating is beneficial in preventing surface charging but does not eliminate entirely the influence of the electrical field due to charge trapping. The recommended strategy for analysis of oxide multilayer structures is to apply a conductive coating and use as low a voltage as possible, compatible with achieving electron penetration into the substrate.

Acknowledgement. The authors wish to thank R. Alani for his assistance in preparation of cross-sectional specimens, and transmission electron microscopy.

References

[1] W. H. Southwell, *Appl. Opt.* **1985**, *24*, 457.
[2] K. H. Guenther, H. K. Pulker, *Appl. Opt.* **1976**, *15*, 2992.
[3] K. H. Guenther, *Appl. Opt.* **1984**, *23*, 3806.
[4] K. Yu-Zhang, G. Boisjolly, J. Rivory, L. Kilian, C. Colliex, *Thin Solid Films* **1994**, *253*, 299.
[5] P. Willich, *Microchim. Acta* **1992**, *[Suppl.] 12*, 1.
[6] J.-L. Pouchou, *Mikrochim. Acta* **1996**, *[Suppl.] 13*, 39.
[7] D. G. Rickerby, J.-F. Thiot, *Mikrochim. Acta* **1994**, *114/115*, 421.
[8] L.-S. Johansson, J. Juhanoja, *Thin Solid Films* **1994**, *238*, 242.
[9] Y. Taga, T. Itoh, *Appl. Opt.* **1989**, *28*, 2690.
[10] G. F. Bastin, H. J. M. Heijligers, in: *Electron Probe Quantitation* (K. F. J. Heinrich, D. E. Newbury, eds.), Plenum, New York, 1991, p. 163.
[11] R. Alani, P. R. Swann, *Mat. Res. Soc. Symp. Proc.* **1992**, *254*, 43.
[12] R. Wepf, M. Amrein, U. Bürkli, H. Gross, *J. Microsc.* **1991**, *163*, 51.
[13] D. G. Rickerby, *Mikrochim. Acta* **1996**, *[Suppl.] 13*, 493.
[14] J.-L. Pouchou, F. Pichoir, *J. Microsc. Spectrosc. Electron.* **1985**, *10*, 279.
[15] J.-L. Pouchou, F. Pichoir, In: *Electron Probe Quantitation* (K. F. J. Heinrich, D. E. Newbury, eds.), Plenum, New York, 1991, p. 31.
[16] M. Hüppauff, K. Bange, B. Lengeler, *Thin Solid Films* **1993**, *230*, 191.
[17] M. K. Smit, G. A. Acket, C. J. van der Laan, *Thin Solid Films* **1986**, *138*, 171.
[18] M. Ritala, M. Leskelä, E. Nykänen, P. Soininen, L. Niinistö, *Thin Solid Films* **1993**, *225*, 288.
[19] J. Cazaux, *J. Appl. Phys.* **1986**, *59*, 1418.
[20] V. I. Arkhipov, A. I. Rudenko, G. M. Sessler, *J. Phys. D: Appl. Phys.* **1991**, *24*, 731.
[21] M. Kotera, H. Suga, *J. Appl. Phys.* **1988**, *63*, 261.
[22] B. Gross, J. Dow, S. V. Nablo, *J. Appl. Phys.* **1973**, *44*, 2459.
[23] J. W. Colby, *Adv. X-ray Anal.* **1968**, *11*, 287.
[24] R. H. Packwood, G. Redmond, P. H. Holloway, *Surf. Interface Anal.* **1988**, *11*, 127.
[25] P. Klein, K. Röhrbacher, M. Andrae, J. Wernisch, *Mikrochim. Acta* **1996**, *[Suppl.] 13*, 377.
[26] P. Willich, K. Schiffmann, *Mikrochim. Acta* **1992**, *[Suppl.] 12*, 221.

Mikrochim. Acta [Suppl.] 15, 155–161 (1998)

Surface Ionization of Thin Films on Substrates: Measurement and Simulation

Xavier Llovet[1,*], **Claude Merlet**[2], and **Francesc Francesc Salvat**[3,4]

[1] Serveis Científico-Tècnics, Universitat de Barcelona, Lluís Solé i Sabarís, 1–3, E-08028 Barcelona, Spain
[2] ISTEEM, FU 160, CNRS, Université de Montpellier II, Sciences et Techniques du Languedoc, Pl. E. Bataillon, 34095 Montpellier Cedex 5, France
[3] Department of Physics and Astronomy, The University of Toledo, 2801 W. Bancroft, Toledo, Ohio 43606, U.S.A.
[4] Facultat de Física (ECM), Universitat de Barcelona, Diagonal 647, E-08028 Barcelona, Spain

Abstract. In an attempt to improve the accuracy of quantitative analysis of thin films with EPMA, experimental measurements and Monte Carlo simulations of the surface ionization have been performed (for Ni K_α x-rays) on Cu films of various thickness deposited onto a variety of substrates and for varying accelerating voltages. Monte Carlo calculations have been carried out using the simulation package PENELOPE for electron transport, together with ionization cross sections computed from an optical data model. In spite of the uncertainties arising from sample preparation and from the smallness of the peak-to-background ratio, the results from the experiments and the Monte Carlo simulations are found to agree to within 5%.

Key words: electron probe microanalysis, Monte Carlo simulation, inner-shell ionization, electron transport.

In electron probe microanalysis (EPMA) and Auger electron spectroscopy (AES), knowledge of the surface ionization ($\Phi(0)$ in EPMA or the R in AES) is an important requirement for quantitative procedures. Castaing and Descamps in 1955 [1], defined the surface ionization for the x-ray emission of an element as the ratio of primary x-ray intensities of that element emitted by an infinitely thin film deposited on a substrate material to the same, unsupported thin film.

In EPMA, the surface ionization is a fundamental parameter that appears in the majority of analytical expressions used for approximating the depth distribution of ionizations. This distribution is required by existing quantitative procedures and it is conventionally expressed as the $\Phi(\rho z)$ function, which gives the density of ionizations caused by the impinging electrons as a function of mass depth ρz (see e.g. [2]).

In an analytical form, the surface ionization can be expressed as [3]

$$\Phi(0) = 1 + \frac{1}{\sigma(E_0)} \int_0^{\pi/2} \int_{E_i}^{E_0} \sigma(E) \frac{\mathrm{d}^2\eta}{\mathrm{d}E\mathrm{d}\Omega} \mathrm{d}E \frac{1}{\cos\theta} \mathrm{d}\Omega \tag{1}$$

where $\mathrm{d}^2\eta/\mathrm{d}E\mathrm{d}\Omega$ is the energy and angular distribution of backscattered electrons, E_0 is the energy of the incident electron beam, E_i is the ionization energy of the considered atomic shell, $\sigma(E)$ is the corresponding ionization cross section and $1/\cos\theta$ is the path elongation factor for directions within the solid angle element $\mathrm{d}\Omega = 2\pi \sin\theta \, \mathrm{d}\theta$. This expression shows that the surface ionization depends mainly on the distribution of backscattered electrons, which is determined by the composition and geometrical structure of the sample. As a consequence, for film-over-substrate specimens, the surface ionization will vary with the film thickness and the substrate composition. For homogeneous samples, a variety of simple models for calculating $\Phi(0)$ have been proposed by several authors ([4]–[12]). The accuracy of the different models has been checked experimentally on massive substrates (see e.g. [11] and references therein), since the surface ionization can be measured in a direct way. However, measurements for complex structures, and specifically for thin films deposited on massive

* To whom correspondence should be addressed

substrates, are still vary scarce. Owing to this lack of information, the generalization of the existing models of $\Phi(0)$ to thin films deposited on substrates is still not firmly established.

Since direct measurement of the surface ionization is costly, time consuming and difficult, accurate theoretical methods for calculating the surface ionization are desirable. To date, Monte Carlo simulation provides the most reliable description of electron penetration in matter. Monte Carlo methods have reached a high degree of sophistication; they can incorporate realistic interaction cross sections and can be applied to complex geometries (e.g multilayer samples, tilted specimens, etc..). However, they suffer from the drawback of requiring very large computing times and thus they cannot be used in practical applications, which frequently involve iterative procedures where parameters of the geometry and/or of the scattering model must be varied. For this reason, the use of Monte Carlo simulation has been restricted to guide and test the development of more simple analytical methods of analysis.

In this work we present experimental measurements and Monte Carlo simulations of the Ni K-shell surface ionization for specimens consisting of thin films of various thicknesses deposited onto a variety of substrates, and for varying accelerating voltages. The simulation of electron transport is the target has been performed by means of the general-purpose simulation package PENELOPE together with semi empirical ionization cross sections obtained from an optical data model. The aim of this work has been to obtain new experimental data for $\Phi(0)$ and to show that the simulation code PENELOPE, combined with realistic ionization cross sections, yields reliable values of the surface ionization. As simulation results are free from some experimental uncertainties, they provide a convenient basis to derive empirical formulae.

Experimental

Cu films have been deposited by thermal evaporation on different polished pure substrates and simultaneously, for mass thickness determination, on four thin aluminum leafs of known area and weight, situated at the same distance from the evaporation source. Ni tracer films were then deposited, also by thermal evaporation, in the previously prepared targets. In the same runs, the tracer films have been deposited on a 30 nm self supporting film of carbon. Mass thicknesses of the evaporated films have been measured by direct weighing of the evaporated material on the aluminium leafs with a microbalance [1, 14] and by electron microprobe measurements in Cu/Zn and Cu/Ni targets at various accelerating voltages (10, 15, 20, 25 kV). The accuracy of the weighing method is estimated to be about 5% for a mass thickness of 100 $\mu g/cm^2$, mainly limited by the precision of the microbalance. This accuracy could not be improved by using a larger area aluminum leaf owing to the inhomogeneity induced by geometrical effects in the evaporating system. For the electron probe measurements, the XPHI multilayer procedure [28] was used. As the atomic number of the film and the substrate selected are similar, electron microprobe gives an accuracy better than 2% for the thickness of the Cu film. Both methods for mass thickness determination have been found to agree to within 5%.

The particular choice of tracer element (Ni) and film composition (Cu) has been motivated by the need to minimize experimental uncertainties. Elements Ni and Cu with atomic numbers 28 and 29 respectively, have similar scattering effects, moreover Cu cannot enhance Ni peaks by characteristic fluorescence, and finally the Cu film reduces the characteristic and continuum fluorescence of the tracer from the substrate.

Using the tracer method, the surface ionization is obtained by comparing the number of characteristic x-rays emitted by the tracer film deposited onto the target to the number emitted by the same, self supporting, tracer film. Measurements have been carried out at 10, 15, 20, 25 and 30 kV on a CAMECA SX-50 electron microprobe using a wavelength dispersive spectrometer, with a LiF crystal. Care has been exercised in selecting pulse height analyzer settings that avoid high order Bragg reflections from the substrate. To minimize the bias in the determination of the surface ionization due to the thickness of the tracer film, the Ni layer must be relatively thin. On the other hand, it must not be too thin, since a small peak-to-background ratio would imply unacceptably large statistical uncertainties in the peak intensity. Ni layers of 2 and 1.5 $\mu g/cm^2$ seem to represent a good compromise between the thickness bias and the statistical requirements (see Fig. 1). Background intensity has been estimated by linear interpolation of measurements at both sides of the characteristic line. In order to minimize sample contamination during measurements, an anti-

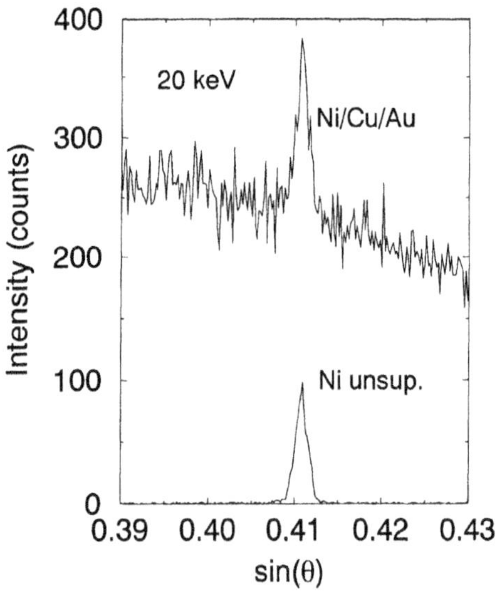

Fig. 1. Experimental wavelength dispersive x-ray spectra of Ni self-supported tracer film and a 196 nm thick Cu film on Au, at 20 keV incident energy with a LiF crystal

contamination device has been used. For each target and each voltage, at least five measurements at differents positions on the specimen were made. The statistical uncertainties, associated to the limited number of counts, were less than 5%. As an example, Fig. 1 shows the experimental spectra of Ni K_α emitted by the unsupported tracer film and by the Cu film deposited on a Au substrate.

The Simulation Code PENELOPE

PENELOPE (an acronym for "penetration and energy loss of positrons and electrons") is a self-contained Monte Carlo simulation code developed to generate electron-photon showers in materials systems consisting of a number of homogeneous bodies, with arbitrary compositions, limited by well defined surfaces. PENELOPE is applicable to electrons and positrons with energies higher than few hundred eV and to photons with energies above 1 keV [16].

Monte Carlo simulation keeps track of the evolution of the history of an electron, which can be considered as an alternate succession of interactions and free flights. In each interaction the electron changes its energy and direction of movement and eventually produces secondary electrons. The kind of interaction is sampled according to the probabilities of occurrence of the possible interaction events. Electrons are followed until their energy falls below a preselected value or leave the materials system. Quantities of interest are stored in appropriate counters.

Electrons are assumed to interact in the medium via elastic collisions, and inelastic interactions. In the energy range of interest, Bremsstrahlung emission has a negligible stopping power and can be disregarded when simulating electron tracks. Although accurate differential cross sections (DCS) for elastic and inelastic scattering are currently available, PENELOPE uses approximate cross sections, given by analytical formulae, that largely simplify the numerical calculations. Elastic scattering is simulated by means of a combination of the Wentzel (screened Rutherford) DCS and a fixed-angle scattering process [18]. This analytical DCSs contains three parameters that are determined by requiring that the mean free path between collisions and the mean and the variance of the angular distribution in single elastic collisions coincide with the values obtained with a realistic numerical DCS. Tables of these quantities have been obtained from partial wave calculations and can be generated by using the program material of the PENELOPE code system. Although the analytical DCSs implemented in PENELOPE differ in detail from the numerical ones, they lead to essentially the same simulation results, provided only that the average number of collisions along an electron track is large enough (of the order of 20 or larger) [19]. The DCS for inelastic collisions is based on a simple generalized oscillator strength model proposed by Liljequist [15]. In this model the ionization of each atomic shell is described by a single oscillator with strength f_i equal to the number of electron in the shell, and resonance energy $W_i = aE_i$, where E_i is the ionization energy of the shell and a is an adjustable parameter. Excitations of valence electrons (with binding energy less than 30 eV) are accounted for by a single oscillator. The value of a is obtained in such a way that the mean excitation energy I recommended by Berger and Seltzer [17] is exactly reproduced.

In PENELOPE, electron tracks are simulated by means of a mixed procedure. Interactions of the electrons in the medium are classified as hard events and soft events. Hard events, in which the angle deflection and the energy loss are greater than a certain cutoff, are simulated in detail, i.e. in chronological succession. The effect of the usually many soft interactions that occur between a pair of consecutive hard events is described as a single "artificial" event in which the electron is deflected and loses energy according to a multiple scattering theory [16].

PENELOPE has been developed to provide an accurate description of the penetration and slowing down of low-energy electrons, which produce, on average, a very small number of characteristic x-rays by direct impact. In fact, the emission of characteristic x-rays is not accounted for in PENELOPE; only the average effect of ionizing collisions on the projectile is considered. However, it is possible to obtain K-shell ionization distributions and thus the surface ionization by using the following procedure. We use PENELOPE to generate electron tracks and, in each free flight, we calculate the expected number of ionizations n produced per unit path length using a more realistic ionization cross section. This number of ionizations may be written as $n = N\sigma_K(E)$ where N is the density of atoms of the ionized element per unit volume, σ_K is the total ionization cross section and E is the electron energy. The surface ionization is therefore obtained by determining the expected number of ionizations produced in a infinitesimal layer situated at the target surface. In the simulations reported here, we have used ionization cross sections computed from an optical data model proposed by Mayol and Salvat [20], which give reliable results when comparing simulated K-shell ionization distributions with experimental data taken from the literature [21].

Simulation Results and Discussion

Prior to the measurements of the surface ionization we made sure that, for the element of interest, the optical-data model correctly reproduces the measured energy-dependence of the ionization cross section. This dependence can be easily determined experimentally, since the number of characteristic x-rays emitted by the self supported film of Ni is proportional to the ionization cross section. We have thus performed a series of mea-surements of the number of characteristic x-rays emitted for different incident electron energies, from threshold to 40 keV. The experimental results, rescaled so as to match the calculation at about 20 keV, are compared in Fig. 2 with the optical-data cross section and with four different simple analytical formulae of common use in EPMA, namely, those of Bethe-Powell [22], Worthington and Tomlin [24], Gryzinsky [23] and de la Ripelle [25]. Owing to the relatively large experimental uncertainties, which are estimated to be of the order of 5%, the optical-data cross section seems to yield results slightly closer to the experiment, as could be expected from its physical content.

Values of the surface Ni K-shell ionization produced by electron beams with energies of 10, 15, 20, 25 and 30 keV at normal incidence, for Cu films of three different thicknesses (67, 100 and 196 nm) over substrates of 9 elements, which span the periodic table, have been measured and computed. In the

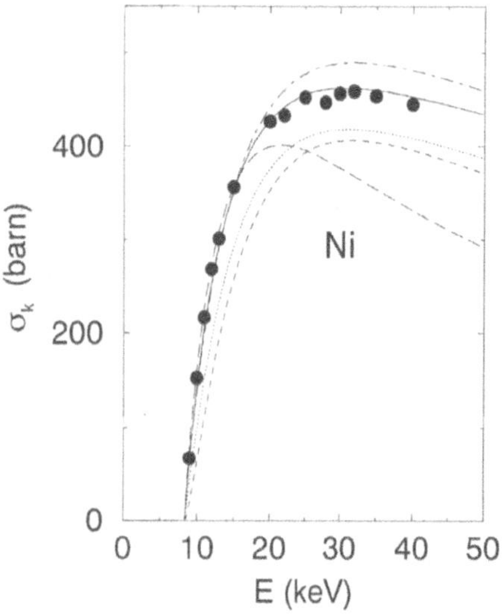

Fig. 2. Nickel K-Shell ionization cross section. Cricles are measured data, lines are the cross section computed from the optical-data model of Mayol and Salvat [20] (continuous) and the analytical formulae of Bethe-Powell [22] (dotted), Gryzinsky [23] (dashed), Worthington-Tomlin [24] (long-dashed) and de la Ripelle [25] (dot-dashed)

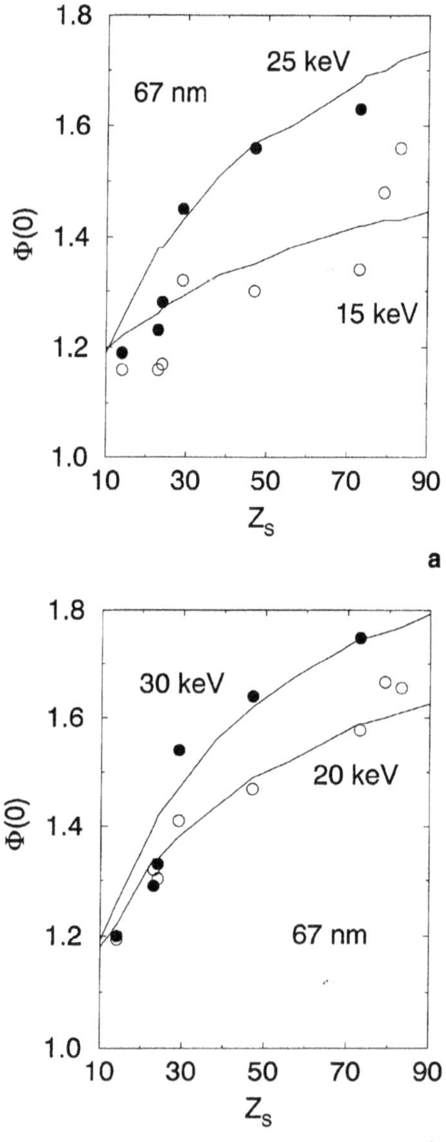

Fig. 3. Surface ionization for 67 nm thick Cu film deposited onto differents substrates as a function of the atomic number of the substrate, at 15 keV and 25 keV (**a**) and at 20 keV and 30 keV (**b**)

simulations, 200,000 electron tracks were generated in each run, which is a large enough number to make sure that the statistical uncertainty (three standard deviations) of the calculated surface ionization is less than 1%. The results of measurements and simulations, as functions of the atomic number for the different bombarding energies, are compared in Figs. 3–5. In spite of the uncertainties arising from sample preparation and from the smallness of the peak-to-background ratio, experimental results and Monte Carlo simulations are found to agree to within about 5%.

From Figs. 3–5, it is seen that the surface film ionization increases with the atomic number of the substrate. This is because the heavier the substrate the larger the number of electrons backscattered from the substrate, which produce more ionizations in the film and at the surface. Accordingly, the dependence of $\Phi(0)$ on the atomic number of the substrate reduces when the thickness of the overlayer increases and/or

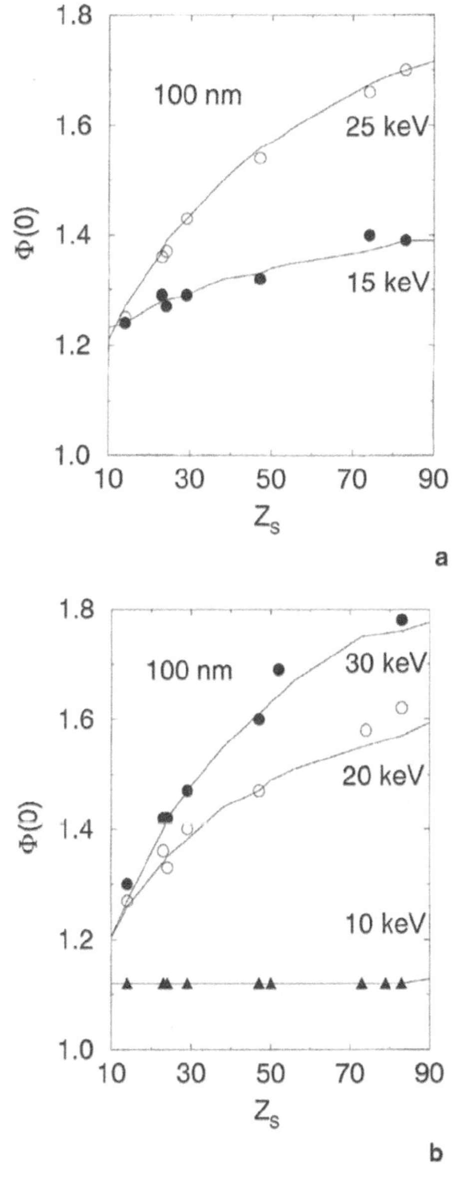

Fig. 4. Surface ionization for 100 nm thick Cu film deposited on differents substrates as a function of the atomic number of the substrate, at 15 keV and 25 keV (**a**) and at 10 keV, 20 keV and 30 keV (**b**)

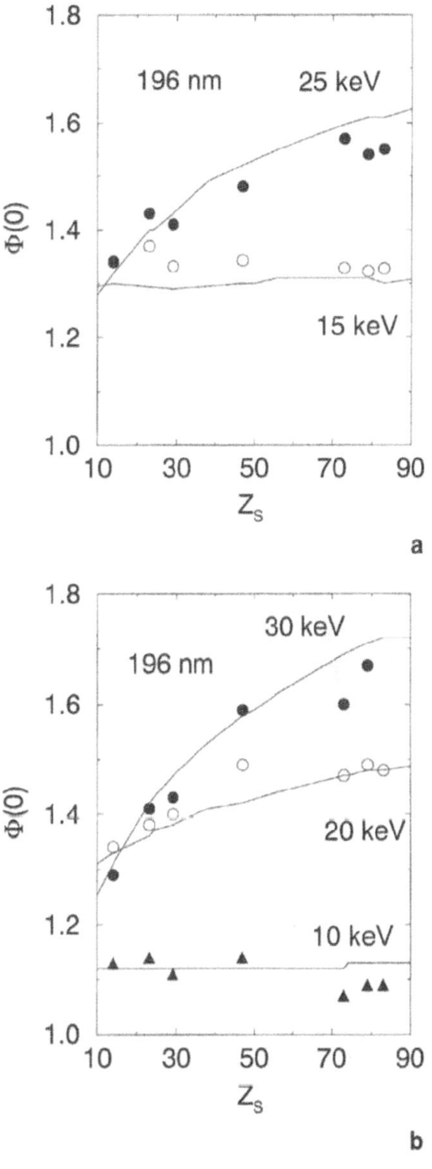

Fig. 5. Surface ionization for 196 nm thick Cu film deposited onto differents substrates as a function of the atomic number of the substrate, at 15 keV and 25 keV (**a**) and at 10 keV, 20 keV and 30 keV (**b**)

when the electron beam energy decreases (i.e. when the number of electrons that reach the substrate with enough energy to return to the surface diminishes). The surface ionization, specially for film-substrate systems, represents a stringent global test for the accuracy of the whole simulation algorithm.

As simulation results are free from some of the experimental uncertainties, they can be used to investigate the validity of analytical empirical formulae. A convenient starting point to establish such formulae is provided by the following approximate relation

$$\Phi^N(0) = \left[\frac{\Phi_{FS}(0) - \Phi_S(0)}{\Phi_F(0) - \Phi_S(0)}\right], \qquad (2)$$

Where $\Phi_F(0)$ and $\Phi_S(0)$ are the surface ionizations for bulk samples of the film and substrate materials, respectively, and $\Phi_{FS}(0)$ is the value for the film-substrate system. This relation was used previously by Hunger and Rogaschewski [26] in a study of the backscattering coefficient of film-substrates systems. The dependence of $\Phi^N(0)$ on the incident electron beam energy E_0 can be largely reduced by expressing the film thickness t in units of the effective range R of electrons in the film. R is defined as the average path length that the electrons travel before slowing down to an energy equal to the ionization energy E_i of the considered atomic shell, i.e. R is the difference between the Bethe ranges of electrons with energies E_0 and E_i in the film material.

Figure 6 shows simulated surface ionization as a function of reduced film thickness, $s = t/R$, calculated for different substrates, film thicknesses and electron energies. It is observed that the normalized surface ionizations $\Phi^N(0)$ obtained from the simulations does scale with R. The continuous curve in this figure represents the function

$$\Phi^N(0) = \tanh(As + Bs^2), \qquad (3)$$

with parameters $A = 3.412$ and $B = 11.216$ determined by least-squares fitting of the simulated data. This function has been previously used by Sogard [27] and Hunger and Rogaschewski [26] for describing the backscattering coefficient and by Merlet [28] and Lee et al. [29] for describing the surface ionization. These authors, however, use a different thickness scaling. Our results thus confirm the validity of the assumed scaling properties, which provide a convenient basis to derive more general empirical formulae. Work along these lines is in progress.

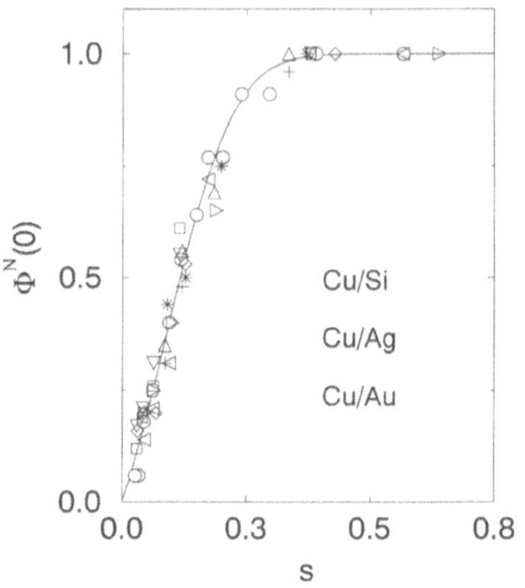

Fig. 6. Scaled surface ionization as a function of the reduced depth for different substrates, film thicknesses and bombarding energies. Symbols are simulation results: Cu/Ag, circles, squares and up triangles; Cu/Au, down triangles, plus, left triangles; Cu/Si, right triangles, diamonds, stars. The continuous curve represents the function (3) with the parameter values indicated in the text

In conclusion, comparison between experimental and Monte Carlo simulation results demonstrates the applicability of the simulation code system PENELOPE for EPMA studies of layered structures. The results presented here clearly indicate that, with the aid of this simulation tool, the uncertainties of exsisting analytical quantitative procedures for layered systems can be significantly reduced.

Acknowledgement. This work has been partially supported by Comisión para la Investigación Científica y Técnica (Spain), project no. PB95-0271-C02-01.

References

[1] R. Castaing, J. Descamps, *J. Phys. Radium* **1955**, *16*, 304.
[2] L. Reimer, *Scanning Electron Microscopy*, Springer, Berlin Heidelberg New York Tokyo, 1985.
[3] P. Klein, M. Andrae, K. Röhrbacher, J. Wernisch, *Mikrochim. Acta* **1996**, *[Suppl.] 13*, 363.
[4] W. Reuter, *Proceedings of the 6th International Congress on X-Ray Optics and Microanalysis* (G. Shinoda, K. Kohra, T. Ichinokawa, eds.) University of Tokio Press, Tokyo, 1972.
[5] G. Love, M. G. Cox, V. D. Scott, *J. Phys. D* **1978**, *11*, 23.
[6] R. H. Packwood, J. D. Brown, *X-Ray Spectrom.* **1981**, *10*, 138.
[7] S. Ichimura, R. Shimizu, J. P. Langeron, *Surf. Sci.* **1983**, *124*, 49.
[8] J. H. Tirira Saa, J. A. Riveros, *X-Ray Spectrom.* **1987**, *16*, 27.
[9] W. Rehbach, P. Karduck, In: *Microbeam Analysis* (D. E. Newbury, ed.), San Francisco Press, San Francisco, 1988.

[10] J. Cazaux, *Mikrochim. Acta* **1992**, *B9*, 107.

[11] C. Merlet, *X-Ray Spectrom.* **1992**, *21*, 229.

[12] S. Segui, J. Trincavelli, G. Castellano, J. A. Riveros, *X-Ray Spectrom.* **1996**, *25*, 110.

[13] G. F. Bastin, J. M. Dijkstra, H. J. M. Heijligers, *Proceedings of Microscopy and Microanalysis* (E. G. W. Bailey, J. M. Corbett, R. V. W. Dimlich, J. R. Michael, N. J. Zaluzec, eds.), San Francisco Press, San Francisco, 1996.

[14] J. D. Brown, L. Parobek, *X-Ray Spectrom.* **1976**, *5*, 36.

[15] D. Liljiequist, *J. Phys. D: Appl. Phys.* **1983**, *16*, 1567.

[16] J. Baró, J. Sempau, J. M. Fernández-Varea, F. Salvat, *Nucl. Instr. Meth. B* **1995**, *100*, 31.

[17] M. J. Berger, S. M. Seltzer, National Bureau of Standards Report NBSIR 82-2550, Washington, 1982; also available as ICRU Report 37, 1984.

[18] J. Baró, J. Sempau, J. M. Fernández-Varea, F. Salvat, *Nucl. Instr. Meth. B* **1994**, *84*, 465.

[19] J. M. Fernández-Varea, R. Mayol, J. Baró, F. Salvat, *Nucl. Instr. Meth. B* **1993**, *73*, 447.

[20] R. Mayol, F. Salvat, *J. Phys. B: At. Mol. Opt. Phys.* **1990**, *23*, 2117.

[21] X. Llovet, J. A. Riveros, F. Salvat, *Mikrochim. Acta* **1996**, *[Suppl.] 13*, 409.

[22] C. J. Powell, *Electron Impact Ionization* (T. D. Mark, G. H. Dunn, eds.), Springer, Wien, New York, 1985.

[23] M. Gryzinsky, *Phys. Rev. A* **1965**, *138*, 336.

[24] C. R. Worthington, S. G. Tomlin, *Phys. Rev.* **1956**, *43*, 839.

[25] F. M. de la Ripelle, *J. Physique* **1949**, *10*, 319.

[26] H. J. Hunger, S. Rogaschewski, *Scanning* **1986**, *8*, 257.

[27] M. R. Sogard, *J. Appl. Phys.* **1980**, *51*, 4412.

[28] C. Merlet, *Proceedings of Microbeam Analysis*, (E. S. Etz, ed.), VCH 1995.

[29] C. L. Lee, K. Y. Kong, H. Gong, C. K. Ong, *Surf. Interface Anal.* **1996**, *24*, 15.

Mikrochim. Acta [Suppl.] 15, 163–170 (1998)

Comparison of Different Methods to Characterize Thin a-Si:H Films

H. G. Esser[1,*], P. Karduck[2], M. Rubel[3], N. Almqvist[4], L. Grobusch[1], J. von Seggern[1], F. Weschenfelder[1], and P. Wienhold

[1] Institute of Plasma Physics, Forschungszentrum Jülich, Association EURATOM-KFA, D-52425 Jülich, Federal Republic of Germany
[2] Central Facility for Electron Microscopy, Aachen University of Technology, D-52056 Aachen, Federal Republic of Germany
[3] Physics Department-Frescati, Royal Institute of Technology, Association EURATOM-NFR, Frescativägen, 24, S-104 05 Stockholm, Sweden
[4] Physics Department, Lulea University of Technology, S-971 87 Lulea, Sweden

Abstract. The objective of this work is to determine the materials properties of thin layers typically deposited for wall conditioning in fusion devices. As reference material, a sequence of thin amorphous hydrogenated silicon layers with different thicknesses were produced on polished glassy carbon by Plasma Chemical Vapour Deposition PCVD. The films were characterised using AES, AFM, EPMA, ellipsometry, interference fringe analysis and RBS. A second aim was to study the accuracy and reliability of EPMA and RBS measurements. The techniques applied provided values for refractive index, absorption coefficient, roughness, thickness and elemental areal density. From the latter two, the density of the film could be determined. It turned out that the density derived by combining the results from different techniques was in good agreement. For the range of thickness investigated, the combination of ellipsometry and EPMA gave the most consistent values for the density with a maximum uncertainty of $\pm 5\%$.

Key words: a-Si:H thin films, EPMA, ellipsometry, interference fringe analysis, RBS, wall conditioning.

1. Introduction and Motivation

Radio frequency assisted plasma chemical vapour deposition (PCVD) of thin hydrogen containing films of carbon, boron or silicon, has proved to be an efficient method for producing layers on plasma facing components in controlled fusion devices. Such films improve the plasma performance significantly [1, 2]. In order to optimise the layer deposition process and to achieve a better understanding of the plasma surface interaction, relating to (i) wall erosion by physical and chemical sputtering and (ii) uncontrolled layer formation by sputtered and transported particles, it is essential to study materials properties (e.g. composition, thickness, density and depth profile) of layers before and after exposure to the plasma. Online investigations during and after the plasma discharges, e.g. optical spectroscopy for layer erosion and particle transport or quadrupole mass spectroscopy of desorbed neutral gas following TOKAMAK discharges, provide useful information. Nevertheless, an ex situ investigation of eroded material surfaces and redeposited layers using surface analysis techniques is essential to gain a better understanding of the interaction processes between plasma and wall. This contribution reports on the production and ex situ characterization of a thin amorphous hydrogen containing silicon film (a-Si:H). The film was prepared specially as a reference layer for the purpose of comparing different analysis methods. The measurements performed, a kind of a cross calibration of some micro properties, were obtained by the routinely used and complementary analytical techniques of Electron Probe X-Ray

* To whom correspondence should be addressed

Microanalysis (EPMA), Ellipsometry, Interference Fringe Analysis, Atomic Force Microscopy (AFM), Rutherford Backscattering (RBS) and Depth Profiling Auger (AES). The independent analytical methods are discussed and compared in terms of reliability, conformity and accuracy.

2. Experimental

a-Si:H Layer Production

For a reliable comparison of the results from the different analysis methods, a thin coverage was produced by RF assisted DC plasma deposition in a vacuum chamber at the laboratory of the Institute of Plasmaphysics, Forschungszentrum-Jülich. A silicon layer consisting of 4 different thickness steps was deposited onto a plain surface of an otherwise irregular shaped glassy carbon sample by exposing, for certain time periods, adjacent areas to the plasma consecutively.

After deposition the thickness steps of the layers could be observed from the interference colours. Fig. 1 shows areas of the different thickness steps, labelled I–IV, surrounded by contour lines. The deposition parameters of the PCVD process were:

- Gas composition: 40%Helium + 60%Silane
- Total pressure: 2×10^{-4} Pa
- RF-Power: 100 Watt (forward-reflected)
- DC-Voltage: 400 V

An average deposition rate of ~ 4.3 nm/min was deduced from results of the analysis methods described below. Most of the analytical measurements presented here were taken along the dashed analysis line AB.

3. Analysis Methods

3.1 Atomic Force Microscopy (AFM)

The AFM technique is based on the measurement of the force between a probing tip and a sample surface [3]. Therefore, the technique is applicable also on insulators. The images recorded are related to surface electronic energies up to the Fermi level, i.e. to the total charge density. The main use of AFM is for topographic studies, but certain mechanical properties of surfaces e.g. friction can be traced.

Experimental

Fine features of surface structure were studied using a Nanoscope II (Digital Instruments) AFM. The scans were made using standard Si_3N_4 cantilevers with a spring constant $k = 0.58\,Nm^{-1}$. The images were recorded with care to avoid tip-connected artefacts, i.e. where surface structures sharper than the AFM tip become distorted. The micrographs presented here are representative of the surfaces under investigation.

Results

Figure 2a shows the AFM image (constant force mode) of a non-exposed glassy carbon. The surface is fairly smooth with some small hillocks 5–10 nm in height and up to 40 nm in diameter. The deposition of

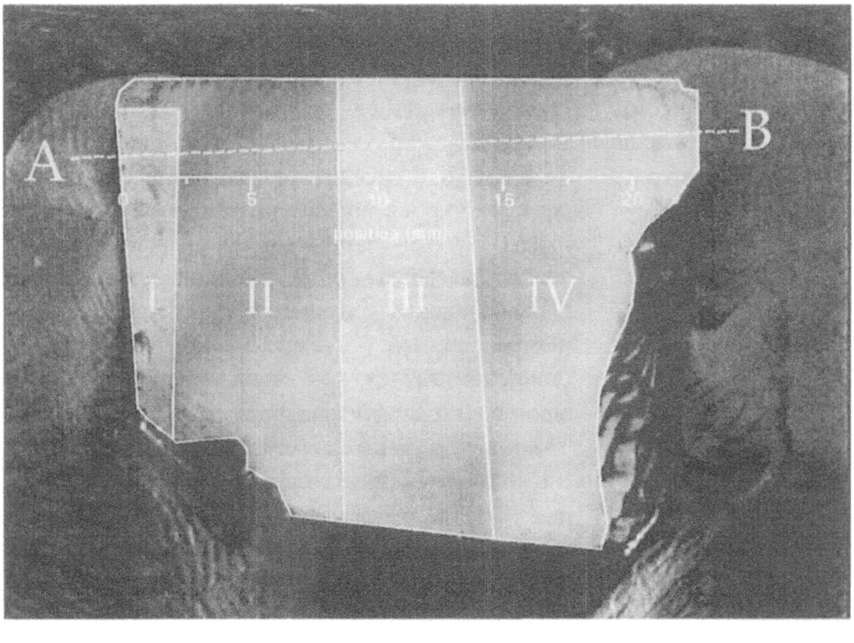

Fig. 1. View of the a-Si:H layer; thickness steps surrounded by contour lines; location of analyses, line AB

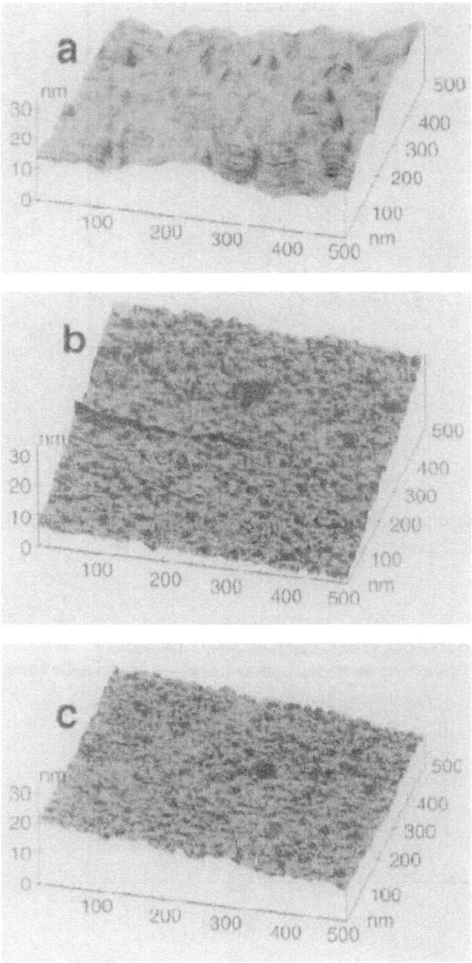

Fig. 2. AFM analysis of **a** the uncovered glassy carbon, **b** the a-Si:H layer thickness step I and **c** thickness step III

the a-Si:H film changes the surface structure as shown in Fig. 2b and 2c, layers on step I and step III respectively. The images indicate that the layers, independently of their thickness, are uniform with small (approximately 10 nm in diameter) grain-like features. At greater magnifications it may be noticed that small sized grains build up larger sized islands. A typical step height between these islands is about 2.5 nm and some porous structures with "holes" about 8 nm deep are found.

3.2 Auger Electron Spectroscopy

To confirm the homogeneity of the layer with depth, which is of great importance for the interpretation of the measured layer properties, depth profiles were measured by means of Sputter Auger Spectroscopy.

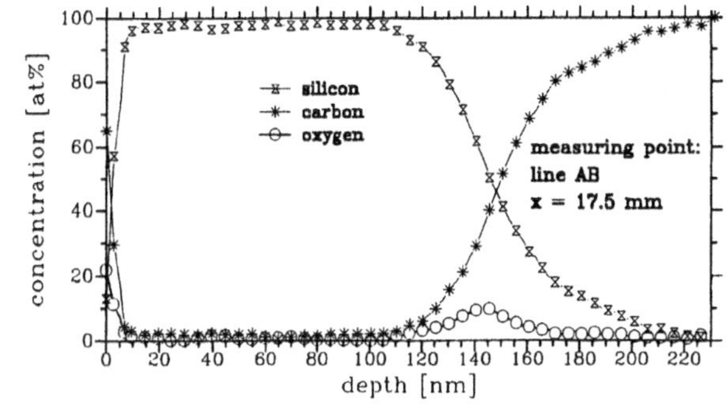

Fig. 3. Sputter Auger depth profile of the a-Si:H film; position of measurement: area IV, position x = 17.5 mm on line AB

Experimental. The sample has been investigated by Auger depth profiling [4–6]. Elemental distributions within every one of the sample areas I to IV were obtained by simultaneously eroding the surface with 3 keV Xe^{+} ions and investigating the composition by AES. A Leybold ion source and Varian Auger spectrometer were used for this purpose. The depth profiles were taken at 1.5 mm, 5.5 mm, 12.5 mm and 17.5 mm along the line AB. The conversion of the sputter time to depth was performed by applying experimentally determined sputter rates established previously.

Results

Figure 3 shows, as a representative example, the measured depth profile in the thickest layer at area IV, x = 17.5 mm. Common to all of the thickness steps is their homogeneous composition with depth (nearly 100% silicon with little carbon and no oxygen). However, the first 5–6 nm of the layers show an enhanced, though not unexpected, concentration of carbon as well as a significant amount of oxygen. These impurities originate from contamination in air. This, apparently thick contamination layer, may originate in compositional mixing under the influence of the incident ion beam (mean range of 3 keV Xe^{+} in a-Si:H is about 6.5 nm). Oxygen around the interface is also a simple contaminant. On the basis of the film thickness determined by ellipsometry, the sputter rate for the 3 keV Xe^{+} beam was calculated to be ≈2.5 nm/min.

3.3 Ellipsometry

Ellipsometry was used to determine the optical constants and thickness of the deposited film. The method is based on the alteration of the polarization state of light upon reflection from a plain or coated surface. After reflection the light will be elliptically polarized in most cases, which explains the name of the technique. If the nature of polarization of the incident light is known then analysis of the polarization state of the reflected light can be used to determine the refractive index (n) and the extinction coefficient (k) of a pure substrate; additionally, for a coated surface the thickness (d) of the film can be deduced from the measurement. (In this paper the extinction coefficient k will be defined positive). Polarization can be divided into two linearly polarized parts: light which is polarized parallel (||) and light which is polarized perpendicular (⊥) to the plane of incidence (defined by the direction of the incident light and the surface normal). Ellipsometry measures the phase shift and change of light intensity of the two polarizations produced by the reflection.

Theoretically this behaviour can be described by two complex reflection coefficients r|| and r⊥ which are functions of the optical constants of the film, the substrate and surrounding medium, the film thickness, the wavelength and the angle of incidence. Reflection coefficients can be calculated using, for example, the Fresnel equations [7]. Assuming that the two electric field vectors of the incident light have the same intensity and phase angle, the reflection can be described by the equation [8, 9].

$$\tan\Psi \cdot e^{i\cdot\Delta} = \frac{r_{\|}}{r_{\perp}}$$

tan Ψ is the quotient of the intensities E of the two components of the reflected light (E||/E⊥) and Δ is the phase shift between these two components. The measured ellipsometric angles ψ and Δ are compared with the calculated values of ψ and Δ to determine n, k and d.

Experimental. For the measurements described here an ellipsometer with a rotating analyser and a single wavelength of 632.8 nm was used; the angle of incidence was 70°.

Results

The sample described above has the advantage that its optical constants n, k and film thickness may be determined with one single wavelength. Assuming that the film properties do not change during the deposition process, one theoretically derived curve in a (ψ,Δ) plot should fit all the measured pairs of ψ and Δ Fig. 4 shows the result of the measurement together with a fitted curve; the measured points are plotted as triangles with the measurement numbers next to them. The optical constants of the glassy carbon substrate were determined as n = 2.0, k = 0.67. The best fit of the theoretical curve to the measured points was

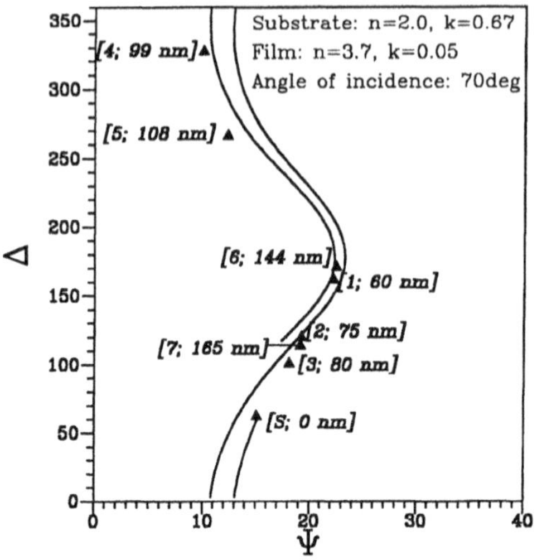

Fig. 4. Ellipsometric measure of layer thickness; theoretical fit line and measured points

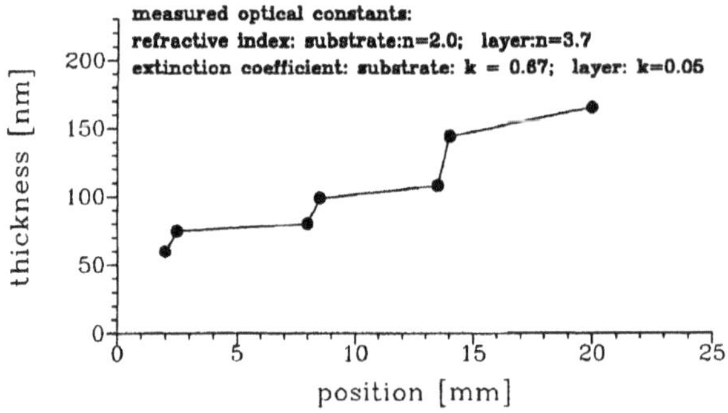

Fig. 5. Ellipsometric thickness determination of the different thickness steps of the a-Si:H film

achieved with n = 3.7 and k = 0.05 as optical constants for the a-Si:H film. From the value of these constants the film thickness could be determined for each point. The results are plotted next to the measurement numbers and Fig. 5 shows the layer thicknesses deduced for all measured points, plotted as a function of the position on the analysis line AB. The positions of the points were chosen to be as close as possible to the thickness steps between the areas I–IV. The thicknesses obtained vary between 60 nm in area I to 165 nm in area IV.

3.4 Interference Fringe Analysis (Colorimetry of the Interference Colours)

Colorimetry measures the colour coordinates R, G and B of the interference colours of thin transparent coatings on a substrate illuminated by light [10]. The coordinates are determined by the film thickness d, but the refractive and extinction constants n, k of both film and substrate, the spectral distribution of the illuminating light $\varphi_i(\lambda)$ and the angle of incidence play a part as well, and influence the reflected spectral distribution $\varphi_r(\lambda)$ according to the interference laws. On the other hand, any distribution $\varphi_r(\lambda)$ represents a particular colour tone which can be expressed as hue angle h with $\tan h = (y-yo)/(x\text{-}xo)$ and $x = R/(R+G+B)$, $y = G/(R+G+B)$. As described elsewhere [11], for simple film/substrate systems the quantity h is uniquely related to the thickness d within each order of interference. It circumvents the colour circle if d increases, thereby passing the yellow (Y), purple (P), blue (B) and green (G) sections. If a CCD colour camera is used $xo = yo = 1/3$, but for the case of a human observer and daylight illumination as applies here, $xo = 0.313$ and $yo = 0.329$. Interference fringe analysis distinguishes only the most pronounced colour impressions (Y, P, B, G) and hence the thickness ranges of about 20–40 nm, which is sufficient in many cases. The corresponding film thicknesses were determined by means of a relation h(d) which was calculated using the ellipsometrically measured optical constants n and k.

Results

Figure 6 shows the spatial ranges of the observed colour fringes.

Since each of the fringes corresponds to a particular thickness range they form a chain of boxes in the diagram. They start with a thickness of about 60 nm at position zero and increase stepwise up to about 170 nm at position 23 mm, thereby passing the first and second order of interference.

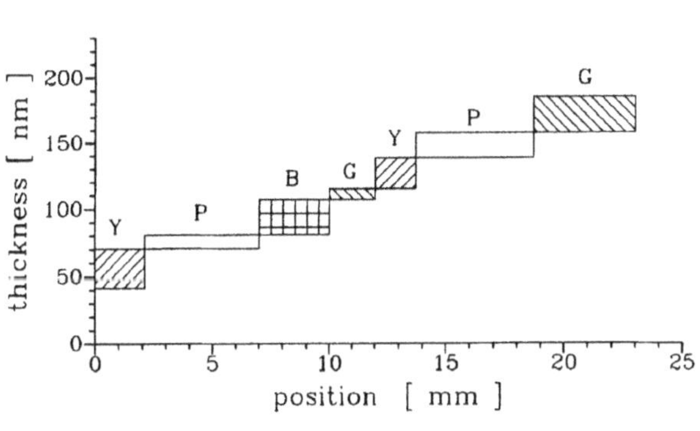

Fig. 6. Layer thickness determination by interference fringe analysis; absorption coefficient k and refractive index n taken from ellipsometry

3.5 Rutherford Backscattering Spectroscopy (RBS)

The technique is based on classical scattering (in a central force field) of a collimated beam of charged particles by nuclei of target atoms [12], the collisions being governed by Coulomb repulsion. In practice, a solid target is usually irradiated with a beam of $^4He^+$ ions in the MeV range and an energy analysis of ions backscattered from the outermost layer and deeper region of the target is performed. The kinematics of collisions and the scattering cross section are independent of chemical bonding, so that no information is obtained on the electronic configuration of species in the target. However, the method allows a sensitive qualitative and quantitative analysis of elemental composition to be made throughout the irradiated layer.

Experimental. RBS was used to determine the areal concentration of silicon and impurity atoms in the analysed film. The analyses were performed using a $^4He^+$ beam with energy of 1500 keV and a beam spot diameter of 0.8 mm. The energy of 1500 keV meant that the whole layer thickness was probed since the information depth for the Si film was below 400 nm. The accelerator in use was an AN 2000 Van de Graff from High Voltage Engineering Corporation. Backscattered ions were detected at a scattering angle of 170° using a surface barrier silicon detector (Ortec). Several RBS scans were carried out on the sample: along the centre line as well as 10 mm apart on the left and right hand sides of it.

Results

Figure 7 shows the results obtained with RBS for a scan performed along the analysis line AB on the sample. Four distinct steps are noticed in the areal concentration of Si atoms. The concentrations vary from $2.4 \times 10^{17}\,cm^{-2}$, the lowest value in area I, to $8.3 \times 10^{17}\,cm^{-2}$, the highest value in area IV. In

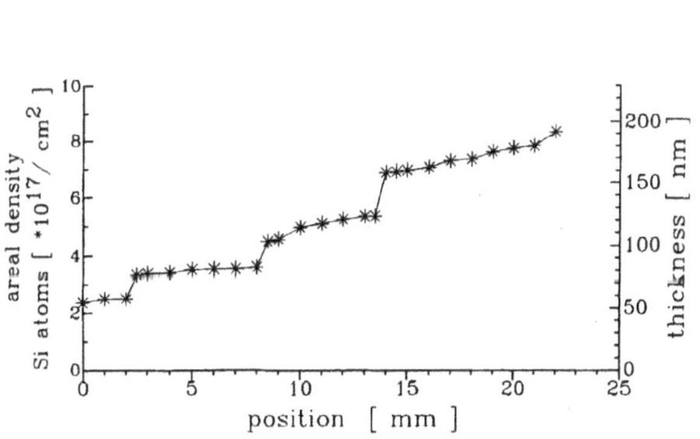

Fig. 7. Line scan of the areal densities of Si atoms along the analysis line AB measured with RBS

addition, the RBS measurements revealed the presence of impurities in the films.

These were mainly elements from stainless steel, originating from the deposition chamber walls, but tungsten impurities were also detected. Both impurities were distributed through the silicon film and the concentrations of stainless steel components and tungsten were in the range 2–8 $\times 10^{14}$ cm^{-2} and 1–4 $\times 10^{14}$ cm^{-2}, respectively.

3.6 Electron Probe X-Ray Microanalysis (EPMA)

For more than a decade a variant of EPMA called thin film EPMA has been a well established technique for the quantitative analysis of thin layers on substrates when the layer thickness, t_F, is in the range 0.1 nm $< t_F < 1$ µm [13–19]. For x-ray intensities calibrated with respect to bulk standard samples the technique is able to calculate the partial coverage of elements present in the layer (areal density of atoms), and also the composition of the film, provided that it is homogeneous with depth. Problems arise in multilayer samples or if elements are present in both film and substrate; for further details we refer to the extensive literature on this subject. The technique applied here to single films presupposes no elements being present in both the film and substrate.

Experimental.

The measurements were carried out using a Camebax SX 50 (CAMECA) microprobe analyser with wavelength dispersive x-ray spectrometers (WDS). To establish proof of consistency several different measuring conditions were chosen. X-ray intensities of Si Kα as an element of the film and C Kα as an element of the substrate were recorded at three different electron beam energies 5.0, 7.5 and 10 keV. These energies correspond to the depths of Si-Kα excitation shown in Table 1. The measured x-ray intensities were calibrated by using reference measurements on pure bulk silicon and pure bulk glassy carbon, respectively. X-ray intensities were measured along a straight line, in steps of 100 µm, across the ledges with different film thickness; the beam diameter was chosen to be 10 µm. Relative x-ray intensities from each point were converted into a silicon coverage (µg/cm^2 or atoms/cm^{-2}) by applying software according to the model of Pouchou and Pichoir [13], which was modified to convert line-scan data automatically into profiles of coverage versus length or distance on a film. The accuracy and precision of this technique was reported to be between 2% and 10% relative, depending on the complexity and thickness of the film(s).

Table 1. Values for the maximum mass depth of Si Kα x-ray generation and most probable mass depth of Si Kα x-ray generation in a Si:H film for beam energies of 5, 7.5 and 10 keV

Beam energy	5.0 keV	7.5 keV	10.0 keV
Max. mass depth of ionisation	75 µg/cm^2	175 µg/cm^2	300 µg/cm^2
Mass depth of max. ionisation probability	10 µg/cm^2	22 µg/cm^2	55 µg/cm^2

Results

Figure 8 presents the results of the three measurements in terms of areal density of silicon atoms along the analysis line trace, AB.

The secondary ordinate gives values of film thickness calculated from areal densities assuming a film density of 2.0 g/cm^3. With high sensitivity and good lateral resolution EPMA is able to trace the variation of film thickness across the sample. Steps, as well as very shallow gradients, can be established. Thickness sensitivity can be deduced from the statistics associated with the measured intensity ratios k and it is proportional to σ(k)/k where σ as the standard deviation of the calibrated intensity. In the present case this sensitivity was determined to be around 1 nm. Thus, the gradient in film thickness, which in the marked area amounts to 1 nm/600 µm, can be detected.

Curves obtained at different conditions of electron excitation do not match exactly, the deviations are systematic and result in an uncertainty of $\pm$ 2–5%. Whilst at 7.5 and 10 keV the lower thickness results are nearly identical the values obtained at 5 keV seem to be underestimated over the whole range of film thickness. A similar tendency appears at larger thickness for the curve of 7.5 keV. This type of deviation is due to the fact that to convert x-ray data to coverage the depth distribution of the ionisation

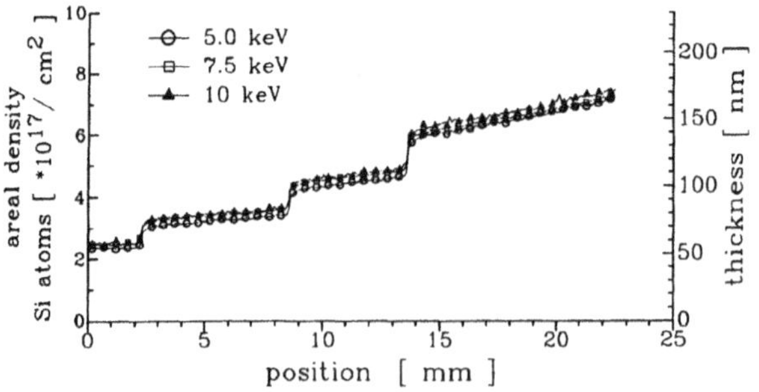

Fig. 8. Line scan of the areal densities of Si atoms along the analysis line AB measured with EPMA

probability, called $\phi(\rho z)$, needs to be known as accurately as possible. In the PAP model used here it is only an approximation to the real distribution, especially for systems where film and substrate differ in mean atomic number. Since the x-ray intensity calculated in the model contains an integral over the ionisation depth distribution, deviations between it and the real distribution result directly in elemental coverage errors.

Nevertheless although systematic, the overall uncertainty of the EPMA data is rather small compared to the results of other techniques.

4. Conclusion

For EPMA and RBS, homogeneity of film composition with depth and low sub-critical surface roughness are requirements. These properties were confirmed by Auger depth profiling and AFM for all layer thickness steps in the specimen studied here.

The techniques applied provide layer information which cannot be compared directly. Ellipsometry and interference fringe analysis give values for the thickness whereas areal densities or mass coverage are obtained by EPMA and RBS. These different layer

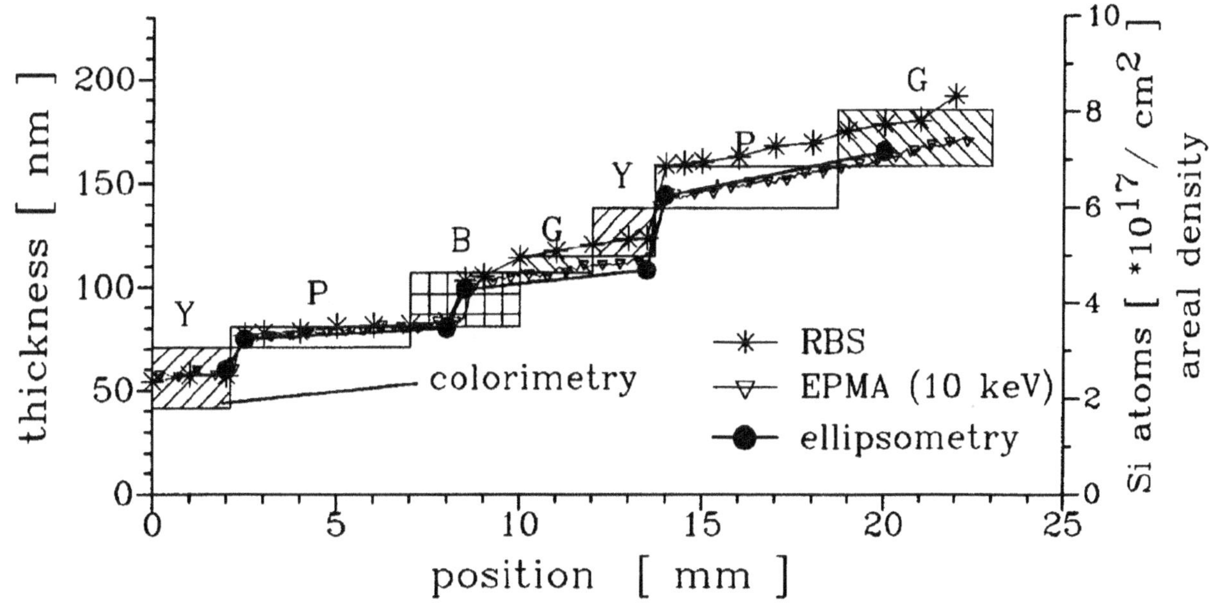

Fig. 9. Comparison of the measurements of ellipsometry, RBS, EPMA and interference fringe analysis

Table 2. Comparison of density determination by different analysis techniques

Analysis method	Optical layer constants				Material properties					
	Refractive index n		Absorption coefficient k		Areal density [Si atoms × $10^{17}/cm^2$]		Thickness [nm]		Density ρ[g/cm^3]	
	step I position 1.5 mm	step IV position 18.5 mm	step I position 1.5 mm	step IV position 18.5 mm	step I position 1.5 mm	step IV position 18.5 mm	step I position 1.5 mm	step IV position 18.5 mm	step I position 1.5 mm	step IV position 18.5 mm
Ellipsometry	3.7	3.7	0.05	0.05	2.5	6.8	60 ± 3	158 ± 5	1.94 ± 0.1	2.0 ± 0.1
Colorimetry	3.7	3.7	0.05	0.05	2.5	6.8	57 ± 15	148 ± 10	2.0 ± 0.5	2.1 ± 0.3
EPMA					2.5	6.8	60 ± 3	158 ± 5	1.94 ± 0.1	2.0 ± 0.1
RBS					2.5	7.5	60 ± 3	158 ± 5	1.94 ± 0.1	2.2 ± 0.2

AES = pure silicon layer, homogeneous in depth; AFM = surface uniform on a scale of a few nanometer.

(shaded) directly measured by corresponding method.
(unshaded) values required for evaluation of the thickness or density.

parameters are linearly related to the density of the material. Thus, for each thickness step the density of the layer can be determined from the results obtained by complementary techniques. Although ellipsometry and interference fringe analysis are in good agreement (Fig. 9), because of its higher accuracy only ellipsometry was used for density determination. Results of measurements and evaluation of layer density are summarised in Table 2 for the thinnest (step I) and the thickest layer (step IV). In this table the results directly deduced from the corresponding measurement are shaded.

Using results from ellipsometry and EPMA an average density of 1.97 ± 0.1 g/cm^3 was found for all thickness steps and no systematic change with layer thickness was evident. However, the density determined from ellipsometry and RBS systematically increased with thickness of the layer, from 1.94 g/cm^3 at 60 nm to 2.2 g/cm^3 at 158 nm. The small density difference can be explained by the amorphous structure of the layer, for comparison the density of Si is 2.33 g/cm. It may be seen that there is closer agreement between ellipsometry and EPMA than there is between ellipsometry and RBS. However, in general, the results obtained by means of the different techniques are highly consistent as illustrated in Fig. 9, which summarizes the results of all the measurements performed.

References

[1] J. Winter, H. G. Esser, L. Könen et al., *J. Nucl. Mater.* **1989**, *162*, 713.
[2] H. G. Esser, J. Winter, V. Philipps et al., *J. Nucl. Mater.* **1995**, *220*, 457.
[3] G. Binnig, C. F. Quate, C. Gerber, *Phys. Rev. Lett.* **1986**, *56*, 930.
[4] C. Smith, *Surface Analysis by Electron Spectroscopy*, Plenum, New York, 1994.
[5] T. E. Gallon, J. A. D. Matthew, *Rev. Phys. Technol.* **1992**, *3*, 31.
[6] P. M. Hall, J. M. Morabito, *Surf. Sci.* **1979**, *83*, 391.
[7] M. Born, E. Wolf, *Principles of Optics*, *6th Ed.* Pergamon, Oxford.
[8] R. M. A. Azzam, N. M. Bashara, *Ellipsometry and Polarized Light*, North-Holland, Amsterdam.
[9] P. Drude, *Ann. Physik* **1889**, *272*, 532; *Ann. Physik* **1889**, *272*, 865; *Ann. Physik* **1890**, *272*, 481;
[10] P. Wienhold, F. Weschenfelder, J. Winter, *Nucl. Instr. Methods Phys. Res.* B **1994**, *94*, 503.
[11] P. Wienhold, U. Littmark, *E-MRS Symp. Proc.*, *Vol. XVII*, Straßbourg, France, 1987, p. 441.
[12] W. K. Chu, J. W. Mayer, M. A. Nicolet, *Backscattering Spectroscopy*, Academic Press, New York, 1978.
[13] J. L. Pouchou, F. Pichoir, *Rech. Aérospat.* **1984–85**, 47.
[14] H. J. Hunger,. W. Baumann, S. Schulze, *Cryst. Res. Technol.* **1985**, *20*, 1427.
[15] R. A. Waldo, *Microbeam Analysis-1988* (D.E. Newbury ed.), San Francisco Press, San Francisco, pp. 310–314.
[16] G. F. Bastin, J. M. Dijkstra, H. J. M. Heijligers, D. Klepper, *Mikrochim. Acta [Suppl.]* **1992**, *12*, 93.
[17] N. Ammann, P. Karduck, *Microbeam Analysis* **1990**, 150.
[18] J. L. Pouchou, F. Pichoir, in: *Electron Probe Quantitation* (K. F. J. Heinrich, D. E. Newbury, eds.), Plenum, New York, 1991, pp. 31–75.
[19] P. Karduck, *Mikrochim. Acta [Suppl.]* **1998**, *15*, 109.

Mikrochim. Acta [Suppl.] 15, 171–176 (1998)

EPMA Studies of the Growth of Thin Surface Coatings Produced by Evaporation

Hildegard Hammer[1,*], **Sabine Kulow**[1], **Christine Schmidt**[2], **Gerhard Schmidt**[1], and **Peter Zanel**[1]

[1] Institut für Angewandte Physik, Heinrich-Heine-Universität Düsseldorf, Universitätsstrasse 1, D-40225 Düsseldorf, Federal Republic of Germany

[2] Institut für Physikalische Chemie und Elektrochemie, Heinrich-Heine-Universität Düsseldorf, Universitätsstrasse 1, D-40225 Düsseldorf, Federal Republic of Germany

Abstract. We have studied by electron probe microanalysis (EPMA) the growth of thin surface layers prepared by evaporation and have compared the experimental *k*(*d*) curves before and after heating the samples with those calculated by Monte Carlo simulation.

When studying systems of substrate and layer materials forming homogeneous flat films, the correspondence of the experimental and calculated *k*(*d*) curves (the normalized intensities of the characteristic x-ray lines of film and substrate material, depending on film thickness *d*) is quite good. However, when forming alloys of these materials, for example by heating the sample, the experimental *k*(*d*) curve is drastically changed.

Key words: EPMA, film growth, alloying, copper, gold, silver.

The method of studying thin coating films by electron probe microanalysis (EPMA) is well known. We have carried out some special experiments [1] to look at the film growth layer by layer and the alloying between the materials of film and substrate. The principal aim of these experiments has been to get some fundamental data so that the application of the EPMA technique to thin films, prepared during external experiments, becomes possible. In contrast to surface-sensitive techniques, as for instance Auger electron spectroscopy, the EPMA technique is applicable to those samples which may be destroyed by sputter depth profiling or those which are not compatible with ultra-high vacuum conditions. A rather good test of the layer-by-layer growth of coatings is to excite the samples by electrons of different energies (10 keV, 30 keV) and to compare the experimental *k*(*d*) curves with those calculated by Monte Carlo simulations for samples consisting of a single element substrate and a thin coating of another element.

Experimental

The film and substrate materials were copper, gold and silver because these show large thermal diffusivity values [2] and form ideal alloys at higher temperatures [3]. Also the films' growth during the coating process by evaporation is layer by layer [4; 1]. Additionally, diffusion or alloying is observable with the naked eye due to the different colours of the materials and of the alloys.

The EPMA experiments have been performed in a scanning electron microscope ISI 60. The x-ray spectra have been measured by an energy dispersive system (ORTEC, Si(Li)-detector), the primary electron energies were 10 keV and 30 keV respectively. Fig. 1 shows the corresponding spectra of a copper, gold and silver foil as standard sample.

The films have been prepared by evaporation and heating under vacuum. The temperatures were about 450 °C for about 60 minutes for the systems Cu/Au and Cu/Ag and 30 minutes for the system Au/Ag, respectively, from consideration of the phase diagrams of these binary solutions [3]. The film thicknesses have been measured in situ by optical transmission and additionally by interference microscopy.

* To whom correspondence should be addressed

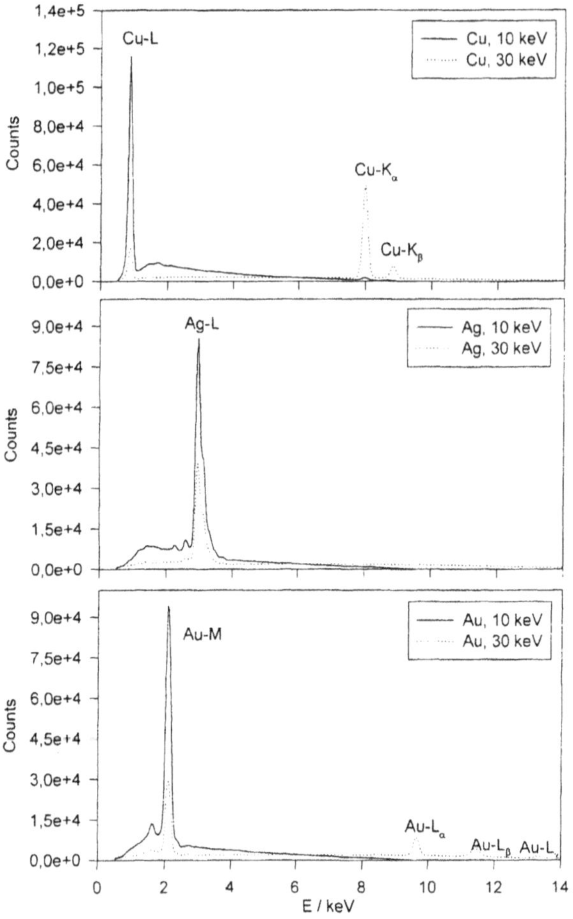

Fig. 1. Spectra of the bulk materials copper, silver, gold (top to bottom)

The results are presented as $k(d)$ curves verses d [5]:

$$k = \frac{I_{AB}}{I_A} = \frac{\int_0^d \Phi(\rho z) e^{-\chi \rho z} d(\rho z)}{\int_0^R \Phi(\rho z) e^{-\chi \rho z} d(\rho z)}$$

I_{AB} : x-ray intensity of the film (material A) on a substrate (material B)
I_A : x-ray intensity of the bulk material A
d : film thickness
R : effective depth of x-ray excitation
χ : $\chi = \mu/\rho \csc(\theta)$
θ take-off angle of the analyzed x-ray intensity
μ/ρ mass absorption coefficient of the characteristic x-rays

The experimental intensity values are recorded by measuring the x-ray peak area in the spectra of the coated samples (I_{AB}) divided by the peak area of the corresponding standard spectra (I_A). The simulated $k(d)$ curves assume single elemental substrates and films, and also ideal layer by layer growth of the coating, i.e. no alloys or compounds or interdiffusion is considered. We are still working on simulations for these problems. The procedure for the calculated $k(d)$ curves by Monte Carlo simulation is a modification of Joy's single scattering method [6]. The x-ray intensity is calculated from knowledge of the ionisation cross section. To get the $\Phi(\rho z)$ curves a fixed maximum depth is defined which is determined by the lightest element used and this is divided into 256 or 512 parts of equal width. The x-ray intensity generated in each depth interval is summed, resulting in $\Phi(\rho z)$ curves which are comparable with regard to their yield scale. The $\Phi(\rho z)$ curves are calculated for bulk standards of all materials used, i.e. films and substrates. The standard intensity is gained for each material from the corresponding $\Phi(\rho z)$ curve by integrating the curve with respect to the depth. $\Phi(\rho z)$ curves are calculated and integrated with respect to the depth for both the coating and the substrate in systems with increasing film thicknesses. These data are divided by the corresponding results from bulk standards to gain $k(d)$ curves. The $k(d)$ values of the substrate are corrected with regard to the x-ray absorption occuring in the film.

Results

The Figs. 2, 3, and 4 show from top to bottom the calculated $k(d)$ curves, the experimental $k(d)$ curves taken directly after deposition, and those after heating the films of silver, gold and copper, respectively.

The simulated $k(d)$ curves for the ideal layered structure demonstrate differences in the values for the film and for the substrate for 10 keV and 30 keV primary electron energies. These differences should be qualitatively seen in the experimental curves whenever the film growth is layer by layer. Comparing the calculated and experimental $k(d)$ values directly after deposition for 10 keV, the agreement is quite good, for 30 keV less good for all systems. At 30 keV, the Monte Carlo predictions of the film $k(d)$ values are lower than the measured values and those for the substrates are higher. These discrepancies are not yet well understood. The $k(d)$ curves are changed by the heating process for the silver films on copper (Fig. 2 left) and on gold (Fig. 2 right) the $k(d)$ curves appear shifted to higher film thicknesses by the diffusion of some silver into the substrate materials copper and gold. At larger thicknesses, the differences between 10 keV and 30 keV values for the layered structures are still observable. Regarding the heated samples, the very thin silver films have nearly totally vanished from the surface.

Looking at Fig. 3 and the gold films on copper (left) and silver (right), the $k(d)$ curves change drastically by heating: the $k(d)$ curves of 10 keV and 30 keV are

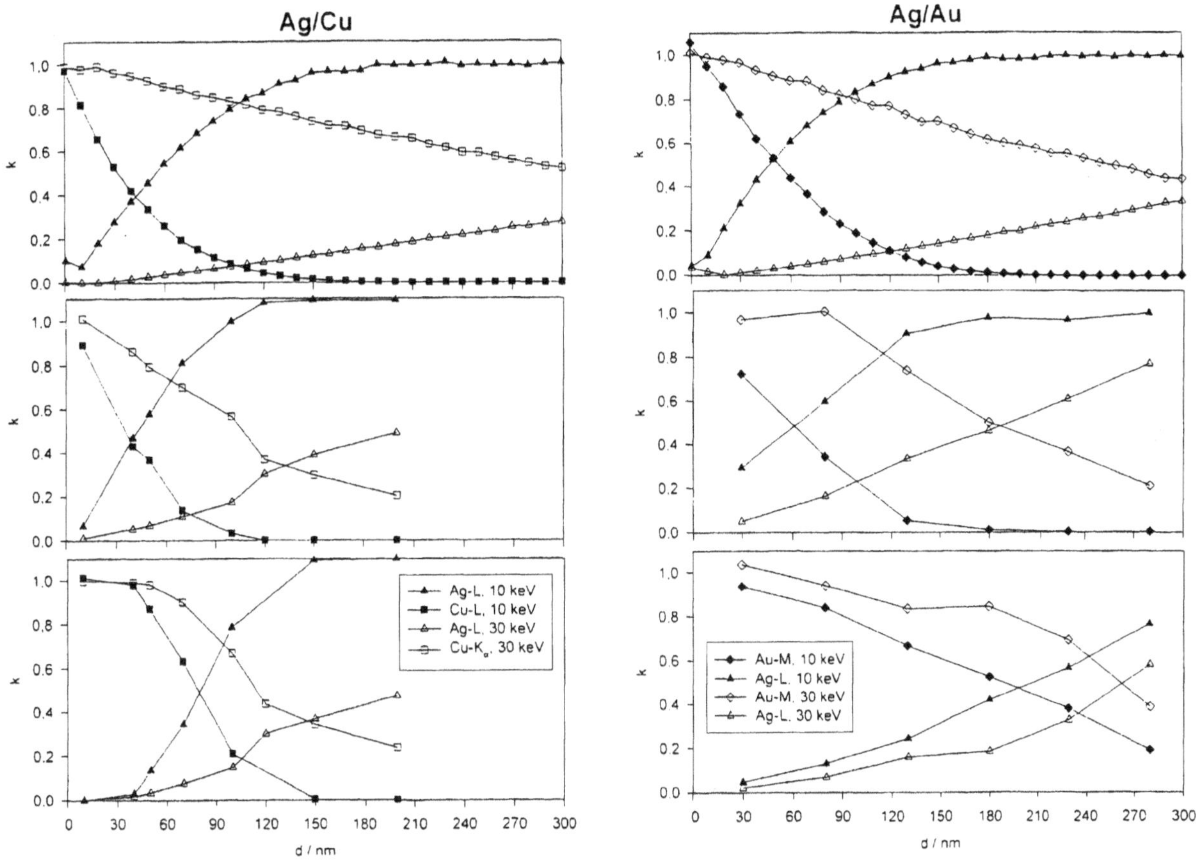

Fig. 2. *k*(*d*) curves for silver films on copper (left series) and on gold (right series). Top: results of Monte Carlo simulations. Middle: experimental results for films directly after deposition. Bottom: experimental results after heating (450 °C, 60 minutes (left) and 30 minutes (right))

nearly identical showing a weak dependence of the film thickness and it would appear that Au/Cu and Au/Ag alloys have formed in the surface region of the sample.

By heating the copper films on silver (Fig. 4 left) and on gold (Fig. 4 right), the copper signal totally vanishes. The copper must have diffused far into the substrate material, because the temperature of about 450 °C is not high enough to evaporate the copper from the samples.

The well known effect of forming by diffusion, a very thin silver film on top of a gold coating on a silver substrate is detectable by EPMA too when using electrons with 5 keV primary energy. The calculated *k*(*d*) curve for 5 keV (Fig. 5) shows zero intensity of silver for films thicker than 30 nm when the electron energy is too low to excite the silver L shell. But the experimental curve for silver shows non zero intensities hinting at a thin layer of silver on top of the gold films.

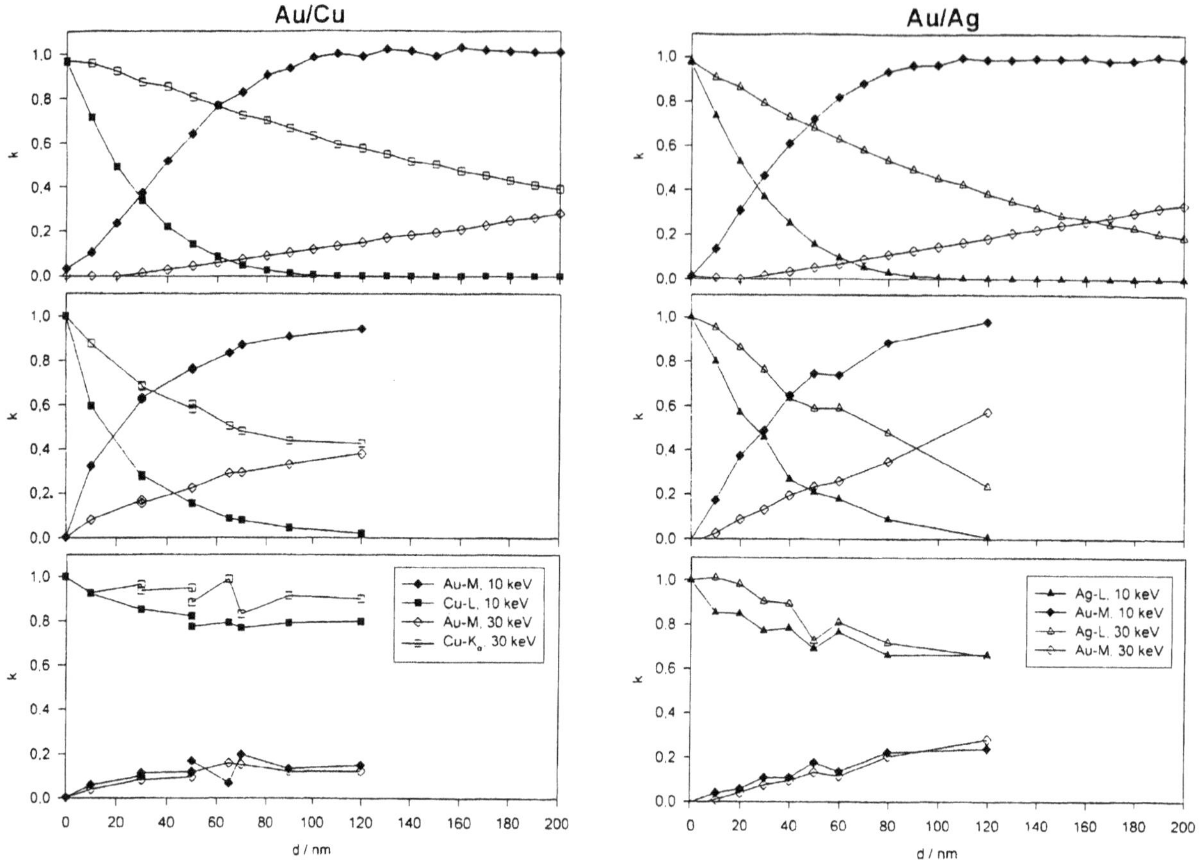

Fig. 3. $k(d)$ curves for gold films on copper (left series) and on silver (right series). Top: results of Monte Carlo simulations. Middle: experimental results for films directly after deposition. Bottom: experimental results after heating (450 °C, 60 minutes (left) and 30 minutes (right))

All these results have been verified by depth profiling and Auger electron spectroscopy.

Conclusion

The following systems have been studied: copper and silver films on a gold substrate, copper and gold films on a silver substrate, silver and gold films on a copper substrate.

The $k(d)$ curves have been measured under two conditions: excitation by 10 keV electrons and by 30 keV electrons before and after heating the samples.

It was possible to point out on the one hand a nearly ideal film growth and on the other hand the alloy transformations in these systems. The degree of alloying has been more dependent on the material of the film than on the combination of

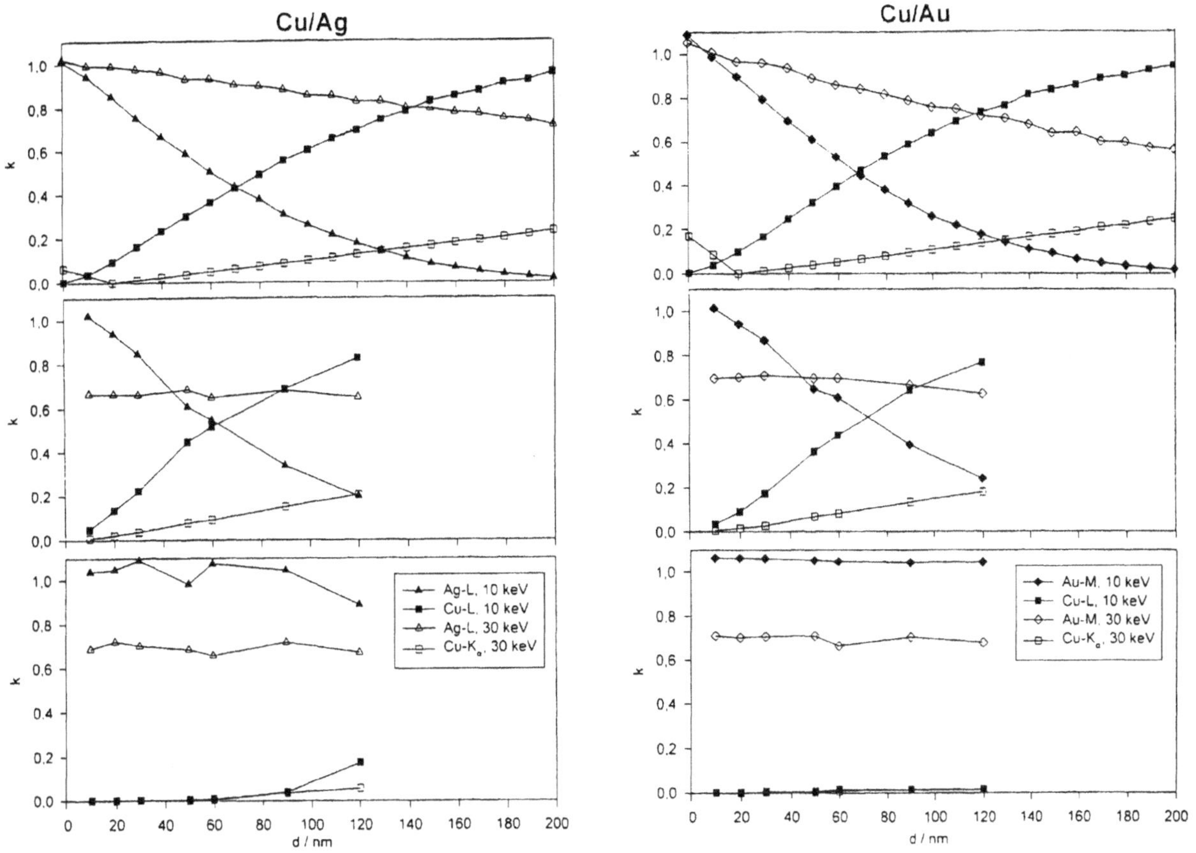

Fig. 4. *k*(*d*) curves for copper films on silver (left series) and on gold (right series). Top: results of Monte Carlo simulations. Middle: experimental results for films directly after deposition. Bottom: experimental results after heating (450 °C, 60 minutes)

materials. The silver, gold and copper films are alloying in an increasing degree on heating the samples.

The results, mentioned above, show, that the study of coatings by the EPMA technique at different primary electron energies in combination with Monte Carlo simulation give some information about the coated system without destroying the coating by sputtering as is normally done when using surface sensitive methods:

i) it is possible to distinguish between an ideal layered coating and the formation of interdiffused or alloyed layers

ii) in the case of ideal layered coatings it is possible to determine the thickness of the coating by comparison with relevant *k*(*d*) standard curves.

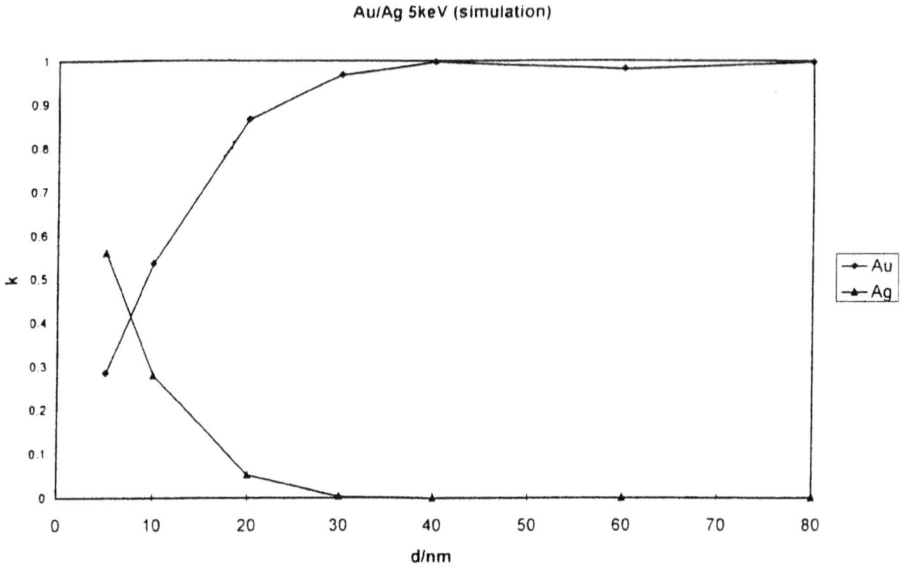

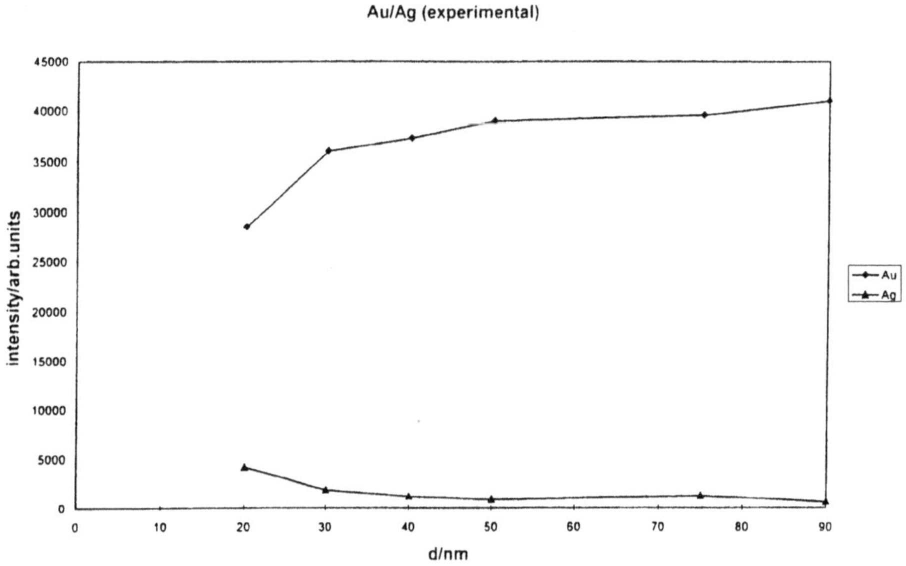

Fig. 5. $k(d)$ curves for gold films on silver excited by 5 keV electrons. Top: results of Monte Carlo simulation. Bottom: experimental results directly after deposition (here intensities rather than k values are plotted)

Acknowledgements. The authors would like to thank Prof. Kisker for some financial support.

References

[1] C. Schmidt, *Diploma Thesis*, Duesseldorf, 1995.

[2] Y. S. Touloukian (ed.), *Thermophysical Properties of Matter Vol. 10: Thermal Diffusivity; IIFI*, Plenum, New York, 1973.

[3] M. Hansen, *Constitution of Binary Alloys*, McGraw Hill, New York, 1958.

[4] S. Kulow, *Master Thesis*, Duesseldorf, 1994.

[5] J. I. Goldstein, A. D. Romig, Jr., D. E. Newbury, C. E. Lyman, P. Echlin, C. Fiori, D. C. Joy, E. Lifshin, *Scanning Electron Microscopy and X-Ray Microanalysis, 2. ed.*, Plenum, New York, 1992.

[6] D. C. Joy, in: *Quantitative Microbeam Analysis* (A. G. Fitzgerald, B. E. Strey, D. Fabian, eds.), 10P Publishing Ltd, London, 1993, p. 213.

Mikrochim. Acta [Suppl.] 15, 177–180 (1998)

Analysis of Thin Films with Slightly Rough Boundaries

Ivan Ohlídal[1,*], **Daniel Franta**[1], **Jaroslav Hora**[1], **Karel Navrátil**[1], **Jan Weber**[2], and **Pavel Janda**[2]

[1] Department of Solid State Physics, Faculty of Sciences, Masaryk University, Kotlářská 2, 611 37 Brno, Czech Republic
[2] J. Heyrovský Institute of Physical Chemistry, Dolejškova 3, 182 23 Prague 8, Czech Republic

Abstract. Technological procedures often produce thin films with rough boundaries as for example, semiconductor films prepared by various chemical methods. If roughness of the boundaries of the films is very small, i.e. if the rms values of the heights of boundary irregularities are equal to several nanometers, it is very difficult to carry out a detailed analysis of the films. Spectroscopic ellipsometry and atomic force microscopy represent two experimental techniques enabling us to determine some important parameters of these films (e.g. thickness) together with the main statistical quantities characterizing slight (nanometric) roughness of their boundaries.

In this paper a method will be presented based on a combination of variable angle spectroscopic ellipsometry, spectroscopic reflectometry and atomic force microscopy enabling us to determine the optical parameters and the most significant statistical parameters of the rough boundaries characterizing the slightly rough single layers.

Within the method the generalized Rayleigh-Rice theory published elsewhere will be used for interpreting experimental data obtained from both the optical techniques, i.e. variable angle spectroscopic ellipsometry and spectroscopic reflectometry.

The method will be illustrated by presenting experimental results achieved by applying the method to slightly rough surfaces of single crystals of GaAs covered by very thin surface layers (i.e. by native oxide layers).

Key words: Spectroscopic Ellipsometry, Spectroscopic Reflectometry, Atomic Force Microscopy.

* To whom correspondence should be addressed

In this paper the optical methods, i.e. variable angle spectroscopic ellipsometry (VASE) and spectroscopic reflectometry (SR), will be used for analysing very thin films with slightly rough boundaries forming on a GaAs single crystal substrate. It will be shown that thickness and spectral dependence of the refractive index of these films together with the basic statistical quantities characterizing random roughness of their boundaries can be evaluated with a relatively high accuracy using the optical methods mentioned. Some of the statistical quantities determined by the optical methods will be confirmed by means of atomic force microscopy (AFM). Moreover, it will be shown that both the optical methods and AFM can be combined in an effective way within the complete analysis of the films mentioned.

Preparation of Samples and Experimental Arrangement

The samples of the GaAs surfaces covered by the rough, very thin, films were prepared by thermal oxidation of GaAs wafers (intrinsic type, (100) orientation, $\rho = (1.0\text{–}1.5)10^{18}\Omega$cm) at a temperature of 400 °C in air. This was followed by dissolution of the oxide layers originated (dissolution of these films was carried out using a mixture of H_2O and HF). It is known that the GaAs surfaces subjected to the procedures mentioned are more or less rough [1, 2]. Because of the fact that the rough GaAs surfaces were placed in air before measurement very thin native oxide films originated on these surfaces.

Spectral dependences of the ellipsometric parameters Ψ (azimuth) and Δ (phase change) for several angles of incidence were measured by a rotating analyser type ellipsometer constructed on the basis of the Gaertner L119 XUV ellipsometer. A halogen bulb with a monochromator and *Si* photodiode were used as the light source and the detector of the ellipsometer, respectively.

The spectral dependence of the reflectance R was measured at near normal incidence (the angle of incidence Θ_0 was about 11°).

For measuring the reflectance a commercial spectrophotometer, Specord 42, was employed.

Roughness of the samples under consideration was also quantitatively studied using an atomic force microscopy (Discoverer, TopoMetrix Inc).

Data Processing

All the values of the ellipsometric parameters and reflectance corresponding to the spectral dependences measured were simultaneously treated by least-squares methods (LSM). Theoretical values of Ψ, Δ and R needed for applying the LSM were calculated by means of a generalized Rayleigh-Rice theory [3] assuming that the films investigated were represented by the identical thin film (ITF). The ITF is a rough film whose boundaries are identical from both the statistical and geometrical points of view. These boundaries are statistically described by the autocorrelation function $\rho(\vec{\tau})$ defined as follows:

$$\rho(\vec{\tau}) = \langle \zeta(\vec{r})\zeta(\vec{r}+\vec{r}) \rangle, \tag{1}$$

where ζ denotes a random function, the vectors $\vec{r}$ and $\vec{\tau}$ determine the position of the points in the mean planes of the boundaries and the symobls $\langle \rangle$ denote the statistical mean values written inside these symbols. In a forthcoming paper it will be shown that the ellipsometric parameters and reflectance of the ITF with slightly rough boundaries are given as follows (for the slightly rough boundaries it is the case that $\sigma \ll \lambda$, where σ is the standard deviation of the height of irregularities and λ is the wavelength of incident light):

$$\tan \Psi e^{i\Delta} = \frac{\hat{r}_p}{\hat{r}_s} \quad \text{and} \quad R = \frac{|\hat{r}_p|^2 + |\hat{r}_s|^2}{2}, \tag{2}$$

where $\hat{r}_p$ and $\hat{r}_s$ denote the Fresnel coefficients of the ITF for p and s polarization, respectively. $\hat{r}_p$ and $\hat{r}_s$ are expressed in the following way:

$$\hat{r}_q = \hat{r}_q^{(s)} + \int_{-\infty}^{+\infty} \int_{-\infty}^{+\infty} f_q(K_x, K_y) W(K_x, K_y) dK_x \, dK_y, \tag{3}$$

where $q = p, s$, W denotes the spectral density of spatial frequencies of roughness. W is also the Fourier transform of the autocorrelation function, ρ, of roughness and f_q is a complex function (the mathematical expression of f_q is extremly complicated and therefore it will be published elsewhere). $\hat{r}_q^{(s)}$ are Fresnel coefficients of the smooth film. K_x and K_y are variable quantities in reciprocal space.

Note that the foregoing equations were derived under the assumption that the slopes of the irregularities are extremely small (i.e. when the standard deviation of the slopes tan β_0 fulfils the following relation: tan $\beta_0 \lesssim 0.1$).

Four models of the ITF were taken into account in treating the experimental data by the LSM. These models are as follows:

1. Model S: The ideally smooth film (i.e. the ITF with smooth boundaries).
2. Model G: The ITF whose boundaries are described with the function W in the Gaussian form, i.e.:

$$W = \frac{\sigma^2 T^2}{4} e^{-(K_x^2+K_y^2)T^2/4}, \tag{4}$$

where T is autocorrelation length. Then the autocorrelation function is expressed in the following form:

$$\rho(\tau) = \sigma^2 e^{-\tau^2/T^2}. \tag{5}$$

3. Model E: The ITF whose boundaries are described with the function W as follows:

$$W = \frac{\sigma_E^2 T_E^2}{4} e^{[-(K_x^2+K_y^2)T_E^2/4]^h}, \tag{6}$$

where σ_E, T_E and h are parameters. In this case the autocorrelation function must be calculated in a numerical way.

4. Model D: The ITF whose boundaries are described with the function W in the following form:

$$W = \frac{\sigma_1^2 T_1^2}{4} e^{-(K_x^2+K_y^2)T_1^2/4} + \frac{\sigma_2^2 T_2^2}{4} e^{-(K_x^2+K_y^2)T_2^2/4}. \tag{7}$$

Thus, within this model roughness of the boundaries are formed by two Gaussian components. The models E and D are taken into account because of the fact that the function W need not generally be in Gaussian form. Within all the models outlined above it was assumed that the spectral dependence of the refractive index of the ITF is given by the Cauchy formula:

$$n_1 = n_{10} + n_{11}/\lambda^2. \tag{8}$$

Thus by means of the LSM the values of n_{10}, n_{11} and d (thickness of the ITF) together with the values of the statistical parameters corresponding to the selected model of the ITF had to be determined. The simultaneous treatment of both the ellipsometric and reflectometric data enabled us to remove or reduce a correlation between the parameters sought by the LSM.

A family of the heights of the irregularities evaluated using the atomic force microscope mentioned was utilized for determining the autocorrelation function $\rho(\tau)$ of boundary roughness. The following formula was employed for this purpose:

$$\rho(\tau) = \frac{1}{M-m}\sum_{n=0}^{M-m-1} \zeta(n\tau_0)\zeta(n\tau_0 + m\tau_0), \qquad (9)$$

where $\tau = m\tau_0$, τ_0 is the distance between the neighbouring points in the mean plane, $m = 0,1,2,3,\ldots, M-1$ and M denote the number of all the points employed for determining $\rho(\tau)$ by Eq. (9). $\sigma = \sqrt{\rho(0)}$ is fulfilled.

Experimental Results

In this paper results are presented of the analysis of a chosen sample of the GaAs surface covered by a very thin oxide film with slightly rough boundaries. The period of thermal oxidation of the corresponding GaAs wafer was 10 hours.

For the optical analysis of all the samples the optical constants of the GaAs substrate were taken from the values published earlier [4].

The results of the analysis of the sample specified above obtained by the optical methods (i.e. by VASE and SR) are summarized for all the models considered in Table 1. The values of σ corresponding to models E and D are calculated by equations $\sigma = \sqrt{\rho(0)}$ and $\sigma = \sqrt{\sigma_1^2 + \sigma_2^2}$, respectively (for model E one must calculate the function $\rho(\tau)$ in a numerical way). From this table one can see that the best fit of the experimental data is obtained for model D (see the values of the minimum of the sum of the squares χ^2). In Figs. 1, 2, and 3 the agreement between both the experimental and theoretical data for this model is presented. The remaining parameters determining the spectral densities of spatial frequencies W of the individual models have been evaluated as follows: $T = (50.0\pm0.8)$ nm (model G), $\sigma_E = (1.2\pm0.4)$ nm, $T_E = (610\pm320)$ nm and h = 0.28±0.03 (model E) and $\sigma_1 = (2.0\pm0.1)$ nm, $T_1 = (12\pm1)$ nm, $\sigma_2 = (3.89\pm0.08)$ nm and $T_2 = (63\pm2)$ nm (model D).

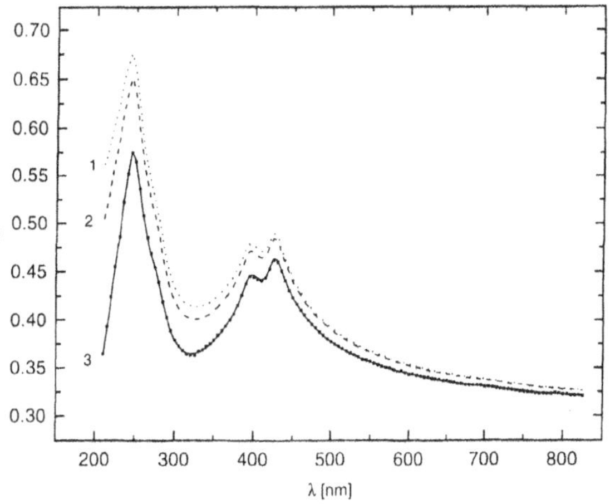

Fig. 1. The spectral dependences of the reflectance: curve 1-theoretical values of the smooth surface GaAs calculated according to [4]; curve 2-theoretical values of the reference sample (the smooth GaAs substrate covered with native oxide layer: $d = 3.65$ nm, refractive index of the layer is taken according to [2]); curve 3-theoretical values calculated by means of the best fit for the sample studied; the points represents the experimental values of the reflectance measured

By Fourier transform of the functions W belonging to models G, E and D the corresponding autocorrelation functions $\rho(\tau)$ are found (see Fig. 4). In this figure the autocorrelation function determined by AFM (see Eq. (9)) is also plotted. The value of σ evaluated by the AFM measurements is as follows: $\sigma = (3.7\pm1.6)$ nm. The value of σ and its error have been calculated in the standard statistical procedures on the basis of the values of the heights corresponding to 6 scans (in one scan 300 × 300 points have been recorded for an area of 5 × 5 μm). One can see that there is moderate agreement between the values of σ and $\rho(\tau)$ determined by the optical methods and by AFM. The differences existing between values of $\rho(\tau)$ are caused by the fact that when using AFM it is not

Table 1. The values of the optical parameters and the standard deivation σ determined for the sample studied

Model	σ [nm]	d [nm]	n_{10}	n_{11} [nm^2]	χ^2
S	–	7.03±0.09	1.86±0.02	10100±500	0.1596
G	4.24±0.06	6.57±0.05	1.69±0.01	13700±300	0.0243
E	4.43±0.06	6.53±0.06	1.59±0.01	17100±400	0.0198
D	4.38±0.06	5.97±0.13	1.64±0.02	15700±700	0.0178

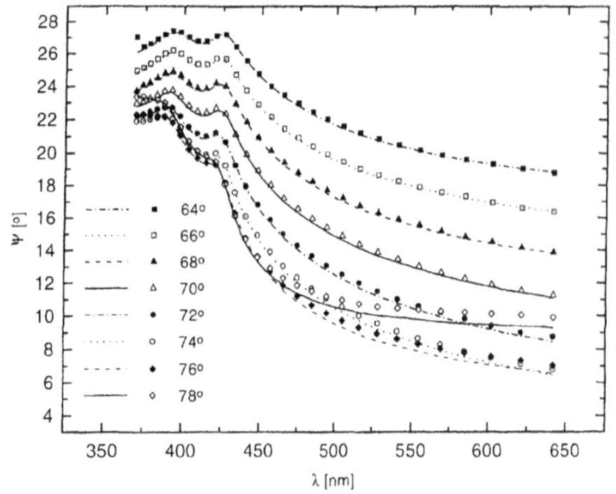

Fig. 2. The spectral dependence of Ψ of the sample studied for different angles of incidence specified in the figure legend (the point and/or curves denote the experimental and/or theoretical values)

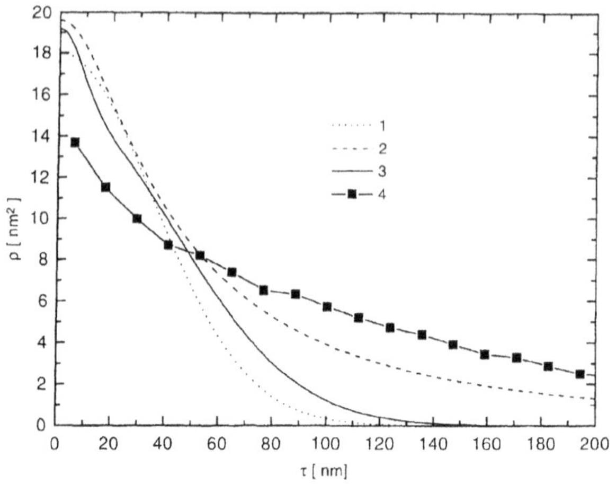

Fig. 4. The autocorrelation functions of the sample studied: curve 1-model G; curve 2-model E; curve 3-model D; curve 4-the AFM measurement

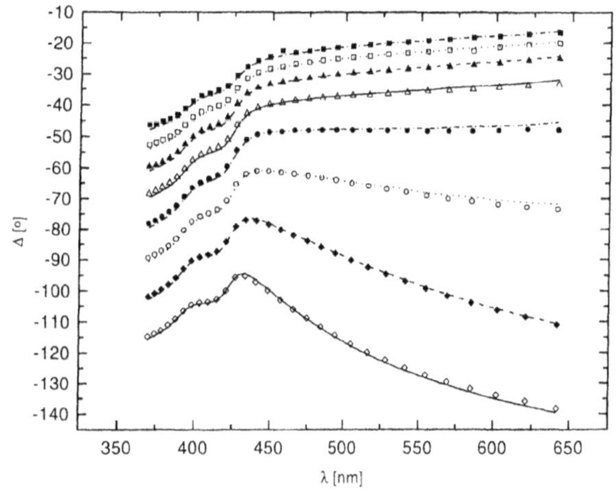

Fig. 3. The spectral dependence of Δ of the sample, studied for different angles of incidence (everything is the same as in Fig. 2)

possible to record the components of roughness characterized by higher spatial frequencies. This statement is supported by the results corresponding to model D (see the agreement between the value of σ determined by AFM and the value of σ_2 evaluated by the optical method within the model D). This problem will be discussed in detail elsewhere. From the foregoing results it is possible to imply that in contrast to AFM the optical methods utilized are sensitive to both the higher and lower spatial frequencies of roughness.

It is also evident that the optical methods can be combined with AFM within the analysis of such the films. If the value of σ_2 from model D is used with the data evaluated by AFM one obtains the following results; $\sigma_1=(2.25\pm0.08)$ nm, $T_1=(15\pm1)$ nm, $T_2=(67\pm2)$ nm, $d=(6.15\pm0.10)$ nm, $n_{10}=(1.62\pm0.02)$ nm, $n_{11}=(16800\pm500)\,\mathrm{nm}^2$, and $\chi^2=0.0181$. By means of this combination the correlation between the remaining parameters sought using LSM is reduced and the accuracy in their determination is increased.

Conclusion

In this paper it is shown that optical methods (i.e. VASE and SR) can be employed for analysing very thin films with slightly rough boundaries. This conclusion has been illustrated by means of the analysis of native oxide layers formed on a rough GaAs surface. The results achieved by the optical methods have been supported by those obtained using AFM. Moreover it has been shown that within the analysis of the films mentioned, optical methods can be usefully combined with AFM.

Detailed results concerning the analysis of the films under consideration will be presented in a forthcoming paper.

References

[1] K. Navrátil, I. Ohlídal, F. Lukeš, *Thin Solid Films* **1979**, *56*, 163.
[2] I. Ohlídal, F. Vižda, M. Ohlídal, *Opt. Engn.* **1995**, *34*, 1761.
[3] S. O. Rice, *Commun. Pure Appl. Math.* **1951**, *4*, 351.
[4] D. E. Aspnes, A. A. Studna, *Phys. Rev. B* **1983**, *27*, 985.

Mikrochim. Acta [Suppl.] 15, 181–186 (1998)

Effect of Chromium Substrate Pretreatment on Diamond Growth by the Chemical Vapour Deposition Method

Olga Glozman[1,3]*, **Alexander Berner**[2], **Dan Shechtman**[2,3], and **Alon Hoffman**[1,3]

[1] Chemistry Department, Technion, Haifa 32000, Israel
[2] Material Engineering Department, Technion, Haifa 32000, Israel
[3] Wolfson Centre for Interface Science, Technion, Haifa 32000, Israel

Abstract. The effects of nitriding and carburizing chromium substrates on the nucleation of CVD diamond were studied. Surface morphology, incubation time and diamond particle density at the initial stages of deposition were examined using high resolution scanning electron microscopy (HRSEM) as a function of the substrate pretreatment. The presence of amorphous carbon, as well as residual stress in the deposited films were determined by Raman spectroscopy.

The composition and structure of the near surface region of the samples were investigated by quantitative electron probe microbeam analysis (EPMA) as a function of the various pretreatments. The elemental profile across the near-surface layer was determined by peak deconvolution with the instrumental function of the electron probe. The function was approximated by a radial Gaussian distribution, whose parameters were determined using a Faraday cup and Monte-Carlo simulations. X-ray diffraction (XRD) was used as a complementary technique for the determination of substrates' phase composition following heat pretreatment and diamond deposition. EPMA appeared to be a suitable method for the investigation of carbon diffusion into the substrate during the diamond CVD process.

Key words: HRSEM, EPMA, Raman spectroscopy, CVD diamond growth.

The unique properties of diamond, such as its high hardness and thermal conductivity at room temperature, and its good resistance to chemicals, heat and radiation are so attractive that it has been named the ceramic coating of the future. Interest in diamond entered a new phase when the production of polycrystalline diamond films became feasible by a wide variety of chemical vapor deposition (CVD) techniques. These processes are characterised by the activation of a mixture of a hydrocarbon gas with excess hydrogen at low-pressure and low-temperature [1, 2]. Since diamond is metastable under these conditions, its nucleation and growth are kinetically driven and strongly affected by deposition conditions and properties of the substrate [3].

CVD growth of diamond on a pre-existing diamond substrate has been reported and its mechanism is relatively well understood. However, knowledge of diamond deposition on non-diamond substrates is poor, mainly because diamond formation is complicated by the need for a nucleation stage. Additionally, the chemical interaction between the substrate material and the activated gas mixture can lead to the formation of a non-diamond carbon phase and of carbide phases [4–7].

Of special interest is the use of CVD diamond films as hard coatings on cutting tools. We have previously reported the use of a chromium nitrided coating as an interlayer for diamond deposition on steel substrates [8, 9]. Using this interlayer, continuous diamond films with good adhesion to the substrates were deposited. Further study of the diamond CVD growth, on steel substrates coated by nitrided Cr films and on nitrided polycrystalline Cr substrates, demonstrated that the phase transformation, recrystallization and diffusion processes in the near-surface regions of substrates

* To whom correspondence should be addressed

resulted in a very stable chemical bonding and good adhesion of the diamond film to the substrates [10].

In this article we report on the effect of nitriding and carburizing of Cr substrates on the initial stages of CVD diamond deposition. The formation and growth of the diamond films, and the composition and structure of the near-surface region of the Cr substrates were investigated as a function of substrate pretreatment, using HRSEM, EDS, XRD and Raman spectroscopy.

Experimental

Substrate Preparation

Mirror polished polycrystalline Cr, supplied by MERCK (catalogue num. 112095), was used as substrates. In the pretreatment step, the substrates were nitrided and carburized. Nitriding was performed in a 100 sccm (cm^3 per second) ammonia flow at a substrate temperature of 900 °C and a system pressure of 200 Torr for 1 h. Carburizing was performed in a hot filament reactor at a substrate temperature of 800 °C for 1 h, under standard CVD conditions (see below). Cr substrates with no pretreatment were used as a reference for diamond CVD growth. Prior to diamond deposition, all substrates were ultrasonically abraded for 20 minutes in a suspension of 0.25 g of diamond powder (1–3 μm) in 10 ml of ethyl alcohol.

Diamond deposition was performed using a hot filament CVD reactor described previously [9, 10]. A methane to hydrogen gas mixture of 1:99 at.% and a working pressure of 50 Torr were used for deposition. The substrates were placed 8 mm below a carburized tungsten filament. During deposition the sample and filament temperatures were 800 °C and 2000 °C, respectively.

Characterization Techniques

The surface morphology of the substrates, the diamond particle density and the morphology of the deposited diamond films were examined by a high resolution field emission gun SEM. The high resolution SEM study was carried out on a DSEM LEO-982 type microscope with an accelerating voltage of 1 to 3 keV. It was found that under these conditions the electron probe does not affect the specimen surface and that image quality is not damaged as a result of charging. HRSEM images were recorded using the vertical in-lens detector of secondary electrons.

The presence of different solid carbon phases in the deposited films and determination of residual stress in the diamond films were determined by Raman Spectroscopy. The Raman measurements were carried out using an Ar laser (514.5 nm) with an output power of 5 m W and a laser beam diameter of 2 μm (DILOR DXY). The Raman spectra were gathered in the 1200–1700 cm^{-1} range. It is known that the Raman peak of diamond at 1332 cm^{-1}, attributed to the zone-centre phonon, is triply degenerated in unstressed diamond. This peak may split and shift from its normal position at 1332 cm^{-1} as result of the stress in the film [11–13]. Compressive stress in the deposited film results in the shift of the diamond peak to a higher wavenumber.

The phase composition in the near surface region of the Cr substrates and the deposited film were studied by x-ray diffraction using a Bragg-Brentano type diffractometer with a long-focus Cu x-ray tube using Cu Kα radiation (wave length 1.54 Å).

The elemental distribution in the near-surface layer of cross-sectioned samples was measured by energy dispersive spectroscopy using a JEOL 840 SEM equipped with energy dispersive spectrometer LINK ISIS (Oxford Instruments, England). Cross sections of the substrates were prepared for compositional analysis using standard metallographic techniques. Quantitative analysis was performed at an incident electron energy of 10 keV and 1 nA beam current using Cr Kα, C Kα, and N Kα analytical lines. Pure Cr, low alloy steel (British Chemical Standard, No 402) for C, and AlN for nitrogen were used as standards. A conventional ZAF correction from the LINK ISIS software package was used to quantify the results of measurements. Accuracy of the quantitative determinations was verified with the help of bulk standards: Cr_3C_2 and CrN. It has been shown that discrepancies were within the limits of statistical error.

Due to the small thickness of some layers formed in the near-surface region, the finite size of the electron probe (an instrumental function) can significantly reduce the precision of quantitative EDS analysis. The measured concentration profile exhibits, in fact, a convolution of a real elemental distribution with the electron distribution within the excitation volume. In an effort to separate the effect of the instrumental function, deconvolution was carried out. We have assumed that the elemental distribution within a thin layer is a right-angled step with a definite width and height.

The electron distribution in the electron beam was measured by the Faraday cup technique [14] and was found to fit a normal distribution of width $\sigma_1 = 233$ nm (under the indicated analytical conditions). The electron distribution in the sample resulting from the scattering of electrons was evaluated by a Monte Carlo simulation [15]. Work [15] has shown that this distribution can be fitted to a Gaussian function and its width, σ_2, was determined for each element of interest. Values of σ_2 calculated for Cr Kα, C Kα and N Kα for different carbides and nitrides of chromium fall in the range from 40 to 70 nm. The instrumental function is a Gaussian distribution with a width σ, where

$$\sigma^2 = \sigma_1^2 + \sigma_2^2. \quad (1)$$

Under the indicated analytical conditions (10 keV) $\sigma_2 \ll \sigma_1$, and σ was taken as $\sigma = \sigma_1 = 233$ nm.

Results and Discussion

Fig. 1 shows HRSEM micrographs of diamond films deposited on differently pretreated Cr substrates for various deposition times. As can be seen in this figure, diamond particle growth takes place on the rough substrate surface in all cases. Formation of diamond grains with characteristic faceting was observed on previously nitrided (Fig. 1a) and carburized (Fig. 1c) Cr substrates following 15 minutes of deposition, the particle density being lower in the latter case. After 30 minutes of deposition, the diamond particle density on both substrates was similar (Fig.1b, d) but the size distribution was different. Secondary nucleation of diamond was observed in both cases. Continuous diamond films were obtained after 60 minutes of deposition on these substrates. When non-pretreated

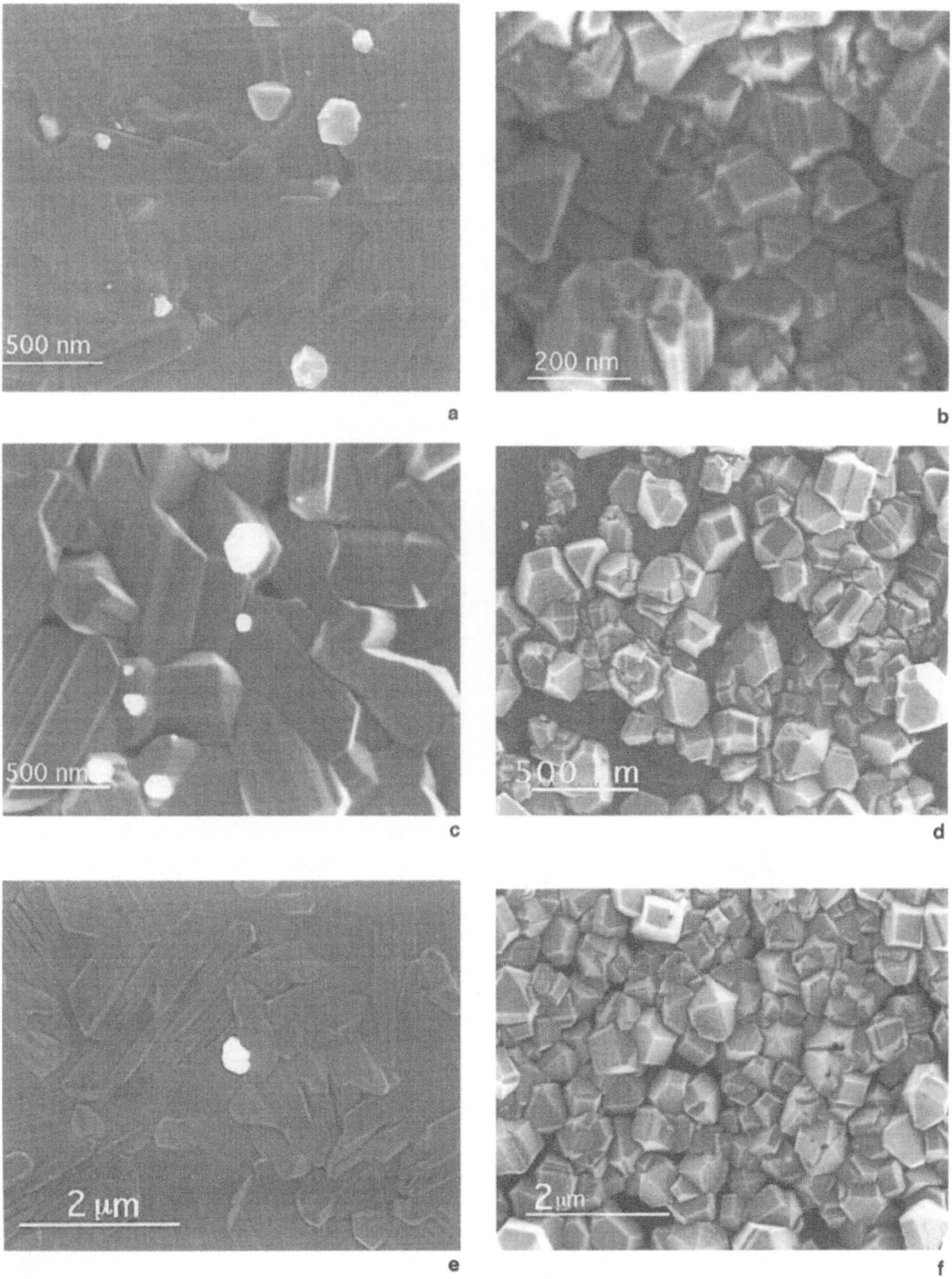

Fig. 1. HRSEM micrographs of diamond films after **a** 15 min. and **b** 30 min. of deposition on nitrided Cr substrates; after **c** 15 min. and **d** 30 min. of deposition on carburized Cr substrates; after **e** 60 min. and **f** 120 min. of deposition on non-pretreated Cr substrates

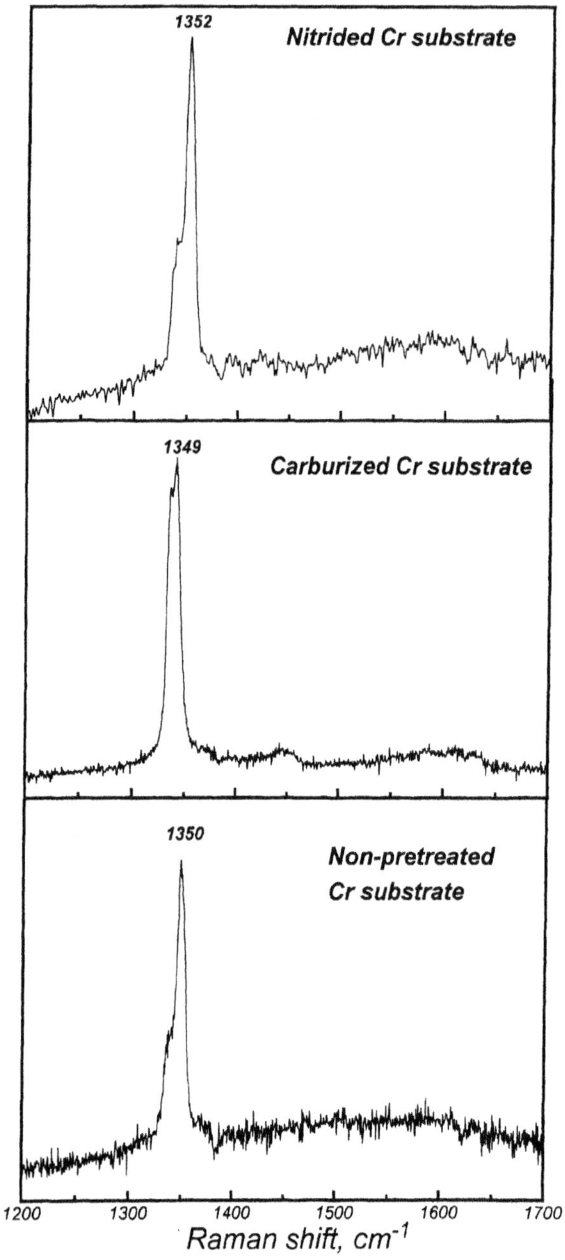

Fig. 2. Raman spectra of diamond films deposited on different substrates during 120 min

Cr substrates were used, diamond particle density was extremely low even after 60 minutes of deposition (Fig. 1e) and a semi-continuous diamond film was obtained only after 120 minutes.

The quality of the diamond films after 120 minutes of deposition was examined by Raman spectroscopy. Typical Raman spectra of films deposited on differently pretreated Cr substrates are shown in Fig. 2. The intense peak at about 1350 cm^{-1} is associated with the diamond Raman line which was split and shifted from its normal position at 1332 cm^{-1}. In addition, a low intensity and broad peak centred at ~1550 cm^{-1} was observed. This peak is associated with sp^2-bonded carbon which is present in the deposited films. The shift of the diamond peak in the Raman spectra to higher wavenumber is caused by residual compressive stress in the diamond films which results from a mismatch between the thermal expansion coefficients of diamond and the upper carbide layer of the substrate. Its value is about 7 GPa for continuous diamond films [13].

Using XRD, it was determined that the carburization of the Cr substrates prior to diamond deposition, as well as the substrate carburization which took place during deposition, resulted in the formation of chromium carbide Cr_3C_2 and Cr_7C_3 phases in the near-surface layer. Nitriding the Cr substrates for 1 hr at 900 °C prior to diamond deposition resulted in the formation of two chromium nitride phases, Cr_2N and CrN. Carburization of nitrided Cr substrates commenced at the beginning of the diamond deposition process. XRD measurements proved the formation of the Cr_3C_2 phase and a small amount of Cr_7C_3 phase after 15 minutes of deposition. In addition, a considerable amount of chromium nitride was detected in this sample. Chromium carbide, Cr_3C_2, and nitride, Cr_2N, were detected following 120 min. of diamond deposition.

Elemental distributions obtained with EDS/SEM by line scanning of cross-sectioned samples, are consistent with the results obtained by XRD. Examples of these results are presented in Figs. 3–4. Fig. 3 shows C and N elemental distributions in the near surface

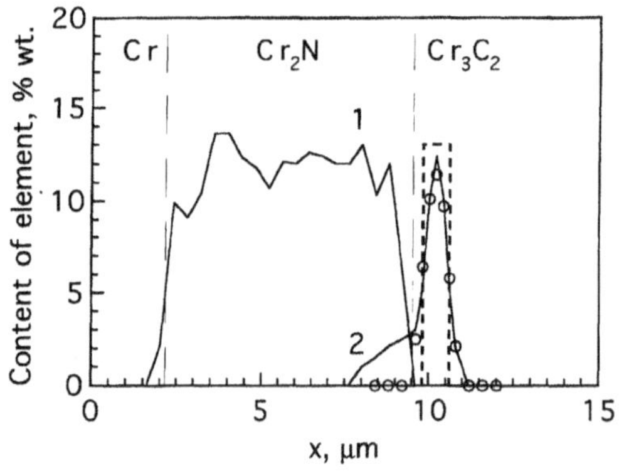

Fig. 3. Nitrogen (*1*) and carbon (*2*) distributions in the near surface region of the nitrided Cr substrate after 15 min. of diamond deposition. Circles present the simulated concentration distribution for the ideal right-angled step (dashed curve)

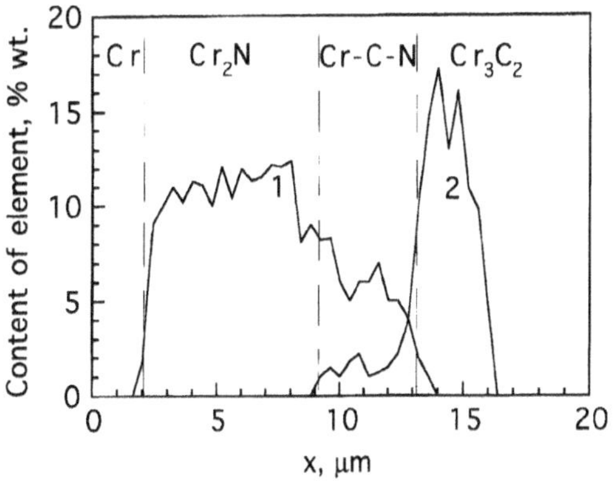

Fig. 4. Nitrogen *(1)* and carbon *(2)* distributions in the near surface region of the nitrided Cr substrate after 120 min. of diamond deposition.

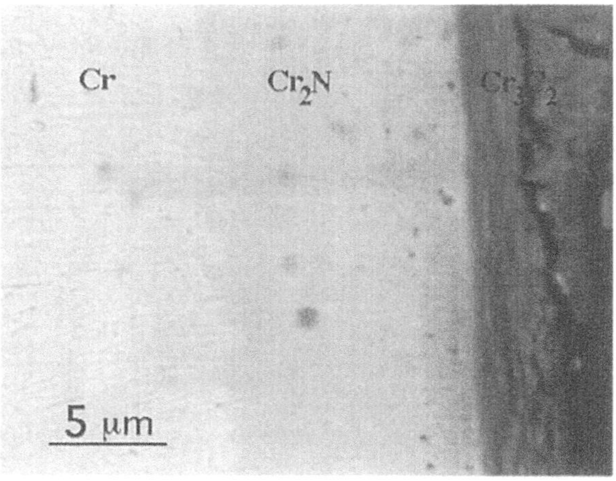

Fig. 5. Nitrided Cr substrate after 120 min. of diamond deposition (cross section, SEM)

region of a nitrided Cr substrate after 15 minutes of diamond deposition. In order to estimate the composition and the thickness of the thin Cr-C layer on the substrate surface, the above-mentioned deconvolution was applied to the measured C profile (curve 2). It was shown that the carbon distribution can be approximated by the right-angled step function (dashed line) with the content of carbon equal to 13.2 ± 1.5% wt. and 0.8 μm thick. This composition can be attributed to the Cr_3C_2 phase (13.3 %wt. carbon). Figs. 4 and 5 show the elemental distributions and the SEM micrograph of the nitrided Cr substrate after 120 min. of diamond deposition. It can be seen that the width of the Cr_3C_2 phase reached 3–4 μm, and the carbon penetration depth was about 7 μm. A summary of the EDS results of carbon penetration depth after 120 minutes of deposition for the various pretreated

Table 1. Carbon penetration depth after 120 min. of deposition and incubation time of diamond formation for the various pretreated substrates

Pretreatment	C penetration depth, μm	Incubation time
Cr with no pretreatment	20	more than 60 min
Carburized Cr	8–10	less than 15 min.
Nitrided Cr	6–7	less than 15 min.

Cr substrates, and the corresponding incubation time for diamond formation is given in Table 1.

These results show that there is a correlation between the incubation time for diamond nucleation and the diffusion of carbon into the substrate. The larger the carbon penetration depth, the longer the incubation time for diamond particle formation. In earlier studies of diamond growth on refractory metals it has been suggested that diamond nucleation requires a minimum carbon concentration at the substrate surface [4]. This concentration cannot be reached during the very early stages of the CVD process as most of the available carbon is consumed by forming carbide phases. As the thickness of the carbide layer increases, the carbon transport rate slows down and the carbon concentration at the substrate surface rises to the critical level of the carbon required for diamond nucleation. The time necessary to reach this critical carbon concentration (incubation time of diamond formation) is determined by the carbon diffusivity into the substrate and by the carburization rate of the substrate surface and near surface region. It is easy to understand that formation of a metal carbide layer near the substrate surface, prior to diamond deposition, slows down the diffusion of carbon, thus decreasing significantly the incubation time.

The results obtained for the Cr substrate demonstrate that formation of a Cr nitride layer prior to diamond deposition is also a useful method for creating a diffusion barrier for carbon and decreasing the incubation time for diamond. As a result, an increased nucleation density of diamond and an enhanced formation of continuous diamond films took place over a short deposition time. The quality of the diamond films deposited for longer periods of time (120 min.) did not depend on the Cr substrate pretreatment, and neither did the residual stress in these films.

Conclusions

In the present work, the effects of nitriding and carburizing Cr substrates on the incubation time for diamond growth and particle density at the initial stage of the CVD process were studied. It was demonstrated that the formation of a nitride or a carbide layer in the near surface region of the substrate prior to diamond deposition, created a diffusion barrier for carbon and significantly decreased the incubation time for diamond formation. EPMA was used for the investigation of the elemental distributions following various substrate pretreatments and at different stages of the diamond CVD process.

Acknowledgements. We would like to acknowledge the Israeli Ministry of Science and Art (grant No 100004), and the Technion Fund for Research financial support during this research work. Part of this study was performed at the Wolfson Centre for Interface Science of the Technion.

References

[1] B. Lux, R. Haubner, *RM HM* **1989**, *8*, 158.
[2] Y. Hirose, Y. Teresawa, *Jpn. J. Appl. Phys.* **1989**, *25(6)*, L519.
[3] V. Spitsyn, L. L. Bouilov, B. V. Derjaguin, *J. Cryst. Growth.* **1991**, *52*, 219/16.
[4] P.-O. Joffreau, R. Haubner, B. Lux, *RM HM* **1988**, *12*, 186.
[5] W. Zhu, P. C. Yang, J. T. Glass, F. Arezzo, *J. Mater. Res.* **1995**, *10(6)*, 1455.
[6] T. P. Ong, R. P. H. Chang, *Appl. Phys. Lett.* **1991**, *58(4)*, 358.
[7] M. Kavarada, *Diam. Rel. Mat.* **1993**, *2*, 1083.
[8] A. Fayer, O. Glozman, A. Hoffman, *Appl. Phys. Lett.* **1995**, *67*, 2299.
[9] O. Glozman, A. Hoffman, *Diam. Rel. Mat.* **1997**, *6*, 796.
[10] O. Glozman, A. Hoffman, *Diam. Rel. Mat.* **1997**, *6*, 1847.
[11] D. S. Knight, W. B. White, *J. Mater. Res.* **1989**, *2*, 385.
[12] S.-A. Stuart, S. Prawer, S. P. Weiser, *Appl. Phys. Lett.* **1993**, *62*, 1227.
[13] J. W. Ager III, M. D. Drory, *Phys. Rev. B.* **1993**, *48*, 2601.
[14] J. I. Goldstein, H. Yakowitz, eds. *Practical Scanning Electron Microscopy*, Plenum, New York, 1975.
[15] A. I. Berner, O. P. Sideleva, B. N. Yurchenko, *High-Purity Subst.* **1990**, *4*, 151.

Mikrochim. Acta [Suppl.] 15, 187–189 (1998)

EPMA Determination of Arsenic Excess in Low Temperature Grown GaAs

Vladimir V. Chaldyshev and **Vladimir V. Tretyakov***

A.F. Ioffe Physical Technical Institute, 194021 St.-Petersburg, Russia

Abstract. Arsenic excess in low temperature grown GaAs (LT GaAs) was determined by EPMA. This excess varied from 0.4 to 1.5 at.% for LT GaAs films grown under various conditions. The EPMA data were compared with independent results from x-ray diffraction and optical studies. The most important errors of EPMA were investigated.

Key words: electron probe microanalysis, low temperature grown GaAs, MBE.

GaAs layers grown by molecular beam epitaxy (MBE) at low temperatures (LT GaAs) have attracted a great deal of attention during the last few years [1–3]. When the growth temperature is as low as 200 °C, a high (~ 1 at.%) arsenic excess is incorporated into the growing crystal. After subsequent annealing, the excess As forms nanoscale clusters within the GaAs matrix and strongly affects the materials properties. LT GaAs exhibits an extremely short carrier lifetime and high electrical resistivity and it has found many device applications, such as active layers in ultra-fast photodetectors, buffer layers in GaAs HEMTs and integrated circuits, etc.

Since the arsenic excess is the most important feature of LT GaAs, an evaluation of this parameter is one of the main problems in the material's characterisation. For this purpose some indirect techniques can be employed, such as x-ray diffraction and near-infrared absorption measurements (NIRA) [4], however, they require a reliable knowledge of the parameters of excess-arsenic-related defects. Much more direct techniques are based on the analysis of x-ray emission induced by accelerated H ions [5] or electron beams (EPMA) [3].

In this paper we analyse the prospects and limitations of EPMA for measurements of arsenic excess in LT GaAs.

Experimental

The LT MBE GaAs layers were grown in a dual-chamber MBE system on undoped semi-insulating 2-inch GaAs substrates with (001) orientation. An 85 nm thick buffer layer of undoped GaAs was grown on the substrate at 580 °C. Then the substrate temperature was reduced to 200 °C and LT GaAs layers were grown. The growth rate was 1 μm/h and the layer thickness was 0.5–1 μm. Each sample was divided into several parts. One part was kept as-grown, the others were annealed at various temperatures in the region 400–700 °C.

The x-ray diffraction study was carried out using an asymmetrical Ge (001) monochromator adjusted for the (004) reflection of Cu $K_{\alpha 1}$ radiation. Optical absorption measurements were performed at 300 °K in the spectral region 0.8–1.2 μm. The transmission electron microscope (a Philips EM420 instrument operating at an accelerating voltage of 120 kV) was employed for observation of As clusters in the annealed samples. Measurements of arsenic excess were carried out using a Camebax microprobe equipped with wavelength spectrometers (take off angle $\theta = 40°$) and gas flow counter of low pressure (< 0.1 bar).

Results and Discussion

The main problem with the EPMA measurements is that only a small variation of the x-ray emission related to excess As occurs, compared with the strong signal from the large content of Ga and As at their stoichiometric concentrations. Detection requires good statistical accuracy obtained under stable microanalyser operating conditions. To increase the precision of the measurements a special regime was created to perform background measurements without

* To whom correspondence should be addressed

Table 1. Growth conditions and composition characteristics of epitaxial LT GaAs films

Sample number	P_{As}, 10^{-5} Pa	T_s, °C	a_l–a_s, 10^{-4} Å	X-ray [As]-[Ga] at.%	Absorption at 1 μm (cm^{-1})	Optical [As]-[Ga] at.%	k-ratio Ga La	k-ratio As La	EPMA [As]-[Ga] at.%
94	1.1	150	85	1.5			0.98	1.006	1.2
93	1.1	200	60	1.1	8800	0.6	0.984	1.008	1.1
95	1.1	250	20	0.35	3300	0.2	0.996	1.00	<0.3
96	1.8	150	83	1.5	16000	1.1	0.973	1.022	1.4
97	1.8	200	65	1.1	9700	0.7	0.987	1.011	1.0
98	1.8	250	27	0.5	4100	0.3	0.9942	1.002	~0.4

spectrometer realignment. We used a conventional stoichiometric GaAs wafer as a reference sample and the deviation of Ga Lα and As Lα lines intensities for LT GaAs samples from those for the reference was determined. The acceleration voltage was 10 kV (limited by the layer thickness) and the computer controlled beam current was 15 nA. In order to increase the measurement precision, we used a long count time (200 sec.) and monitored the random distribution of counts. A defocused electron beam (up to 50 μm) was employed to reduce the influence of contamination by carbon deposition. The analysis program was adapted to obtain the deviation of Ga and As related signals from the standard. Possible errors due to the instability of measurement process were checked by periodic measurements of the standard GaAs sample. Taking into account the Poisson distribution of x-ray emission [6], the precision of As and Ga signal measurements was estimated as ± 0.003.

Table 1 shows the EPMA results for different epitaxial LT GaAs films grown at various arsenic pressures P_{As} and substrate temperatures T_s. One can see that the arsenic excess[As]-[Ga] is detected in the LT GaAs samples. It increases with decreasing growth temperature (T_s) and depends slightly on arsenic pressure at low temperature. The highest arsenic excess detected is 1.4 ± 0.2 at.%. The limit of detection was found to be ≤ 0.4 at.%.

In order to elucidate the accuracy of the EPMA method, the data were compared to those obtained from x-ray diffraction studies and optical measurements. In x-ray diffraction, the arsenic excess was calculated from the experimentally determined expansion of the lattice parameter a_l–a_s in the as-grown LT GaAs films. The calibration by Liu et al. [4] was used in this calculation.

Strong absorption related to arsenic anti-site defects was detected in near infrared spectra of the LT GaAs samples. The concentration of arsenic anti-site defects and corresponding values of excess arsenic were calculated from the absorption coefficient at 1.0 μm using Martin's calibration [7]. The data obtained from the x-ray diffraction and optical absorption studies are shown in Table 1. One can see that these data and the EPMA measurements of arsenic excess are reasonably consistent.

Thus the EPMA technique allowed us to measure the small arsenic excess in LT GaAs films. Such precise measurements require a careful analysis of possible errors and development of the most appropriate conditions.

One might expect that the precision and detection limit could be enhanced by measuring at higher electron beam currents to improve the counting statistics. However, we found an increase in the confidence interval for the data obtained at high beam currents. This phenomenon can be attributed to poor stability of the gas flow counter at high x-ray intensities (10000 c/sec.).

One can see from the EPMA data in Table 1 that the arsenic excess in LT GaAs manifests itself by a decrease in the Ga Lα k-ratio and an increase in the As Lα k-ratio. The changes in the Ga and As k-ratios are not equal, the reason being the different absorption of Ga and As x-rays. Using x-ray distributions with depth [8] this effect can be accurately taken into account for any take-off angle. However, a small tilting of the sample always takes place but the angle of tilt is unknown. Table 2 shows the results of calculating Ga Lα and As Lα intensity variations

Table 2. Model calculation of Ga Lα As Lα k-ratios at 10 kV for various tilts of a GaAs sample in the vicinity of its exact position in the Camebax microanalyser ($\theta = 40°$)

	Tilt angle (°)					
Line	+3	+2	+1	−1	−2	−3
Ga La	0.9814	0.9904	0.9953	1.0044	1.0087	1.0086
As La	0.9696	0.9802	0.9903	1.0097	1.0183	1.0269

Table 3. Experimental Ga Lα and As Lα k-ratios obtained for different LT GaAs films in the initial position and after rotation by 180°. The data demonstrate the effect of small sample tilts on EPMA precision

Orientation	Samples 1		2		3	
	Ga Lα	As Lα	Ga Lα	As Lα	Ga Lα	As Lα
Initial	0.9832	1.004	0.9841	1.005	0.9768	1.018
rotated by 180°	1.000	1.0114	0.9982	1.011	0.9692	1.020

from a GaAs sample in the Camebax machines as a function of tilt about the exact sample position. One can see that the error originating from a small sample tilt is comparable with that from x-ray emission statistics.

In order to reveal the tilt effect experimentally, we performed EPMA measurements on several LT GaAs samples in an initial position and after rotation of the holder by 180° [9]. The data obtained are shown in Table 3. One can see a pronounced difference in the k-ratios for both Ga Lα and As Lα in the initial positions and after rotation. Thus, our calculation and experimental data showed that a small sample tilt was an important limitation and affected the precision of EPMA measurements.

Conclusions

We have shown that the EPMA technique can be employed for the determination of arsenic excess in LT GaAs. The detection limit of these measurements was $\sim$0.4 at.%. The EPMA results were found to show good consistency with the data obtained from independent x-ray studies and optical measurements on samples where the arsenic excess was in the range 0.4–1.5 %. The most important points of such precise EPMA measurements were considered, including x-ray statistics, long-term stability of the analyser and influence of small sample tilts. EPMA data from the small arsenic excess in LT GaAs are a convenient measure of the quality of the instrument performance.

Acknowledgement. The research was supported by Russian Foundation for Basic Research. The authors are grateful to Dr. V. V. Preobrazhenskii and Dr. B. R. Semyagin for LT GaAs samples, Dr. A. E. Kunitsyn for optical measurements, Dr. M. P. Yakubenya and Dr. N. N. Faleev for x-ray diffraction data.

References

[1] F. W. Smith, A. R. Calana, C. L. Chen, M. J. Manfra, L. J. Mahoney, *IEEE Electron. Lett.* **1988**, *9*, 77.

[2] M. R. Melloch, K. Mahalingam, N. Otsuka, J. M. Woodall, A. C. Warren, *J. Crystal Growth* **1991**, *111*, 39.

[3] N. A. Bert, A. I. Veinger, M. D. Vilisova, S. I. Goloschapov, I. V. Ivinin, S. V. Kozyrev, A. E. Kunitsyn, L. G. Lavren'eva, D. I. Lubishev, V. V. Preobrazhenskii, B. R. Semyagin, V. V. Tretyakov, V. V. Chaldyshev, M. P. Yakubenya, *Phys. Solid State* **1993**, *35(10)*, 1289.

[4] X. Liu, A. Passad, J. Nishio, E. R. Weber, Z. Liliental-Weber, Walukiewicz. *J. Appl. Lett.* **1995**, *67*, 279.

[5] K. M. Yu, Z. Liliental-Weber, *Mat. Res. Symp. Proc.* **1992**, *v241*, 145.

[6] K. F. G. Heinrich, *Electron Beam X-ray Microanalysis*, Reinhold, New York, 1981, 196.

[7] G. M. Martin, *Appl. Phys. Lett.* **1981**, *39*, 747.

[8] J. L. Pouchou, F. Pichoir, *Rech. Aerosp.* **1984**, *3*, 13.

[9] A. P. Mackenzie. *Proc. XIIth Int. Congr. Elect. Microsc*, San Francisco Press, San Francisco, 1990, p. 220.

Mikrochim. Acta [Suppl.] 15, 191–200 (1998)

EPMA of Melted UO_2 Fuel Rods from the Phebus-FP Reactor Accident Experiment

Paul D. W. Bottomley, François Montigny, Achilleas D. Stalios, and **Clive T. Walker***

Institute for Transuranium Elements, European Commission, Joint Research Centre, Postfach 2340, D-76125 Karlsruhe, Federal Republic of Germany

Abstract. The paper reports electron probe microanalysis and optical microscopy results obtained on a disc from the melted fuel bundle of the first Phebus test (FPT0). In this test, 20 lightly irradiated UO_2 fuel rods and a central Ag-In-Cd control rod were heated by fission power to an estimated 3000 °C. The disc was cut from the bottom of the molten pool in the lower part of the test assembly. The corium matrix of the solidified molten pool was a single homogeneous phase of composition $(U_{0.5}Zr_{0.47}Fe_{0.02}Y_{0.01})O_2$. The molten pool had reached a temperature of at least 2900 °C, but at the axial location of the disc it had cooled from a lower temperature. The rate of cooling had been sufficient to inhibit separation of the Zr-rich tetragonal and monoclinic (U, Zr)O_2 phases. Inclusions in the corium were ReIr alloys, W oxide and sesquioxides based on Cr and ferric iron. Attack of the ZrO_2 thermal shrouds by liquid corium was increased by the presence of Fe, Ni and Cr in the melt. Local concentrations of Ag and In from the central control rod were found only in a region of the disc where limited melting had occurred. They were discovered combined with Zr in two intermetallic compounds. In the same region, severe steam oxidation of Zircaloy had resulted in the rejection of the alloying element tin which was also a component of the two intermetallic compounds. Cadmium, the control rod constituent with the lowest concentration, was not detected.

Key words: phebus project, reactor accident, corium, electron probe microanalysis.

* To whom correspondence should be addressed

The international Phebus-FP (Fission Product) project is managed by the Institut de Protection et Sureté Nucleaire, Cadarache, France. It is designed to measure the source-term (mass or activity of fission products, activation products and actinides released from a nuclear fuel in a reactor accident) and to study the degradation of irradiated UO_2 fuel in conditions typical of a severe loss of coolant accident in a pressurised water reactor (PWR). In the first test (FPT0), performed in December 1993, a bundle of 20, 1.1 m long, lightly irradiated UO_2 fuel rods and a central Ag-In-Cd control rod were heated by fission power in the Phebus reactor at Cadarache in France to an estimated 3000 °C over a period of six hours.

At the Institute for Transuranium Elements a section from the bottom of the molten pool in the lower part of the test-assembly has been examined by optical microscopy and EPMA. The aims of the EPMA investigation were threefold. First, determination of the chemical composition of the solidified melt, which was a solid solution of UO_2 and ZrO_2 referred to as corium. Second, identification of metallic and oxide inclusions in the corium matrix. Third, characterisation of the chemical interaction between the molten corium and (1) the fuel rods, (2) the Zircaloy stiffeners and the outer zirconia thermal shroud, (3) the ThO_2 ultrasonic thermometer and (4) the W/WRe thermocouple wires.

The FPT0 Test

The test assembly (Fig. 1) was placed in the hollow centre of the "driver" core of the Phebus reactor. The

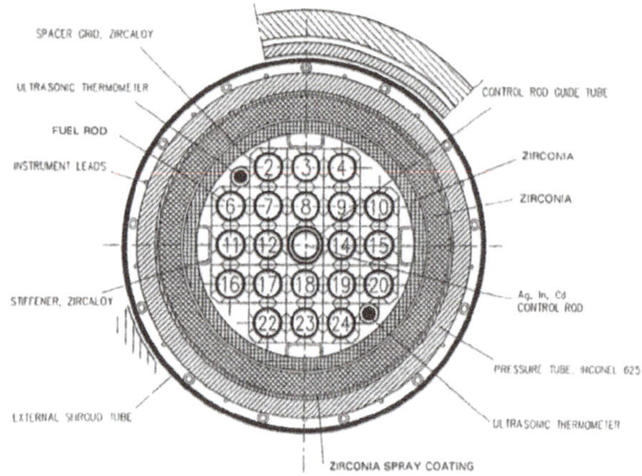

Fig. 1. Section through the FPT0 test-assembly

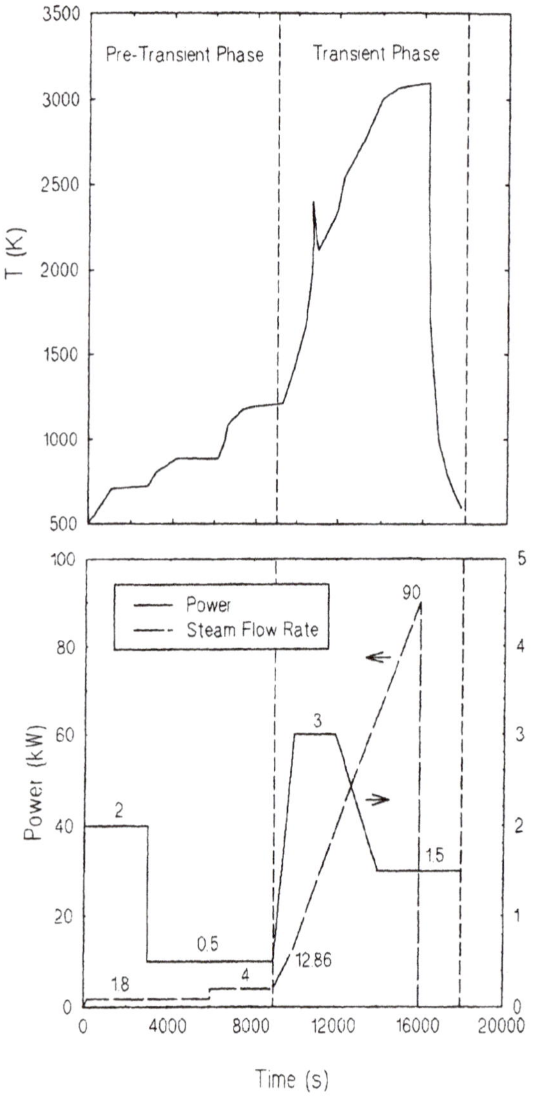

Fig. 2. The power schedule, steam mass flow rate and the temperature in the fuel bundle during the FPT0 test

power schedule, steam mass flow rate and the temperature in the fuel bundle during the test are shown in Fig. 2. In the first phase of the test, the temperature of the bundle was increased from 230 °C (outer temperature of the fuel cladding in a PWR) to 900 °C under conditions of low reactor power (4 kW max.) and modest steam flow rates (0.5 gs^{-1}). In the second, or transient phase, the power was first raised together with the steam flow rate until this reached 3.0 gs^{-1}, then after about 30 min. the steam flow was reduced to 1.5 gs^{-1}. The power ramp was continued until about 20% of the fuel bundle had melted producing a pool of molten material in the lower part of the assembly. At this point the reactor was shut down.

Figure 3 is a false colour x-ray radiograph from IPSN, Cadarache showing the condition of the fuel bundle at the end of the test. The material density; that is to say, attenuation of the x-rays, increases in the order blue, green, yellow, red. It is seen from the

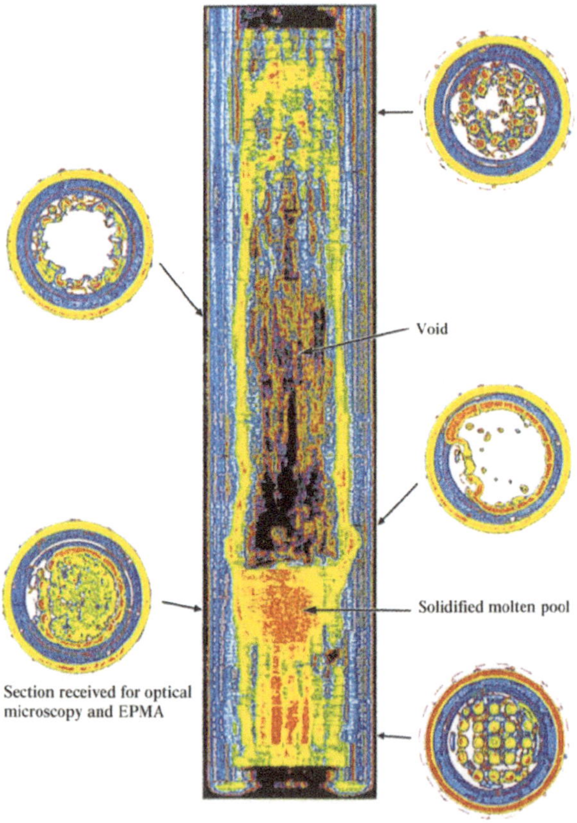

Fig. 3. False colour X-ray radiograph and computed tomography at five axial positions showing the condition of the FPT0 fuel bundle at the end of the test. The material density increases in the order blue, green, yellow, red

Table 1. Nominal composition of the main components of the FPT0 test-assembly

Component	Material
Fuel	UO_2
Cladding	Zircaloy 4[a]
Absorber rod	Ag-15In-5Cd
Cladding	304 Austenitic stainless steel[b]
Guide tube	Zircaloy 4[a]
Spacer grids	Zircaloy 4[a]
Grid clips	Inconel[c]
Thermocouples	
Sheath	Re (plated with 25 μm Ir)
Wires	W/W-10Re
Internal spacers	HfO_2
Ultrasonic thermometer	
Outer sheath	ZrO_2
Inner sheath	ThO_2
Sensor wire	W
Inner thermal shroud	ZrO_2, 5% Y_2O_3 (coated with 0.4 mm CeO_2)

[a] Zr-1.6Sn-0.2Fe-0.13Cr.
[b] Fe-19Cr-9Ni-2Mn-1Si.
[c] Ni-14Cr-6Fe.

Table 2. Approximate masses of the major elements in the FPT0 test-assembly

Element	Mass, g
U	9610
Zr	2260
Ag	471
In	88
Fe	77
Sn	39
Cd	30
Cr	25
Ni	13
Ce	0.24
Y	0.05

After ref. [1].
Re and W from the thermocouples and Th from the ultrasonic thermometers are omitted.

figure that in the middle section of the test-assembly little of the fuel bundle remained. In the upper part of this section some of the outer rods remained, while in the lower part all the rods had melted. At this location the melt had eaten into the zirconia thermal shroud leaving a peripheral crust of corium. At a height of 250 mm from the bottom of the fuel bundle a pool of solidified corium about 130 mm deep had formed. Below the pool the fuel rods were mostly in position. Likewise, at the top of the test-assembly partial melting only had occurred and most of the fuel rods had survived.

The nominal compositions of the main components in the test assembly are given in Table 1 and the approximate masses of the major elements, excluding Th, Re, Ir and W which were constituents of the thermocouples and the ultrasonic thermometers, are listed in Table 2. It is seen that the materials inventory is dominated by the UO_2 fuel and the Zircaloy cladding. The next most abundant elements after U and Zr are Ag and In from the absorber control rod followed by Fe from stainless steel and inconel.

The Sample

The sample was cut from the solidified molten pool at a height of 149–172 mm from the bottom of the fuel bundle and identified as disc 3. The appearance of the upper surface examined by optical microscopy and EPMA is shown in Fig. 4. As seen, except for a thin segment on the left, the molten pool extends over the whole cross-section. This exhibits an outer shell of columnar grains – typical of fast cooling. Within the pool 17 thermocouples can be counted. Most of these are believed to have been carried down with the molten corium from higher axial positions in the fuel bundle. On the left, outside of the molten pool, a

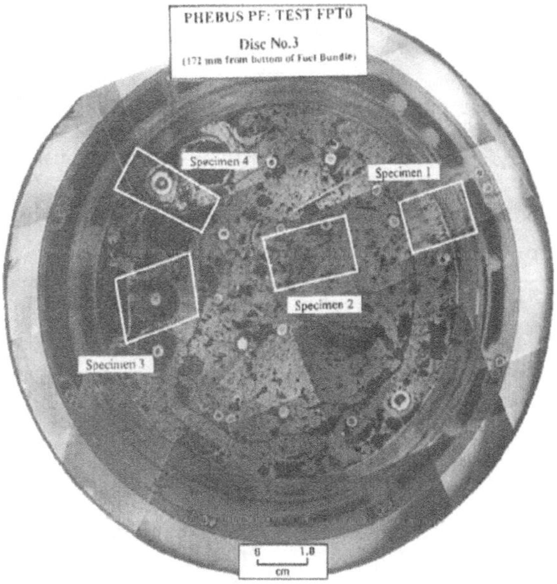

Fig. 4. Optical macrograph of the section from the fuel bundle sent for post-test analysis at the Institute for Transuranium Elements. The locations of the specimens cut for optical microscopy and EPMA are shown

Zircaloy stiffener and two fuel rods containing thermocouples are visible. The stiffener and the cladding of both fuel rods are severely oxidised and in some places have melted.

From the disc four specimens were cut for optical microscopy and EPMA. The location of these are marked in Fig. 4. Specimen 1 included a region where the molten pool had attacked the inner zirconia shroud. Specimen 2 from the centre of the molten pool contained a thermocouple and specimen 4 from the edge of the pool an ultrasonic thermometer. Specimen 3 from the opposite side of the disc to sample 1 comprised part of the stiffener and one of the fuel rods that lay outside of the molten pool.

Experimental

Specimen Preparation

Specimens were removed from the disc using a SiC cutting wheel, vacuum impregnated with Araldite and mounted in a special aluminium holder to facilitate handling by manipulators. After grinding on 400 and 600 grade SiC papers with lapping oil from Buehler as lubricant, the specimens were polished using a water free suspension of 3 μm diamond powder (Kulzer, type MM 140). Grinding and polishing were repeated until a surface finish of good metallographic quality was obtained.

Optical Microscopy

The polished specimens were examined on a Reichert Telatom optical microscope installed in a hot cell. Immediately, prior to optical microscopy the polished surface of the specimen was ion etched to improve optical contrast which was further enhanced by the use of oblique illumination. The ion etching equipment was custom built. Etching was accomplished using a low energy, low intensity beam of Ar^+ ions (acceleration voltage 8 kV; beam current, 1 mA). The angle of incidence was 45° and three traverses of the specimen surface were made, each of 15 min. duration. Generally, areas of interest were examined at magnifications of 200 and 500×.

Electron Probe Microanalysis

EPMA was carried out using a Cameca MS46 electron microprobe specially shielded with tungsten and lead to permit the analysis of irradiated nuclear fuel [2]. Analysis was performed on the same specimens used for optical microscopy which were re-polished to remove surface relief produced by ion etching and coated with a conducting film of aluminium about 20 nm thick to avoid charging effects. The analysis conditions used for the major elements of interest are summarised in Table 3. With the exception of oxygen, all elements were analysed at an electron acceleration potential (E_0) of 15 kV and a beam current of 100 nA. Oxygen was analysed at 10 kV and 100 nA. The change in lateral resolution caused by switching from 15 to 10 kV did not significantly affect the results because the phases analysed were homogeneous. Since a standard was not available, the concentration of Ir in the binary IrRe inclusions in the corium and in the IrRe interaction layer on the outer surface of the thermocouple sheaths was determined by difference. k-values were calculated either from five peak and five background determinations or from three peak and three background determinations depending on the element concentration level. The matrix correction was made using the PAP [3] option in the XMAS® x-ray microanalysis program marketed by SAMx, France. This program is designed to accept elements analysed at different acceleration voltages.

Results

The Corium Matrix of the Soldified Molten Pool

Point analysis results for the corium matrix are given in Table 4. It is seen that its composition was reasonably uniform. In specimen 2 the element concentrations approximated to a chemical composition of $(U_{0.5}Zr_{0.47}Fe_{0.02}Y_{0.01})O_{1.73}$. In specimen 3, however, the oxygen to metal ratio appeared to be

Table 3. Conditions used for quantitative electron probe microanalysis

Element	X-ray Line	E_0 (kV)	Beam Current (nA)	Counting time Peak (s)	Background (s)	Diffracting crystal	Standard
Oxygen	$K\alpha_1$	10	100	50	20	Ni-C multilayer	SiO_2
Chromium	$K\alpha_1$	15	100	50	20	LiF	Cr
Iron	$K\alpha_1$	15	100	50	20	LiF	Fe
Nickel	$K\alpha_1$	15	100	50	20	LiF	Ni
Yttrium	$L\alpha_1$	15	100	50	50	PET	Y
Zirconium	$L\alpha_1$	15	100	50	50	PET	Zr
Silver	$L\alpha_1$	15	100	50	50	PET	Ag
Indium	$L\alpha_1$	15	100	50	50	PET	In
Tin	$L\alpha_1$	15	100	50	50	Quartz 1011	Sn
Iridium[a]							
Rhenium	$M\alpha_1$	15	100	50	20	PET	Re
Uranium	$M\alpha_1$	15	100	50	50	Quartz 1011	UO_2

[a] Determined by difference.

Table 4. EPMA results for the corium matrix of the solidified molten pool

Location	Concentration, wt%				
	Uranium	Zirconium	Yttrium	Iron	Oxygen
Specimen 1					
1	63.4 (2.3)	21.5 (2.3)	...[a]	0.8 (0.2)	...
Specimen 2					
1	62.8 (1.4)	22.1 (0.4)	0.5 (0.1)	0.6 (0.1)	14.1 (0.5)
2	62.4 (1.7)	22.0 (1.2)	0.7 (0.2)	0.6 (0.1)	14.3 (0.4)
3	61.7 (1.2)	22.0 (0.7)	0.7 (0.1)	0.6 (0.1)	15.0 (0.2)
4	62.6 (1.0)	22.3 (0.2)	...	0.5 (0.1)	...
Specimen 3					
1	60.9 (0.7)	21.9 (0.7)	0.4 (0.3)	0.7 (0.1)	16.1 (0.5)
2[b]	62.5 (1.2)	21.1 (1.3)	0.0	0.5 (0.1)	15.8 (0.3)
3[b]	61.0 (1.2)	23.1 (1.4)	0.0	0.4 (0.1)	16.0 (0.4)

Standard deviation is shown in parentheses.
[a] Not determined.
[b] Adjacent to melted Zircaloy cladding.

close to the stoichiometric value of two (1.96 and 1.97). Further, in specimen 3, Y was not detected close to where the fuel cladding had melted.

Inclusions in the Solidified Corium Matrix

Both oxide and metallic inclusions were present in the corium melt. Two types of oxide inclusions were detected; one containing Cr, Fe and Ni; the other W. The metallic inclusions were ReIr binary alloys of varying composition (in specimen 2 an inclusion with 23 wt% Re and another with 39 wt% Re were analysed).

The electron absorption micrograph (specimen current micrograph) and x-ray distribution maps in Fig. 5 show a cluster of inclusions in specimen 2. The oxide inclusions containing Cr, Fe and Ni were of two different compositions. Inclusion ① contained high concentrations of all three transition elements whereas inclusion ② contained less than 1 wt% Cr (see Table 5). Both, however, are considered to be oxides of the type A_2O_3. As to the W inclusions, it is evident from the phase contrast in electron absorption micrograph in Fig. 5 as well as from the absence of lustre in the optical microscope that these are also oxides.

Interaction Between the Corium Melt and the Zirconia Thermal Shroud

Specimen 1 incorporated a section of the thermal shroud that had been attacked by the molten corium. The distribution in the interaction zone of the main constituents of the corium melt (U, Zr, Fe and Y), as well as Ce, a constituent of the plasma-spray coating on the inner surface of the shroud, is shown in Fig. 6. It is seen that the width of the interaction zone, defined by the distance needed for the U concentration to fall from its level in the corium melt (around 62 wt%) to zero, was about 1.25 mm. Cerium, although initially concentrated in a thin layer 400 μm thick, had spread throughout the interaction zone and had even diffused about 600 μm into the corium and zirconia. Most importantly, however, it is seen that Fe from the corium melt had penetrated the zirconia shroud to a depth of at least 1 mm. The concentration profile for this element exhibits a peak just in advance of the peak in the Ce concentration (approaching from the corium side). This may be interpreted as Ce hindering the diffusion of Fe to the zirconia thus causing a slight build-up in its concentration in the outer part of the zone.

Thermocouple in Specimen 2

During the test, molten corium containing Cr, Fe and Ni had penetrated the Re sheath of the thermocouple and had reacted with the thermocouple wires and the HfO_2 ceramic spacer pellets. Inside the sheath, the remains of a thermocouple wire could be discerned surrounded by a mass of material that had clearly been molten. It emerged that this was the W-10Re wire. The accompanying W wire had been completely

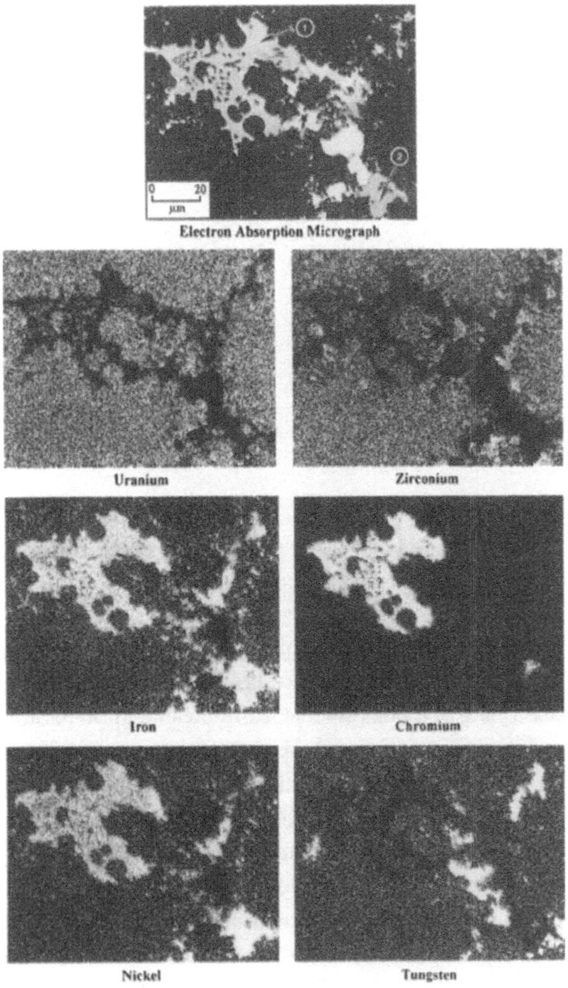

Fig. 5. Element specific x-ray maps showing the composition of a cluster of inclusions in the corium in specimen 2. Both inclusions ① and ② are considered to be oxides of the type A_2O_3. Inclusion ① contains nearly 40 wt% Cr, whereas inclusion ② contains less than 1 wt% Cr

Table 5. Composition of the A_2O_3 inclusions in the corium in specimen 2

Phase	Concentration, wt%			
	Chromium	Iron	Nickel	Oxygen[a]
1	36.0 (0.8)	25.5 (0.7)	7.4 (0.1)	31.1
2	0.4 (0.1)	48.9 (0.6)	16.8 (0.1)	33.9

Standard deviation is shown in parentheses.
[a] Determined by difference.

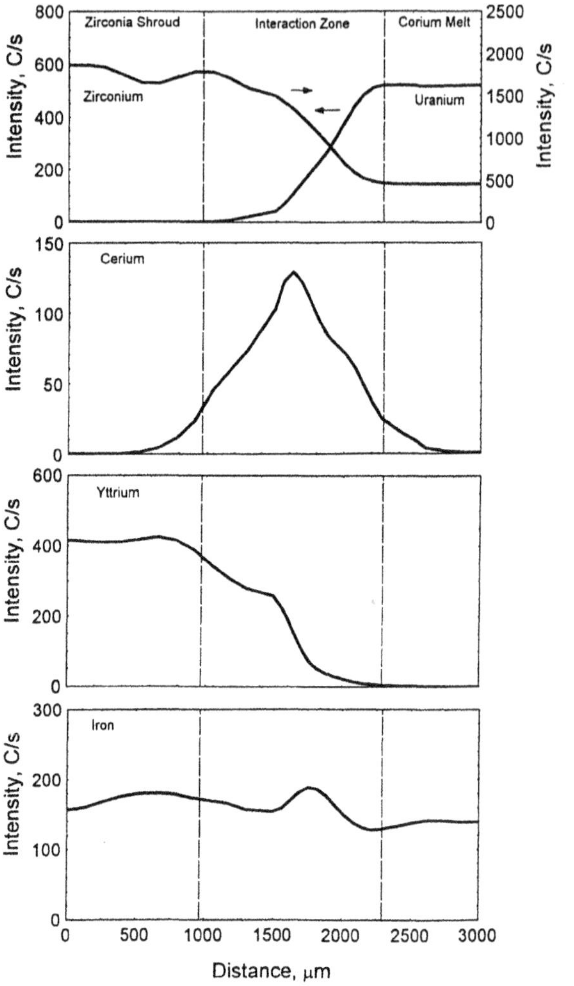

Fig. 6. x-ray line scans showing the distribution of U, Zr, Ce, Fe and Y across the interface between the thermal shroud and the corium melt in a region where chemical attack had occurred in specimen 1. The Ce from the plasma-spray coating on the inner shroud surface has spread throughout the interaction zone and Fe from the corium melt has penetrated the shroud to a depth of 600 μm

consumed by the molten corium. As to the sheath itself, the Ir coating on the outer surface had interacted with the underlying Re to a depth of about 80 μm forming an alloy of containing 59 wt% Ir.

As seen from Fig. 7 the thermocouple wire exhibited a fine eutectic type structure composed of lamellae of W oxide and ReNi. At grain boundaries within this structure, three relatively coarse phases; one rich in W, another rich in U and a further one rich in Hf and Zr, were identified. Around the periphery of the eutectic phase were numerous inclusions of Fe and the odd inclusion of Ni.

The mass of material surrounding the wire also consisted of a U rich phase, a phase rich in Zr and Hf, and a W rich phase. These phases were probably oxides (UO_2, (Hf, Zr)O_2 and possibly W oxide) and are most likely the same ones found at grain boundaries in the

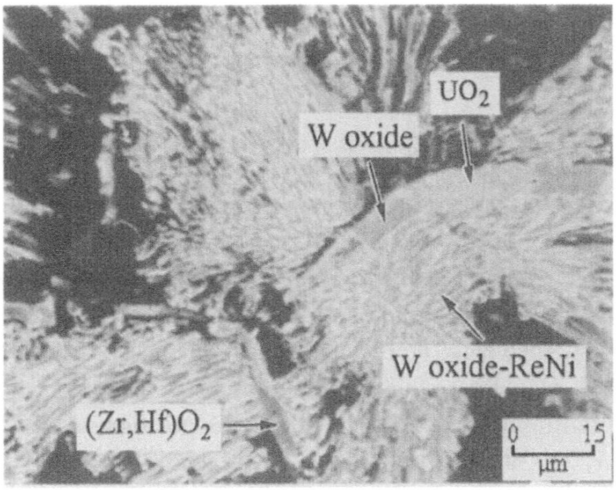

Fig. 7. Degraded W-10Re wire in the thermocouple in specimen 2. The fine eutectic type structure is composed of lamellae of W oxide and ReNi. The coarse phases at grain boundaries are probably UO_2, (Hf, Zr)O_2 and possibly W oxide

degraded thermocouple wire. Spots showing concentrations of Fe were pervasive. Chromium was also detected. In the region examined its presence centred on a cavity in the material. The W contained in the material very likely came from the lost W wire.

Ultrasonic Thermometer in Specimen 4

Optical microscopy revealed extensive intergranular cracking in the zirconia shroud and that a second phase was present on many grain boundaries. Further, extraneous material was observed on the inner surface of the shroud. Clearly, this had entered in the molten state through a breach at a higher level in the assembly and had flowed down the inside of the tube. Electron absorption micrographs of this material and of the microstructure at the centre of the shroud wall are shown in Fig. 8. From micrograph 8a it is seen that the material attached to the inside of the sheath was a mixture of corium and an oxide of Cr, Fe and Ni. It appears that Fe and U from the corium had penetrated the grains and grain boundaries of the zirconia equally to a depth of about 50 μm, but beyond this depth the intrusion of Fe and U was restricted to the grain boundaries. As is evident from micrograph 8b, even at the centre of the shroud wall the grain boundaries had been greatly altered by the invasion of U, Cr, Fe and Ni. Grain boundary regions that are delineated by a light band contained concentrations of U and Fe from the corium, while dark areas denote a second phase identified as an oxide of Cr, Ni and Fe.

Fuel Rod in Specimen 3

Molten material had penetrated radial cracks in the fuel and had accumulated in the space between the inner fuel surface and the Re sheath of the central thermocouple as seen in Fig. 9. The molten material was shown to be either ZrO_2 or αZr(O) containing a small amount of UO_2 (about 12 wt% equal to approximately 2 at %). In this material between the fuel and the thermocouple sheath Re inclusions containing about 1 wt% each of Ni and Fe had collected. A number of the larger inclusions examined exhibited a

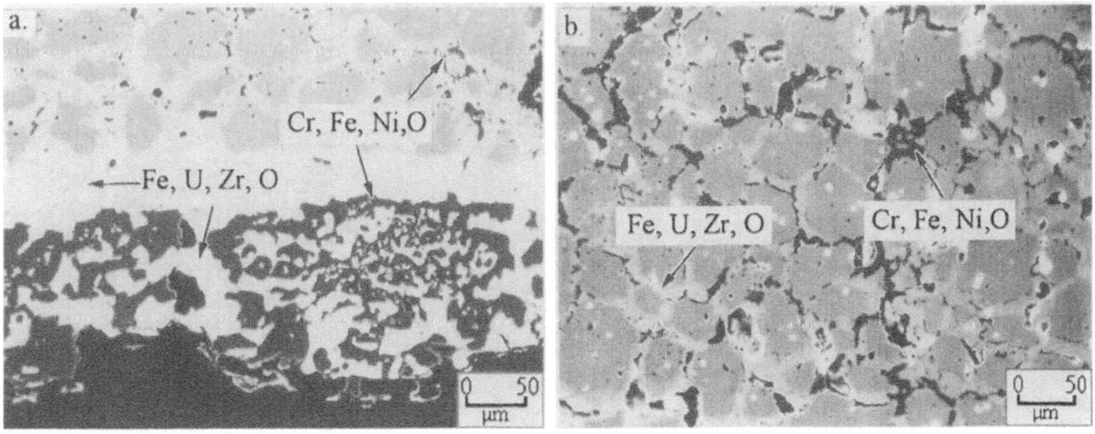

Fig. 8. Electron absorption micrographs of the zirconia thermal shroud from of the ultrasonic thermometer in specimen 4. **a** Solidified molten material on the inner surface of the shroud; **b** intergranular attack at the centre of the shroud cross-section associated with the invasion of U, Cr, Fe and Ni from molten material

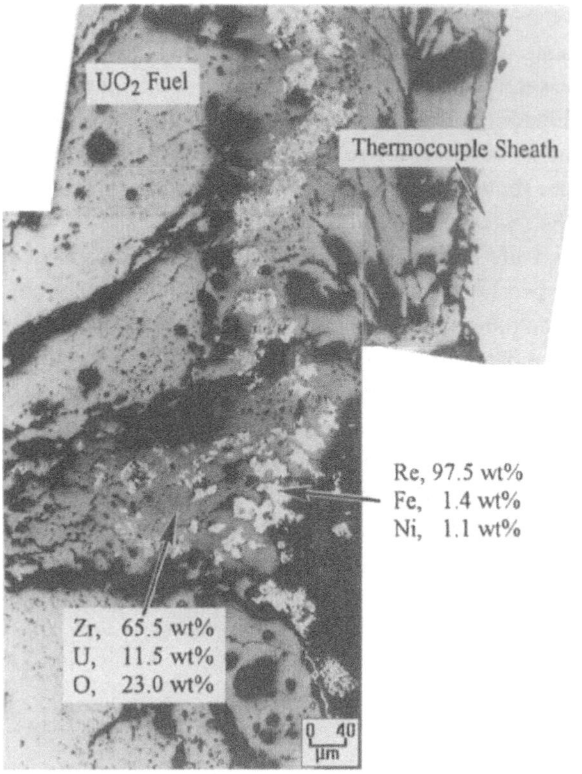

Fig. 9. ZrO_2 or $\alpha Zr(O)$ in the space between the inner fuel surface and the rhenium sheath at the location of a radial crack in the fuel rod in specimen 3. The oxygen saturated Zr contains a small amount of U. Metallic inclusions in the material are essentially Re from the thermocouple sheath

fine covering of Sn. In addition to these, the material also contained numerous small CrFe inclusions.

Zircaloy Stiffener in Specimen 3

Both surfaces of stiffener were severely oxidised, but at the centre of the stiffener cross-section optical microscopy revealed the presence of a metallic phase (Fig. 10, micrograph a). Closer examination using the absorbed electron image revealed this phase to consist of two intermetallic alloys rich in Zr (Fig. 10, micrograph b). The chemical composition of phase 1, the lighter of the two phases in micrograph b, corresponded to $Zr_3(Sn_{0.87}In_{0.10}Ni_{0.03})_2$ whereas the composition of phase 2 corresponded to $Zr_2(Fe, Ni, Cr)_{0.74}(Ag, Sn, In)_{0.26}$. Interspersed between these two metallic phases were islands of Zircaloy. At free surfaces throughout the microstructure, concentrations of Ag and In were detected.

Discussion

The corium matrix of the soldified molten pool was reasonably uniform in composition indicating that it had been fully molten with strong convection currents leading to near complete mixing. It is judged to be an almost stoichiometric oxide. The O/M ratio of 1.73 obtained from analyses on specimen 2 is attributed to

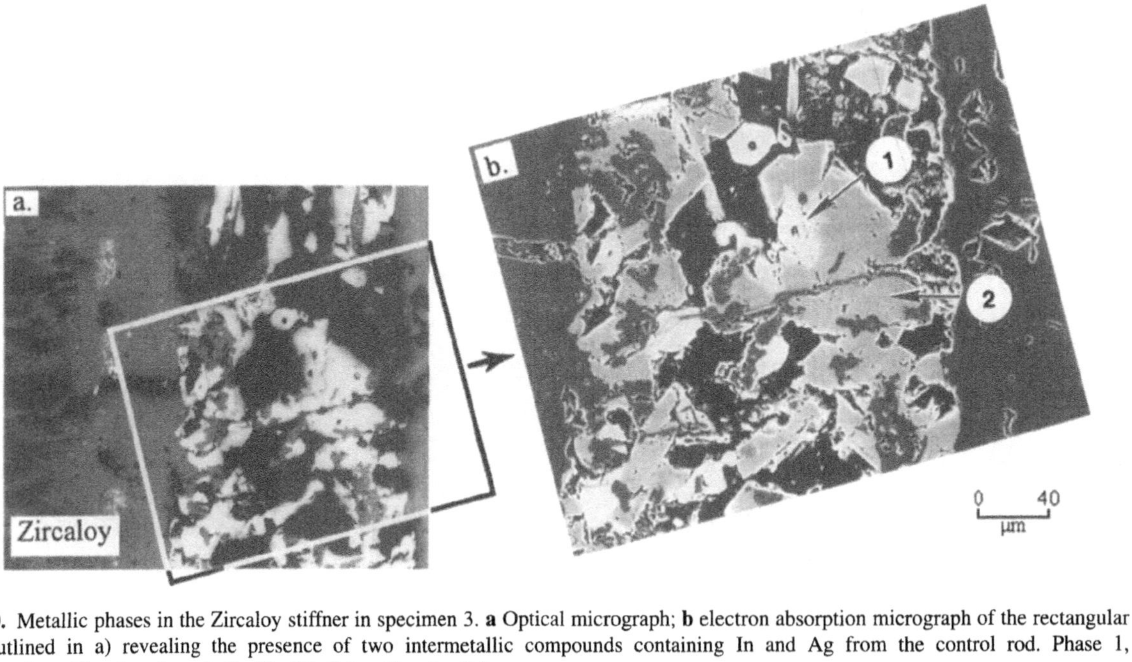

Fig. 10. Metallic phases in the Zircaloy stiffner in specimen 3. **a** Optical micrograph; **b** electron absorption micrograph of the rectangular area outlined in a) revealing the presence of two intermetallic compounds containing In and Ag from the control rod. Phase 1, $Zr_3(Sn_{0.87}In_{0.10}Ni_{0.03})_2$; phase 2, $Zr_2(Fe, Ni, Cr)_{0.74}(Ag, Sn, In)_{0.26}$

a defective counter window in the x-ray spectrometer. The corium contained approximately 62 wt% U, 22 wt% Zr, 0.6 wt% Fe and 0.6 wt% Y. Its Zr concentration is therefore higher than the typical PWR inventory of 16 wt% [4]. This is almost certainly due to the addition of ZrO_2 from the thermal shroud as shown by the presence in the corium of a small amount of Y.

The corium melt had probably reached a temperature of at least 2900 °C. This is mainly deduced from the fact that above the molten pool the corium had been sufficiently hot to consume almost three-quarters of the 11.5 mm wall thickness of the ZrO_2 thermal shroud with a melting point of 2700 °C (see Fig. 3). Moreover, since the solidified corium was uniform in composition it can further be deduced that the cooling rate had been sufficiently fast to retain the high temperature cubic (U, Zr)O_2 phase and to avoid the separation of the Zr-rich tetragonal and monoclinic (U, Zr)O_2 phases (see Fig. 11). This was confirmed by x-ray diffraction which established that the solidified corium melt was a single phase with the FCC fluorite structure of UO_2 and a lattice parameter of 0.5282 nm. Further, the solidified melt exhibited a cast structure consisting of an outer shell of columnar grains and a core of equiaxed grains. Such structures generally result when the temperature of the melt is only slightly above the freezing point. It is pointed out that the Zr-rich monoclinic phase was observed in samples taken from the melted core of the Three Mile Island (TMI) reactor [5]. Thus it appears that in the TMI accident the molten core cooled more slowly.

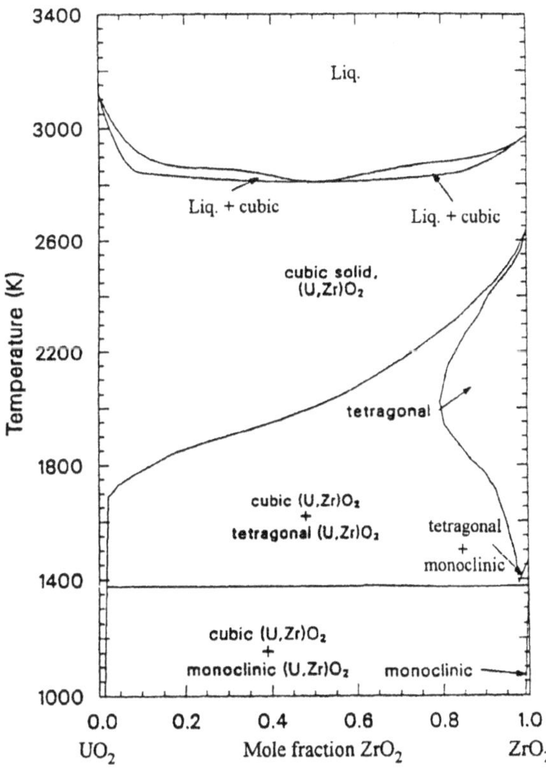

Fig. 11. The UO_2-ZrO_2 binary phase diagram

Three main types of inclusions were detected in the solidified corium melt in disc 3. Metallic ReIr binary alloys of varying composition, W evidently as an oxide, and sequioxides based on chromium and ferric iron, also of varying composition. Clearly, these inclusions separated when the corium melt solidified at about 2550°C (see Fig. 11). Since one of the ReIr inclusions analysed was of similar composition to the ReIr alloy at the surface of the thermocouple sheath in specimen 2 it appears that the origin of these inclusions was the dissolution of this alloy which has a melting point of 2850 °C [6]. The discovery of W oxide inclusions was not foreseen. The melting point of W is 3410°C which is at least 300 °C higher than the highest temperature reached by the corium melt. Nevertheless, the finding is consistent with the knowledge that this element has a solubility of several percent in liquid UO_2 (C. Ronchi, Institute for Transuranium Elements, personal communication). It is assumed that W metal separated from the melt and that this oxidised as cooling advanced. Consideration of the melting point of the ReIr inclusions and of W leads to the assumption that the A_2O_3 oxide inclusions with a melting point close to 1600 °C (m.p. of Fe_2O_3 is 1565 °C [7]) were the last to solidify.

The zirconia components in disc 3 showed attack not only by the molten corium but also interaction with Cr, Fe and Ni. For example, in specimen 1, Fe from the corium had diffused into the inner zirconia thermal shroud in advance of the dissolution front to a depth of least 1 mm (see Fig. 6). Further, in specimen 4, corium containing Ni and Cr and additional Fe had penetrated along the grain boundaries of the thermal sheath of the ultrasonic thermometer and on cooling had formed a separate (Cr, Fe, Ni) oxide phase (probably (Fe, Cr, Ni)O_2) leaving a broad zone enriched in U and Fe on each side of the grain boundaries (see Fig. 8). The tendency of Cr, Fe and Ni (presumably as liquid oxides) to attack zirconia suggests that in future Phebus tests a different material may need to be considered for the thermal shrouds.

The thermocouple wires and spacer pellets were also susceptible to attack by the corium melt and the molten steel components. For example, in the case of the thermocouple in specimen 2, Ni had alloyed with the Re in the W-10Re wire which had transformed from a solid solution to a two-phase eutectic type mixture composed of ReNi and W oxide (see Fig. 7). Also, Zr from the corium had combined with HfO_2 from spacer pellet leaving behind UO_2.

Silver and In from the central control rod were found only in specimen 3 which was taken from a region where only limited melting had occurred and the molten pool had evidently not penetrated (see Fig. 4). Both elements were constituents of two intermetallic alloys that contained Zr, Sn together with Cr, Fe and Ni and which were paradoxically located at the centre of the stiffener cross-section. Cadmium, the control rod constituent with the lowest concentration (5 wt%) was not detected.

The Sn in the intermetallic alloys referred to above apparently came from the surrounding Zircaloy stiffener. Tin is barely soluble in ZrO_2 and $\alpha Zr(O)$ and consequently it was rejected by the Zircaloy when the stiffener oxidised. In the same specimen, Sn was also detected at the centre of the fuel rod where it formed a coating on metallic Re inclusions in the space around the thermocouple. Evidently, the Sn had separated from the oxidised Zr in which the Re inclusions were found. The origin of this material was Zircaloy cladding which had melted and flowed down radial cracks in the fuel. Since the cladding had only partially melted it appears that the temperature in this region of disc 3 had barely exceeded the melting point of Zr which is approximately 1850°C. This implies that oxidation of the Zr occurred after the Zircaloy had melted and when it was already in place in the cracks and in the space around the thermocouple.

Conclusions

The corium matrix of the soldified molten pool was a single homogeneous phase of composition $(U_{0.5}Zr_{0.47}Fe_{0.02}Y_{0.01})O_2$ with a melting point of approximately 2550 °C. The corium melt had reached a temperature of at least 2900 °C, but at the axial location of Disc 3 it had cooled from a lower temperature. The rate of cooling had been sufficient to retain the high temperature cubic phase and to inhibit separation of the Zr-rich tetragonal and monoclinic $(U, Zr)O_2$ phases. Inclusions in the corium contained constituents from structural materials (stainless steel cladding of the control rod, inconel grid clips, etc.) and thermocouple components. The inclusions detected in disc 3 were of three types; namely, ReIr binary alloys of varying composition, W oxide, and sesquioxides based on Cr and ferric iron, also of varying composition. These had formed early in the cooling process when the corium melt solidified. Liquid corium acted aggressively towards the ZrO_2 thermal shrouds and resulted in the intergranular attack of these components. This attack was increased by the presence of Fe, Ni and Cr in the melt. Local concentrations of Ag and In from the central control rod were found only in sample 3 where the temperature had not exceeded 2000°C. They were found combined with Zr in two intermetallic compounds. Also in sample 3, severe steam oxidation of Zircaloy had resulted in the rejection of the alloying element tin. This element was found together with Ag and In in the intermetallic compounds and also as a coating on Re inclusions embedded in the oxidised Zr (either ZrO_2 or $\alpha Zr(O)$) at the centre of the fuel rod. Cadmium, the control rod constituent with the lowest concentration was not detected.

References

[1] A. Tani, P. von der Hardt, *Phebus FPT0 Expected Sampling Mass and Activity*, Report Phebus TG176, CEA Cadarache, 1993, Annex 3.

[2] C. T. Walker, in: *Proc. 2nd Regional, EMAS Workshop on Electron Prob Microanalysis of Materials Today – Practical Aspects* (J. L. Lábár, E. Heikinheimo, P. Nicholson, eds.), Kossuth Lajos University, Debrecen Hungary, 1996, p. 102.

[3] J. L. Pouchou, F. Pichoir, *La Recherche Aerospatiale*, **1984a**, *5*, 167.

[4] *In-Vessel Core Degradation in LWR Severe Accidents – A State of the Art Report to CSNI*, Report OECD/NEA/CSNI/R(91), 1991.

[5] A. Braun, G. J. McIntyre, C. Gräslund, *Nucl. Technol.* **1989**, *87*, 137.

[6] T. B. Massalski, H. Okamoto, P. R. Subramanian, L. Kacpzak (eds.), *Binary Alloy Phase Diagrams, Vol. 3.*, ASM International, Materials Park, Ohio, 1990, p. 2346.

[7] D. R. Lide, H. P. R. Frederukse (eds.), *Handbook of Chemistry and Physics 1993–1994, 74th Ed*, CRC, Boca Raton, 1993, pp. 4–36.

Mikrochim. Acta [Suppl.] 15, 201–205 (1998)

Steels, Carbon Concentration, and Microhardness

P. Poelt* and **A. Fian**

Research Institute of Electron Microscopy, Technical University of Graz, Steyrergasse 17, A-8010 Graz, Austria

Abstract. A simplified method for the quantitative analysis of carbon in steels is presented. It is based on the use of two steel standards and a linear interpolation between their carbon concentrations. With this method only x-ray intensity measurements at the C-K_α wavelength are necessary and background measurements are superfluous.

The method is used to measure carbon concentration profiles in case hardened steels and to correlate these with corresponding microhardness profiles.

Key words: EPMA, x-ray microanalysis, carbon, steels, microhardness.

The mechanical behaviour of steels depends strongly on their carbon content. A typical example is case hardened steels, where case carburizing results in a hard wear resistant surface and a soft interior. Whereas the average bulk carbon concentration can easily be determined by chemical analysis, for the measurement of carbon concentration profiles other techniques have to be used. One of the most common is electron microprobe analysis (EPMA). But carbon quantification in steels by EPMA is a difficult task. The main problems are carbon contamination of the specimens and interference from higher order lines of chromium and nickel. Also the generally inhomogeneous dispersion of carbon in steels makes the use of even a sophisticated $\phi(\rho z)$ correction method rather doubtful. A concise description of all these difficulties and how to cope with them can be found in [1].

To avoid most of these problems, the use of two steel standards and linear interpolation in the region 0 to 2 wt% is suggested in this paper. This makes background measurements superfluous and the only measurements that have to be performed are x-ray intensity measurements at the carbon K_α wavelength.

In the following sections proof will be given that with this method results with an accuracy better than 0.05 wt% for the carbon concentration can be achieved. Subsequently, this method for carbon quantification will be used to establish some correlation between carbon concentration and microhardness for case hardened steels.

Experimental

The steel samples were embedded in resin. After polishing with alumina the resin was removed and the samples were cleaned with acetone in an ultrasonic bath and then dried. The carbon measurements were performed using a WDX-2A spectrometer from Microspec attached to a Philips SEM 505. An oxygen jet served as an anticontamination device. Fluctuations of the probe current, measured using a Faraday cup and the built-in specimen current meter, were below one per cent. For the calculation of theoretical k-ratios the program PROZA, based on the work of Bastin et al. [2] and distributed by TECHAN B. V., was used. Microhardness measurements were performed by a μ-indenter 5 microhardness tester from Paar KG. This device can be operated inside the specimen chamber of the SEM. The indenter was a Vickers pyramid.

Results and Discussion

Generally in quantitative electron microprobe analysis, for every element of the unknown sample the intensity of a specific x-ray line is measured. Additionally the same measurements have to be performed on standards with well-known compositions. For the subsequent calculation of chemical concentrations, programs are used that correct for electron backscattering, absorption of x-ray lines in

* To whom correspondence should be addressed

the specimen itself and for x-ray fluorescence. Using wavelength dispersive spectrometry this can be very time consuming, especially for a specimen with many elements. Additionally, in the case of light element analysis peak area factors have to be determined if peak shape alterations between unknown and standards occur [2]. It is to be questioned whether most of this tedious work could not be avoided in cases where only the concentration of one element of the specimen is of interest; for example that of carbon in steels. Possibly by the use of two appropriate steel standards a linear interpolation between their carbon concentrations would suffice to determine the concentration in an unknown with sufficient accuracy. The practicability of the method will depend on the linear relationship between k-ratios and concentrations in the region of interest, the influence of the overall composition of standards and the unknown sample on the results and, as a consequence, how closely the standards have to resemble the unknown.

Linearity k-Ratios – Carbon Concentrations

In Table 1 four steels currently in service with their different compositions are presented. The abbreviations are arbitrary and only used for shortness in the text. Fig. 1 shows a nearly perfect linear relationship between the k-ratios of carbon and the corresponding concentrations in the range 0 to 3 wt% carbon. This linear relationship has already been shown by Andersen [7] by measurements of a series of different concentrations of carbon in iron. To obtain an overall concentration of 100 wt%, in the calculations a change in the C-concentration was always compensated by an equivalent change in the Fe-concentration. Another important result from Fig. 1 is, that the graphs for steels with different composition coincide nearly completely. Even for the steel CS4, with its high tungsten content, the deviations are smaller than 10%

Table 1. Composition (elements with concentrations greater 1 wt% only) of customary steels. Typical examples for the arbitrary abbreviations used are: CSI: 20 MnCr5; CS2: X 2 CrNiMo 17 13 2; CS3: X 48 CrMoV 8 11; CS4: X 50 WNiCrVCo 12 12

	wt%								
Abbrev.	Si	V	Cr	Mn	Fe	Co	Ni	Mo	W
CS1	–	–	1.2	1.2	97.6	–	–	–	–
CS2	–	–	16.8	1.3	68.6	–	11.2	2.1	–
CS3	–	1.5	7.5	–	89.5	–	–	1.5	–
CS4	1.5	1.0	4.0	–	68.0	1.5	11.5	–	12.5

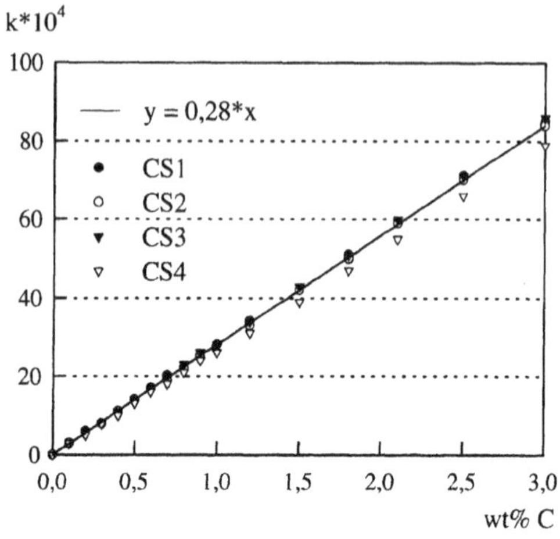

Fig. 1. Variation of the k-ratios of carbon as a function of the carbon concentration dependence on the composition of the steels (calculated by PROZA, 15 keV, m.a.c.'s: default values)

Table 2. Mean values of the mass absorption coefficient of C-K_α, the atomic number Z, the atomic weight A and the density ρ for the steels of Table 1

	Mean values					
Steel	mac[3]	mac[4]	mac[5]	Z	A	ρ
CS1	16,208	13,832	14,043	26.0	55.8	7.8
CS2	16,185	13,848	14,245	26.2	56.4	7.9
CS3	15,953	13,614	13,978	26.0	55.3	7.8
CS4	17,181	15,193	15,039	32.0	71.6	9.3

relative. The reason for this is clear from Table 2, where the mean values for atomic number, atomic weight, density and mass absorption coefficient (m.a.c) of carbon are shown. These values which strongly influence electron backscattering, x-ray generation and x-ray absorption, are very similar for the steels CSI-CS3 of Table 1. Thus no great differences in the k-ratios of carbon can be expected. Also the background intensity depends mainly on the atomic number and thus background variations between steels of different composition should not be significant.

Table 3 gives a survey of the m.a.c.s used for the calculation of the mean values in Table 2 and in Fig. 2 the variation of the calculated k-ratios dependence on different m.a.c. sets is shown. It is evident from a comparison with Fig. 1 that there is a greater dependence of the results on the m.a.c.s than on the overall composition of the steels.

In the conventional procedure a close similarity between standard and unknown makes the results

Table 3. Mass absorption coefficients for Carbon K_α x-rays from various sources

	Mass absorption coefficients (cm^2/g)			
Absorber	Henke 93[3]	Henke 82[4]	Heinrich [5]	PROZA [6]
Si	35,900	36,800	34,881	36,000
V	10,100	8,840	9,403	9,700
Cr	11,900	10,600	10,978	10,700
Mn	13,000	11,500	12,257	12,330
Fe	16,300	13,900	14,103	13,500
Co	15,500	15,600	15,527	14,730
Ni	21,300	18,200	18,009	17,270
Mo	21,400	16,400	26,159	20,600
W	18,400	18,800	16,706	18,600

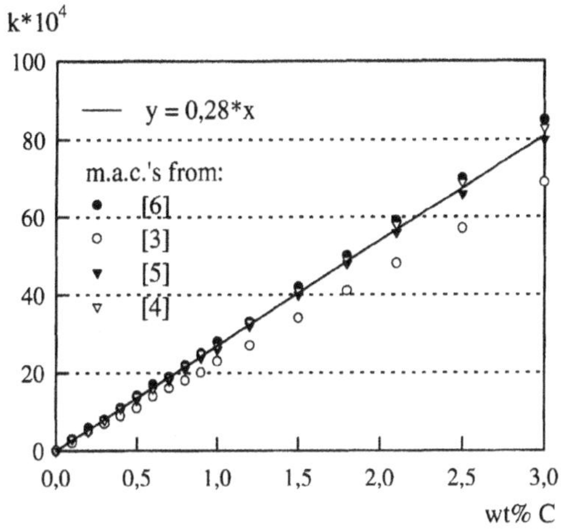

Fig. 2. Variation of the k-ratios of carbon as a function of the carbon concentration dependence on the set of m.a.c.s (CS2, calculated by PROZA, 15 keV)

nearly independent on the m.a.c. set chosen. But if, for example, diamond is used as a carbon standard, then the choice of the 'correct' m.a.c. set is critical for the accuracy of the results. In fact, Bastin et al. [6] recommend a specific set of m.a.c.s for the absorption of carbon in different metals; the one they have optimized for their program PROZA.

However, quantitative analysis of carbon in steels by the use of two standards and linear interpolation between their concentrations should be accurate and independent of any m.a.c. The composition of the steel standards can vary in a broad range. The only requirement is that no peak shape alteration should be observable between the carbides in the standards and those in the unknown specimen.

Measurement Procedure

Generally carbon in steels is not dispersed homogenously in the matrix, but part of it is bound in the form of carbides. Thus measurements at a large number of spots have to be performed to get a reliable average value for the carbon content. The area for a single measurement was around 0.001 mm^2. For this magnification no substantial decrease in the x-ray intensity was observed between the centre point and the edge points of that area. Care has to be taken, that the magnification is the same for unknown and standards, because it will also determine the amount of carbon contamination which occurs.

A primary energy of 15 keV has been chosen. Calculations with the program PROZA showed, that this corresponds to a maximum depth of around 0.8 μm for the generated and 0.2 μm for the emitted C-K_α x-rays. Lower primary energies may result in higher x-ray intensities, but the x-rays are also generated closer to the surface, and the portion of the signal originating from the carbon contamination may become more dominant. Thus, in effect, the accuracy of the measurements of the carbon concentration in steels may decrease.

The take-off angle was 35 degrees. Generally a higher take-off angle and a lower primary energy decreases the dependence of the results on the accuracy of the mass absorption coefficients. However the advantage of this method is that the mass absorption coefficients are not required.

Only the x-ray intensity at the C-K_α wavelength was measured (peak + background + contamination). Following each measurement the mean value of all the measured intensities up to the last one was calculated. The number of measurements was found sufficient when additional measurements no longer significantly changed the mean value (see Fig. 3). The recommendation by Ruste [1] was observed; wait around one minute until contamination has stabilized then start the measurement.

Carbon Quantification

The use of two steel standards and linear interpolation makes background measurements unnecessary. One basic assumption is, that the amount of contamination is independent of the composition of the steels. The thickness of the contamination film depends mainly on the accumulated charge density and the surface mobility of the hydrocarbon molecules [8]. Since the former is connected with electrical conductivity, and the latter may not differ too much in steels, the amount of contamination may in fact be rather independent of the type of steel. The dependence of the background intensity on the composition has already been discussed above.

As already mentioned, the carbides in the standards and the unknown sample have to be of the same type, otherwise peak integration would be necessary and this would require background subtraction and therefore corresponding background measurements.

Calculation of the carbon concentration is straightforward. A demonstration is given by use of the intensity values of Table 4, Analysis 1. The carbon concentration in all three specimens had been determined chemically. ST1 and ST2 were chosen as standards, ST3 served as the unknown.

$$\text{counts (ST2)-counts (ST1)} = 1009 \text{ counts,}$$
$$\text{equiv. to } (0,52\text{–}0,025) = 0,495\,\text{wt\% C;}$$
$$\Rightarrow \quad 51 \text{ counts,}$$
$$\text{equiv. to } 0.025\,\text{wt\% C;}$$
$$\text{counts (ST1)-51} = 853 \text{ counts,}$$
$$\text{equiv. to background + contamination;}$$
$$(\text{counts (ST3)-853}) * 0.495/1009 = 0.37,$$
$$\text{equiv. to C-concentration of ST3.}$$

Thus the EPMA result is in excellent agreement with the chemical analysis result of 0.38 wt%. Another result can be found in Table 4, Analysis 2. The deviations between the chemical and the x-ray analyses are around 0.01 wt%. But, of course, the accuracy depends on the difference between the composition of the unknown sample and steel standards (Fig. 1). Thus, as a conservative value for the attainable accuracy, independent of the composition of the standards, a value of 0.05 wt% seems reasonable.

Table 4. x-ray intensity at C-K_α wavelength and carbon conc. by chemical analysis for three steel specimens respectively, two of which served in each case as standards for analysis of the third by EPMA (Analysis 1: 15 keV, LOD crystal, 50 sec meas. time; Analysis 2: 15 keV, ML crystal, 30 sec meas. time)

Analysis 1	Specimen	Carbon conc. (chem. analysis)	Counts C-K_α wavel	Carbon conc. (calculated)
	ST1[a]	0.025	904	(used as Std.)
	ST2[b]	0.52	1,913	(used as Std.)
	ST3[b]	0.38	1,616	0.37
Analysis 2				
	ST1[a]	0.025	6,546	(used as Std.)
	ST3[b]	0.38	13,497	(used as Std.)
	ST4[b]	0.15	8,789	0.14

[a] Equivalent to CS2, [b] equivalent zu CS1.

Microhardness Measurements

On polished cross-sections of case hardened steels both carbon concentration and microhardness profiles have been measured from the edge to the interior. The aim was, to examine the correlation between these two quantities. The procedure for carbon measurements described above was only used for the determination of the carbon concentration in the interior of the sample. In the profiles themselves quantification was performed from single measurements. But several profiles on the same specimen showed only little deviation from one another. Some of the microhardness measurements were performed at the same points as those where the C-K_α x-ray intensities were measured. This was an easy task, because the microhardness tester operates inside the specimen chamber of the SEM [9], and the spots chosen for carbon measurement were clearly visible by the contamination there. Because of the strong scattering of the microhardness values many more microhardness than carbon concentration measurements had to be performed. Both the diameters of the indentations and their distances from the edge were measured in the SEM.

The results are presented in Fig. 4. They clearly show a correlation of carbon concentration and microhardness. Up to a hardness of around 1000 HV both graphs can be brought to nearly complete coincidence. Fig. 4a demonstrates also, that with the decarburization near the surface a decrease in hardness occurs. Despite the strong scattering of the

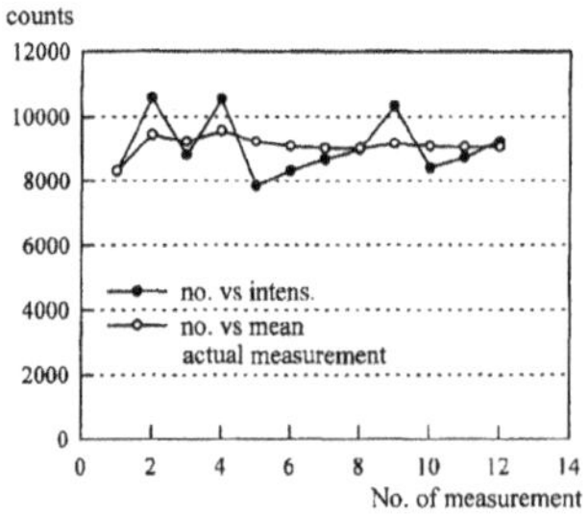

Fig. 3. Variation of the x-ray intensity (Carbon from steel + contamination + background) at different points on the specimen (CS1). Included are the mean values of all the intensities measured up to the last one

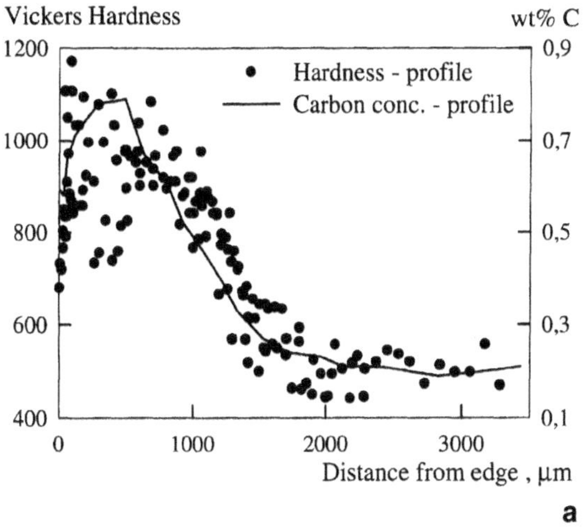

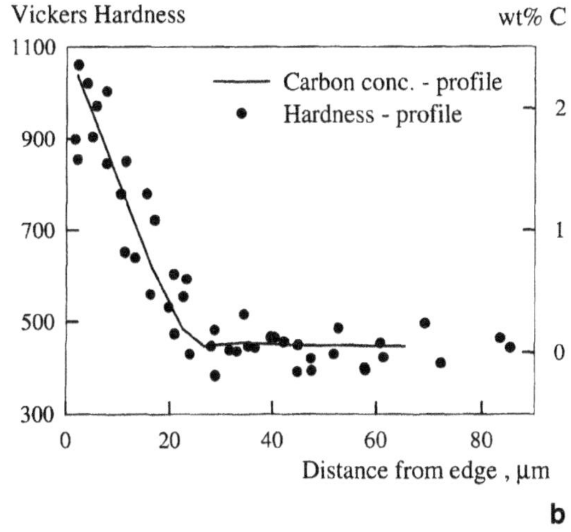

Fig. 4. Carbon concentration and microhardness (load: 2p; indentation speed: 0.07 p/s; dwell time: 10 s) profiles over the cross-sections of two case hardened steels. **a** bulk composition of a steel of equivalent composition to CS2, **b** bulk composition of a steel of equivalent composition to CS1

microhardness values both Figs. 4a and 4b seem to show that the maximum possible hardness at the surface of case carburized steels is around 1000 HV. Because the specimen surfaces were bent and rough, measurements directly at the surface could not be performed.

The case hardened layer in Fig. 4b is only 30 µm thick. Thus carbon measurements very close to the edge were necessary. But because of the maximum depth of 0.2 µm for the emitted C-K_α x-rays up to a distance of around 1 µm near the edge errors should not occur, not even if the edge is turned in the direction of the detector.

Because of the small indentation diameters in microhardness measurements (around 2 µm in the present examples), the average hardness of the crystallites is not obtained in the case of polycrystalline samples. The results are dependent on whether the measurements have been performed in the interior of the grains or in the grain boundary region. They are also dependent on the orientation of the grains [10]. This may partly explain the strong scattering of the measured microhardness values. Also the microhardness values are generally higher than the corresponding macroscopic values [10].

Conclusion

By use of two steel standards and linear interpolation the measurement of the carbon concentration of steels is possible with high accuracy. Only x-ray intensity measurements at the C-K_α wavelength are necessary and therefore this method is especially advantageous in cases where a lot of measurements at different points on the specimen have to be performed. In the present work the method has been used to measure carbon concentration profiles of case hardened steels. Additional microhardness measurements showed a strong coincidence between the profiles of carbon concentration and microhardness.

Acknowledgement. We have to thank the Fa. Anton Paar GmbH, Graz, for sponsoring the Ultramicrohardness tester.

References

[1] J. Ruste, *Mikrochim. Acta [Suppl.]* **1992**, *12*, 247.
[2] G. F. Bastin, H. J. M. Heijligers, in: *Electron Probe Quantitation* (K. F. J. Heinrich, D. E. Newbury, eds.), Plenum, New York, London 1991, p. 145.
[3] B. L. Henke, E. M. Gullison, J. C. Davis, in: *Atomic Data and Nuclear Data Tables Vol. 54, Nr. 2,* (A. Li Scholz, ed.), Academic Press, New York, 1993.
[4] B. L. Henke, P. Lee, T. J. Tanaka, R. L. Shimabukuro, B. K. Fujikawa, in: *Atomic Data and Nuclear Data Tables Vol. 27, Nr. 1,* (K. Way, ed.), Academic Press, New York. 1982.
[5] K. F. J. Heinrich, *Proc. 11th ICXOM, Canada* (J. D. Brown, R. H. Packwood, eds.), University of Western Ontario, London, Canada, 1987.
[6] G. F. Bastin, H. J. M. Heijligers, *X-Ray Spectrometry* **1986**, *15*, 143.
[7] C. A. Andersen in : *The Electron Microprobe* (T. D. McKinley, K. F. J. Heinrich, D. B. Wittry, eds.), Wiley, New York, 1964, p. 58.
[8] L. Reimer, *Scanning Electron Microscopy*, Springer, Berlin Heidelberg New York Tokyo 1985, p. 126.
[9] H. Bangert, X. Cai, A. Wagendristel, A. Kaminitschek, *Scanning Microsc.* **1987**, (*1*), 127.
[10] D. Tabor, *Rev. Phys. Technol.* **1970**, *1*, 145.

Mikrochim. Acta [Suppl.] 15, 207–210 (1998)

Determination of Chemical and Phase Composition of Fly-Ashes by Combined EPMA and XRD Methods

Michał Żelechower[1,*], **Danuta Smołka**[2], **Mariola Jabłońska**[2], and **Andrzej Dytkowicz**[1]

[1] Silesian University of Technology, Chair of Materials Science, Krasinskiego 8, 40-019 Katowice, Poland
[2] Silesian University, Sciences of Earth Faculty, ul. Bedzińska 60, 41-200 Sosnowiec, Poland

Abstract. Fly-ashes collected actively and passively in the Silesia region have been studied by EPMA and XRD methods. The forced collection (using vacuum cleaners) has been applied to gather fly-ashes on cellulose filters. It is well known that a greater part of fly-ash is composed of amorphous materials (glasses) and fine grained compounds, hence the XRD method has limited application. The combination of both EPMA (semi-quantitative analysis) and XRD however, gives acceptable results. Typical oxides (SiO_2, Al_2O_3, Fe_2O_3 – in separate particles) and their mixtures containing additionally potassium, calcium, sulphur and titanium have been found as well as some carbonates and sulphates ($CaCO_3$, $CaSO_4$). Some particles containing considerable amounts of titanium, barium, zinc, magnesium, sodium, phosphor and chlorine were found but total amounts of those elements in all specimens examined were very low. A number of sulphur-rich particles have been observed and the form of sulphur in these has not been determined. Several hypotheses have been rejected {$(NH_4)_2SO_4$, Fe_2S, adsorbed H_2SO_4)} and the question still remains open.

Key words: fly-ashes, EPMA, XRD, sulphur.

Examination of fly-ashes' chemical and phase composition was part of a wider environmental project in the Silesia region. Its aim was to identify emission sources by specific composition of fly-ashes. The main source of fly-ashes is from coal combustion but there are many other "manufacturers" of various kinds of industrial aerosols in Silesia (for instance steel plants or zinc and lead manufacturers). Results of studies of chemical and phase composition of fly-ashes reported in the following sections are in good agreement with results given by other authors [3–7]. However, there is one exception – a certain number of sulphur-rich particles have been discovered in which the form of sulphur could not be determined and we would like to pay special attention to this problem.

Results of investigations on fly-ashes and aerosols in many countries have been reported by several authors and both aerosols and fly-ashes from various sources have been studied, namely rural [1], volcanic [2] and industrial [3–7, 9, 10]. Fly ashes have been examined by various techniques such as ESCA (electron spectroscopy for chemical analysis) [3], SEM (scanning electron microscopy) and EPMA (electron probe microanalysis) [1], [2], [4], [6]. Usually the investigations were focused only on the heavy element content in aerosols, although some attempts were undertaken to identify phase composition of fly-ashes [10, 12]. Phase composition of fly-ashes can be specific to certain regions and type of emission source e.g. natural, industrial or municipal. Certain authors observed in fly-ashes and aerosols a number of particles with a very high content of sulphur. In particular, Rietmeijer [2] reported sulphur-rich particles in volcanic fly-ashes as droplets of sulphuric acid adsorbed on surface grains. Novakov et al. [3] discussed the role of carbon (soot) particles in the catalytic transition of sulphur dioxide into sulphates. Results of examinations of sulphur-rich particles reported below allow us to exclude both sulphuric acid droplets and sulphates.

* To whom correspondence should be addressed

Experimental

Sampling Procedures

All samples have been collected according to the principles usually applied in environmental projects. Special devices (types of vacuum cleaners) were applied to actively collect fly-ashes. Three independent devices, equipped with cellulose filters, have been located in three spots around the power station near Katowice (Silesian coal basin). The fly-ash collection time was 24 hours (rainless day). Open containers were also placed (under the same conditions) in order to collect the so-called fly-ash fall (passive collection). It should be noted that although measuring devices were located around a power station (coal fired), other sites (steel plant, zinc and lead plant) not far away should also be considered as possible fly ash sources.

X-Ray Diffraction

Fly-ashes collected passively (the fly-ash fall) were examined with the use of XRD (X-ray diffraction). In the XRD method an X-ray tube with a copper anode (voltage 22.5 kV, current 10 mA) was applied as a radiation source and a scintillation counter as a detector (JDX7 X-ray diffractometer). Diffraction patterns were interpreted with a use of FAZAN 1.0 software [8] and JCPDS catalogues as the reference data base. Fly-ashes collected actively on cellulose filters were not examined by XRD due to their very low volume.

SEM Observations and Electron Probe X-Ray Analysis

Fly-ashes collected actively on cellulose filters and collected passively (fly-ash fall) were examined using SEM and EPMA. In order to eliminate overlap of characteristic X-ray lines from filter material and fly-ashes, cellulose filters were used instead of mineral ones. Filters with small amounts of fly-ash on their surface were coated in vacuum with a thin film of carbon in order to dissipate electrical charge formed during SEM observations and electron probe X-ray analysis. Samples of fly-ash fall were prepared on plastic plates and also coated with a film of carbon. Carbon coating has been chosen despite high carbon mass absorption coefficients because other possible coating elements (Al, Ti, Cr, Cu) could be present in fly-ashes. Two scanning electron microscopes and two energy dispersive spectrometers were used in X-ray analysis: one was a beryllium window spectrometer (JSM35 scanning electron microscope combined with LINK 290 energy dispersive spectrometer) and the other an ultra thin window detector (S4200 cold emission gun scanning electron microscope combined with VOYAGER energy dispersive spectrometer), which can analyse soft X-ray (C K_α, N K_α, O K_α). During SEM observation and X-ray analysis a voltage of 25 kV and beam current of 2×10^{-10} A were applied. These experimental conditions have been selected to obtain a high X-ray count rate and good quality of SEM micrographs. Although the fly-ashes examined are composed mainly of light elements, some heavy elements may be present (for instance lead, zinc), and the minimum recommended overvoltage ratio $U = E_0/E_{K(L)}$ should equal at least 2 to obtain reasonable characteristic X-ray lines intensities (i.e. Pb – L series or even Zn – K series). Here, E_0 denotes the primary electron beam energy and $E_{K(L)}$ the critical excitation energy of the K or L series. Quantitative analysis of separate particles was not performed for reasons mentioned below but spectra from some pure sulphates (of ammonium, calcium, magnesium and sodium) have been measured in order to compare them qualitatively (via peak intensities ratios) to the spectra of some particles contained in fly-ashes.

Results

XRD Results

The materials collected passively (fly-ash fall) were studied by XRD and typical products of coal combustion were found, namely silicon dioxide (quartz) as well as calcium sulphate (anhydrite), calcium pyrophosphate, and calcium carbonate (calcite). A list of the compounds is shown in Table 1.

The EPMA results show some glassy phases which could not be interpreted by XRD. Additionally it is suspected that some phases present in fly-ashes have either too fine a grain or they are present in quantities below that detectable by XRD.

EPMA Results

Results of qualitative X-ray analysis of fly-ash fall are presented in Table 2 and Fig. 1. X-ray emission from a surface area of 0.1×0.1 mm was measured, hence the results of those measurements can be considered as the volume composition (qualitative). In Table 2 elements are positioned according to the successive decrease of peak intensities; not equivalent to a decreasing concentration level of each element. Since the beryllium window X-ray detector was applied to these measurements, soft X-rays of energy below Na K_α were cut off.

Table 1. Compounds identified by XRD in fly-ash fall

Specimen symbol	Source	Identified phases
A1	fly-ash fall	quartz (SiO_2), anhydrite ($CaSO_4$)
C1	fly-ash fall	quartz (SiO_2), anhydrite ($CaSO_4$), calcium pyrophosphate ($Ca_2P_2O_7$)
D1	fly-ash fall	quartz (SiO_2), anhydrite ($CaSO_4$), calcite ($CaCO_3$)

Table 2. Results of qualitative X-ray analysis of fly-ash fall (surface analysis)

Specimen symbol	Main elements	Low content elements
A1	Si, Al, K, Ca, S, Fe	Ti, Mg, Cl
C1	Si, Al, K, Ca, S, Fe	Na, Mg, P, Cl, Ti, Ba
D1	Si, Al, K, Ca, S, Fe	Mg, P, Cl, Ti, Ba, Na

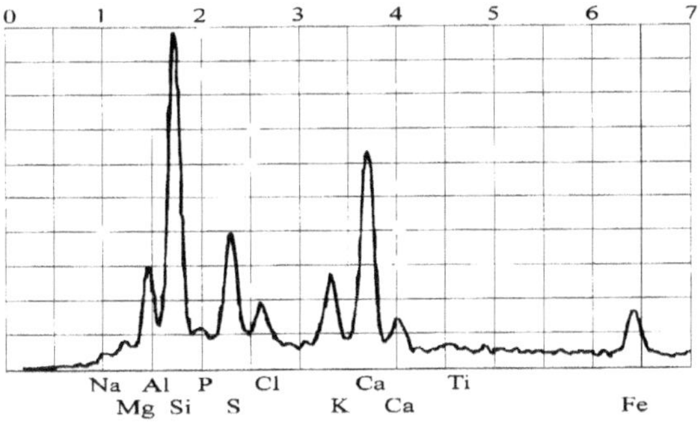

Fig. 1. Surface X-ray anlysis of fly-ash fall (qualitative). Beryllium window X-ray detector

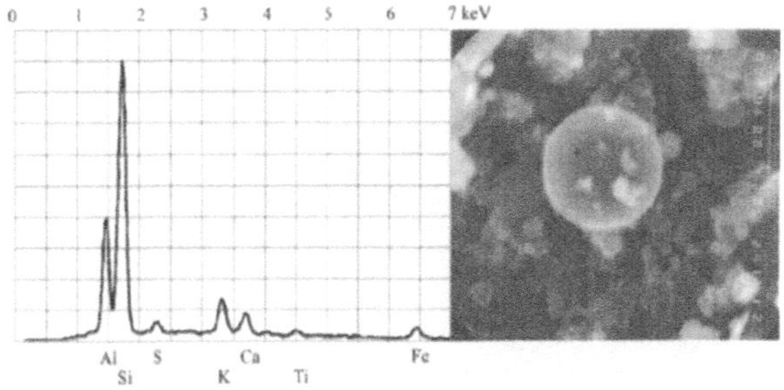

Fig. 4. Particle of glassy mixture of Al, Si, K, Ca, Ti, Fe oxides (microsphere). SEM micrograph and corresponding X-ray spectrum. Beryllium window detector

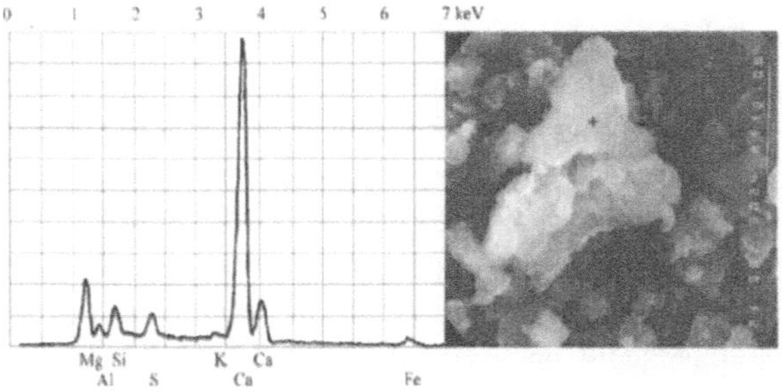

Fig. 2. Particle of calcite ($CaCO_3$). SEM micrograph and corresponding X-ray spectrum. Beryllium window detector

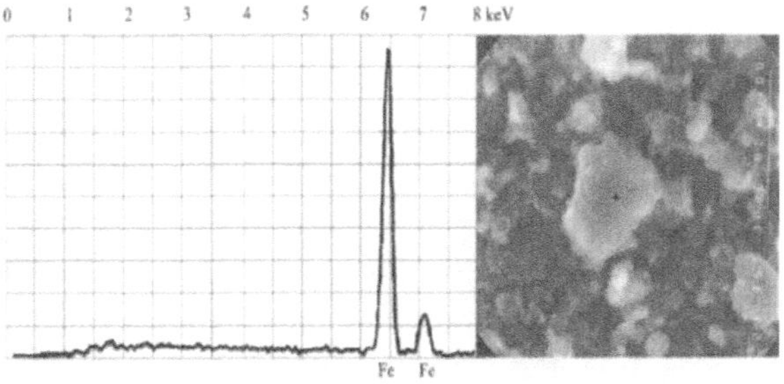

Fig. 5. Particle of iron oxide (Fe_2 O_3). SEM micrograph and corresponding X-ray spectrum. Beryllium window detector

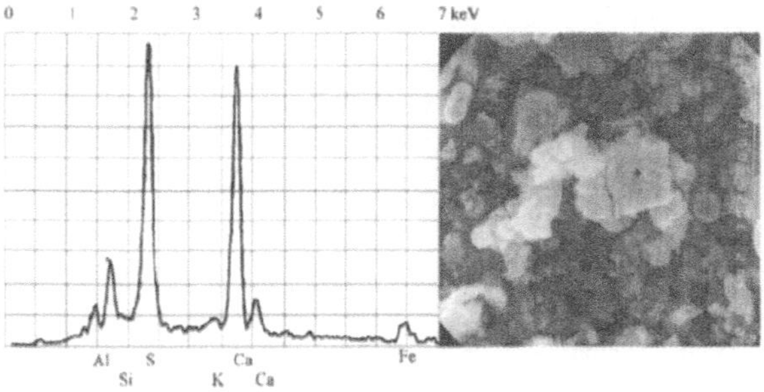

Fig. 3. Particle of anhydrite ($CaSO_4$). SEM micrograph and corresponding X-ray spectrum. Beryllium window detector

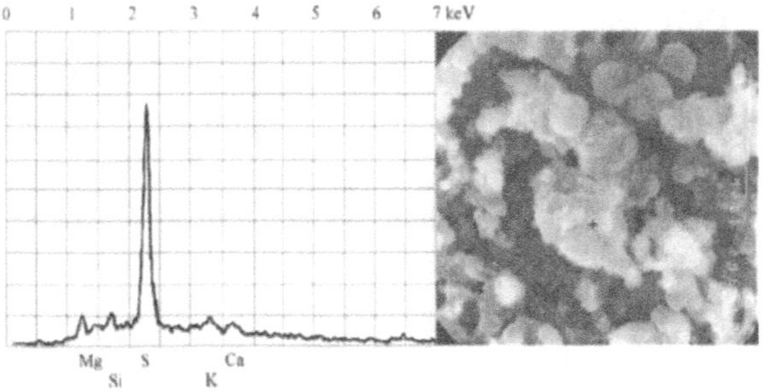

Fig. 6. Sulphur-rich particle. SEM micrograph and corresponding X-ray spectrum. Beryllium window detector

Fly-ashes collected actively on filters were examined with special care and various particles of different morphology were analyzed. The results are shown in Fig. 2–6 (beryllium window detector). Points of qualitative analyses are marked with black crosses (Figures 2–6). The spectra presented correspond to calcium sulphate (anhydrite) and calcium carbonate (calcite) identified by XRD. Additionally an iron oxide particle is seen in Fig. 5. Spherical particles (microspheres) shown in Fig. 4 have an amorphous structure [9, 10] and their composition is a mixture of aluminum oxide, silicon oxide, potassium oxide, and iron oxide, and sometimes also calcium oxide and titanium oxide. A small sulphur peak is observed in the spectrum but its origin is not clear. In some particles zinc has been detected (usually together with chlorine); although it was not identified in surface analysis. It should be noted that several filters were examined and over one hundred particles were observed. Those presented on micrographs are only typical examples.

A number of sulphur-rich particles have been found. One of them is shown in Fig. 6, together with the corresponding X-ray spectrum (the beryllium window detector). In Fig. 7 an X-ray spectrum corresponding to another, similar particle is shown (collected by the

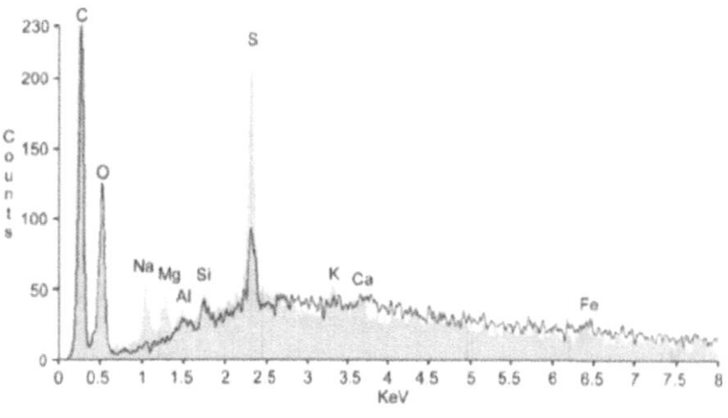

Fig. 7. Sulphur-rich particle (gray) and cellulose filter material (black line) X-ray spectra (SEM micrograph not enclosed). Ultra-thin window detector

ultra-thin window detector). Its electron micrograph is not enclosed. The form of the sulphur in these sulphur-rich particles is not known and this problem is discussed in the next section. Since in Fig. 7 the lower spectrum (line only) corresponds to the composition of the cellulose filter, it cannot be established whether the oxygen peak comes from the particle or from the filter material. The carbon peak probably originates from the coating.

Discussion

Fly-ashes collected in the neighborhood of the power station near Katowice contain, in addition to particles of soot, many fine mineral particles. The latter were composed mainly of light elements and their density should be low. This is a reason for their quantification by EPMA appearing to be uncertain, since it is very difficult to determine the X-ray emission source volume. If the material density 2.5–2.8 g/cm^3 is assumed, then the X-ray emission source diameter should be of the order of 5–8 μm (Monte Carlo simulation [13]). Fig. 3 for instance shows that the X-ray emission source volume contains not only the anhydrite particle (black cross) but also some aluminum, silicon, potassium and iron compounds. For this reason also, detection limits for trace elements have not been presented. However they were estimated (for instance: according to Ziebold expression [11] 0.013 wt% for sulphur in pure ammonium sulphate). The chemical and phase composition of fly-ashes examined is typical for coal combustion products. Theoretically the products of coal combustion are fully identified and have been known for over one hundred years. Most of the results presented agree well with those reported by other authors. However it cannot be definitely established in what form sulphur appears in the observed sulphur-rich particles. Several hypotheses have been considered but none of them can be accepted as being reasonable. In order to exclude some sulphates, X-ray spectra of pure specimens of magnesium sulphate, sodium sulphate, calcium sulphate and ammonium sulphate have been measured (spectra not shown). All these sulphates should be excluded (see Fig. 7); compounds of the type CS, COS, CS_2 are liquid and should quickly evaporate in high vacuum during specimen preparation and during SEM observations; solid SO_3 is very active and transforms instantly into sulphates; SO_2 is volatile. Atomic sulphur could result only as a reduction product, a reaction which is thermodynamically improbable in a combustion chamber. The hypothesis proposed by Rietmeijer [2] that "sulfuric acid droplets boil off rapidly under exposure to the electron probe but *not before a sulphur X-ray signal could be obtained*" is not applicable in our case, since our measurements are repeatable. For the reasons mentioned above, the authors could not establish definitely in what form sulphur appears in sulphur-rich particles and that question still remains open.

Acknowledgments. Authors would like to thank to Dr. J. Szala for helpful assistance in experimental works, and to Dr. S. Serkowski and Dr. M. Sopicka-Lizer for discussion, as well as to K. Rockstein, MSc. for help in preparation of the manuscript.

References

[1] E. Thomas, P. R. Buseck, *Atmospheric Environment* **1983**, *17(11)*, 2299.
[2] F. J. M. Rietmeijer, *J. Volcanol Geothermal Res.* **1988**, *34*, 173.
[3] T. Novakov, S. G. Chang, A. B. Harker, *Science* **1974**, *186*, 259.
[4] M. Manecki, M. Marszalek, *Ecological Problems of Cracow*, Editions of mining Academy, Cracow, 1993 (in Polish).
[5] J. Pastuszka, S. Hlawiczka, K. Willeke, *Atmos. Environ.* **1993**, *278 (1)*, 59.
[6] K. Rockstein, M. Borowska, A. Dytkowicz, M. Zelechower, *Air Protection and Wastes Problems, Vol. XXX, No. 6*, 1996, pp. 200–204 (in Polish).
[7] Dutch Society of Fly-Ashes, Report, 11.03.86, unpublished.
[8] K. Stroz, Private Communication, unpublished software, Katowice, 1979.
[9] S. Pawłowski, A. Gierek, J. Hycnar, S. Serkowski, *Proc. Int. Conf. CIMTEC*, Sant Vincent, Italy, 1980.
[10] G. J. McCarthy, O. E. Manz, D. M. Johansen, S. J. Steinwand, R. J. Stevenson, *Mat. Res. Soc. Proc.* **1986**, *86*, 109.
[11] T. O. Ziebold, *Anal. Chem.* **1967**, *39*, 859.
[12] M. Enders, *Cem. Concr. Res.* **1995**, *25(6)*, 1369.
[13] D. C. Joy, *Monte Carlo Simulation in Turbo Pascal*, Joy, 1995.

Mikrochim. Acta [Suppl.] 15, 211–217 (1998)

EPMA of the Composition of Opal-Based Nanostructured Materials

Vladimir V. Tretyakov*, **Sergei G. Romanov**, **Aleksandr V. Fokin,** and **Vladimir I. Alperovich**

A. F. Ioffe Physical Technical Institute, 194021 St. Petersburg, Russia

Abstract. Electron probe microanalysis (EPMA) was used to study the composition of nanostructured materials based on insulating porous 'hosts' (namely synthetic SiO_2 opal) infilled with different 'guest' materials. In the bare opal a reduction of the *k*-ratio compared with that of bulk amorphous silica was observed. In the case of partially or completely infilled opals this discrepancy was found to be less pronounced. The effect was explained in terms of the space charge induced under the incident electron beam. Obviously, this effect is of importance for all microporous materials if the size scale of their voids matches the ionisation length of the electron beam. We have shown that with the right choice of the accelerating voltage and probe size the EPMA-determined composition can be consistent with the data from wet chemical analysis.

Key words: electron probe microanalysis, nanostructured pores materials, induced space charge

Materials with structural microporosity nowadays attract growing attention as powerful tools for designing nanocomposite solids. If pores of such materials, namely, zeolites, porous glasses, silica-gels, asbestos, opals, etc., are interconnected throughout this allows them to be filled with metals or semiconductors. In this case porous materials are called 'hosts' or templates depending on the purpose for which they are used. The resulting nanocomposites consist of two or more components, the size and the position of 'guest' particles being fully controlled by the topology of the template voids. Making use of crystalline porous templates results in well-ordered nanostructured composite materials, which, potentially, have electronic and photonic energy structures that may be engineered [1]. In contrast to artificial layered structures or 2-dimensional arrays of laterally patterned nanostructures, the characterisation of 3-dimensional arrays is a more difficult problem because their structure and composition varies not only laterally but also normal to the surface.

One of the most important parameters of a nanocomposite is the volume fraction of its components. The composition may be determined by various methods, such as wet chemical analysis, quantitative spectral analysis, etc., but non-destructive methods are more preferable. EPMA seems very suitable as it can resolve the local composition (with 1.5 μm precision) as well as the large-scale distribution (over hundred micrometers) of the 'guest' fraction in a composite. Regardless of whether the infill technology is a physical or a chemical routine, characterisation requires a modification of standard examination techniques since the component contributions to the total signal can change with acceleration voltage and probe size because of the variation in the structure within the volume accessed by the electron beam. Understanding the EPMA data is straightforward if the analysis region is a lot larger or a lot smaller than the size of the composite inhomogeneity, since in both cases the sample under examination can be treated as a bulk target. However, special care must be taken if the space inhomogeneity of the nanocomposite is similar in size to the analysed region. EPMA of porous materials has been carried out from the middle 60s [2], but no general rules have yet been derived. Another complication arises if the template is made

* To whom correspondence should be addressed

from an insulator. Although experimental difficulties related to the analysis of insulators have been comprehensively surveyed [3], a methodological investigation is still necessary in order to be confident of EPMA results from nanocomposites based on insulating templates.

Recently, artificial templates with well-ordered sub-micrometer structural porosity, namely opals, were examined as prospective electronic materials when impregnated with superconductors [4], semiconductors [5, 6] or dielectrics [7]. Being based on the same template and having a similar structure, these materials demonstrate properties that strongly depend on composition. Therefore, precise information about the infill content and its distribution over the sample becomes vitally important for material design. In this paper we demonstrate how the opal-based materials' sub-micrometer structure affects the determination of their composition by EPMA and how the analysis results depend upon the measurement conditions.

Samples

To reveal the effect of porosity on the precision of EPMA data, the porous insulators have to be chosen with account taken of (i) the size of voids – which should be comparable with the ionisation path of electrons, (ii) the well defined topology of voids and (iii) the possibility of extending the results to deal with irregular structures. We used opals since their structure satisfies the above requirements and because of their potential as templates in the design of photonic crystals or in the stabilisation of nanostructure lattices. The structure of the precious opal and its artificial simulation was studied in detail by scanning and transmission electron microscopy as well as by atomic force microscopy (SEM, TEM and AFM) [8–10]. Only the synthetic opals – the product of the colloidal chemistry, were the subject of our interest. The SEM micrograph of the bare opal is shown in Fig. 1a. Opal is assembled from identical silica balls, that are closely packed according to face centred cubic (*fcc*) symmetry. Obviously, such a package contains empty voids between the touching spheres. The correlation between the diameter D of the spheres and the size of structural voids can be derived straightforwardly from the geometry – large voids may be characterised by the diameter of the inscribed sphere $d_1 = 0.41D$ and small voids by $d_2 = 0.23D$ [9]. For example, the size of the large voids in typical opal

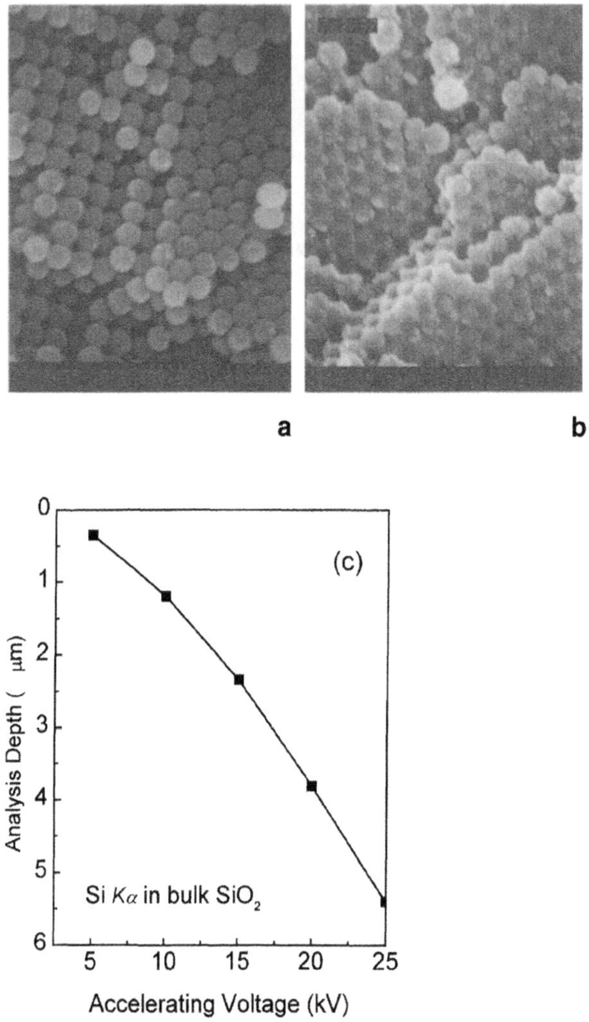

Fig. 1. a SEM image of accidental cleavage of the bare opal with porosity near 30%. Structural voids are clearly seen in both (100) and (111) planes of the *fcc* package. **b** SEM micrograph of the sintered and "glued" opal with porosity near 11%. **c** Analysis depth Si K_α in SiO_2 according to [7], $\rho_{SiO_2} = 2\,g/cm^3$

with $D \approx 200$ nm is approximately 82 nm. Large and small voids follow each other alternately and penetrate each other via bottle-neck-like shaped channels with a minimum diameter of $0.115D$. The total porosity of an ideal *fcc* package is 26%.

The growth process of opal balls includes, particularly, the coagulation of 5 nm subballs, which results in additional porosity; a typical size of the largest void from this subporosity is about 2.5 nm. Further, the hardening of the ball compact can be achieved either by sintering, which increases the diameter of the contact points between spheres, or by fixing the balls with amorphous silica introduced into the compact.

Both procedures result in squeezing the free internal volume of an opal package compared with that of the compact. Opals under investigation were prepared by applying both sintering and overgrowth with silica, a further sintering being eventually used to change the spacing in the package. SEM inspection has revealed that the excess 'silica glue' forms small particles distributed randomly within the space of a void but homogeneously on a scale of several voids (see Fig. 1b). The total internal surface of 'glued' opal increases very much compared with sintered opal, whereas the decrease of overall porosity due to overgrowth is relatively small. To characterise the porosity we have measured two related parameters: the volume of water, that can be accommodated by the opal, and the internal surface of the opal.

The polycrystalline structure of opals (a typical sample consists of 50–100 μm crystallites) brings another complication since the inter-crystallite boundaries break the long-range ordering and change the local porosity. Obviously, these defects may affect the precision of EPMA analysis.

Generally, all samples, except the intentionally disordered opal whose pores are distributed accidentally, possess the same structure but they differ with regard to the fraction of the free internal volume and the conductivity of the infill.

Measurement Technique

All measurements were carried out using a 'Camebax' microprobe equipped with four wavelength and one energy dispersive spectrometers. The wavelength spectrometers were used to perform quantitative analysis and the energy dispersive spectrometer to analyse the content of elements and to record the continuous X-ray background spectra. Composition calculations were based on the X-ray depth distribution of the 'PAP' model [11]. Before measurement all samples were polished with 1 μm diamond powder and cleaned by acetone and alcohol. To avoid charging a thin (100–300Å) aluminium layer was deposited onto the sample surface. The effect of this layer upon the data was taken into account using the following expression:

$$k_i = c_i e^{-\chi_i^f \rho z_f} \frac{\int_{\rho z_f}^{R_x} \varphi_s e^{-\chi_i^s(\rho z - \rho z_f)} d\rho z}{\int_0^{R_x} \varphi_t e^{-\chi_i^t \rho z} d\rho z} \qquad (1)$$

where

c_i-calculated fraction of i-element, k_i-measured k-ratio for i-element, φ-in depth X-ray generation function for sample φ_s and for standard φ_t, $\chi_I = (\mu/\rho)_i \times \operatorname{cosec}\theta$, $(\mu/\rho)_i$-mass absorption of i-element. θ-X-ray take-off angle ($=40°$), ρz-mass depth (g/cm^2), ρz_f-mass thickness of conducting film, R_x-the ultimate depth of the X-ray generation intensity function.

The thicknesses of the conducting films were determined from the average k-ratios obtained for the film at low acceleration voltages by using the film correction technique [12]. The corrections for Si K_α line 0.950, 0.970, 0.974 were calculated from the expression (1) for an aluminium film of 20 nm thickness at accelerating voltages of 10, 15 and 20 kV.

The volume fraction of the 'guest' material was evaluated using the following expression:

$$\Psi_i = \frac{\frac{C_i}{\rho_i}}{\sum \frac{C_i}{\rho_i}} \qquad (2)$$

where ρ_i is the density of i-component.

The integration time for data recording was chosen to provide a statistical error below 0.5%; therefore uncertainlty of data will be caused mainly by the irregularity of the opal package and by defects introduced during sample polishing. To avoid short-term instability of readings and to increase reproducibility, the measurements were carried out with a probe of approximately 10 μm in diameter. The confidence range, that was checked experimentally, does not exceed ± 10% of the relative error. The present investigation was stimulated, in part, by the obvious inconsistency of compositions measured for different opals. The systematic study was performed by applying different accelerating voltages to study bare opals with different porosity and opals impregnated with 'guest' materials. The Si K_α line was used in the analysis of the composition of the bare opals with the reference standards being SiO_2 (coated with a thin aluminium film) and crystalline silicon. In the latter case the shift of the line position was taken into account. The second type of sample was represented by opals whose voids were completely infilled with indium. The opals whose free volume was partially infilled with TiO_2 were considered as another type of composite. The fraction of TiO_2 was verified by analysis of three component samples composed of TiO_2-opal and indium occupying the rest of the free

Table 1. Parameters of bare opals

Sample	Structure	Ball diameter D (nm)	Porosity by water (%)
1	*fcc*	200	35
7	*fcc+*	350	12
9	*fcc*	210	36
A1	disorder	200–300	30
9-5	*fcc*	truncated cube $D=170$	3.85

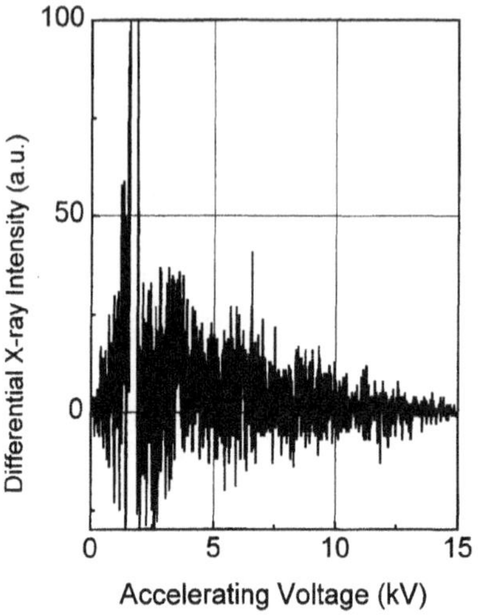

Fig. 3. The differential continuous X-ray intensity spectrum of amorphous silica and the bare opal

space of the sample. Table 1 shows the structural details of the opals examined.

Bare Opals

The relative intensities of the X-ray radiation (k-ratio) are plotted as a function of voltage in Fig. 2 for different opals, with amorphous SiO_2 being used as the standard. It is seen, that $k<1$ applies for all samples. The k-value depends on the porosity of the opal package and decreases with increasing accelerating voltage. Comparison of the k-ratio variation with opal porosity demonstrates a change which is proportional to the ratio of the volume fracton of SiO_2 balls in the package and the penetration depth of incident electrons. The closer the density of opal to that of amorphous SiO_2 – the closer the k-ratio is to unity. It was found that this tendency remains the same regardless of whether the volume fraction of silica is increased by sintering or by overgrowing. We assume, that the surface of the silica-vacuum interfaces remains nearly the same for opals with different porosity, since squeezing of voids is associated with increasing roughness. Therefore, if the number of vaccum gaps (structural voids) is constant, the important parameter is the size of each vacuum gap relative to the penetration depth of electrons. The trend of the k-ratio in opals with decreasing porosity towards that of amorphous SiO_2 points to a similar X-ray generation mechanism in different modifications of silica.

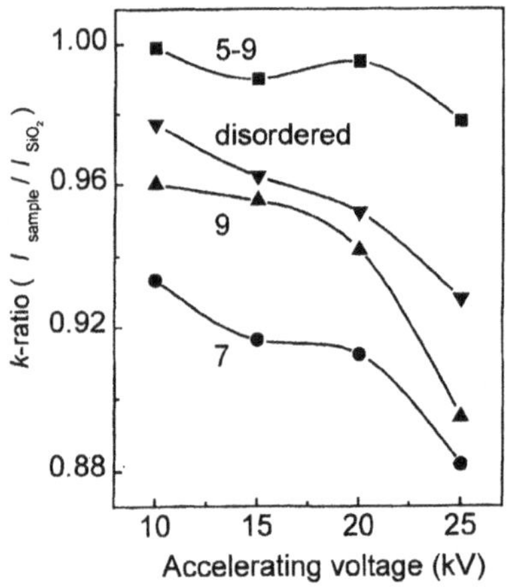

Fig. 2. k-ratios at Si K_α line of opals with regard to amorphous SiO_2 standard upon the accelerating voltage. Numbers identify the samples

Two types of X-radiation are generated along the electron path in the material: characteristic radiation due to excitation of electronic shells and continuous radiation due to the electron deceleration in the field of the nucleus. Fig. 3 shows the differential continuous spectrum of SiO_2 relative to that of opal. It is seen that SiO_2 signal exceeds that from the opal in the whole of the range examined and this difference increases at the low energy end. This effect is the same for all opals at any accelerating voltage. This means that the ratio of the background radiation of opal to that of bulk SiO_2 is below unity, similar to the k-ratio Si K_α; however, its absolute value is higher than the k-ratio. Therefore, we may assume the same origin for both phenomena. Unfortunately, the present level of understanding of the energy relaxation process in the opal structure is not sufficient to determine the correlation between the difference in the intensity of the continuous spectra and the reduction of the k-ratio for characteristic radiation.

The X-radiation may be described as the convolution of the spatial distribution of the energy of electrons $\Phi(\rho,E,\Omega)$ with the relevant X-ray ionisation cross-section $d\sigma/dE$ [13]:

$$\varphi(\rho z) = \int_{4\pi} \int_{Ec}^{Eo} \Phi(r, E, \Omega) \frac{d\sigma_i}{dE} dE d\Omega. \qquad (3)$$

The reduction of the k-ratio may be modelled by varying parameters in this equation to change the scattering efficiency according to the internal structure of opal. It seems reasonable to ascribe this change to the space charge [14] which may be readily built-up in the volume of opal because of worse conditions for charge recombination than that in the bulk SiO_2, assuming that the voids in the opal prevent leakage of a space charge. Recombination of space charge in the porous structure requires either a high electrical field to breakdown the vacuum gaps or time for non-equilibrium charge to complete a path around voids and reach the grounded surface. Therefore, dynamic equilibrium arises under the probe beam. In other words the balance between the rate of the build-up and the relaxation of space charge depends on the kind of structural porosity. As a result, the apparent k-ratio reflects the interplay between the volume under analysis (the volume depends on the accelerating voltage) and the number of vacuum gaps accommodated within the volume (this is the structural parameter). If this is correct, the impregnation of an opal with a conducting material should reduce uncertainty in the determination of the k-ratio in porous materials.

Another source of change in electron scattering conditions is the deep channelling of the electron beam into the material along the voids because of electrostatic focusing of the beam by the space charge accumulating on the void surface. Apparently, this effect becomes significant only if the pore size becomes comparable with the penetration depth of the electron beam. The lower the energy of the electrons the better the focusing, so the effect can change the spatial distribution of the electron density compared with the bulk material and cause a reduction in the continuous radiation from opals.

Completely Infilled Opals

The complete impregnation of voids in the opal with the metal was approached by infilling an opal with the melt under high hydrostatic pressure. This method allows the structure of an opal to remain unaffected during impregnation. The volume fraction of indium in opals with different porosity was determined by wet

Table 2. Porosity of opals and volume fractions of indium according to the wet chemical analysis

Sample	Porosity (%)	In (weight %)
1	37.5%	
9-1	40.5%	68%
10-1	43.0%	73%
7-1	12.0%	28%

Table 3. Volume fraction of indium in completely infilled opals at different accelerating voltages

Sample	10 keV	15 keV	20 keV	25 keV
1	0.6.322	0.6506	0.6495	0.6841
9-1	0.6273	0.6261	0.6303	0.6561
10-1	0.6461	0.6066	0.6303	0.6503
7-1	0.2677	0.2589	0.2736	0.2925

chemical analysis and specific density measurements (Table 2). These data correlate well with the porosity of the bare opals taken before impregnation. Indium forms a 3-dimensional replica of the porous material. This replica exists in the grains coupled by bridges for voids and inter-void channels, respectively. The conductivity of this replica completely removes the problem of the space charge. This allowed us to carry out the EPMA on the monolithic conductive material.

The weight fractions of indium derived from EPMA at different accelerating voltages are given in Table 3. In spite of a less than 1% statistical error, the confidence interval is as high as $\pm 2.5\%$ in absolute magnitude. This uncertainty appears due to the surface defects, that were introduced in the course of polishing, and to inhomogeneity of the indium distribution in an opal.

The deviation of the EPMA-determined indium fraction from the wet chemical analysis data has been observed at the high indium concentration. However this difference does not relate to opal porosity as was shown for opals nos. 1, 9–1 and 10–1. The calculated concentration of indium also increases with accelerating voltage but this can hardly be ascribed to the greater penetration depth and better conditions for averaging over the inhomogeneous medium, since the experimental error does not decrease. Probably, the polishing of a soft metal like indium is intrinsically imperfect. EPMA data referenced to SiO_2 and Si standards correlate one with another confirming the negligible effect of the space charge upon the intensity of the Si K_α line in the SiO_2 standard.

Partially Infilled Opals

Partial infilling of an opal was approached by the chemical vapour deposition of a TiO_2 layer on the inner surface of opal voids [15]. EPMA shows the homogeneity of the TiO_2 distribution on the internally coated opal without detectable inclusions. The data for the partially infilled opals are collected in Table 4. With an increase of the accelerating voltage the calculated volume fraction of SiO_2 decreases but the weight sum keeps close to 100%.

As compared with the bare opals, those coated with TiO_2 demonstrate a 10% higher *k*-ratio for the Si K_α line in contrast with the theoretical model which predicts the opposite. Apparently, the major role in this experiment belongs to the conductivity of the TiO_2 coating. It was found out independently, that coating the opal with a TiO_2 layer leads to an increase in the high-temperature conductivity of the composite by several orders of magnitude over that of bare opals [16]. A similar increase in conductivity is well documented for rutile under vacuum and electron irradiation due to self-doping by the oxygen deficit in its structure. The negligible reduction of the *k*-ratio in the TiO_2-opal composite is in line with our speculation about the role of a space charge effect.

In order to examine the accuracy of the composition analysis the indium was embedded in the TiO_2-opal composite. This procedure let us fill in the vacuum gaps in the composite and repeat the TiO_2 fraction analysis in a monolithic conductive material.

The composition of the In-TiO_2-opal multiple component samples is shown in Table 5. At 20 kV the porosity of opals derived from EPMA of the In-TiO_2-opal nanocomposites corresponds to the capacity of opals with respect to water. Wet chemical analysis confirms the average composition of In-TiO_2-opal for both indium and TiO_2 components. This assumes that no serious complication has occurred in the analysis of the multiple component nanocomposites without structural porosity. However, the penetration depth of the electron beam should be high enough to exceed the length of structural inhomogeneity.

Table 4. Volume fraction of TiO_2 in partially infilled opals at different accelerating voltages

	k-ratio		volume fraction (vol.%)	
keV	TiO_2	SiO_2	TiO_2	SiO_2
10	0.1465	0.9153	0.038	0.962
15	0.1235	0.8913	0.032	0.968
20	0.0918	0.7919	0.03	0.97

Table 5. Composition of three component opal-based composites at different accelerating voltages

	k-ratio			Volume fraction (%)		
keV	In	TiO_2	SiO_2	In	TiO_2	SiO_2
10	0.6355	0.1245	0.4712	0.271	0.042	0.687
15	0.72	0.104	0.3612	0.316	0.04	0.643
20	0.792	0.092	0.3048	0.327	0.04	0.634

Conclusion

The results of this study are instructive for the correct application of compositional analysis by EPMA in the case of an inhomogeneous medium where the length scale of the inhomogeneity is comparable with the ionisation electron path. A reduction in the intensity of X-rays emitted from the porous insulator compared with the intensity from the bulk insulator was observed. This effect depends on the topology of voids and can be associated with an extra space charge due to the deterioration of the electron beam runoff. It is impossible to avoid inconsistency in EPMA data from porous and monolithic insulators by increasing the analysed volume to achieve better averaging. However, with a decrease in the volume fraction of porosity in the body of the insulator the inconsistency gradually disappears.

In the case of partially infilled templates the *k*-ratio depends on the conductivity of the 'guest' material. For conductive nanocomposites no precautions are needed with EPMA, but special care should be taken to obtain satisfactory averaging over the inhomogeneous medium. To obtain consistent composition data from EPMA and wet chemical analysis an accelerating voltage of not less than 20 kV should be applied.

The main result from this investigation is a detailed description of the problem of EPMA analysis in porous insulating structures where the effect of a space charge is enhanced by the presence of the vacuum gaps. The next deeper level of understanding may be approached by applying the Monte-Carlo simulation to model energy losses of electrons as well as the interplay between energy losses and the electron beam channeling along the structural voids.

Acknowledgement. This work was supported in part by Russian Science-Technical Programme "Superconductivity" (grant no.

96099), International Soros Foundation (grant no. R52300) and Russian Foundation for Basic Research (grant 96-02-17963).

References

[1] S. G. Romanov, C. M. Sotomayor Torres, in: *Nanoparticles in Solids and Solutions: Preparation, Characterisation and Utilization* (J. H. Fendler, ed.) VCH, Weinheim, 1997, in press.

[2] T. Ichinokawa, H. Kobayashi, M. Nakjima, *Jpn. J. Appl. Phys* **1967**, *8*, 1563.

[3] G. F. Bastin, H. J. M. Heijligers, in: *Electron Probe Quantification* (K. F. J. Heinrich, D. E. Newbury, ed.) Plenum, New York, 1991, p. 193.

[4] V. N. Bogomolv, Y. A. Kumzerov, S. G. Romanov, V. V. Zhuravlev, *Physica C: Superconductivity* **1993**, *208*, 371.

[5] S. G. Romanov, A. V. Fokin, D. K. Maude, J. C. Portal, *Appl. Phys. Lett.* **1996**, *69*, 2897.

[6] S. G. Romanov, N. P. Johnson, H. M. Yates, M. E. Pemble, V. Y. Butko, C. M. Sotomayor Torres, *Appl. Phys. Lett.* **1997**, *70*, 2091.

[7] S. V. Pankova, V. V. Poborchii, V. G. Soloviev, *J. Phys. Condens. Matter* **1996**, *8*, L203.

[8] J. V. Sanders, *Nature* **1964**, *204* (*4964*), 1151.

[9] V. G. Balakirev, V. N. Bogomolov, V. V. Zhuravlev, Y. A. Kumzerov, V. P. Petranovsky, S. G. Romanov, L. A. Samoilovich, *Crystallogr. Rep.* **1993**, *38*, 348.

[10] R. Mayoral, J. Requena, J. S. Moya, C. Lopez, A. Cintas, Hernan Miguez, F. Meseguer, L. Vazquez, M. Holgado, A. Blanco, *Advanced Materials* **1997**, *9*, 257.

[11] J. L. Pouchou F. Pichoir, *Rech. Aerosp.* **1984**, *3*, 13.

[12] S. V. Kazakov, S. G. Konnikov, V. V. Tretyakov, *Bull. Acad. Sci. USSR, Ser. Phys.* **1991**, *55*, 147.

[13] T. S. Rao-Sahib, D. B. Wittry, *J. Appl. Phys.* **1974**, *45*, 5060.

[14] J. Cazaux, *X-ray Spectrosc.* **1996**, *25*, 256.

[15] S. G. Romanov, A. V. Fokin, V. Y. Butko, V. V. Tretyakov, S. M. Samoilovich C. M. Sotomayor Torres, *Fizika Tverdogo Tela* **1997**, *39*, 3347 (in Russian), (English translation *Phys. Solid State* **1997**, *39*, 727).

[16] V. G. Soloviev, *Private Communication.*

Mikrochim. Acta [Suppl.] 15, 219–225 (1998)

NDIC and EMP Study of Plagioclase Mineral Zoning: An Example from Nea Kameni Lavas

Alba P. Santo* and **Rosa Maria Bomparola**

Dipartimento di Scienze della Terra, Via La Pira, 4, I-50121 Firenze, Italy

Abstract. Compositional zoning of plagioclase in magmatic rocks provides an insight into the crystallisation history, as this mineral is capable of recording even subtle physical and chemical changes in magmatic liquids. In this paper plagioclase phenocrysts in lavas from the island Nea Kameni of the Santorini volcanic complex (Aegean Sea, Greece) have been studied using Nomarski imaging and detailed electron microprobe scans. The combined use of these techniques permitted the reconstruction of accurate compositional zoning profiles. On the basis of textural and chemical characteristics plagioclase phenocrysts can be divided into two main groups: 1) crystals displaying a core composition in the range Anorthite 40–65%, generally characterised by an unzoned or poorly zoned clear core and by variously zoned mantle and rim; 2) crystals showing a very Ca-rich core (Anorthite >75%), often displaying a fine sieve-textured core and an oscillatory-zoned region mantling the core. These latter plagioclase crystals have been interpreted as xenocrysts. Their textural characteristics confirm this hypothesis and represent additional evidence that, in addition to fractional crystallisation, mixing processes played a role in the evolution of the Nea Kameni magmas.

Key words: NDIC, EMP, plagioclase, zoning profiles, mixing.

Chemical zoning, which is of common occurrence in nearly all magmatic minerals, plays a significant role in petrogenetic studies. In fact, compositional zoning reflects changes of chemical and physical conditions during mineral growth, thus providing information on the evolution processes. In particular, plagioclase is found to be extremely useful because it is capable of recording even subtle changes in the magmatic systems [e.g. 1–3] due to its wide possible compositional variation and its high sensitivity to chemical and physical conditions. Plagioclase in magmatic rocks and particularly in orogenic andesites displays a variety of textures including complex compositional zoning and several different morphological features. Several papers [e.g. 4–6] dealt with the description and classification of these features and there is still no definite agreement about their origin.

In mineral zoning studies, an exact representation of the compositional zoning profile is very necessary [6] but the traditional transmitted light polarizing microscope is not always suitable for this purpose. In addition, an inappropriate use of the electron microprobe (EMP) may also contribute to misleading profiles. Indeed, a limited knowledge of the zoning pattern may lead to an erroneous choice of the points to be analysed. The use of the Nomarski Differential Interference Contrast (NDIC) technique combined with accurate electron microprobe analyses may be very helpful in resolving these problems. NDIC imaging is able to reveal very fine textural details, otherwise almost invisible in transmitted light (e.g. Fig. 1), thus permitting points submitted to chemical analysis, to be properly defined.

In this paper the NDIC and EMP techniques have been applied to the study of plagioclase in volcanic rocks from Nea Kameni. This island belongs to the Santorini volcanic complex which is part of the Aegean magmatic arc. Nea Kameni has resulted from

* To whom correspondence should be addressed

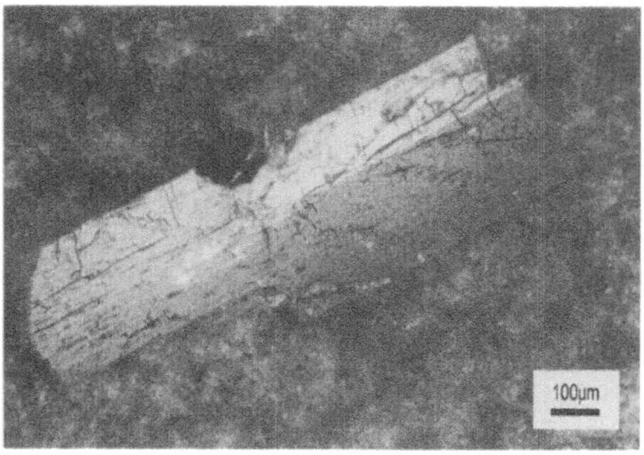

a

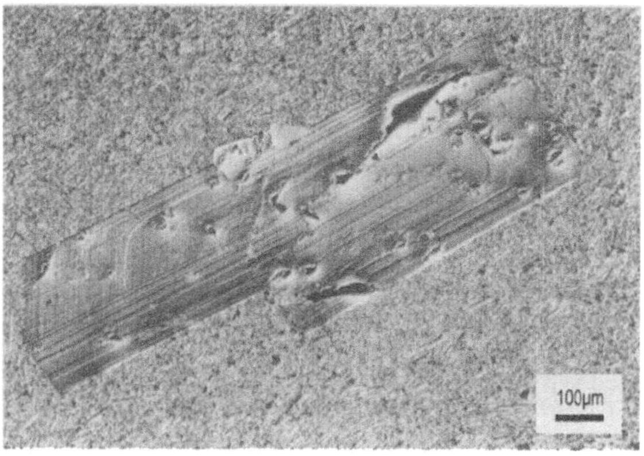

b

Fig. 1. **a** Cross-polarized light image of a plagioclase phenocryst from the Nea Kameni lavas; **b** Nomarski DIC reflected light image of the same crystal in **a**

several intermittent eruptions, starting in 1570 and continuing to the present day [7, 8].

Previous studies [e.g. 9–11] dealing with the geochemical, petrological and mineralogical data of the Nea Kameni rocks provided insights on the evolution of these arc lavas and showed that fractional crystallisation and magma mixing processes played an important role during their evolution. In this paper we present compositional zoning profiles obtained on selected plagioclase crystals from lavas belonging to single eruption cycles, with the purpose of (1) demonstrating effectiveness of combined NDIC and EMP techniques in mineral zoning studies, and (2) using the zoning profiles to obtain independent petrological evidence on the Nea Kameni magma evolution.

Experimental

For the present study, lava samples belonging to all eruption cycles between 1570 and the present were selected. Lavas have dacitic composition and holocrystalline porphyritic texture with a phenocryst mineralogy represented by plagioclase (plg), clinopyroxene, orthopyroxene and Fe-oxides. In addition, olivine is present in rounded and resorbed crystals. Chemical and modal analyses of the studied rocks are given in Table 1.

The Nomarski differential interference contrast imaging technique has been used to observe textural features in selected plagioclase crystals. The crystal selection was done with great care as many of the observed phenocryst characteristics may be due to the orientation of the section. Crystals oriented in order to provide maximum possible information [6, 13] were selected. The NDIC technique involves etching of polished thin sections with

Table 1. Major (wt%) element and modal (volume %) analyses of representative Nea Kameni rocks

Eruption	1570	1707	1866	1925	1940	1940	1950
SiO_2	66.7	67.4	66.3	65.7	66.6	65.7	65.4
TiO_2	0.68	0.74	0.80	0.77	0.76	0.81	0.79
$Al2O_3$	15.3	15.1	14.9	15.4	15.2	15.3	15.3
$Fe2O_3$	4.4	12.0	1.6	1.4	1.4	1.3	1.5
FeO	0.44	3.6	3.3	3.8	3.6	3.7	3.9
MnO	0.14	0.14	0.15	0.15	0.16	0.15	0.16
MgO	1.2	0.83	1.3	1.4	1.2	1.4	1.5
CaO	3.6	3.45	3.7	3.8	3.5	3.6	4.0
Na_2O	5.2	5.21	5.3	5.2	5.2	5.2	5.1
K_2O	1.9	1.87	1.9	1.8	1.9	1.8	1.8
P_2O5	0.13	0.13	0.12	0.15	0.13	0.15	0.12
L.O.I.	0.32	0.40	0.80	0.50	0.38	0.45	1.2
Plg	12.4	5.7	8.2	9.2	7.2	8.8	10.2
Opx	0.5	0.1	0.4	0.6	0.1	0.5	0.2
Cpx	1.8	0.9	1.3	1.6	1.8	1.6	1.9
Ox	1.1	0.8	1.0	0.7	0.4	1.1	0.8
Gdm	84.2	92.7	89.1	87.9	90.5	88.0	86.8

L. O. I. Loss On Ignition; *Plg* plagioclase; *Opx* orthopyroxene; *Cpx* clinopyroxene; *Ox* oxide; *Gdm* groundmass. Analyses of lava from 1707 eruption are from Petrone [12].

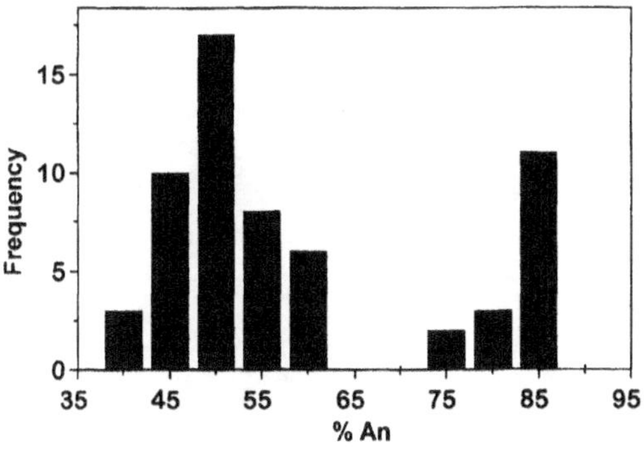

Fig. 2. Frequency distribution and composition of plagioclase phenocryst cores

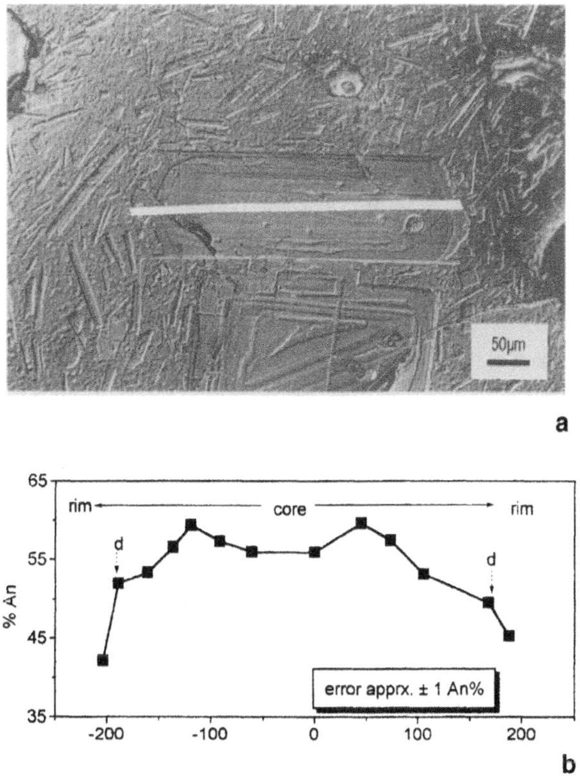

Fig. 3. a Nomarski DIC image of a plagioclase phenocryst from 1950 lavas; **b** Rim-to-rim variation in anorthite content in the above crystal. **d** = dissolution surfaces

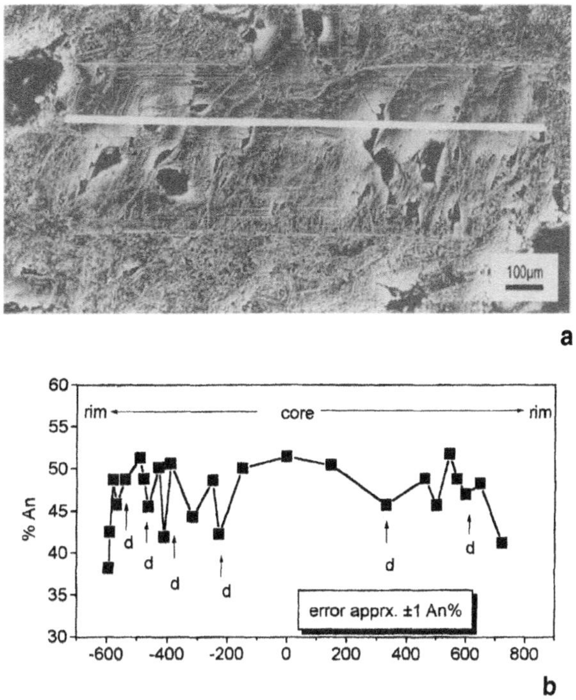

Fig. 4. a Nomarski DIC image of a plagioclase phenocryst from 1570 lavas; **b** Rim-to-rim variation in anorthite content in the above crystal. **d** = dissolution surfaces

concentrated fluoboric acid (HBF_4). The etching time is variable depending on the sample composition. For example, the etching times for a basaltic and an andesitic sample from Aeolian islands (Southern Tyrrhenian Sea) were respectively 20 and 35 sec. (Santo, unpublished data). The acid is then neutralised by immersion in a saturated solution of Na_2CO_3 and the section is rinsed with water and dried [14–16]. The section is finally carbon-coated to increase the reflectivity of the polished surface and to eliminate internal reflection in the crystals before the Nomarski interference observations. Ca-rich plagioclase zones are preferentially leached more deeply than those zones containing less Ca thus producing a microtopographic relief of highs (Na-rich zones) and lows (Ca-rich zones). This relief is generally of about 0.2 to 0.5 µm [16]. In this study polished thin sections were etched for 40 seconds and differential interference contrast images were obtained by using a Nikon Labophot-2 microscope equipped with Nomarski DIC objectives.

Major and minor-element compositions of plagioclase were obtained by a wavelength dispersive system using a JEOL JXA-8600 electron microprobe operating at 15 kV and variable counting time and beam current (10–20 nA). The microprobe is equipped with a Series II Tracor Northern system. Prior to analysis each thin section was repolished in order to eliminate any possible contribution due to the etching [14, 17] and analytical points were located precisely by using photomicrographs of each crystal. Corrections for matrix effects were performed according to the method of Bence and Albee [18]. During analyses, particular care was taken regarding the zone width. Zones displaying very narrow width (< 10 µm) have not been analysed as plagioclase measurements represent an average composition. Compositional zoning profiles were obtained by accurate electron microprobe scans and some representative ones are showed in Figs. 3–7 where the plagioclase composition is given in terms of Anorthite (An)%.

Results and Discussion

Even though emitted during different eruptive cycles, the rocks studied display generally similar character-

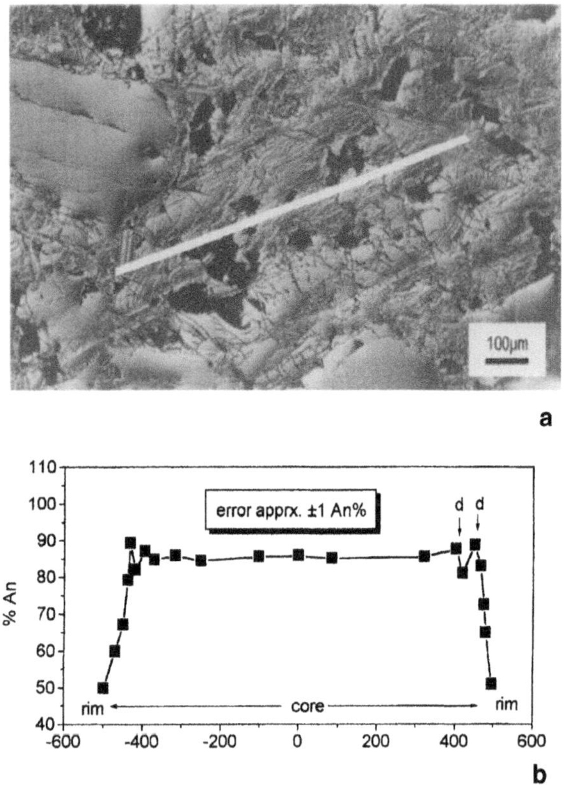

Fig. 5. **a** Nomarski DIC image of a plagioclase phenocryst from 1570 lavas; **b** Rim-to-rim variation in anorthite content in the above crystal. **d** = dissolution surfaces

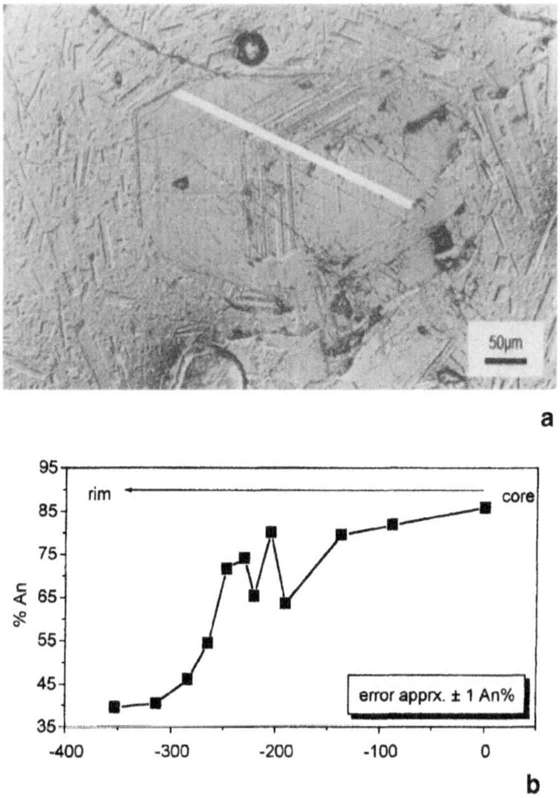

Fig. 6. **a** Nomarski DIC image of a plagioclase phenocryst from 1870 lavas; **b** Rim-to-core variation in anorthite content in the above crystal

istics. Phenocryst content is always less than 20 vol.%. Modal compositions (Table 1) indicate that plagioclase is by far the most abundant phenocryst phase. Plagioclase occurs as euhedral or subhedral and variously twinned crystals. Overall plagioclase compositions vary from An_{32} to An_{89}. Selected plagioclase analyses representing the most and least An-rich zones that were analysed in each crystal are reported in Table 2. Core phenocrysts display a wide compositional variability (An_{40-86}), even though they exhibit a bimodal distribution with a gap existing between An_{62} and An_{72} (Fig. 2). The composition of microphenocryst cores (An_{50-60}) as well as that of plagioclase in groundmass (An_{32-50}) lies in a more restricted range.

Nomarski interference contrast imaging allowed us to observe in plagioclase crystals a variety of interesting features such as fine-scale oscillatory zoning, resorption surfaces, sieved and patchy textures previously undetected with conventional transmitted-light microscopy. These observations, combined with the electron microprobe analyses, permitted accurate profiles of compositional variation across crystals to be obtained. In these profiles three morphological units may be distinguished from the centre outwards: the core, the mantle and the rim. On the basis of textural and compositional characteristics plagioclase phenocrysts can be divided in two main groups: 1) crystals displaying a core composition in the range Anorthite 40–65%, generally characterised by an unzoned or poorly zoned clear core and by variously zoned mantle and rim; 2) crystals showing a very Ca-rich core (usually An > 75%) that commonly exhibits a fine sieved texture.

Group-1 plagioclase crystals occur in all the studied lavas, whereas Group-2 crystals have been found exclusively in lavas of 1570, 1866 and 1950 eruptive cycles.

Group-1 Plagioclase

The common feature of this group of crystals is the occurrence of a large clear core generally euhedral or lightly rounded. The crystal core is usually unzoned or

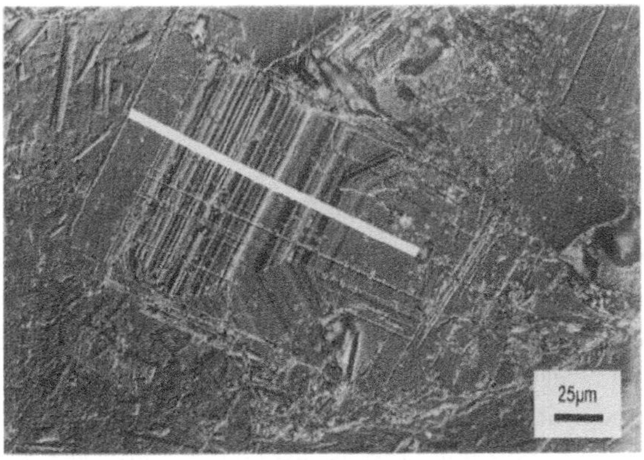

a

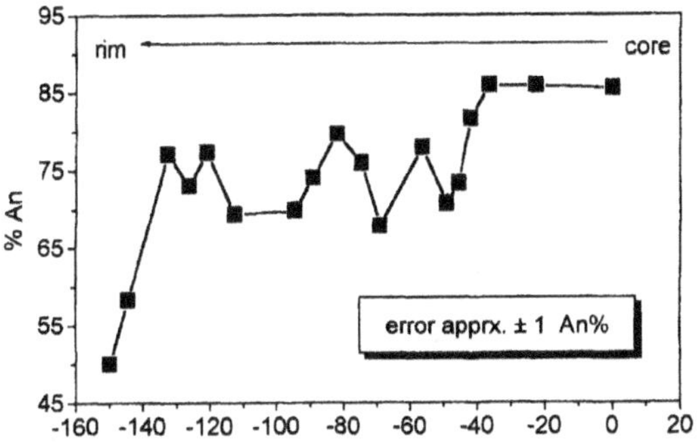

b

Fig. 7. a Nomarski DIC image of a plagioclase phenocryst from 1950 lavas; **b** Rim-to-core variation in anorthite content in the above crystal

exhibits a weak normal zoning. Generally anorthite content decreases continuously from the edge of the core outwards. However, some crystals display an anorthite increase at the edge of the core followed by a decline in the mantle (Fig. 3). A resorption surface associated with a non-significant anorthite decrease (< 10%) generally marks the core-mantle boundary. Mantles display normal, wavy-oscillatory or normal-oscillatory zoning with euhedral conformable dissolution surfaces marking the different zones. The amplitude of this oscillatory zoning is no more than 10% An (Fig. 4). Anorthite content declines rapidly through the rim and lies in the range An_{37-50}.

Group-2 Plagioclase

This plagioclase group is characterised by the high anorthite content of the core that commonly is unzoned (Fig. 5) and shows resorbed boundaries and sieved texture. The mantle region of the majority of these plagioclase crystals displays regular, conformable normal-oscillatory zoning (Fig. 6). The corresponding zoning pattern is characterised by spikes of anorthite content, which is in some cases rather high. The boundary between mantle and rim is marked by a reaction surface and a variable drop in anorthite content that is in some cases very sharp. In general the rim is represented by a normal zoned layer. Some

Table 2. Representative EMP analyses (wt%) of plagioclase phenocrysts from the Nea Kameni rocks

Eruption	1570		1570		1707		1707		1866	
SiO_2	55.4	57.6	45.7	49.6	56.6	56.2	58.9	55.5	56.4	49.6
$Al2O_3$	28.2	26.6	34.3	31.3	27.6	28.1	26.6	27.8	27.7	31.9
FeO	0.43	0.64	0.71	0.67	0.40	0.47	0.41	0.49	0.60	0.48
CaO	10.6	8.5	17.7	14.6	10.0	10.2	8.1	10.4	9.6	16.2
Na_2O	5.4	6.5	1.2	3.0	5.2	5.4	6.1	5.0	5.4	2.2
K_2O	0.15	0.27	0.00	0.03	0.18	0.22	0.23	0.24	0.34	0.06
Sum	100.18	100.11	99.61	99.20	99.98	100.59	100.34	99.43	100.04	100.44
An	51.8	41.2	88.8	72.6	50.8	50.6	41.8	52.5	48.6	79.8
Ab	47.3	57.2	11.2	27.2	48.1	48.1	56.8	46.1	49.4	19.8
Or	0.9	1.6	0.0	0.2	1.1	1.3	1.4	1.4	2.0	0.4
Eruption	1866		1940		1950		1950		1950	
SiO_2	55.9	55.3	54.4	57.0	57.7	53.3	58.7	46.7	53.2	46.7
$Al2O_3$	28.0	27.6	28.7	27.0	26.0	29.3	26.3	34.1	29.0	33.6
FeO	0.43	0.52	0.57	0.52	0.77	0.60	0.36	0.84	0.52	0.62
CaO	10.2	10.4	11.2	8.9	8.5	12.0	8.1	17.0	12.3	17.3
Na_2O	6.3	5.7	5.2	6.3	6.2	4.4	6.6	1.5	4.7	1.6
K_2O	0.21	0.15	0.14	0.27	0.36	0.12	0.37	0.04	0.17	0.06
Sum	101.04	99.67	100.21	99.99	99.53	99.72	100.43	100.18	99.89	99.88
An	46.6	50.0	54.0	43.1	42.1	59.6	39.6	85.9	58.3	85.6
Ab	52.3	49.1	45.2	55.4	55.7	39.7	58.3	13.9	40.7	14.1
Or	1.1	0.9	0.8	1.6	2.1	0.7	2.2	0.2	1.0	0.4

An anorthite; *Ab* albite; *Or* orthoclase.

crystals exhibit a normal-oscillatory zoned rim (Fig. 7). Several resorption surfaces punctuate and truncate the different zones which display only small compositional changes. Each oscillation is in some cases too narrow to be resolved by the microprobe. The anorthite content of the outermost part of rim is similar to that of the other plagioclase group.

Microprobe analyses revealed, as observed in previous studies [9, 10, 19], the occurrence of two populations of plagioclase phenocrysts in the studied lavas. The composition of Group-2 plg cores is too anorthitic to have crystallised from an evolved melt [20]. Indeed, these cores as well as some crystals of clinopyroxene, olivine and orthopyroxene occurring in these rocks [9, 21], and clearly not in equilibrium with the host melt, have been interpreted as xenocrysts deriving from the disaggregation of cognate xenoliths [9]. This hypothesis is consistent with the presence in the Nea Kameni lavas of several mafic xenoliths whose major and trace element mineral composition [9, 19, 21] is similar to that of xenocrysts observed in the host lavas.

In addition to the high-Ca content, several crystals of the Group-2 plagioclase display a fine sieve-textured core. Several mechanisms such as under-cooling, magma mixing and magmatic decompression [22 and references therein] have been proposed to explain the origin of sieve textures in plagioclase. None of these mechanisms can be completely excluded as the compositional and textural plagioclase features reflect their concomitant effect. In general, in the absence of other evidence, it is difficult to understand which mechanism played the main role.

In the case under study, a process of magma mixing has been largely invoked by previous authors [10, 11, 23] as adequately explaining petrography and trace element and isotope variations of lavas. The proposed models require the existence, beneath the Kameni volcano, of a shallow stratified magma chamber where periodic arrivals of hot mafic magmas can determine thermal convection and/or mixing with a fractionating silicic magma. In the light of this model, the dissolution surfaces and the rounded cores of Group-1 plagioclase might reflect the temperature increase in the chamber due to influxes of new hot magma. However, textural evidence, such as the small shift in anorthite content corresponding to many dissolution surfaces, indicates that magma mixing was not very extensive in the volcanic system under study. Furthermore, the chemical trends of the discontinuously zoned plagioclase of Group-2 require the involvement of another process in addition to magma mixing. A mechanism capable of producing the chemical and textural discontinuities occurring across most of Group-2 plagioclase crystals might be one of the type proposed by Blundy and Shimizu [24], i.e. a crystal-liquid mixing in which calcic plagioclase crystals are simply transferred from one melt to another. Such a process may account for the occurrence of high-Ca plagioclase in silicic magmas through disruption of early cumulates. The presence of these anorthitic crystals displaying similar textural features in lavas from different eruptive cycles also seems to indicate the recurrence of processes active over the entire life of the volcanic system. According to the crystal-liquid mixing, the rim of both groups of plagioclase which display similar composition might have been formed simultaneously after the re-entrainment of calcic cores in the host melt.

Finally, concerning the fine-scale zoning present in almost all zoned crystals, we believe, in agreement with several author [e.g. 3, 6], that it reflects the small-scale perturbations from equilibrium in melt around the growing crystal. This is consistent with the mainly euhedral shape of the zones and the compositional differences of a few An% observed in the Nea Kameni plagioclase crystals.

Conclusion

Accurate microprobe scans of plagioclase crystals permitted the zoning patterns revealed by NDIC observations to be quantified. The compositional profiles obtained clearly show how combined NDIC and EMP techniques represent a valuable tool for studying zoning in minerals and prove to be very helpful in elucidating the crystallisation history. On the basis of the chemical and textural data obtained we suggest that a crystal-liquid mixing process played an important role in determining many of the mineral features of the Nea Kameni lavas.

Acknowledgements. The authors would like to thank prof. P. A. Mando' for critical reading of the manuscript and Dr. F. Olmi for technical assistance during EMP analyses. Financial support was provided by C N. R. "Centro di Studio per la Minerogenesi e la Geochimica Applicata" of Firenze.

References

[1] S. Maaloe, *J. Petrol.* **1976**, *17*, 398.
[2] L. C. Kuo, R. J. Kirkpatrick, *Contrib. Mineral. Petrol.* **1982**, *79*, 13.

[3] T. P. Loomis, P. W. Welber, *Contrib: Mineral. Petrol.* **1982**, *81*, 230.
[4] F. Homma, *SMPM* **1932**, *12*, 345.
[5] J. V. Smith, W. L. Brown, *Feldspar Minerals: Crystal structures, Physical, Chemical, and Microtextural Properties, 2nd Ed.*, Springer, Berlin Heidelberg New York Tokyo, 1988, p. 828.
[6] T. H. Pearce, A. M. Kolinsnik, *Earth-Sci. Rev.* **1990**, *29*, 9.
[7] M. Fytikas, O. Giuliani, F. Innocenti, G. Marinelli, R. Mazzuoli, *Tectonophisics* **1980**, *31*, 29.
[8] G. Ferrara, M. Fytikas, O. Giuliani, G. Marinelli, *Thera and the Aegean World, Vol. II. Publ.* Z 1980, p. 37.
[9] M. Barton, J. P. P. Huijsmans, *Contrib. Mineral. Petrol.* **1986**, *94*, 472.
[10] K. Seymour, D. Vlassopoulos, T. H. Pearce, C. Rice, *Contrib. Mineral. Petrol.* **1990**, *104*, 73.
[11] L. Francalanci, G. Vougioukalakis, G. Eleftheriadis, L. Pinarelli, C. Petrone, P. Manetti, G. Christofides, *Proceeding of the 2nd Workshop of European Laboratory Volcanoes* – May 1996, Santorini Island, Greece.
[12] M. C. Petrone, *Thesis*, University of Florence, 1995.
[13] T. H. Pearce, *Contrib. Mineral. Petrol.* **1984**, *86*, 149.
[14] A. T. Anderson Jr., *Am. Mineral.* **1983**, *68*, 125.
[15] T. H. Pearce, A. H. Clark, *Geology* **1989**, *17*, 757.
[16] T. H. Pearce, P. L. Roeder, I. Wolfson, *Am. Mineral.* **1986**, *71*, 734.
[17] T. H. Pearce, J. K. Russell, I. Wolfson, *Am. Mineral.* **1987**, *72*, 1131.
[18] A. E. Bence, A. L. Albee, *J. Geol.* **1968**, *6*, 382.
[19] S. Conticelli, L. Francalanci, A. P. Santo, C. Petrone, *Proceeding of 2nd Workshop on European Laboratory Volcanoes* – May 1996 Santorini Island, Greece.
[20] J. B. Gill, *Orogenic Andesites and Plate Tectonics*, Springer, Berlin Heidelberg New York, 1981, p. 390.
[21] A. P. Santo, J. D. Mac Arthur, P. Manetti, *Mikrochim. Acta* **1994**, *114/115*, 441.
[22] S. T. Nelson, A. Montana, *Am. Mineral.* **1992**, *77*, 1242.
[23] M. D. Higgins, *J. Volcanol. Geotherm. Res.* **1996**, *70*, 37.
[24] J. D. Blundy, N. Shimizu, *Earth Planet. Sci. Lett.* **1991**, *102*, 178.

Mikrochim. Acta [Suppl.] 15, 227–235 (1998)

Compositional X-Ray Maps of Metamorphic and Magmatic Minerals

Alessandro Borghi[1,*], **Roberto Cossio**[2], **Filippo Olmi**[3], and **Gloria Vaggelli**[3]

[1] Dipartimento di Scienze Mineralogiche e Petrologiche, Via Valperga Caluso 37, Università di Torino, I-10100 Torino, Italy
[2] Dipartimento di Scienze della Terra, Via Valperga Caluso 37, Università di Torino, I-10100 Torino, Italy
[3] C.N.R.-C.S. per la Minerogenesi e la Geochimica Applicata, Via G. La Pira 4, I-50121 Firenze Italy

Abstract. This paper deals with the applications of qualitative and quantitative x-ray compositional maps collected by either an energy-dispersive spectrometer (EDS) or a wavelength-dispersive spectrometer (WDS) on magmatic and metamorphic minerals.

The data from qualitative elemental maps, represented by matrices of counts, are converted to a MS-DOS media format and then processed by using an image analysis program.

The digital acquisition of quantitative EDS and WDS x-ray maps produces matrices of spectra and peak counts, respectively, with known X, Y coordinates. The matrices collected are converted to MS-DOS media-format and then processed by a dedicated program which applies a quantitative matrix correction (ZAF/FLS for EDS and pure ZAF for WDS maps). The results obtained are comparable with conventional quantitative microprobe analyses.

The final quantitative data are displayed as wt% oxide, cation or end-member two-dimensional compositional maps, either in a grey-scale or a false-colour, which can contribute significantly to the knowledge of the P-T history of rock formation. In fact, it is possible to determine the type and shape of the chemical zoning and to evaluate the proportion of each mineral involved in post-growth chemical diffusion processes.

The procedure described has been tested by collecting qualitative and quantitative x-ray maps on metamorphic and magmatic minerals.

Key words: x-ray maps, EDS, WDS, mineral phases.

* To whom correspondence should be addressed

Most magmatic and metamorphic minerals are solid solutions of two or more end-members and are rarely homogeneous in composition but often display chemical zoning.

The chemical composition of rock-forming magmatic and metamorphic minerals reflects their history of crystallisation and provides information on the temperature and pressure conditions during their formation. The chemical and physical changes recorded by minerals during their initial growth or subsequent re-equilibration can be used to reconstruct either the evolutionary trends of magmas [1] or the P-T path of metamorphic rocks [2]. Thus, chemical analysis of the rock-forming minerals is necessary for a complete petrological study.

The electron microprobe [3] is an ideal tool to study these minerals as it provides analysis of selected areas as small as a few microns diameter on the surface of the sample. However, minerals show complex chemical zoning due to various physico-chemical processes during crystallisation. In metamorphic minerals for example, different zoning types can be indicative of either growth or chemical diffusion processes [4]. Thus, a single quantitative line scan across grains may be unsatisfactory in an investigation of the whole petrological process.

The best way to determine carefully the chemical zoning is by using two-dimensional x-ray compositional maps. Previously, such an investigation was performed by an extrapolation of either grids of quantitative spot analyses or by a series of x-ray line scans. Semi-quantitative x-ray maps are now routinely produced for single elements using different commercial EDS microanalytical systems [5], [6] and [7].

Recently, techniques for quantitative x-ray compositional EDS maps have been developed [8].

This work describes an application and comparison of digitally acquired qualitative and quantitative compositional x-ray maps on metamorphic and magmatic minerals using both an energy dispersive microanalytical system and a wavelength dispersive electron microprobe. A Cambridge Stereoscan 360 electron microscope equipped with a Link-Analytical EDS (QX-2000) and a JEOL JXA 8600 electron microprobe, equipped with four wavelength dispersive spectrometers, were used for EDS and WDS compositional maps, respectively.

Experimental

Interaction of an electron beam with the surface of a massive sample leads to the emission of characteristic x-rays, which can be collected by either an energy-dispersive detector or a wavelength-dispersive spectrometer.

The main advantage of EDS is its ability to record simultaneously data from a large number of elements over a wide energy range, whereas in WDS the elements need to be recorded sequentially and the range will be constrained by the number and type of the spectrometers available. In addition, no mechanical movements or geometrical constraints are necessary in EDS mapping; whereas in WDS a large number of stage movements are required to cover the surface area of the sample. In contrast, the detector resolution of EDS is about an order of magnitude worse than that of WDS, although it is still adequate to separate the K-lines of the most common elements forming magmatic and metamorphic minerals [9]. However, deconvolution of spectra for specimens containing, at the same time, elements such as Na and Zn, S and Pb or Ti and Ba, is still prone to some degree of uncertainty. Finally, EDS is less suitable for mapping at low concentrations owing to its poorer peak-to-background ratios compared with WDS.

Qualitative Elemental Maps

Qualitative EDS maps are collected by means of a MDI board (LINK system) which controls remotely the scanning of the electron beam, using an appropriate routine for beam positioning. The intensities within selected energy windows, corresponding to individual elements, are acquired by an ADM program (LINK system) and stored to disk. Disk storage of maps is performed in a standard LINK format, with up to 24 images acquired simultaneously.

Each matrix value, representing one pixel of information, gives the concentration of the respective element for that spot but also includes x-ray counts of the background below the peak and therefore with matrix correction applied, produces only qualitative maps.

Qualitative WDS maps are acquired by a LAMAP program (TRACOR-NORTHERN system) which uses stage automation in conjuction with the mapping routine. Stage movement is necessary to preserve the correct diffraction geometry of the spectrometers when a large area of the sample is mapped. So, whilst the beam remains stationary, a specific routine of the LAMAP program moves the stage in a matrix pattern. The total number of x-rays maps acquired at the same time is equal to the number of spectrometers present, as each spectrometer records only one element at a time in the mapping process.

In both EDS and WDS systems the scanning area can be selected by the SEM monitor from the secondary or backscattered electron image. The size of the analysed area can be varied, as can the resolution of the maps (from 16×16 to 512×512 pixels). The dwell time (from 10 ms up to 1 s) for each pixel can be selected too. The digital data is converted to a MS-DOS media format, exported to a PC and then processed by using an image analysis program. In fact, an important advantage of digital recording is the ability to carry out manipulation of the maps off-line from the recording process.

Quantitative Elemental Maps

Quantitative EDS maps are collected using the DIGIPOINT software (LINK system) which automatically carries out fully matrix-corrected spot analyses at different stage positions. This program provides a method for creating and storing analysis positions, such as grids, lines and random points, from a stored image. For each position, x-ray spectra can be automatically acquired and stored on hard disk.

Quantitative EDS maps are collected under quite different instrumental working conditions from those normally used during conventional (spot) microprobe analyses. In particular, a short pulse processing time constant of 15 µs (instead of 40 µs) is used, in order to increase the probe current up to ca. 1.5 nA (compared with ca. 400 pA) and to achieve statistically significant counts. Under these conditions, 12, 000 counts/pixel can be reached with a dwell time $= 3$ s. However, very long acquisition times are necessary to obtain good spatial resolution (large number of pixels) and statistical precision (counts/pixel) in the maps. Quantitative x-ray maps require high stability of the electron beam current for long acquisition times and are susceptible to probe current drift, which must be measured and corrected. Thus, the drift of the probe current was experimentally determined using a reference standard of cobalt. Numerous independent measurements for our microprobe yielded an average linear regression when applying the total area method. Therefore, a cobalt spectrum acquisition before and after map collection is sufficient to evaluate the initial and final probe current.

Every EDS map consists of a square matrix of *nxn* ($64 \leq n \leq 512$) EDS spectra of known X, Y coordinates, collected on the microprobe hard disk. Then, the map is converted to MS-DOS media format and exported to a PC. Here, each spectrum of the matrix is separately processed by means of a dedicated program called PETROMAP [9]. Filtered least squares processing [10] and a quantitative ZAF correction (e.g. [11] and many other more recent formulations) for each point on the map are automatically calculated.

Quantitative WDS maps are obtained with a network of peak counts from each point using remote stage movement and the spectrometers at fixed positions. The initial X, Y coordinates, the number of peak counts per point, the step distance (in microns), and the acquisition time need to be selected. Thus, a network of *nxn* ($64 \leq n \leq 512$) points may be obtained according to the resolution desired.

The result is a set of matrices, one matrix for each element selected, containing the x-ray counts measured at the element's peak position. Net peak counts are obtained by subtracting the background counts measured on reference standards under the same analytical conditions. The results are exported to a PC and then corrected using a dedicated ZAF program compiled by one of the authors (R.C.), which determines the quantitative wt% of the elements in the matrix.

Both (EDS and WDS) ZAF correction programs permit calculation of the cationic formula of the mineral based upon stoichiometry. During this calculation all analytical points are rejected which do not correspond to the stoichiometric constraints relating to the selected mineral. The final matrices can be imported into commercial spreadsheets and graphic programs to give false-coloured or grey-scale two-dimensional maps where each colour or grey tone represents a selected concentration range for the analysed element. Additionally, the end-members of the analysed minerals may be displayed as a two-dimensional quantitative map.

Results and Discussion

The procedure described to obtain qualitative and quantitative X-ray compositional maps is applied here to a metamorphic garnet from the Stilo unit (Calabrian-Peloritan Arc, see [12] for a detailed geological framework), and to a magmatic clinopyroxene from the last historical 1944 A.D. Vesuvius eruption [13]. Specimens were prepared as polished geological thin sections, mounted on a glass slide and carbon coated.

Sections as close as possible to the core of the garnet porphyroblast (maximum size 1.3 mm) and of the clinopyroxene phenocryst (maximum size 0.7 mm.) were chosen (Fig. 1). In this way, the whole metamorphic and magmatic history recorded by the garnet and the clinopyroxene, could be observed. Surface areas of ca. 2.5 mm^2 and of ca. 0.5 mm^2 (for garnet and clinopyroxene respectively) were selected for mapping. An accelerating potential of 15 kV and 1.5 μA beam current were used.

The *qualitative maps* of garnet and clinopyroxene were acquired using a very short dwell time of 120 ms for both EDS and WDS systems and at a high resolution of 512×512 pixels. Thus, each set of maps requires an acquisition time of ca. 12 hours (which includes a detector dead time ca. 30%) in the EDS system and ca. 7 hours in the WDS system. However, for WDS the maximum number of elemental maps in each acquisition set is limited to the number of available spectrometers.

Although qualitative maps provide a detailed image of the compositional variations within the minerals, no information on the absolute elemental content can be derived from such images.

Figures 2 and 3 show a selection of EDS qualitative x-ray elemental maps for the garnet and the clinopyroxene, respectively. We chose to present qualitative EDS compositional maps instead of the WDS maps as no appreciable differences were observed between the two types.

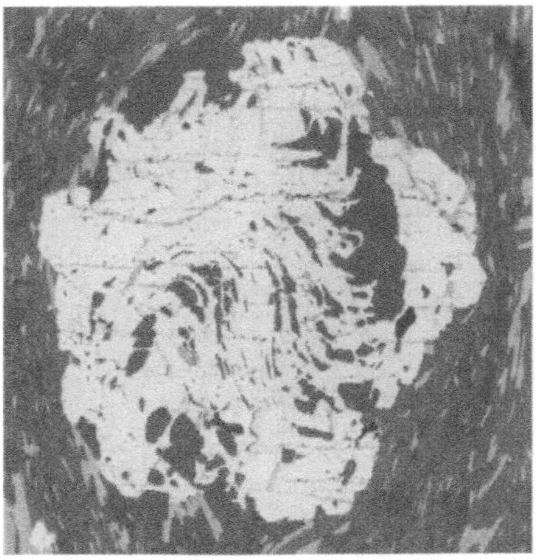

Garnet backscattered image (512 X 512 pixel)

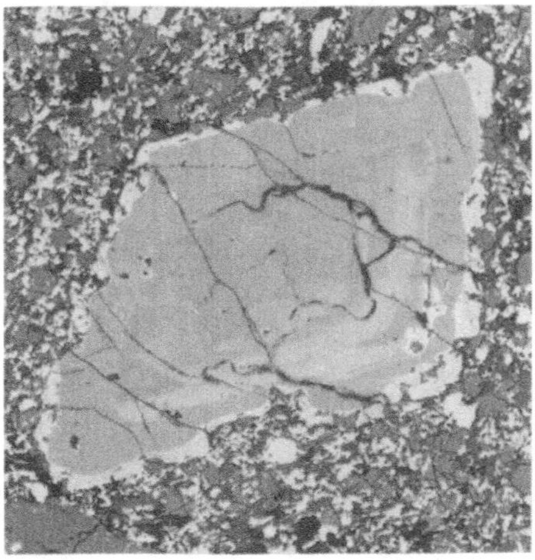

Pyroxene backscattered image (512X512 pixel)

Fig. 1. Back scattered images of the garnet and clinopyroxene mapped crystals. Their maximum size is 1.3 mm and 0.7 mm, respectively

For the garnet and clinopyroxene *quantitative maps* lower spatial resolutions of 128×128 and 64×64 were selected respectively because of the longer acquisition times for each point. Notwithstanding this, the distance between each analytical point (pixel) is ca. 15 micron. Thus, as the diameter of the excitation sphere is 2–8 micron, the selected area will effectively be fully scanned. A dwell time of 3 s was used for EDS maps in order to collect sufficient counts in each pixel for digital filtering and background correction. Thus, a map composed of 128×128 pixels needs ca.

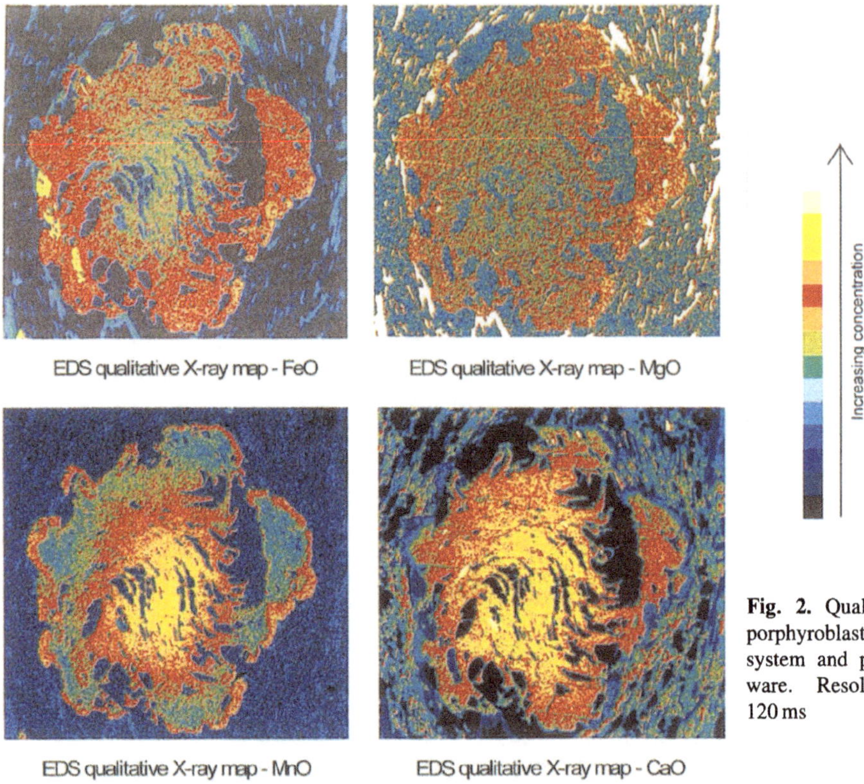

Fig. 2. Qualitative X-ray maps for the garnet porphyroblast. The image was collected by an EDS system and processed with image analysis software. Resolution = 512×512; dwell time = 120 ms

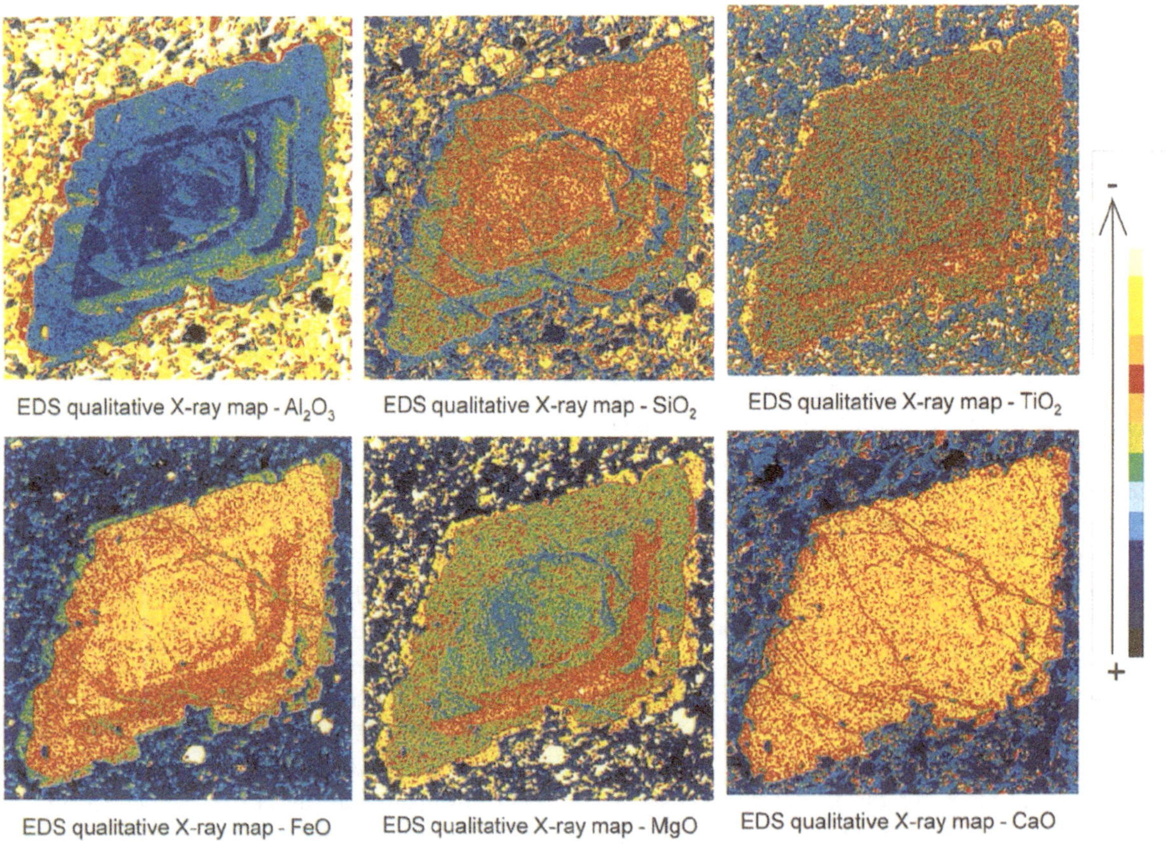

Fig. 3. Qualitative X-ray maps for the clinopyroxene phenocryst. The image was collected by an EDS system and processed with image analysis software (Photostyler®). Resolution = 512×512; dwell time = 120 ms

14 hours (with a dead time approximately 30%). For WDS maps, an acquisition time of 1 second was sufficient to collect statistically significant peak counts.

After the processing procedure, quantitative maps gave much poorer detail compared to the qualitative maps even though they required a longer acquisition time. However, in the quantitative maps (Figs. 4 and 5), each pixel represents a ZAF-processed EDS spectrum or WDS peak counts, with its own precision and accuracy. Thus, the compositional ranges – between the maximum and the minimum value – may be divided into a selected number of gray or false-colour classes where each class corresponds to a well defined and narrow range of oxide concentrations.

The absolute (σ) and the relative (%) errors for each element collected, referred to a representative analytical point on the EDS and WDS maps for both garnet and clinopyroxene, are reported in Tables 1 and 2, respectively. It is evident that the absolute errors in EDS and WDS systems are comparable as are the relative errors for concentrations higher than 1–2 wt%. The effects for lower concentrations are illustrated in the garnet Mg map (Figs. 4A and 4B) where the EDS elemental map is more homogeneous and less detailed than the WDS one. In all other cases, comparison between EDS and WDS quantitative maps reveals no significant differences in the distribution of the major chemical elements.

Garnet Elemental Maps

In high grade metapelites, garnet generally shows a strong compositional zoning due to several overlapping processes (fractional crystallization, homogenisation and diffusion) acting at different times during the metamorphic history of the host rock. Growth zoning occurs when a new shell of different composition is added onto a growing crystal and reflects a changing of P-T conditions or a change in the local bulk composition of the rock. Diffusion zoning is a modification of a pre-existing composition by the process of volume diffusion [14]. Growth and diffusion zoning are two end-members cases and naturally zoned samples are likely to have been influenced by both processes. The identification and interpretation of such composite zoning is extremely difficult if microstructural studies or traditional electron microprobe analyses are performed only. Thus, detailed compositional maps are important for observing and interpreting the compositional zoning in both metamorphic and magmatic minerals.

Figures 4a and 4b show quantitative 128×128 WDS and EDS elemental maps for the garnet. Output is represented by false-coloured maps of weight % oxides. The four elements (Fe, Mg, Mn and Ca) involved in the exchange processes within the octahedral site of the garnet are considered here. The mapped garnet crystal shows well defined compositional zoning, characterised by a gradual and concentric variation of all elements from core to rim. This elemental distribution represents a zoning growth developed in greenschist towards amphibolite metamorphic facies rocks (see [14], with ref.). In particular, this type of Mn zoning has been described in the literature as "bell shaped zoning" and is typical of zoning growth in medium-grade metamorphic rocks.

The outer rim of the garnet shows a large increase in Mn content balanced by a decrease in Fe and Ca. This thin portion of the crystal (better displayed in Fig. 2) is affected by intercrystalline diffusion processes, which modify the pre-existing pattern composition.

Clinopyroxene Elemental Maps

In general, clinopyroxene from Vesuvius historical eruptions (between 1631–1944 A. D.; [15]) is mainly diopside or salite which show normal, reverse or oscillatory zoning [16].

Figure 5 shows the quantitative 64×64 WDS and EDS elemental maps for the selected clinopyroxene. The output is represented by pseudo-coloured maps of weight % oxides. The two exchanging elements in the tetrahedral (Si–Al) and in the octahedral (Fe–Mg) sites are considered here. The mapped crystal shows normal and concentric oscillatory zoning. In particular, the decrease of Si is followed by an increase of Al and is accompanied by an enrichment of Fe, the latter substituting for Mg in the octahaedral site. Ti also increases whereas Ca remains mostly constant (Fig. 3). The outer rim of clinopyroxene is characterised by the highest Fe and Al contents and has the same composition as the small clinopyroxene in the groundmass. Thus, the clinopyroxene outer rim is the only portion in equilibrium with the final magmatic liquid.

The compositional variations described here for the clinopyroxene suggest the occurrence of different evolutionary processes during the recent activity of

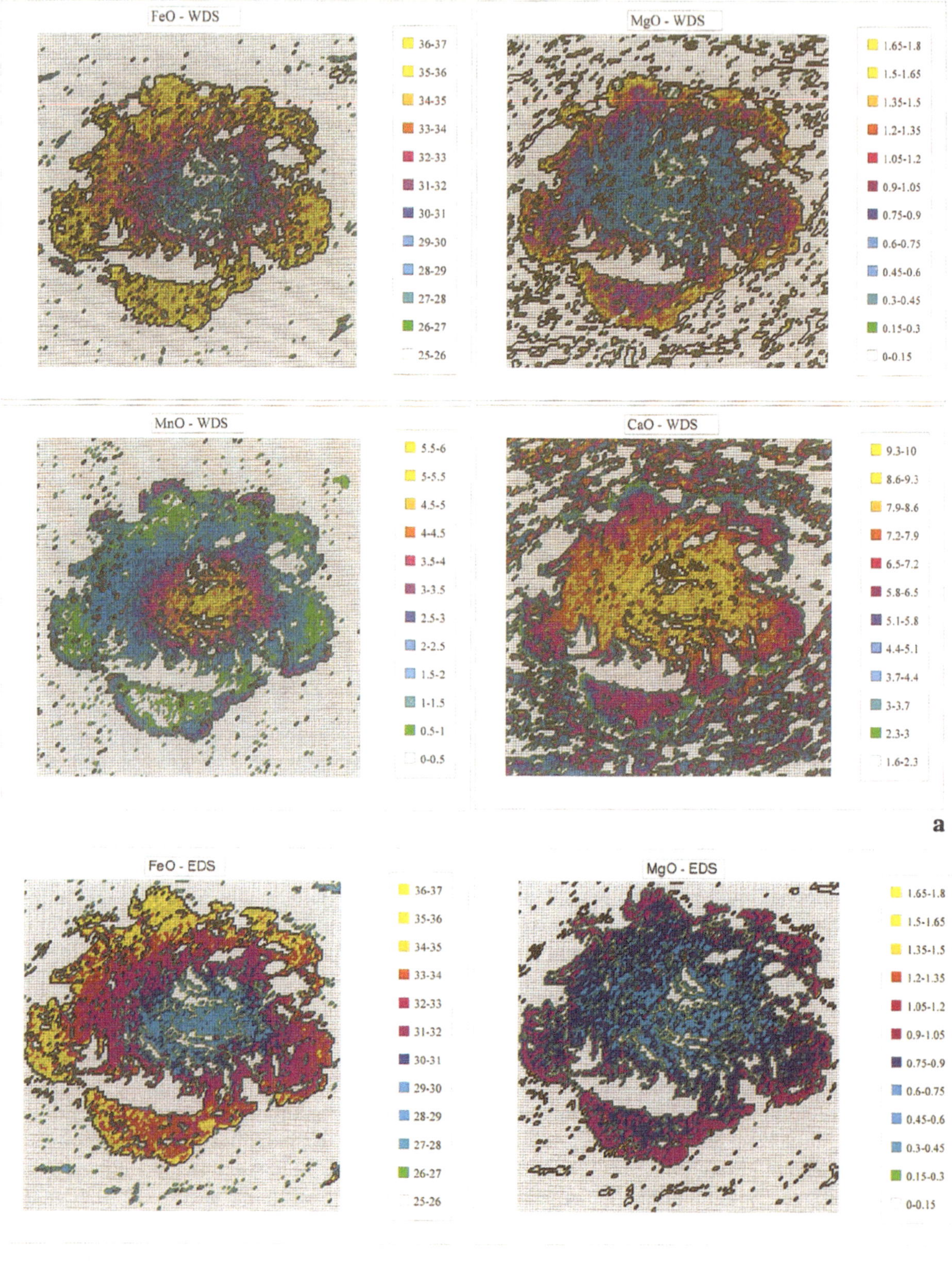

Fig. 4a. WDS quantitative compositional maps of four main exchanging chemical elements (Fe, Mg, Ca, Mn) expressed as weight % oxide for the metamorphic garnet. **b** A selection (Fe and Mg) of EDS quantitative elemental maps are also shown for comparison. For each map the range between the maximum and the minimum values is divided into 12 pseudo-coloured classes

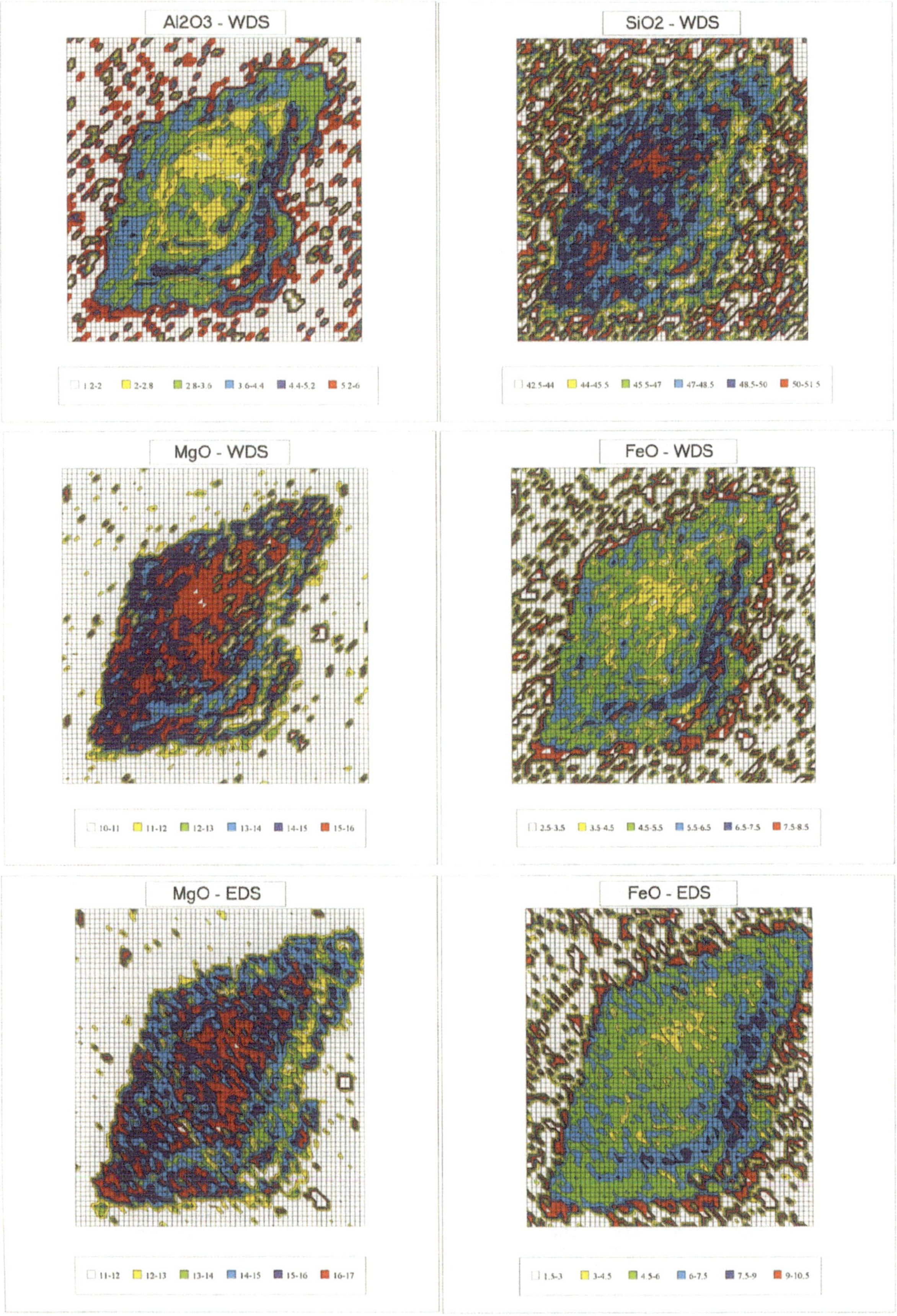

Fig. 5. WDS quantitative compositional maps of four main exchanging chemical elements (Al, Si, Mg, Fe) expressed as weight % oxide for the magmatic clinopyroxene. A selection (Mg and Fe) of EDS quantitative elemental maps are also shown for comparison. For each map the range between the maximum and the minimum values is divided into 6 pseudo-coloured classes

Table 1. Microchemical EDS and WDS analyses, error (relative and absolute) and cationic formula (based on 12 oxygen) for the analysis of (100, 64) coordinates of the garnet X-ray maps. Dwell time = 3 s and = 1 s for EDS and WDS analysis, respectively

Oxide	Weight %		Absolute error (σ)		Relative error (%)		Cationic formula	
	EDS	WDS	EDS	WDS	EDS	WDS	EDS	WDS
SiO_2	37.00	37.36	0.96	0.75	2.6	2.0	2.995	3.034
Al_2O_3	20.14	19.59	0.79	0.55	3.8	2.8	2.035	1.875
FeO	30.19	30.59	1.43	1.32	4.6	4.3	1.959	2.077
MgO	1.11	0.64	0.49	0.16	44.3	25.0	0.133	0.078
MnO	3.44	3.15	0.58	0.48	16.7	15.2	0.236	0.217
CaO	7.76	8.41	0.44	0.38	5.6	4.5	0.673	0.732
Total	99.64	99.74					8.031	8.013

Table 2. Microchemical EDS and WDS analyses, error (relative and absolute) and cationic formula (based on 6 oxygen) for the analysis of (27, 30) coordinates of the clinopyroxene X-ray maps. Dwell time = 3 s and 1 s for EDS and WDS analysis, respectively

Oxide	Weight %		Absolute error (σ)		Relative error (%)		Cationic formula	
	EDS	WDS	EDS	WDS	EDS	WDS	EDS	WDS
SiO_2	50.68	51.38	1.01	0.82	2.0	1.6	1.917	1.907
TiO_2	0.49	0.59	0.43	0.15	87.8	25.4	–	0.016
Al_2O_3	2.75	2.92	0.60	0.24	21.8	8.2	0.123	0.128
FeO	5.00	4.66	0.75	0.48	15.0	10.3	0.158	0.145
MgO	15.01	15.69	0.72	0.47	4.8	3.0	0.846	0.868
CaO	24.14	23.52	0.70	0.66	2.9	2.8	0.978	0.935
NaO	–	0.20	–	0.04	–	20.0	–	0.014
Total	98.07	98.96					4.022	4.013

Vesuvius. In particular, both fractional crystallization and periodic episodes of magmatic refilling may be responsible for the zoning observed and this is supported by the study of clinopyroxene-hosted silicate-melt inclusions [17].

Conclusion

Compositional x-ray maps can play an important role in petrological studies. They allow for a detailed assessment of the distribution of several elements within a single mineral phase and can provide useful indications on the P-T history of the host rock. In particular, the distribution of the Mg/Fe ratio (an important parameter in petrogenesis) can be determined for ferro-magnesian phases. This ratio is important when selecting analytical points to use in the thermobarometric calculation. In general, the type (growth or diffusion) and the existence of chemical zoning can be detected. In the retrograded portions of rock, the intensity and the extent of the diffusion processes can be appreciated.

Acknowledgements. This study was carried out with the financial support of M. U. R. S. T., grant 60% to A. Borghi, of the CNR – C. S. Geodinamica Catena Collisionali – Torino, of the CNR-C.S. per la Minerogenesi e la Geochimica Applicata – Firenze, and of the CNR – Gruppo Nazionale per la Vulcanologia. Two anonymous referees are also thanked for reviewing the manuscript and for providing substantial improvements to the English.

References

[1] M. Shore, A. D. Fowler, *Can. Mineral.* **1996**, *34*, 1109.
[2] F. S. Spear, in: Evolution of Metamorphic Belts (J. F. Daly, R. A. Cliff, B. W. D. Yardley, eds) Geological Society of London, Special Publication, 1989, pp. 43, 63.
[3] R. Castaing, *Thesis*, University of Paris, Paris, 1951.
[4] F. S. Spear, *J. Metamorphic Geol.* **1991**, *9*, 379.
[5] T. Ikeda, *Lithos* **1993**, *30*, 109.
[6] F. P. Florence, F. S. Spear, *Earth Planet. Sci. Lett.* **1995**, *134*, 329.
[7] F. S. Spear, M. J. Kohn, S. Paetzold, *Am. Mineral.* **1995**, *80*, 361.

[8] S. J. B. Reed, in: Microprobe Technique in the Earth Sciences. (P. J. Potts, J. F. W. Bowers, S. J. B. Reed, M. R. Cave, eds.) Chapman and Hall, London, 1995, 49.
[9] R. Cossio, A. Borghi, *Comput. Geosci.* **1998**, in press.
[10] P. J. Statham, *Anal. Chem.* **1977**, *49*, 2149.
[11] D. R. Beaman, J. A. Isasi, *ASTM Special Technical Publication* **1972**, *506*, 80.
[12] G. Bonardi, A. Messina, V. Perrone, S. Russo, A. Zuppetta, *Mem. Soc. Geol. It.* **1984**, *17*, 1.
[13] H. E. Belkin, R. J. Kilburn, B. De Vivo, *J. Volcanol. Geotherm. Res.* **1993**, *58*, 273.
[14] F. S. Spear, in: *Mineralogical Society of America, Monograph* 1, Washington, 1993, 799 pp.
[15] R. Santacroce, *J. Volcanol. Geotherm. Res.* **1983**, *17*, 237.
[16] G. Vaggelli, *Doctoral Thesis*, University of Firenze, Firenze, 1992.
[17] G. Vaggelli, B. De Vivo R., Trigila, *J. Volcanol. Geotherm. Res* **1993**, *58*, 367.

Mikrochim. Acta [Suppl.] 15, 237–246 (1998)

Chemical Mapping of Weathering Stages in Laterites

Wiebke Hachmann and **Gerd Tietz***

Geologisch-Paläontologisches Institut und Museum der Universität Hamburg, Abteilung Sedimentologie und Anorganische Geochemie, Universität Hamburg, D-20146 Hamburg, Federal Republic of Germany

Abstract. In thin sections from laterites, studies of large parts of rock structures are often hampered by browning from secondary iron mobilization. Halos around Fe-containing minerals change translucent areas into almost opaque zones which cannot be directly observed in a normal petrographic microscope. With the method of chemical mapping, a combination of SEM and EDX, the true element distribution within the area studied can be established because, under electron beam irradiation, the behaviour of opaque and translucent minerals is similar.

The examples presented here are from a profile through (almost) unweathered to entirely modified biotite – hornblende schist (Ilesha, SW-NIGERIA). The thin sections were first coated with C and then chemically mapped. The mapping technique creates element distribution patterns from a preselected area of interest but final interpretation of data is based on a combination of chemical mapping and numerous quantitative EDX-spot analyses.

Some parts of almost fresh ilmenite crystals show iron deficits and in the parent rock small secondary element accumulations (Churchite) were detected. The central area of the profile still contains primary minerals, but many of them were already affected by mobilization of elements (mainly Fe, Mn, P). The top of the profile is dominated by a matrix of kaolinite impregnated with goethite but some primary minerals such as ilmenite and quartz are still present. The mobilized elements were well separated. Iron formed broad halos around its ilmenite source whilst manganese was fixed in small crusts. The precipitated MnO_2 became a shelter for many trace elements such as K, Ca, Ce, P and S.

Key words: weathering stages; laterites/chemical mapping/EDX-SPOT analyses

In geology and soil sciences, chemical analysis on thin sections or on rock slabs by an EDX-system attached to a scanning electron microscope (SEM) is a standard technique for rock investigation (Potts et al., 1995). For many years a vigorous and sometimes controversial debate has been taking place among geoscientists on the reliability of quantitative data obtained by EDX-analyses. Meanwhile, rapid developments in computer technology have made possible a new application of the EDX-method in chemical rock investigations: sophisticated combinations of multi-coloured maps, representing elemental distributions within the polished and C-coated rock surface, are replacing the old x-ray dot mapping technique (cf. many papers in: Submicroscopy of soils and weathered rocks, 1981). The aim of mapping is to interpret element localisation in a two-dimensional area which relates to the petrographic composition of the rock sample. Unlike a standard petrographic thin section, the technique is not sensitive to whether the mineral is translucent or opaque. The numerous quantitative EDX-spot analyses involved in this type of chemical petrography may give us a better understanding of the micro-patchwork in chemical weathering of crystalline rocks.

Experimental

Due to the preparation problems in resin-stabilised soft rock samples (i.e. in obtaining a flat polished surface for microprobe

* To whom correspondence should be addressed

analysis), we applied the technique of SEM+EDX unit to locate distinct secondary element accumulations. Any polishing of lateritic soft rocks will not result in a well defined plain surface but several surface levels develop due to the specific hardness of the minerals present. When analyzing within an area containing marked mineral-related topography, the precise geometrical alignment between surface feature and detector is lost and the correction methods employed in quantitative analysis will fail.

During sample preparation using a smooth grinding process no mineral related relief will develop and – at a low magnification range (< 1000x) – one may find many small areas geometrically well aligned.

When using these smooth surfaces, the generally higher collection efficiency for x-rays [3] favours EDS rather than WDS.

From an extensive comparison of element maps we finally chose the points for quantitative EDX analyses and in our interpretation we combined both evaluations for "chemical petrography".

The method of chemical mapping depends on two prerequisites: (i) the stability of the scanning microscope (i.e. CamScan-microscope) and (ii) the quality of the EDX software (EDITOR-EDX unit).

For area selection on a rock slab/thin section we used, in the first instance, a BEI facility. In the next step, a number of spot analyses were performed to check for the elements present. Then, these elements (or some of them) were selected for mapping and, by switching to the spot mode of the SEM, control was transferred to the EDX-unit. In a special map mode, the SEM spot was directed to scan the area at a certain resolution (640×512 pixels). A pre-selected dwell time (in millisec) limited the data acquisition period. At a count rate of about 2200–2500 cnts/sec the dead time for the x-ray detector was in the order of 25–30%, and the calibrated resolution was 144.2 eV at Mn Kα. With a 100 millisec life-time the entire mapping is finished after about 15 hours. Doubling the livetime doubles the run-time, but the colour density of the element maps is higher and the data from elements with low concentrations is much improved. The colour depth has to be selected visually, black indicating the absence of an element and white its presence at high concentration. Data are stored as one map per element and maps may be recalled and combined to form a set of four for visual comparison (see the examples below), either on the monitor screen or as hardcopy. Alternatively, a distinct colour may be allocated to an element and a series of coloured maps may be amalgamated into one multi-coloured map. A multi-coloured map is not a pile of single element layers with a mixture of colours, the EDITOR-system calculates which element has the highest concentration at each point and the colour of this element shows in the map. Combining Ca with Na, K or even Al in a feldspathic rock will create an impressive map of plagioclase and microcline intergrowth. Depending on the scale of the map, one may even recognize internal zoning in plagioclase crystals – or one may recognise different goethite generations distinguished by their Al incorporation.

We do not conceal a typical problem that requires careful consideration: the limited energy resolution of a qualitative EDX analysis frequently causes peak overlap problems. In general, it is very difficult to deconvolute two peaks separated by 50 eV – no matter which peak-stripping method is used. In quantitative EDX analysis, a long livetime (>150 sec) gives a good statistical base for separating peaks and distinguishing them from the background. During the mapping procedure the dwell time is generally in the order of 100–200 msec and this very short counting period does not allow any statistical evaluation of peak overlap. Therefore, the recording of maps from certain elements can prove difficult and a number of spot analyses have to be carefully performed before the relevant maps may be taken as a reliable representation of the

Table 1. Some of the most seriously interfering elements in EDX analysis

Element	Interfering X-ray line	interferes with	X-ray line interfered with
Ti	Kβ	V	Kα
V	Kβ	Cr	Kα
Cr	Kβ	Mn	Kα
Mn	Kβ	Fe	Kα
Fe	Kβ	Co	Kα
P	Kα	Zr	Lα

Fig. 1. Sections from sample KP8–11 to show the simulation of V in ilmenite as result of poor statistical stripping due to short dwell time during signal recording. Picture width is 0.673 mm. The upper picture is for Ti and demonstrates the localization of Ilmenite (light grey). The middle picture gives the Fe distribution for: Biotite: middle-grey; Magnetite: very light grey; Ilmenite: grey. The lower picture shows the recorded V-distribution: traces in Biotite, slightly more in Magnetite and highest in Ilmenite

element's concentration. The most difficult elements to map are presented in Table 1 (Goldstein et al., 1992). The Fe-Ti-V maps in a magnetite/ilmenite integrowth (from sample KP8-11) may illustrate the Ti and V interference (Fig. 1). This portrays a very good correspondence of V with the Ti-map, but a comparison of spot analyses clearly demonstrates the exact opposite: in magnetite the V-concentrations are slightly higher than in ilmenite. At the short dwell times used in the mapping procedure there was no chance to distinguish between Ti-Kβ and V-Kα, and Ti-Kβ was identified as V-Kα. During Ti mapping at low concentrations a certain fraction of the V counts were also recorded giving rise to the misleading Ti and V correspondence observed.

Based upon the element localisation within chemical mapping routine, we selected some spots for distinctive and higly reliable quantitative analyses. When matrix effects are corrected using mineral standards (VSCF values in EDITOR), highly reliable results may easily be obtained in a modern EDX-system.

Results and Discussion

The basis of our investigation is a typical section of a biotite-hornblende-schist (sample KP 8-3, Figs. 2, 3) with a mineral assemblage as reported in Table 2. These minerals were identified from their characteristic chemical composition as established from quantitative EDX-analyses. Additionally, local concentrations of REE (Rare Earth Elements) were recorded but not yet identified. The shape of mineral grains/crystals (see Figs. 2, 3) and their relevant analyses do not give any hint of being influenced by weathering. However, from a few spots (white circle on Fig. 3, P-map) we are aware of the presence of $YPO_4 \times 2H_2O$ (churchite), probably an early weathering product derived from the onset of decomposition of apatite and the suggested REE-minerals mentioned above. The two plates Figs. 2 and 3 represent the composition of the parent rock, arranged in Fig. 2 to Si+Fe+Mn+Ti to emphasize the contrast between the iron minerals magnetite and ilmenite and the silicates from quartz to Mn-amphiboles. In Fig. 3 one may recognize the apatite field (Ca+P-map) and the strong separation of plagioclases (Na-map) compared to biotites (K-map).

The next step in our investigation was the evaluation of minerals in the saprolite sample KP 8–11. The recorded area contained both opaque and translucent minerals. Translucent minerals are represented in the Si-map (Fig. 4) with the conspicuous white fields of quartz, different shades of grey represent biotite and plagioclases while the dark areas (= no Si) indicate opaque minerals. These dark areas change to different shades of white and grey in the Fe-map (opposite to the Si-map in Fig. 4) which represents the distribution of opaque minerals. Only the dark-grey fields of the Fe-containing biotite correspond to the dark-grey fields within the Si-map, indicating the biotites as silicates. The identification of biotite is supported by the regional distribution of Mg and K on Fig. 4 too. We used these biotites for discussing early signs of chemical weathering. The first step in chemical weathering (lateritisation) of biotite is the mobilization of Fe^{2+} and the subsequent oxidation of Fe^{2+} to Fe^{3+}. The relevant areas in the K and Mg maps are black, thus indicating these elements absence. Simultaneously with iron mobilization the depletion and migration of potassium occurs and the first stage of the destruction of biotite begins. In chemical maps it is easy to recognize gradual

Table 2. Chemical analyses for distinct minerals recorded by chemical mapping of an almost fresh biotite-hornblende-schist (KP 8-3)

	Oligoc. – Andes.	Biotite	Amphibole	Ilmenite	Magnetite	Apatite
SiO_2	62.01	41.17	40.08	1.72	1.47	1.61[a]
TiO_2	1	1.84	0.33	46.24	6.28	0
Al_2O_3	22.46	16.57	19.11	1.27	0.92	1.01[a]
Fe_2O_3	0	17.94	26.12	45.10	86.88	0
MnO	0	0	6.46	2.57	0.66	0
MgO	0	12.54	2.38	1.43	0.43	0
CaO	7.43	0.39	4.49	0.36	2.72	55.72
Na_2O	8.10	0	0.65	0	0	0
K_2O	0	9.55	0.38	0	0.20	0.28[a]
P_2O_5	0	0	0	0	0	39.07
PdO	0	0	0	0.70[b]	0	0
SO_2	0	0	0	0.48[b]	0.44	0
F	0	0	0	0	0	2.31

[a] From silicatic inclusions/impregnations

[b] Out of 8 spots from ilmenite recorded PdO in a range from 0.5 to 0.87%. The association with SiO_2 corresponds to sulphide inclusions too small to be identified.

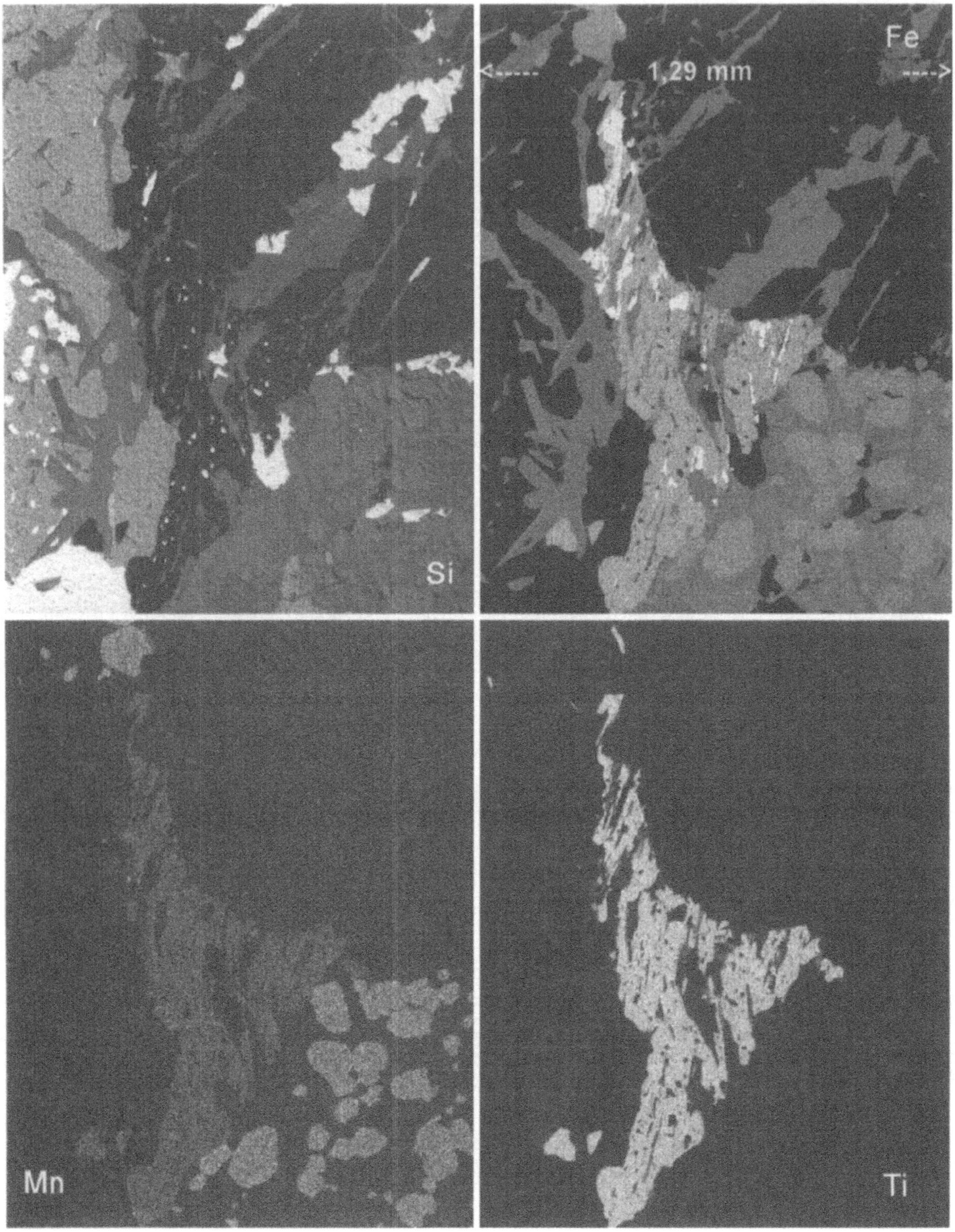

Fig. 2. Chemical mapping Biotite-Hornblende-Schist (KP 8-3) from Ilesha/SW-Nigeria

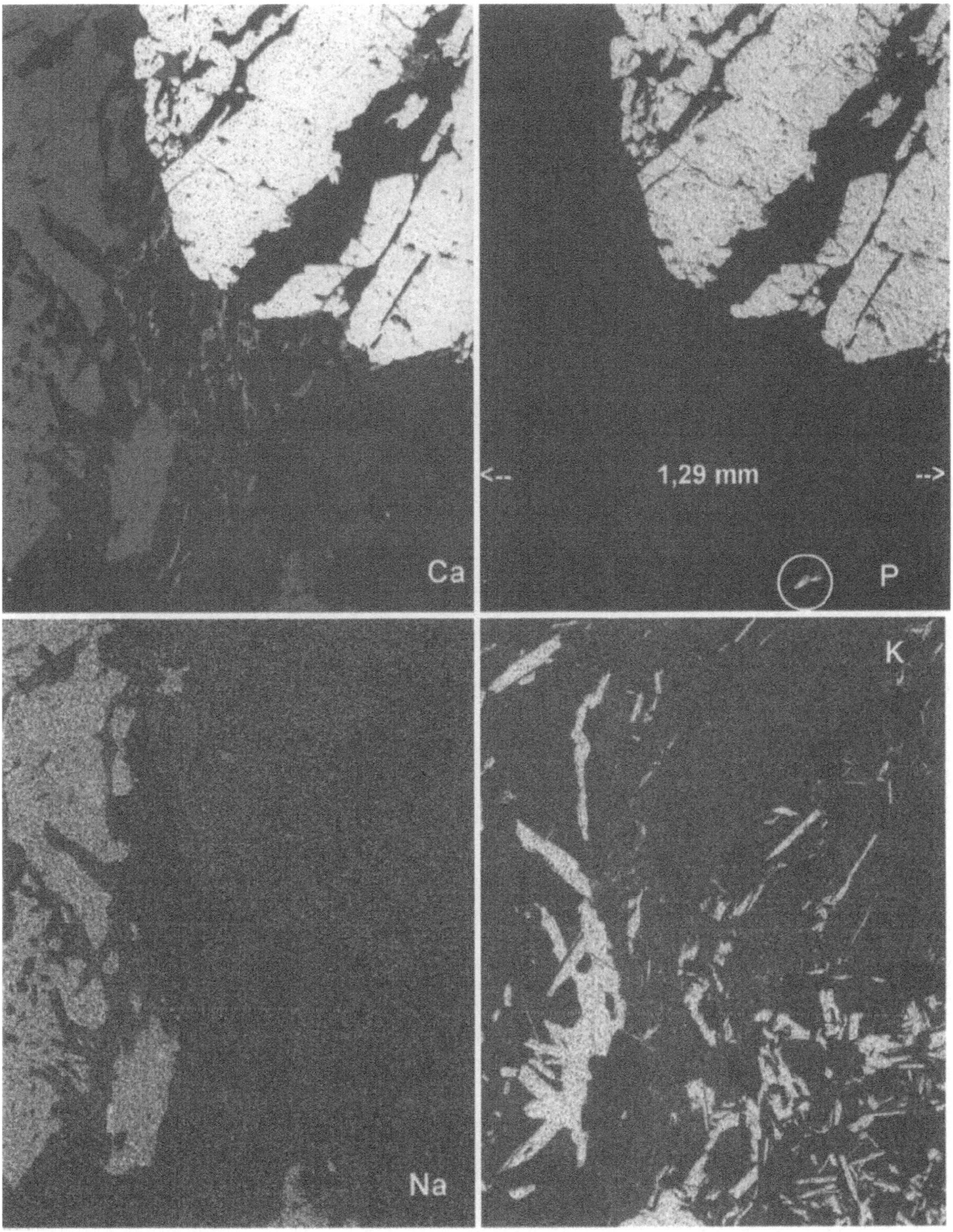

Fig. 3. Chemical mapping of Biotite-Hornblende-Schist (KP 8-3) from Ilesha/SW-Nigeria

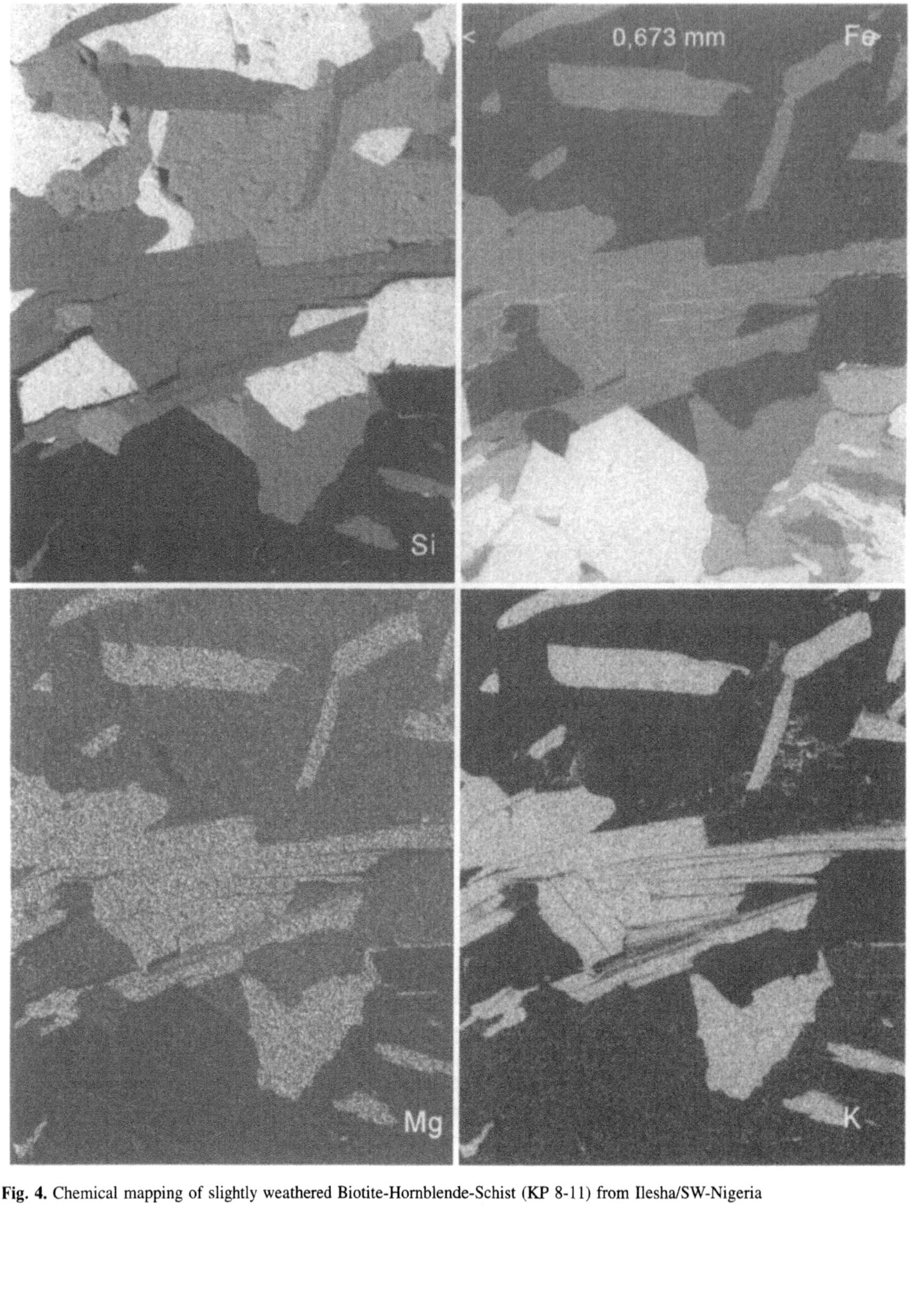

Fig. 4. Chemical mapping of slightly weathered Biotite-Hornblende-Schist (KP 8-11) from Ilesha/SW-Nigeria

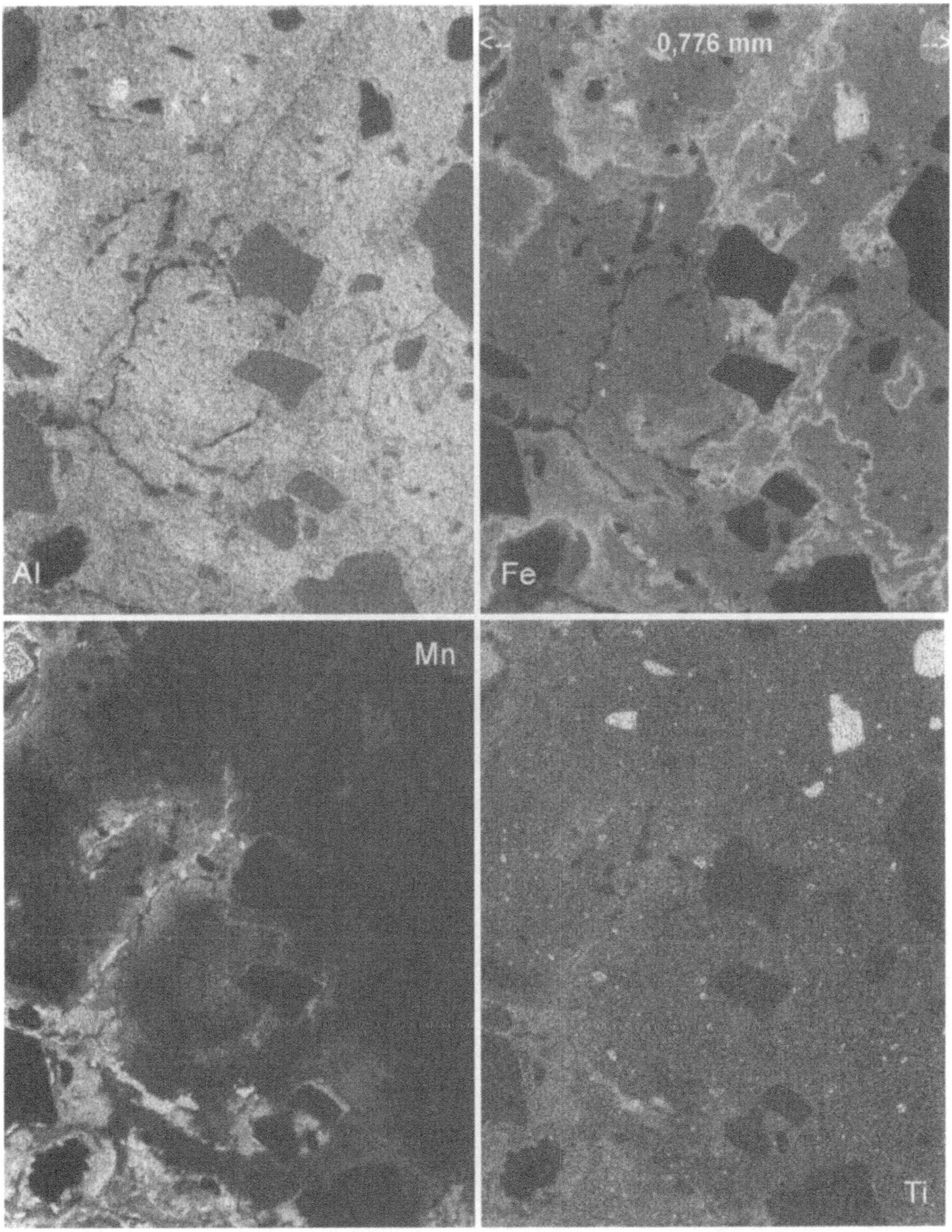

Fig. 5. Chemical mapping of entirely weathered Biotite-Hornblende-Schist (KP 8-12 Matrix) from Ilesha/SW-Nigeria

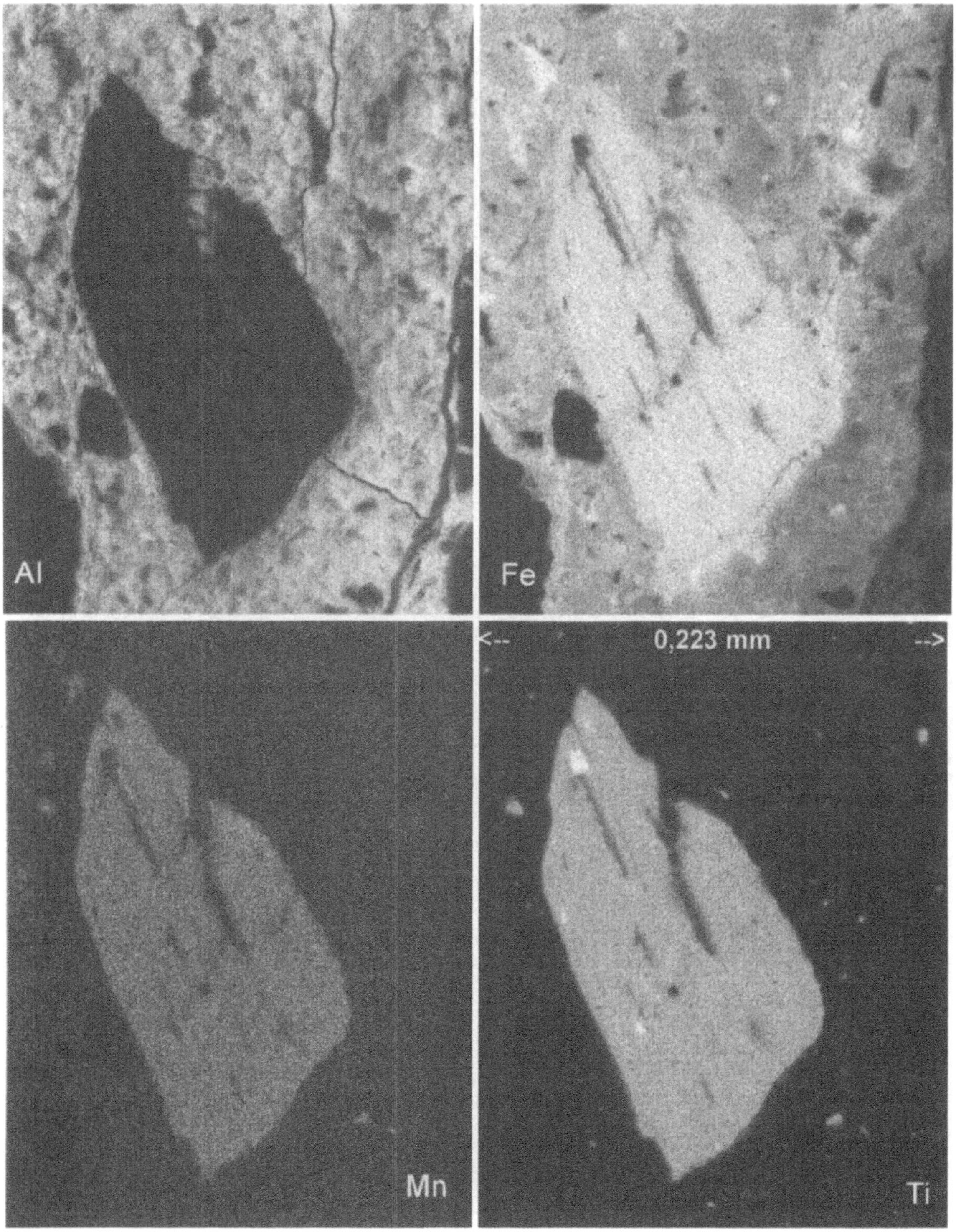

Fig. 6. Chemical mapping of entirely weathered Biotite-Hornblende-Schist (KP 8-12 Ilm.xx) from Ilesha/SW-Nigeria

Table 3. Chemical analyses of weathered Biotites from KP 8-11

	Biotite 1	Biotite 2	Biotite 3
SiO_2	37.96	36.17	35.97
TiO_2	1.34	1.78	1.53
Al_2O_3	16.14	15.33	15.26
Fe_2O_3	25.12	33.26	31.40
MnO	0.26	0.36	0.22
MgO	12.93	8.36	10.19
CaO	0.47	0.94	0.83
K_2O	5.63	3.61	4.45
V_2O_3	0.14	0.17	0.15

Table 4. Significant element ratios for fresh and weathered Biotites

Factors for Biotites						
KP 8-3						
	Si:Al	Al:Mg	Al:K	Al:(Mg+K)	Fe:(Mg+K)	Al:Fe
Spot 19	2.198	0.976	1.597	0.606	0.419	1.444
Spot 18	2.2	0.998	1.786	0.64	0.379	1.688
Spot 13	2.108	1.044	1.604	0.632	0.437	1.447
Spot 11	2.136	1.047	1.6	0.633	0.457	1.386
KP 8-11						
Spot 1	1.996	0.987	2.65	0.719	0.715	1.007
Spot 2	1.998	1.456	3.944	1.064	1.468	0.725
Spot 3	1.997	1.188	3.179	0.865	1.132	0.764

depletion of the relevant elements (K/Mg/Si) by a decreasing brightness of their grey-shades: K > Mg > Si. From the change of grey, one may be able to locate these weathering zones precisely for further spot analyses as presented in Table 3. To emphasise the degree of weathering, we compared significant element ratios from biotites in the parent rock (sample KP 8-3) with those from the saprolite sample (KP 8–11), see Table 4 and Fig. 7. The Si:Al ratio indicates a slight decrease in Si, but the weathering influence appears striking when comparing the magnesium and potassium ratios. Usually, magnesium is one of the most mobile elements under tropical weathering conditions (Millot 1964). The special structure of the biotite lattice as sheet silicate is responsible for easy depletion of potassium which is positioned between two Si-tetrahedron layers. However, magnesium is relatively stable in its position at the centre of the $Mg_3(OH)_6$ octahedrons (Heim, 1990). From the iron ratios it is obvious that some 'fresh' iron has become additionally incorporated into the slightly defective biotite lattice at the position of silicon and magnesium.

The completely lateritised sample KP 8–12 does not contain any biotite. The entire sample is made up of a mixture of secondary minerals which are characteristic of lateritised soft rocks: poorly crystallised kaolinite and goethite (Fig. 5, Al-map and Fe-map). The black fields from both maps represent relic quartz crystals, their shape is subrounded compared to the angular shape of quartzes in Fig. 2, thus indicating weathering attacks on quartz surfaces. The main feature of this sample is the impregnation of the kaolinitic-goethitic matrix by tertiary iron and manganese accumulations (Fig. 5, Fe-map and Mn-map). The impregnation of a pre-existing matrix is substantiated from the calculated gross mineralogical composition:

normal matrix:
- Kaolinite: 82%
- Goethite: 17% — Traces: 1%

impregnated with iron:
- Kaolinite: 50%
- Goethite: 48% — Traces: 2%

impregnated with manganese:
- Kaolinite: 43%
- Goethite: 24% — Traces: 9%
- MnO_2: 24%

In the parent rock iron and magnanese were combined in the same minerals; due to their individual weathering mobility both elements migrated different distances but manganese attracted more trace ele-

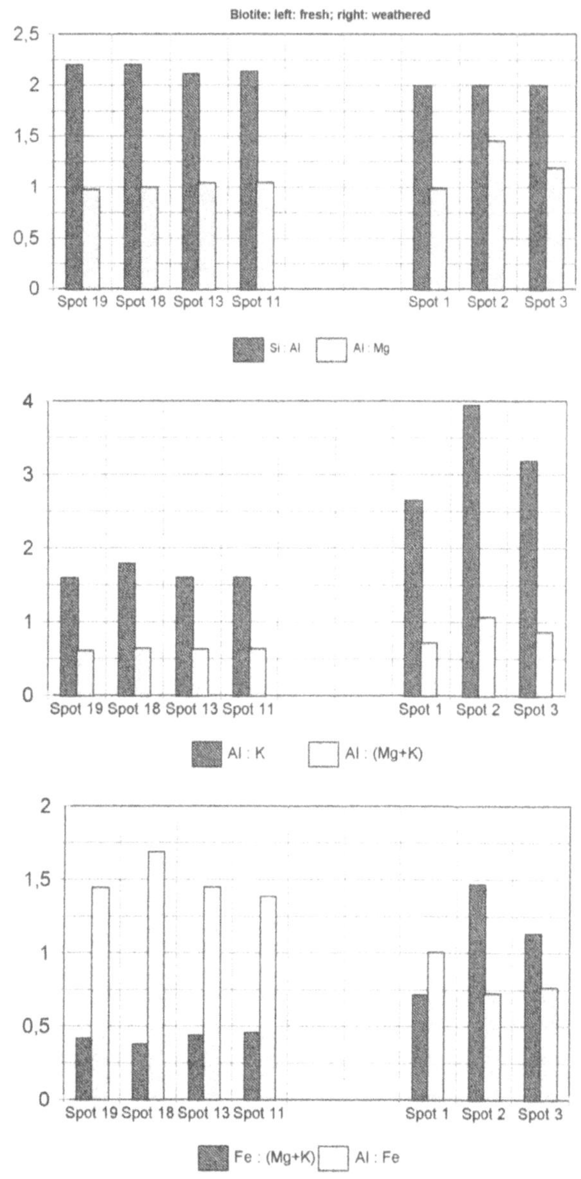

Fig. 7. Significant element ratios for fresh (KP 8-3) and weathered (KP 8-11) Biotite

Table 5. Characteristic Element ratios of fresh to weathered Ilmenites

		Fe:Ti	Fe:Mn
n = 6	KP 8-3	0.951	14.64
n = 3	KP 8-11	0.927	13.65
n = 12	KP 8-12	0.929	11.95

ments than iron. Even titanium is now incorporated within the manganese crust (Fig. 5, Mn-map and Ti-map). When looking for the stable minerals, in this lateritic sample (KP 8–12) there are still many ilmenite grains/crystals. One was selected for chemical mapping (Fig. 6): the different single element maps demonstrate a distinct migration of iron into the Al-rich matrix (kaolinitic) as indicated by a pronounced iron halo around the ilmenite crystal (Fig. 6, Fe-map), while titanium and manganese are still restricted to the crystal area (Fig. 6, Ti-map. Mn-map). Some black fields in all the crystal mappings indicate serious decomposition of parts of the stable ilmenite, which is most probably related to internal structural defects. As for biotite, the rate of decomposition may be derived from element ratios (Table 5). Titanium is more stable than iron as seen from the slight increase in Fe:Ti, while manganese, when present as MnO_2, is much more immobile than iron. This suggests that the manganese in the crust may originate from Mn-amphiboles (as present in KP 8-3) and Mn-containing biotites but not ilmenites.

Finally, certain spots indicate the presence of secondary/tertiary REE minerals (i.e. florencite, Ce-Al phosphate-hydrate).

Conclusions

In this paper we have demonstrated the scientific value of chemical mapping for the detailed study of rock/mineral changes. It is one of the first attempts to trace mineral decomposition in natural rocks that have had a complex and long lasting weathering history. From the general situation of laterites in Nigeria, the profile we studied may have developed on the African Surface (DeSwardt, 1956) about 30–50 Ma ago.

References

[1] E. B. A. Bisdom (ed.) 1st Workshop of the International Working Group on Submicroscopy of Undisturbed Soil Materials (IWGSUSM) 1980, PUDOC, Wageningen, The Netherlands, 1981, 320 pp.

[2] A. M. J. DeSwardt, *Proc. 3rd Internat. W. African Conf. 1949*, pp. 180–186.

[3] J. I. Goldstein, D. E. Newbury, P. Echlin, D. C. Joy, Jr. A. D. Romig, C. E. Lyman, C. Fiori, E. Lifshin (eds.), *Scanning Electron Microscopy and X-Ray Microanalysis*, Plenum, New York, 1992, 820 pp.

[4] D. Heim, *Tone und Tonminerale*, Enke, Stuttgart 1990, 157 pp.

[5] G. Millot, *Géologie des argiles*, Masson, Paris, 1964, 499 pp.

[6] P.J. Potts, F.W. Bowles, S. J. B. Reed, M. R. Cave (eds.), *Microprobe techniques in the Earth Sciences*, The Mineralogical Society Series No. 6, Chapman and Hall, London, 1995, 419 pp.

[7] A. A. G. Schieferdecker (ed.), *Geological Nomenclature*, Gorinchem. Royal Geological and Mining Society of the Netherlands, 1959, 523 pp.

Mikrochim. Acta [Suppl.] 15, 247–252 (1998)

Electron Microprobe Determination of Minor and Trace Concentrations of Gold and Platinum Group Elements in Sulphides and Sulpharsenides: Problems, Solutions, and Applications

Irene P. Laputina[1,*] and **Vadim A. Batyrev**[2]

[1] Russian Academy of Sciences, Institute of Geology of Ore Deposits (IGEM), Staromonetny per. 35, 109017 Moscow, Russia
[2] Federal Scientific Center "TCNIITMASH", Moscow, Russia

Abstract. EPMA, as a complementary technique to standard analytical methods, is reviewed in the context of microprobe determination of spatially resolved trace amounts of platinum group elements (PGEs), gold and silver in sulphide and sulpharsenide minerals. Attention is focused on the methods for introducing the peak-overlap correction and background modelling with the use of two-background standards. Precise background determination is achieved by using a modified form of the Kramers expression (in accord with the approach proposed by C. Merlet and J.-L. Bodinier). Simultaneous measurement of major and trace elements (up to 20) makes is possible to monitor the specimen stability and the instrumental drifts. The limits of detection of Pd, Rh, Ru, and Ag (90–100 ppm) and Pt, Ir, Os, and Au (220–280 ppm) in sulphides and sulpharsenides achieved with a probe current of 200 nA and a counting time of 100 s can be improved to 40–100 ppm in some cases by increasing the current and counting time. The method also permits the required analysis time to be reduced by a factor of two. It is applicable not only to the analysis of minor and trace elements, but also to the determination of major elements.

Key words: precious elements, overlap, background modelling.

Precious metals may occur either as natural minerals or as minor and trace constituents of other, usually sulphide and sulpharsenide, minerals. The first determinations of isomorphous Pd in pentlandite (Norilsk deposits, Russia) have been carried out with the EPMA technique (CAMECA MS-46 microprobe) in ores of various types within the concentration range 40–120 ppm [5, 7, 12]. More recently, these results have been obtained on the same specimens using proton-microprobe analysis [3]. Minor and trace-element analysis is essential for several reasons: the knowledge of element localization provides information on crystallochemical, physical and geochemical conditions of mineral formation; from the economic standpoint it is necessary to obtain an accurate balance of element content within microscopic mineral phases present in the ores so that the appropriate method of ore processing may be chosen. EPMA is quicker and less expensive than other methods of determining minor and trace amounts of valuable elements but it is less sensitive, however. A few hundred ppm is a commonly accepted detection limit (DL) with EPMA of some trace elements in sulphides: for example, DL = 270 ppm for silver-bearing chalcopyrite [10, 14] and DL = 400 ppm for gold in arsenopyrite [8].

Correct measurement of peak and background intensities in the EPMA of PGEs presents difficulties because of the many interfering lines which require a careful choice to be made of spectrometer settings for background measurements. To overcome these problems it is appropriate to determine the background intensities by calculation and to use a special computer program for removing overlaps before applying ZAF matrix corrections.

* To whom correspondence should be addressed

A variety of models have been described in the literature for determination and calculation of the background intensity as well as for the calculation of the peak-overlap correction, particularly in the analysis of rare earth elements (REEs), but also in others [1, 2, 6, 11, 15, 16].

Merlet and Bodinier [13] have proposed an efficient approach to the modelling of background values (B_l) with the background response of the wavelength of the analytical line l at the fixed accelerating voltage. In the case of silicate minerals this approach is based on the equation:

$$B_l = T_l\,(\bar{Z}_{sp}^{m(Z,E)} f_l / \bar{Z}_{st}^{m(Z,E)} k_l) \qquad (1)$$

where T_l is the background response of the lth wavelength for pure standard;

$$m = 1.271 - 0.0011\bar{Z} + 0.007\,E \qquad (2)$$

k_l and f_l are the correction factors (accounting for absorption and backscattering) for the standard and the specimen, respectively; $\bar{Z}_{st}$ and $\bar{Z}_{sp}$ are the average atomic numbers of the standard and the specimen; $\bar{Z} = \sum(C_i Z_i)$, where C_i is the concentration of the ith element in the specimen; m is the coefficient of deviation of the $B_l\,(\bar{Z})$ dependence from the linear law described by the Kramers expression.

In this work an attempt is made to apply the Merlet and Bodinier approach to background modelling for analyzing PGEs, Au and Ag minerals, and for determining minor and trace precious elements in sulphides and sulpharsenides. Our improved WDS technique also includes a program for the peak-overlap correction which is introduced before applying the standard ZAF matrix corrections.

Experimental

Measurements were carried out with a CAMEBAX-MICRO-BEAM electron microprobe equipped with three WD X-ray spectrometers (one tilted and two vertical) and a PDP 11/23 computer. The operating voltage was fixed at 20 kV. When analyzing PGEs, Au, and Ag minerals, simultaneous measurement of major, minor and trace elements (up to 20) was possible with the use of OsMα, IrMα, PtMα, AuMα, PbMα, BiMα, RuLα, RhLα, PdLα, AgLα, SnLα, SbLα, TeLα, SKα (with a PET crystal), AsKα, SeKα, FeKα, CoKα, NiKα, and CuKα (with a LiF crystal) X-ray lines. When Hg determination was also needed, one of the least important elements was excluded from the data file. Note that only 15 elements (Sn, Te, Se, Pb, Bi are ruled out) are to be determined in analysis of sulphides and sulpharsenides.

Background Modelling

To check the precision of the background calculation with the aid of Eq. (1) in the analysis of PGE-containing minerals, as well as in the determination of minor and trace amounts of precious elements in sulphides and sulpharsenides, we carried out a series of measurements of the background intensity for all 20 analysed elements on 40 standards (pure elements or two- and three-component minerals of known composition) in the wavelength range from 0.10573 to 0.96709 nm (AsKβ – AsLα lines).

To reduce statistical error in the background measurements, we optimized the measurement conditions with respect to the counting time and probe current. Since the specimens and standards of interest are thermally stable, the measurements were made with a beam current of 200 nA and a counting time of 100 s. The measured background intensities from pure elements or binary compounds (standards of known composition) were corrected for absorption and backscattering only (the main background components) with the aid of the f factor ($f = f_l/k_l$) and the resulting values were plotted on the $B_l\,(\bar{Z})$ graph. We constructed the graphs for the CoKα radiation, in the narrow wavelength region (from the RhLα line to the AgLα line, i.e. in the range 0.459743–0.415443 nm, see Fig. 1), and in the broad region, from the AsKβ line to the AsLα line (0.10573–0.96709 nm), where the graphs were reduced to a single scale. It follows from the data obtained that the Z-dependence of background intensity agrees with the Kramers expression for Z < 50.

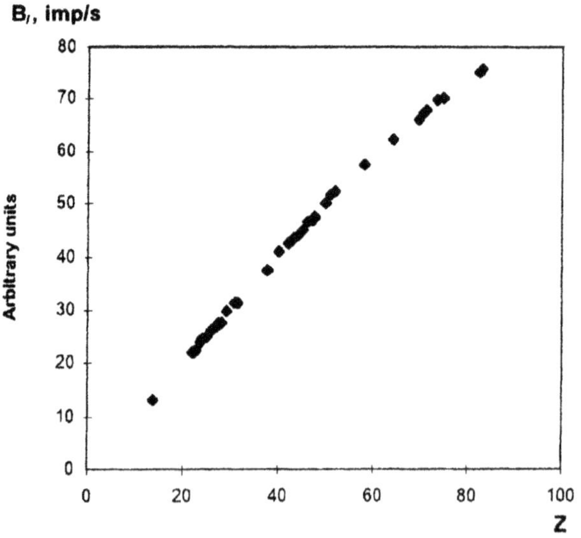

Fig. 1. The Z-dependence of the measured background intensity corrected for absorption and backscattering. Formulae (1) and (3) in the paper is based on this dependence

Some deviation from this dependence observed for the samples with Z > 50 is particularly pronounced for longer wavelengths ($\lambda > 0.5$ nm). Using the least squares method, we approximated the m values for two wavelength regions, $\lambda < 0.5$ nm and $\lambda > 0.5$ nm, by the following expressions:

$$m = 1.0092 + 0.000046\bar{Z} - 0.0000036\bar{Z}^2 \quad \text{for } \lambda < 0.5\,\text{nm},$$
$$m = 0.9953 + 0.0008\bar{Z} - 0.000015\bar{Z}^2 \quad \text{for } \lambda > 0.5\,\text{nm} \qquad (3)$$

Upon substitution in equation (1), these m values were used to calculate the background intensity in the analysis of standards and the samples containing PGEs, as well as to determine minor and trace concentrations of precious elements in sulphides and sulpharsenides. Since the character of the B_l ($\bar{Z}$) curve changes insignificantly over a wide wavelength range, particularly at the shorter wavelengths (<0.5 nm), all the measurements can be performed in the integrated mode without the risk of adding higher order reflections of short wavelength to the total background intensity. Comparative background measurements made in the integrated and differential modes indicate the same character of B_l ($\bar{Z}$) dependence for these two modes.

It was necessary to choose as a standard for the background calculations a pure element (or a compound) that produced no interfering lines for the elements of interest and did not contain the analyzed element. We used metallic Co for all elements, except for Os, S, and Te, the analytical lines of which overlap with very weak high-reflection orders of CoKα radiation. For these elements and for Co we used Ni as a background standard. A comparison of the measured background intensities and those calculated from formula (1) with the m values estimated from Eq. (3) showed good agreement (Fig. 2). The absolute background intensities for a given wavelength over a wide range of $\bar{Z}$ are calculated with a precision of about $\pm$2–8% but for $\bar{Z}$ between 22 and 35, characteristic of all natural sulphides and sulpharsenides, the precision at all wavelength is better than $\pm$2%.

Peak-Overlap Corrections

For the peak-overlap corrections we used the simple expression:

$$KO_{i,j} = DKO_{i,j}\,K_j \qquad (4)$$

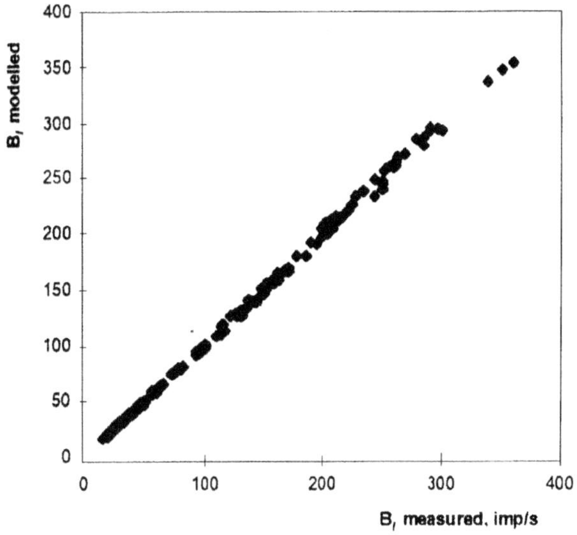

Fig. 2. The relation between measured and modelled background (formulae 1 and 3) for different Z

where $KO_{i,j}$ (the relative overlapping intensity; subscript i denotes the analysed element) is the ratio of the intensity of the overlapping line (or several lines) of element j at the spectrometer setting of the analytical line of element i (absent in the sample) to the intensity of the analytical line from the pure element i; $DKO_{i,j}$ is the same relative intensity measured on the pure element j; and K_j is the ratio of the intensity of the analytical line of element j in the sample to the intensity of the same line from the pure element j (the original measured quantity in the determination of the quantitative content of element j). Eq. (4) is sufficiently rigorous if the overlapping and analytical lines of element j refer to the same series of the X-ray spectrum. Taking this fact into account, the analytical lines in the group of minerals considered were chosen so as to satisfy the above condition at least in cases when $DKO_{i,j}$ exceeds 0.01, while the mass fraction of element j was usually lower than 5%. The entire set of $DKO_{i,j}$ for the elements in the group under consideration was obtained on the basis of careful measurements. Since the set of PGEs is characterised by many mutual interferences, the correct spectrometer positions for background measurements on pure elements were found after careful examination of the spectral intensity distribution by the preliminary recording on the plotter of the intensity distribution in a broad spectral range, the aim being to reveal all possible overlaps near analytical lines.

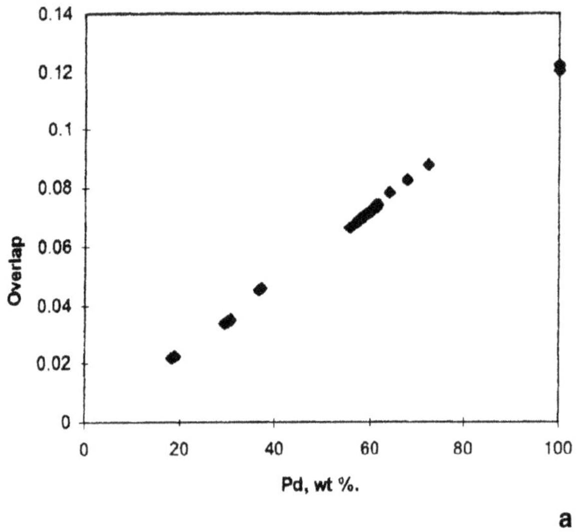

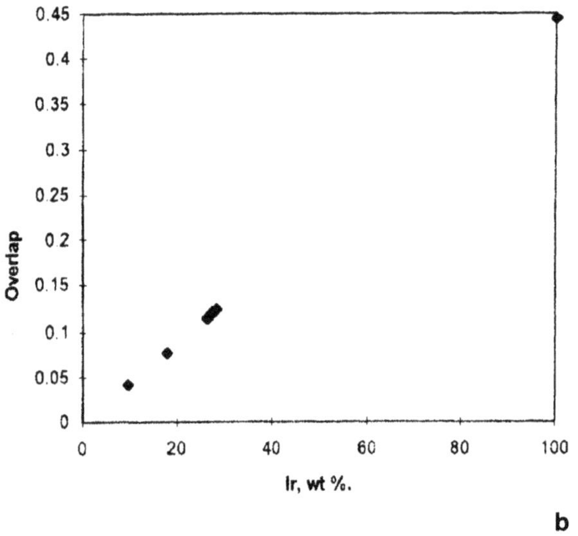

Fig. 3. The peak-overlap corrections calculated from formula (4) for the AgLα line in the Pd-Ag system (**a**) and for the PtMα line in Ir-Pt system (**b**)

When analyzing elements present in the system under study, we took into account 32 cases of peak overlap, the principal ones being: the overlap of the AsKα with Pb, Bi, Os and Ir X-ray lines; the SeKα peak with Pt, Ir and Au lines; the RuLα peak with the Bi line; the RhLα peak with Ru and Pt lines; the PdLα peak with Ru and Rh lines; the AgLα peak with the Pd line; the SnLα peak with the Sb line; the SbLα peak with Te and Sn lines; the TeLα peak with the Sn line; the OsMα peak with Au lines; the IrMα peak with Os and Au lines; the PtMα peak with Ir and Sb lines; the AuMα peak with the Pt line, and so on. All these overlaps were recorded on standards of pure elements except for the AsKα line where the overlaps (NiAs and FeAsS standards) were measured on PbS and PbSe compounds. To estimate the validity of calculations made with the use of formula (4) we carried out the measurements of element concentrations in synthesised samples for the systems Ag-Pd (30 determinations) and Ir-Pt (10 determinations) (Figs. 3a and 3b). The results obtained demonstrate the high quality of the overlap correction.

Software

The computational subroutine for determination of background intensity (Eq. (1) with the m values from (3)), taking the peak-overlap corrections into account (see formula (4)), was incorporated in our basic program for the quantitative analysis of 20 elements with the aid of the CAMEBAX-MICROBEAM microanalyzer. When creating the file of initial data, the coefficients for recalculating the background intensity measured on the background standard must be introduced in the corresponding background intensities of pure elements for all analytical lines of the analyzed elements. In our case, we obtained the following coefficients: 2.5306 for OsMα, 2.4361 for IrMα, 2.3650 for PtMα, 2.3041 for AuMα, 2.1508 for PbMα, 2.1088 for BiMα, 1.6618 for RuLα, 1.6352 for RhLα, 1.6173 for PdLα, 1.6051 for AgLα, 1.5963 for SnLα, 1.6014 for SbLα, 1.6082 for TeLα, 0.9725 for FeKα, 1.0000 for CoKα, 1.0288 for NiKα, 1.1798 for CuKα, 1.2359 for AsKα, 1.2590 for SeKα, and 0.9614 for SKα.

The necessity of including in the computer program all the overlaps present called for simultaneous measurements of major and trace elements in each particular case. The results obtained for the major components make it possible to monitor the instrumental drift and the long-term stability of specimens in the course of measurements. This makes information on the content of trace elements more reliable.

Prior to the use of the standard ZAF correction iterative procedure, we subtract the possible overlaps from interfering lines (which requires 2 or 3 iterations) and calculate the background intensities for each microprobe analysis. The completely corrected relative intensities K_j are then introduced in the ZAF correction procedure.

The correctness of determination of major elements with the aid of the above program was verified by using the two-component specimens of known composition: PdPb, PdBi, $PdBi_2$, PdSb, Pd_3Sb, Pd_2As, Pd_3Sn_2, PtRh (5 compositions), Ag_2Te, $AuSb_2$, $PtSb_2$, $PtAs_2$, PbTe, PbSe, OsIr, SnTe, Sb_2Sb_3, Bi_2S_3, Bi_2Te_3. Good agreement between the known and calculated compositions was obtained.

The precision of determination for minor concentrations was checked on synthesised sulfides: cobalt-pentlandite $(Co, Fe, Ni, Rh)_{1.125}S$ with 1800 ppm Rh, and nickel-pentlandite $(Ni, Fe, Co, Pd)_{1.125}S$ with 1000 ppm Pd. Ten measurements were repeated for each specimen under the experimental conditions used. The statistical accuracy of intensity measurements and the standard deviation of the data was of the same order of magnitude, indicating the presence of Rh and Pd solid solutions in both pentlandites. The average rhodium concentration was 1790 ± 130 ppm for the cobalt-pentlandite and 990 ± 160 ppm for the nickel-pentlandite sample, respectively. In some cases, the intensities of the spectral lines of minor elements were recorded on the plotter in order to control the accuracy of data obtained.

The 3σ criterion was used to calculate the detection limit

$$DL = 3(\sqrt{2}\sqrt{B_l}\,t/I_{st}t)C_{st}F_{sp},$$

where t is the measurement time and F_{sp} is the ZAF correction factor for the analytical line in the sample. The detection limit for Ru, Rh, Pd, and Ag in sulphides and sulpharsenides was estimated to be 90–100 ppm, whereas for Os, Ir, Pt, and Au in the same minerals we have 220–280 ppm due to the great F_{sp} values (1.5–2.2 for $i_0 = 200$ nA and $t = 100$ s). With increasing probe current and counting time, the detection limit for elements of the first group can be reduced to 40 ppm and for elements of the second group, to 100 ppm.

Applications

We have analyzed many natural specimens from different deposits using the present technique. The heterogeneous distribution of valuable elements was established in sulphides and sulpharsenides. Concentrations varied within a wide range. Some separate blocks of sulpharsenides from low-sulfide disseminated Cu-Ni-PGE mineralisation from the Mt. Generalskaya (the Kola Peninsula, Russia) [9] contain Ru with concentrations from 90 ppm to 6800 ppm, Rh from 200 ppm to 26.3 wt%, Pd from 0.63 to 0.67 wt%, Os from 0.17 to 12.6 wt%, Ir from 400 ppm to 14.9 wt%, and Pt from 900 ppm to 10.3 wt%, giving a continuous series of (Co, Ni, Fe)AsS – (PGE)AsS minerals from Rh,Pd-cobaltite-gersdorfite (CoAsS-NiAsS) to Ni, Co-hollingworthite (RhAsS).

Some pentlandites from high-temperature sulphides, i.e. quenched solid solutions, of the Norilsk region contain Pd in the range 1.9–3.05 wt%, which is the highest concentration found in pentlandites of the world's copper-nickel deposits [4]. However, Pd is distributed very irregularly among the pentlandite

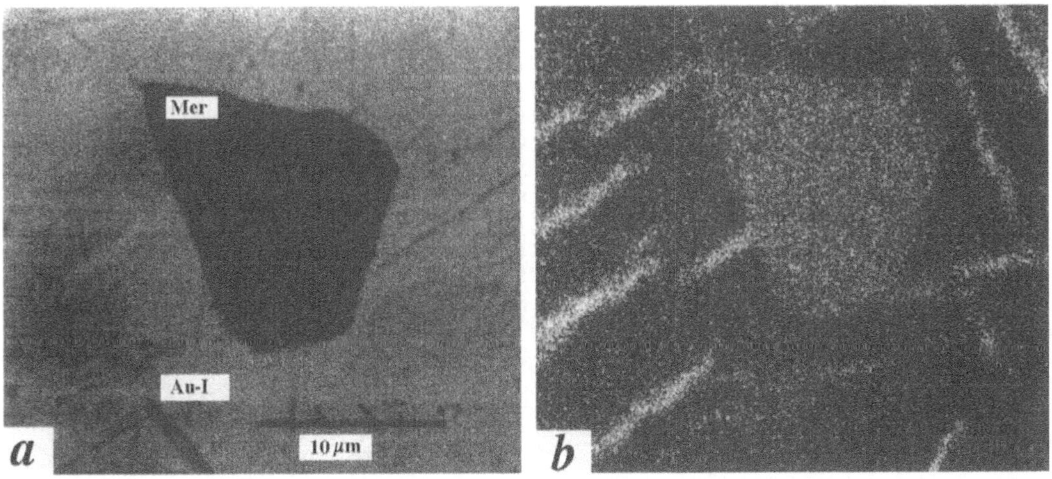

Fig. 4. The intergrowth of native gold 1: **a**) electron back-scattered image, scale bar 10 μm, **b**) characteristic CuKα radiation. Cu-enriched lamellae in Au 1 are clearly seen but are not discernable in the back-scattered image

crystals of the same composition. Thus, many pentlandite grains contain from 90 ppm to 5700 ppm (0.57 wt%) Pd. In Pd-rich pentlandites the presence of Rh and Ir from 90 to 1200 ppm is also detected.

The ores of Circumpolar Ural contain as Ag-, Cu-, Hg- and Pd-bearing native Au I and high-grade Pd-bearing Au II (Pd up to 1.9 wt%) [17]. The internal structure of native Au I is complicated by the exsolution structure, which consists of the Au + Ag + Cu + Hg + Pd-matrix (84.1–87.1 wt.% Au, 9.8–12.1 wt.% Ag, 1.6–1.9 wt.% Cu, 0.4–0.8 wt.% Pd, 0.7–1.3 wt.% Hg) and Au + Cu + Pd-laminae (82–84.1 wt.% Au, 5.0–10.4 wt.% Ag, 5.9–10.5 wt.% Cu, 0.6–1.1 wt.% Pd (Fig. 4). The improvement of the WDS analytical technique achieved in the present work made it possible to obtain equally good results in routine analyses of intrinsic PGE, Au and Ag minerals, as well as in determination of trace amounts of these elements in sulphides and sulpharsenides. The method described is convenient for routine analyses of inhomogenous distributions of precious elements in minerals.

Conclusion

The efficiency of the proposed analytical method improves the WDS technique for routine microprobe determinations of major and trace platinum group elements, gold and silver in their own minerals and in sulphide and sulpharsenides. High quality measurements are achieved by (1) the precise peak-overlap correction preceding the standard ZAF correction procedure; (2) precise determination of the background intensity by means of a modified form of the Kramers expression; (3) simultaneous measurements of major and trace elements (up to 20), which allow the instrumental drift and specimen stability to be monitored in the course of measurements. Application of this method to the determination of precious elements in sulphide and sulpharsenide minerals permitted the minimum detection limit to be reduced to 90–280 ppm, depending on the particular element, with a probe current of 200 nA and a counting time of 100 s: this can be improved further to 40–100 ppm by increasing the current and counting time. The use of the approach described in this paper also lowers by a factor of two the required analysis time. The results obtained were also checked by the conventional EPMA technique.

Acknowledgments. The authors would like to thank the referees and editors for their helpful comments. The authors are also grateful to Dr. T. L. Grokhovskaya, Dr. A. D. Genkin, S. F. Sluzhenikin, and Dr. G. Moralev (the Institute of Ore Deposits Geology, the Russian Academy of Sciences) for placing at their disposal a large number of minerals used in this study, and to Dr. A. Yu. Malevsky (the Institute of Mineralogy and Geochemistry of Rare Earth Elements) for providing most of the synthesised specimens. This study was supported in part by the Russian Foundation for Basic Research (project no. 95-05-14970).

References

[1] R. Amli, W. L. Griffin, *Am. Mineral.* **1975**, *60*, 599.
[2] J. Boecker, T. Hehenkamp, *Mikrochimica Acta. Suppl.* **1977**, *7*, 209.
[3] L. J. Cabri, J. L. Campbell, J. H. G. Laflamme, R. G. Leigh, J. A. Maxwell, S. D. Scott, *Can. Mineral.* **1985**, *23*, 133.
[4] V. V. Distler, E. A. Kulagov, S. F. Sluzhenikin, I. P. Laputina, *Geology of Ore Deposits* **1996**, *38*, 35.
[5] V. V. Distler, A. Yu. Malevsky, I. P. Laputina, *Geokhimiya* **1977**, *11*, 1646.
[6] J. J. Donovan, D. A. Snyder, M. L. Rivers, *Microbeam. Anal.* **1993**, *2*, 23.
[7] A. D. Genkin, V. V. Distler, I. P. Laputina, A. A. Filimonova, *Geokhimiya* **1973**, *9*, 1336.
[8] A. D. Genkin, L. J. Cabri, N. S. Bortnikov, G. McMahon, C. J. Stanley, F. E. Wagner, J. Friedl, Yu. G. Safonov, G. N. Gamyanin, *Proc. Global Exploitation of Heap Leachable Deposits* (D. M. Hausen, ed.), TMS Annual Meeting, Warrendale, PA, 1997, pp. 129–140.
[9] T. L. Grochovskaya, I. P. Laputina, G. S. Kuznetsov, G. N. Muravitskaya, S. V. Sokolov, V. A. Tel′nov, A. N. Bol′shakov, *Geology of Ore Deposits* **1996**, *38*, 183.
[10] D. C. Harris, L. J. Cabri, R. Nobiling, *Can. Mineral.* **1984**, *22*, 493.
[11] M. Fialin, *X-ray Spectrom.* **1992**, *21*, 171.
[12] I. P. Laputina, *IGEM Scientists,USSR Acad. Sci.* 1972, pp. 107–108 (Abstracts).
[13] C. Merlet, J. L. Bodinier, *Chem. Geol.* **1990**, *83*, 55.
[14] G. Remond, F. Cesbron, K. Traxel, J. L. Campbell, L. J. Cabri, *Scan. Microsc.* **1987**, *1*, 1017.
[15] G. Remond, P. Coutures, C. Gilles, D. Massiot, *Scan. Microsc.* **1989**, *3*, 1059.
[16] P. L. Roeder, *Can. Mineral.* **1985**, *23*, 263.
[17] M. B. Tarabaev, S. K. Kuznetsov, G. V. Moralev, A. A. Soboleva, I. P. Laputina, *Geology of Ore Deposits* **1996**, *38*, 11.

Mikrochim. Acta [Suppl.] 15, 253–267 (1998)

Composition of 15–17th Century Archaeological Glass Vessels Excavated in Antwerp, Belgium

Koen H. Janssens[1,*], **Ine Deraedt**[1], **Olivier Schalm**[1], and **Johan Veeckman**[2]

[1] Department of Chemistry, University of Antwerp, Universiteitsplein 1, B-2610 Antwerpen, Belgium
[2] Excavation Department of the City of Antwerp, Godefriduskaai 36, B-2000 Antwerpen, Belgium

Abstract. In the beginning of the 16th century, Antwerp (Belgium) was one of the largest ports of Europe and an important centre for the manufacture and import of glassware, especially for Venetian and *façon-de-Venise* glass vessels. As a result, archaeological excavations in the historical centre of Antwerp have yielded a number of important glass finds from the 15th until the 17th century. Within this extensive collection, several typological categories of glass vessels can be distinguished. The major and trace composition of a series of 96 representative glass objects was determined by means of EPXMA (electron probe X-ray micro analysis) and μ-SRXRF (synchroton radiation induced X-ray fluorescence analysis). Four major compositional types could be distinguished. The correlation between object composition, colour and type is discussed. Special emphasis is placed on the composition of the *façon-de-Venise* or Venetian glass objects found in Antwerp and on the comparison of this composition to that of glass of the same type found in other locations in Europe.

Key words: archaeological glass, Venetian glass, façon-de-Venise glass, electron probe micro-analysis, synchrotron radiation induced X-ray fluorescence analysis.

During the first thirty years of the 16th century, the city of Antwerp became one of the most important ports on the European continent. One of the materials which were being imported into the low countries via the port of Antwerp from Northern Italy (e.g., from Venice) was glass; next to trade in ready-made glass objects (especially in high quality and costly luxury pieces), glass vessels of more ordinary quality were also manufactured locally. The earliest historical documents referring to local glassmakers taking up residence in Antwerp date from 1537 [1]; until well into the 17th century, glassmakers and glassmerchants, usually of Italian origin, are known to have been active in the city and were granted monopolistic priviliges by the city council or higher authorities to ensure the success of their business. During this period, the production of *façon-de-Venise* glass (FdV) became very important in Antwerp. As well as the latter category of colourless glass, objects made from green 'forest-glass' or from colourless 'fern-glass' were found in important quantities [2].

As a result, archaeological excavations in the historical city centre of Antwerp, mostly of basements and cesspits of dwellings from that period, have yielded an extensive number of glass vessels of different sizes, shapes and quality. Although Antwerp is regarded in that period as being one of the most important centres of glass manufacture and trade in Western Europe, a systematic investigation of the composition of these glass finds has never been undertaken until now. Nevertheless such an investigation might shed light on several aspects of the economic and social life in Antwerp in the above-mentioned period.

With the aim of (a) making an inventory of the different typological/compositional categories of archaeological glass vessels found in Antwerp, (b) establishing what proportion of the glass finds

* To whom correspondence should be addressed

correspond to local production and (c) obtaining a better insight into the possible origin of the various types of glassware which were imported into the city during the 15–17th century, the major and trace composition of a series of 96 archaeological glass vessels, dating from the end of the 15th century to the 17th century was determined. These objects were selected as representative of the various typological categories of objects which were recovered during the excavations. For the major element analysis, EPXMA (electron probe X-ray micro analysis) [3] was employed. This information was useful to group the objects into several broad compositional categories, each featuring a significantly different glass composition. By means of the trace element signatures of the glass vessels, as obtained by means of μ-SRXRF (synchrotron radiation induced X-ray fluorescence analysis) [4], these broad categories may be further subdivided.

In what follows, after a brief description of the sample preparation procedures and analytical methods which were employed, a concise description of the history and typology of the series of glass vessels which were analysed is given. Then, the various major-element groups which could be distinguished within this series are discussed, in relation to the typological category of the objects they contain. Also, and especially for the *façon-de-Venise* type of vessels, this major composition data is critically compared to that of similar finds at other archaeological sites in Europe and to the published composition of the various types of Venetian glass produced during the 15–17th century. Finally, the subdivision of the major-element groups by means of trace-element information is described and the usefulness of the trace-element data in this context is discussed.

Experimental

Sample preparation. From each of the glass objects, small samples were taken from one or more fragments by means of a diamond-coated saw. Most glass samples of 1–10 mm^2 in size and, in groups of about 10 pieces were embedded into resin blocks. These were subsequently cut to expose a cross-section perpendicular to the original glass surface and polished using diamond paste (1 μm grain size). For the EPXMA measurements, the polished surfaces were carbon coated. Prior to the μ-SRXRF measurements, the topmost part of each resin block was sawed-off by means of a low-speed saw, yielding a 300–500 μm thick resin disk containing a number of glass slices of easily measurable thickness. This reduction in specimen thickness was performed to reduce the scatter intensity in the SRXRF spectra and to bring the detected X-ray fluorescence count rate within usable limits.

Major element composition determination. EPXMA measurements were performed at the University of Antwerp, Belgium, on a JEOL JSM 6300 Scanning Electron Microscope, equipped with an energy-dispersive X-ray detector. For the measurements, an electron beam current of 1 nA and a voltage of 20 kV were employed; a low magnification setting of the instrument (700 x) and a limited analysis time (100 s) were used to insure that no significant migration of Na occured during the irradiation. The X-ray intensities obtained were quantified by means of a standardless ZAF scheme [5]. Various glass standards of known composition from NIST (National Institute of Standards and Technology) and from an institute specialising in Archaeological glass analysis were used to evaluate the accuracy of the analysis (see below).

Trace element determinations. Trace element determinations in silicate-type materials such as glass are usually performed by means of destructive techniques such as AAS (Atomic Absorption Spectrometry) [6], ICP-MS (Inductively Coupled Plasma-Mass Spectrometry) [7] and ICP-OES (Inductively Coupled Plasma-Optical Emission Spectrometry) [8]. Since these techniques require acid dissolution of the samples being analysed, the trace analysis of glass samples invariably involves the (partial) destruction of a number of historically valuable glass fragments. In this study, preference was given to another method of elemental analysis which (a) leaves the sample material intact, (b) only requires minute samples and (c) can yield data on trace element concentrations with high quantitative reliability. μ-SRXRF measurements only differ from conventional EPXMA measurements in the sense that an intense (micro) beam of polarised X-ray photons is employed as a means of sample irradiation instead of an electron beam [4]. μ-SRXRF measurements were executed at the Hasylab (Hamburger Synchrotron Labor, Hamburg, Germany) beamline L using a $20 \times 20\,\mu m^2$ white synchroton X-ray beam. A spectrum collection time of 300 s was used per sample. Quantification of the μ-SRXRF spectra obtained was performed using a fundamental parameter method [9]. The accuracy of the method was checked as with the EPXMA measurements by means of a series of glass standards with known trace element composition.

Analysed Glass Samples

The excavated glass vessels which were analysed in the framework of this study mainly originate from three archaeological sites in the centre of Antwerp. Most objects originate from the site in the *Zwartzustersstraat* [2] where an important collection of glass vessels (67 objects) were found, dating from around 1500 (35 objects, mainly green coloured and colourless beakers) and the middle of the 17th century or somewhat later (32 objects, beakers and goblets of colourless glass, assumed to be soda glass [2]). A second site was located in the *Kaasstraat* and yielded a wide typological range of glass vessels dated to the end of the 16th/beginning of the 17th century (imported forest glass and *façon-de-Venise* objects, assumed to be local products [1, 10]). Remains of very luxurious glass vessels were found during restoration activities in the palace of the Bishop of Antwerp at the *Schoenmarkt*. A large collection (of about 40 objects) of Venetian and/or *façon-de-Venise* vessels was among the finds. In addition, some of the analysed glass vessels were found during various emergency excavations carried out at different locations in the historic centre of the city.

In order to obtain an overview of the composition of the material the different types of Antwerp glass finds are made of, a limited number (e.g., 4–10) of examples from each stylistic group were analysed. Table 1 provides an overview of each group of analysed vessels, the time period the vessels are dated to, the number of analysed vessels per group and the type of glass (forest glass, fern glass, soda glass) the vessels were assumed to be made of in the literature [11] based on their appearance and colour, i.e., prior to the analyses. The distinction between the three above-mentioned types of glass is based on the type of ash which is used during its manufacture [12]: forest glass is made with Ca and Fe rich ash of (beech) trees, yielding a dark green glass; fern glass (made from K-rich ashes from fern plants) usually is almost colourless or straw-coloured. In the period investigated here, soda glass was made by means of the ashes of particular saline plants, rich in Na_2CO_3, imported for this purpose, e.g. from Syria. Nearly colourless soda glass can be produced if care is taken to employ very pure raw materials (mostly quartz pebbles and ash).

During the analyses, special emphasis was placed on the *Façon-de-Venise* group of glasses, which are very abundant among the glass vessels excavated in Antwerp. Accordingly, a larger number of these objects (33) were analysed in an attempt to distinguish local production from imported wares.

As can be seen from Table 1 the glass vessels which were analysed span the period from the beginning of the 15th to the end of the 17th century. The periods before (the 13th century and to a lesser extent the 14th century) are much poorer with respect to glass finds: in Western Europe the use of glass as drinking implements was then restricted to the nobility and high clergy, making glass tableware a rarity before 1400. However, in the period 1450–1550, simple drinking vessels such as *maigeleins* and *maigel beakers* made from green 'forest glass' were again reaching a wider public in Western Europe. These are generally assumed to have been imported from Germany where the basic materials for glass making (sand, chalk) as well as wood for the ovens (and as a source of alkali) were quite plentiful. Next to this first group of (green) forest glass object, a second group comprises all the uncoloured glasses which sometimes are denoted as 'soda-glass' and sometimes as 'fern-glass'. Within the group of colourless glass vessels, one can make a further subdivision among the late 15th/early 16th century glasses and the late 16th/17th century objects. Below, the different typological categories indicated in Table 1 are briefly described.

Green (forest) glass. The earliest analysed objects are the *maigeleins and maigel-beakers.* Low cups (*maigeleins*) and beakers (*maigelbecher*) both commonly of green glass and sometimes decorated with a ridge-pattern were blown in large numbers in Germany [13]. The *Krautstrunk* (cabbage stem), a green beaker decorated with pointed prunts, was also in demand in the Low Countries. These vessels evolved into *prunted beakers* and *berkemeiers* (a type of conical beaker, see Table 1) [14]. An important step in the evolution during the period 1550–1650 was the transition of the *berkemeier* into the *roemer*, which proved to be a wine-glass much favoured by citizens during the 17th century ([11], p. 190). *Roemers* differ from *berkemeiers* by having a convex bowl. Probably *roemers* were made in Germany as well in the Netherlands.

Colourless glass. Within this group two large categories can be distinguished. The first consists of ribbed beakers and beakers on folded foot. Both types date from the end of the 15th until the beginning of the 16th century. The second group comprises all colourless glass from the second half of the 16th and the 17th century.

Colourless Glass Vessels from the End of 15th/ Beginning of 16th Century

Mould-blown *ribbed beakers* are the continuation of the 14th–15th century vessels of similar shape; by the end of the 16th century, this vessel type has disappeared. These cylindrical beakers are thought to be blown from fern glass and are assumed to result from the French glassmaking tradition ([11], pp. 93–94); they were probably made in the Southern Netherlands or Northern France. Most excavated examples are thin-walled and colourless or very lightly tinted (straw-coloured). The *beakers on folded foot* were probably also produced in the same period and geographical area as the ribbed beakers. This type of vessel is made from one piece of glass: the folded foot is obtained by pushing in the constricted lower part of a (softened) glass bowl while evacuating the

Table 1. Overview of the typological categories of glass vessels

Typical shape	Category	Time period	Assumed glass type	Analysed glass type	Number of objects
	Maigeleins Maigel-beakers	late 15th, early 16th	forest	calcic	5 2
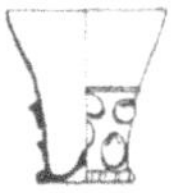	Berkemeiers	1st half 16th	forest	calcic	5
	Prunted beakers	1st half 16th	forest	calcic	4
	Ribbed beakers	early 16th	fern	sodic	12
	Beakers on folded foot	1st half 16th	fern	sodic	13
Various shapes	Façon-de-Venise vessels Filigrain objects	16th	soda	sodic	33
	Roemers	1st half 17th	forest	calcic	4
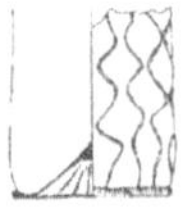	Mesh-work beaker	1st half 17th	soda	potassic	1
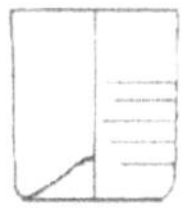	Stackable beakers	17th	soda	potassic	8
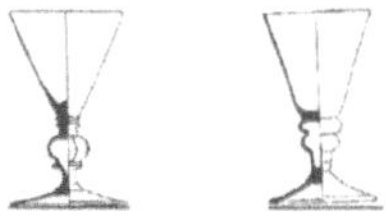	Façon-de-Venise goblets	17th	soda	potassic	8

air inside ([11], p. 96). This form appears both in the French and the German tradition [15]: the beakers made from forest glass are assumed to be produced in Germany; the French beakers are probably made of fern glass and are either colourless or lightly straw-coloured and manufactured in Momignies-Beauwelz, near the French-Belgian border [16]. Some of these beakers are decorated with enamel paintings.

Colourless Glass Vessels from the End of the 16th and of the 17th Century

Although there are indications that in the beginning of the 16th century, Italian glassmakers started producing glass in Antwerp similar to that manufactured in the North of Italy (see above), the type and composition of the early glass objects they produced are unknown. Finds of the second half of the 16th century include beakers and goblets, with or without decorations. In this period, it is typologically difficult to distinguish the genuine Venetian products from those of local origin. The decoration techniques used with the *façon-de-Venise* (FdV) glass were the same as those previously developed in Venice, such as the use of applied lion head masks or masks of Neptune, possibly combined with the application of decorative threads (*vetro a filigrano*, VaF), often of different colours ([11], p. 170). In the 17th century the excavated *goblets* acquired a heavier look and the delicate and elegant Venetian or *façon de Venise goblets* evolved into unrefined drinking glasses. A last category of analysed objects comprises *stackable beakers* and *meshwork beakers*, both in colourless glass from the 17th century: on the one hand beakers with horizontal ridges, fitting into one another [1, 17] and on the other hand beakers with a decorative mesh-work pattern. Both the above-mentioned goblets and beakers are assumed to be made of soda glass and remained in use until the end of the century when they were supplanted by vessels in English lead glass and Bohemian potash-lime glass.

Results and Discussion

Accuracy of Major Element Analyses by Means of EPXMA

Various NIST glass standards were used to validate the ZAF-based quantitative procedure (see Table 2). Since NIST SRM's (standard reference materials) all have Na^+ as a major cation, a number of glass standards available from the Corning Museum of Glass (CMG), with higher K and Ca levels, were also analysed. For concentration levels above 1 percent, deviations of the order of only a few percent are noted for the most part; larger errors are observed when concentration levels approach the detection limit of the method (below the 0.1% level). Also for heavier elements such as Zn, Pb, Sn and Ba a poorer accuracy is obtained. It can be concluded that the accuracy of the method is suitable for an overall analysis of the major and minor constituents of the glass. However, in order to detect more subtle differences among compositional categories of glass vessels with sufficient confidence, the complementary use of a more sensitive method, yielding more reliable data on the elements present at lower concentrations, appears appropriate.

Accuracy of Trace Element Determinations by Means of μ-SRXRF

A method complementary to EPXMA, that provides information on the major and minor constituents of the glass (in this case: Na, Al, Si, Cl, K, Ca, Mn, Fe), is μ-SRXRF, which can yield concentration information on minor and trace constituents such as Cu, Zn, Pb, Br, Rb, Sr, Mo, Zr, Sn, Sb and Ba [4]. In order to evaluate the accuracy with which the concentrations of these minor and trace constituents can be determined, a series of geological glasses with known trace element composition was analysed [18]. In Table 3, some of the results obtained are compared to the certified concentration levels. The accuracy in most cases is of the order of the uncertainty caused by the statistical noise in the spectral data. It can be concluded that in addition to having a much higher sensitivity for heavy trace elements (at the 1–10 ppm level), μ-SRXRF also permits the quantitative determination of the concentration levels of the traces element in glass matrices with an accuracy of about 10% [4, 18].

Major Element Composition of the Glass Vessels

As expected, the major composition of all analysed glass vessels is dominated by four components: SiO_2, Na_2O, K_2O and CaO. In Fig. 1, the characteristic ratio $CaO/(CaO + K_2O)$ is plotted against the Na_2O concentration for all glass vessels. Clearly, four distinct compositional categories can be discerned. The largest group is the sodic glass group, containing 11–18% Na_2O, which comprises several typological categories of glass vessels (see Table 5c). Next to this large cluster, a group of Na-poor and Ca-rich vessels can be identified (calcic group), which also contain relatively little K. In addition, two groups (potassic and potasso-calcic glass), showing intermediate abundances of Na, K and Ca can be identified.

In Tables 4a–c, the average glass composition of the various typological groups is presented. On the whole, most type-groups proved to have a reasonably homogeneous composition, although in a few groups, 'outlier' objects were found to have a different composition, usually similar to the average composition of another typological group. The average values

Table 2. Quantitative EPXMA Analysis results of various glass standards (w/w %)

	NIST SRM 620 Soda lime flat glass		NIST SRM 1830 Soda lime float glass		NIST SRM 1831 Soda lime sheet glass		NIST SRM 621 Soda lime container glass	
	Certified	Measured	Certified	Measured	Certified	Measured	Certified	Measured
Na	10.68	10.75	10.2	10.63	9.88	9.78	9.45	9.77
Mg	2.23	2.24	2.35	2.34	2.12	2.02	0.16	0.04
Al	0.95	1.03	0.06	0.32	0.64	0.78	1.46	1.52
Si	33.7	33.78	34.16	33.93	34.16	34.28	33.25	33.39
S	0.11	0.06	0.1	0.08	0.1	0.07		
K	0.34	0.36	0.03	0.05	0.27	0.32	1.67	1.70
Ca	5.08	4.95	6.12	5.88	5.86	5.74	7.65	7.40
Ti	0.01	0.01	0.007	n.d.	0.011	n.d.		
Fe	0.03	0.06	0.11	0.10	0.061	0.07		
Ba							0.11	0.06

	CMG standard A		CMG standard B		CMG standard D	
	Recomm.	Measured	Recomm.	Measured	Recomm.	Measured
Na	10.77	10.57	12.81	13.04	0.98	n.d.
Mg	1.69	1.63	0.72	0.39	2.47	2.60
Al	0.53	0.68	2.23	2.05	2.87	2.63
Si	31.12	30.92	28.77	28.89	25.82	25.94
P			0.37	0.40	1.75	1.85
S			0.22	0.37	0.12	0.11
Cl	0.1	0.06	0.2	0.15	0.4	0.10
K	2.43	2.76	0.91	0.91	9.51	9.45
Ca	3.79	3.93	6.23	5.96	10.76	10.91
Ti	0.48	0.50			0.24	0.22
Mn	0.75	0.75	0.18	0.17	0.36	0.38
Fe	0.76	0.83	0.24	0.24	0.35	0.34
Co	0.12	0.16				
Cu	0.97	1.06	2.16	2.29	0.32	0.26
Zn			0.16	0.15		
Sn	0.22	0.26			0.1	0.37
Sb	1.29	1.17	0.35	0.22	0.72	0.97
Ba	0.48	0.33	0.13	0.13	0.3	0.26
Pb			0.37	0.60	0.25	0.29

Recomm. recommended value.
CMG: Corning Museum of Glass.

in Table 4 are obtained after removal of these outlier objects.

Calcic and potasso-calcic groups (Table 4a). The calcic group of vessels ($\sim$22% CaO, 5–6% K_2O, $Na_2O<2\%$) consists of two type groups: maigeleins and roemers; similarly the potasso-calcic group is comprised of prunted beakers and berkemeiers ($\sim$15% CaO, 7–11% K_2O, 5–6% Na_2O). It is remarkable to note that three of the four roemers, although typologically evolving from the berkemeiers, show an overall composition resembling that of the maigeleins; the composition of the fourth roemer is similar to that of the average prunted beaker/berkemeier composition. In Table 5a, another group of vessels with calcic composition of miscellaneous type is listed. This sub-group contains one '*façon-de-Venise*' vessel, two beakers on folded foot and one straight beaker. It can be concluded that throughout the period investigated, calcic objects of various shapes are found in Antwerp; these are usually assumed to be imported wares, originating from the German 'Waldglas' tradition.

Without exception, all of the green 'forest glass' objects (see Table 1) turned out to be of calcic or potasso-calcic composition (see below). However, also a few colourless or straw-coloured objects of different shapes (see Table 4a, miscellaneous group) were found to have a similar calcic composition. These objects may have been locally manufactured by recycling (broken) forest glass objects or they may be imitation products, directly manufactured in the same (German) glass workshops that were producing green glassware, after suitable decolourisation of the wood-ash based glass, probably with MnO.

Table 3. Analysis results of three geological glasses obtained by μ-SRXRF (from [18]). The indicated 1 s standard deviation refers to the variation in composition at different locations on the (homageneous) sample

	ATHO		KL2		ST HS6/80	
Element	Literature w% (ppm)	Measured w% (ppm)	Literature w% (ppm)	Measured w% (ppm)	Literature w% (ppm)	Measured w% (ppm)
K (%)	2.34	1.64 ± 0.17	0.37	0.53 ± 0.16	1.11	1.0 ± 0.1
Ca (%)	1.15	0.94 ± 0.09	7.8	7.1 ± 0.2	3.73	3.2 ± 0.1
Ti (%)	0.15	0.11 ± 0.02	1.6	1.4 ± 0.1	0.43	0.38 ± 0.05
Cr	180	177 ± 34	205	417 ± 148	89	59 ± 53
Mn (%)	0.08	0.10 ± 0.02	0.16	0.18 ± 0.03	740	720 ± 44
Fe (%)	2.76	reference	8.32	reference	3.4	reference
Ni			115	124 ± 25		
Zn	169	153 ± 10	120	123 ± 14	59	60 ± 8
Ga	25.6	28 ± 4	20	23 ± 10	18	20 ± 9
Rb	59.4	65 ± 4	9.4	10 ± 4	29.2	31 ± 4
Sr	92.3	99 ± 5	363	391 ± 33	480	472 ± 17
Y	103	105 ± 4	24.4	22 ± 6		
Zr	567	557 ± 15	163	177 ± 29	120	120 ± 6
Nb	65.5	52 ± 3	17	13 ± 5	–	4 ± 2
Ba	557	592 ± 16	124	153 ± 43	300	297 ± 8
La	58.3	57 ± 3	13.7	12 ± 7	11	11 ± 3
Ce	126	125 ± 5	33.7	34 ± 13	23	23 ± 3
Pr	14.9	13 ± 2	4.71	5 ± 4		
Nd	58.9	72 ± 3	22.3	29 ± 11	14	14 ± 3
Sm	13	19 ± 3	7.8	13 ± 12		
Eu	2.13	6 ± 4				
Gd	19.2	16 ± 4				
Dy	15.4	13 ± 5				
Er	9.4	14 ± 5				

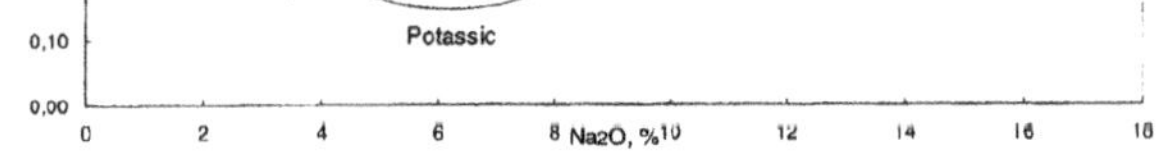

Fig. 1. The ratio $Cao/K_2O + CaO$ plotted against the Na_2O concentration for all analysed vessels

The calcic glass vessels show $CaO/(CaO + K_2O)$ ratios in the range 0.75–0.85 (see Fig. 2) which is very close to the value of 0.8 typical for ash derived from beech wood burning [19]. Similar compositions were also obtained by other authors who have analysed late medieval vessels. Golebiewski [20] analysed Stangengläzer, prunted beakers and roemers from the North of Poland and found vessels with either a very high CaO content (19–21%), almost no Na_2O and a K_2O level between 4 and 11%, and objects with a more comparable CaO and K_2O levels (respectively in the range 16–17% and 8–16%) but did not encounter objects which also contained up to 5% Na_2O as is the case here. Wedepohl analysed about 150 late-medieval glass objects from different locations in Germany and distinguishes 6 compositional types [21]; one of these (the wood-ash chalk glass type) having $23 \pm 2\%$ CaO and $5 \pm 1\%$ K_2O is similar to that of the maigeleins and roemers; however, in this extensive survey of German forest glass, nothing similar to the composition of the berkemeirs and roemers was found. Since the type of vessels concerned here (prunted beakers and berkemeiers from the first half of the 16th century) were not only manufactured east of the Rhine but also west of it [22] and in Beauwelz (close to what is now the Belgian-French border) [16, 11 (p. 191)], a possible explanation for the origin of the objects with potasso-calcic composition could be that these vessels were made in the western part of France instead of in Germany.

Table 4a. Composition of Ca-rich glass vessels. in mass % (Na_2O-Fe_2O_3) and in ppm (Ni–Bi). (n.d. = not detected) The group 'miscellaneous' consists of 1 straight beaker. 1 ribbed beaker and 2 Façon-de-Venise vessels

	Maigeleins n = 7		Prunted Beakers n = 4		Berkemeiers n = 5		Roemers n = 3		Miscellaneous n = 4		Roemer* n = 1
	Mean StDev	Min–Max	Mean StDev	Min–Max	Mean StDev	Min–Max	Mean StDev	Min–Max	Mean StDev	Min–Max	
Na_2O	1.35 ± 0.26	0.99–1.70	4.48 ± 0.70	4.07–5.53	5.38 ± 0.49	4.81–5.89	1.57 ± 0.41	1.21–2.01	0.03 ± 0.03	0.00–0.07	4.35
MgO	4.07 ± 0.52	3.57–5.05	3.48 ± 0.09	3.35–3.54	3.51 ± 0.20	3.20–3.67	3.20 ± 0.49	2.87–3.77	3.67 ± 0.30	3.36–3.96	1.80
Al_2O_3	3.84 ± 0.73	2.67–4.74	2.28 ± 0.06	2.19–2.34	2.73 ± 0.47	2.30–3.25	3.67 ± 0.23	3.51–3.93	2.18 ± 0.67	1.45–3.00	2.48
SiO_2	56.10 ± 2.19	53.69–59.30	60.01 ± 0.46	59.66–60.69	61.64 ± 1.21	60.01–63.34	60.43 ± 1.49	58.72–61.45	60.89 ± 2.24	57.90–62.89	64.59
P_2O_5	3.39 ± 0.27	3.04–3.79	1.18 ± 0.04	1.13–1.22	1.15 ± 0.07	1.08–1.26	1.87 ± 0.15	1.75–2.04	2.50 ± 0.77	1.83–3.45	1.06
SO_3	0.12 ± 0.04	0.06–0.18	0.15 ± 0.02	0.13–0.17	0.17 ± 0.10	0.05–0.30	0.07 ± 0.04	0.04–0.11	0.32 ± 0.09	0.20–0.40	0.10
Cl	0.27 ± 0.06	0.17–0.34	0.11 ± 0.03	0.08–0.14	0.21 ± 0.06	0.14–0.26	0.42 ± 0.07	0.36–0.49	0.00 ± 0.00	0.00–0.00	0.33
K_2O	5.96 ± 0.65	5.07–6.85	11.49 ± 0.81	10.30–12.04	7.97 ± 2.44	5.48–10.56	4.10 ± 0.68	3.49–4.84	6.09 ± 0.28	5.78–6.45	9.29
CaO	21.80 ± 1.86	18.70–23.64	14.66 ± 0.37	14.12–14.89	14.81 ± 0.62	14.31–15.54	22.42 ± 0.82	21.61–23.26	21.94 ± 2.17	20.04–25.07	14.50
MnO	1.78 ± 0.62	0.83–2.36	0.91 ± 0.03	0.88–0.94	0.91 ± 0.11	0.72–1.00	1.07 ± 0.04	1.05–1.11	1.40 ± 0.32	1.01–1.74	0.70
Fe_2O_3	0.79 ± 0.16	0.62–1.13	0.67 ± 0.04	0.63–0.72	0.74 ± 0.09	0.61–0.81	0.61 ± 0.03	0.57–0.62	0.43 ± 0.06	0.35–0.51	0.44
Ni	30 ± 10	17–47	38 ± 6	32–44	40 ± 8	29–47	47 ± 15	36–64	42 ± 18	25–67	16
Cu	127 ± 101	62–352	165 ± 8	159–175	176 ± 27	144–205	38 ± 9	28–43	81 ± 21	55–100	37
Zn	183 ± 59	151–315	255 ± 10	246–266	202 ± 18	182–230	64 ± 10	53–71	236 ± 93	143–359	71
Br	4 ± 2	2–8	12 ± 3	9–15	20 ± 5	14–26	7 ± 0	6–7	n.d.		6
Rb	143 ± 21	122–168	520 ± 44	473–561	300 ± 142	144–420	82 ± 18	65–101	94 ± 44	45–132	151
Sr	1388 ± 303	710–1547	792 ± 36	754–825	770 ± 40	701–799	1488 ± 119	1410–1625	985 ± 155	829–1164	950
Y	13 ± 3	11–19	34 ± 2	32–36	31 ± 6	23–38	11 ± 2	9–12	9 ± 4	5–13	12
Zr	184 ± 21	169–231	118 ± 4	115–122	116 ± 16	92–134	194 ± 30	171–228	131 ± 45	70–175	183
Cd	n.d.		6 ± 0	6–7	7 ± 3	5–12	n.d.		3 ± 2	0–6	7
Sn	28 ± 68	0–182	765 ± 122	685–905	897 ± 164	655–1067	7 ± 7	3–15	126 ± 222	1–459	n.d.
Sb	n.d.		4 ± 0	4–5	3 ± 2	1–6	n.d.		n.d.		n.d.
Ba	4784 ± 1358	1978–6178	2659 ± 62	2587–2698	2375 ± 463	1812–3081	5031 ± 2181	3219–7455	5252 ± 2703	3734–9297	2931
La	21 ± 4	18–29	17 ± 0	17–17	15 ± 7	9–25	24 ± 3	21–27	15 ± 3	12–19	27
Ce	15 ± 9	8–35	18 ± 2	16–19	18 ± 7	11–26	18 ± 5	12–21	10 ± 6	4–16	13
Pb	272 ± 560	26–1539	3380 ± 585	2834–3997	5040 ± 3991	1835–9605	37 ± 13	27–52	227 ± 261	91–618	33
Bi	9 ± 20	0–55	65 ± 17	53–85	71 ± 66	16–162	n.d.		17 ± 8	10–26	n.d.

Table 4b. Composition of K-rich glass vessels. in mass % for Na_2O to Fe_2O_3 and in ppm for Ni to Bi. (n.d. = not detected); FdV: façon-de-Venise

	Stacked Beakers n = 8		FdV Goblets n = 8		Network Beaker n = 1
	Mean StDev	Min–Max	Mean StDev	Min–Max	
Na_2O	5.78 ± 1.43	3.52–8.14	6.19 ± 0.74	5.27–7.31	6.31
MgO	1.35 ± 0.43	0.77–1.95	1.59 ± 0.17	1.34–1.78	1.66
Al_2O_3	1.33 ± 0.13	1.13–1.47	1.25 ± 0.05	1.19–1.31	1.41
SiO_2	69.21 ± 2.35	64.43–71.24	68.24 ± 2.29	64.35–71.38	65.39
P_205	0.11 ± 0.05	0.02–0.19	0.09 ± 0.05	0.03–0.14	0.08
SO_3	0.09 ± 0.05	0.00–0.19	0.09 ± 0.04	0.02–0.13	0.06
Cl	0.41 ± 0.07	0.30–0.53	0.42 ± 0.04	0.37–0.48	0.38
K_2O	14.32 ± 2.17	10.30–17.03	13.54 ± 1.48	10.34–15.43	12.86
CaO	4.42 ± 0.79	3.58–5.51	5.21 ± 0.55	4.41–6.24	6.47
MnO	0.35 ± 0.11	0.20–0.50	0.24 ± 0.07	0.14–0.36	0.38
Fe_2O_3	0.36 ± 0.07	0.26–0.43	0.26 ± 0.03	0.21–0.31	0.37
Ni	57 ± 48	8–131	70 ± 43	11–149	92
Cu	87 ± 37	36–144	78 ± 25	32–111	63
Zn	84 ± 40	27–118	47 ± 13	32–69	44
Br	40 ± 22	22–80	49 ± 28	26–111	72
Rb	203 ± 106	82–362	154 ± 82	44–305	109
Sr	269 ± 64	196–392	318 ± 80	237–447	282
Y	28 ± 14	8–48	37 ± 17	13–58	54
Zr	45 ± 10	29–58	42 ± 11	22–63	35
Cd	8 ± 11	0–26	14 ± 13	0–39	16
Sn	5 ± 5	0–11	5 ± 3	1–10	5
Sb	551 ± 497	36–1317	508 ± 728	18–2042	1405
Ba	425 ± 221	194–786	257 ± 82	131–367	231
La	5 ± 2	4–8	7 ± 2	4–10	5
Ce	7 ± 2	5–11	9 ± 1	7–11	8
Pb	18023 ± 19760	908–51020	23902 ± 16112	1743–52840	35520
Bi	243 ± 279	18–730	321 ± 252	23–795	507

Indeed, Barrera and Velde [19], in their very extensive survey of the composition of archaeological glass vessels from various locations in France dating from the 10th to the 18th century, for the period 1600–1700 identify a type of potasso-calcic glass having a ratio of $CaO/(CaO + K_2O)$ in the range 0.4–0.8, a Na_2O content of 0.5–3.4% and an MgO content of 2.6–8.0% as being typical for the west of France. Since the composition of this glass is intermediate between that of calcic forest glass and of fern glass (see below), they suggest that this glass type might have been manufactured by using mixtures of (Ca-rich) beech and (K-rich) fern ash.

Sodic-glass groups (Table 4c). As with the calcic glass vessels, within the series of sodic glass vessels, the K_2O content can be used to differentiate between on the one hand the beakers on folded foot and ribbed beakers ($2.5 \pm 0.5\%$ K_2O) and on the other hand (most of) the *façon de Venise* vessels (>3% K_2O). All Na-rich subgroups contain 13–16% Na_2O and 8–10% CaO. The major composition of the two beaker types are very similar to each other. The *façon de Venise* and filigrain vessels can be further sub-divided into three groups: one containing $7.5 \pm 0.5\%$ K_2O (subgroup FdV-I) and another featuring $5 \pm 1\%$ K_2O (subgroup FdV-II). Apart from these two groups, which both feature MgO concentrations of about 3%, a third small group of three VaF objects can be considered which show a lower content of K_2O ($2.2 \pm 0.1\%$) and a higher MgO abundance ($4.2 \pm 0.1\%$) (subgroup VaF-I).

The distinction between the two FdV subgroups and the VaF subgroup becomes clearer when the K_2O content of all the Na-rich glasses is plotted against their Na_2O level (see Fig. 2). The datapoints of the two FdV groups are roughly oriented along a straight line, indicating a negative correlation between Na_2O and K_2O. A similar correlation was found by Barrera and Velde [19] for the soda glass of the South of France. The VaF-I subgroup does not follow this trend; here, the Na_2O concentration appears to vary independently from the (low) K_2O level. Next to the 3 objects that make up the VaF-I subgroup, two other

Table 4c. Major composition of Na-rich glass vessels. in mass % for Na_2O to Fe_2O_3 and in ppm for Ni to Bi. (n.d. = not detected); BFF: beaker on folded foot; RB: ribbed beaker; FdV: façon-de-Venise; VaF: Vetro-a-fili

	BFF n = 11		RB n = 12		FdV-I n = 8		FdV-II n = 17		VaF-I n = 3		VaF-II		VaF-II		
	Mean StDev	Min–Max	Mean StDev	Min–Max	Mean StDev	Min–Max	Mean StDev	Min–Max	Mean StDev	Min–Max	n = 1	n = 1	n = 1	n = 1	n = 1
Na_2O	14.39 ± 0.85	12.94–15.66	14.73 ± 1.19	13.44–17.17	13.15 ± 0.73	11.83–14.19	15.71 ± 0.60	14.68–17.02	14.96 ± 0.67	14.23–15.55	15.01	15.54	13.54	17.43	11.42
MgO	2.36 ± 0.37	1.65–3.03	2.02 ± 0.31	1.42–2.32	2.99 ± 0.10	2.87–3.13	2.90 ± 0.57	1.98–3.77	4.24 ± 0.08	4.17–4.32	3.44	2.51	2.11	1.97	2.89
Al_2O3	1.39 ± 0.23	1.07–2.01	1.20 ± 0.24	0.97–1.71	1.51 ± 0.06	1.39–1.57	1.38 ± 0.08	1.25–1.54	1.34 ± 0.14	1.24–1.50	1.49	1.55	4.45	1.43	1.85
SiO_2	66.15 ± 2.81	63.11–70.64	68.18 ± 1.22	65.93–70.19	62.74 ± 0.44	62.09–63.44	63.31 ± 2.09	59.99–65.88	65.33 ± 1.00	64.67–66.47	62.07	62.76	58.62	68.01	64.22
P_2O_5	0.86 ± 0.26	0.21–1.20	0.70 ± 0.11	0.51–0.84	0.31 ± 0.07	0.22–0.43	0.30 ± 0.05	0.20–0.35	0.33 ± 0.05	0.29–0.38	0.28	0.32	5.53	0.27	0.39
SO_3	0.08 ± 0.08	0.02–0.31	0.05 ± 0.03	0.01–0.10	0.10 ± 0.03	0.07–0.15	0.16 ± 0.05	0.07–0.26	0.25 ± 0.03	0.22–0.27	0.07	0.13	0.26	0.15	0.12
Cl	0.90 ± 0.09	0.68–0.99	0.96 ± 0.08	0.83–1.09	0.57 ± 0.04	0.52–0.62	0.69 ± 0.09	0.57–0.82	0.65 ± 0.04	0.61–0.69	0.50	0.68	0.77	1.01	0.47
K_2O	2.85 ± 0.37	2.33–3.40	2.44 ± 0.53	1.82–3.47	7.35 ± 0.47	6.83–8.31	4.79 ± 1.00	2.78–6.28	2.18 ± 0.09	2.08–2.25	6.49	4.59	2.40	3.38	6.87
CaO	9.66 ± 2.33	6.16–12.99	8.53 ± 1.10	6.36–9.72	10.42 ± 0.58	9.69–11.40	9.81 ± 0.99	8.27–11.33	10.12 ± 0.50	9.54–10.48	9.71	8.05	6.85	5.17	10.52
MnO	0.79 ± 0.32	0.21–1.24	0.77 ± 0.43	0.29–1.29	0.28 ± 0.04	0.23–0.35	0.38 ± 0.18	0.16–0.72	0.23 ± 0.01	0.22–0.24	0.41	0.31	0.28	0.38	0.45
Fe_2O3	0.41 ± 0.12	0.24–0.65	0.30 ± 0.11	0.19–0.53	0.42 ± 0.03	0.37–0.48	0.40 ± 0.07	0.29–0.53	0.32 ± 0.08	0.26–0.41	0.38	0.45	0.29	0.42	0.60
Ni	6 ± 4	0–13	3 ± 2	1–7	6 ± 2	3–9	9 ± 5	0–18	12 ± 2	11–14	10	8	18	5	15
Cu	29 ± 31	10–116	12 ± 3	7–17	22 ± 4	15–28	39 ± 25	13–119	53 ± 36	30–94	28	25	101	156	36
Zn	44 ± 24	1–73	54 ± 15	35–82	46 ± 9	32–56	46 ± 18	20–78	37 ± 5	33–41	35	34	36	70	51
Br	42 ± 5	33–53	41 ± 5	33–50	37 ± 6	32–49	44 ± 21	28–110	51 ± 9	41–59	30	40	60	62	42
Rb	23 ± 8	16–37	21 ± 4	14–27	74 ± 5	67–82	36 ± 15	18–61	18 ± 2	16–20	54	39	17	19	71
Sr	314 ± 60	220–402	263 ± 37	197–332	771 ± 61	690–882	664 ± 137	390–850	982 ± 135	843–1114	675	565	684	524	607
Y	4 ± 1	2–6	3 ± 1	2–4	6 ± 1	4–8	8 ± 5	4–18	4 ± 4	2–9	10	6	11	4	8
Zr	54 ± 44	17–184	35 ± 4	29–44	99 ± 17	74–115	80 ± 20	33–105	32 ± 2	29–33	105	71	24	82	79
Cd	6 ± 3	1–10	6 ± 4	1–11	2 ± 1	0–4	5 ± 4	0–13	2 ± 4	0–6	5	6	11	4	7
Sn	88 ± 70	17–188	58 ± 24	25–88	283 ± 315	38–825	1197 ± 1260	56–4509	471 ± 732	35–1317	1122	2464	3158	362	1267
Sb	n.d.		n.d.		n.d.		n.d.		4 ± 3	1–6	n.d.	n.d.	n.d.	n.d.	n.d.
Ba	530 ± 195	121–819	570 ± 314	249–1129	323 ± 41	245–381	291 ± 81	165–394	254 ± 98	189–366	252	300	368	338	632
La	8 ± 3	2–12	7 ± 3	3–14	6 ± 2	5–9	6 ± 2	1–10	4 ± 2	2–5	6	7	5	4	10
Ce	9 ± 5	4–20	7 ± 3	4–14	10 ± 1	9–11	9 ± 2	6–12	4 ± 1	3–5	9	8	5	8	13
Pb	104 ± 74	34–226	68 ± 31	0–103	217 ± 137	114–484	900 ± 904	98–2493	480 ± 652	86–1233	1152	557	2375	256	482
Bi	n.d.		n.d.		22 ± 4	17–27	27 ± 15	7–53	19 ± 11	11–31	29	26	42	14	38

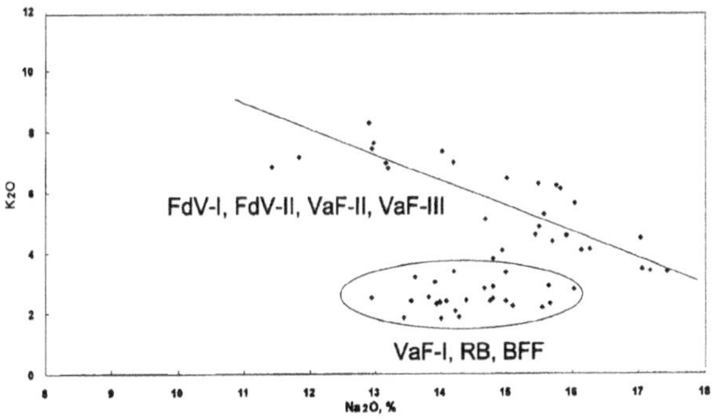

Fig. 2. K_2O vs. Na_2O abundance plot of the Na-rich glass vessels

vetro-a-fili objects show a composition similar to that of subgroups FdV-I and FdV-II (group VaF-II, see Table 4c). Also the data pertaining to three 'outlier' *vetro-a-fili* vessels are listed (VaF-III) showing respectively uncharacteristically high Al_2O_3/P_2O_5 (both 5% vs. normally 1.4% for Al_2O_3 and 0.3% for P_2O_5) and Na_2O/Cl levels (17%/1% vs. 15%/0.5%). In Fig. 2, the data corresponding to the beakers on folded foot (BFF) and the ribbed beakers (RB) are also shown and overlap with the VaF-I datapoints. However, the low MgO content of the BFF and the RB (ca. $2.0 \pm 0.4\%$) clearly distinguishes them from the VaF-I objects in which MgO is present at a fairly high level ($4.2 \pm 1\%$).

In order to explain the overwhelming presence of soda-objects in 16th century Antwerp (Table 4c), either soda or sodium-rich ash (and possibly the other raw materials for glass making), soda-glass in bulk and/or finished glass vessels must have been imported in considerable quantities from the Mediterranean (either from Northern Italy, Spain or more eastern locations). In view of the presence of the port, the cheapest and safest transportation route from, e.g., Venice, Alicante or the Middle-East to Antwerp undoubtedly would have been by sea. The import of Levantine ash ('lume') from different locations in the Middle East (in Syria and Egypt) by the Venetians is well documented [12]; Genoese traders in some cases also transported the soda-rich ash directly to Flanders and England [23]. An alternative source area of (lower quality) ash might have been the region of Alicante, Cartagena and Malaga, Spain [12, 24].

Potassic glass groups (Table 4b). Although they are listed as two separate sub-groups in Table 4b, the major composition of the 17th century stackable beakers and of the *façon-de-Venise* goblets does not differ significantly. The composition of the meshwork beaker is also very similar, showing intermediate values for the Na_2O and CaO abundances (resp. ca. $6 \pm 1\%$ and $5 \pm 2\%$) and a K_2O concentration of $13 \pm 1\%$.

Whereas the majority of the 16th century colourless glass vessels (both tableware for daily use and luxury items) are made of soda-glass (see previous section), in the 17th century, only potassic glass appears to have been used in Antwerp, even for vessel types such as the goblets which are assumed to have evolved from 'façon-de-Venise' or Venetian 16th century predecessors. When one considers the ratio CaO/(CaO + K_2O) for the potassic glass (see Fig. 1), values in the range 0.17–0.33 are obtained which are reasonably close to the corresponding value of 0.20 for fern ash [19]. It is tempting to attribute this discontinuity in glass composition which appears to have occurred around 1600 to the economical and political crisis which took place in Antwerp and surroundings as a result of the invasion of Spanish troops, their siege and conquest of the city in 1585 and the separation of the (catholic) Southern Low Countries from the (protestant) United Provinces. Since after 1585, the newly established Dutch state closed the River Scheldt for all naval traffic, it can be surmised also that the import by sea of ash, soda and/or soda glass into Antwerp ceased and that the local glass workshops were forced to find alternative (and more local) sources of raw materials. On the other hand, the changeover may have occurred simply because it became possible in the 17th the century to produce colourless glass using the much cheaper fern ash rather than imported soda-rich ashes. It is interesting to note that several of the analysed objects made from this 'new' type of potassic glass also contain appreciable amounts of PbO (>1%) in contrast to the calcic- and sodic glass vessels (see Table 4).

Trace Element Data

As can be seen in Table 4, within each of the large compositional groups, (sometimes small) variations in the minor constituents appear to be present. Most of these (more subtle) compositional distinctions appear to coincide with differences in the shape of the objects and therefore are probably meaningful. However, because of the variation in the data within each subgroup, it is sometimes difficult to objectively evaluate whether all of these typological groups have a significantly different composition or not. To establish whether some of the differences are due to erratic variations in the glass manufacturing process or are caused by differences in the (origin of the) raw materials which were used, it is relevant to consider also the pattern of trace element abundances in each of the glass objects. As an example, Fig. 3 shows a plot of the strontium content of all analysed objects against their rubidium abundance. The scatter plot of these trace elements emphasises the differences and similarities in composition of the various typological categories.

From this plot (and from a consideration of the other trace element abundance levels) it can be concluded that all 17th century objects (i.e., the goblets 'à-la-façon-de-Venise' and the stackable beakers) were produced from the same raw materials since they feature not only a very similar major and minor composition, but also nearly identical trace element signatures. In this respect it is interesting to mention the elevated level of Sb which found in all the 17th century potassic glass vessels; the use of Sb, next to Mn, as decolourising agent ('glass-makers soap') is well documented, especially in the Roman period [25]. Whereas in the sodic and calcic vessels, the Sb abundance is always below the detection limit ($\sim$5 ppm in this case), for the K-rich vessels shown in Table 2b, an average concentration of ca. 550 ppm is found. This Sb abundance is quite variable however: within the set of 17 potassic glasses, it varies from 20 ppm up to 2000 ppm in some cases. The presence of Sb in the potassic glass may explain how these glass vessels acquired their nearly colourless outlook and were assumed to be made of soda-glass.

The two categories of 16th century glassware (ribbed beakers and beakers on folded foot) have a similar composition of major elements and also show the same trace elemental pattern, indicating the common origin of both glass types; however, as can be seen in Fig. 3, this latter trace element pattern is distinctly different from that of the FdV and VaF groups. Although they show (slightly) different major element compositions, all the 'façon-de-Venise' objects (FdV-I and FdV-II) as well as the VaF-II and VaF-III objects (see Table 2c), form one large cluster in Fig. 3. This group, however, does not include the three filigrain objects (VaF-I) which feature a definitely higher Sr abundance (800–1200 ppm). The difference in Sr abundance between the VaF-I objects on the one hand and the ribbed beakers on folded foot

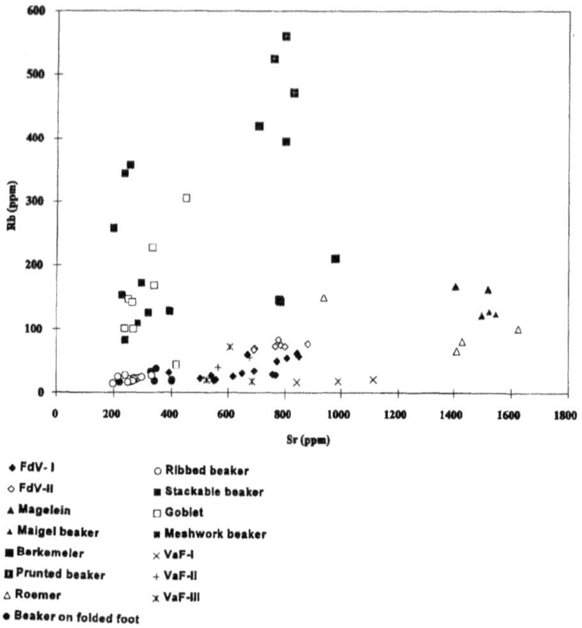

Fig. 3. Rb vs. Sr scatter plot of all analysed glass objects

on the other hand emphasises the different origin of these vessel types, even if they show nearly the same major composition (see Table 4c). A similar distinction can be made by means of the Sn abundance: the two beaker types show average concentrations of respectively 60 and 80 ppm while the FdV-I and FdV-II groups feature average concentrations at the 300 and 1200 ppm level. The VaF-I objects contain ca. 500 ppm Sn. The elevated concentrations of Sn in the FdV and VaF groups is not surprising since Sn is used to produce opacity in the manufacture of white opaque glass ('lattimo') which is frequently used to decorate these types of glass vessels.

Within the high-Ca group of glass vessels, similarities or differences in composition between the three (17th century) roemers and the (16th century) maigeleins with calcic composition can be evaluated in greater detail. As well as having a (nearly) identical major composition to the maigeleins, the roemers also feature the same trace element pattern (except for their copper and zinc abundance), although they are separated from the maigeleins in time by about a century. This may be an indication that the 17th century roemers were manufactured from recycled 16th century (or earlier) forest glass similar (or even identical) to the maigeleins. As can be seen in Fig. 2, the difference in origin between the berkemeiers and prunted beakers on the one hand and the maigeleins, maichel beakers and roemers on the other hand is confirmed by the trace element data. Fig. 2 also appears to indicate that the group of 5 berkemeiers might be divided further into two subtypes having Rb concentrations of 400–600 ppm (such as the prunted beakers) and 200 ppm (such as the fourth roemer, see last column in Table 4a). However, a larger series of vessels of this type should be analysed in order to substantiate such a subdivision.

Comparison of the Antwerp Façon-de-Venise and Filigrain Compositions to Literature Data

In view of the above, it is interesting to compare the compositional data from the Antwerp filigrain and 'façon-de-Venise' glass objects with analysis results of glass of the same type and period and of well-known origin. Table 5 provides an overview of published compositions of Venetian and 'façon-de-Venise, glass from different locations in Europe and compares it to the FDV-I, FdV-II and VaF-I average compositions.

The first comparison group is '*cristallo*', i.e., a transparent and colourless soda-glass which was fabricated in the middle of the 15th century in Venice. Its name was derived from rock crystal, to which it resembled in purity, brightness and homogeneity. The composition of Cristallo was reputed to be very constant with time and was kept a secret. Another group consists of Venetian objects made of '*Vitrum Blanchum*', which are dated from the 14th to the 16th century. This is also soda glass but of inferior quality, probably due to the lower purity of the raw materials employed for its production [26]. Brill analysed fragments of six Venetian glass objects excavated from the (16th century) Gnalic shipwreck and belonging to the collections of the Narodni Museum in Zadnar [27]. About 20 goblets dating from the late 16th to early 17th century from the Lido of Venice were analysed by H. Bronk (personal communica-

Table 5. Comparison with other Venetian and 'façon-de-Venise' glass of definite origin

	Cristallo[a] Venice		Venetian[b] Gnalic Wreck		FaV-I Antwerp		Vitrum Blanchum[a] Venice		Venetian[c] Lido/Venice		Venetian[c] Lido/Venice	
	Mean	Range	Mean	Range	Mean	Range	Mean	Range	Mean	Range	Mean	Range
Na_2O	16.90	14.30–19.20	12.90	12.30–13.70	14.96	14.23–15.55	13.00	11.2–14.90	13.24	10.50–15.30	19.55	18.90–20.50
MgO	1.80	1.10–2.35	2.20	1.76–2.58	4.24	4.17–4.32	3.40	1.15–5.10	2.80	1.57–3.40	2.10	1.02–3.18
Al_2O_3	0.70	0.48–0.90	1.30	1.76–2.58	1.34	1.24–1.50	1.10	0.70–1.95	0.98	0.65–1.38	0.87	0.60–1.22
SiO_2	70.90	68.50–73.00	71.20	71.00–72.00	65.33	64.67–66.47	67.50	63.4–70.00	68.54	65.40–73.60	65.45	62.10–69.80
P_2O_5	0.14	0.09–0.25	–	–	0.33	0.29–0.38	0.36	0.20–0.50	0.73	0.39–0.97	0.48	0.35–0.57
SO_3	0.31	0.23–0.42	–	–	0.25	0.22–0.27	0.27	0.15–0.40	–	–	–	–
Cl	1.00	0.90–1.20	–	–	0.65	0.61–0.69	0.82	0.75–1.00	–	–	–	–
K_2O	2.80	2.30–3.25	2.60	2.36–2.88	2.18	2.08–2.25	2.50	1.45–3.00	3.60	2.41–6.03	2.18	0.96–3.34
CaO	4.90	3.90–6.40	7.80	6.53–8.86	10.12	9.54–10.48	9.50	8.20–11.90	9.04	5.08–10.60	8.30	6.35–10.30
MnO	0.30	0.21–0.58	0.60	0.50–0.94	0.23	0.22–0.24	0.50	0.25–0.95	0.61	0.33–0.82	0.52	0.25–0.73
Fe_2O_3	0.24	0.17–0.38	0.60	0.39–0.75	0.32	0.26–0.41	0.36	0.22–0.48	0.44	0.25–0.71	0.35	0.21–0.60
	FdV[d] Grez–Doiceau		FdV I Antwerp		FdV II Antwerp		Veste Coburg[e] Hall/Innsbruck		Veste Coburg[e] Spain I		Veste Coburg[e] Spain II	
			Mean	Range	Mean	Range	Mean	Range	Mean	Range	Mean	Range
Na_2O	13.70		15.71	14.68–17.02	13.15	11.83–14.19	12.70	12.00–13.20	11.70	10.60–11.90	10.29	9.50–11.60
MgO	1.00		2.90	1.98–3.77	2.99	2.87–3.13	3.10	1.25–4.13	3.40	2.70–3.85	1.10	0.91–1.75
Al_2O_3	1.30		1.38	1.25–1.54	1.51	1.39–1.57	1.00	0.76–1.67	1.70	1.01–2.41	1.10	0.74–2.51
SiO_2	63.90		63.31	59.99–65.88	62.74	62.09–63.44	68.00	65.00–79.50	67.40	65.50–69.20	67.30	63.40–69.30
P_2O_5	–		0.30	0.20–0.35	0.31	0.22–0.43	0.46	0.29–0.61	0.60	0.53–0.87	1.00	0.83–1.20
SO_3	–		0.16	0.07–0.26	0.10	0.07–0.15	0.34	0.26–0.42	0.31	0.21–0.36	0.23	0.16–0.31
Cl	1.00		0.69	0.57–0.82	0.57	0.52–0.62	–	–	–	–	–	–
K_2O	5.50		4.79	2.78–6.28	7.35	6.83–8.31	3.00	2.21–5.86	3.50	3.10–6.75	6.90	4.92–9.02
CaO	10.60		9.81	8.27–11.33	10.42	9.69–11.40	10.20	5.39–12.46	9.80	7.97–12.45	9.80	8.91–12.72
MnO	1.00		0.38	0.16–0.72	0.28	0.23–0.35	0.46	0.17–0.74	0.60	0.46–0.96	1.40	0.72–1.61
Fe_2O_3	0.85		0.40	0.29–0.53	0.42	0.37–0.48	0.43	0.30–0.63	0.70	0.43–0.96	0.60	0.48–0.65

[a–e] See refs. [24], [27], [28], [29] and [30] resp.

tion); within this series, a distinction can be made between a group containing about 14% Na_2O, resembling *Vitrum Blachum* and a few objects containing up to 19% Na_2O. Closer to Antwerp, in Grez-Doiceau, Belgium a small oil/vinegar bottle dating to the late 16th century was found, made 'à-la-façon-de-Venise' and decorated with filigrain [28]. Ultizka analysed 15th to 19th century Venetian and 'façon-de-Venise' glass vessels from the collection of the '*Veste Coburg*', Austria [29]. He distinguished three compositional groups, two of Spanish origin and one with objects from Hall and Innsbruk, Austria. A latter group shows an average composition resembling that of *Vitrum blanchum.* It is historically proven that the Venetian glassmakers in Innsbruck were allowed to bring along or import their raw materials from the North of Italy. The 'façon-de-Venise' objects of Spanish origin probably were made from glass manufactured using ash from local saline plants as the sodium source; these are reputed to contain much less Na than their counterparts growing in the Middle East ([12], p. 493). This might explain the relatively low Na_2O abundance in these vessels (10–12%) and the higher K_2O level (7%).

The two Antwerp 'façon-de-Venise' groups (FdV-I and FdV-II) show no immediate resemblance to any of the comparison groups. Although the FdV-I and II groups feature a Na_2O content which is bracketed by the Na_2O ranges or respectively 'Cristallo' and 'Vitrum Blanchum', the abundance of K_2O in both groups (and of CaO in the case of FdV-I) is too high relative to that of the corresponding Venetian groups. The K_2O levels found for the two Antwerp groups (5–7%) are also quite high relative to the other groups; approximately the same level is found in the object from Grez-Doiceau and in the group Spain-II of Ulitzka. This could be an indication that the local glass production in Antwerp was making use of soda of variable Na/K content or of a mixture of two distinct sources of soda, one with high Na_2O and low K_2O content (maybe of Levantine origin, as used in Murano) and one of lower Na_2O and higher K_2O abundance (possibly Spanish in origin). The use of this kind of alkali source, in a fixed proportion to the amount of sand and chalk used, might explain the negative Na_2O/K_2O correlation shown in Fig. 2. In contrast to the FdV groups, the VaF-I group composition appears to resemble that of the *Vitrum Blanchum* glass quite well; it also is similar to the first group of goblets from the Lido of Venice. Considering the complex shape of these objects, it may be that they are the only ones actually imported as finished products from Venice within the analysed series.

Conclusions

The chemical analysis of 96 15th to 17th century glass objects excavated in Antwerp revealed the existence of three glass compositions. The first group covers glass from the late 15th until the early 17th centry; maigeleins, berkemeiers, prunted beakers and roemers. All these glasses are characterised by a high content of calcium oxide. There is a distinction between these typological groups when the minor elements and the amount of CaO are considered. It should be noticed that the roemers, although typologically evolving from the berkemeiers, have a similar major and trace composition as the maigeleins and differ from the berkemeiers and prunted beakers. The second group are sodic glass objects. This type includes all 16th century objects, i.e., the ribbed beakers, the beakers on folded foot and the 'façon-de-Venise' and filigrain glass vessels. The first two typological groups feature similar major as well as trace compositions. The 'façon-de-Venise' objects can be further subdivided in terms of their major constituents, but no such distinction is revealed in their trace composition. Among these soda glass objects, only a few filigrain glass objects have a composition resembling that of the Venetian 'Vitrum Blachum' glass. A third group consists of all 17th century objects (a meshwork beaker, stackable beakers and goblets 'à-la-façon-de-Venise'). All these objects have a similar major and trace composition and are made in potassium-rich glass decolourised with Sb, although in the literature, they are expected to consist of soda glass. The abrupt changeover from the sodic glass of the 16th century to the potassic glass of the 17th century objects may be connected with the blockade of the River Scheldt at the end of the 16th century or may be the result of the introduction of the use of Sb as a decolourizing agent for glass made from fern ash.

Special emphasis was placed on the 16th century 'façon-de-Venise' objects. The central question in this respect is whether there was local production of these kind of glass vessels in Antwerp and how this local production can be distinguished from 'real' Venetial wares. The analysis results of the 'façon-de-Venise' group of vessels were therefore compared with compositional data on glass of well-known origin. It appears that, except for three filigrain objects, the Antwerp glass vessels show no resemblance to the other groups. On the other hand, the filigrain objects have a composition similar to *vitrum blanchum*, a typical product from Venice. In order to obtain a clearer insight into the significance of these differences and to establish with a higher degree of confidence whether most of the 'façon-de-Venise' objects show a typically 'Antwerp' kind of soda-glass composition, the extension of the present programme of analyses is planned. On the one hand, this extension will involve the analysis of a larger number of 'façon-de-Venise' glass objects excavated in Antwerp, but on the other hand will also involve comparison of the 'Antwerp' soda glass composition with that found at other sites in Western Europe.

Acknowledgements. The authors would like to thank H. Wouters and Ch. Fontaine for providing them with the CMG standards and to L. Vincze for helping with the quantification of the XRF data; the authors are also indebted to H. Henkes for the interesting discussions on glass provenance and history in the Southern and Northern Netherlands during the 15–17th century. The SRXRF measurements were performed under Hasylab project nr. II-95–05 EC and with the help of M. Haller. M. Radkte and A. Knöchel (University of Hamburg). We also would like to thank Mrs. H. Bronk of the TU Berlin for letting us reproduce her analysis results on the glass from the Lido of Venice in Table 5.

References

[1] S. Denissen, *III: Het Glas, Bulletin* van de *Antwerpse Vereniging voor Bodem- en Grotonderzoek*, **1984**, *5*, 113.
[2] J. Veeckman, *Berichten en Rapporten over het Antwerps Bodemonderzoek en Monumentenzorg 1*, 1996, pp. 11–37.
[3] S. J. B. Reed, *Electron Microprobe Analysis, 2nd Ed.*, Cambridge University Press, Cambridge, 1993.
[4] K. Janssens, L. Vincze, B. Vekemans, A. Aerts, F. Adams, K. W. Jones, A. Knöchel, *Microchim. Acta [Suppl.]*, **1996**, *13*, 87.
[5] O. Schalm, K. Janssens, F. Adams, *Proc. EMAS'97, Torquay, UK, May 11–15*, 1997, p, 378 (Abstract).
[6] R. H. Brill, in: *Excavations at Jalame, Site of a Glass Factory in Late Roman Palestine*, University of Missouri Press, 1988, Chapter 8.
[7] M. C. Calvi, M. Tornati, M. L. Scandellari, in: *I Vetri Romani del Museo de Acquileia*, 1968; P. Mirti, A. Casoli, L. Appolonia, in: *Archaeometry* **1993**, *25*, 225.
[8] C. M. Jackson, J. R. Hunter, S. E. Warren, H. E. M. Cool, in: *Archaeometry '90*, Birkhauser, Basel, 1991, pp. 295–304.
[9] L. Vincze, *PhD Dissertation*, University of Anwerp, .
[10] S. Dennissen, in: *Het glas in Belgie van de oorsprong tot heden* (L. Engen, ed.), Antwerpen, 1989.
[11] H. E. Henkes, *Rotterdam Papers* 9, 1994.
[12] E. Ashtor, G. Cevidalli, Levantine Alkali Ashes, European Industries, *J. Eur. Hist.* **1993**, *12*, 475.
[13] E. Baumgartner, I. Krueger, *Phoenix aus Sand und Asche, Glas des Mittelalters*, München, 1988.
[14] S. Denissen, *Van nederzetting tot Metropool: Archeologisch-historisch onderzoek in de Antwerpse binnenstad, Exhibition Catalogue*, Antwerp, 1982, pp. 113–181.
[15] E. Baumgartner, I. Krueger, *Bonner Jahrbücher*, **1985**, *185*, 363.
[16] R. Chambon, *L'histoire de la verrerie en Belgique du IIème siècle à nos jours*, Bruxelles, 1955, p. 65 and note on p. 307.
[17] S. Denissen, *Bulletin van de Antwerpse Vereniging voor Bodem- en Grotonderzoek* **1985**, *5*, 9.
[18] K. P. Jochum, A. W. Hoffmann, M. Haller, M. Radtke, A. Knöchel, L. Vincze, K. Jansens, *Hasylab Jahresbericht, Vol. II*, Hamburg, Germany, 1995, pp. 1003–1004.
[19] J. Barrera, B. Velde, *Archéologie Médiévale*, **1989**, *XIX*, 81; J. Barrera, B. Velde, *J. Glas Studies* **1989**, *31*, 48.
[20] A. Golebiewski, *Zeitschrift für Archäeologie des Mitellalters* **1993**, *21*, 107.
[21] K. H. Wedepohl, *Die Herstellung mittelalterlicher und antiken Gläser*, Steiner, Stuttgart, 1993.
[22] P. Klingenfus, *Verrerie de l'Est de la France, XIIIe–XVIIIe siècles*, Dijon, 1990, pp. 93–103.
[23] E. Ashtor, in: *Saggi e Documenti* I, Civico Instituto Colombiano, Genova, 1987, p. 432.
[24] M. Verita, T. Toninato, *Riv. della Staz. Sper. Vetro* **1990**, *4*, 169.
[25] E. V. Sayre, *Science* **1961**, 1824–1826.
[26] M. Verità, *Rivista della Stazione Sperimentale del Vetro* **1985**, *1*, 17.
[27] H. R. Brill, *J. Glass Studies* **1973**, *15*, 93.
[28] Ch. Fontaine, *VIIIème Journées des Conservateurs-Restaurateures en Archéologie*, Bordeaux, 1992, pp. 35–42.
[29] S. Ulitzka, in: *Venezianisches Glas der Kunstsammlungen der Veste Coburg* (A.-E. Theuerkauff-Liederwald, ed.), Luca, Lingen, 1994.
[30] H. Bronk, personal communication.

Mikrochim. Acta [Suppl.] 15, 269–272 (1998)

Potassium Migration in Silica Glass During Electron Beam Irradiation

Karel Jurek[1,*], **Ondrej Gedeon**[2], and **Václav Hulínský**[2]

[1] Institute of Physics AV CR, Cukrovarnická 10, 162 53 Praha 6, Czech Republic
[2] Institute of Chemical Technology, Technická 5, 166 28 Praha 6, Czech Republic

Abstract. Potassium glass was irradiated by 10, 30 and 50 keV electron beams of various diameters and the time decay of potassium x-radiation was measured. These measurements were complemented by SEM observation of the etched cross sections of exposed areas and potassium depth concentration profiles in order to observe structural changes in the glass. Whilst the temperature increase remained low, the length of the incubation period was inversely proportional to the specimen current density and a possible explanation of this phenomenon was suggested.

Key words: Potassium migration, silica glass, electron beam, x-ray microanalysis.

It has been already proved [1] that if silica glass is exposed to an electron beam, alkaline ions migrate from the grounded surface into the specimen, driven by an electric field created by the negative charge injection [2]. If the charged volume is of a cylindrical shape and a uniform distribution of this charge is assumed [3], the electric field is maximum at the surface and tends linearly to zero at a depth approximately equivalent to the electron range. Alkali ions accumulate at about half or two thirds of this electron range probably in the form of aggregates of various size (10–200 nm) [1]. In spite of the fact that the electric field reaches its stationary value in microseconds [4, 5] and the localised specimen temperature increase occurs in few seconds [6, 7], the time dependence of the alkaline elements x-ray intensity shows some delay, called the incubation period, before it starts to decrease [8].

* To whom correspondence should be addressed

Experimental

Potassium glass was chosen because of the relatively low absorption of the potassium x-radiation in the glass (about 40 percent of its intensity is emitted from a depth of 10 μm). The composition of the glass investigated was (weight percentage): 79.5 SiO_2, 8.5 CaO, 12 K_2O. The block of polished glass was coated by a layer of carbon approximately 30 nm thick and irradiated by electron beams of 50, 100, 200 and 400 μm diameter. The broad electron beam was used to simplify the geometry of the experiment and to facilitate the preparation of cross sections. The energies of the primary electrons were 10, 30 and 50 keV and the specimen current was in the range of 15 nA to 4000 nA. Cross sectioned glass was either coated by carbon for EPMA or etched by 1% HF for about 20 s and then coated by gold for SEM. (For more details see [1]). All observations and measurements were performed with a JXA-733 X-ray microanalyser of JEOL to which was attached a DELTA V ED spectrometer of KEVEX (Fisons). The high x-ray intensities were measured with WD spectrometers using the K_β line to prevent over saturation of the counting electronics, while concentration profiles were measured with an ED spectrometer under moderate beam conditions (15 keV, 2 nA).

Results

The time dependency of potassium x-ray intensity variation can be divided into two typical groups as regards the shape of the intensity versus time curves. Representatives of both groups are presented on Figs. 1 and 2, respectively. Fig. 1 is typical of a low temperature increase occurring in experiments performed at 50 and 30 keV using relatively low beam currents, i.e. the intensity began to decrease rapidly after a fixed period of time during which it remained constant. SEM photomicrographs of exposed areas and corresponding depth concentration profiles of potassium taken at specific time intervals are also shown in Fig. 1. After 30 s exposure nothing could be seen in the SEM image but the concentration profile shows a slight decrease in the potassium concentration

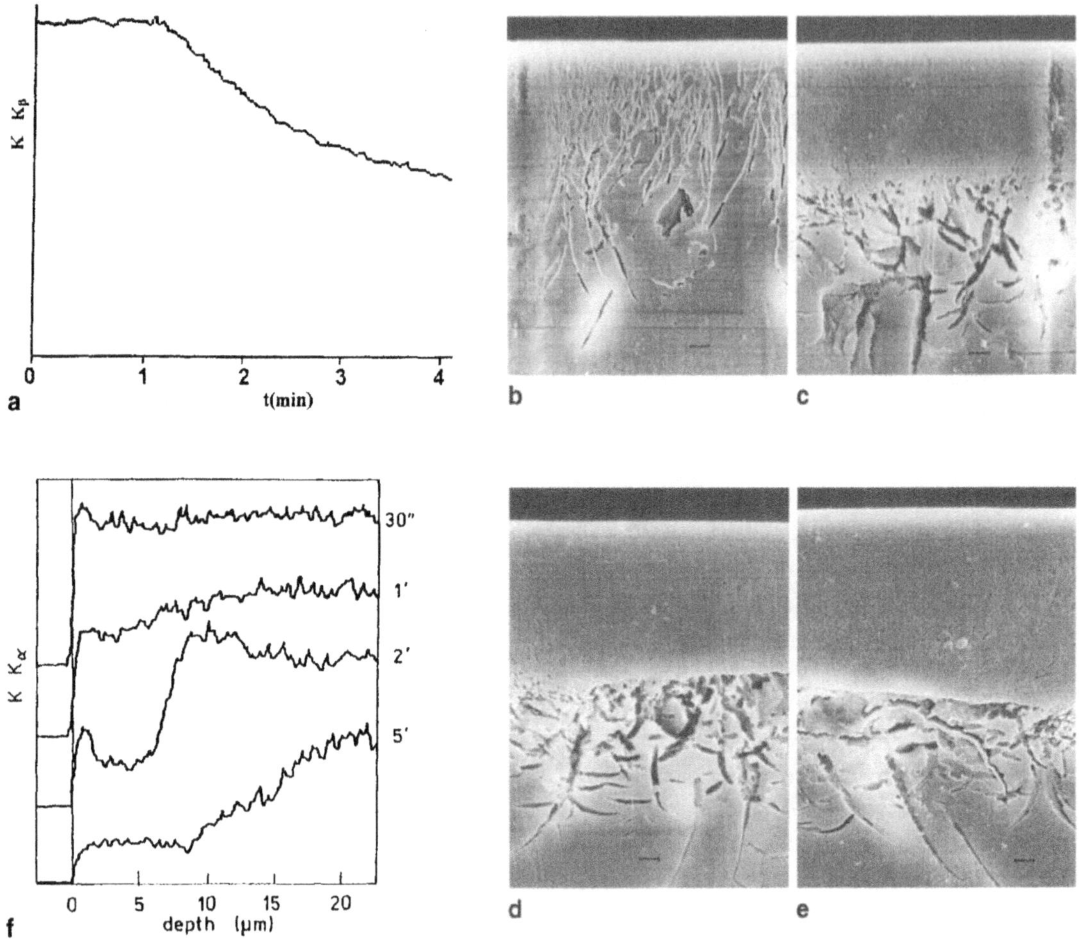

Fig. 1. a Time dependence of the intensity of K_β line. Beam diameter 50 μm, $E_0 = 50$ keV, specimen current 252 nA. Current density 12.8 mA/cm^2, $T_{max} = 385$ K. **b** SEM of cross-section after 70 s exposure, **c** 2 min. exposure, **d** 3 min. exposure, **e** 5 min. exposure, **f** corresponding depth concentration line profiles labelled with appropriate exposure times. The bar on photomicrographs is 1 μm. (trace after line profile measurement can be seen on **b** and **c**

at depths between 2 and 7 μm. After a 1 minute exposure, a number of "channels" oriented approximately perpendicularly to the surface could be observed and the concentration profile already began to exhibit a significant deficiency of potassium near to the surface. After 2 minutes exposure the density of the channels rapidly increased and large craters occurred deeper in the specimen. The concentration profile showed a large depletion of potassium at the surface, followed by an increase to a depth of about 10 μm. After 3 minutes exposure a rather compact surface layer about 7 μm thick was created, followed by a heavily damaged area of glass. The situation did not change much after 5 minutes, except that the depleted layer became thicker (nearly 10 μm) and had a lower concentration of potassium.

At higher specimen currents (Fig. 2) the intensity of K_β line of potassium is shown to decrease slowly immediately following the onset of the irradiation and after about 5 minutes begins to decay rapidly. A small decrease of potassium concentration under the surface could also be seen even in the early stage, after 2 minutes exposure. The corresponding SEM image shows short (1 or 2 μm) channels. These channels lengthened after 3 minutes, their density increased and the decrease in potassium became more distinct. After 5 minutes exposure the first stage of the surface layer and the large holes under it are revealed. The concentration profiles show significant depletion in potassium in the surface up to about 9 μm, but a peak of higher concentration is located very close to the surface. A compact, potassium depleted surface layer

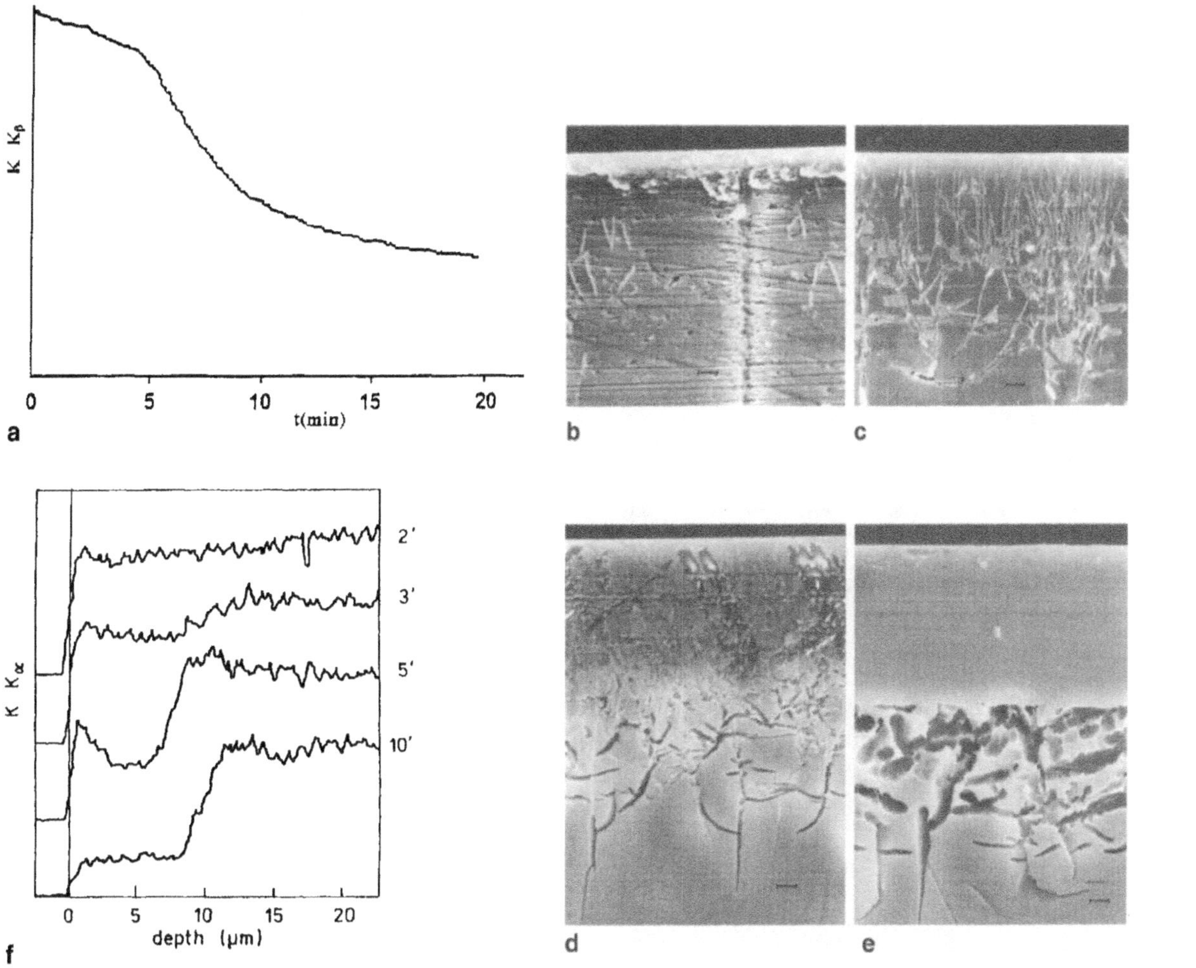

Fig. 2. a Time dependence of the intensity of K_{β} line. Beam diameter 200 μm, $E_0 = 50$ keV, specimen current 1070 nA. Current density 3.4 mA/cm^2, $T_{max} = 422$ K. **b** SEM of cross-section after 2 min. exposure (the trace after line profile measurement can also be seen), **c** 3 min. exposure, **d** 5 min. exposure, **e** 10 min. exposure, **f** corresponding depth concentration line profiles labelled with appropriate exposure times. The bar on photomicrographs is 1 μm

about 8 μm thick has been created after 10 minutes exposure. This layer is sharply separated from the heavily etched lower area. It was impossible, however, to determine the true concentration of potassium in this area because the small aggregates of metallic potassium or glass particles with extremely high content of potassium were completely dissolved during the etching (in some cases previously, during the polishing) so that only empty holes remained.

At 10 keV, on the other hand, the incubation period, if any, was very short, and due to the rather slow decay rate, it was sometimes hard to determine.

To establish the dependence of the incubation period on the absorbed current density, the incubation period was measured under various conditions and the results are presented in Fig. 3.

Discussion

The existence of the incubation period was proved and its length was measured. The formation of the individual channels, starting with some of very short in length, seems to be especially important. The results suggest, that the migration produced by the electric field, starts creating random individual channels and that this can be a rather slow process. After the number of channels increases they can interconnect making the diffusion easier. The diffusion mean free path of this process depends on the intensity of the electric field, as calculated in [9]. The penetration of primary electrons is quite small at low energies (at 10 keV about 1.5 μm and in Auger spectroscopy even shorter), so that the diffusion path

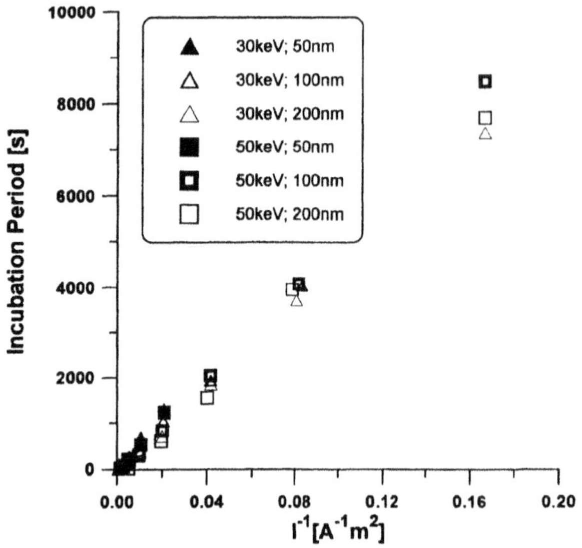

Fig. 3. Dependence of the incubation period on the inverse of the absorbed current density

length may exceed it. In this case the interconnection of the diffusion channels cannot influence the transport and an incubation period cannot be observed. Thus the absence of an incubation period, as reported in [10], can be explained.

Figure 3 shows that the incubation period is inversely proportional to the specimen current density (i.e. to the number of incident electrons) and at high currents becomes shorter. This can be caused by the temperature increase, which stimulates electrodiffusion (see [11]). The temperature increase depends strongly on the shape of the heated volume – in our case on the ratio of the height of the cylinder to the diameter of its base [6], which is especially unfavourable at low primary electron energies. The instant intensity decrease in Fig. 2 could be explained by this temperature increase. Maximum temperatures in our experiments were calculated according to [12].

It can be deduced from the time dependence of the concentration profiles, that the electric-field-stimulated diffusion begins at some depth beneath the surface of the specimen, where the intensity of the electric field is still strong and the number of ionisations and therefore the bonding breakdown is the highest (i.e. where the ionisation function is a maximum – see [9]). This idea is supported by the creation of the valley in the concentration profile, i.e. ions at some depth begin to migrate first, creating small diffusion channels and thus enabling transport of other ions. At this moment practically no change in the generated x-ray intensity can be observed. When the number of channels increases, so that they interconnect, a way for massive ion transport is open until the increase in the chemical potential stops this diffusion.

Conclusion

The existence of the incubation period was experimentally proved. It was shown that at low temperatures it is inversely proportional to the absorbed current density. An explanation of this phenomenon was suggested.

References

[1] K. Jurek, V. Hulínský, O. Gedeon, *Mikrochim. Acta* **1996** [*Suppl.*] *13*, 339.
[2] A. Miotello, P. Mazzoldi, *J. Phys. C* **1982**, *15*, 5612.
[3] J. Cazaux, *J. Appl. Phys.* **1986**, *59*, 1418.
[4] A. Miotello, *Phys. Lett.* **1984**, *103A*, 279.
[5] A. Miotello, P. Mazzoldi, *Phys. Rev. Lett.* **1985**, *54*, 1615.
[6] H. J. Dudek, *Z. Angew. Phys.* **1971**, *31*, 331.
[7] K. Závěta, K. Jurek, V. Kamberský, P. Král, *J. Magn. Magn. Mater.* **1990**, *86*, 254.
[8] L. F. Vassamilet, V. E. Caldwell, *J. Appl. Phys.* **1969**, *40*, 1637.
[9] O. Gedeon, V. Hulínský, K. Jurek, *Mikrochim. Acta* **1998** [*Suppl.*] *15*, 301.
[10] O. Jbara, J. Cazaux, P. Trebbia, *J. Appl. Phys.* **1995**, *78*, 868.
[11] P. Regnier, Y. Serruys, A. Zemskoff, *Phys. Chem. Glasses* **1986**, *27*, 185.
[12] V. Hulínský, K. Jurek, O. Gedeon, *Microchim. Acta* **1996**, [*Suppl.*] *13*, 325.

Mikrochim. Acta [Suppl.] 15, 273–282 (1998)

X-Ray Microanalysis of Frozen-Hydrated Biological Bulk Samples

Alan T. Marshall

Analytical Electron Microscopy Laboratory, School of Zoology, La Trobe University, Bundoora, Victoria 3083, Australia

Abstract. The x-ray microanalysis of biological samples is best carried out in the frozen-hydrated state. Frozen bulk samples are the most stable form and the most resistant to radiation damage. A 20 nm thick metal coat appears to prevent the development of an internal space charge. A fracture surface through this type of sample reveals morphological detail but requires careful selection of the analysis sites, using backscattered electron images, for quantitative analysis. If the sample is planed at low temperature to provide a polished surface no morphological detail is visible in electron images. This disadvantage may be overcome by subliming (etching) the ice matrix from the sample surface to reveal the embedded organic structures. Such a process, however, must necessarily compromise quantitative analyses. It is shown that quantitative elemental x-ray imaging, and the selection of analysis sites by the rapid acquisition of qualitative oxygen and carbon x-ray images to reveal morphological detail, is a feasible alternative. This method does not require etching of the sample and electron images are not needed. X-ray images are processed using phi-rho-zed matrix corrections to yield concentration images. Analyses of model systems of albumin and salts show that x-ray images produce accurate quantitative results. Quantitative oxygen images can be used to estimate cell water concentrations.

Key words: x-ray microanalysis, x-ray imaging, biological frozen-hydrated, phi-rho-zed model, insect Malpighian tubules.

Biological samples consist of a number of compartments, or phases, varying in size from mm to nm. The composition of these compartments is very labile and subject to change when a sample is removed from an intact living organism. This is because living organisms are collections of interacting compartments which depend on steady state processes for their structural and compositional integrity. The size of a compartment to be analysed by x-ray microanalysis depends on the nature of the physiological question being addressed and on the type of sample preparation and the instruments available. In the case of bulk biological samples the cell and the major subcellular divisions of nucleus and cytoplasm are the minimum size of compartment that can be analysed in a conventional scanning electron microscope (SEM). There may be a possibility of analysing rather smaller compartments in a field emission SEM (FESEM) at low voltage but the difficulties are considerable.

Cells consist largely of H, C, N and O. Typically 94% of the oxygen present in a cell is a component of water. The other light elements are components of organic compounds, primarily proteins. Oxygen is the most interesting of the light elements, from an analytical point of view, as it may be a possible indicator of water content. Differences in, or gradients of, oxygen concentration in a biological sample will indicate differences in, or gradients of, water concentration since the average organic oxygen content probably varies little in cells unless considerable lipid is present in a particular region. Sulphur and phosphorus are present in much lower concentrations and are probably associated largely with organic compounds although some inorganic phosphates will be present. The elements of physiological interest that are present in quantitites measurable by x-ray microanalysis include Ca and Mg which are largely associated with organic compounds and Na, K and Cl which are generally considered to be mostly present as ions in cell water. Water itself is an important physiological variable and it is therefore desirable to know both the concentration of ions in water in a compartment and the concentration of water, since changes in ion concentration can be a consequence of ion shifts or water shifts or both. If it is not possible to determine water concentration then elemental concentration in terms of wet mass is much more meaningful than concentrations in terms of dry mass. For this reason the analysis of hydrated samples of biological material is preferable to the analysis of dried samples. The concentrations of the "physiological" elements in a hydrated sample are extremely low and for x-ray microanalysis qualify as trace elements. Sodium concentration in an animal cell would be typically about 0.023% whilst potassium would be about 0.39%.

Sample Preparation

Freezing

Since the composition of biological samples is labile the samples must be rapidly stabilised by freezing when removed from the intact organism. Freezing may also be carried out in some cases prior to removal from the organism [1]. An intermediate step may involve incubating the sample in a physiological medium, in which a new steady state is achieved, prior to stabilisation of the sample. Stabilisation involves rapid freezing. A variety of devices are available for this purpose [2] but all have limitations in terms of the depth of sample from the surface at which water can be transformed into amorphous ice. This depth varies from 10–20 μm in the case of the majority of devices to possibly 150–600 μm in the case of high pressure freezers [3]. Below this depth hexagonal ice is formed with crystallites of increasing size with increasing depth. For the analysis of bulk samples the ice crystals can be in the submicron range, since the excitation volume will be 4–5 μm^3, without compromising the analysis.

Fracturing and Planing

The frozen samples may be treated in different ways:

a) Samples can be fractured in vacuum at low temperature (100 K) and metal coated [4, 5] to reveal surface morphology.
b) After freezing the samples can be planed in a cryo-ultramicrotome at 153 K, cryo transferred to a low temperature coating device where they are etched and metal coated. Alternatively the etching process, i.e. sublimation of surface ice, is accomplished by raising the temperature of the sample in a microscope column, whilst viewing the sublimation process, prior to coating [6, 7].

Coating

Since ice is an excellent insulator and has a very low conductivity below 203 K [8] frozen-hydrated samples must be coated to prevent charging of the sample. A carbon coating is inadequate for preventing charge build up in frozen-hydrated samples but metal coatings of Be, Al or Cr may be used. Beryllium coating requires care owing to the highly toxic nature of this metal and evaporation devices should be modified accordingly [9]. The thermal and electrical conductivities of Be at low temperature are superior to those of Al and Cr (Table 1). Chromium absorbs Na and Mg radiations strongly (Table 2) and has an L line coinciding with the OKα line which will interfere with O analysis. Generally a 20 nm coat of Al is adequate to prevent charging. The work of Fuchs et al. [10] on uncoated frozen-hydrated samples showed that an internal space charge developed which resulted in a pancake-shaped electron interaction volume below the surface of the sample. However, with a highly conductive metal coat charge storage in the sample

Table 1. Thermal and electrical conductivities of coating metals at 100 K

	Resistivity (μOhm cm)	Thermal conductivity ($W cm^{-1} K^{-1}$)
Be	0.09	10
Al	0.47	3
Cr	1.6	1.5

Table 2. Percentage of x-ray absorption by metal coats, assuming an x-ray emergence angle of 40° to the plane of a 20 nm-thick coat

Emitter	Absorber		
	Be	Al	Cr
O	2.23	5.49	6.78
Na	0.30	0.89	13.97
Mg	0.17	0.54	9.07
Cl	0.02	0.51	1.34
K	0.01	0.49	0.75

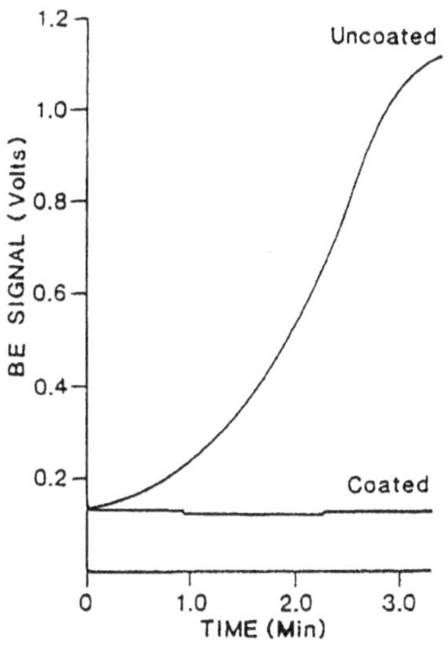

Fig. 1. Amplified backscattered electron signal at constant beam current from coated and uncoated frozen-hydrated gelatine gels (from Marshall and Condron, 1985)

may be negligible at the beam currents used for biological analyses (10^{-10}–10^{-9}A). A number of lines of evidence support this [11], a) the continuum cut off or Duane-Hunt limit is equal to the accelerating voltage, b) the backscattered electron yield remains constant in a coated sample whereas it increases dramatically in uncoated samples (Fig. 1), c) the measured analytical depth resolutions in frozen-hydrated samples match those expected from $\phi(\rho z)$ depth distribution curve calculations and the measured lateral resolution is not consistent with a pancake-shaped interaction volume.

Quantitation

Quantitative x-ray microanalysis of biological bulk samples can be carried out using either matching samples, which are constituted from an organic compound such as protein together with various salts, or microprobe standards of well defined stoichiometry. The latter have the advantages that they can be polished and they are stable. If non-matching standards are used then matrix corrections become necessary to account for the effects of differences in composition of sample and standard on x-ray generation and absorption. These are the so-called ZAF (atomic number, absorption and fluorescence) corrections.

Matching Standards

The quantitation of biological bulk samples has been largely based on the use of matching standards e.g. [12]. This has been done to overcome the problem of defining the composition of the light element matrix. A matching standard should obviate the necessity of applying any matrix corrections. There are major problems with matching standards for frozen-hydrated biological samples in that the composition is difficult to control precisely and they are labile. Nevertheless, they continue to be used e.g. [7, 13].

Peak to Background Ratios

The peak to local background model has been used successfully for the analysis of fractured frozen-hydrated samples which have an irregular surface [14, 15]. Since the absorption of characteristic radiation and continuum radiation in the same energy band is very similar, the use of the P/B ratio diminishes the effects of sample topography on absorption. This principle seems to have first been applied to biological samples by Millner and Cobet [16]. Subsequently it was developed for particle analysis [17, 18] with the inclusion of matrix corrections. The concentration of an element is calculated as:

$$C_{sp} = \frac{P/B_{sp}}{P/B_{st}} \cdot \frac{ZAF_{st}}{ZAF_{sp}} \cdot \frac{G_{sp}}{G_{st}} \cdot C_{st}$$

Where *P/B* is peak to background ratio, *ZAF* are the matrix corrections, *G* is mean Z^2/A and the subscripts *sp* and *st* represent sample and standard respectively. The matrix composition must be "guesstimated" and light elements (C, N and O) can not be analysed because of the difficulty of estimating background under the peaks for these elements. It may also be the case that the assumption of approximate equivalence in generation and absorption of characteristic and continuum radiation does not hold at these low energies. It should be noted that a disadvantage of the *P/B* method is that it is dependent on the elimination of all non-sample sources of background radiation. Marshall and Condron [19] showed that unless changes in local tilt angle are large (> ±20° for K and >±10° for Na) there seems to be little advantage in using *P/B* ratios for biological samples since the effect on the detected x-ray counts is slight. It was also demonstrated that O intensities could be corrected for changes in local tilt angle by normalising with the backscattered electron signal.

Conventional ZAF

Attempts to apply conventional *ZAF* corrections to biological samples have met with little success in the past [20] (reviewed by Roomans [21]). This may be attributed to the limitations of commercially available software which has not contained algorithms appropriate for corrections in matrices consisting mostly of the light elements H, C, N and O (see Fig. 2). Although more recent versions of *ZAF* software may be satisfactory for higher atomic number elements in an organic matrix they may still fail with light element analysis. No recent attempts appear to have been made to apply modern formulations of *ZAF* corrections such as that of Farthing et al. [22].

ϕ (ρz) model

A promising approach is the $\phi(\rho z)$ model. The ionization function or $\phi(\rho z)$ curve describes the x-ray depth generation during electron excitation in a solid target. This function was originally modelled by two sets of empirical equations derived by Parobek and Brown [23] and Brown and Robinson [24] and subsequently modified [25]. Further changes have been carried out by Bastin et al. [26, 27] and a variety of other modifications and models exist [28]. The $\phi(\rho z)$ model has been developed for biological analyses by Marshall [29–32]. The model allows the absorption of generated x-rays by the matrix to be readily calculated and the combined atomic number and absorption correction obtained from the integrated value of the total emitted x-rays. This correction can be applied to x-ray intensity data to obtain elemental concentrations in biological bulk samples using inorganic standards.

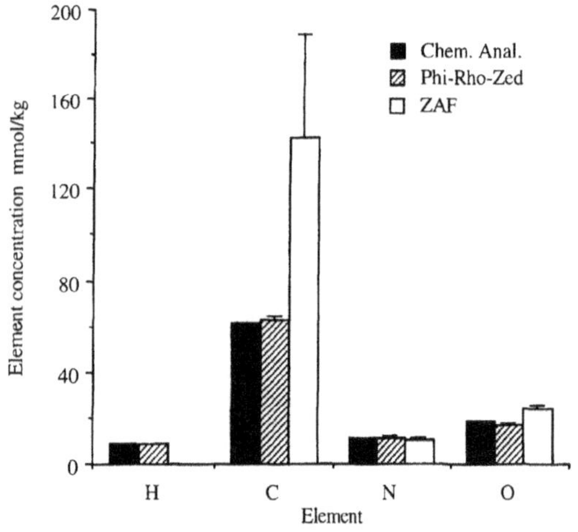

Fig. 2. Quantitative analysis of Nylon using the $\phi(\rho z)$ model and a commercial classical ZAF model

Table 3. Correction factors for elements in a protein/water matrix with varying % concentrations of protein

Element	% Emitted intensity 20%	10%	5%
Na	49.8	48.5	47.9
K	97.8	97.7	97.7
Cl	95.8	95.6	95.5
S	93.9	93.7	93.6
P	91.1	90.7	90.5

The element concentration in the sample is calculated from:

$$C_{sp} = \frac{I_{sp}}{I_{st}} \cdot \frac{\int_0^\infty \phi(\rho z)\exp(-\mu\rho z \operatorname{cosec}\theta)d\rho z_{st}}{\int_0^\infty \phi(\rho z)\exp(-\mu\rho z \operatorname{cosec}\theta)d\rho z_{sp}} \cdot C_{st}$$

where C is the concentration of the sample (*sp*) and standard (*st*), I is the measured x-ray intensity, $\phi(\rho z)$ is the depth distribution of characterstic x-ray production, ρ is density, z is depth, θ is the take off angle and μ is the mass absorption coefficient. The organic matrix of a biological sample can be modelled by a solution of a generalised protein in water. Variations in protein concentration over a range which would exceed that usually occuring in biological samples has virtually no effect on the $\phi(\rho z)$ corrections for the usual elements of interest ($Z = 11$–20). Some improvement can be obtained for Na and Mg if the variation in percentage organic matrix present is considered but even for these elements the changes in the corrections are small (Table 3).

If an ultra thin window EDS detector is used then light elements (C, N, O) can be detected and quantitatively analysed using the $\phi(\rho z)$ model (Fig. 2) [33, 34]. This provides the possibility of determining O concentrations in frozen-hydrated bulk samples and subsequently either relative or absolute water concentrations [34].

Analytical Resolution

The depth resolution for the elements of interest in frozen-hydrated biological samples has been calculated from $\phi(\rho z)$ curves [29]. At 15 kV the depth resolution is close to 2 μm for all elements, including the light elements. A number of experiments have substantiated this finding [11]. Using a modified sandwich technique the depth resolution of Br L and Ag L x-rays was measured to be 1.5–2 μm in frozen-hydrated photographic film (30% protein and 70% water) which approximates a biological sample. Calculated depth resolution for Br $L\alpha$ emission was 2.4 μm. Frozen-hydrated sections of 16% gelatin, cut at various thicknesses and placed on aluminium grids, were used to measure the depth resolution for Al $K\alpha$ x-rays and gave similar results to Br. Lateral resolution was also measured in frozen-hydrated photographic film and was approximately 2 μm.

Selection of Analysis Sites

If a frozen-hydrated biological sample is fractured then the topography of the fracture surface reveals morphological detail in the sample. However, an uneven surface can present problems for analysis and the nature of the irregular surface is not readily discernable in a secondary electron image (Fig. 3a). The surface topography can be readily interpreted in a backscattered electron image (Fig. 3b) obtained from an annular backscattered electron detector attached to the x-ray detector [35]. Sites can be selected which have a clear line of sight to the x-ray detector and which have minimal tilt.

If the frozen-hydrated sample is planed then the image is featureless in both secondary and back-scattered electron images. The strategy that has been most commonly adopted to render morphological detail visible in such samples is to induce sublimation of surface ice by raising the temperature of the sample whilst under vacuum. This process is generally referred to as etching. Recent examples of this method include Morgan et al. [13] and Canny [36]. This procedure must neccessarily compromise the analysis. Although Wyness et at. [12] maintained that there were no differences between etched and unetched standards of frozen gelatine solutions containing salts Marshall [35] showed both an increase in count rates for various elements, after etching similar samples, and an increase in the coefficient of variation for replicated analyses. The effects of etching will clearly depend on the depth of the etch but this is extremely difficult to control and is probably not reproducible [2]. It seems certain that the rate of sublimation of ice will also vary with the composition of the organic matrix.

A planed surface for the analysis of frozen-hydrated biological samples is highly desirable but etching the surface will compromise the analysis. A possible solution to the problem is to forgo the requirement for electron images and rely on x-ray images to reveal the morphology of the sample. A review of experiments investigating this approach [37] is described in the next section.

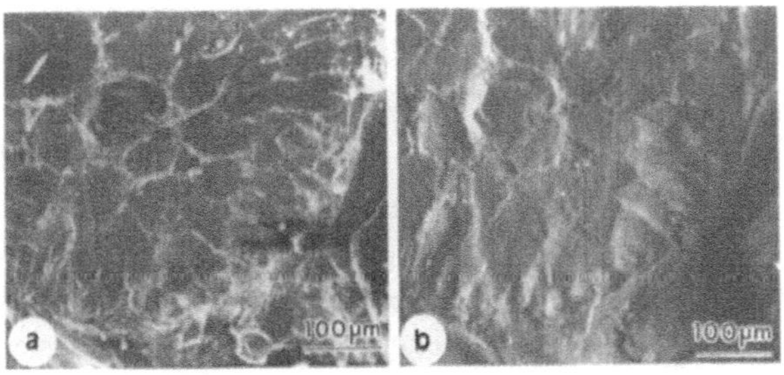

Fig. 3. a Secondary electron image from the fracture surface of a frozen-hydrated lachrymal salt gland from a sea turtle. **b** Back-scattered electron image from the same sample. The BE detector was mounted on the x-ray detector

Experimental

Sample Preparation

Samples of animal tissue and model test specimens of dialysed albumin solutions containing known concentrations of salts were frozen on aluminium pins for the Reichert FC4E cryo ultra-microtome. All samples were frozen by plunging into liquid propane at 83 K using a Reichert KF80 freezing device. The frozen samples were planed with a diamond trimming knife in a Reichert FC4E cryo-ultramicrotome at 163 K. Following planing the samples were cryo-transferred to a low temperature coating unit and coated with 20 nm Al at 83 K. After coating the samples were cryo-transferred to a Jeol JSM 840A low temperature scanning electron microscope and analysed at 100 K using a Link exL analyser with LZ4 turreted detector having an ultrathin window and a Be window.

Analytical Model

Quantitative data were obtained by applying matrix corrections derived from the Phi-rho-zed model to the x-ray intensities from samples and standards [29, 31, 32]. The quantitative model used in this investigation was that provided in the Link exL software which is based on the model developed by Bastin and colleagues [26, 27].

The standards used for elements $Z > 10$ were microprobe standards (Biorad). For C, N and O microprobe standards were used as primary standards for peak fitting but frozen-hydrated samples of 20% albumin (w/w) were used as secondary standards to obtain concentrations of these elements. Albumin samples were chemically analysed for C, N and O content. In applying matrix corrections it was assumed that the biological matrix is sufficiently well modelled by a protein solution and that non-analysed elements could be represented by the concentrations calculated for a protein solution i.e. H in the case of ultra thin window (UTW) analyses and H, C, N and O in the case of Be window analyses. For biological samples the composition of a generalised protein was used [38] and for albumin samples the composition as determined by chemical analysis was used.

Water concentration was derived from oxygen concentrations by reference to the linear relationship existing between them in solutions of protein in water [29].

X-Ray imaging

All analyses were made using an accelerating voltage of 15 kV. The Link x-ray imaging software for quantitative mapping calculates peak integrals, using multiple least squares fitting of library peaks to the spectrum for each pixel, and apparent concentrations i.e. concentrations which have not been corrected for matrix effects. Apparent concentration images were corrected by applying matrix corrections derived from static beam analyses of the same samples. Since the corrections vary very little with protein concentration or changes in elemental concentration over the normal biological range [29, 31] this is a reasonable procedure. Quantitative data were extracted from selected regions of images by means of the Link program Featurescan.

Results and Discussion

Validation of Quantitation by X-Ray Imaging

A series of samples of 20% dialysed albumin containing various salt concentrations was analysed for 100 s by a rapidly scanning raster over an area of approximately 66×66 µm at a beam current of 1×10^{-9}A. These are referred to as "static beam" analyses and each sample was analysed in three different regions. Similar areas on the same samples were analysed as x-ray images using the same beam current and a matrix of 32×32 pixels with a dwell time of 6 s. A comparison of the analyses of potassium and chlorine (Figs. 4 and 5) shows that the two analytical methods give identical results. Although the dwell time per pixel in the x-ray image is only 6s the mean element concentration for each sample is derived from 1024 pixels. These data indicate that averaging concentrations from pixels with small numbers of counts provides values comparable to those obtained from spectra accumulated over a much longer time and containing many more counts than the spectra from each pixel. This is consistent with the findings of Fiori et al. [39] and LeFurgey et al. [40] that small numbers of counts per pixel give valid quantitative data when averaged or summed over regions.

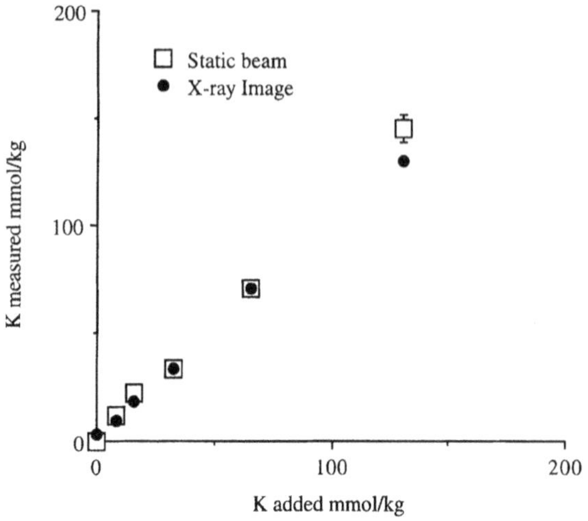

Fig. 4. Comparison of K concentrations in frozen-hydrated samples of 20% albumin measured by "static beam" analyses (Mean of 3 analyses per sample±SE) and by averaging pixel values in 32×32 pixel images

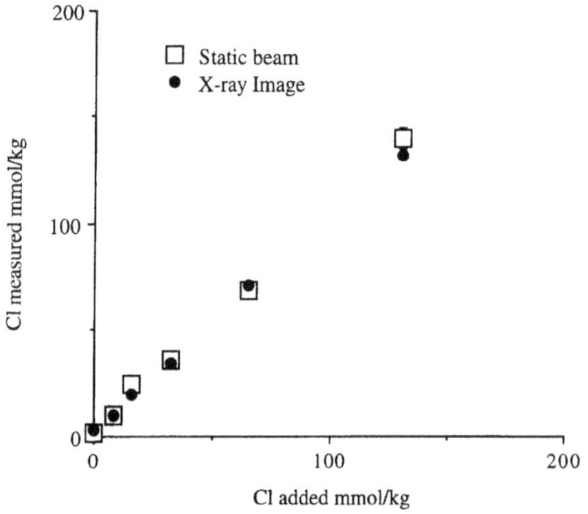

Fig. 5. Comparison of Cl concentrations in frozen-hydrated samples of 20% albumin measured by "static beam" analyses (Mean of 3 analyses per sample ± SE) and by averaging pixel values in 32×32 pixel images

A similar analysis was carried out on frozen-hydrated albumin samples containing Na. Although the albumin was dialysed, flame emission spectrophotometry showed that a 20% solution contained 34 mmol kg^{-1} Na. The results from x-ray imaging are shown in Fig. 6 where measured Na is plotted against total Na. A regression line fitted to the data points fails to go through the origin and indicates that the minimum detection limit is approximately 11 mmol kg^{-1} wet weight. The Na image was processed by a simultaneously collected standard deviation image [41] so that all pixels in the Na image

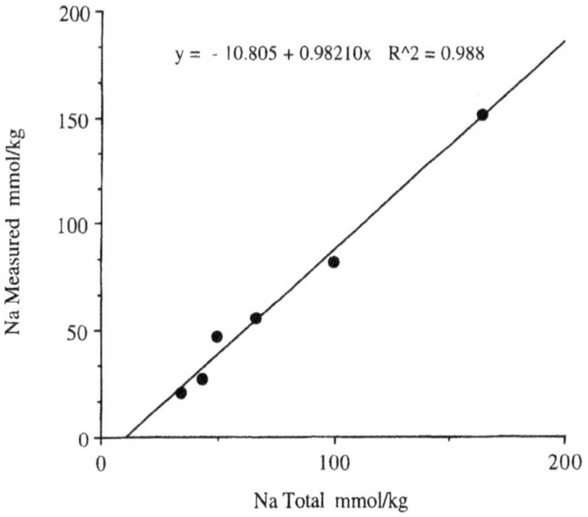

Fig. 6. Na concentrations in frozen-hydrated samples of 20% albumin measured by averaging pixel values in 32×32 pixel images

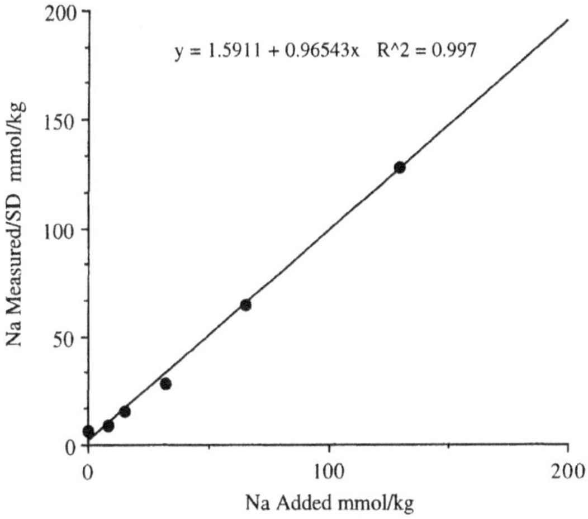

Fig. 7. Na concentrations in frozen-hydrated samples of 20% albumin measured by averaging pixel values in 32×32 pixel images after setting to zero all pixel values less than 1.5×SD

with counts less than 1.5 times the standard deviation were set to zero. When the measured concentration values are plotted against the concentrations of added Na, not total Na, an excellent agreement is obtained (Fig. 7). This indicates that the minimum detectable concentration is close to the amount of Na present in the dialysed albumin before addition of Na i.e. 34 mmol kg^{-1}. An important deduction with respect to the measurement of Na concentrations in frozen-hydrated bulk samples can be made. A precise estimation of Na concentrations is not possible if an

Table 4. Comparison of O analyses by "static beam" and x-ray images in frozen-hydrated samples of 20% albumin

Theoretical O%	Static beam O%	X-ray image O%
76.6	74.5±1	87.2±2.7

Mean±SE, n = 6.

x-ray image contains pixels with Na concentration values between zero and 34 mmol kg^{-1}. In this range true Na counts cannot be discriminated from statistical noise. The application of a standard deviation image to remove pixels of non-zero intensity which are not significant will inevitably lead to underestimates of Na concentration in the circumscribed region. Nevertheless, accurate relative estimates of Na concentration in samples are still possible. Improvement in Na detection in frozen-hydrated samples by use of an ultrathin window detector is not possible due to the pulse pile-up of O counts to form a sum peak at the position of the Na peak [34].

The analysis of oxygen in frozen-hydrated samples is not straightforward. Oxygen concentrations were measured in six frozen-hydrated samples of 20% albumin by static beam analyses using a rapidly scanning raster covering an area of 15625 μm^2 for 100 s. Three areas in each sample were analysed using a beam current of 1×10^{-10} A. An x-ray image was obtained from each of the same samples for a similar area at a resolution of 64×64 pixels and with the same beam current. The results (Table 4) show that the static beam analyses were more accurate than the x-ray images. After conversion of O concentration to water concentration static beam analyses underestimated the theoretical water concentration by 4% whilst x-ray images overestimated by 16%. A possible explanation of this discrepancy is that the considerably lower electron dose during x-ray imaging results in reduced mass loss or radiation damage. The problem may be overcome by the use of standards data obtained by x-ray imaging, i.e. the conditions of standards analysis must match those of analysis of unknown samples in this respect.

The application of a $\phi(\rho z)$ correction factor for oxygen to all pixels in an x-ray image produces inaccuracies when the dry mass (protein concentration) in different regions differs substantially from that at which the correction factors were measured. This is because the correction factor for oxygen varies markedly with protein concentration. This is illu-

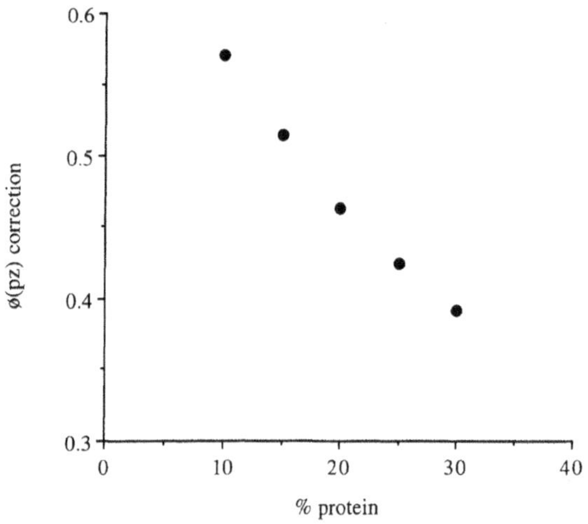

Fig. 8. Calculated change in $\phi(\rho z)$ correction factors at 15 kV for O with changes in protein concentration in solutions of protein in water

strated in Fig. 8 in which concentration factors are calculated using a $\phi(\rho z)$ model based on the equations of Packwood [42].

A difference of ±5% in protein or dry weight concentration leads to an inaccuracy of approximately ±8% in oxygen concentration. Accurate measurements can only be made by either obtaining a fully corrected analysis at each pixel, a procedure which is not supported by the software currently available, or by conducting the analysis at low voltage. At 5 kV the corrections are considerably smaller than at 15 kV and inaccuracies would be greatly reduced [43]. But a disadvantage is that the depth of the interaction volume would be considerably reduced also. This means that the depth resolution of analysis for oxygen would be different from that of the other analysed elements of higher atomic number. At 15 kV the depth resolution for all elements of interest is similar [29] and any estimate of water content can be used for the calculation of dissolved element concentrations in the same analysed volume.

In spite of the foregoing reservations reasonably accurate measurements, from x-ray images, of oxygen and water concentration can be made in cells and accurate estimates can be obtained of any change in water concentration that may occur.

Application of Quantitative X-Ray Imaging

The method has been applied to the analysis of ion transport mechanisms in the excretory organs (Malpighian tubules) of an insect (*Teleogryllus oceanicus*) (Xu and Marshall, in preparation). Approximately 100–150 of these tubular organs lie free in the body cavity. They have a considerable advantage for physiological investigations in that they can be kept alive in a functioning state *in vivo* for many hours.

As discussed previously no morphological detail is visible in the electron image of a planed frozen-hydrated sample. In order to visualise the structures to be analysed an O image was obtained at a magnification of 200× and a resolution of 64×64 pixels with a dwell time of 0.3 s and a beam current of 0.5×10^{-10}A. Because the oxygen concentration in a sample of this type is very high an O image in which the sample structure is recognisable can be obtained with these operating conditions in approximately 30 min. This image was used to select a region at 400× magnification for quantitative x-ray imaging. Positioning the beam to analyse this region was facilitated by matching the O image with a SE or BE image. Particles of frost or planing debris were usually sufficient to provide coordinates, with the aid of computer generated grids on the images, for region location on the otherwise featureless electron image. Areas for static beam analysis, primarily for obtaining matrix corrections, were also located in this manner.

A region selected for detailed x-ray imaging was analysed at 64×64 pixels, a dwell time of 1 s and a beam current of 1×10^{-10}A for light element analysis, principally oxygen, with a UTW detector. This type of image required 1–2 h (Fig. 9a). For the other elements of interest an x-ray image from the same region was obtained with a Be window at a dwell time of 6 s and a beam current of 1×10^{-9}A. This type of image required 10–12 h to acquire (Fig. 9c, d).

The x-ray images obtained in this way were apparent concentration images and were processed using the matrix corrections obtained from static beam analyses to obtain corrected concentration images. Concentrations in selected regions were averaged and extracted using the Link program Featurescan. Binary masks were made for the single cell thick epithelia of the Malpighian tubules by circumscribing the S containing regions of the tubules from the S image. In the case of the lumen the O image was used for this purpose. These masks were then applied to all the element x-ray images to obtain mean concentrations per tubule from the averaged pixel intensities. In the case of sodium, images were

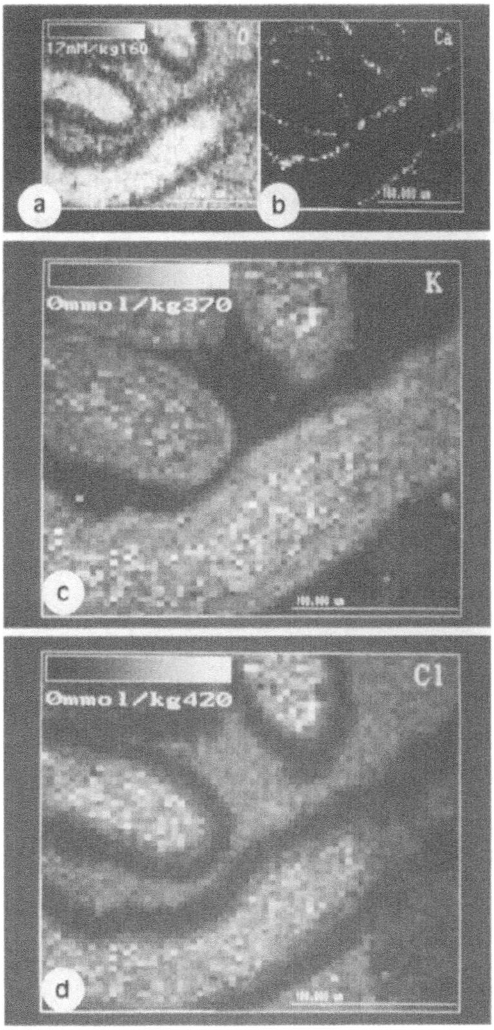

Fig. 9. X-ray images of planed frozen-hydrated Malpighian tubules. **a** O concentration image, **b** qualitative Ca image to show the distribution of calcium phosphate inclusions, **c** K concentration image, **d** Cl concentration image

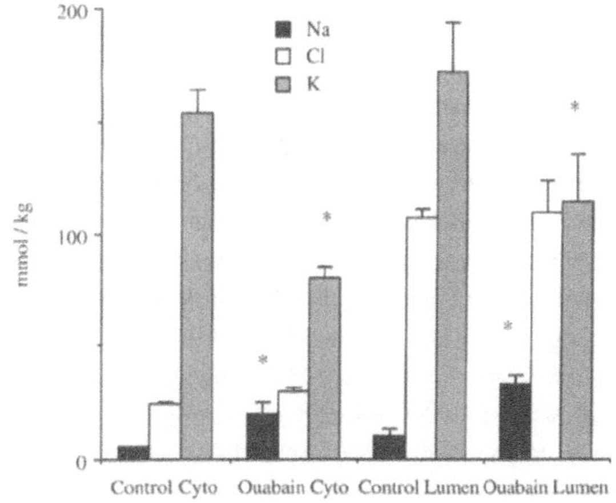

Fig. 10. Comparison of element concentrations in the cell (Cyto) and lumina (Lumen) of control and ouabain treated Malpighian tubules. Mean ± SE N_1 (insects) = 3, N_2 = 9,15 (tubules), N_3 (Cyto pixels) = 4440, 4568, N_4 (Lumen pixels) = 1155, 987, *P<0.05

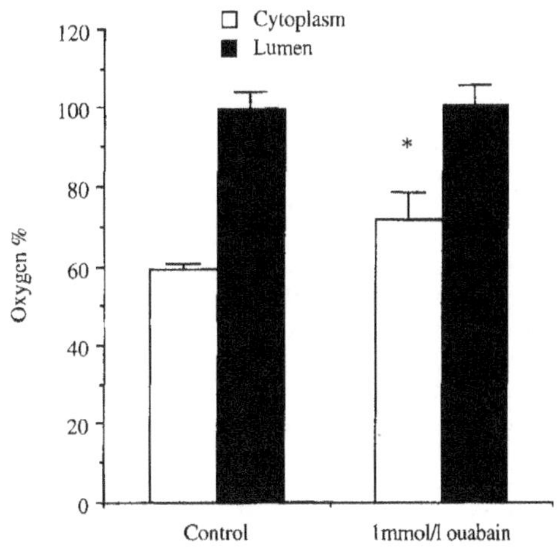

Fig. 11. Comparison of O concentration in cell cytoplasm and lumen of control and ouabain treated Malpighian tubules. Mean ± SE, N as in Fig. 10. *P<0.05

first processed with a 1.5×SD image to set all pixels below this value to zero.

The results of an analysis of control tubules and tubules treated with ouabain, which inhibits Na K ATPase in the basal cell membranes, shown in Fig. 10. Clear differences in Na and K concentration are seen in both the cells and the tubule lumina. Differences in oxygen concentration in the epithelial cells in relation to treatment were also evident (Figure 11). The difference is due entirely to an increase in cell water content as a result of the ouabain treatment and translates to an approximately 34% increase in water content. The mean water content of these cells is lower than that normally expected for biological samples because the cells contain a large number of calcium phosphate inclusions (Fig. 9b).

Conclusion

The quantitative analysis of fully frozen-hydrated biological bulk samples is possible using fractured samples, if care is exercised in the selection of analysis sites, using either a $\phi(\rho z)$ model or a *P/B* approach. The use of samples which have been planed at low temperature overcomes the problems which are inherent for quantitative analysis of samples with irregular surfaces. If, however, the morphology of the sample is exposed by sublimation of ice, i.e.

etching, the accuracy of the analysis must inevitably be compromised to some extent and O concentrations, from which water concentration can be determined, cannot be accurately measured. X-ray imaging, using O images to delineate the sample morphology, yields accurate quantitative data for all higher atomic number elements of biological interest. Although the detection limit for precise estimation of sodium is in the region of 30–35 $mmol\,kg^{-1}$ wet weight much smaller relative changes in Na concentration can be measured. More accurate, but less precise, measurements of sodium can be made with a detection limit of approximately 11 $mmol\,kg^{-1}$ wet weight. The detection limits for other elements is of the order of 5 $mmol\,kg^{-1}$ wet weight.

Acknowledgements. The work described in this paper was carried out with the aid of grants from the Australian Research Council.

References

[1] C. C. Freudenrich, D. Hockett, P. Ingram, A. LeFurgey, *J. Struct. Biol.* **1994**, *112*, 173.
[2] P. Echlin, *Low-Temperature Microscopy and Analysis*, Plenum, New York, 1992, pp. 1–539.
[3] H. Moor, in: *Cryotechniques in Biological Electron Microscopy* (R. A. Steinbrecht, K. Zierold, eds), Springer, Berlin Heidelberg New York Tokyo, 1987, pp. 175–191.
[4] A. T. Marshall, *Scanning Electron Microsc.* **1980**, 335.
[5] A. T. Marshall, in: *Cryotechniques in Biological Electron Microscopy* (R. A. Steinbrecht, K. Zeirold, eds), Springer, Berlin Heidelberg New York Tokyo, 1987, pp. 240–257.
[6] W. T. W. Potts, K. Oates, *J. Exp. Zool.* **1983**, *227*, 349.
[7] M. J. Canny, C. X. Huang, *Physiol. Plant.* **1993**, *87*, 561.
[8] M. Durand, M. Deleplanque, A. Kahane, *Sol. Stat. Comm.* **1967**, *5*, 759.
[9] A. T. Marshall, D. Carde, M. J. Kent, *J. Microsc.* **1985**, *139*, 335.
[10] W. Fuchs, J. D. Brombach, W. Trosch, *J. Microsc.* **1978**, *112*, 63.
[11] A. T. Marshall, R. J. Condron, *J. Microsc.* **1985**, *140*, 109.
[12] L. E. Wyness, J. A. Morris, K. Oates, W. G. Staff, H. Huddart, *J. Pathol.* **1987**, *153*, 61.
[13] I. J. Morgan, W. T. W. Potts, K. Oates, *J. Exp. Biol.* **1994**, *194*, 139.
[14] A. Boekstein, E. v. Kerkhof-Peters, A. Stadhouders, *Electron Microscopy* (P. Brederoo, V. Cosslett, eds), Seventh European Congress on Electron Microscopy Foundation, Leiden, 1980, pp. 74–75.
[15] P. Echlin, S. E. Taylor, *J. Microsc.* **1986**, *141*, 329.
[16] R. Millner, U. Cobet, *Phys. Med. Biol.* **1972**, *17*, 736.
[17] J. Small, K. Heinrich, C. Fiori, D. E. Newbury, R. L. Mykelbust, *Microbeam Analysis* **1978**, 13, 56A.
[18] P. Statham, J. B. Pawley, *Scanning Electron Microsc.* **1978**, *1*, 469.
[19] A. T. Marshall, R. J. Condron, *J. Microsc.* **1985**, *140*, 99.
[20] A. Boekestein, A. M. Stadhouders, A. L. H. Stols, G. M. Roomans, *Scanning Electron Microsc.* **1983**, 725.
[21] G. M. Roomans, *J. Electron Microsc. Technique* **1988**, *9*, 19.
[22] I. Farthing, G. Love, V. D. Scott, C. T. Walker, *Mikrochim. Acta [Suppl.]* **1992**, *12*, 117.
[23] L. Parobek, J. D. Brown, *X-Ray Spectrom*, **1978**, *7*, 26.
[24] J. D. Brown, W. H. Robinson, *Microbeam Analysis* **1979**, 238.
[25] J. D. Brown, R. H. Packwood, *X-Ray Spectrom.* **1982**, *11*, 187.
[26] G. F. Bastin, F. J. J. van Loo, H. J. M. Heijligers, *X-Ray Spectrom.* **1984**, *13*, 91.
[27] G. F. Bastin, H. J. M. Heijligers, F. J. J. van Loo, *Scanning* **1986**, *8*, 45.
[28] J. A. Riveros, G. E. Castellano, J. C. Trincavelli, *Mikrochim. Acta Suppl.* **1992**, *12*, 99.
[29] A. T. Marshall, *Scanning Electron Microsc.* **1982**, 243.
[30] A. T. Marshall, *Scanning Electron Microsc.* **1984**, 49.
[31] A. T. Marshall, R. J. Condron, *Micron Microsc. Acta* **1987**, *18*, 23.
[32] A. T. Marshall, R. J. Condron, *Microbeam Analysis* **1987**, 269.
[33] A. T. Marshall, A. Patak, *Scanning Electron Microsc.* **1993**, *7*, 677.
[34] A. T. Marshall, *Scanning Microscopy [Suppl.]* **1994**, *8*, 187.
[35] A. T. Marshall, *Scanning Electron Microsc.* **1981**, 327.
[36] M. J. Canny, *Phil. Trans. R. Soc. Lond. B* **1995**, *348*, 457.
[37] A. T. Marshall, W. Xu, *J. Microsc.*, **1998**, in press.
[38] A. Engström, in: *Physical Techniques in Biological Research* (A. W. Pollister, ed.), Academic Press, New York, 1966, pp. 87–171.
[39] C. E. Fiori, R. D. Leapman, C. R. Swyt, S. B. Andrews, *Ultramicroscopy* **1988**, *24*, 237.
[40] A. LeFurgey, S. D. Davilla, D. A. Kopf, J. R. Sommer, P. Ingram, *J. Microsc.* **1992**, *165*, 191.
[41] P. J. Statham, *Scanning* **1988**, 10, 245.
[42] R. Packwood, *Microbeam Analysis* **1986**, 268.
[43] E. D. Boyes, et al., *Proc. Ann. Meeting Electron Microsc. Soc. America* (G. W. Bailey, J. Bentley , J. A. Small, eds.), San Francisco Press, San Francisco, 1992, pp. 1630–1637.

Mikrochim. Acta [Suppl.] 15, 283–293 (1998)

Environmental SEM and X-Ray Microanalysis of Biological Materials

David C. Sigee

School of Biological Sciences, University of Manchester, 1.800 Stopford Building, Oxford Road, Manchester M13 9PT, U.K.

Abstract. Environmental scanning electron microscopy (ESEM) permits the examination and analysis of uncoated biological specimens in the fully-hydrated wet state. The application of x-ray microanalysis (XRMA) to ESEM encounters two major problems – atmospheric effects and specimen damage. The presence of a gaseous atmosphere results in a loss of spatial XRMA resolution due to beam skirting, an x-ray contribution from the atmosphere and a reduced overall x-ray count. Operational parameters can be optimised and corrections made to reduce these effects. Beam damage to the specimen involves mass loss, fine-structural changes, cross-linking of polymers and oxidation/hydrolysis reactions. Migration of cations and anions in relation to the probe spot may be particularly important for XRMA.

Key words: environmental microscopy, ESEM, x-ray microanalysis.

Conventional scanning electron microscopy is now firmly established as a routine tool for the examination and analysis of a wide variety of biological and materials science specimens. Although this technique has proved extremely useful, the need to operate the whole column (plus specimen chamber) at high vacuum imposes a number of limitations – including the avoidance of samples containing volatile constituents (such as water) and a requirement to coat specimens with a conducting material (normally gold or carbon). The requirements for high vacuum also limit the use of this instrument for carrying out dynamic experiments within the specimen chamber.

The development of a scanning electron microscope that allows specimens to be surrounded by a gaseous environment (hence the term environmental scanning electron microscopy, or ESEM) provides a completely new approach to scanning microscopy and circumvents some of the limitations noted previously. The flexibility that this instrument brings in terms of specimen preparation and chamber environment also brings opportunities for a more dynamic approach to microscopy, with possibilities for experimentation *in situ*. General accounts of ESEM and its applications have recently been published [1–3] together with a bibliography [4].

A. The ESEM Concept – Development and Instrumentation

The concept of environmental scanning electron microscopy was first developed by Danilatos and essentially involves the use of a scanning probe with specimens surrounded by a gaseous environment, such as water vapour. The gaseous environment is important both in terms of the type of specimen that can be imaged and also instrumental design and operation, and clearly distinguishes environmental scanning electron microscopy (ESEM) from the conventional scanning electron microscopy (SEM) – which operates only at high vacuum.

The distinction between ESEM and the SEM needs to be more clearly defined, since even at high vacuum conventional scanning electron microscopes have some gas molecules inside their specimen chamber. Danilatos [5] has suggested that the major criterion of ESEM should be the ability to observe specimens in their wet state. This requires a minimum water vapour pressure (at 0 °C) of 4.579 torr, to maintain water in its liquid phase. On this basis, Danilatos [5] has defined the environmental scanning electron microscope as an instrument which can operate up to the specimen chamber pressure required to observe liquid distilled water. This definition distinguishes ESEM from both conventional SEM and low vacuum SEM. In practice ESEM can obtain specimen images over a wide range of specimen chamber pressures, ranging from high vacuum to about 20 torr.

Instrument Design and Operation

The use of electron optics with a gaseous atmosphere clearly has problems in terms of electron penetration. This has implications both for beam penetration (requiring radical changes in the design of the microscope column) and electron detection (requiring the development of a completely new type of detector). Full details of the general principles and practical details of the design of the environmental scanning electron microscope have been given elsewhere [1].

(1) The Microscope column. The problem of beam penetration of the gaseous atmosphere has been overcome by keeping the distance travelled within the chamber atmosphere as short as possible (minimal distance between specimen and objective lens) and maintaining pressure in the rest of the electron column as low as possible using a differential pumping system.

The design of the electron optics column has a number of novel features [1], including a large specimen chamber (allowing introduction of large specimens and optimal control of the chamber atmosphere) and a differential pumping system in the main part of the column. The latter involves a number of pumps and valves, maintaining a sequence of atmospheric stages at 10^{-7}, 10^{-6}, 10^{-4}, 10^{-1} and 0–20 (specimen chamber) torr. Gaseous diffusion between stages is restricted by pressure-limiting apertures. The objective lens has two limiting apertures within a single aperture holder (referred to as the 'bullet' or 'objective aperture assembly'), which restricts vertical gas diffusion but has lateral openings for free lateral diffusion.

Passage of the beam through the chamber atmosphere causes a certain amount of electron scattering. At very low gas pressure (e.g. 10^{-6} torr – typical of that in a conventional scanning electron microscope), there is hardly any scattering and the focussed probe has a gaussian profile in the plane of the specimen. Introduction of increasing levels of gas into the specimen chamber results in a progressive increase in electron scattering and a corresponding reduction in the intensity of the central probe. Under normal operating conditions for biological specimens (up to about 10 torr) only part of the beam is scattered (oligo-scattering) and the beam profile has the appearance of a skirted probe. At higher chamber pressures (20 torr and above), all of the beam is scattered (plural scattering), and the beam profile appears as a broadened probe. Under oligo-scattering conditions, the beam skirt does not affect the spatial resolution of the microscope (quoted maximum value of 4–5 nm) because the central intense part of the beam profile is not affected.

(2) Detection of secondary and environmental electrons. The limited ability of secondary electrons, emitted from the specimen surface, to penetrate the chamber atmosphere means that detection and imaging with a conventional scanning electron microscope (Everhart-Thornley) detector is not possible, though a conventional backscattered detector may be used in ESEM without modification [2]. This apparent problem with the chamber atmosphere can be turned to advantage if the gas itself is used as a detection medium, amplifying the secondary electron signal at the specimen surface via a cascade of gaseous ionisation and excitation within the specimen chamber.

Detection of the amplified signal involves a novel type of detector, which has been variously referred to as a gas detection device (GDD), environmental secondary detector (ESD) or gaseous secondary electron detector (GSED). This was originally designed and developed by Danilatos [6], and is positioned around the final pressure-limiting aperture. Application of a voltage (up to several hundred volts) between the specimen surface and the GSED, accelerates both secondary and derived electrons towards the detector. Collisions between secondary electrons and gas molecules are of three main types – elastic, ionising (creating electron-ion pairs) and excitatory (transferring energy to gas molecules), all of which are important in the amplification process. The creation of electron-ion pairs by ionising collisions leads to a continuation of the electron signal (as environmental electrons) in the chamber atmosphere, with a residual cloud of positive ions around the sample. Monte-Carlo simulation of the ion-cascade within the chamber atmosphere [7] has revealed that excitation collisions are the most important factor in determining amplification characteristics, and provides potentially useful information for improving the design and selectivity of future environmental detectors.

B. Use of ESEM with Biological Samples

The presence of a gaseous atmosphere in the specimen chamber has two main advantages over conventional SEM in relation to biological specimens. Firstly they require minimal preparation, with no need for a conductive coating, and secondly they can be examined and maintained in a fully or partially hydrated (wet) state.

(1) Direct Observation of Uncoated Specimens

In a conventional SEM, surface charging requires coating with a conductive layer of material – normally gold or carbon. This is not necessary in ESEM because air in the specimen chamber is ionised by secondary electrons close to the specimen surface, resulting in the generation of anions which continually neutralise surface charge. Although ESEM is useful for direct observation of uncoated specimens, neutralisation of surface charge by gas molecules does not require the high chamber pressures (1–20 torr) normally used in this instrument. Much lower gas pressures are adequate for this, and the low vacuum scanning electron microscope (normally 10^{-1}–10^{-2} torr) is widely used for the examination of uncoated specimens.

In many cases, such as museum specimens, the uncoated preparation is completely dry. This technique has been used, for example, in the examination of fossil bryozoans [8], where scanning of specimens without preparation, damage or alteration could be routinely carried out. Valuable type specimens in particular, where gold coating is generally prohibited, can now be examined by ESEM and low vacuum SEM.

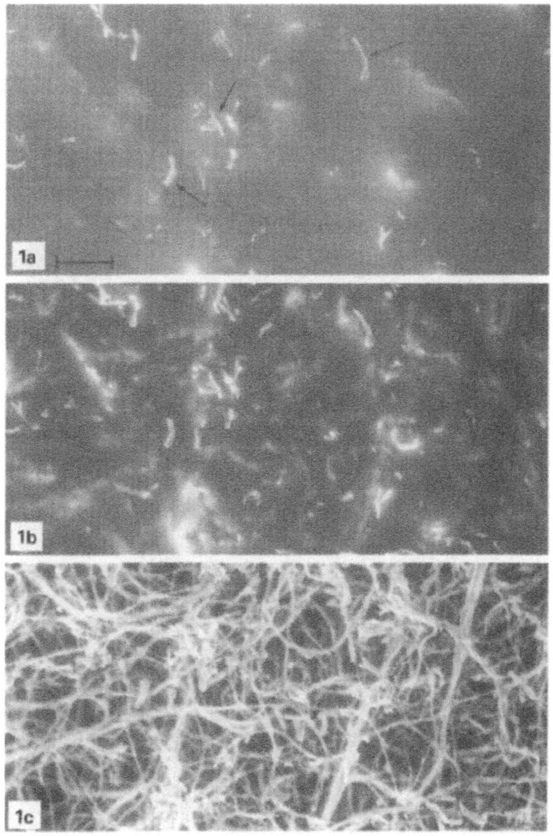

Fig. 1. Progressive removal of surface water film from wet biological specimen. **a**. View of surface water film laying over mycelium of the fungus *Cladosporium cladosporioides*. Fungal sporangia (arrows) can be seen poking above the surface of the water. The image was obtained shortly after the preparation was flooded during specimen chamber pump-down. Preparation maintained at a vapour pressure of 5 torr. **b** View of same field after preparation held for 5 minutes at 3.9 torr. Some free water has been removed from the surface film, and the top part of the fungal mycelium is becoming visible. **c** View of same field after preparation held for a further 5 min at 3.6 torr. Most of the free surface water has now been removed and the fungal mycelium can be clearly seen. The preparation was maintained at 4 °C throughout the sequence. Bar scale in Fig. 1a = 50 µm, also applies to (**b**) and (**c**). (Photographs taken in collaboration with P. Chareonsudjai and R. Bardgett)

(2) Observation of Hydrated Specimens

The use of ESEM with wet specimens is particularly relevant to biological samples, which typically have a high water content.

Operation of the Environmental Scanning Electron Microscope

In order to maintain the specimen in a wet state, the surrounding water vapour should be at saturation point. At a particular temperature, higher water vapour pressures than this result in water deposition, while lower pressures result in dehydration.

In practice, it is difficult to adjust water vapour pressure in the specimen chamber to exactly match saturation. It is particularly important not to dehydrate the specimen at the onset of the investigation during evacuation of the specimen chamber, and it is advisable to flood the chamber with water vapour initially and then gradually reduce the water vapour pressure until the surface of the specimen is exposed. A typical sequence is shown in Fig. 1, which shows the appearance of a mycelium of *Cladosporium cladosporioides* under progressively reduced water vapour pressures at 4 °C. Saturation pressure at this temperature is 6.1 torr. Initial flooding of the specimen to maintain full hydration results in a surface water film, with only the spore-bearing hyphae poking above the liquid surface (Fig. 1a). Reduction of the vapour pressure to 3.9 torr over a 5 min period causes loss of free water, with some exposure of the underlying mycelium (Fig. 1b). A further 5 min at 3.6 torr results in the removal of most of the water film, revealing the appearance of the mycelium in a wet state (Fig. 1C). Further removal of water by maintaining the specimen at this vapour pressure (or lower) results in partial dehydration, with the development of increased specimen contrast and some distortion of the hyphae due to shrinkage.

Applications of ESEM to Different Types of Wet Specimen

The ability to examine specimens in a complete or partially hydrated (wet) state has two major applications for biological samples – they can be observed in the native state and experiments can also be carried out in relation to variations in humidity.

Observation of Biological Specimens in the Wet State

The maintenance of specimens in a wet state means that delicate objects, such as micro-organisms, biofilms and gels, can be imaged without potentially destructive cryopreservation or dehydration procedures. ESEM thus combines the possibility of viewing biological specimens close to the native (living) state without the need to use preparation techniques that may be time-consuming and may generate artefacts.

The advantages of using ESEM with a range of micro-organisms, including bacteria, fungi, protozoa and micro-algae – many of which are damaged by conventional preparation techniques, have been discussed by Collins et al. [9].

ESEM has potential for the examination of biofilms, as noted by Sutton et al. [10]. These authors compared the varied appearance of biofilms of *Streptococcus crista* using conventional scanning electron microscope preparation techniques (critical point dring, freeze-drying), low temperature SEM (LTSEM) and direct observation by ESEM. Critical point drying showed single or paired cocci in the biofilm, with complete disappearance of the extracellular polysaccharide (EPS) matrix. Freeze-drying caused loss of matrix and detachment of the biofilm from the underlying surface. By comparison, LTSEM showed improved preservation of EPS but the biofilm exhibited shrinkage artefacts and had discontinuities in the matrix. Direct imaging of the biofilm under ESEM was the only approach that preserved the integrity of the EPS matrix with no shrinkage artefacts. Individual bacterial cells were observed as indistinct shapes within the wet matrix.

Superior imaging of biological specimens in the wet state using ESEM has also been shown for various plant and animal tissues. Uwins et al. [11] studied the fine structure of potato tuber tissue and demonstrated that fewer artefacts were introduced with ESEM compared to other types of microscopy. As an example of the recent use of ESEM with animal tissues, Peters [12] has investigated the structure of the air-interface of lung tissue, providing information on the 3-D structure and surface matrix of the alveolar lining.

Experimentation within ESEM – the Effects of Varying Humidity

In addition to the direct imaging of wet biological specimens in the native state, ESEM has also been used to study the response of different specimens to variations in humidity. Dynamic events related to humidity changes have been reported [13] for various botanical samples. These include filaments of Myxomycetes, hydration of which leads to rearrangements and spore dispersal, and leaf hair cells – which also show movements in relation to humidity changes.

ESEM has also been used [14] to study a range of pharmaceutical formulations at different states of hydration and to relate these to drug release. The ability of ESEM to visualize wet preparations has been useful in 'pseudo-dissolution' studies of various drug substances, and has also proved to be a valuable tool in the assessment of certain drug-delivery formulations (micro-particulate systems, biodegradable matrices and tablet formulations). Previous SEM investigations had suggested that drug release by polyanhydride micro-spheres was mediated by entry of water into surface cracks in the polymer. Recent ESEM studies have shown, however, that these cracks do not develop until dehydration occurs – so they cannot be involved in drug release in the wet state. Other potential mechanisms of drug release have been successfully investigated by ESEM [15], including the use of ultrasound to promote erosion of the matrix of polyanhyride carriers. Although ESEM has now been used with a wide range of pharmaceutical preparations, it should be noted that certain samples – such as emulsions, do not lend themselves to analysis by this technique.

C. X-Ray Microanalysis with ESEM

Conventional use of x-ray microanalysis with either scanning or transmission electron microscopy in biology typically involves the use of either freeze-dried or hydrated frozen specimens [16]. The development of ESEM raises the possibility that XRMA may be used to determine the elemental composition of biological specimens in the fully hydrated wet state, possibly even extending this to analysis of living cells. There are two major problems encountered, however, in the application of XRMA to ESEM – atmospheric effects and beam damage to the specimen.

(1) Atmospheric Effects

The presence of a gaseous atmosphere in the specimen chamber presents serious limitations for the use of XRMA with ESEM [17, 18] due to three major effects – a loss of spatial resolution arising from beam skirting, an x-ray contribution from the gas atmosphere and a reduced overall x-ray count.

Loss of Spatial Resolution Due to Beam Skirting

The dominant effect of specimen chamber pressure on XRMA spatial resolution has been investigated by various workers [17–19] and can be demonstrated using a simple experimental model such as that shown in Fig. 2. In this system, an electron beam is directed onto a layer of carbon DAG, with a copper (TEM) grid placed at varying distances from the primary probe area. The skirting effect of the beam within a water vapour atmosphere is shown by the ability to pick up a Cu signal of varying intensity (Cu P/B ration) at different distances of the Cu target from the central probe area (Fig. 3). Reduction in accelerating voltage causes an increase in the skirting effect, with an increased Cu P/B ratio at set target distances on changing from 20 kv to 10 kv.

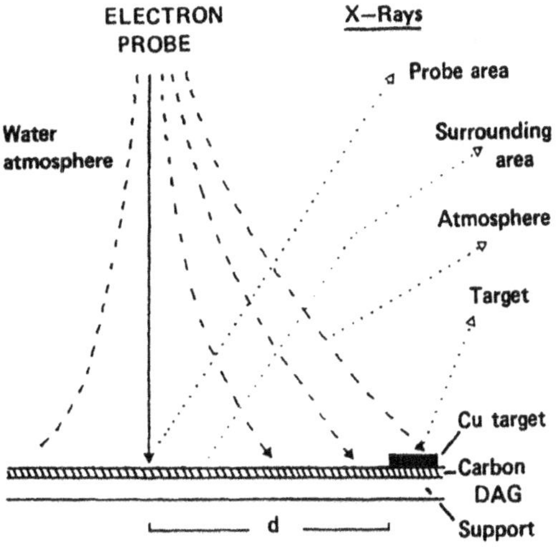

Fig. 2. Demonstration of beam skirting using a Cu target. Primary probe – continuous line, skirting electrons – broken lines. The support layer is Whatman filter membrane. Derivation of x-rays is shown as dotted lines. (Taken, with permission, from Sigee and Gilpin [17])

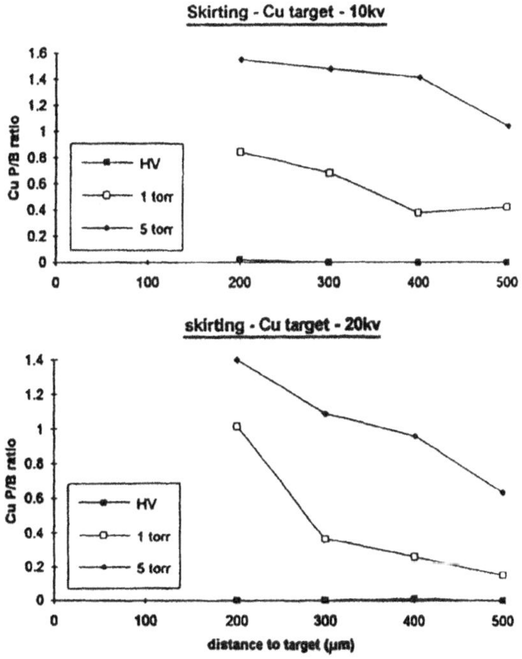

Fig. 3. Beam skirting in a water vapour atmosphere at 10 kV and 20 kV. Graphs show P/B ratios of Cu, obtained during two traverses (10 kV and 20 kV), at varying distances from the Cu target. With each distance, x-rays were collected over 200 µm^2 areas at high vacuum (HV), 1 torr and 5 torr atmospheric pressure. (Taken, with permission, from Sigee and Gilpin [17])

The effect of beam skirting with biological specimens can be demonstrated both in terms of x-ray

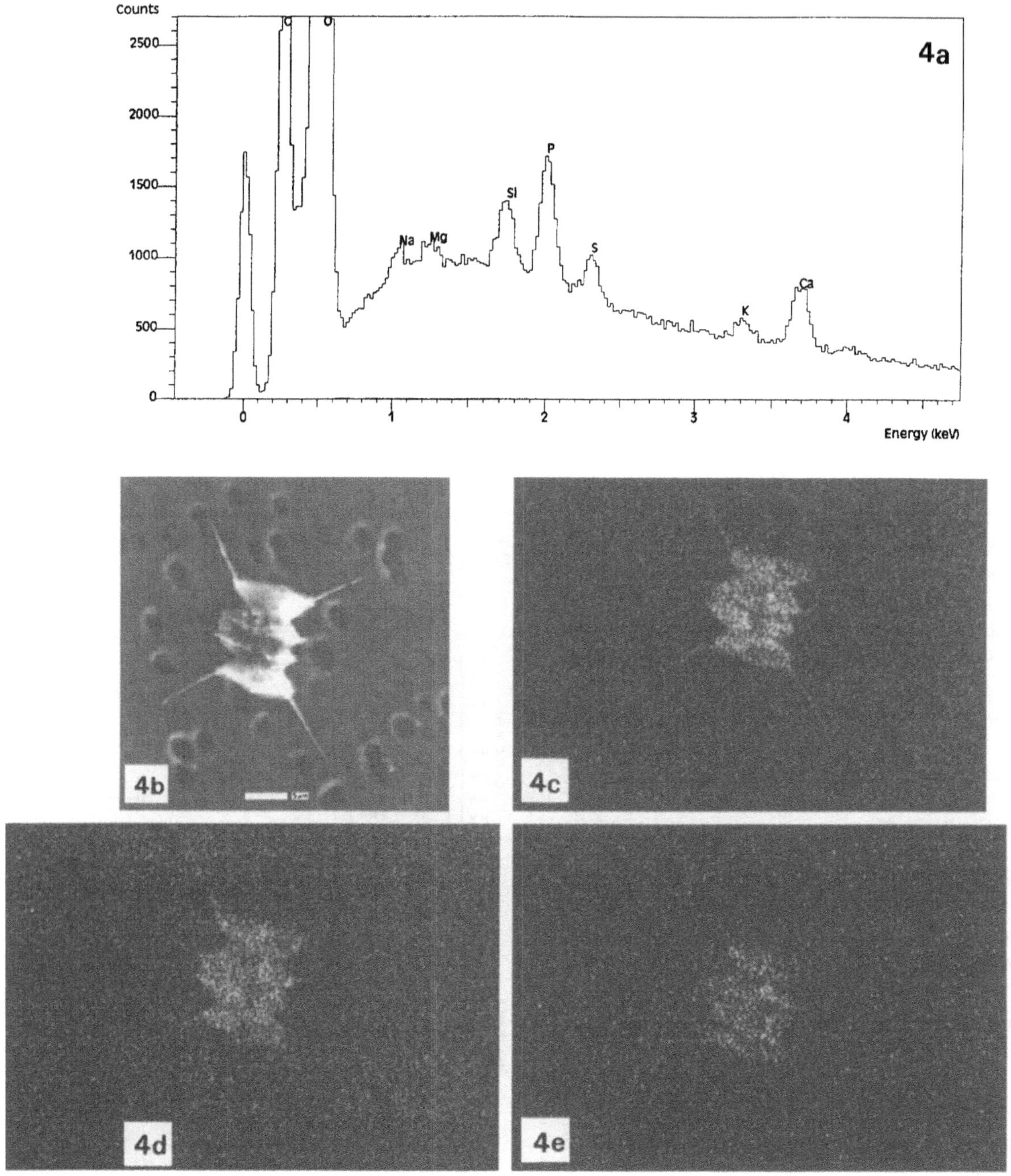

Fig. 4. Effect of specimen chamber atmospheric pressure on spatial resolution of x-ray emission mapping. **a** x-ray emission spectrum taken from cell of *Staurastrum* (**b**) under conditions of high vacuum. **b** SEM image of single freeze-dried cell of the alga *Staurastrum*. Bar scale = 5 μm. **c** Phosphorus map of (**b**), obtained at a chamber pressure of 1 torr. This element occurs throughout the cell, and the spatial image has sufficient resolution to clearly define the cell outline, including the cell processes. **d** Phosphorus map of (**b**), obtained at 2 torr. Cell outline less clearly defined. **e** Phosphorus map of (**b**), obtained at 5 torr. Cell outline poorly defined

elemental mapping and by collection of x-ray emission spectra. Fig. 4 shows the effects of increased atmospheric levels on the mapping resolution of a single cell of the unicellular alga (desmid) *Staurastrum*. Mapping of P (one of the major elements – Fig. 4a) at high vacuum shows a clear image of the alga, consistent with the element being distributed throughout the whole of the cell (compare Figs. 4b and 4c). Increase in chamber pressure causes a progressive deterioration of the x-ray map (Figs. 4d and 4e) since

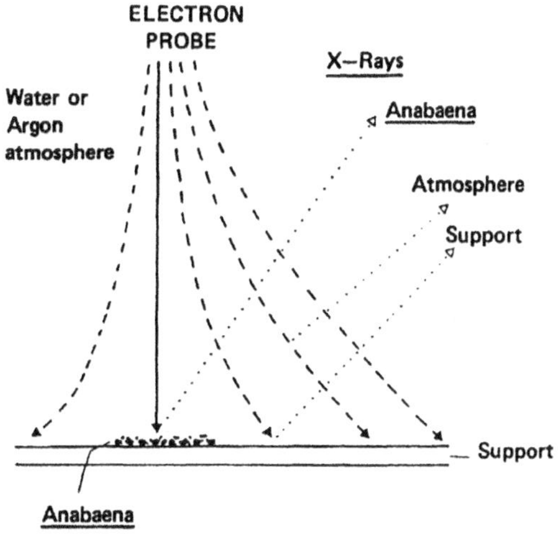

Fig. 5. Analysis of a discrete sample of *Anabaena* using a water vapour or argon atmosphere. Format as with Figure 2. (Taken, with permission, from Sigee and Gilpin [17])

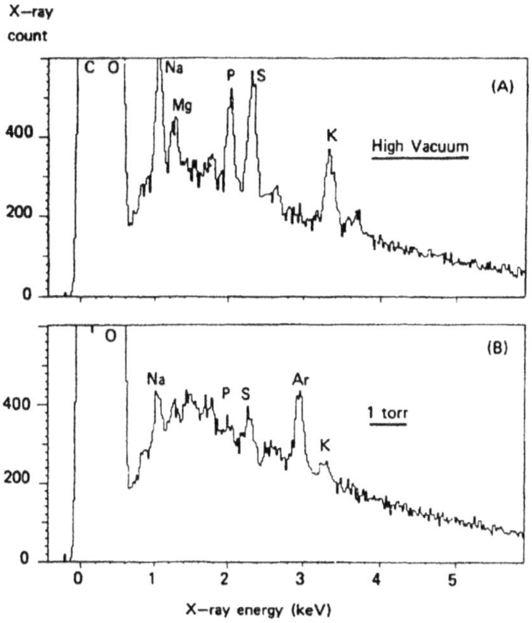

Fig. 6A, B. X-ray emission spectra from a sample of *Anabaena* at high vacuum and in an argon atmosphere. XRMA spectra taken from a 1000 μm^2 area within the discrete *Anabaena* sample (see Fig. 5) show a clear Ar Kα peak in the argon atmosphere, with depressed peaks of other elements. (Taken, with permission, from Sigee and Gilpin [17])

spectra from adjacent areas of formvar also pick up elements from the cell due to beam skirting.

In the above example, beam skirting is demonstrated for a single cell. Similar atmospheric effects can be demonstrated for larger samples, as shown in Figs. 5 and 6, where a discrete zone of the blue–green alga *Anabaena* has been deposited onto a support film (Fig. 5). The typical spectrum from a freeze-dried specimen at high vacuum (no beam skirting) has clear peaks of Na, Mg, P, S and K (Fig. 6A, in agreement with previous studies on this alga). Introduction of an argon atmosphere causes a dramatic reduction in the characteristic peaks of the above elements since the beam now spreads over the surrounding area of carbon support film, reducing the net contribution of specimen constituents (Fig. 6B).

Corrections and Reduction in Beam Skirting for X-Ray Microanalysis

Beam skirting clearly limits the spatial resolution of x-ray microanalysis. The effect can be reduced in various ways, including use of optimal operation parameters, implementation of correction procedures, use of extended specimens and software corrections in x-ray mapping.

Operational Parameters

Various operational parameters can be optimised to reduce beam skirting and increase XRMA resolution – including atmospheric pressure, nature of atmospheric gas, accelerating voltage and working distance.

Atmospheric pressure. Chamber pressure should be kept as low as possible. If water vapour is being used to maintain hydration then it would be advantageous to reduce specimen temperature to near freezing so that the saturation vapour pressure is minimal.

Nature of atmospheric gas. For most biological specimens, the major biological purpose of the chamber atmosphere is to maintain the specimen in a wet state, and would thus be water vapour. Circumstances might arise, however, where other gases would be used either separately or in combination with water vapour. Commonly available gases that would provide an inert or reducing atmosphere include helium, nitrogen, carbon dioxide and argon. These differ considerably in the atomic mass and resulting ability of their molecules to scatter electrons.

Using a discrete freeze-dried biological specimen, Sigee and Gilpin [17] demonstrated a much greater skirting effect with an argon (atomic mass 40) compared to a water vapour (atomic mass 18) atmosphere.

Accelerating voltage. Although this parameter should be kept as high as possible (Fig. 3), there

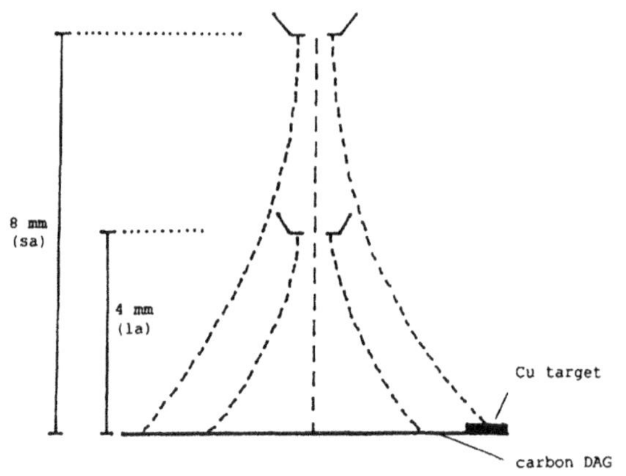

Fig. 7. Experimental model – effect of different detector assemblies on beam skirting. Diagrammatic comparison of beam skirting with short detector assembly (*sa*, specimen distance 8 mm) and long detector assembly (*la*, specimen distance 4 mm) in relation to a copper grid target on a layer of carbon DAG. (Taken, with permission, from Gilpin and Sigee [18])

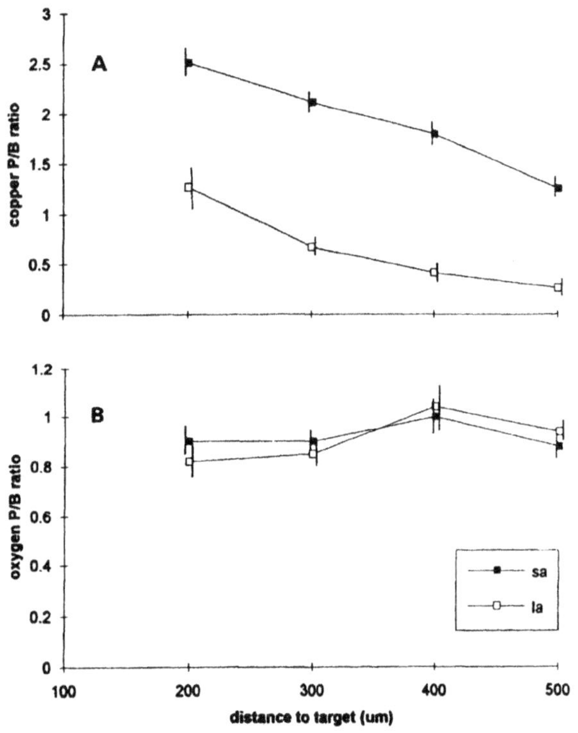

Fig. 8. Beam skirting and specimen distance. **A** Beam skirting, monitored as pickup of the Cu Lα peak (see Fig. 7), is shown for short (*sa*) and long (*la*) detector assemblies, over a range of specimen (probe target) distances, at a water vapour pressure in the specimen chamber of 5.2 torr. **B** Atmosphere contribution to x-ray count. The oxygen signal (P/B ratio) is shown for both detector assemblies over a range of specimen distances. Data in (**A**) and (**B**) are derived from the same experiment. (Taken, with permission, from Gilpin and Sigee [18])

may be a practical limitation with certain preparations due to beam over-penetration. In the case of fully hydrated specimens, however, over-penetration will not normally be a problem. Marshall [20] and Marshall and Condron [21] have shown that the maximum depth from which x-rays are emitted in frozen-hydrated bulk preparations is 2–3 μm at an accelerating voltage of 15 kV.

Working distance. Decreasing the distance travelled by the electron probe within the chamber atmosphere decreases the skirting effect and increases the spatial resolution of XRMA. This was demonstrated by Gilpin and Sigee [18] using an extended detector assembly which reduced the working or specimen distance (between specimen and environmental detector) from 8 to 4 mm (Fig. 7). Change to the extended assembly (reduced working distance) resulted in a much reduced Cu signal at particular distances between the probe and Cu target (Fig. 8A). The extent to which working distance can be reduced ultimately depends on the spatial constraints, with limitations imposed by detector access within the probe area.

'Beam Stop' and 'Pressure Variation' Correction Procedures

Two basically different correction methods have been described by Bilde-Sørensen and Appel [22, 23], referred to as the 'beam stop' and 'pressure variation' procedures.

Beam stop method. One version of this procedure involves the use of a needle (composed of a known element not detected in the sample) that can be inserted over the specimen to act as a beam stop for the central (unscattered) electrons but not the peripheral beam skirt. The spectrum obtained with the beam stop in position contains x-rays derived from the known element (beam stop) plus the area covered by the beam skirt. The pure spectrum from the central probe area can be obtained by removing the characteristic peaks of the known element from the first spectrum and subtracting the remaining spectrum from the second.

Pressure variation method. This approach is based on the fact that intensity of skirt varies with chamber pressure, and thus correction for electron scattering can be made by obtaining x-ray emission spectra at different chamber pressures. According to Danilatos [24] at any particular pressure the fraction of unscattered electrons is given by the equation:

$$I/I_o = \exp(-p\sigma L/kT) \quad (1)$$

where p = pressure, σ = the scattering cross section of the gas, L = distance from the last pressure limiting aperture to the sample, k = the Boltzmann constant, and T = absolute temperature.

The above expression can be used to relate the measured count rate (C_T) from a particular element in the sample to count rates at zero scattering (C_U) and complete scattering (C_S). According to Bilde-Søren-sen and Appel (1996a, b):

$$C_T = C_U \cdot \exp(-pm) + C_S \cdot (1 - \exp(-pm)) \quad (2)$$

where $m = \sigma L/kT$.

The factor C_S is unknown, but will be a constant provided that pm is sufficiently low for multiple scattering of electrons to be ignored. Under these conditions, C_U can be derived from two measurements of C_T at different pressures, where p and m are known.

Use of Extended (Continuous) Samples

With some biological preparations, the investigator may simply wish to determine the mean elemental composition of the sample, without the need for high resolution analysis of individual components. In this situation, the microsample can be analysed as an extended preparation, where beam skirting simply results in a wider area of analysis.

This approach was used by Sigee and Gilpin [17] to determine the mean elemental composition of the alga *Anabaena*, deposited as a continuous monolayer on a filter membrane. Increase in the chamber water vapour pressure from 1–5 torr, with increase in beam skirting, did not result in any major changes in the x-ray emission spectra taken from the wet sample. Mean elemental concentrations (P/B ratios) typically showed only minor changes with increase in atmospheric pressure to 5 torr, though some of these were statistically significant.

Improvement of Spatial Resolution in X-Ray Mapping

The definition of x-ray maps is severely limited at water vapour pressures required to maintain the specimen in a wet state, as shown in Fig. 4. Although this can be improved by analysing under optimal conditions (see previously), further corrections are clearly desirable.

Improvements in element mapping at high chamber pressures have been reported by Griffin et al. [25]. Scattered beam profile data were modelled as the sum of two gaussian peaks, leading to the derivation of an algorithm for the calculation of effective spatial resolution under different conditions of chamber pressure and accelerating voltage. A gaussian filter (based on the above profile measurements) was subsequently applied to digitised binary x-ray dot maps, followed by segmentation of the image using image analysis software. These modifications resulted in the production of acceptable x-ray maps of wet specimens in ESEM.

X-Ray Contribution from the Chamber Atmosphere

Passage of the electron beam through the specimen chamber atmosphere leads to the generation of characteristic and continuum x-rays from the gas molecules.

Contribution of characteristic x-rays to the overall emission spectrum is shown in Fig. 6B, where the introduction of argon into the chamber results in a clear elemental peak. Using a non-biological specimen, Gilpin and Sigee [18] demonstrated an approximate linear increase in the P/B ratio for argon (from argon gas) and oxygen (from water vapour) with increase in the atmospheric pressure up to 10 torr. The presence of an atmospheric x-ray contribution clearly presents problems for the determination of oxygen levels in a wet biological specimen.

Contribution of continuum x-rays from the chamber atmosphere results in depressed P/B ratios and reduced detectability for specimen elements. Using continuous biological and non-biological specimens to avoid complications from beam skirting, Sigee and Gilpin [17], and Gilpin and Sigee [18] showed that water vapour has relatively little effect on specimen element detectibility, while argon causes major decreases in P/B ratios.

It might be expected that reduction in the distance travelled through the chamber atmosphere by the electron beam (i.e. the working distance) would reduce the atmospheric contribution to the overall x-ray emission spectrum. Using both water vapour and argon atmospheres, however, Gilpin and Sigee [18] showed that this was not the case, and that the atmospheric P/B ratio does not vary with working distance. This lack of correlation between gas signal and distance travelled within the chamber atmosphere arises because the gas P/B ratio reflects the concentration of gas molecules within the whole volume of x-ray generation (gas and specimen). The geometry of the excitation volume, and thus the proportion occupied by the atmosphere, does no change with reduction in working distance (Fig. 7). Because of this proportionality, the distance travelled by the electron beam in the chamber atmosphere has no demonstrable effect on the depression of specimen elemental P/B ratios by atmospheric continuum. Close similarity in oxygen P/B values with different detector assemblies is shown in Fig. 8B, where the values are obtained at different Cu-target distances.

Atmospheric x-ray contribution thus varies with type of gas and atmospheric pressure, but not working distance [18].

Reduced X-Ray Count

Introduction of an atmosphere into the specimen chamber can lead to a marked fall in the overall x-ray count. This was demonstrated by Gilpin and Sigee [18] using a continuous carbon specimen, where the overall X-ray count was monitored as the continuum under the carbon peak. These studies demonstrated that water vapour had relatively little effect on x-ray count up to a pressure of 10 torr, but argon caused a marked decrease (leading to a 75% reduction in count rate at 10 torr).

The effects of chamber atmosphere on x-ray counts may arise for several reasons, including reduced electron beam penetration to the specimen, x-ray absorption by gas molecules prior to detection and x-ray generation/scatter out of the detection angle.

Some indication of the effects of gas pressure and molecular weight on beam penetration can be obtained from physical modelling of electron scatter using a standard Monte Carlo simulation (Table 1). The results obtained from this should only be regarded as semi-quantitative, since the simulation provides a theoretical description of electron penetration in solids rather than gases and does not take into account such factors as Brownian motion and lack of molecular homogeneity. The data obtained do, however, provide an interesting comparison of water vapour and argon gases, and provide a clear demonstration of the effect of gas pressure on beam penetration. At each vapour pressure, the water vapour/argon penetration ratio was 2.6. Such a difference in beam penetration may be a major factor in the observed differences between these atmospheres in terms of x-ray count.

Table 1. Maximum penetration (μm) of a 10-kV electron beam in water-vapour and argon atmospheres (Monte Carlo simulation)

Gas	1 torr	2 torr	5 torr	100 torr
Water vapour	4.52×10^6	2.26×10^6	9.03×10^5	4.52×10^4
Argon	1.75×10^6	8.75×10^5	3.5×10^5	1.75×10

(2) Beam Damage to the Specimen in ESEM

Specimen beam damage in ESEM presents a potentially major problem. This is particularly the case with XRMA, where it is necessary to deliver at least 20 nC of irradiation to the analysed region in order to obtain adequate x-ray emission [26]. The question of beam damage in ESEM can best be approached by initial consideration of the effect of radiation damage in conventional XRMA.

Beam Damage in Freeze-Dried and Frozen Hydrated Specimens

Radiation effects can be divided into two major categories [27] – relocation artefacts and mass loss. Relocation artefacts (also referred to as dislocation damage) involve breakdown of chemical bonds, production of molecular fragments and dispersal of these form the primary radiation site in a cascade of reactions. Since the mean free path for highly reactive species (such as free radicals) is rather short, this type of damage will be of minimal importance if the environmental scanning electron microscopy XRMA is carried out at low resolution. Mass loss during XRMA has been investigated particularly in freeze-dried and frozen-hydrated specimens analysed at low temperatures [27]. With frozen hydrated samples, substantial etching is normally observed at radiation doses required for XRMA. For both dry mass and frozen-hydrated specimens, element-specific losses have been shown to occur – particularly for light elements C, O, N, Cl and S.

Beam Damage in ESEM

With conventional preparations, the two major measures used to limit beam damage are the use of dry specimens and the application of low (sub 0 °C) temperatures, neither of which are relevant for the analysis of wet biological specimens in ESEM.

Although there is relatively little information on the problem of beam damage in ESEM, a number of investigators have revealed that significant beam damage occurs when the specimen is wet [5]. The surface of cellulose fibres, for example, appears unexpectedly smooth when examined in ESEM during wetting and drying cycles [28] – probably due to the production of hydrogen and free hydroxy radicals which attack the surface organic material of the sample. Recent studies by Kitching and Donald [29] on electron-beam damage in thin polypropylene films, using infra-red spectroscopy, have demonstrated mass loss and cross-linking in the polymer. Prolonged beam exposure lead to a brittle polymer with a high carbon content, which is stable and resistant to further change. In addition to the direct effects of beam damage on the specimen, the presence of water vapour in the specimen chamber leads to oxidation and hydrolysis of the sample.

XRMA studies by Egerton-Warburton and Griffin [30] have investigated the possibility of ion migration within wet biological samples in relation to the primary electron excitation volume. Using a range of hydrated specimens (liquid droplets, wet filter paper and methylcellulose gels), with XRMA in spot mode, these authors demonstrated clear changes in the elemental composition of the sample. Over a livetime of 250s, there was an increase in the concentration of K, and decreases in S and P in the probe area. The results obtained were interpreted as reflecting ion migration in the sample resulting from an electrochemical bias in and around the probe volume. This bias would be negative in the electron interaction volume and positive in the surrounding region, leading to the observed inward migration of cations and exit of anions.

These studies [30] provide the first evidence of the effects of beam damage during XRMA in ESEM, and indicate that considerable caution must be exercised in the interpretation of data – particularly in relation to diffusible elements. The extent to which this effect varies with operating conditions and different types of specimen awaits future investigation.

Acknowledgement. The author wishes to thank the Editors of the Journal of Microscopy and Scanning Microscopy for permission to publish figures and tables from copyright material.

References

[1] G. D. Danilatos, *Microsc. Res. Tech.* **1993**, *25*, 354.
[2] R. E. Cameron, *Microsc. Anal.* **1994**, *41*, 11.
[3] C. Mathieu, *Microsc. Anal.* **1996**, *55*, 19.
[4] G. D. Danilatos, *Microsc. Res. Tech.* **1993**, *25*, 529.
[5] G. D. Danilatos, *Adv. Electron. Electron. Phys.* **1988**, *71*, 109.
[6] G. D. Danilatos, *Adv. Electron. Electron. Phys.* **1990**, *78*, 1.
[7] I. C. Bache, B. L. Thiel, A. M. Donald, *Proceedings of the XI European Congress on Microscopy*, Dublin, Ireland, 1996, in press.
[8] P. D. Taylor, C. G. Jones, *Microsc. Anal.* **1996**, *52*, 27.
[9] S. P. Collins et al., *Microsc. Res. Tech.* **1993**, *25*, 398.
[10] N. A. Sutton, N. Hughes, P. S. Handle, *J. Appl. Bacteriol.* **1994**, *76*, 448.
[11] P. J. Uwins et al., *Microsc. Res. Tech.* **1993**, *25*, 412.
[12] K. R. Peters, *Scanning* **1990**, *12*, 40.
[13] R. Pope, R. W. Scheetz, *Scanning* **1990**, *12*, 38.
[14] A. D' Emanuuele, *Pharm. J.* **1992**, *June 13*, 785.
[15] A. D' Emanuuele, *Macromolecules* **1992**, *25*, 511.
[16] D. C. Sigee, A. J. Morgan, A. T. Sumner, A. Warley, *X-Ray Microanalysis in Biology: Experimental Techniques and Applications*, Cambridge University Press, Cambridge, 1993.
[17] D. C. Sigee, C. Gilpin, *Scanning Microsc.* **1994**, *[Suppl. 8]*, 219.
[18] C. Gilpin, D. C. Sigee, *J. Microsc.* **1995**, *179*, 22.
[19] B. J. Griffin, R. L. Trautman, J. Coffey, *Proceedings of the Microbeam Analysis Society Annual Meeting, USA*, 1993 pp. 7–8.
[20] A. T. Marshall, *Scanning Electron Microsc.* **1982**, *1*, 243.
[21] A. T. Marshall, R. J. Condron, *J. Microsc.* **1985**, *140*, 109.
[22] J. B. Bilde-Sørensen, C. C. Appel, *Proceedings of the XI European Congress on Microscopy, Dublin, Ireland*, 1996, in press.
[23] J. B. Bilde-Sørensen, C. C. Appel, *Preceedings of the 48th Annual Meeting of the Scandinavian Society for Electron Microscopy*, 1996, pp. 4–5.
[24] G. D. Danilatos, Mikrochim Acta **1994**, *114/115*, 143.
[25] B. J. Griffin, A. van Riessen, L. Egerton-Warburton, J. Hatch, R. L. Trautman, *Proceedings of the 18-th Annual MAS Meeting, USA*, 1994, p. 245.
[26] T. A. Hall, B. L. Gupta, *J. Microsc.* **1984**, *136*, 193.
[27] T. Von Zglinicki, in: *X-Ray Microanalysis in Biology: Experimental Techniques and Appplications.* (D. C. Sigee, A. J. Morgan, A. T. Sumner, A. Warley, eds.) Cambridge University Press, Cambridge, 1993.
[28] J. G. Sheehan, *Proceedings of the 49th Annual Meeting of the Electron Microscopy Society of America , USA*, 1991, p. 1130.
[29] S. Kitching, A. M. Donald, *Proceedings of the XI European Congress on Microscopy, Dublin, Ireland*, 1996, in press.
[30] L. M. Egerton-Warbuton, B. J. Griffin, *Micron* **1994**, *25*, 607.

Mikrochim. Acta [Suppl.] 15, 295–300 (1998)

Effects of Electron-Beam/Gas Interactions on X-Ray Microanalysis in the Variable Pressure SEM

Christian Mathieu

L.P.C.I.A. Université d'Artois, SP 18, 62307 Lens Cedex, France

Abstract. The objective of this investigation is to evaluate the effects induced by the electron-beam/gas interaction on x-ray microanalysis in the variable pressure scanning electron microscope. The occurrence of an electron beam skirting effect implies a loss of x-ray spatial resolution. Skirting is first studied by current measurement, the conduction current being a useful parameter for its evaluation. To minimise the skirting effect, the experimental conditions must be in the pressure range 1 to 20 Pa, and a low working distance and a high accelerating voltage should be used. The atmospheric effect, which includes a direct x-ray contribution from the atmosphere, produces an increase in the oxygen signal when the pressure of air inside the chamber increases.

Key words: variable pressure SEM, XRMA, ESEM.

A variable pressure scanning electron microscope (VP-SEM) may be defined as a scanning electron microscope (SEM) that can operate from high vacuum up to a pressure level equal to 270 Pa. The implications of this capability are enormous. A gaseous environment acts to dissipate charge. The collisions of the electron beam with gas molecules produce positive ions near the specimen surface. These ions can then combine with excess electronic charges on the sample surface and in this way neutralise surface charge. The VP-SEM also offers a variety of new capabilities for characterising materials [1]. Most important is the ability to study insulators without applying a conductive coating. Unfortunately, directing an electron beam into a gas creates several problems for microanalysis: degradation of x-ray spatial resolution due to electron scattering, and changes in beam current with pressure and working distance. The electron-beam/gas effect [2–4] and the consequences for x-ray microanalysis [5–8] are described in the literature on the environmental scanning electron microscope ESEM. The purpose of this investigation is to determine some of the optimum operating conditions for x-ray measurements in the variable pressure scanning electron microscope.

Experimental

Studies were carried out using a VP-SEM (Hitachi S2460N) fitted with a KEVEX atmospheric thin window detector. Specimen chamber pressure ranged from high vacuum (10^{-3} Pa) to 270 Pa with an air atmosphere. The microscope was also equipped with a probe current meter and the variations in the conduction current flowing to earth through an aluminium SEM specimen were measured.

Specimens were prepared by mounting a carbon stick onto an aluminium stub and the specimen working distance was 25 mm when the microscope was set up for analysis. X-ray microanalysis was carried out with a 5 and 15 kV electron probe for a live time of 100 seconds. Specimens were analysed in spot mode and the count rate was typically $1-2\times10^3$ cps. Electron skirting around the central probe at different atmospheric levels was tested by placing the probe at different distances from a Cu target and measuring the Cu signal.

Results and Discussion

Gas Beam Interaction

In the conventional specimen current mode in high vacuum (10^{-3} Pa), the specimen current is given by the well-known equation [9].

$$I_{sc} = I_{pe} - I_{se} - I_{bse} \qquad (1)$$

where I_{sc} = specimen current in vacuum, I_{pe} = primary beam current in vacuum, I_{se} = secondary electron current in vacuum, I_{bse} = back-scattered electron current in vacuum.

Equation (1) can be written as

$$I_{sc} = I_{pe}(1 - \eta - \delta),$$

where δ = coefficient of secondary electron emission, η = coefficient of back-scattered electron emission. Therefore, the probe current and conduction current are only functions of the specimen coefficients and remain constant in the high vacuum mode.

Figure 1 shows probe current and conduction current variations versus the pressure at 15 kV. With increase of pressure, the probe current falls and the specimen current rises. These variations are characteristic of the influence of a gaseous environment. For an aluminium sample, the back-scattered coefficient, η, is 0.15 and the secondary coefficient, δ, is 0.1 so that the ratio I_{sc}/I_{pe} will be equal to 0.75. The variation of I_{sc}/I_{pe} versus the pressure (Fig. 2) is a good criterion to evaluate the influence of the pressure. Between 1 and 20, the ratio remains constant but when the pressure is higher, an increase is observed. Beam broadening or the skirting effect begins at 20 Pa. so that a change in I_{sc}/I_{pe} indicates the onset of the effect.

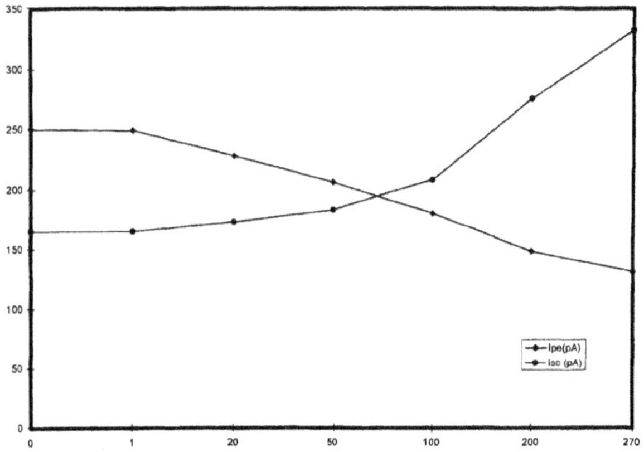

Fig. 1. Variation in probe current and conduction current with the pressure at 15 kV

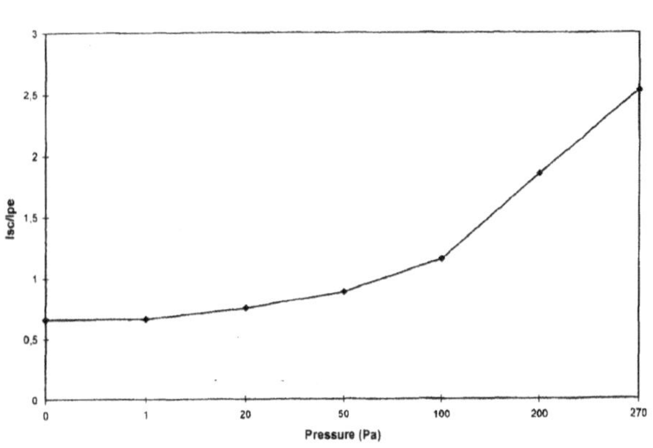

Fig. 2. I_{sc}/I_{pe} versus the pressure

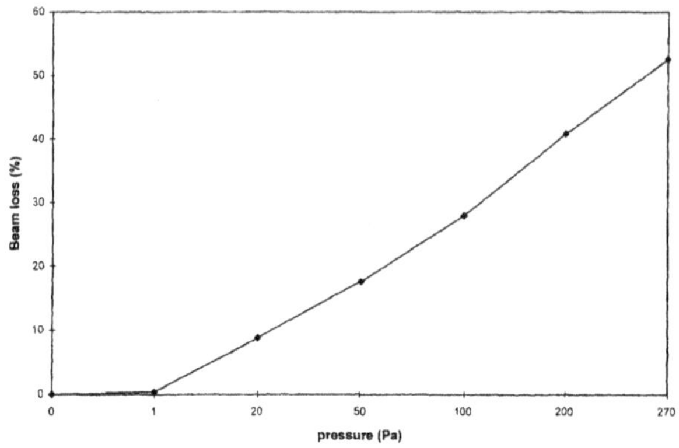

Fig. 3. Percent of the beam lost (15 kV Wd = 11 mm)

The percentage of the beam lost can be estimated. Indeed, if the intensity of electron beam in high vacuum is Io and the fraction of beam transferred intact (i.e. wholly unscattered) at the measured pressure is I, the percent of beam lost is given by (1-*I*/*I*o). Results are shown (Fig. 3) for a working distance of 11 mm and a probe voltage of 15 kV, from which the beam loss can be estimated as 10% at 20 Pa and 50% at 270 Pa.

In order to investigate the influence of the working distance and the accelerating voltage, the probe current is maintained at a fixed value equal to 250 pA and the change in the conduction current with pressure is measured. The conduction current increases when the accelerating voltage decreases (Fig. 4) and the electron-beam/gas effects are more important when the accelerating voltage is lower. This result corresponds to the variation of the ionisation efficiency of the electrons with accelerating voltage in nitrogen and water vapour atmospheres [10]. The ion current generated by the beam depends on the ionisation efficiency of the electrons, which is defined as the number of ion pairs produced per unit path length and is directly related to the inelastic scattering cross section (Fig. 5).

The conduction current increases when the working distance increases (Fig. 6), a variation that can be explained if the average number of collisions per electron, m, [4] is considered.

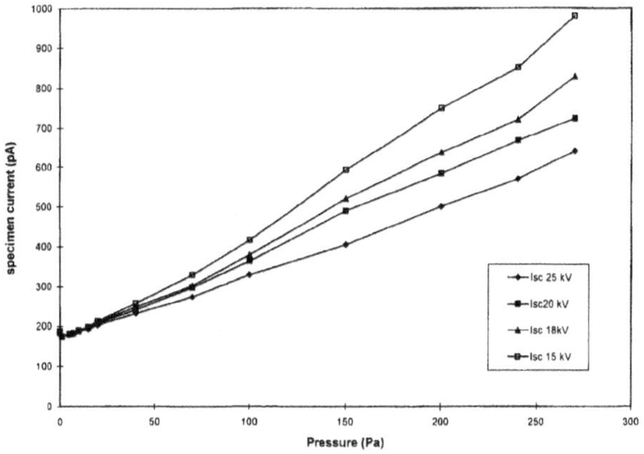

Fig. 4. Influence of the accelerating voltage ($I_{pe} = 250$ pA)

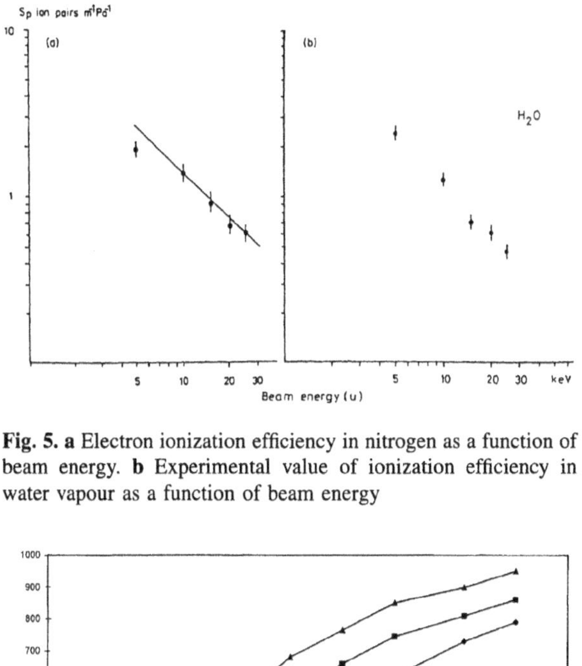

Fig. 5. a Electron ionization efficiency in nitrogen as a function of beam energy. **b** Experimental value of ionization efficiency in water vapour as a function of beam energy

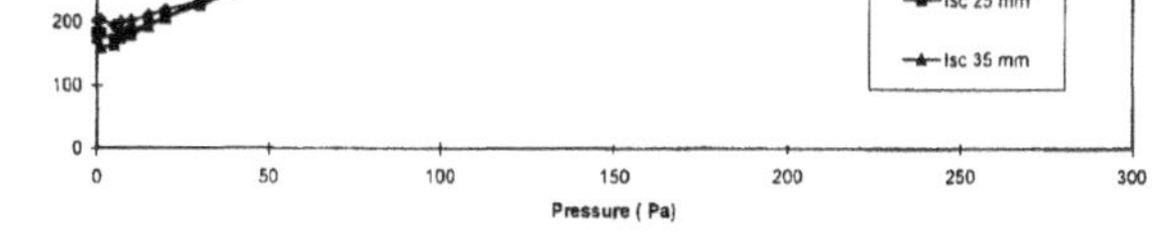

Fig. 6. Influence of the working distance ($I_{pe} = 250$ pA)

This number is defined by

$$m = \sigma p L / kT$$

where σ scattering cross section for the particular gas used, p pressure, L working distance, T temperature.

The relationship indicates that the number of collisions increases with the working distance, thereby corroborating our results.

The general guidelines to reduce beam broadening in the VP-SEM is to choose experimental conditions in the range from 1 to 20 Pa, and to use short working distances and a high accelerating voltage.

X-Ray Microanalysis Results

The presence of a gaseous environment in the specimen chamber presents serious limitations for the use of XRMA with VP-SEM due to three major effects – loss of spatial resolution because of beam skirting, an x-ray contribution from the gas atmosphere and a reduced x-ray count.

Loss of Spatial Resolution Due to Beam Skirting

With the ESEM, the dominant effect of specimen chamber pressure on XRMA spatial resolution has been investigated by several authors [5–8]. Sigee et al. [5–6] proposed a simple experiment to demonstrate the beam skirting which had been reproduced in the VPSEM. An electron beam is directed onto a layer of carbon with a copper target placed at varying distances from the primary probe area. The skirting effect of the beam with an air atmosphere is shown by the ability to pick up a Cu signal of varying intensity from the central probe area. Most data were obtained at distances of 100–1000 µm from the Cu target. At high vacuum, no Cu peak was detected at these distances.

At a distance equal to 1 mm, the copper content varied from 2 to 9% when the pressure ranged from 1 to 270 Pa (Fig. 8a and b). The Cu peaks were always

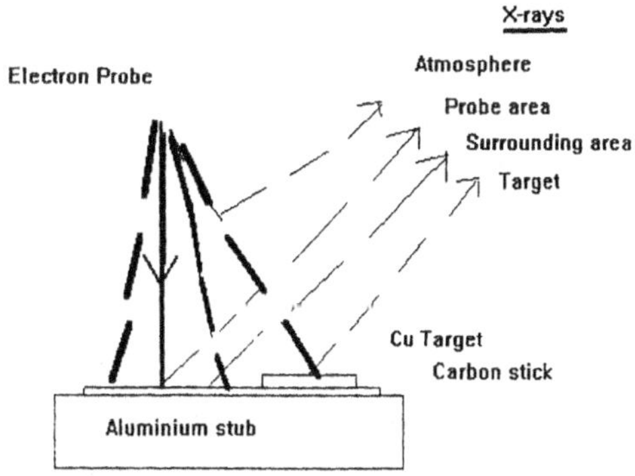

Fig. 7. Demonstration of beam skirting using a Cu target (from ref. [6])

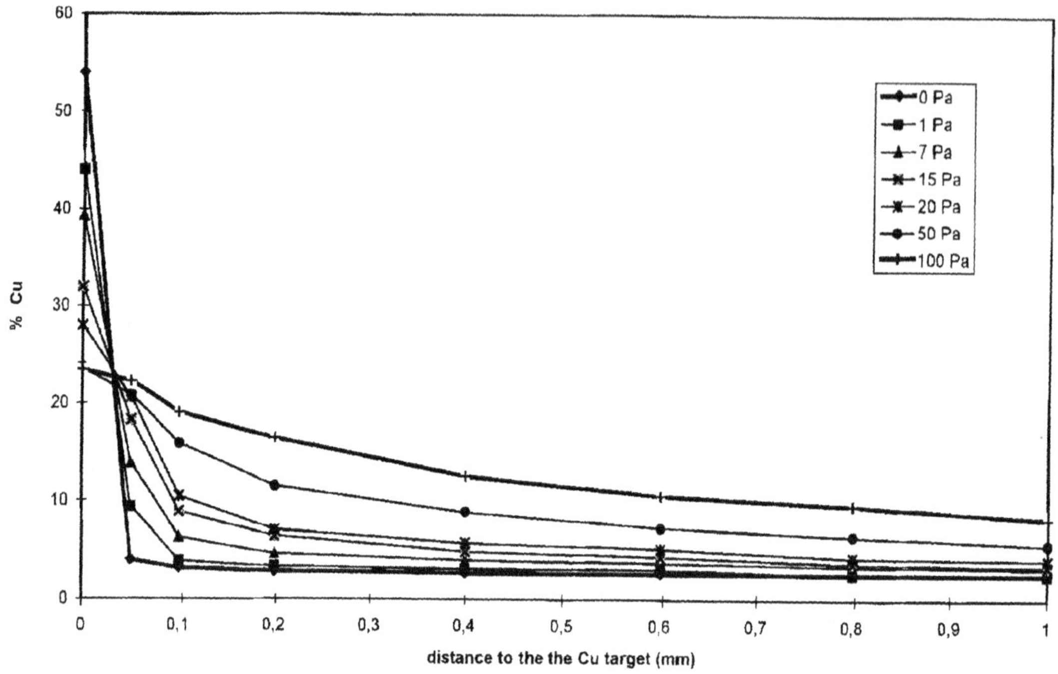

(a)

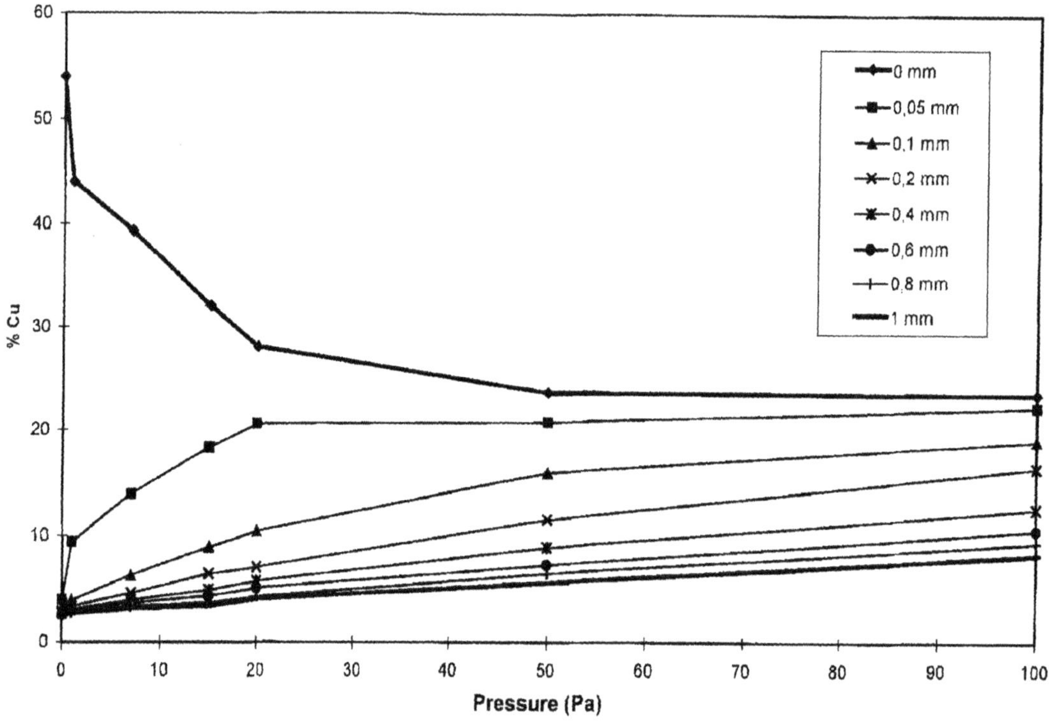

(b)

Fig. 8. a Variation of the Cu content versus the pressure. **b** Variation of the Cu content with the distance to the Cu target

detected over the 100–1000 μm range between 1 to 270, indicating a clear skirting effect in the presence of an atmosphere. In the pressure range from 1 to 20 Pa, the atmospheric level was sufficient to induce perturbations on x-ray microanalysis. Beam skirting clearly limits the spatial resolution of x-ray microanalysis but various operational parameters, including atmospheric pressure, type of gas atmosphere, accelerating voltage and working distance, can be optimised to reduce its effect and to increase XRMA resolution.

Atmospheric pressure: Chamber pressure should be kept as low as possible

Type of gas atmosphere: Using a discrete freeze dried biological sample, Sigee and Gilpin [5] demonstrated a much greater skirting effect with an atmosphere of argon (atomic mass 40) than with water vapour (atomic mass 18). In the VPSEM, the air atmosphere has an average atomic mass 29 and therefore a choice of gas with a low atomic mass such as hydrogen or helium could provide a solution.

Accelerating voltage: This parameter should be kept as high as possible.

Working distance: Reducing the distance travelled by the electron probe within the chamber atmosphere decreases the skirting effect and increases the spatial resolution of XRMA. This was demonstrated by Gilpin and Sigee and is confirmed by current measurement in the previous section.

X-Ray Contribution from the Chamber Atmosphere

Passage of the electron beam through the specimen chamber atmosphere leads to the generation of characteristic and continuum x-rays from the gas molecules. In the ESEM, Sigee and Gilpin [6] demonstrated an approximately linear increase in the P/B ratio for argon (from argon gas) and oxygen (from water vapour) with increase in atmospheric pressure up to 1330 Pa.

A sample (a carbon stick) which contained only C, O, Na was used to determine the effects of the atmospheric contribution in the VPSEM. The analysis was performed at 5 kV and, using the same excitation conditions, the effect of introducing increased amounts of air was investigated. X-ray emission spectra at high vacuum indicated only peaks of C, O, Na and S but introduction of air into the specimen chamber resulted in a rise in the oxygen signal and a corresponding decrease in the levels of C, Na and S.

Table 1. The atmospheric effect. Variation of the sample composition and the total counts with the pressure at 5 kV

P(Pa) %	C	O	Na	Al	S	Nt(cps)
1	90.66	7.9	0.66	0.06	0.66	69724
10	84.88	13.74	0.7	0.37	0.44	71402
20	82.62	15.99	0.43	0.6	0.36	71522
50	77.32	20.10	0.45	1.81	0.31	67785
100	75	20.79	0.35	3.58	0.27	61114
200	70.89	21	0.31	7.53	0.26	43960
270	69.62	21.5	0.25	9.31	0.2	32612
1	90.7	7.91	0.66	0	0.7	
10	85.11	13.87	0.62	0	0.4	
20	82.97	16.24	0.43	0	0.36	
50	78.22	20.98	0.45	0	0.33	
100	76.72	22.61	0.39	0	0.29	
200	74.36	24.94	0.37	0	0.31	
270	74	25.35	0.33	0	0.24	
ref.	92.36	6.33	0.66	0	0.66	

Moreover, the aluminium signal increased with the chamber pressure because of the skirting effect. However, with a homogeneous sample such as this, the skirting effect can be eliminated and the sample composition calculated (see Table 1).

The air atmosphere in the specimen chamber produced an oxygen signal which increased with atmospheric pressure and its presence clearly creates problems for the determination of oxygen content in any specimen analysed. To minimise such effects it is recommended that the chamber pressure is kept as low as possible and that a gas such as H_2 or He is employed so that any atmospheric contribution will not be detected by XRMA.

Reduced X-Ray Count

Introduction of an atmosphere into the specimen chamber can lead to a marked fall in the total counts. In the ESEM, Gilpin and Sigee [5–6] demonstrated that water vapour had relatively little effect on the x-ray count rate up to a pressure of 1330 Pa, but argon caused a marked decrease (leading to a 75% reduction in count rate at 1330 Pa). In the VPSEM, with the same excitation conditions, the air atmosphere has an important effect on the total x-ray counts and the decrease observed can be explained by increased electron scattering which, in turn, reduces electron beam penetration into the specimen.

Some indication of the effects of gas pressure and molecular weight on beam penetration can be obtained from physical modelling of electron scatter-

Table 2. Maximum penetration (μm) of a 10 kV electron beam in water vapour and an argon atmosphere (Monte Carlo Simulation) (from ref [6])

Gas	133 Pa	266 Pa	670 Pa	13300 Pa
Water vapour	$4.52\ 10^6$	$2.26\ 10^6$	$9.03\ 10^5$	$4.52\ 10^4$
Argon	$1.75\ 10^6$	$8.75\ 10^5$	$3.5\ 10^5$	$1.75\ 10^4$

ing using a standard Monte Carlo simulation (Table 2 from ref. [6]). The results obtained from this should only be regarded as semi-quantitative, since the simulation provides a theoretical description of electron penetration in solids rather than gases and does not take into account such factors as Brownian motion and lack of molecular homogeneity. The data obtained do, however, provide an interesting comparison of scattering in water vapour and argon gases, and demonstrate the effect of gas pressure on beam penetration. At each vapour pressure, the water vapour/argon penetration ratio was 2/6. In the VPSEM, with an air atmosphere, the beam penetration is somewhere between water vapour and argon gas and it may be a major factor in terms of x-ray count reduction. In order to reduce this effect, the choice of gas with a low atomic mass such as hydrogen or helium may provide a solution.

Conclusion

The aim of this study was to investigate the effect of electron-beam/gas interactions in the variable pressure scanning electron microscope and determine their influence on quantitative microanalysis results. The beam gas interaction was firstly studied by current measurement as the conduction current appeared a useful parameter to evaluate the skirting effect. To minimise the skirting effect the range of gas pressure in the specimen chamber should be between 1 and 20 Pa, and the instrument should be operated at a low working distance and a high accelerating voltage. The atmospheric effect, which includes an x-ray contribution from the atmosphere, shows an increase in oxygen signal with air pressure inside the chamber. In order to eliminate this, a practical solution is to choose a gas with a very low atomic mass like hydrogen or helium.

Future quantitative x-ray microanalysis in the VP-SEM may involve the introduction of correction factors to take into account some of the effects discussed here.

References

[1] C. Mathieu, *European Microscopy and Analysis* **1996**, *43*, 13.
[2] G. D. Danilatos, *Adv. Electron. Electron Phys.* **1988**, *71*, 109.
[3] G. D. Danilatos, *Scanning Microsc.* **1990**, *4*, 799.
[4] G. D. Danilatos, *Mikrochim. Acta* **1994**, *114/115*, 143.
[5] D. C. Sigee, C. Gilpin, *Scanning Microsc.* **1994**, *8*, 219.
[6] C. Gilpin, D. C. Sigee, *J. Microsc.* **1995**, *179*, 22.
[7] R. B. Bolon, *Microbeam Anal.* (D. G. Howitt, ed.), San Francisco Press, San Francisco, 1991, p 201.
[8] R. Carlton, *Scanning* **1997**, *19*, 85.
[9] D. E. Newbury, *Scanning Electron Microsc.* **1976**, *1*, 112.
[10] A. N. Farley, J. S. Shah, *J. Microsc.* **1990**, *158*, 379.
[11] D. A. Moncrieff, V. N. E. Robinson, *J. Phys. D: Appl. Phys.* **1978**, *11*, 2315.
[12] D. A. Moncrieff, P. R. Barker, V. N. E. Robinson, *J. Phys. D: Appl. Phys.* **1979**, *12*, 481.

Mikrochim. Acta [Suppl.] 15, 301–305 (1998)

The Analytical Signal in EPMA and the Influence of the Electric Field Created by the Primary Beam

Ondrej Gedeon[1,*], **Václav Hulínský**[1], and **Karel Jurek**[2]

[1] Institute of Chemical Technology, Department of Glass and Ceramics, 166 28 Prague, Czech Republic
[2] Institute of Physics, Academy of Sciences of Czech Repubic, Prague, Czech Republic

Abstract. The electric field created by the primary electron beam inside an insulator may change the analysed X-ray signal compared to that from a conductor. It is assumed that homogeneous charge build-up occurs inside the excited volume. Distributions of slowed down primary electrons are calculated by the Monte Carlo method for light and heavy element samples at both 10 keV and 50 keV. Because the incoming electron will be affected by the Bethe stopping power and the electric field it is advantageous to express the former in V/m to explain the qualitative features of the electric field's influence. A few distribution functions were calculated to estimate the lowest value of electric field that would be necessary in the sample to influence the analytical signal.

Key words: EPMA of insulators, electric field in insulators, Monte Carlo, analytical signal.

A primary electron beam injected into an insulator creates an electric field within the excited volume. The field exists even if the specimen surface is coated by a conductive layer and it may produce structural changes in the sample by causing bonds to break and then producing subsequent migration of certain chemical species [7]. The other possible effect [4] is modification of the analytical signal caused by primary electron deflections and by changes in primary electron energy loss process due to retardation by the field. A change in the analysed signal coming from an insulator compared to a conductor is one of the most interesting problems in quantitative analysis using EPMA [1].

This paper discusses the influence on the detected signal of an electric field created inside the insulator. We assume an insulator target coated by a conductive thin layer set at a ground potential. We also neglect the influence of the upper, positively charged, layer created by secondary electron emission [4] which is much smaller than the electron range. The steady state charge distribution takes place in a time much shorter than that required for analyses so that we do not need to take into account the dynamics of the process.

Electrons inside the sample can, at any instant in time, be divided into three groups. Primary electrons, not yet completely slowed down, represent the first group and their energy ranges from the incident beam energy down to a few eV. These electrons are affected by the insulating properties of a specimen only at high overvoltages where the relaxation time is long, or when a focused beam and high current are used.

Electrons captured in electronic structure defects of the target form the second group. These are localised, having their energy below the bottom of the conductivity band.

The third groups consists of electrons which have slowed down and are now moving in the conduction band. These are either electrons previously captured and then released or electrons that have never fallen into an energy trap; they have energy of a few eV.

As a first approximation, the charge density of the electrons from the first group is independent of the

* To whom correspondence should be addressed

sample state, i.e. it is not sensitive to whether the sample is conductive or nonconductive. This is unlike the charge density of the electrons forming the second and the third groups.

It is quite reasonable to expect that the ratio between the number of electrons in the second and the third groups will be high for glass samples, where there are more defects in the electronic structure than is the case for crystals. This means that an excellent insulator does not guarantee that a higher electric field will be created in the sample because there may be a lack of traps inside.

Results and Discussion

As outlined, a detailed picture of the electron density distribution requires knowledge of electronic structure of the sample which can, itself, be influenced by incoming electrons. A minimum requirement is to calculate the charge distribution of slowed down primary electrons, to serve as a source of the relaxed charge distribution. To achieve this we have used the Monte Carlo algorithm described elsewhere [5] and slightly modified it for our purpose. The calculations were performed without the presence of an electric field. Figures 1 and 2 show charge density distributions for lithium and gold, elements representative of a light and heavy matrix respectively, and for 10 keV and 50 keV. Differences between the light and heavy matrices are similar for both energies: the distribution in the heavy matrix is shifted toward the surface of the sample due to higher elastic scattering. These distributions are close to reality if the slowed down electron is instantly captured and its lifetime in the localized state is much longer than the time between its release from the trap and its escape from the sample. This is equivalent to a sample in which there are a relatively small number of energetically deep defects. A comparison with the homogeneous charge distribution is also shown in the figures.

The real overall charge distribution involving all three groups is different thanks to reasons mentioned above. As a first approximation we will assume a homogeneous distribution of charge inside the sample. This implies for the electric field E_E

$$E_E(z) = \frac{\rho}{\varepsilon}(z - R), \qquad (1)$$

where ρ is the charge density, ε is the permittivity of the sample and R is the electron range. Under these assumptions the electric field is a maximum at the surface and decreases linearly to zero at the electron range.

The electrons entering the sample are driven by two forces. The first is electric in nature and its strength is expressed by Eq. (1). The second is the effective quantum-mechanical force resulting from the interaction of the primary electron with the sample. The force decelerating an electron with energy E can be described by the Bethe stopping power [3] over the electron range relating to EPMA and its corresponding intensity is

$$E_B = \frac{2\pi e^2}{(4\pi\varepsilon_0)^2}\frac{Z}{A}\frac{\kappa}{E}\ln\left(1.116\frac{E}{J}\right) \quad \text{in} \quad [V/m], \qquad (2)$$

where the ionisation potential J is given by [2]

$$J = 9.76Z + 58.8Z^{-.19.}, \quad Z > 6 \qquad (3a)$$

$$J = 11.5Z; \quad Z \leq 6. \qquad (3b)$$

The symbols have following meaning: e-Elemental charge; Z-Atomic number; A-Relative atomic weight; k-Density of the sample.

E_B is directly proportional to sample density and inversely proportional to electron energy (neglecting the logarithmic term). This means that denser matrices and lower energies increase E_B and that, in these circumstances, the influence of any electric field is likely to be less.

It is advantageous to express E_B in V/m (Eq. (2)) so that we can compare it directly with the electric field of a particular strength created in the sample by electrons. Fig. 3 shows E_B versus energy for a hypothetical sample ($Z/A = 1/2$, $\kappa = 1\,\mathrm{g/cm^3}$). The figure instantly proves that no influence of the electric field on primary electrons can be expected for the low electron energy region because E_B substantially exceeds the electric breakdown limit.

Expressing E_B in terms of V/m is quite instructive but as the electron penetrates into the sample its energy decreases and therefore the influence of the electric field increases. However, in the above consideration the effects of the field on elastic scattering have been ignored together with the result that the electric field situated deeper in the sample is weaker. This is why we can obtain a better estimation of the influence of the electric field by calculating the distribution functions. To do this we have used the Monte Carlo algorithm again. In the model we have

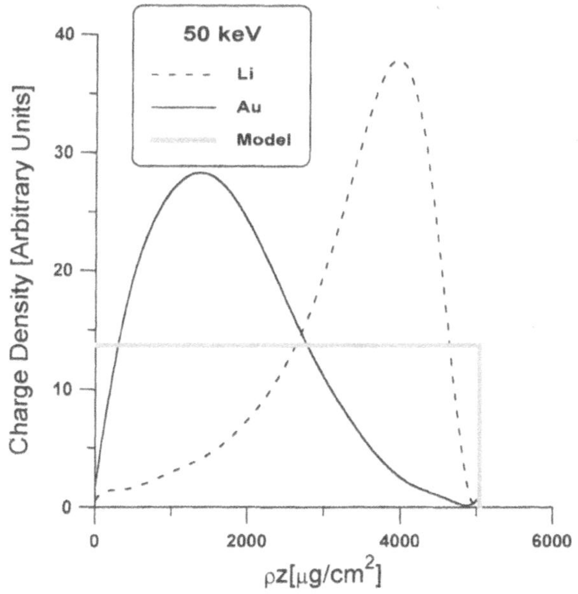

Fig. 1. Charge density of slowed down primary electrons for a 50 keV primary beam in lithium and gold matrices. The model assuming a homogeneous charge density is also depicted

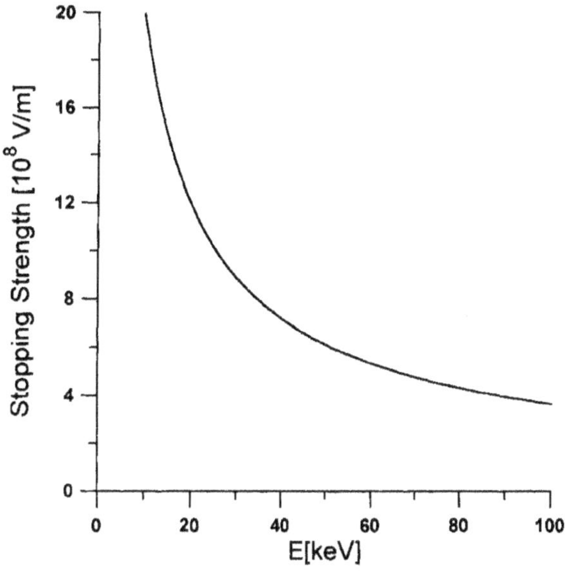

Fig. 3. Bethe stopping intensity, E_B, for a hypothetical sample ($Z = 1$, $A = 2$ and $\rho = 1\ \mathrm{g/cm^3}$) expressed in V/m

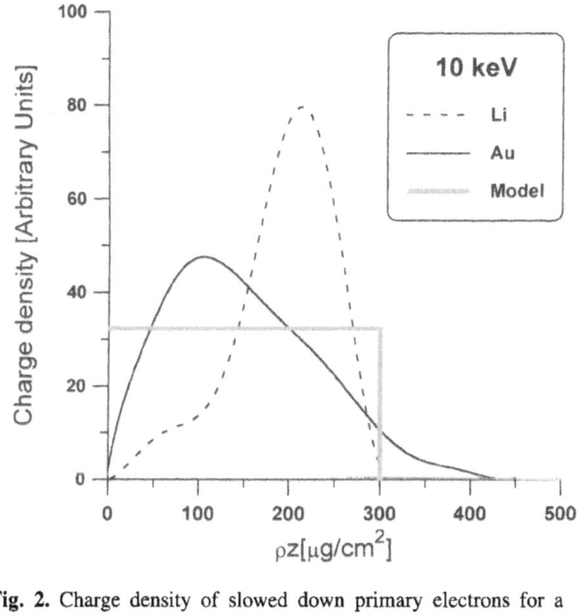

Fig. 2. Charge density of slowed down primary electrons for a 10 keV primary beam in lithium and gold matrices. The model assuming a homogeneous charge density is also depicted

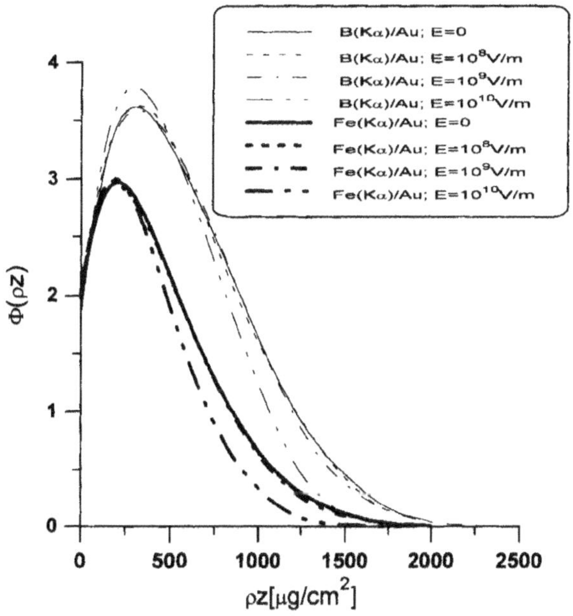

Fig. 4. x-ray depth distribution functions of Kα lines for boron and iron in a gold matrix at a primary energy of 30 keV showing the effects of different values of the maximum electric field

assumed a homogeneous charge distribution as depicted in Figs. 1 and 2. The lithium matrix and gold matrix represent two extreme cases. The lithium sample represents a matrix with low density and low atomic number. The low atomic number implies a deeper electron penetration into the sample and a greater influence of the electric field on the electron due to its longer path. The low density matrix reduces

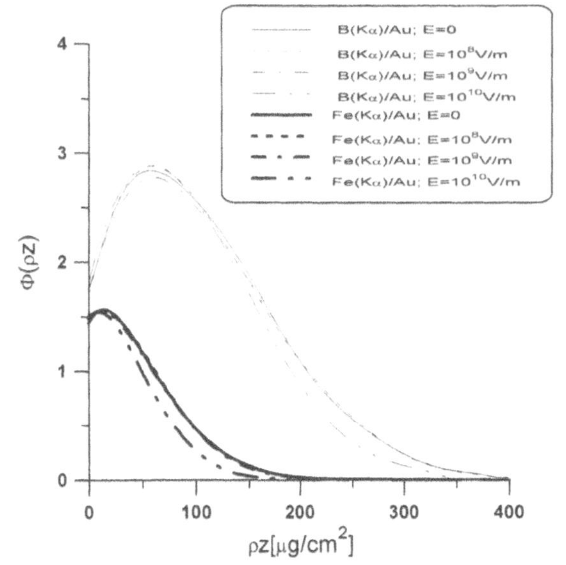

Fig. 5. x-ray depth distribution functions of $K\alpha$ lines for boron and iron in a gold matrix at a primary energy of 10 keV showing the effects of different values of the maximum electric field

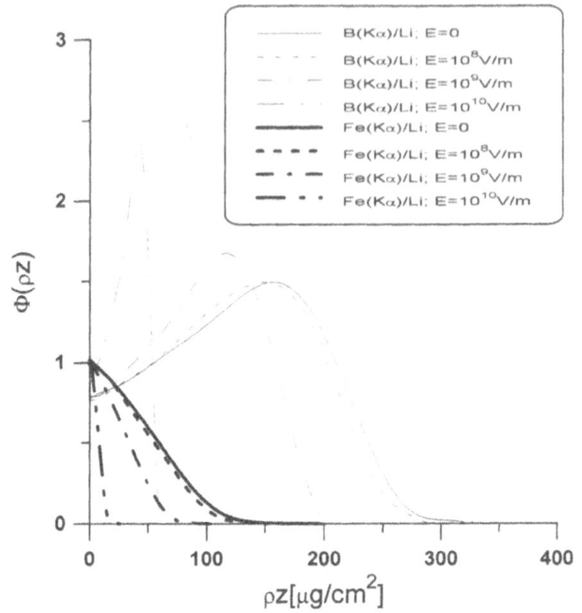

Fig. 7. x-ray depth distribution functions of $K\alpha$ lines for boron and iron in a lithium matrix at a primary energy of 10 keV showing the effects of different values of the maximum electric field

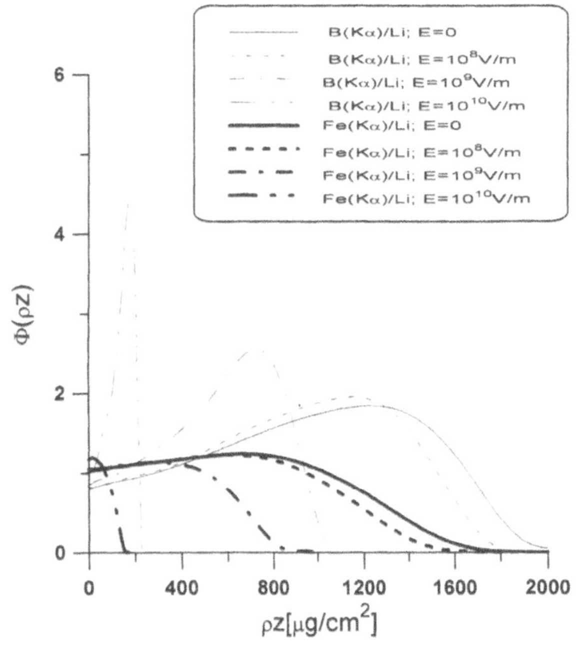

Fig. 6. x-ray depth distribution functions of $K\alpha$ lines for boron and iron in a lithium matrix at a primary energy of 30 keV showing the effects of different values of the maximum electric field

E_B and emphasizes the influence of the electric field. Therefore, we can expect the lithium-like matrix to be more sensitive to a particular electric field than the gold-like matrix.

The electric field intensities assumed at the surface were varied. The electric field affected the electron between successive interactions with atoms in two ways: changing its energy (accelerating or decelerating the electron) and altering its movement (by deflecting the electron from its original path). The results for the gold matrix are presented in Figs. 4 and 5 for 30 keV and 10 keV, respectively. Distribution functions for the gold matrix explicitly prove the predominant role of the dense matrix where a field higher than 10^9 V/m at the surface is necessary to influence the primary electron and the analysed signal. The results for the lithium matrix are presented in Figs. 6 and 7 for 30 keV and 10 keV, respectively. For this matrix (approximately 40 times lighter than gold matrix) the field is about an order lower and the influence of the primary energy is stronger. From this, we can conclude that the electric field intensity at the surface should be at least approximately 10^8 V/m to influence the analytical signal and affect any quantitative analysis of the sample. Estimation of intensity by studying alkali diffusion in soda-silica [8] and soda-lime-silica glass [6] gives 10^6V/m; a value substantially lower than that which we have found necessary to influence the x-ray signals.

Conclusion

Assuming a homogeneous distribution of electric charge inside the sample we have shown, by calculating distribution functions, that an electric field strength of at least 10^8V/m must be created to influence primary electrons and the resulting x-ray signal. This is the case for a lithium-like matrix with low density and low atomic number. For denser samples with higher average atomic numbers the intensity of electric field must be even greater. It is still an open question as to whether such electric fields can be created except in special samples.

References

[1] G. F. Bastin, H. J. M. Hijligers, in: *Electron Probe Quantification* (K. F. J. Heinrich, D. E. Newbury, eds.), Plenum, New York, 1991, p. 193.

[2] M. J. Berger, S. M. Seltzer, *Nat. Res. Council. Publ. 1133, Washington: Nat. Acad. Sciences*, 1964, p. 252.

[3] H. Bethe, *Ann. Physics* **1930**, *5*, 325.

[4] J. Cazaux, *X-Ray Spec.* **1996**, *25*, 265.

[5] O. Gedeon, V. Hulínský, *Mikrochim. Acta* **1994**, *114/115*, 305.

[6] O. Gedeon, K. Jurek, V. Hulínský, in preparation.

[7] K. Jurek, O. Gedeon, V. Hulínský, *Mikrochim. Acta*, submitted for publication.

[8] A. Miotelo, *J. Phys. C: Solid State Physics* **1986**, *19*, 445.

Mikrochim. Acta [Suppl.] 15, 307–316 (1998)

Standardless Analysis

Johann Wernisch* and **Kurt Röhrbacher**

Institute of Applied and Technical Physics, Technical University of Vienna, Wiedner Hauptstrasse 8–10, A-1040 Wien, Austria

Abstract. Three totally different methods for standardless analysis are reviewed and a new method is presented. Advantages, restrictions and estimated errors of these methods are discussed. For the theoretical prediction of standard net intensities it is shown that the fundamental parameters for K X-rays imply errors at least to the same extent as those for L radiation.

Key words: standardless analysis, EDS, ionisation cross section, fluorescence yield, TWIX.

When G. Bastin, a recognised expert in the field, was asked about standardless analysis his ironic reaction was "standardless analysis – the next step will be sampleless analysis and still at the highest precision!". It is now more than 20 years since the first standardless methods for composition determination by EPMA were developed, but none of these or subsequent methods have found any acceptance amongst electron microscopists. Much larger errors are reported than occur in conventional analysis with standards and to cite another famous researcher "is it worth the risk?" [1]. The aim of this work is to give a review of several standardless methods including their respective advantages and limitations.

Theory

Formerly, the net X-ray intensity from a sample under electron impact at an acceleration energy E_0 and X-ray take-off angle ψ was equated to the product of three factors dependent upon atomic number Z, X-ray absorption, A, $(=f(\chi)=F(\chi)/F(0))$ and, if necessary, secondary fluorescence F (eq. 1a).

$$I = n^{e-} \cdot \frac{\Omega}{4\pi} \cdot \frac{N_0}{a} \cdot c \cdot \omega \cdot p \cdot \varepsilon(E) \cdot R \cdot \int_{E_0}^{E_c} \frac{Q(E)}{dE/d\rho z} \cdot dE \cdot f(\chi) \cdot F \quad (1a)$$

$$= n^{e-} \cdot \frac{\Omega}{4\pi} \cdot \frac{N_0}{a} \cdot c \cdot \omega \cdot p \cdot \varepsilon(E) \cdot \int \varphi(\rho z) \cdot \exp\left(-\frac{\mu}{\rho} \cdot \frac{\rho z}{\sin\psi}\right) \cdot d\rho z \cdot F \quad (1b)$$

This so-called ZAF approach has now been replaced by a more appropriate $\varphi(\rho z)$ approach (Eq. 1b), which incorporates atomic number and absorption corrections simultaneously. Here the depth distribution function $\varphi(\rho z)$ represents the amount of ionisation occurring in mass depth ρz and $\exp(-\chi \cdot \rho z)$ allows for the attenuation of the generated X-rays in the sample. Both expressions (1a) and (1b) are equivalent by the correlation $F(0) \cdot Q(E_0) = R/S$ [see Ref. 2].

Standardless Methods

1. Conventional Standardless Analysis (Method of Intensity Calculation)

In conventional analysis *with* standards the so-called k-ratio is utilised, defined as the ratio of the net intensity of the X-ray line of interest from the specimen to that from a reference sample. Several factors such as the number of electrons n^{e-}, the detector entrance angle $\Omega/4\pi$, the fluorescence yield ω, the relative line weight p, and the detector response $\varepsilon(E)$ are cancelled out of Eqs. (1a) and (1b). In addition, any errors from an inaccurate take-off angle

* To whom correspondence should be addressed

or kilovoltage setting and even from spectrum processing during background subtraction or peak deconvolution (in cases of peak overlap) generally compensate when the same analysis routines are applied to a reference specimen. This is not the case when there is no standard, as occurs for inert gases, highly radioactive elements or non-vacuum resistant compounds, and the net intensity of the standard has to be calculated theoretically to obtain the required k-ratio. Since the integrals in Eq. (1) both for the ZAF and for the $\varphi(\rho z)$ approaches are functions of the sample's composition, the weight concentrations of all matrix elements must be calculated iteratively. Several different iteration algorithms have been reported (simple iteration, hyperbolic iteration, Wegstein iteration [3], Ebel iteration [4]). Hyperbolic iteration is probably the most popular currently but, even for large correction factors, all of them typically converge within five iteration steps. If no standard is involved at all, the weight fractions c_i of the matrix elements require renormalising by Eq. (2a) after each iteration step. This condition aids convergence, but redistributes the sum of errors in the concentrations determined, smaller concentrations being more sensitive to any error [3, 5].

$$c_i = \frac{c_i}{\sum_j c_j} \tag{2a}$$

$$k_i = \frac{k_i}{\sum_j k_j} \tag{2b}$$

While n^{e-} and $\Omega/4\pi$ can be eliminated by setting the sum of all experimental k-ratios to one (Eq. 2b), all other relevant parameters have to be inserted in Eq. (1). The crucial problem is deciding which of the large theoretical, semi-empirical and experimental data sets to choose, because all parametrizations scatter largely as a function of atomic number and energy. Newbury et al. [6] have carried out careful measurements and analyses on 23 pure elements and compounds to simulate the net $K\alpha$ intensity from a (fictitious) argon standard and to enable an estimation of the resulting error. They point out that spectral simulation should be preceded by a systematic comparison of measured with theoretical peaks for the shell and energy range of interest to find the best choice of parameters. This procedure reduces anticipated errors significantly, but we wish to state here that choosing parameters found in such an extensive study may not necessarily turn out to give the best results when measurement conditions are different (beam energy, take-off, thin films, ...).

Russ [7] has taken the ionisation cross section of Green and Cosslett [8] and the energy loss law of Thomson and Whiddington [9] to enable analytical integration of Eq. (1a) and obtained relative deviations from nominal concentrations in steel samples of less than 5% when $c > 1$ wt% and less than 60% when $c < 1$ wt%. Nasir [10] employed the more accurate Bethe energy loss law [11] and used Eq. (1a) to construct an intensity ratio from two matrix elements to eliminate any dependence on atomic number. Both authors report significant errors if the sample contains elements with widely differing atomic numbers and in many articles the uncertainties in the fundamental parameters of L lines are blamed for causing such difficulties. In this paper we show that the specific values and data fits for the K lines, which are usually employed in standardless programs, introduce errors at least to the same extent as those for L lines. For this purpose we plot ratios of different ionisation cross-sections and ratios of different fluorescence yields against atomic number. In these plots it can be seen directly where deviations occur if a change from one data set to another is performed during the analysis procedure.

1a. Ionisation Cross Sections and Fluorescence Yields

After electron impact the general expression for ionisation of atomic shells with quantum numbers n and 1 has the form

$$Q_{nl} \cdot E_C^2 = 6.51 \cdot 10^{-14} \cdot Z_{nl} \cdot b_{nl} \cdot \ln(c_{nl} \cdot U_{nl}) / U_{nl}^m \ [\mathrm{cm^2\,eV^2}] \tag{3}$$

where $c_{nl} \neq 1$ and $m = 1$ corresponds to the Bethe formula [11]. The term U_{nl} stands for the overvoltage ratio $U = E/E_C$ with E_C as the binding energy of the relevant shell, Z_{nl} describes the number of electrons in the shell and b_{nl} and c_{nl} are parameters specific to the shell and element. In a so-called Fano plot, experimental data $Q \cdot E_C^2 \cdot U$ are plotted against $\ln(U)$ allowing a convenient evaluation of the Bethe parameters b_{nl} and c_{nl}. Eq. (3) has been derived by use of the first Born approximation assuming that U_{nl} is sufficiently large [12]. Hence, especially for $c_{nl} < 1$, this expression becomes meaningless near the threshold for ionisation $U \sim 1$. This is why a number of

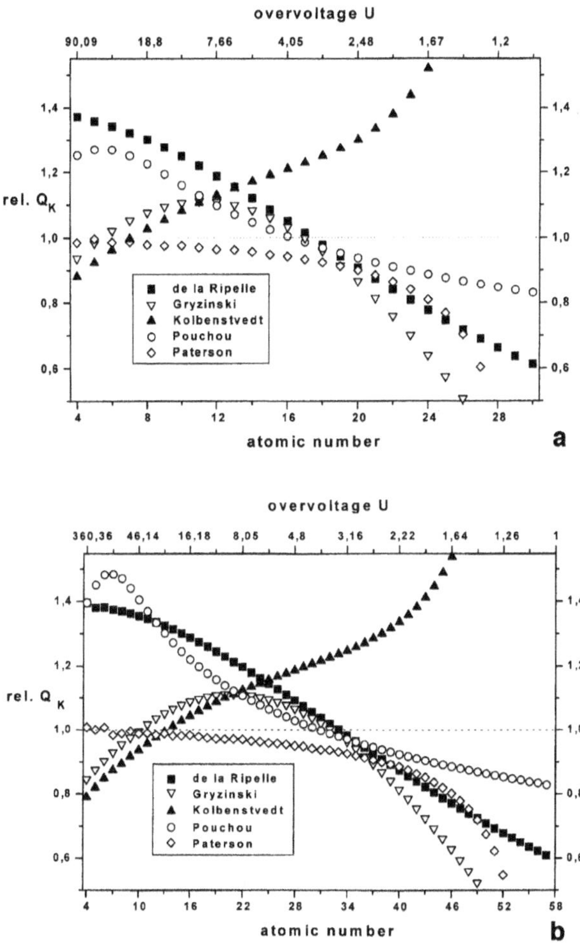

Fig. 1. Several expressions for the cross section of K shell ionisation relative to that of Green and Cosslett are plotted versus the atomic number. For a fixed acceleration energy of $E_0 = 10\,\text{keV}$ (**a**) or $E_0 = 40\,\text{keV}$ (**b**) this is equivalent to plotting the ratio versus overvoltage $U_0 = E_0/E_C$. Deviations of up to 40% resulting from the different parametrizations enter directly into the analytical results in a TEM. In the case of bulk analysis the area under each curve from the right hand axis up to the applied overvoltage U_0 gives approximately the amount of ionisation in the sample relative to that of the Green expression

authors returned to the idea of Webster [13] in selecting $c_{nl} = 1$ and $m \neq 1$ [14, 15].

In Fig. 1a and 1b several expressions for the ionisation cross section of the K-shell [15–17] are ratioed to that of Green and Cosslett [8] and plotted against atomic number, Z, for an acceleration energy, E_0, of 10 and 40 keV. At a fixed value of E_0 this corresponds to a plot of Q_K versus decreasing overvoltage, U, since binding energy increases with atomic number. The second plot is identical to the first except for the expression of Pouchou [15], which is a function of the atomic number, and the fact that the Z labels have an offset according to the respective overvoltage. The expression of Green and Cosslett has been chosen as a reference because of its broad popularity. As can be seen all ionisation cross sections are almost identical at an overvoltage of $U \sim 8$ (corresponding to $Z \sim 12$ at 10 keV and $Z \sim 22$ at 40 keV) but differ from each other by up to 40% when overvoltages and atomic numbers are high or low. Of course a comparison like this can be applied directly to estimate analytical results only in a TEM, where for ultra thin foils the integral in Eq. (1a) reduces to the simple term $Q(U)$; in a SEM the integral has to be fully executed. Using the Thomson-Whiddington law $dE/d\rho z \propto -b/E$ as an approximation for the stopping power, reduces the integral to $\int Q(U) \cdot dU$. The ratio of ionisations in the sample at a certain overvoltage U to those according to Green-Cosslett is then equivalent to the area under the respective curve from the right hand ordinate up to the overvoltage in question. As can be seen, all Q_K parametrizations except that of Kolbenstvedt [18] lead to less ionisation (negative area) compared to Green and Cosslett up to $U \sim 3$ and then partially compensate at higher overvoltages; an exception to this is Paterson [17] which remains smaller for all values of U.

Therefore deviations of up to 80% (Gryzinski, Ref. [19]) for overvoltages, U_0, less than 3 or 4 can be expected depending on the expression employed in the standardless procedure. The Kolbenstvedt expression leads to significantly greater ionisation up to $U_0 \sim 8$ compared to all others and we recommend its use only in a TEM where overvoltages less than 8 do not arise because electrons pass through the foil before they have been decelerated to low energies. Poor agreement of Kolbenstvedt's formula with experimental data for $U < 5$ has also been documented in a review of inner shell ionisation cross section by Powell [16].

In the case of analysis using L lines the factor $(1 + g_{ck})$ induced by radiationless Coster-Kronig transitions f_{ij} has to be taken into account, which enhances the amount of primary L_3 subshell ionisations N_3 by

$$(1 + g_{ck}) = 1 + [N_1 \cdot (f_{12} \cdot f_{23} + f_{13}) + N_2 \cdot f_{23}]/N_3 \tag{4}$$

As has been reported previously [5], Krause's semi-empirical data set of g_{ck} [20] differs by 30% from the theoretical predictions of McGuire ([21], approximate Hartree-Slater theory) and Chen et al. ([22], Dirac-Hartree-Slater theory). We prefer to examine the

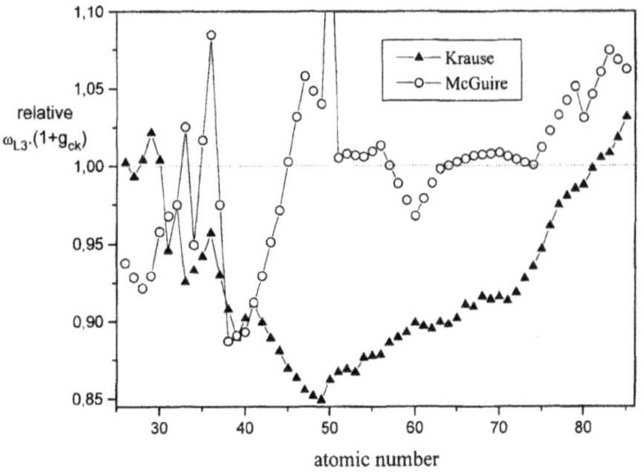

Fig. 2. The product of the fluorescence yield ω_{L_3} of the L_3 subshell with the enhancement factor $(1+g_{ck})$ of the L_3 subshell ionisation cross section of McGuire ([21], AHS theory) and Krause ([20], semi-empirical) relative to that of Chen et al. ([22], DHS theory) versus the atomic number. Krause's overestimation of ω_{L_3} brings his underestimation of g_{ck} at $Z=29$ in excellent accord with the theoretical data (see text)

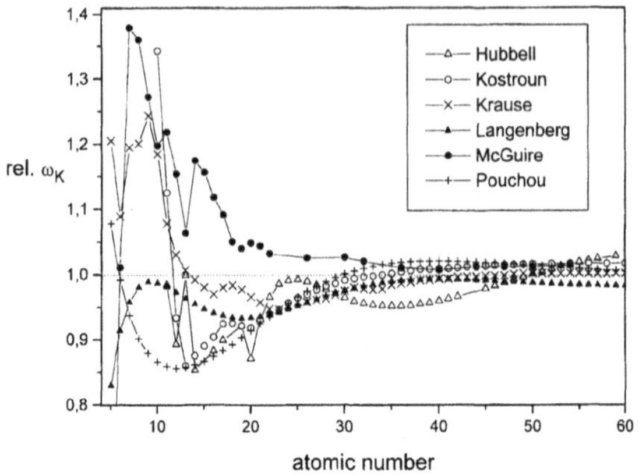

Fig. 3. The fluorescence yield ω_K after K shell ionisation of several authors relative to that of Bambynek's fit of "most reliable experimental data" [26]. Good agreement between all parametrizations is given only for atomic numbers $Z>21$, while deviations at lower atomic numbers, e.g. between the data of Pouchou and Krause at $Z\sim 8$, exceed 35%

product of $(1+g_{ck})$ and the fluorescence yield ω_{L_3}, which is also included in these three tables, because we think that Krause partially overestimated ω_{L_3} values. In fact, Krause's semi-empirical g_{ck} values differ most (29%) from the theoretical predictions at $Z=29$, but a 11% overestimation of ω_{L_3} in copper brings the product $\omega_{L_3}\cdot(1+g_{ck})$ in excellent accord with the theoretical data of Chen et al. (Fig. 2). Only this product multiplied by the respective cross section is physically observable as it represents the total number of generated X-rays originating from the L_3 subshell. Nevertheless, between atomic numbers $Z=40$ and $Z=73$, a difference of 10–15% remains between theory and Krause's semi-empirical data for $\omega_{L_3}\cdot(1+g_{ck})$. These deviations enter directly into Eq. (1) and can affect analytical results severely. What has not been recognised until now is that published data on *K* shell fluorescence yields [see 15, 20, 23, 44] imply even greater uncertainties. Fig. 3 shows some well known expressions for ω_K with different Wentzel parameters following Wentzel parametrization [25]

$$\frac{\omega_K}{1-\omega_K}\propto Z^4$$

or

$$\omega_K=\frac{\sum_{i=0}^{4}B_i\cdot Z^i}{1+\sum_{i=0}^{4}B_i\cdot Z^i} \tag{5}$$

relative to that of Bambynek [26], which represents a fit of "most reliable experimental data". Good accord between all sets can be observed for $Z>21$ but significant deviations (up to 40%) are visible for $Z<15$; e.g. between the data of Krause and Pouchou [15]. From this point of view it is clear that good results can be obtained if K lines are involved from elements with neighbouring atomic numbers greater than 20 (e.g. steel), but severe errors can occur when K line elements with widely differing atomic numbers are analysed.

1b. Relative Transition Probability *p*

This parameter in Eqs. (1a) and (1b) describes the probability of a particular X-ray line being generated compared to the probability of generating any X-ray line from the same subshell. For $K\alpha$ lines this is

$$p(K\alpha)=\frac{I_{K\alpha}}{I_{K\alpha}+I_{K\beta}}$$

and

$$p(L\alpha)=\frac{I_{L\alpha}}{I_{L\alpha}+I_{L1}+I_{L\beta2,15}+I_{L\beta6}+\cdots} \tag{6}$$

for $L\alpha$ lines. Theoretical values for most elements have been tabulated by Scofield [27] and have proved to be highly reliable [28].

This parameter must not be confused with the relative line weights a_K and a_L which describe, for example, the fraction of $L\alpha$ photons in the whole emitted L peak. These line weights are not elemental constants but depend upon the acceleration energy [29]. As has been shown recently, this effect results from the different mass attenuation coefficients

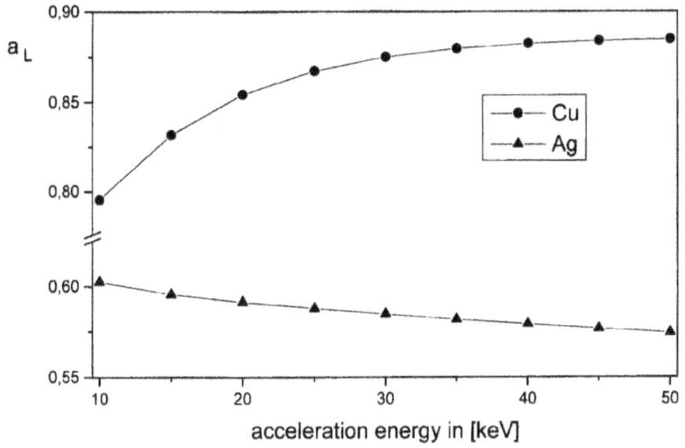

Fig. 4. The relative intensity factor a_L for copper and silver versus acceleration energy. a_L decreases from 10 to 50 keV by 4.4% relatively, while a_L for copper shows a strong energy dependence and increases by 10.2% relatively due to mass attenuation effects. According to the data fit of Heinrich [37] $(\mu/\rho)_{\mathrm{Cu}\,L\beta} \approx 5.22 \cdot (\mu/\rho)_{\mathrm{Cu}\,L\alpha}$ in Cu, $(\mu/\rho)_{\mathrm{Ag}\,L\beta} \approx 0.87 \cdot (\mu/\rho)_{\mathrm{Ag}\,L\alpha}$ in Ag

associated with the X-ray lines comprising the L peak [30]. Particularly for atomic numbers less than 33 the energy of the strongest characteristic $L\beta$ line, $L\beta_1$, lies above the excitation energy for L_3 subshell ionisation and thus self-attenuation of $L\beta_1$ lines is 5 times higher than the respective $L\alpha$ line. With increasing beam energy and generation depth of L X-rays the fraction of $L\beta$ lines in the emitted L peak decreases. Fig. 4 shows the dependence of the relative line weight a_L in pure copper and silver on the acceleration energy E_0 at a take-off angle of 35°.

Since a specific line of element i is attenuated in all elements j of the matrix by

$$(\mu/\rho)_i = \sum_j c_j \cdot (\mu/\rho)_j \tag{7}$$

a_L is also a function of the composition of the investigated sample. Hence a general formula for a_L as a function of E_0 and c_i is not possible.

For this reason we recommend using the total L peak for analysis instead of deconvoluting when peak overlap arises ($Z < 50$). Consequently the theoretical intensities for $L\alpha$ and $L\beta$ radiation have to be added to obtain the total measured L counts

$$\begin{aligned} I_L^{\mathrm{exp}} &\propto c \cdot Q_L \cdot \omega_L \cdot p_{L\alpha} \cdot \varepsilon_{L\alpha} \\ &\quad \cdot \int \varphi(\rho z) \cdot \exp\left(-\left(\frac{\mu}{\rho}\right)_{L\alpha} \cdot \frac{\rho z}{\sin\psi}\right) \cdot d\rho z \\ &\quad + c \cdot Q_L \cdot \omega_L \cdot p_{L\beta} \cdot \varepsilon_{L\beta} \\ &\quad \cdot \int \varphi(\rho z) \cdot \exp\left(-\left(\frac{\mu}{\rho}\right)_{L\beta} \cdot \frac{\rho z}{\sin\psi}\right) \cdot d\rho z \end{aligned} \tag{8}$$

where $Q_L \cdot \omega_L$ is the total fluorescence yield after L shell ionisation. Here the parameters $p_{L\alpha}$ and $p_{L\beta}$ are constants specific to each element but independent of experimental conditions. Furthermore it can be shown that, by developing the exponential functions and assuming that $p_{L\beta} = 1 - p_{L\alpha}$, Eq. (8) can be rewritten as

$$I_L^{\mathrm{exp}} \propto c \cdot Q_L \cdot \omega_L \cdot \varepsilon \cdot \int \varphi(\rho z) \cdot \exp\left(\frac{\bar{\mu}}{\rho} \cdot \frac{\rho z}{\sin\psi}\right) \cdot d\rho z \tag{9}$$

with

$$\frac{\bar{\mu}}{\rho} = p_{L\alpha} \cdot \left(\frac{\mu}{\rho}\right)_{L\alpha} + p_{L\beta} \cdot \left(\frac{\mu}{\rho}\right)_{L\beta} \tag{9a}$$

This is still correct for first order ρz (higher orders can be neglected since $\rho z \sim 10^{-4}$) and is equivalent to Eq. (1b) but for one (imaginary) L line with a mass attenuation coefficient which is weighted by the generation probabilities $p_{\mathrm{L}\alpha}$ and $p_{L\beta}$ [30].

The variation of the relative line weight a_L with the acceleration energy influences analysis especially if L lines from germanium or below are investigated. We think that this is the reason why Russ found his analytical results agreeing better with the nominal concentrations when he applied total L line count rates in his procedure [31].

1c. Detector Efficiency

The detector response term $\varepsilon(E)$ in Eq. (1) is responsible for the fact that conventional standardless analysis is not possible with wavelength dispersive spectrometers because of alignment problems associated with the crystals. In an energy dispersive system on the other hand the detector sensitivity is determined by the beryllium window, the gold conductive layer, the silicon dead-layer and a contamination layer that the generated X-radiation has to pass through before being recorded. Hence the response function can be characterised by

$$\begin{aligned} \varepsilon(E) = \frac{\Omega}{4\pi} \cdot \prod_{i=\mathrm{Be,Au,Sin,cont.}} \exp\left[-\left(\frac{\mu}{\rho}\right)_i \cdot \rho d_i\right] \\ \cdot \left[1 - \exp\left(-\left(\frac{\mu}{\rho}\right)_{\mathrm{Si}} \cdot \rho d_{\mathrm{Si}}^{\mathrm{active}}\right)\right] \end{aligned} \tag{10}$$

where the last term allows for some photons passing through the active silicon volume without being detected. The function $\varepsilon(E)$ is almost constant for K lines of Ti and above and has only a small influence if radiation is investigated with a characteristic energy higher than 4 keV. However, while the Be window and the Au conductive layer can be regarded as approximately 9 µm and 20 nm thick respectively, the Si dead layer and the contamination layer are undefined functions of time, reducing the detector efficiency rapidly for low energy X-rays. (As an illustration please see Ref. [32], Fig. 1 also in this volume.). Consequently, $\varepsilon(E)$ affects the analysis especially of high and low energy X-rays, e.g. from elements with widely differing atomic numbers.

The uncertainty associated with detector characteristics has been recognised for a long time and can be utilised (intended or not) to compensate for errors produced by some of the other parameters discussed above. But this single function can only compensate for errors from one family of X-radiation – K, L or M X-rays. If two families are examined at low energies simultaneously, e.g. Al K and Ga L or Cu L and Au M, the error becomes evident.

2. *Hybrid Method (Method of Linear Interpolation)*

As proposed by Barbi [33] it is possible to predict the net intensity from the standard by linear interpolation from neighbouring elements. Fig. 5 shows the dependence of $K\alpha$ and $L\alpha$ X-ray intensities emitted from pure elements on the respective characteristic energy, which we found during our investigations of detector efficiency at an acceleration energy of 30 keV, a take-off angle of 35° and a counting time of 400 seconds [32]. While the emitted $K\alpha$ intensities vary almost linearly with characteristic energy (and atomic number), the emitted $L\alpha$ intensities show a smooth deviation from the linear dependence including an offset between $Z=49$ and $Z=50$ ($E \approx 3.4$ keV) and between $Z=74$ and $Z=79$ ($E \approx 8.3$ keV) proportional to the offset in the enhancement factor $(1+g_{ck})$ discussed in a previous section. When detected X-radiation intensities instead of emitted ones are interpolated and employed to build the k-ratio, additional offsets must be expected at the binding energies of the materials forming the detector unit (Si or Ge; Be or compounds of C, B, N and O for the window; Au or Al for front layers; C and O for contamination layers).

If applied to standardless analysis a large number of calibration curves are required, one for each acceleration energy and take-off angle [33]. Since linear dependence is valid for the emitted, but not for the measured intensities, an analysis necessitates the experimental data being corrected for detector efficiency. As pointed out earlier, the detector response, $\varepsilon(E)$, is a function of time so one simple analysis implies the necessity of its new validation. This can be circumvented by the next two methods.

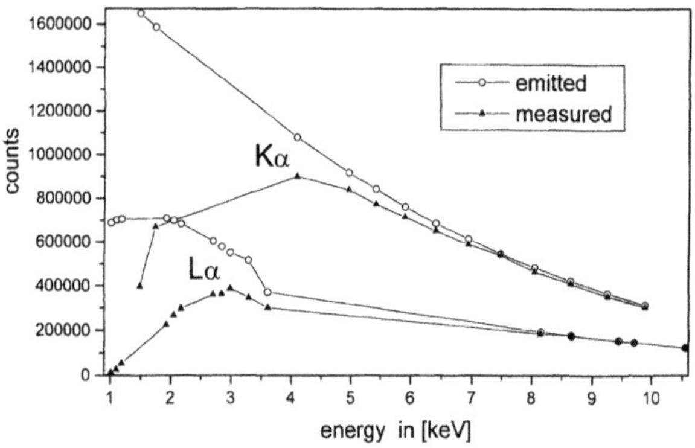

Fig. 5. Net intensities of $K\alpha$ radiation from pure elements depend almost linearly upon atomic number and those of $L\alpha$ radiation vary smoothly with Z. (for a better presentation the characteristic energies have been chosen for the abscissa.) The Offsets between $Z=49$ and $Z=50$ ($E \approx 3.4$ keV) and between $Z=74$ and $Z=79$ ($E \approx 8.3$ keV) are proportional to the offset in the enhancement factor $(1+g_{ck})$ discussed in the text. Emitted intensities are obtained from division of measured intensities by the detector efficiency $\varepsilon(E)$ presented in [32]. $E_0 = 30$ keV, Take-off angle = 35°, live time 400 seconds

3. *Method of Peak-to-Background Ratios*

Firstly presented by Statham [34] this method has been refined by Heckel and Jugelt [35]. The integrated characteristic line intensity has to be related to the background counts in one channel under the maximum of the characteristic peak. This eliminates the detector efficiency $\varepsilon(E)$, but introduces atomic number, absorption and fluorescence correction factors for the continuous radiation, which are derived empirically in their work. Provided that all P/B ratios can be determined experimentally neither relative k-ratios (Eq. 2b) nor normalisation of weight concentrations (Eq. 2a) is necessary. The latter feature is advantageous because the analytical results can then be checked by summing all weight fractions of the matrix elements in the specimen. Investigating 14 binary samples the authors obtained relative errors of less than 5% when $c \geq 10$ wt%, with a mean relative error of 1.7%. By calculating an expression for the expected relative concentration error they find that it is more advantageous to start from the P/B ratio of the element with the lower atomic number when analysing binary samples with high atomic number difference. Unfortunately, the method's usefulness is limited considerably by the difficulty in quantifying the background with sufficient accuracy, particularly in the low energy region (less than 1.5 keV), where the background is curved.

4. A New Method: TWIX (Method of Peak-to-Peak Ratios)

A completely new method in which the investigated sample itself is used as a reference is now presented by our group [36]. Instead of P/B ratios we measure the characteristic peak intensity twice (thus the name): at a fixed incidence angle, but at two different take-off angles. This is most conveniently achieved by placing the sample on a wedge and after the first measurement rotating the wedge by 180° for the second measurement (Figs. 6a and 6b). This is equivalent to tilting the sample by the angle and replacing the take-off angle ψ in Eq. (1b) by ($\psi\pm$) for the two measurements. The tilt angle, depth distribution function and the number of emitted photons are identical in both cases; only the

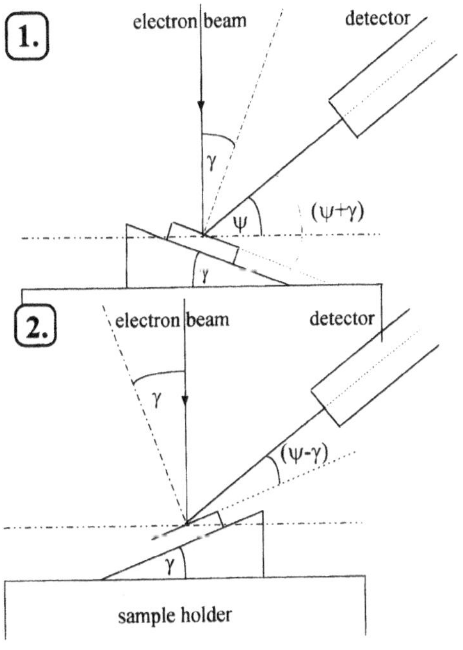

Fig. 6. Experimental arrangement for the TWIX-method. The first measurement (*1*) is followed by rotation of the sample holder by 180° for the second measurement (*2*)

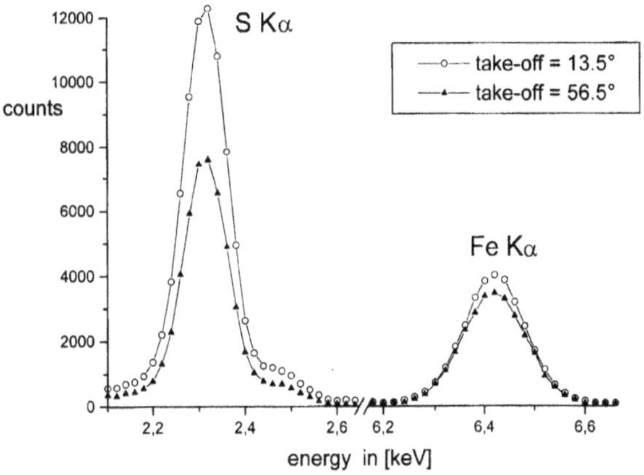

Fig. 7. The spectra of S $K\alpha$ and Fe $K\alpha$ lines recorded at two different take-off angles $\psi\pm$ according to Figs. (6a) and (6b). $E_0 = 20$ keV, $\psi = 35°$, tilt angle $= 21.5°$. Due to mass attenuation effects the S $K\alpha$ signal from the first measurement is 1.6 times higher than that from the second measurement ($k_{twix} \sim 1.6$), while the Fe $K\alpha$ signal reduces by only 1.1

path length of the generated radiation in the sample and therefore mass attenuation effects determine the net intensities from the two measurements. Consequently this ratio is significant (and analysis is even possible with reliable precision) if:

- the mass attenuation of the measured line in the other elements of the sample is high
- the mass attenuation of the measured line in the other elements of the sample is much higher than the self-attenuation coefficient.

Measured spectra from FeS_2 at two different take-off angles plotted in Fig. 7 illustrate the importance of these two conditions for the new method: The ratio of mass attenuation coefficients for S $K\alpha$ radiation in Fe and S is $(\mu/\rho)_{Fe}/(\mu/\rho)_s \approx 4.57$ and for Fe $K\alpha$ radiation in S and Fe is $(\mu/\rho)_s/(\mu/\rho)_{Fe} \approx 2.53$ according to the data fit by Heinrich [37]. Hence the twix-ratio and thus the analytical significance is high for S ($k_{twix} \sim 1.6$) and low for Fe ($k_{twix} \sim 1.1$). This fact can be seen directly in Fig. 7, where k_{twix} is equal to the ratio of the S $K\alpha$ and Fe $K\alpha$ peak from the first to the second measurement. The significant decrease of the S $K\alpha$ peak by varying the take-off angle according to Figs. 6a and 6b enables a reliable deduction of the weight fraction of S in FeS_2 from the two spectra.

For determining the composition of samples the weight fractions c_i of the matrix elements have to be varied until the theoretical twix-ratio (Eq. 11a) equals the experimental ratio (Eq. 11b).

$$K_{\text{twix}}^{\text{theoret}} = \frac{\int \Phi(\rho z) \cdot \exp\left[\left(-\sum_i c_i \cdot \left(\frac{\mu}{\rho}\right)_i\right) \cdot \frac{\rho z}{\sin(\psi+\gamma)}\right] \cdot d\rho z}{\int \Phi(\rho z) \cdot \exp\left[\left(-\sum_i c_i \cdot \left(\frac{\mu}{\rho}\right)_i\right) \cdot \frac{\rho z}{\sin(\psi-\gamma)}\right] \cdot d\rho z} \tag{11a}$$

where

$$k_{\text{twix}}^{\text{exp}} = \frac{\psi_+}{\psi_-} \tag{11b}$$

For determining film thickness the mass attenuation of the measured substrate radiation in the film should be high. In the case of a pure element film and substrate

$$K_{\text{twix}}^{\text{theor.}} = \frac{\int_d^\infty \Phi(\rho z) \cdot \exp\left[-\left(\frac{\mu}{\rho}\right)_{\text{Layer}} \cdot \frac{\rho d}{\sin(\psi+\gamma)} - \left(\frac{\mu}{\rho}\right)_{\text{Substrate}} \cdot \frac{\rho z - \rho d}{\sin(\psi+\gamma)}\right] \cdot d\rho z}{\int_d^\infty \Phi(\rho z) \cdot \exp\left[-\left(\frac{\mu}{\rho}\right)_{\text{Layer}} \cdot \frac{\rho z}{\sin(\psi-\gamma)} - \left(\frac{\mu}{\rho}\right)_{\text{Substrate}} \cdot \frac{\rho z - \rho d}{\sin(\psi-\gamma)}\right] \cdot d\rho z} \tag{12}$$

where the depth distribution function and the mass attenuation coefficients correspond to the substrate signal and d denotes the unknown film thickness, which has to be used in the iteration procedure. Nine stoichiometric binary alloys and thirteen layered samples were analysed at an acceleration energy of 20 keV, a tilt angle of 21.5 degrees and take-off angles of 13.5 and 56.5 degrees. Errors of less than 10% relative have been obtained for concentrations and film thickness, but these are expected to drop significantly with further improvement of the method [36].

Special care has to be taken in determining accurately the tilt and the take-off angles. As we found during our investigations the last is strongly influenced by the vertical position of the sample surface. This problem is not present in wavelength dispersive systems, where the sample surface is held mechanically exactly on the Rowland circle. In an energy dispersive system the problem can be solved by bringing the electron beam on the sample surface into focus using a light microscope prior to each measurement [36].

The method presented offers the same advantages as the Peak-to-Background method which are:

- elimination of the fundamental parameters in Eqs. (1a) and (1b), especially the detector efficiency $\varepsilon(E)$
- no necessity for relative k-ratios (Eq. 2b) or normalisation of concentration (Eq. 2a)
- applicability to wavelength dispersive spectrometers
- independence of sample surface roughness

Furthermore: (1) The method can be easily applied to the analysis of thin films on a substrate. This is not possible with the P/B method, because variation of the continuous radiation with the sample composition is not nearly significant enough: the background from e.g. a thin chromium layer on iron is almost identical with that from a pure iron bulk and independent of beam energy, tilt or the take-off angle, but the characteristic chromium X-ray intensity increases strongly with increasing tilt angle or decreasing beam energy. (2) Spectra can be treated as in conventional analysis with standards (background subtraction, peak deconvolution, ...) with no need to calculate the continuous radiation at the high precision required for the P/B method. Consequently: (3) Standardless investigations can be easily extended down to 1 keV, where the background is hard to quantify accurately, but characteristic peaks – here used for the reference intensity – can still be determined with high statistical significance. (4) By comparing calculated and measured line intensities from pure elements or simple stoichiometric compounds the validity of tabulated mass attenuation coefficients can be tested. (5) Secondary fluorescence both from characteristic peaks and from the continuum should be much less significant when compared to analysis with standards, because here the amount of additionally generated ionisation is equal for both measurements and mainly cancels out of the twix-ratio. (6) No empirical formulae for atomic number or absorption correction factors for continuous radiation are necessary if – and this is the crucial point – a $\varphi(\rho z)$-model, which predicts the amount of ionisation at an *oblique* angle of incidence, is available.

Some parameterizations for the depth distribution function $\varphi(\rho z)$ at oblique angle of incidence have been presented recently [38–40]. Until now the demand for such models has arisen only in the case of ultra thin films on a substrate in order to enhance the X-ray yield from the layer relative to the background. Therefore the main attraction is still focused on developing and improving depth distribution functions for normal electron incidence. Since there is now a third application of such $\varphi(\rho z)$-models, namely the verification of mass attenuation coefficients in solids by a comparison of calculated and measured line intensities, we hope that further refinement of models will improve the applicability and accuracy of the new standardless analysis method, TWIX.

Conclusion

Summarising the methods described and their limitations we come to the conclusion that standardless analysis can not be recommended at all for inexperienced analysts. This is especially so if the standardless procedure is part of a "black box" computer program, where the user does not know either the method or the parameters employed and then errors of up to 150% relative can occur [1]. If at all possible a stoichiometric compound should be used as a reference. Nevertheless such a standard is sometimes impractical as in the case of inert gases, highly radioactive elements or compounds incompatible with vacuum systems. If the computer program employed allows an interchange of parameters the standardless analysis should be preceded by a systematic study as shown by Newbury et al. [6] to find the best choice of parameters and hence to reduce the expected errors. However, the need to do this makes nonsense of the often cited speed advantage of standardless analysis. In the other case of fixed, unknown parameters (a real black box) one should at least check the possible accuracy of the procedure. As discussed in detail in this work the greatest problems arise when analysing high and low characteristic energies simultaneously in one sample. A very simple check is thus to measure two lines of different radiation families from a pure element, e.g. Cu $K\alpha$ and Cu L or Au $L\alpha$ and Au M. The standardless program should then of course respond with 100 wt% (or 50 wt% if a normalisation condition (Eq. 2a) is used in the program) for each investigated line. For guidance of how to assess a particular computer program the inexperienced user is refered to [5], where procedural steps for such a test are described in detail.

Standardless analysis requires an expert in the method applied to achieve errors less than 10% relative, but the method is still prone to unacceptable errors especially for low energy X-rays below 1.5 keV. Furthermore we think that data fits for several fundamental parameters (ionisation cross section, fluorescence yield) should not be applied at all. With fast computers and huge disk space it should be possible to store all relevant data specific to each element on disk, since theoretical calculations of McGuire [41, 42] as well as statistical considerations of Langenberg and van Eck [43] indicate systematic deviations from the general Bethe (Eq. 3) and Wentzel (Eq. 5) expressions respectively due to the electron configuration. The deviations arise because in these equations, derived more than 60 years ago, neither screening nor relativistic effects were considered. Consequently as long as such equations are employed instead of element specific values errors will inevitably be inherent in any procedure.

Acknowledgements: We acknowledge the support given by the Austrian „Fonds zur Förderung der wissenschaftlichen Forschung" (Project N° P10415-PHY).

References

[1] D. E. Newbury, C. R. Swyt, R. L. Myklebust, *Microbeam Analysis, Proceedings of the 28th Annual MAS Meeting* (J. Friel, ed.), VCH, New York, 1994, pp. 29–30.
[2] J. L. Pouchou, F. Pichoir, *Electron Probe Quantitation* (K. F. J. Heinrich, D. E. Newbury, eds.), Pleunum, New York, 1991, pp. 1–75.
[3] V. D. Scott, G. Love, *Quantitative Electron-Probe Microanalysis, Ellis Horwood Series: Physics*, Department of Physics, University College of Swansea, 1983, pp. 234–62.
[4] K. Röhrbacher, *Thesis*, Technical University of Vienna, 1994, pp. 84–85.
[5] J. L. Lábár, *Microbeam Analysis* **1995**, *4*, 65.
[6] D. E. Newbury, C. R. Swyt, R. L. Myklebust, *Microbeam Analysis, Proceedings of the 28th Annual MAS Meeting* (J. Friel, ed.), VCH, New York, 1994, pp. 231–32.
[7] J. C. Russ, *Proceedings of the 9th Annual Conf. of the Microbeam Analysis Society* (Goldstein, ed.), Ottawa, Ontario, 1974, pp. 22A–22C.
[8] M. Green, V. E. Cosslett, *Proc. Phys. Soc.* **1961**, *78*, 1206.
[9] R. Whiddington, *Proc. Roy. Soc.* **1912**, *A86*, 360.
[10] M. J. Nasir, *J. Microsc.*, 1976, *108*, 79.
[11] H. Bethe, *Ann. Phys.* **1930**, *5*, 325.
[12] M. Inokuti, *Rev. Mod. Phys.* **1971**, *43*, 297.
[13] D. I. Webster, H. Clark, W. W. Hansen, *Phys. Rev.* **1931**, *37*, 115.
[14] G. A. Hutchins, *Characterisation of Solid Surfaces* (P. F. Kane, G. B. Larrabee, eds.) Plenum, New York, **1974**, p. 441.
[15] J.-L. Pouchou, *Microchim. Acta*, **1994**, *114/115*, 33.
[16] C. J. Powell, *Electron Impact Ionisation* (T. D. Märk, G. H. Dunn, eds.) Springer, Wien New York, 1985, pp. 199–231.
[17] J. H. Paterson, J. N. Chapman, W. A. P. Nicholson, J. M. Titchmarch, *J. Microscopy*, **1989**, *154(1)*, 1.
[18] H. Kolbenstvedt, *J. Appl. Phys.* **1967**, *38*, 4785.
[19] M. Gryzinski, *Phys. Rev.* **1965**, *138*, A336.
[20] M. O. Krause, *J. Phys. Chem. Ref. Data.* **1979**, *8(2)*, 307.
[21] E. J. McGuire, *Phys. Rev.* **1971**, *A3(2)*, 587.
[22] M. H. Chen, B. Crasemann, H. Mark, *Phys. Rev. A*, **1981**, *24(1)*, 177.
[23] W. Bambynek, B. Crasemann, R. W. Fink, H.-U, Freund, H. Mark, C. D. Swift, R. E. Price, P. Venugopala Rao, *Rev. Mod. Phys.*, **1972**, *44(4)*, 793.
[24] J. H. Hubbell, P. N. Trehan, N. Singh, B. Chand, D. Mehta, M. L. Garg, R. R. Garg, S. Singh, S. Puri, *J. Phys. Chem. Ref. Data*, **1994**, *23(2)*, 339.
[25] G. Wentzel, *Z. Phys.* **1927**, *43*, 524.

[26] W. Bambynek, *X-Ray and Inner Shell Processes in Atoms, Molecules and Solids*, Leipzig, 1984, pp. 1–2.
[27] J. H. Scofield, *Phys. Rev. A*, **1974**, *9*(3), 1041.
[28] M. R. Khan, M. Karimi, *X-Ray Spectrom.* **1980**, *9*, 32.
[29] J. I. Goldstein, D. E. Newbury, P. Echlin, D. C. Joy, A. D. Romig, C. E. Lyman, C. Fiori, E. Lifshin, *Schanning Electron Microscopy and X-Ray Microanalysis, 2nd* Ed., Plenum, New York, 1992, p. 1–27.
[30] K. Röhrbacher, M. Andrae, P. Klein, J. Wernisch, *Microchim. Acta [Suppl.]*, **1996**, *13*, 507.
[31] J. C. Russ, *Proc. of the 13th Ann. Conf. of the Microeam Analysis Soc.*, Miami, Florida, 1978, paper 46.
[32] K. Röhrbacher, M. Andrae, M. Völkerer, J. Wernish, *Mikrochim. Acta [Suppl.]* **1998**, *15*, 21.
[33] N. C. Barbi, D. P. Skinner, S. Blinder, *Proc. of the 11th Ann Conf. of the Microbeam Analysis Soc.*, Miami, Florida, 1976, Paper 8.
[34] P. J. Statham, *Mikrochim. Acta [Suppl.]* **1979**, *8*, 229.
[35] J. Hechel, P. Jugelt, *X-Ray Spectrom* **1984**, *13(4)*, 159.
[36] M. Völkerer, M. Andrae, K. Röhrbacher, J. Wernisch, *Mikrochim. Acta [Suppl.]* **1998**, *15*, 317.
[37] K. F. J. Heinrich, *11th International Congress on X-Ray Optics and Microanalysis IXCOM 11*, August 1986, London, Canada, 1986, p. 67.
[38] M. Andrae, K. Röhrbacher, P. Klein, J. Wernisch, *Mikrochim. Acta [Suppl.]* **1996**, *13*, 225.
[39] C. Merlet, *Mikrochim. Acta* **1994**, *114/115*, 363.
[40] J.-L. Pouchou, F. Pichoir, D. Boivin, *Proc. of the 12th Int. Congress on X-ray Optics and Microanalysis*, 1989, pp. 105–143.
[41] E. J. McGuire, *Phys. Rev.* **1977**, *A16*, 73.
[42] E. J. McGuire, *Phys. Rev.* **1979**, *A20*, 445.
[43] A. Langenberg, J. van Eck, J. Phys. B, *Atom. Molec. Phys.* **1979**, *12*, 1331.

Mikrochim. Acta [Suppl.] 15, 317–320 (1998)

A New Technique for Standardless Analysis by EPMA-TWIX

Martin Völkerer*, Michael Andrae, Kurt Röhrbacher, and **Johann Wernisch**

Institute for Applied and Technical Physics, Technical University of Vienna, Wiedner Hauptstrasse 8-10, A-1040 Wien, Austria

Abstract. A new standardless method (TWIX) for quantification of thin films and bulk materials has been developed. Based on a correction model for oblique angle of incidence the effect of attenuation of the detected x-ray line resulting from path length elongation in tilted samples has been used for thickness determination of thin films and composition of binary samples. It should be noted that this method is also applicable on WDS systems. Results with an average deviation of less than 10 per cent are obtained.

Key words: EPMA, standardless analysis, thin films, bulk concentrations, TWIX.

Until now standardless analysis has been merely of academic interest because the knowledge of the physical parameters is still insufficiently accurate. Therefore results with an accuracy of 10 per cent can be considered as rather good [1]. Moreover standardless analysis has been generally restricted to EDS systems because in WDS the spectrometer efficiency function of each crystal has to be determined using standards [2]. Using the new method, TWIX, described in this work standardless analysis is possible even with WDS systems without this effort; restrictions will be discussed later. TWIX can be called a semi-standardless method of analysis because the sample itself is used as a reference. For this reason all unknown parameters cancel out of the equation used for analysis. Thus it is possible to achieve results with an accuracy within the range obtained when using standards.

* To whom correspondence should be addressed

Experimental

Theory

The physical basis of this method is the effect of attenuation on the detected radiation in the sample. Since the detected intensities depend on the path length of the x-ray line of interest in the sample, the count rate is significantly changed by a variation of the take-off angle. The angle between electron gun and detector of most electron microscopes is fixed and therefore tilting the sample is the easiest way to achieve this variation.

Two measurements of the same tilted sample at various take-off angles have to be made (Fig. 1, Fig. 2). Two different spectra are obtained by a horizontal rotation of the centered, tilted sample round 180°. In standardless analysis with a $\Phi(\rho z)$ approach the intensities are calculated using

$$I_{\text{theor}} = \frac{\Omega}{4\pi} \cdot \frac{N_0}{A} \cdot i \cdot t \cdot Q \cdot c \cdot \Omega \cdot \varepsilon \cdot p \cdot (1 + f_c) \cdot (1 + g_{ck}) \cdot \int \varphi(\rho z) \cdot e^{-\chi \rho z} \cdot d\rho z \tag{1}$$

Notation

$\Omega/4\pi$	detector solid angle
N_0	Avogadro's number
A	atomic weight
i	electron probe current
t	time of measurement
Q	ionization cross-section
c	weight% of the measured element
ω	fluorescence yield
ε	detector efficiency
p	relative x-ray line intensity
$(1+f_c)$	fluorescence by continuum
$(1+g_{ck})$	Coster-Kronig transitions (g_{ck} is zero for K lines)
$\int \varphi(\rho z) \cdot e^{-\chi \rho z} \cdot d\rho z$	intensity integral

The main problem is the uncertainty of some of the parameters that occur in this equation. Since these parameters are equal for the two measurements, forming a k-ratio between the net intensities of the detected characteristic x-ray line leads to a simple mathematical algorithm for the analysis. For the calculation of concentrations the composition of the sample is altered until the theoretical k-ratio in Eq. (2) equals the measured k-ratio in Eq. (3). In the case of a thin

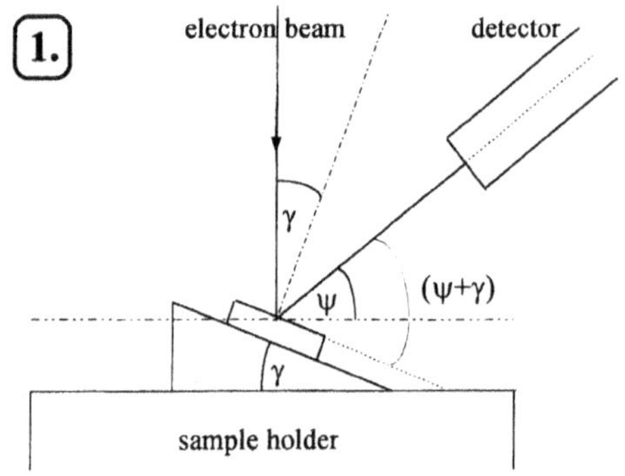

Fig. 1. Experimental arrangement part 1

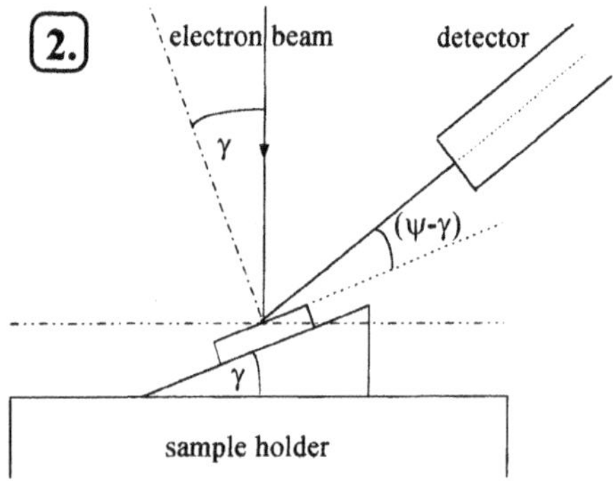

Fig. 2. Experimental arrangement part 2

layer thickness determination the layer thickness in Eq. (4) is varied through the calculation. Eq. (4) is for a pure element layer on a pure element substrate.

$$K_{\text{twix}}^{\text{theor.}} = \frac{\int \Phi(\rho z) * \exp\left[\left(-\sum_i \left(\frac{\mu}{\rho}\right)_i * c_i\right) * \frac{\rho z}{\sin(\psi + \gamma)}\right] d\rho z}{\int \Phi(\rho z) * \exp\left[\left(-\sum_i \left(\frac{\mu}{\rho}\right)_i * c_i\right) * \frac{\rho z}{\sin(\psi - \gamma)}\right] d\rho z} \quad (2)$$

$$K_{\text{twix}}^{\text{exp}} = \frac{I_{\psi+\gamma}}{I_{\psi-\gamma}} \quad (3)$$

$$K_{\text{twix}}^{\text{theor.}} = \frac{\int_d^\infty \Phi(\rho z) \cdot \exp\left[-\left(\frac{\mu}{\rho}\right)_{\text{Layer}} \cdot \frac{\rho d}{\sin(\psi+\gamma)} - \left(\frac{\mu}{\rho}\right)_{\text{Substrate}} \cdot \frac{\rho z - \rho d}{\sin(\psi+\gamma)}\right] \cdot d\rho z}{\int_d^\infty \Phi(\rho z) \cdot \exp\left[-\left(\frac{\mu}{\rho}\right)_{\text{Layer}} \cdot \frac{\rho z}{\sin(\psi-\gamma)} - \left(\frac{\mu}{\rho}\right)_{\text{Substrate}} \cdot \frac{\rho z - \rho d}{\sin(\psi-\gamma)}\right] \cdot d\rho z} \quad (4)$$

Notation

$\Phi(\rho z)$ depth distribution function for tilt angle γ
c_i concentration of element i
$(\mu/\rho)_i$ mass attenuation in element i
$(\mu/\rho)_{\text{Layer/Substrate}}$ mass attenuation of substrate material in the layer/substrate
ρz mass depth
ψ take-off angle
γ tilt angle
I_+ measured intensity for take-off angle $(\psi + \gamma)$
I measured intensity for take-off angle $(\psi - \gamma)$

Since the depth distribution function $\varphi(\rho z)$ is equal for both geometries the k-ratio, defined in Eq. (2) and (4), is determined only by the mass attenuation coefficient (MAC) of the detected x-ray line. Two conditions were found, which influence the accuracy and make analysis possible: firstly, the attenuation of the investigated line in the other elements has to be high and secondly, the ratio between the self attenuation coefficient and the MAC in the other elements has to be significant. Whereas for the determination of concentrations of binary alloy it is sufficient that these conditions are true for the MAC's of one element in the other, the determination of thickness requires these conditions to apply to the substrate material only. This is because alteration of the layer signal with take-off angle merely depends on self attenuation which is normally rather weak.

Naturally these two conditions restrict this method to combinations of elements with appropriate MAC's.

Instruments and Materials

The scanning electron microscope (SEM) used was a JEOL JSM T330A equipped with an energy dispersive x-ray spectrometer. The angle between the electron gun and the detector is fixed at 55° (take-off angle for normal electron incidence $\psi = 35°$). Measurements were taken at an acceleration voltage of 20 kV and a sample tilt angle of $\gamma = 21.5°$, determined optically with a laser. Since the attenuation effect is better the more the sample is tilted but the intensity decreases, we decided to choose a $\gamma \approx 20°$. Analysis has been carried out using a multiple scattering model first published by Cosslett and Thomas [3], later refined by Lantto [4] and August et al. [5], adapted for oblique incidence by Andrae et al. [6].

The MAC's used in the analysis procedure are fits based on data from Mc Master et al. [7] and Henke et al. [8].

The measured samples are standard etalons from Micro-Analysis Consultants Ltd..

Results and Discussion

The experimental results are summarized in Tables 1 and 2. The relative errors for the determination of layer thicknesses as well as of concentrations are

Table 1. Results of the layer thickness determination

Sample El1 on El2	Thickness [nm]	Rel. error [%]	Line	(μ/ρ) in substrate	(μ/ρ) in layer	K_{TWIX}
120 nm Cu on Si	121.4	1.01	Si-K	354	3252	3.588
50 nm Cu on Si	51.2	1.02	Si-K	354	3252	1.978
26 nm Cu on Si	25.4	2.3	Si-K	354	3252	1.618
10 nm Cu on Si	9.7	3	Si-K	354	3252	1.419
100 nm Au on Ag	101.5	1.02	Ag-K	525	2157	4.714
50 nm Au on Ag	49.6	0.8	Ag-K	525	2157	2.582
20 nm Au on Si	21.8	9	Si-K	354	1572	1.527
3 nm Au on Si	3.1	3.33	Si-K	354	1572	1.336
67 nm Al on Si	67.9	1.34	Si-K	354	3257	1.560
42 nm Al on Si	40.6	3.33	Si-K	354	3257	1.455
8.5 nm Al on Si	8.4	1.18	Si-K	354	3257	1.340
100 nm Co on Si	101.2	1.2	Si-K	354	2629	2.553
90 nm Pd on Si	88.7	1.44	Si-K	354	1890	2.241

Table 2. Results of the composition determination of binary samples

	Nominal [wt. %]	Calc. C [%]	Rel. err. [%]	Line	(μ/ρ) in 2nd element	R	K_{TWIX}
InAs	60.5% In	58.58	3.2	In-L	863	1.8	1.539
FeS_2	53.4% S	53.04	0.7	S-K	1105	4.5	1.606
PbSe	72.4% Pb	76.3	5.4	Pb-M	2323	2.6	1.972
HgTe	61.5% Hg	60.82	1.1	Hg-M	1622	1.7	1.963
$TbSi_2$	26.1% Si	28.45	9	Si-K	5014	14	2.360
CdSe	41.04% Se	37.79	7.9	Se-L	3948	3.5	2.639
CdSe	58.96% Cd	62.23	5.55	Cd-L	1066	2.1	1.602
TlI	61.7% Tl	53.58	13.2	Tl-M	1600	1.7	1.976
Ag_2Te	37.2% Te	35.48	4.6	Te-L	1340	3.2	1.816

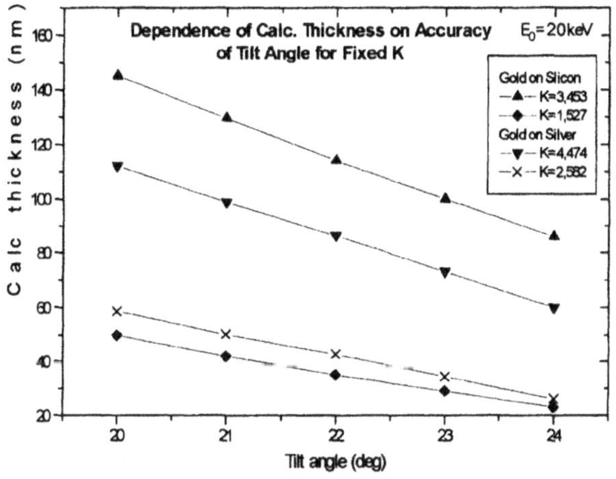

Fig. 3. Dependence on the tilt angle

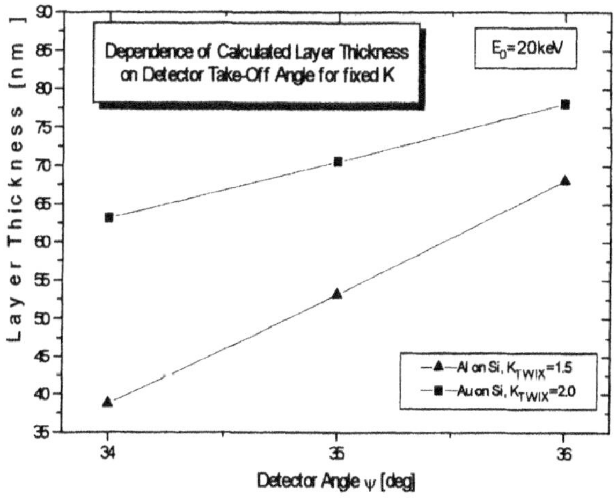

Fig. 4. Dependence on the detector take-off angle

within a range of ten percent. It has to be stated that combinations of elements were chosen such that the two MAC conditions mentioned previously are fulfilled.

The change of the signal with the tilt angle as the determining factor for TWIX has to be looked at from two aspects: on the one hand the sensitivity on the angle is exactly what we need for analysis, but on the other the sensitivity to tilt angle (Fig. 3) and to detector angle (Fig. 4) requires very exact positioning of the sample. Slightly inexact positioning of the sample leads to significant errors in the results. The take-off angle is especially difficult to adjust in EDS [9]. The surface of the sample has to be at the same

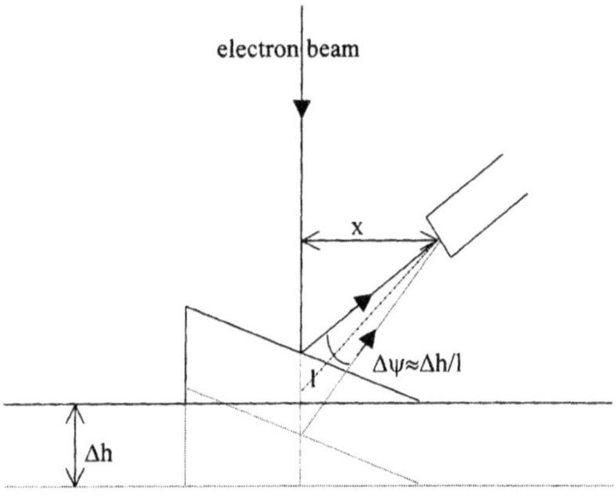

Fig. 5. Variation of the detector take-off angle

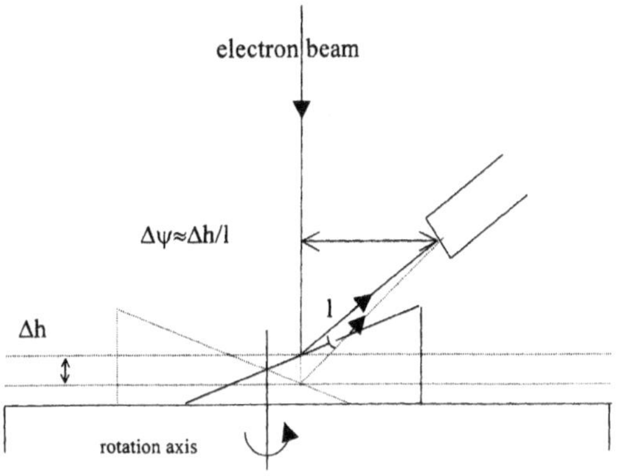

Fig. 6. Errors due to the eccentricity of rotation axis

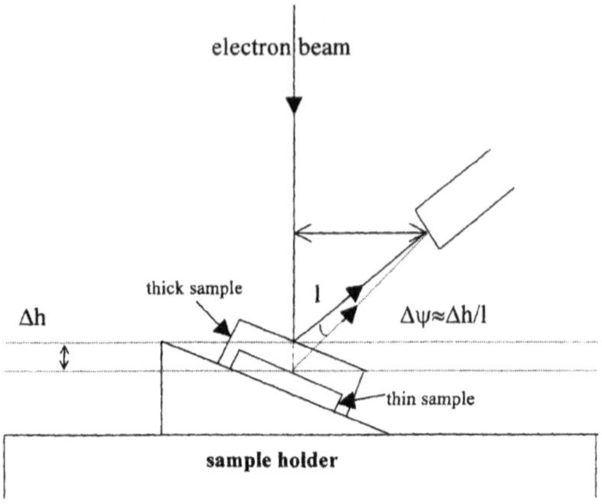

Fig. 7. Influence of the sample thickness on the take-off angle

height for each measurement, otherwise the detector angle changes as shown in Fig. 5. Hence, great care has to be taken that the sample holder is rotated round the axis of electron incidence (Fig. 6). We found that the mechanical positioning of samples with different thicknesses is very difficult (Fig. 7) and therefore a light microscope is advantageous for exact sample positioning. If the image is focused at the height, where the surface of the sample has to be for correct geometry, then the specimen's position can be adjusted by varying the height until the surface is sharply focused.

In WDS this problem of positioning is omitted because the sample is automatically adjusted on the Rowland circle and correct geometry is provided.

Conclusion

TWIX is a powerful method for standardless analysis. When no standards are in stock it can be a real alternative to analysis with standards. Results with realtive errors of less than 10 per cent are possible. For the first time standardless analysis with high reliability and with little expense is not restricted to EDS systems because TWIX is applicable to WDS too. Further investigations on the reliability and applicability will be necessary for industrial use.

Acknowledgements. We acknowledge the support given by the Austrian „Fonds zur Förderung der wissenschaftlichen Forschung" (Project N° P10415-PHY).

References

[1] J. L. Lábár, *X-Ray Spectrom.* **1987**, *16*, 33.
[2] J. Wernisch, *X-Ray Spectrom.* **1985**, *14*, 109.
[3] V. E. Cosslett, R. N. Thomas, *Br. J. Appl. Phys.* **1965**, *16*, 779.
[4] V. Lantto, *J. Phys. D* **1979**, *12*, 1181.
[5] H. J. August, J. Wernisch, *X-Ray Spectrom.* **1991**, *20*, 131.
[6] M. Andrae, K. Röhrbacher, P. Klein, J. Wernisch, *Microchim. Acta* **1996**, *[Suppl.] 13*, 225.
[7] W. H. McMaster, N. K. Del Grande, J. H. Mallet, J. H. Hubbell, *Rep. UCRL-50174, Sec II*, 1969.
[8] B. L. Henke, E. S. Ebisu, *Adv. X-Ray Anal.* **1974**, *17*, 150.
[9] J. L. Pouchou, *Mikrochim. Acta* **1994**, *114/115*, 33.

Mikrochim. Acta [Suppl.] 15, 321–324 (1998)

Stopping Power Factor for Standardless QEPMA

Birman I. Losic and **Fuad A. Tabak***

Department of Physical Metallurgy, Metallurgical Institute "Kemal Kapetanovic," 7200 Zenica, Bosnia-Herzegovina

Abstract. Based on semi-empirical modelling a flexible mathematical expression for the stopping power factor in standardless quantitative electron microanalysis has been obtained. Parameters in the model have been adjusted using a non-linear fitting procedure to match relative x-ray intensity data with that given by experiment and Monte Carlo simulations. When appropriate parameters for the ionization cross-section are used the model is shown to give a satisfactory performance over a wide range of atomic numbers, overvoltage ratios and primary electron energies.

Key words: stopping power factor, standardless QEPMA, Monte Carlo calculation.

Quantitative electron probe microanalysis (QEPMA) is a standard application of electron microprobe analysis (EPMA) and a well established technique for quantitative chemical analysis of material microconstituents in bulk samples. The key difference between conventional and standardless methods is that the experimental x-ray intensity from the standard is substituted by a calculated value. For reliable standardless quantification of the element-specific x-ray intensity, a fundamental requirement is to account satisfactorily for the energy loss of the impinging primary electrons (the stopping power factor). The lack of accuracy of most standardless analysis is mainly caused by uncertainties in the value of basic parameters. However, due to the very complex physical nature of ionization and the mechanism of energy loss, errors in calculated concentrations also result from using inappropriate analytical expressions for the stopping power factor.

* To whom correspondence should be addressed

Further development and improvement of correction factors can be achieved by models that describe physical processes as accurately as possible by use of flexible mathematical expressions. By experimental measurements and Monte Carlo (MC) calculations of relative line intensities from pure elements, the constants and parameters of such a model can be adjusted and its range of applicability tested.

Stopping Power Factor

The stopping power factor is defined by [1]

$$1/S_j = \int Q_j(E)/[dE/d(\rho_j z)]dE \qquad (1)$$

where $Q_j(E)$ is the ionization cross-section of element j, ρ_j is the density of element j and the integration limits are the energy of the impinging electrons and the critical excitation energy of the element-specific line.

Based on semi-empirical modelling of the ionization cross-section and electron energy loss a generalised expression for stopping power factor is obtained:

$$1/S_j = \text{const} \cdot U^b/w[\alpha \cdot (J/E)^{1/2} \cdot (1 - 1/U^a - a \cdot \ln U) + \beta \cdot (1 - 1/U^b - b \cdot \ln U)] \qquad (2)$$

where const, α and β are the constants of modelling, a and b are the parameters of modelling, E is the energy of impinging electrons (keV), E_{cj} is the critical excitation energy (keV), $U(=E/E_{cj})$ is the overvoltage ratio, J is the mean ionization potential as described by Love et al. [2], A_j is the atomic weight, Z_j is the atomic number, c_j is the concentration expressed as a mass fraction and $w = \Sigma_j c_j \cdot Z_j/A_j$. In the derivation of this expression a modified version of the equation of Bethe for the ionization cross-section is used, as described by Hutchins [3]:

$$Q_j(E) = 0.7\pi \cdot e^4 \cdot (1/E_{cj}^2) \cdot \ln U/U^P \qquad (3)$$

where e is the electron charge and p is a parameter.

Also used is the electron energy loss equation given by Love et al. [4], expressed in generalised form:

$$dE/d\rho_j z = (1/J) \cdot (Z_j/A_j) \cdot \{1/[\alpha_0 \cdot (E/J)^c + \beta_0 \cdot (E/J)^d]\} \qquad (4)$$

where in the original paper $\alpha_0 = 1.18 \times 10^{-5}$, $\beta_0 = 1.47 \times 10^{-6}$, $c = 1/2$ and $d = 1$.

Absorption and Backscattering Factor

The absorption factor expression developed by Love and Scott [2] is used with an error no greater than 5% even when applied to soft x-ray element-specific lines [5]. The backscattering coefficient of Hunger and Kuechler [6] and mass absorption coefficients determined by Heinrich [7] are included in the x-ray absorption factor calculation. The backscattering factor, developed by August and Wernisch [8] can be calculated with an error of less than 1% for acceleration energies between 4 keV and 30 keV.

The fundamental parameters used as multiplying factors (x-ray fluorescence yields, transition probabilities etc.) in the expression of the absolute line intensity have only Z dependence and are not included in the calculation of relative line intensities. Thus, calculated relative intensities are not influenced by uncertainties incorporated in these fundamental parameters making it easier to compare the model calculation with experimental results and MC values.

Monte Carlo Calculations

The Monte Carlo values are calculated using a version of the Reimer-Krefting algorithm [9], slightly modified by the choice of an ionization cross-section suggested by Hutchins. The algorithm takes into account single elastic electron scattering (Mott cross section), multiple elastic small angle scattering and single inelastic electron-electron scattering, including the production of fast secondary electrons up to the third generation. Electron energy loss is calculated using the Bethe formula from which losses by single inelastic scattering are subtracted. To avoid an "energy gain", the energy loss calculation for $E<6.338\,J$ is performed using the equation of Rao-Sahib and Wittry [10].

$$dE/d(\rho_j Z) = -62360 \cdot (Z_j/A_j) \cdot [1/(E \cdot J)^{1/2}]. \quad (5)$$

The critical aspect of the calculations, the x-ray production, is strongly influenced by the choice of an ionization cross section. From the experimental results for K_α radiation collected by Powell [12] the modified expression of Bethe (see equation 3) is selected, where p is set at 0.9. Random numbers are used in the MC program to determine whether an elastic or inelastic scattering event takes place and also in calculating the length of the free path of electrons between interactions.

Experimental

The measurements were carried out using a JSM 35C scanning electron microscope, equipped with beam stabilizer and fitted with a TN-5500 energy-dispersive system. For x-ray energies below 3 keV a windowless Si(Li) detector was used to avoid x-ray attenuation in the beryllium window or in any ice layer. Due to the high risk of detector contamination the measurements were performed only for six element-specific lines. To reduce the influence of pulse pile-up effects on the measured spectra a beam current below 10^{-10} A was applied (giving a low input count rate).

Pure-element standards were carefully prepared by polishing using well-known metallographycal techniques [11]. Spectra were acquired and processed for 22 integrated $K\alpha$ and $L\alpha$ element-specific lines.

Results and Discussion

The results of measurements and MC calculations for CK_α, Si K_α, and Cu K_α lines are illustrated in Figs. 1–3. The relative intensities calculated by Monte Carlo are in very good agreement with measured ones. It should be mentioned that MC values are computed without adjusting any additional parameter. The discrepancies between the fit curve and both the experimental results and MC values in Fig. 1 can be explained as follows. In the case of the $C\ K_\alpha$ line, the fitting curve cannot reach the higher overvoltage points without a large change in standard deviation. This indicates that the discrepancies are mainly the result of uncertainties in the absorption factor (mass absorption coefficients, x-ray take-off angle), which is to be expected from the susceptibility of soft x-rays to errors in mass absorption coefficients and other parameters. By varying the mass absorption coefficients and/or the x-ray take-off angle within 15%

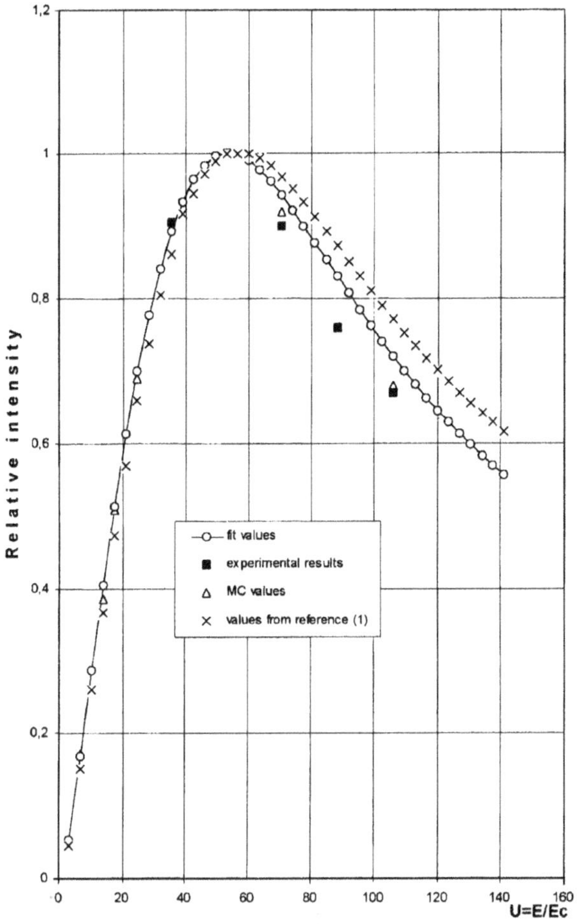

Fig. 1. Relative intensity versus overvoltage ratio for CKα

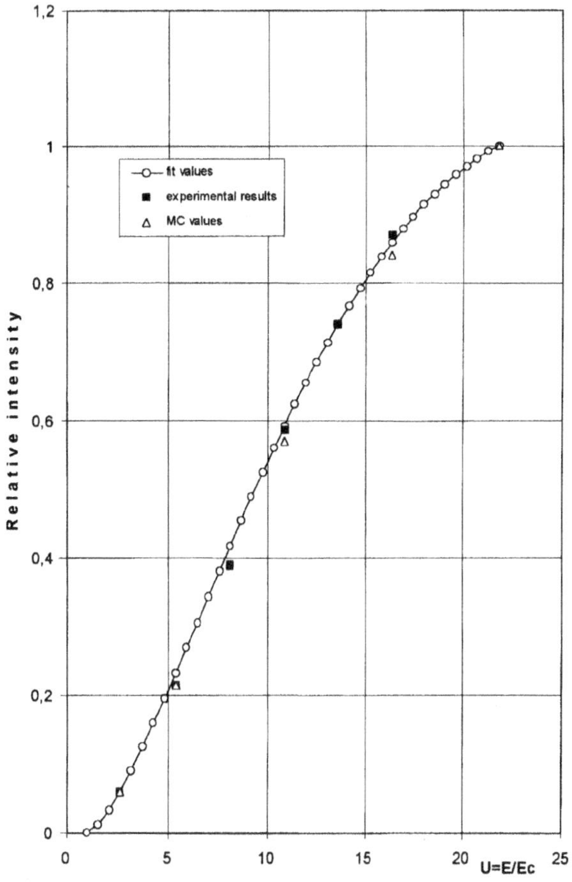

Fig. 2. Relative intensity versus overvoltage ratio for SiKα

very good agreement can be achieved between the fitted curve and both the experimental and MC points.

Non-linear fitting procedures on experimental results and MC values give for constants α and β the ratio of $\alpha/\beta = 25.67$ and for parameters a and b values ranging from 0.56 to 0.62 and from 1.08 to 1.13, respectively. The calculated ratio of model constants emphasises the great importance of using the energy loss term determined by Rao-Sahib and Wittry (5) at low electron energies rather than that of Love et al. (4), where the ratio $\alpha/\beta = 3.73$ is obtained. In our case the contribution of the second term in the denominator is below 5% for $E < 6.338J$, in contrast to greater than 30% for previous modelling, and in better agreement with the asymptotic character of (5).

The model parameters show slight Z dependence but a satisfactory fit is obtained for $a = 0.6$ and $b = 1.1$ (fit values in all illustrated figures) in agreement with considerations of Hutchins [3] and in accordance with experimental ionization cross sections for K_α radiation of Powell [12].

The constant multiplier in the derived expression of importance in the absolute intensity calculation can be estimated using the relations (3) and (5):

$$\mathrm{Const} \approx 2.88 \times 10^{-4} \cdot \pi \cdot e^4.$$

Due to the similarity in the physical processes responsible for the energy loss the same model is applicable to L radiation.

Additional testing, with reference samples, is required to check the stopping power factor in

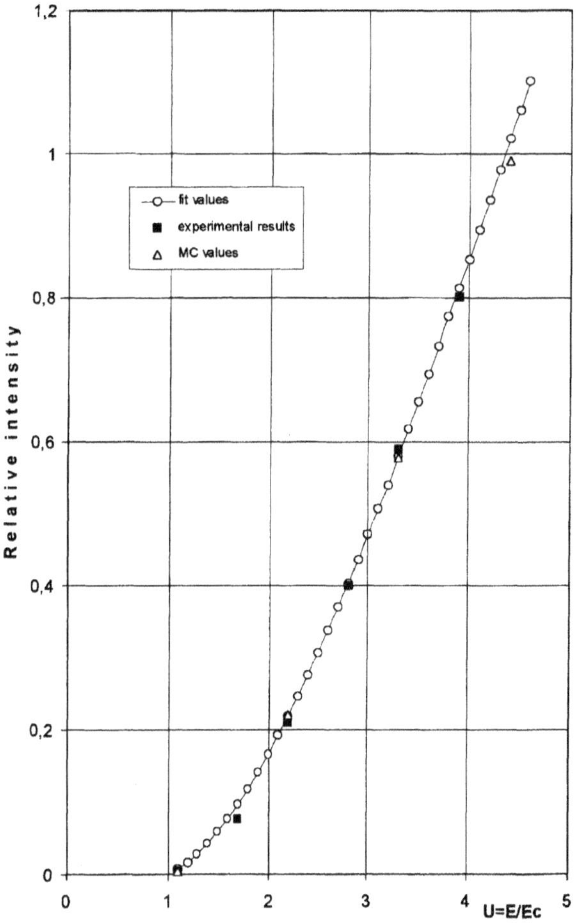

Fig. 3. Relative intensity versus overvoltage ratio for CuKα

practice and well-konwn statistical procedures (histogram, standard deviation etc.) should be used for this purpose.

Conclusions

We have derived and tested a flexible expression for the stopping power factor as a correction term in the conversion of net x-ray intensities into concentrations (iteration procedure) in standardless QEPMA. After a non-linear fitting procedure, comparison of the model calculation with experimental results and MC values for pure element relative x-ray intensities has shown that: (a) the model expression is fully applicable and shows clear interdependence of experimental results, MC values and the physical model; (b) the discrepancies in the model prediction are mainly influenced by uncertainties associated with the absorption factor (mas absorption coefficients, x-ray take-off angle); (c) in comparison with previous modelling it gives better fits to experimental data and MC values over a wide range of atomic numbers, overvoltage ratios and primary electron energies (up to 40 keV) especially in the case of high overvoltage ratio (soft characteristic x-rays, low Z); (d) the values of fitting parameters are in excellent accordance with the ionization cross-section measurements.

For future work the same model will be applied to *L* radiation. After that, the practical aspects of the model application can be checked using a statistically relevant number of standard samples (alloys) with well defined composition.

Acknowledgements. Grateful acknowledgement is made for the support of this work from The Soros Foundations, Open Society Fund Bosnia-Herzegovina and "Zenicatrans" D. D., Zenica.

References

[1] J. Wernisch, *X-ray Spectrom.* **1985**, *14*, 109.
[2] G. Love, M. G. Cox, V. D. Scott, *J. Phys. D* **1977**, *10*, 7.
[3] G. A. Hutchins, in: *Characterization of Solid Surfaces* (P. F. Kane, G. B. Larrabee, eds.), Plenum, New York, 1974, p. 441.
[4] G. Love, M. G. Cox, V. D. Scott, *J. Phys. D* **1978**, *11*, 7.
[5] G. Love, V. D. Scott, *J. Phys. D* **1980**, *13*, 995.
[6] H. J. Hunger, I. Kuechler, *Phys. Stat. Sol.* **1979**, *56*, K45.
[7] K. F. J. Heinrich, in: *Proceedings 11th ICXOM* (J. D. Brown, R. H. Packwood, eds), London, Canada, 1986, 67.
[8] H. J. August J. Wernisch, *Radex-Rundschau* **1988**, *2/3*, 628.
[9] L. Reimer, D. Stelter, *Scanning* **1980**, *8*, 265.
[10] T. S. Rao Sahib, D. W. Wittry, *J. Apply. Phys.* **1974**, *45*, 5060.
[11] Metals Handbook, *Metallography and Microstructures, Vol. 9, 9th* Ed., American Society of Metals, Metals Park, Ohio, 1985, p. 163.
[12] C. J. Powell in: *Electron Beam Interaction with Solids for Microscopy, Microanalysis and Microlithography* (D. F. Kyser, H. Niedrig, D. E. Newbury, R. Shimizu, eds.), SEM, Amf O'Hare, Chicago, 1982, 19.

Mikrochim. Acta [Suppl.] 15, 325–331 (1998)

On the Measurement of the Backscattering Coefficient for Low Energy Electrons

Mohamed M. El-Gomati* and **Ahmad M. D. Assa'd**

Department of Electronics, University of York, Heslington, York YO1 5DD, U.K.

Abstract. An electron detector for the measurement of backscattering and secondary electron coefficients based on that of Reimer and Tollkamp has been devised and built. The detector is used in ultra high vacuum and allows one to use in-situ ion bombardment of the sample to remove surface contaminants before data collection. New experimental measurements of the backscattering and secondary electron coefficients in the energy range 0.5–6 keV are reported for C, Al, Si, Cu, Ge, Mo, Ag and Au. These are collected from both argon ion cleaned and as-inserted surfaces. The backscattering coefficients of the argon ion cleaned surfaces monotonically increases for elements of atomic number $Z > 28$ whilst it decreases for elements of $Z < 28$ for all energies used. This is in contrast to the as-inserted surfaces that show a more complex pattern below 1 keV incident energies.

Key words: backscattering coefficients, electron scattering, LVSEM, electron detector.

Low voltage scanning electron microscopy (LVSEM) is being increasingly used to complement the advances made by conventional scanning electron microscopes (SEM) operated between 10–40 keV. In spite of the fact that the early SEM developments were in the low voltage region, it is only recently that the full advantages of LVSEM have been highlighted, as a result of progress in field emission based SEMs. These advantages include; reduced charging and radiation damage of the samples inspected, increased topographic contrast as a result of reduced edge artefacts caused by energetic backscattered electrons, and the higher secondary electron coefficient of the low energy electrons. For example, it is the combination of negligible backscattered electron interference and improved specimen contrast at low energy that has recently allowed the detection of small concentrations of dopants in semiconductor materials [1] and the imaging of superlattices [2].

Imaging in LVSEM is mostly carried out using either backscattered or secondary electrons. The image contrast is normally related to the backscattered and secondary electron coefficients, which are both functions of the target atomic number, and also depend on the incident electron energy and the angle the electron beam makes with the surface normal. The backscattered electron coefficient (η) is defined as the ratio of the number of electrons that leave the sample with energies greater than 50 eV to the total number of incident electrons. Secondary electrons are those electrons that leave the sample with energy less than 50 eV, and their coefficient (δ) is defined as the ratio of the total secondary electrons leaving the sample to the total number of incident electrons.

While there is a steady increase in the number of studies involving LVSEM, many of these activities, particularly those involving material's contrast, are qualitative in nature. In addition, there appears to be some contrast reversal of backscattered images associated with a change in the incident electron energy [3, 4]. Inspections of the published data on the backscattering and secondary electron coefficients for low energy electrons show a wide variation for a given element [5–11]. For example, the data show a large difference in the behaviour of the backscattering

* To whom correspondence should be addressed

coefficient with respect to the target atomic number between low incident electron energies (< 2keV) and incident energies greater than 5 keV [10–12]. While in the latter case a monotonic increase of the backscattering factor with an increase in the target atomic number is obtained, the low energy regime can not be represented by a simple function. Such a picture makes it difficult to quantify the LVSEM results, and hinders progress on image interpretation. Reported theoretical estimates of η and δ at low incident energies also show a different behaviour between the low and high energy regions, and they differ too with respect to the experimental measurement [12–17].

It is known that the backscattered electron signal is dominated by the ratio of elastic to inelastic scattering of electrons by the target material. At high energies (>5KeV) this ratio is almost independent of energy and varies monotonically with the target atomic number. As the beam energy becomes comparable with the binding energy of the electrons in the target we must expect variations from this smooth behaviour. Inelastic interaction is a strong function of a number of losses; plasmon excitation, band transitions and core excitations. All of these are dependent on target and incident electron energy in a complex way that does not reflect the simple atomic number dependence associated with high energy incident electrons. However, there may be still some systematic behaviour across the periodic table but this can only be discovered with data obtained from well characterised surfaces.

The aim of the present study is to report on new experimental measurements of backscattering and secondary electron coefficients for a number of elements spanning the periodic table excited by incident electrons in the energy range 0.5–6 keV. The difference between the present set of results and those reported elsewhere is that target surfaces have been subjected to energetic ions to remove any contaminants prior to measurements. Data from the "as-inserted" sample surfaces and those subjected to ion bombardment are compared with other published data. A new electron detector design that allows for in-situ target cleaning by ion bombardment is also reported.

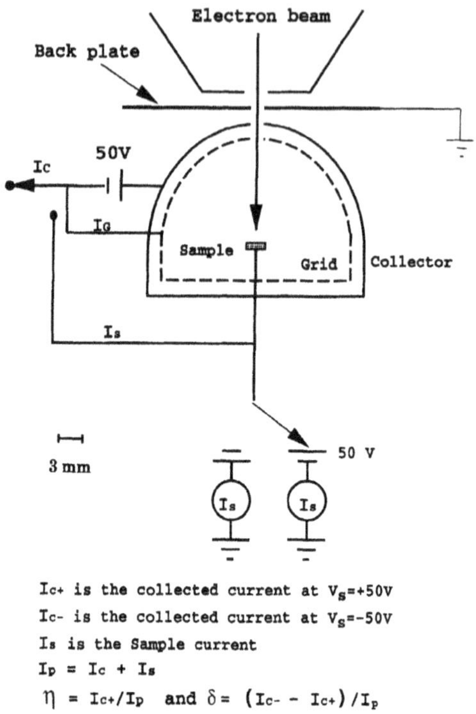

Fig. 1. Schematic drawing of the electron detector and the experimental set-up

Experiment

A backscattered electron detector based on that reported by Reimer and Tollkamp [11] has been designed and built for the present study. Fig. 1 is a schematic drawing of the detector and the experimental set-up used. The main features of the present design are as follows:

(1) The collector electrode is made out of molybdenum in the shape of a hemisphere of radius 15-mm. A 1-mm hole is drilled at the apex of this electrode for the entrance of the electrons that impinge on the target, placed at the centre of the detector. This geometry makes a collection angle for the backscattered electrons of 2πsr. The hemisphere extends 3 mm into a cylindrical shape and is connected to a 30-mm diameter stainless steel plate that carries the target. This ensures that a uniform field exists around the sample and that electrons that backscatter from the collector at low angles to the target surface are captured. The collector can be tilted on its mount to angles of up to 60° with respect to the normal. Appropriate 1 mm holes, corresponding to these angles, are drilled on the collector for the entrance of the primary electrons in these cases.
(2) The target is electrically isolated from the rest of the system. The sample can be moved in situ from the detection position to face an energetic ion beam for surface cleaning. Up to a total of four different samples can be loaded into a carousel type sample manipulator so that measurements are carried out without the need to break the vacuum seal.
(3) An earthed plate is placed between the electron column and the collector electrode to prevent stray electrons generated in the electron column from reaching the collector electrode.

The grid electrode (G) is made out of a woven stainless steel mesh of 80% transparency. The function of this electrode, together with the collector (C), is to prevent secondary electrons generated by the backscattered electrons from leaving the collector. We have found that electrode G in particular is a major source of tertiaries. These electrons can alter the backscattering coefficients by more than 50% if the grid is not properly designed and treated. It was found necessary to use a high transparency grid and to coat it and all the inner surfaces of the collector C with a low secondary electron yield carbon compound [19]. In addition, all the detector components and the targets investigated were subjected to normal ultra high vacuum (UHV) cleaning procedures [20] before assembly and insertion into the vacuum system.

The experimental set-up used in this study is part of a cylindrical mirror analyser (CMA) based Auger electron spectrometer (AES), the details of which are given elsewhere [20]. However, only brief mention will be made here of features relevant to the present work. The main part is an all metal vacuum system bakeable to 160 °C, where a base pressure in the region of $2–5\times10^{-10}$ mbar is normally achieved after 24 hrs of baking. A variable energy electron beam in the range 0.5–10 keV with beam current of typically $0.1–1\times10^{-6}$ A in a spot 10–30 μm diameter is used. The system is equipped with a sample current imaging facility for selecting the electron position on the target surface. Ion beam cleaning conditions were $2–4\times10^{-6}$ A at 3 keV for about 30–60 minutes. These ion beam parameters have been used previously in cleaning similarly prepared surfaces for AES applications [20]. The samples were made of thin foils of high purity materials (99.999%) and were typically 0.5 mm thick and 2–3 mm in diameter. The carbon target was made out of a graphite rod used as the electrode for carbon coating. The beam current stability was better than 0.05%/hr and all electron current measurements were made with a Keithley 604 picoammeter. The measurements were repeated at least twice and η values were found to be reproducible with a standard deviation of better than 1%. The detector was mounted on the entrance aperture of the CMA.

The calculation of η and δ are determined as follows. With the target positively biased (50 V), electrons of energy less than 50 eV will be prevented from leaving it and a current I_{c^+} is measured at the collector. Conventionally electrons of energy less than 50 eV are considered to be true secondary electrons. Therefore, I_{c^+} is considered to consist of only backscattered electrons, inspite of the fact that there is a small percentage of secondaries of energy greater than 50 eV included in that signal. If the target is negatively biased, a current I_{c^-} is measured which contains both backscattered and secondary electrons. The incident primary current, I_p is measured by connecting the target, grid and collector electrode together. η and δ are therefore determined as follows:

$$\eta = I_{c^+}/I_p \quad \text{and} \quad \delta = (I_{c^-} + I_{c^+})/I_p$$

Although the "cut off" value of secondary electrons has been chosen at 50 eV for consistency with the literature, a better measure of this voltage can be obtained by establishing a retarding field curve. This is obtained by measuring the collector current as a function of the target potential (−120 V +120 V), as depicted in Fig. 2. In the present design, we found that the value of 50 eV is perhaps on the low side and would have the effect that our measured η values may therefore be higher up to 1–2% than they actually would be if a value of 80 eV was used.

Following the arguments given by Reimer and Tollkamp, we estimate that the total systematic errors in the measurement of the backscattering coefficient to be less than 1%. The main contributors to this value are electrons that scatter back from the collector and grid onto the target as well as electrons that leave the entrance aperture, both effects reduce I_{c^-} and I_{c^+}.

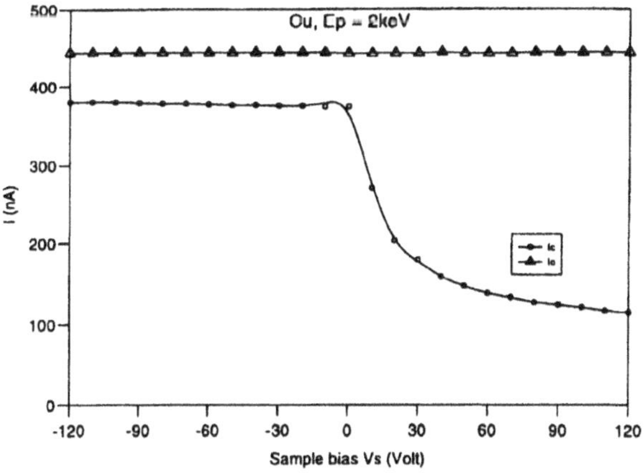

Fig. 2. Retarding field curve from a copper target obtained at 2 keV incident energy

Results and Discussion

Backscattering coefficients η were measured in the energy range 0.5–6 keV for C, Al, Si, Cu, Ge, Mo, Ag and Au at normal incidence. Fig. 3 shows the data of the "as-inserted" and argon ion cleaned surfaces of C, Al, Cu, Ag and Au from the present study along with the data of Bongeler et al. [12]. The latter data set also uses as-inserted samples and was chosen for comparison because it is the most recent report and covers a large number of elements. Inspection of Fig. 3 shows that with the exception of carbon and aluminum, all three data sets show a similar pattern of low η values at low energies rising with different degrees of steepness as E_p is increased up to 3–4 keV after which the increase becomes smaller.

It is clearly seen from these results that the η values obtained at low electron energy on the cleaned samples in the present study are higher than the as-inserted (uncleaned) ones. The following observations are made: η values for elements of atomic number greater than 28 with clean surfaces increase only slowly as a function of the incident beam energy, by a maximum of about 20% between 0.6 keV and 6 keV (for Au, at 0.6 keV $\eta = 0.37$, and at 6 keV $\eta = 0.47$). This is in contrast to the elements with uncleaned surfaces which show an increase by more than 60% (for Au this corresponds to $\eta = 0.17$ at 0.6 keV to $\eta = 0.46$ at 6 keV). The η values at 6 keV are closely similar for both the uncleaned and cleaned surfaces.

There is also a significant difference between the η values of our 'as-inserted' targets and those of Bongeler et al., where perhaps one would have expected closer agreement between the two sets of data. The reason for such discrepancy could be due

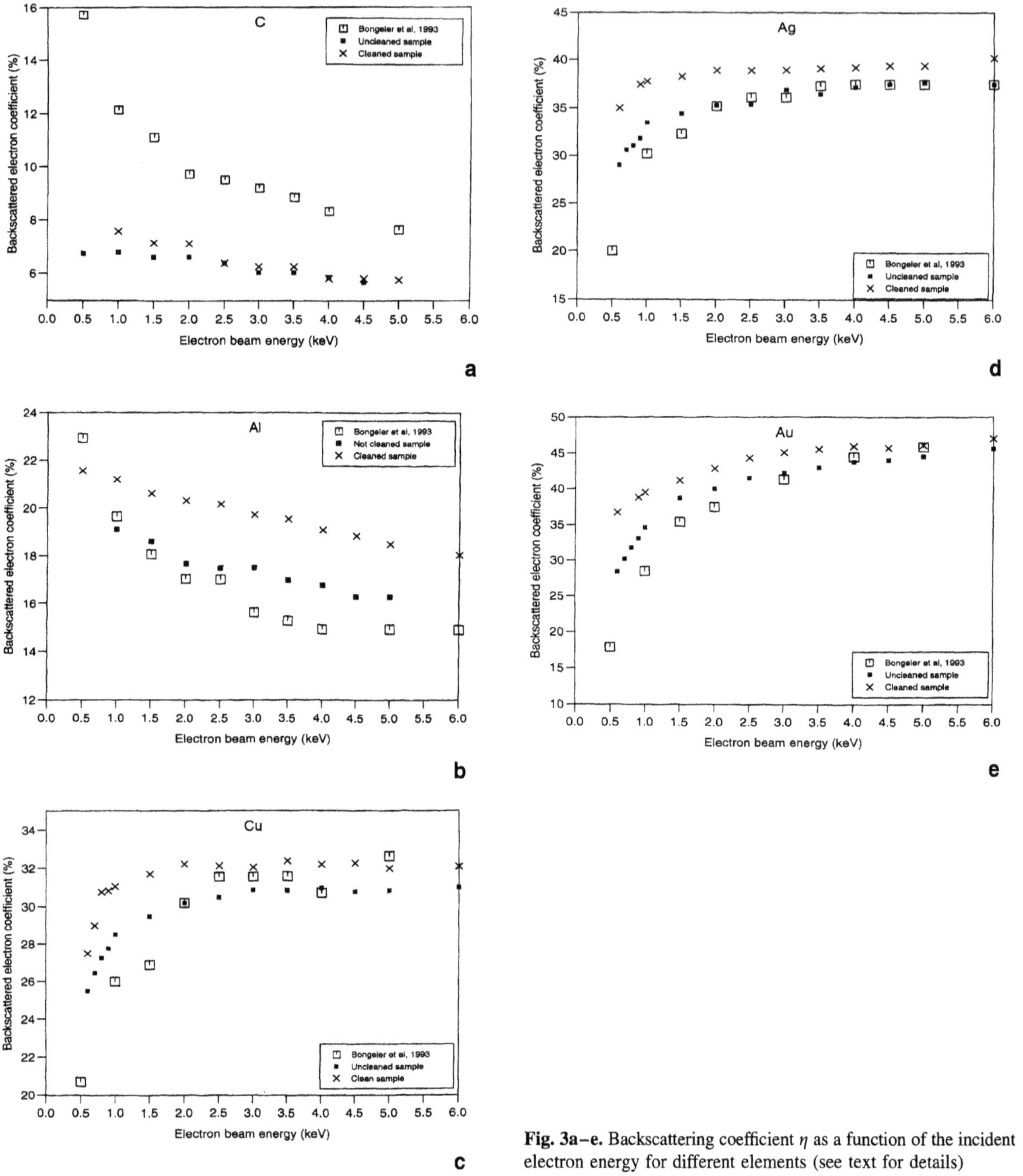

Fig. 3a–e. Backscattering coefficient η as a function of the incident electron energy for different elements (see text for details)

to different preparation of the two sample surfaces. Our data were always collected under UHV conditions, and therefore any electron beam surface deposition of hydrocarbons is eliminated. This is supported by the observation from another experiment showing that the measured η values at 2 keV from the cleaned surfaces change very little under prolonged electron bombardment. This confirms the surface cleanliness of the ion-bombarded targets as well as the suitability of the vacuum environment used.

The above explanation does not apply to the C and Al data of our as-inserted targets. It is likely that the discrepancy between the present set of results and those of Bongeler et al. is due to contamination films on the two surfaces being different in thickness and composition. For carbon it may be that a compound of higher average atomic number has formed on the target measured by Bongeler et al., while in our case a surface layer containing hydrogen may have formed to give a lower η value than clean carbon; for the Al target of Bongeler et al. any surface deposit would be of lower atomic number. If a film of low Z material is deposited on a substrate of high Z, then the interaction volume of the incident electrons with the target will be greatly influenced by the surface film thickness and its composition, particularly at low primary energy. Simple Monte Carlo simulations [21, 22] of a 10 nm thick carbon film on a gold substrate give backscattering coefficients of 0.212 and 0.5 for 2 keV and 20 keV, respectively. These are compared with values of 0.39 and 0.505 in the case of a pure gold target at the same energies. It is important to note that the film thickness affects the absolute value obtained for a given beam energy. We have confirmed this pattern by collecting η at various stages during the cleaning process of Cu and Au. While the η value at 5 keV remained approximately the same, the measured values for 1–2 keV incident electrons increased as the cleaning progressed. Browning et al. [18] calculated η for a gold target and one that had on it 20 nm thick film of oil-like material, and have arrived at similar conclusions to our own. They found that at 0.4 keV, η for the two systems changes in value from 0.17 to 0.27, while at 3–5 keV it remains almost the same to within 0.01.

In Fig. 4 we present a comparison of η values versus the target atomic number (Z) for a wide range

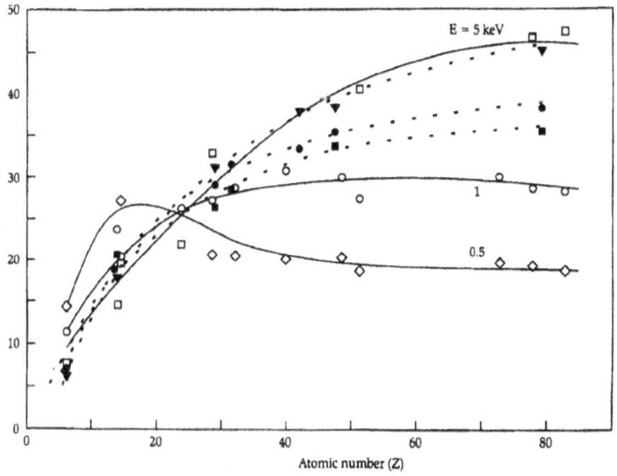

Fig. 4. The backscattering electron coefficient η as a function of the atomic number at different electron energies for the surface cleaned targets of the present study (dashed lines) and the as-inserted targets (solid lines) from Bongeler et al. [12]

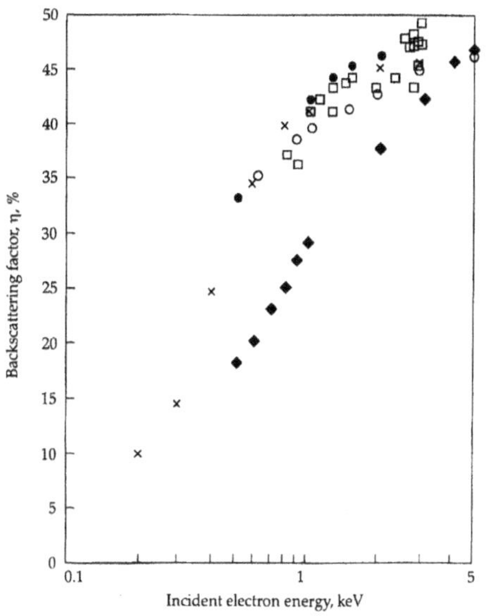

Fig. 5. Experimental measurements of the backscattering coefficient η from Au as a function of energy. × Bronshtein and Faiman [6], ● Thomas and Pattinson [7], □ Schou and Sorensen [10], ◆ Bongeler et al. [12], and ○ the present study

of materials at normal electron incidence. It is clearly seen that the behaviour of η versus Z at lower E_p is quite different for the clean and the as-inserted samples. For example, η from the clean surfaces (dashed lines) increases with Z, irrespective of the incident electron energy. This is unlike the as-inserted data (solid lines) which shows a complex pattern with respect to Z and E_p. In addition, for electrons of 0.5 keV incident energy, as-inserted low Z materials give, in general, a higher η value than for higher Z materials. Further, for the 0.5 keV incident electrons and for $Z > 28$, η appears to decrease slowly.

Fig. 5 is a comparison of experimentally measured η in the case of gold as a function of E_p. There is a large scatter of to up 100% at about 500 eV, but similar values at 5 keV. Fortunately, however, the situation can be simplified as the data set showing the largest difference appears to be obtained under conventional vacuum conditions, i.e. $>10^{-5}$ mbar. Further, the data by Fainman and Bronshtein were collected from evaporated films, while those by Thomas and Pattinson were from in-situ heated samples. Together with the present work, all have been collected under UHV conditions and appear to be in closer agreement with each other. Schou and Sorensen's data show a large scatter from one energy to another even for $E_p > 2$ keV. The data of Bongeler et al. show a monotonic increase in η as a function of

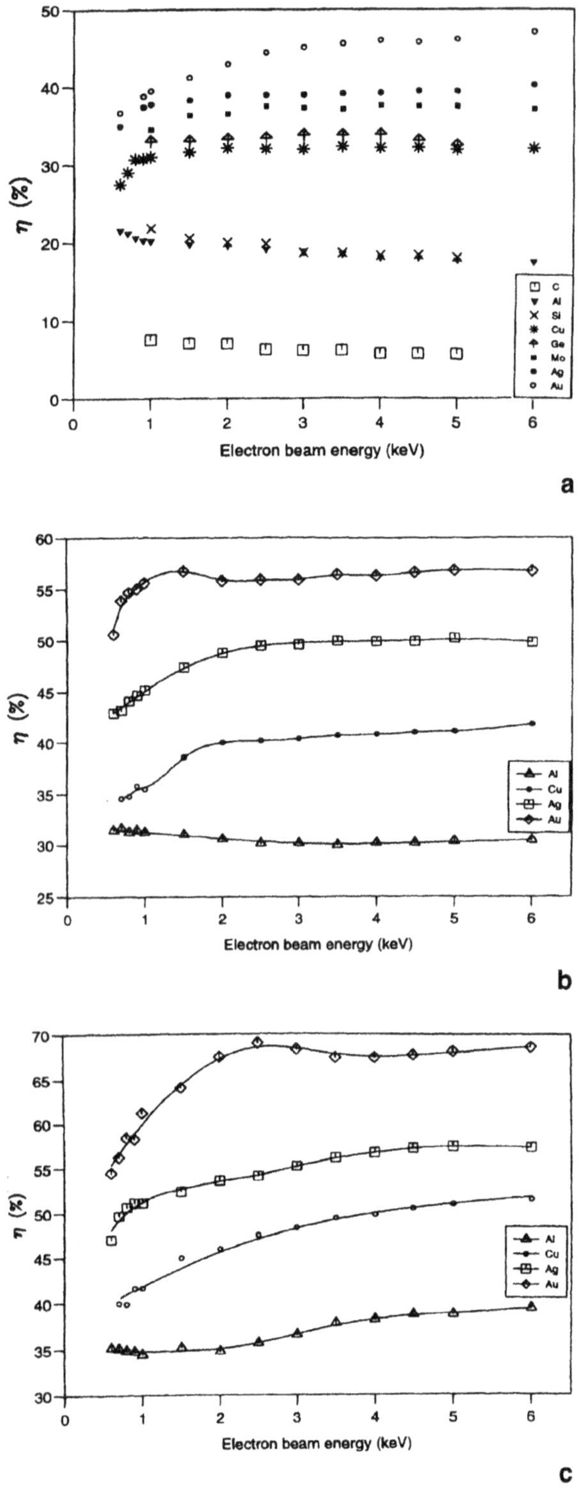

Fig. 6a–c. Measurements of the backscattering coefficient η as a function of the incident electron energy E_p of the cleaned surfaces at: **a** normal electron incidence, **b** 45° incidence, and **c** 60° incidence

Z at an increased incident electron energy, in agreement with the suggestion made here that there might be a carbon film on these targets.

Fig. 6 shows the backscattering coefficient from the cleaned surfaces as a function of E_p for a number of elements of varying atomic numbers and with different angles of electron incidence. In the case of oblique angles of incidence, Fig. 6b and 6c, the collector was tilted to intercept all the forward scattered electrons in this geometry as described in the experimental section. The general observation from these data is that at normal beam incidence, η for electron energies greater than 3 keV is approximately constant for all elements studied. At electron energies below 3 keV, η from heavy elements decreases as E_p decreases, while the reverse is true for light elements. As the angle of incidence is increased, η also increases as a function of E_p, albeit with different rates, for all elements, Fig. 6c. The fine details of the changes for high and low atomic number materials will depend on the balance between elastic and inelastic scattering of electrons inside the reduced interaction volume at these angles.

Generally speaking, however, there appears to be a clear separation in the value of η between the different atomic number materials studied, irrespective of the incident electron energy. This pattern is encouraging as it suggests that a similar atomic number contrast to the case of conventional SEM operated in the energy range 10 to 40 keV may also be obtained in LVSEM. The cost of such unambiguity in contrast lies in sample preparation and the vacuum environment used. In the case of clean surfaces a monotonic increase of η with E_p is obtained irrespective of the incident electron energy. This is unlike the data from the as-inserted surfaces, which show a complicated pattern for low energy electrons. It is suggested that the reason for such behaviour is due to the surface conditions of the samples investigated as well as to the vacuum environment. It is well known that most surfaces grow a natural contamination layer the thickness and composition of which is a function of the material's chemical reactivity. For example, inspection of the data by Bongeler et al. [12], shows an increase in η for elements of $22<Z<35$. This group of elements is known to be highly reactive and grows oxides of thickness up to 10 nm. On the other hand elements like gold have a thin carbon layer deposited on their surface. Under conventional vacuum conditions, found in most electron microscopes, the incident electron beam is known to deposit a film of carbon on the sample surface. It is also possible that this film may contain an oxygen compound of the target material, depending on its chemical reactivity. This may explain why our aluminum data appear lower than those of Bongeler et al., while η obtained for gold is higher. Compounded with this explanation is the fact that the interaction volume of low energy electrons (100 eV) is only about 10–20 nm and the contamination film could occupy a large proportion of this volume. It may be necessary therefore to conduct such measurements with a detector capable of surface characterisation, as in AES. Such modifications are under way in the author's laboratory.

The pattern of behaviour of the backscattering coefficients reported here agrees qualitatively with recent studies of the electron transport properties [23–26] which are influenced by two basic parameters, χ, the ratio of the inelastic mean free path (λ_i) to the transport mean free path (λ_t), and σ, the ratio of the electron range (R) to λ_t. For electron energies of the order of 1000 eV, it is found that χ is higher for higher atomic number materials implying the pattern of atomic number dependence found here. The picture only reverses for electrons of energy less than 400 eV or so and for elements such as gold this ratio shows a minimum at about 200 eV. Further experiments in the energy range 200 eV to 1000 eV are planned.

Conclusion

A new set of measurements of the backscattering coefficients in the energy range 0.5–6 keV is reported. The general conclusion from this study is that atomic number contrast in LVSEM may be quantified as in a conventional SEM operated above 10 keV. However, two vital conditions are necessary to obtain such a clear pattern; surface cleanliness and ultra high vacuum conditions. While the first requirement may be difficult to achieve for some samples, the second requirement should be easier to satisfy with electron microscopes employing field electron emission sources.

Acknowledgment. The authors would like to thank Professor Matthew for helpful discussion. The financial support of the EPSRC and the EU grant PC 12283 are acknowledged.

References

[1] D. Venables, D. M. Maher, *J. Vac. Sci. Tech.* B **1996**, *14(1)*, 421.
[2] K. Ogura, *JEOL News* **1991**, *29E* (2), 1.
[3] K. Ogura, M. M. Kersker, *46th Annual EMSA Meeting, San Francisco Press, San Francisco*, 1988, 204.
[4] D. D. Perovic, M. R. Castell, A. Howie, C. Lavoie, J. S. W. Cole, *Ultra Microsc.* **1995**, *58*, 104.
[5] H. Bishop, *J. Phys. D*, **1967** 18, 703.
[6] I. M. Bronshtein, B. S. Faiman, *Secondary Electron Emission*, Nauka-Press, Moscow, 1969.
[7] S. Thomas, E. B. Pattinson, *J. Phys. D: Appl. Phys.* **1970**, *3*, 349.
[8] E. H. Darlington, V. E. Cosslett, **1972**, *J. Phys. D*, *5*, 1969.
[9] H. J. Fitting, *Phys. Stat. Sol. (a)*, **1974**, *26*, 525.
[10] J. Shou, H. Sorensen, *J. Appl. Phys.* **1978**, *49*, 816.
[11] L. Reimer, C. Tollkamp, *Scanning*, **1980**, Vol. 3, 35–39.
[12] R. Bongeler, U. Golla, M. Kassens, L. Reimer, B. Schindeler, R. Senkel, M. Spranck, *Scanning*, **1993**, *15*, 1.
[13] M. Kotera, K. Murata, K. Nagami, *J. Appl. Phys.* **1981**, *52*, 997, 7403.
[14] S. Ichimura, R. Shimizu, *Surf. Sci.* **1981**, *112*, 386.
[15] D. Joy, *J. Microsc.* **1987**, *147*, 51.
[16] H. J. Fitting, C. H. Hinkforth, H. Glafeke, J. Ch. Kuhr, *Phys. Stat. Soli. (a)*, **1991**, *126*, 85.
[17] K. Janssen, A. Walker, *Surf. Sci.* **1993**, *292*, 83.
[18] R. Browning, T. Z. Li, B. Chui, Jun Ye, R. F. W. Pease, Z. Czyzewski, D. Joy, *Scanning*, **1995**, *17*, 250.
[19] *GRAPHIT* 33, Kontakt Chemie, D-7557, Iffezheim, Germany.
[20] T. El Bakush, *D Phil Thesis*, The University of York, 1995.
[21] A. M. D Assa'd, *D Phil Thesis*, The University of York, 1995.
[22] A. M. D Assa'd, M. M. El Gomati, *Scanning Microsc.* **1998**, in press.
[23] J. A. D. Matthew, A. R. Jackson, M. M. El Gomati, *J. Electron Spect. Relat.* **1997**, *Phen. 85*, 205–219.
[24] P. Cumpson, *Surf. Interface Anal.*, **1997**, *25*, 277.
[25] I. S. Tilinin, W. S. M. Werner, *Mikrochimica Acta.* **1994**, 485.
[26] R. Mayol, F. Salvat, *Atomic Data Nucl. Data Tab.*, **1997**, *65*, 55.

Mikrochim. Acta [Suppl.] 15, 333–339 (1998)

Monte Carlo Simulations of Edge Artefacts in MULSAM Images

Mohamed M. El-Gomati* and **Ahmad M. D. Assa'd**

Department of Electronics, University of York, Heslington, York YO1 5DD, UK

Abstract. A fast Monte Carlo model for the simulation of electron-solid interactions in small dimensional systems such as raised and buried structures and thin multilayer films is reported. It is based on single scattering and uses a modified Rutherford scattering cross-section. The model is used to develop a new method for the detection and correction of edge enhancement encountered in high resolution Auger imaging of topographical structures and is applied to the case of a 1.4 µm thick aluminum overlayer on a silicon substrate. Secondary, backscattered, Auger and low energy loss electrons and characteristic x-rays are produced. These are simulated for collection in an array of detectors around the azimuth. In this method the low energy loss electrons are used to detect the edge position while backscattered electrons of energies ≥ 0.75 of the incident electron beam energy are used for the correction of the Auger edge enhancement. The correction of the substrate enhancement from a bevelled multilayered sample made of 12 alternating 10 nm thick films of cobalt and platinum on a silicon substrate is also reported. Use here is made of the relationship of the sample current with the backscattering coefficient for the correction of the substrate backscattering effects.

Key words: Monte Carlo simulations, electron scattering, Rutherford cross-section, edge artefacts.

Auger electron spectroscopy (AES) and microscopy (SAM) have been established for over two decades as "mature" techniques in surface analysis. This is because the underlying physics of electron-solid interactions, as used in these techniques and in electron microprobe analysis (EPMA) is well understood. It is now possible in AES to estimate the surface concentration of an element in a multi-component alloy or a compound to within few atomic percent [1–2]. The use of AES, with energetic ions of inert gases to depth profile multi-elemental samples, has underpinned many of the technological advancements made in microelectronics and materials science, to name but two disciplines. Auger electron imaging has, on the other hand, allowed one to map, at high spatial resolution, the surface distribution of elements on solid surfaces [3–4]. Imaging with Auger electrons made it easy to correlate the chemical information with the sample structure and hence dynamic type surface studies have become possible.

However, many of these developments have been only successful in a rather limited range of systems that mainly exhibit flat surfaces. Quantification at high spatial resolution has remained a problem in AES and SAM applications [5–7]. In such cases, there are several factors that can, if not detected and corrected, lead to false conclusions being drawn about a given sample. Solids exhibiting surfaces with varying degrees of topography and chemical inhomogeneties may yield a number of analysis artefacts. The origin of these resides in a combination of sample properties and composition as well as the geometrical arrangement of the instrument used [8]. The analyser collection angle, the incident electron energy with respect to the energy of the Auger transition used, and the direction of scanning the incident electrons relative to the position of a given chemical or topographical edge are factors known to affect the collected Auger signal [8].

* To whom correspondence should be addressed

In AES, both the sample topography and the substrate properties have been recognised and their effects included in estimating the surface concentration of a given material [1, 2]. However, allowing for these effects in data processing and/or interpretation depends on a prior knowledge of the sample under investigation and the experimental conditions used. Extension of such a treatment to the case of high resolution SAM imaging has proved to be non-trivial if not an impossible exercise in some cases. For example if a sample made of one element exhibits surface topography variations of more than $\pm 5°$ from one place to another with respect to the incident electron direction, then these variations will be reflected in estimates of the surface concentration of the element being mapped. If, in addition, the sample being analysed is heterogeneous in surface composition as well as in depth then the analysis exercise will be further complicated.

The recent developments at the University of York of the multi-spectral Auger analysis (MULSAM) technique [9–12] has further advanced the use of quantitative Auger electron spectroscopy and imaging to chemically inhomogeneous solids and non-flat surfaces. This is carried out through the use of a combination of a number of energy analysed signals/images. The present MULSAM instrument can be used to collect simultaneously up to 20 energy analysed images in addition to specimen current and characteristic x-ray images. Of the energy analysed images, four are backscattered and collected from a quadrant of p-n type detectors. The use of the backscattered electron signals from this detector has allowed the simultaneous correction of sample topography and substrate effects [13–14]. It is now possible to map the surface concentration of a given element in multi-component, inhomogeneously distributed (both on the surface and in the bulk) and topographically sharp samples of the type one frequently encounters with technological systems. A number of different examples demonstrating the power of MULSAM have been reported recently [15–16]. However, in spite of these successes, there remains one imaging artefact that needs correction; the edge enhancement artefact from sharp topographical edges [7–8] caused by energetic electrons that are transported through the sidewall of an overlayer. On their way out, these electrons will have a lower energy than the primary energy and may have a more grazing angle of incidence with respect to their original directions. The effects act to increase the Auger electron yield of both the overlayer and the adjacent substrate. It is thus possible that, if the sidewall of the overlayer is facing the Auger electron analyser collection angle, then the Auger yield of the overlayer and its adjacent substrate will be greatly enhanced, sometimes by more than a factor of two with respect to those obtained from flat surfaces.

The present study is a Monte Carlo simulation of edge and substrate enhancement effects in SAM imaging. The aim is to suggest correction mechanisms for these effects. In this respect, use of energetic backscattered electrons is proposed to correct both types of artefacts. Backscattered electrons of energy greater than 75% of the incident electron energy are used in the correction of the edge enhancement. Following the suggestion by Cazaux [17] of relating the specimen current to the Auger electron backscattering factor, the latter is used for the correction of the substrate enhancement. Simulation examples demonstrating both methods are given together with a suggested experimental set-up using a new arrangement of a low energy electron loss detector for the correction of the edge artefact.

The Monte Carlo Model

The Monte Carlo model used in this study has been developed by the authors and is based on single scattering of electrons to suit thin layers and small dimension structures. Details of the model can be found elsewhere [18] and will therefore not be repeated here. One difference between the current model and the earlier one is in the treatment of elastic scattering where use is made now of a modified version of the screened Rutherford scattering cross-section [19]. This has the following form:

$$d\sigma(\varphi)/d\Omega = e^4 Z^2/[16E^2(\sin^2(\varphi/2) + (\theta_0^2/4)] \quad (1)$$

where φ is the scattering angle, E is the electron energy in keV, Z is the atomic number of the atom in question, and $\theta_0^2/4 = \alpha$ is the screening parameter. α is also expressed as

$$\alpha = KZ^{2/3}/E \quad (E \text{ in } eV) \quad (2)$$

where K is a numerical constant chosen to fit the Monte Carlo calculations to experimental data on backscattering coefficients (η). We have calculated η for a large number of samples of varying atomic number, angle of incidence and electron beam energy.

The value of K was adjusted to force the calculation to yield the experimental backscattering coefficients in the energy range 5–25 keV measured by Bongeler et al. [19] and by Asaa'd and El Gomati in the energy range 0.5–6 keV [20]. A polynomial fit was then used to obtain a new K value which is expressed as a function of the atomic number (Z), the incident electron energy (E_P) in (keV), and the density (ρ) in (g/cm^3). This has the form:

$$K = a + \exp^{bEp+c} \tag{3}$$

where,
$a = 3.706 + 8.784 \times 10^{-3}Z + 0.2613\rho$,
$b = -0.283 - 6.82 \times 10^{-4}Z + 7.191 \times 10^{-3}\rho$,
$c = 1.641 + 3.595 \times 10^{-2}Z - 2.776 \times 10^{-3}\rho$.

A comparison of the differential cross section of the Rutherford formula using the above screening parameter expression and another suggested by Bishop [21] and the Mott cross section, given by Czyzewski et al. [22] are shown in Figs. 1 and 2. It is clear from these figures that there is a good agreement for the higher energy data between all three expressions while at the lower energies and for heavy elements the discrepancies are much larger. In the case of Au at 2 keV the reduction in the differential cross-section at lower angles is very large, but this is not significant because such small angle scattering processes have little influence on the electron transport. A comparison of the angular region where elastic scattering is dominant (i.e. between 50° and 120° and for 200 eV and 1200 eV electrons) shows that the present modification averages that of the Mott cross section.

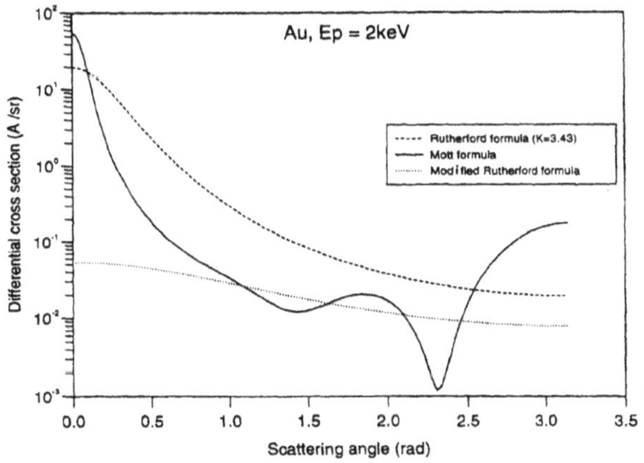

Fig. 1. A comparison of the differential scattering cross-section of the unmodified Rutherford cross section using Bishop K = 3.43, the modified Rutherford formula with the present K value and the Mott cross-section [22] for a Au target at 2 keV

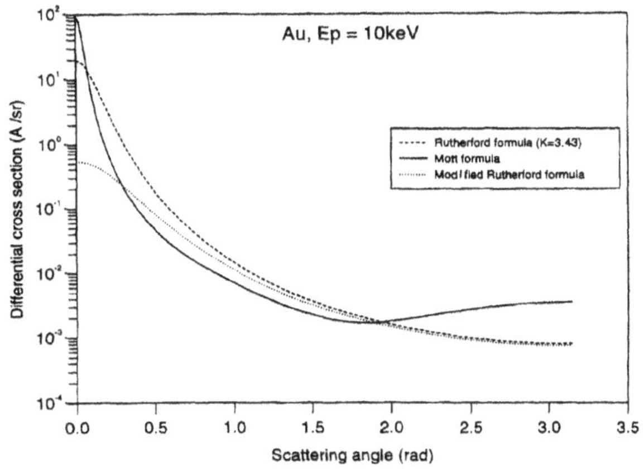

Fig. 2. As in Fig. 1, but for 10 keV electrons

Further, discrepancies between the two models are largely compensated for by fitting the calculated η to the experimental values. The net effect is that the present modifications, although crude in comparison to the more exact Mott cross-section, minimise the amount of numerical data to be fed into the program.

Results

The model is first used to calculate the backscattering coefficients (η) for a number of elements at various incident electron energies. Results obtained using the un-modified and the modified Rutherford cross-sections and those measured by Bongeler et al. [19] are shown in Fig. 3. There is a good agreement between the measured values and those calculated

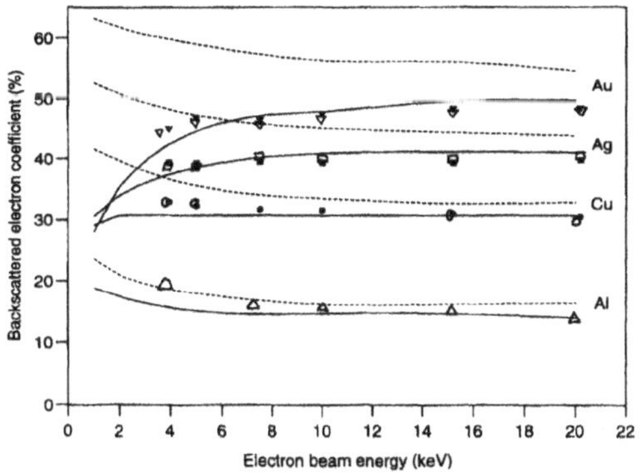

Fig. 3. The backscattering electron coefficient for a number of elements as a function of electron energy using the present modification (solid lines), the unmodified Rutherford formula (dashed lines) and the experimental measurements of Bongeler et al. [19], solid symbols and of Hunger et al. [27] open symbols. Note that in the case of Al, the Hunger et al. data are those of Si

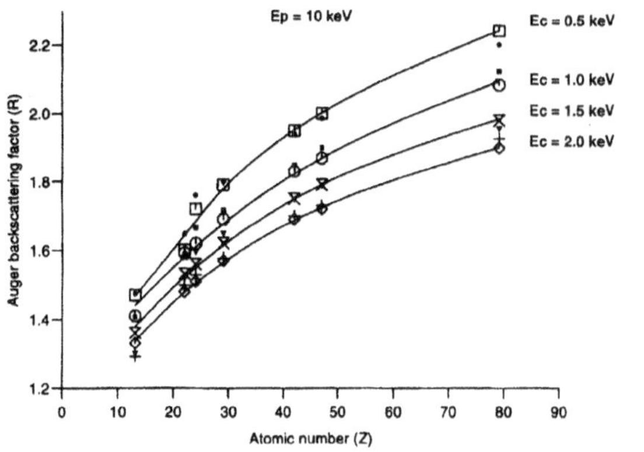

Fig. 4. A comparison of the Auger backscattering factor as a function of the target atomic number calculated in the present work (solid symbols) with those of Ichimura and Shimizu (open symbols)

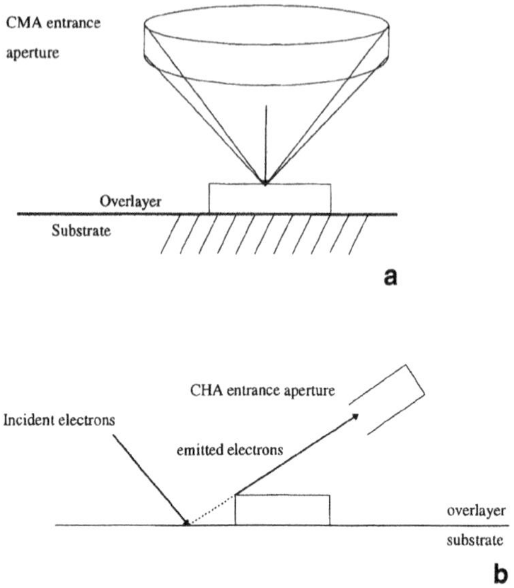

Fig. 5. Schematic of the simulated experimental geometries, **a** the CMA with normal incidence electrons and **b** the CHA geometry of the MULSAM

with the modified formula. Below 5 keV, the difference between the two formulae is large, not surprisingly so since the conventional Rutherford formula breaks down at these energies. In a separate study [23], the modified formula was used in the simulation of the spectral background in the energy range 200–5000 eV, and good agreement with experimental data from a metrology spectrometer [24] was obtained. As a further test of the accuracy of the model, the Auger backscattering factor was calculated for a number of elements and at different incident electron energies. Fig. 4, a comparison between the results obtained using the present model and those reported by Ichimura and Shimizu [25], shows a very good agreement.

The above tests have ascertained the suitability of the model to simulate the interaction of energetic electrons with solids. The present model has been applied for the simulation of complex sample structures. Two examples will be given below; the edge enhancement artefact and the substrate enhancement artefact. The simulated experimental conditions are depicted in Fig. 5a for the CMA geometry and 5-b for the CHA geometry.

a) The Correction of the Edge Enhancement

In the MULSAM arrangement, the incident electron beam makes an angle of 64° with respect to the axis of the input lens of the hemispherical type (CHA) electron energy analyser. In this geometry, the sample is positioned 10 mm from the input lens of the CHA and normally analysis is carried out at about 32° with the surface normal. We have also simulated normal electron incidence as is used in cylindrical mirror analyser (CMA) based instruments with an integral electron gun. The simulations of the edge effects presented here are all carried out for an incident beam energy of 20 keV and for an electron beam of FWHM = 200 nm impinging upon a 1.4 micron Al overlayer on a silicon substrate. It is possible in the present model to simulate a number of signals including, Auger electrons, backscattered electrons of different energies, secondary electrons and characteristic x-rays.

The backscattered electrons, in the present simulation, are collected according to their energy and exit angle with respect to the surface normal. A comparison of the various combinations of energy and angles with the profile of the Auger linescan of the Al on Si edge shows a close similarity between backscattered electrons of energy >75% of the incident primary energy with those of the Auger profile (artefact), as shown in Fig. 6a for the CMA geometry and 6-b for the MULSAM geometry. It was also found, as shown in Fig. 6, that there is a close similarity between the profile of both the total number of energetic electrons of energies >75% of E_p collected in 2πsr and those of the same energy but collected only in an annulus between 20°–30°. The latter was therefore chosen in the present simulation for the correction of the edge

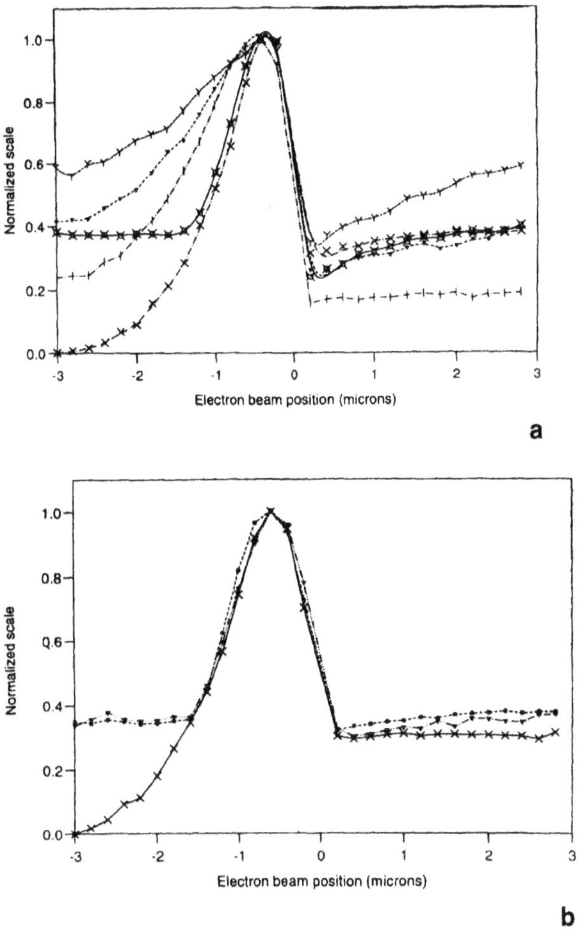

Fig. 6. The profile of energetic backscattered electrons of energies >75% of the incident electron energy on an Al film 1.4 µm thick positioned at point 0.0 in the x-axis on a Si substrate. **a** The case of 20 keV incident electrons at normal incidence as in the CMA geometry with a co-axial gun. The symbols represent the following conditions: × = Si KLL Auger signal, ⊠ = BSE of energies > 0.75% of incident energy, + = BSE emitted between 10° and 20°, ▼ = BSE emitted between 20° and 30° and Y = BSE emitted between 30° and 40°. **b** The case of 20 keV electrons incident at 32° with respect to the surface normal of the Al on Si films as in the MULSAM geometry, × represents Si KLL Auger electrons, ● = total BSE of energies >0.75% of E_p and ▼ is for BSE emitted between 20° and 30°

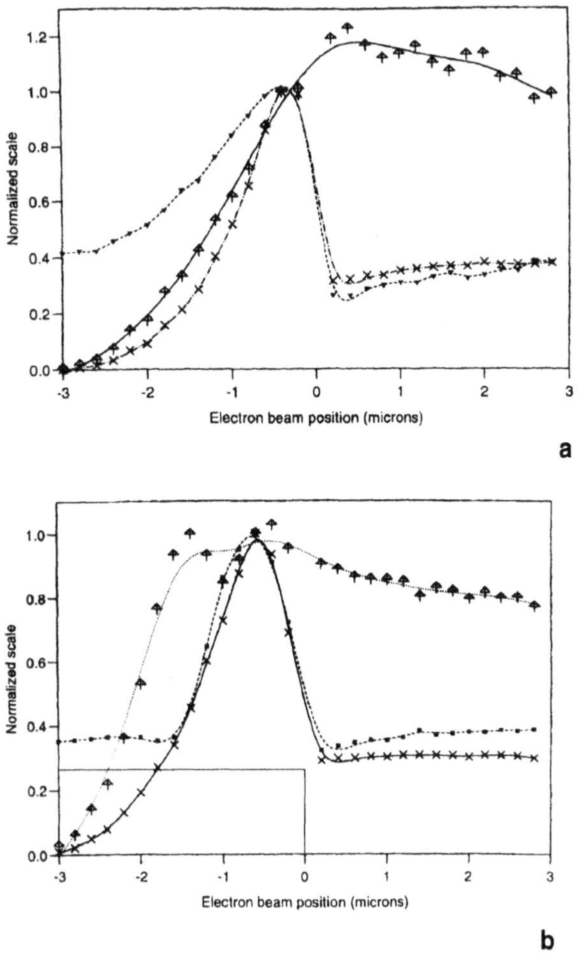

Fig. 7. Line scan profiles of the edge of Fig. 6 above using the same experimental conditions and showing the Si KLL Auger signal (× symbols) and the energetic backscattered electron signal (▼ symbol) for electrons emitted between 20° and 30°, together with the corrected Auger signal (⍋ symbol). **a** CMA geometry. **b** MULSAM geometry

enhancement artefact as these will be most suitable for a CMA geometry. The correction of the edge artefact is shown in Fig. 7a, and 7b for both the CMA and the MULSAM geometries, respectively.

Close inspection of the two cases shown in Fig. 7 shows that the correction in the MULSAM geometry causes a shift in the position of the Si edge; placed at point zero in the figure by about 1.5 µm. This causes an apparent shrinkage of the Al overlayer under study. The edge in the case of the CMA geometry appears to shift slightly in the opposite direction to that of the CHA geometry. The cause of this shift effect is that the energy distribution of the backscattered electrons used for the correction is forward peaked due to the non normal incidence of the primary electrons. It is likely also that a similar effect will be obtained in non normal incidence for the CMA geometry. In both cases, the correction appears to reduce the value of the enhancement, from >200% to about 20% with respect to the Si Auger signal from the substrate.

The x-ray analysis of the Al on Si edge simulated above does not show the enhancement effect present in the Auger mode. This is because the x-ray signal comes from a large interaction volume in comparison with the beam diameter, which is the dominant factor

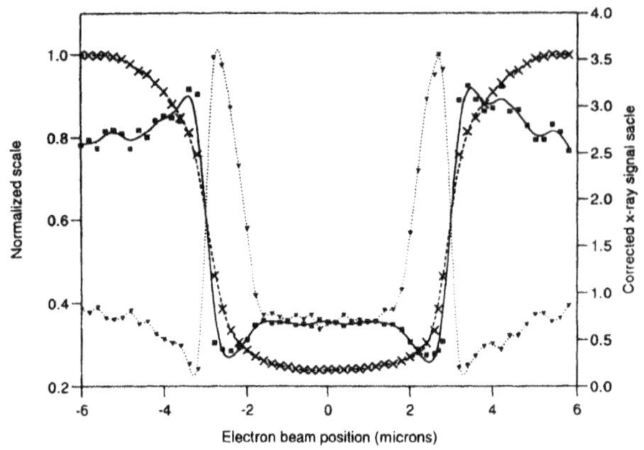

Fig. 8. Simulated Si $K_{\alpha 1}$ x-ray signal (dashed line), the energetic backscattered electrons of energy >75% of the incident electron energy (dotted line) and the corrected x-ray signal (solid line). The x-ray take-off angle is 90° for a 20 keV electrons at normal incidence on the same Al on Si edge of Fig. 6. The edge is placed between −3 and 3 microns in the figure

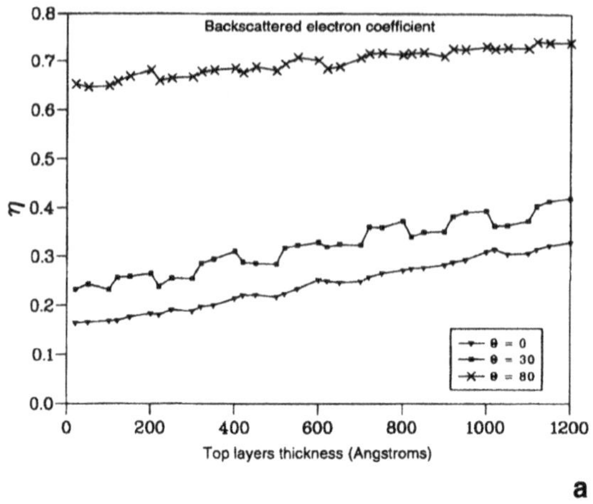

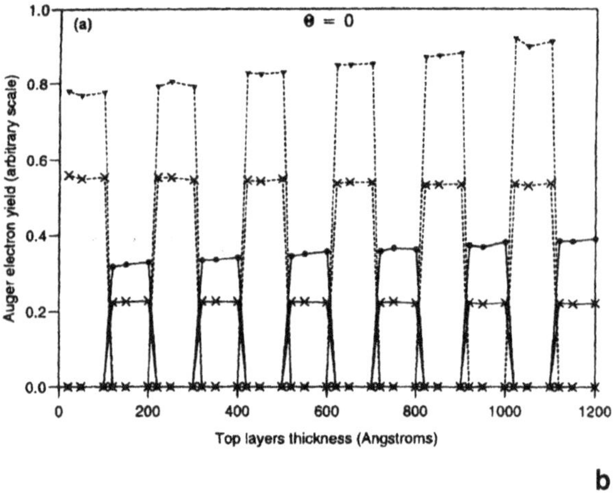

Fig. 9. a Backscattered electron line scans of a bevelled 12 Pt and Co 10 nm thick films on a Si substrate at various angles of electron incidence. **b** The linescans of the KLL Auger electrons (▼, ●) and its corrected signal (×) for Pt (solid line) and Co (dashed line) as a function of the top layer thickness of the bevelled Pt and Co structure

in the Auger mode. This interaction is largely governed by the solid's scattering properties and the energy of the incident electrons, and may be represented by a Gaussian of FWHM equivalent to the backscattering radius of the redifused primary electrons. The convolution of this Gaussian response function with the edge profile gives the rounded edge shape observed in Fig. 8, which for 20 keV electrons gives an edge resolution of about 1 μm. This figure is improved to 0.43 μm by dividing the x-ray signal by the energetic backscattered electrons as in the Auger mode.

b) The Substrate Enhancement Effect

Auger electron spectroscopy combined with depth profiling (AESDP) by inert ions is one of the most popular modes of operation in surface analysis by electron spectroscopy [25]. The disadvantages of this method are that it is sample destructive and also that the energetic ions used, cause a broadening of the interfaces between two chemical layers. Bevelling a sample by energetic ions or chemical acid etching followed by low energy ion bombardment to remove the damaged surface layer has proved to be a successful alternative to conventional AESDP. In this case an electron beam is used to scan the bevelled sample and the resulting line scan is the corresponding depth profile of the sample. In order to quantify the depth profile of either the conventional method or from a bevel, one must correct for the substrate backscattering effects.

In Fig. 9 we show some preliminary results of a system made of ×12 alternating films of platinum (Pt) and cobalt (Co) on a Si substrate. The films are 10 nm thick and are scanned by a beam of 20 keV electrons at different angles of incidence. The effect of the backscattered electrons can be seen in Fig. 9a to double in value from one side of the bevel to the other at normal incidence. Although this is not greatly reflected in the Auger line-scans obtained (Fig. 9b), because of a fortuitous combination of film thickness and weak matrix effects, the correction obtained shows the effectiveness of the method. This should therefore prove useful particularly in systems employing thicker films and stronger backscattering effects.

Conclusion

A fast Monte Carlo model for the simulation of electrons in solids comprising raised and buried structures and thin-multilayer films is reported. The model has been used to develop methods for the correction of imaging artefacts in the analysis of topographical structures. In the case of Auger and x-ray linescan modes the model has been illustrated on a 1.4 μm Al edge on a silicon substrate. The correction proposed is based on using backscattered electrons of energy >75% of the incident electrons. A reduction of the edge artefact observed in the Auger mode by an order of magnitude has been achieved. In the x-ray mode, the correction method improves the spatial resolution by a factor of ∼2. The model has also successfully been used to correct for the substrate effects encountered in Auger electron spectroscopy depth profiling.

Acknowledgements. The authors would like to thank Professor Matthew for helpful discussion. The financial support of the EPSRC and the EU grant PC12283 are acknowledged.

References

[1] M. P. Seah, *Vacuum* **1988**, *36*, 399.
[2] Walker, C. G. H., D. C. Peacock, M. Prutton, M. El Gomati, *Surf. Interface Anal.* **1988**, *11*, 266.
[3] J. Cazaux, J. Chazelas, M. N. Charasse, J. P. Hirtz, *Ultramicroscopy* **1988**, *25*, 31.
[4] J. A. Venables, G. G. Hembree, *Inst. Phys. Conf. Ser. EMAG 91, No. 119*, 1991, pp. 33-38.
[5] M. M. El Gomati, M. Prutton, *Surf. Sci.* **1978**, *72*, 485.
[6] M. M. El Gomati, *Vacuum* **1988**, *38*, 337.
[7] M. M. El Gomati, M. Prutton, B. Lamb, C. G. Tuppen, *Surf. Interface Anal.* **1988**, *11*, 251.
[8] M. M. El Gomati, I. Barkshire, J. C. Greenwood, P. G. Kenny, R. Roberts, M. Prutton, *Microscopy: The Key Research Tool*, 1992, pp. 29–38.
[9] P. G. Kenny, P. G., M. Prutton, M. El Gomati *Inst. Phys. Conf. Ser. EMAG-MICRO 89, No. 98*, Inst. of Physics, London, 1989, pp. 295–298.
[10] M. Prutton, M. El Gomati, P. G. Kenny, *J. Electron Spectros. Rel. Phen.* **1990**, *52*, 197.
[11] M. Prutton, C. G. H. Walker, J. C. Greenwood, P. G. Kenny, J. C. Dee, I. R. Barkshire, R. H. Roberts, M. El Gomati, *Surf. Interface Anal.* **1991**, *17*, 71.
[12] I. R. Barkshire, J. C. Greenwood, P. G. Kenny, M. Prutton, R. H. Roberts, M. M. El Gomati, *Surf. Interface Anal.* **1991**, *17*, 203.
[13] I. R. Barkshire, M. El Gomati, J. C. Greenwood, P. G. Kenny, M. Prutton, R. H. Roberts, *Surf. Interface Anal.* **1991**, *17*, 209.
[14] I. R. Barkshire, R. H. Roberts, J. C. Greenwood, P. G. Kenny, M. Prutton, M. El Gomati, *Inst. Phys. Conf. Ser. No. 119. EMAG 91*, 1991, pp. 185–188.
[15] M. Prutton, I. R. Barkshire, M. Crone, *Ultramicroscopy* **1994**, *59*, 47.
[16] J Cazaux, *Mikrochim. Acta*, **1992**, *107*, 220.
[17] M. M. El Gomati, W. C. C. Ross, J. A. D. Matthew, *Surf. Interface Anal.* **1991**, *17*, 183.
[18] A. M. D. Assa'd, M. M. El Gomati, J. Robinson, *Inst. Phys. Conf. Ser.* **1993**, *130*, 403.
[19] R. Bongeler, U. Golla, M. Kassens, B. Schindeler, R. Senkel, M. Sprank, *Scanning* **1993**, *15*, 1.
[20] A. M. D. Assa'd M. M. El Gomati, *Scanning Microscopy*, 1997, in press.
[21] H. Bishop, *Ph. D. Thesis*, University of Cambridge, 1996.
[22] Z. Czyzewski, D. O. MacCallum, A. Romig, D. Joy, *J. Appl. Phys.* **1990**, *68*, 3066.
[23] J. A. D. Matthew, W. C. C. Ross, M. M. El Gomati, *Inst. Phys. Conf. Ser.* **1992**, *130*, 383.
[24] M. Seah, G. C. Smith, *Surf. Interface Anal.* **1988**, *12*, 105.
[25] S. Ichimura, R. Shimizu *Surf. Sci.* **1981**, *112*, 386.
[26] S. Hofmann, *Surf. Interface Anal.* **1980**, *2*, 148.
[27] H.-J. Hunger, L. Kuchler, *Phys. Stat. Sol. (a)* **1979**, *56*, K45.

Mikrochim. Acta [Suppl.] 15, 341–349 (1998)

Assessment of the Inelastic Scattering Model in Monte-Carlo Simulations

Vladimír Starý

Department of Materials Science, Faculty of Mechanical Engineering, CTU Prague, Karlovo nám. 13, 121 35 Prague, Czech Republic

Abstract. Electron backscattering and transmission through thin films have been simulated by the Monte-Carlo method. Instead of the backscattering coefficient we calculated Niedrig's coefficient $C_N = \eta/(N_1 Z^2 d_{1/2})$, where η is the backscattering coefficient, N_1 is the number of atoms in a volume unit, Z is atomic number, and $d_{1/2}$ is the film thickness giving a value for the backscattering coefficient of half that for a bulk specimen. In the case of electron transmission, the most probable energy loss, W_p, was found. The calculations were carried out for Al, Cu, and Au, the former for 10–100 keV primary energy and in the 50–2000 μg/cm^2 film thickness range, the latter for a primary energy of 20 keV and in the 150–650 μg/cm^2 film thickness range. Tables calculated by means of the Hartree-Fock atom function were used for simulating elastic scattering, and several models of inelastic scattering were employed – the modified Bethe model and models with exponential and hyperbolic shapes of energy losses (with or without electron deflection at scattering). Our findings were compared with published experimental results. In the case of Niedrig's coefficient, an exact comparison was impossible. Nevertheless, for the systems studied best agreement between calculated and experimental values of W_p was reached when using the model containing the hyperbolic shape of energy loss, (usually without deflection during scattering). This result provided an opportunity, when combined with previous assessments of elastic models, to improve the accuracy of calculation for future employment of the Monte-Carlo method.

Key words: Monte-Carlo, electron scattering, thin films, X-ray microanalysis.

A Monte-Carlo method (MC) is currently used for simulating electron scattering in transmission electron microscopy [1], scanning electron microscopy [2] and analytical electron microscopy [3] and for calculating the intensity of X-ray emission in Electron Probe X-ray Microanalysis [4]; extensive reviews are given in [5, 6]. MC makes it possible to calculate and estimate the difficult-to-measure parameters influencing experimental results (e.g., contrast in the image, probe size, X-ray lateral and depth distribution, size of the interaction volume, etc). Nevertheless, the validity of the models and the accuracy of MC results needs to be tested by comparing with some experimentally determined parameters, (e.g., the backscattering coefficient and angular and energy distribution of electrons transmitted and also backscattered by thin films). The MC algorithm is well known, widely used and accepted. However, several different models for elastic and inelastic scattering exist. Sometimes their effect on the reliability of the calculated results and on their agreement with the experiment is not fully checked.

The backscattering coefficient and its dependence on film thickness were used for checking the MC model in [7]. These calculations using the Mott cross-section for elastic scattering and Bethe's continuous slowing down model of inelastic scattering agreed very closely with experiments [8] in the 10–100 keV energy range for Al and Au. Nevertheless, as mentioned in the paper, the backscattering coefficient alone was not very sensitive to the model type. Calculated angular and energy distributions of backscattered electrons from thin films were also compared with experiment [9, 10], but not systematically.

In our previous work [11] we used the quantity C_N (Niedrig's coefficient) defined according Niedrig's review of electron backscattering from thin films [12] as

$$C_N = \frac{\eta}{N_1 Z^2 d_{1/2}} \tag{1}$$

where η is the backscattering coefficient, N_1 is the number of atoms in a volume unit ($N_1 = N_A\rho/A$, N_A is Avogadro's number, ρ is density and A atomic weight), Z is atomic number and $d_{1/2}$ the thickness of the film, where the magnitude of the backscattering coefficient reaches half that of the bulk specimen. The coefficient characterizes the slope of the backscattering coefficient vs. film thickness dependence. We compared experimental values of Niedrig's coefficient with those calculated by our Monte-Carlo program for Al, Cu, and Au. This comparison had three advantages:

1. Values calculated using several different models of elastic scattering cross-sections were checked by defined statistical methods (comparison of the statistical residuals between experimental and calculated values).
2. Using C_N we simultaneously compared data for a wide range of energy (10–100 keV) and film thickness (50–2000 µg/cm^2).
3. We used a large set of experimental data from the literature on electron energy and film thickness depenence of the backscattering coefficient (Seidel [8], Niedrig and Sieber [13], Neubert and Rogaschewski [14] and Cosslett and Thomas [15]), together with data from the unpublished work of Doležel [16].

The tables computed from the Hartree-Fock atomic wave functions [17] for elastic scattering cross sections gave the best agreement with the experiment, though the differences between models for the elements Cu and Au were relatively small.

The main purpose of this work is to compare experimentally measured data with those calculated using different models of inelastic scattering and to decide which model gives the best agreement using a similar method to that described in [11]. Parameters characterizing the energy distribution of transmitted electrons seem to be more sensitive to the model type, because the mechanism of energy loss influences mainly the energy loss distribution of transmitted and backscattered electrons. Thus, for comparison we use not only Niedrig's coefficient but also the value of the most probable energy loss of transmitted electrons, W_p.

Model Used for MC – Inelastic Scattering

The Bethe model for the continuous slowing down of electrons [5–7] has often been used for evaluation of inelastic collisions. In this model, an electron loses its energy continuously and the energy is determined from the length of its path in the sample. The rate of energy loss is given by the stopping power S

$$S = -\frac{dE}{ds}, \tag{2}$$

where E is the electron energy, and s is the path length. Bethe's well known formula for S_B is (e.g., [18])

$$S_B = 2\pi e^4 N_A \frac{\rho Z}{AE} \ln\left(\frac{1.166E}{J_B}\right). \tag{3}$$

where e is the electron charge and J_B is the mean ionization potential. The subscript B indicates that the Bethe mean ionization potential J_B is used in the formula. For J_B we used values according to [19].

Unfortunately, for low electron energies, the classical formula (3) turns negative due to the properties of the logarithmic function. There are two modifications of Bethe's formula to prevent this: the first by Rao-Sahib and Wittry [20] changes Eq. (3) to

$$S_B = C\frac{\rho Z}{A}\frac{1}{\sqrt{J_B E}} \tag{4}$$

from the inflection point of Eq. (3) ($E = 6.338\ J_B$) to $E = 0$, C is a constant ensuring continuous change of S at the inflection point. This formula gives a positive value of S, but S tends to infinity as the energy decreases, which is nonrealistic.

Joy and Luo [21] modified the formula for mean ionization potential, so that it depended on electron energy and at low energy adjusts the value of the logarithm to avoid the occurrence of the anomaly. The mean ionization potential is then

$$J_J = \frac{J_B}{1 + \dfrac{kJ_B}{E}}, \tag{5}$$

where k is a constant (approximately, $k = 0.85$).

The actual energy loss W is established from the path length s between two collisions at distances l_1 and l_2 along the path according to

$$W = \int_{l_1}^{l_2} \left(\frac{dE}{ds}\right) ds. \tag{6}$$

The usual way of calculating energy loss is simply to calculate W as

$$W = S_J(E_1).(l_2 - l_1). \tag{7}$$

Alternatively, integration of (6) from E_o at several steps $(\ldots, E_i,\ E_{i+1}, \ldots)$ to some minimum value of energy (usually to J) may be used. Then the length of the path at these points $(\ldots, l_i,\ l_{i+1}, \ldots)$ must be calculated. An electron scattered at distance l' between points l_i and l_{i+1}, where it had energies E_i and E_{i+1}, respectively, loses energy

$$W = E - E' = E - \left(E_{i+1} + \frac{l_{i+1} - l'}{l_{i+1} - l_i}(E_i - E_{i+1})\right), \tag{8}$$

E being the energy before collision at l'. For dE/ds in Eq. (6), we used formula (3) with corrected mean ionization potential J_J, Eq. (5). Although there was no practical difference between values of W calculated by Eqs. (7) and (8), we employed Eq. (8). Thus, in our MC at all scattering events the deflection is calculated as at the elastic one, but simultaneously, the loss of energy proceeds due to Eq. (8). This model was the first used; it will be referred to as the modified continuous slowing down model – MODCSD.

In the next version, we used a hyperbolic model of energy loss. The shape of the energy loss function is given by the normalized probability density $f(W)$ that is defined as

$$f(W) = \frac{\dfrac{d\sigma(W)}{dW}}{\displaystyle\int_{W_{min}}^{W_{max}} \frac{d\sigma(W)}{dW} dW}, \tag{9}$$

where $d\sigma/dW$ is the differential inelastic cross-section, integrated over the whole solid angle. In our calculation, in accordance with [22], we assumed that

$$f(W) = KW^{-2} \tag{10}$$

in the interval (W_{min}, W_{max}). W_{max} and W_{min} are the limits of electron energy loss in matter, W_{min} usually being taken as an adjustable parameter which, because of the hyperbolic shape of the distribution, must be >0. Due to normalization of $f(W)$ to unity

$$K = \frac{W_{max} W_{min}}{W_{max} - W_{min}} \tag{11}$$

is valid.

In the calculation, we use simply $W_{max} = E$, the actual energy of an electron. Moreover, we assume $W_{min} = 10\,\text{eV}$ and the same shape of $f(W)$ for all elements.

The normalized probability density $f(W)$ is given by

$$f(W) = \frac{EW_{min}}{E - W_{min}} W^{-2} \cong W_{min} W^{-2} \tag{12}$$

$(E \gg W_{min})$. Therefore we can calculate the actual energy loss W at each inelastic event, in a similar manner to that used in the Monte-Carlo codes, by means of the random number R $(0 \leq R \leq 1)$ and the formula

$$\int_0^W f(W')dW' = R. \tag{13}$$

Due to our knowledge of $f(W)$, we can simply calculate W_m, the mean energy loss,

$$W_m = \int_{W_{min}}^{E} Wf(W)dW \cong W_{min} \ln\left(\frac{E}{W_{min}}\right) \tag{14}$$

The mean free path (MFP) for inelastic scattering was calculated, according to [22, 23], as

$$\lambda_{in} = \frac{W_m}{S_B}, \tag{15}$$

where S_B is the Bethe stopping power. The extrapolation proceeds in the program, in accordance with [20].

In the first version of the hyperbolic method, HYPERB, we will suppose that the value of scattering angle θ is given [5] by

$$\sin^2\theta = \frac{W}{E}. \tag{16}$$

In the second version of the model called MODHYP the same method of hyperbolic approach to energy loss is assumed but now no deflection occurs at the inelastic scattering event. The change of direction can be omitted in the case of plasmon excitations [5].

In the last model, following [24], an exponential distribution of electron energy loss is assumed

$$\frac{d\sigma}{dW} = \frac{1}{J_B} \exp\left(-\frac{W}{J_B}\right). \tag{17}$$

To calculate the normalized distribution density function according to Eq. (9), Eq. (17) must be integrated in the interval (W_{min}, W_{max}). Then,

$$f(W) = \frac{\exp\left(-\frac{W}{J_B}\right)}{J_B\left[\exp\left(-\frac{W_{min}}{J_B}\right) - \exp\left(-\frac{E}{J_B}\right)\right]} \quad (18)$$

and the actual energy loss W is determined exactly by means of Eq. (13). The mean energy loss W_m is

$$W_m = \frac{(W_{min}+J_B)\exp\left(-\frac{W_{min}}{J_B}\right) - (E+J_B)\exp\left(-\frac{E}{J_B}\right)}{\exp\left(-\frac{W_{min}}{J_B}\right) - \exp\left(-\frac{E}{J_B}\right)}, \quad (19)$$

and the inelastic MFP is calculated according to Eq. (15). Again we put $W_{max} = E$ and $W_{min} = 10\,\text{eV}$ even though, due to the exponential shape of distribution, $W_{min} = 0$ can be defined. The same angular deflection at an inelastic event was used as in the HYPERB model. In the following we denote this model as EXPONENT.

It can be seen that the last three models of inelastic collisions are not fully equivalent to the continuous slowing down model, but they can be taken more exactly as a single scattering MC i.e., a MC calculation, taking into account each collision individually and dealing separately with elastic and inelastic collisions.

Subsequent MC Program Details

For all models, we calculate the elastic scattering cross-sections by means of tables computed from the Hartree-Fock model of an atom [17]. This model was selected as the best for Al and Au and as a close second best for Cu in the elastic model assessment [11]. In MODCSD, $\lambda_{tot} = \lambda_{el}$, otherwise, after calculation of λ_{in} and the elastic MFP λ_{el}, the total MFP λ_{tot} is

$$\frac{1}{\lambda_{tot}} = \frac{1}{\lambda_{el}} + \frac{1}{\lambda_{in}}, \quad (20)$$

and the actual path s is

$$s = -\lambda_{tot}\ln(R), \quad (21)$$

where R is a random number.

The limits of the calculation were set to 10000 backscattered electrons, or to 20000 transmitted electrons, which means that the calculation was carried out for approximately 30000–100000 impinging electrons. The electron beam is normal to the sample surface. Even though the extrapolation [20] makes the behaviour of the stopping power S_B around 1 keV better, the real dependence of S well below 1 keV is rather different than given by Eq. (3). Thus the value of λ_{in} calculated in HYPERB, MODHYP and EXPONENT according to Eq. (15) is not correct. Moreover, the lowest energy for which the differential cross-sections of elastic scattering are tabulated is 1 keV. For these reasons, the electron path was terminated when the electron energy reached the mean ionization potential J_B (i.e., 0.150, 0.314 and 0.797 keV for Al, Cu and Au, respectively). Below this energy, the electron was regarded as being absorbed. Other details are those usually used in MC codes.

Results and Discussion

Firstly, we calculated by our MC code values of the backscattering coefficient η at the primary electron energies 10, 20, 50, or 100 keV, and the corresponding mass thicknesses 50 $\mu g/cm^2$ (i.e., 185, 56, and 26 nm for Al, Cu, and Au, respectivley), 100 $\mu g/cm^2$, 500 $\mu g/cm^2$, and 2000 $\mu g/cm^2$ (i.e., 7407, 2232 and 1036 nm for Al, Cu, and Au, respectively). Using the values of η, we calculated Niedrig's coefficient C_N Eq. (1). Thicknesses used in determining η were approximately equal to $d_{1/2}$. The calculated values are subject to random errors; their standard deviation was estimated in [11].

MC values of η and of C_N were established for all models that we compared. For each model, the dependence of MC values of C_N on the primary energy of electrons, E_o, was fitted by the functions $C_N = A\ E_o^{-B}$, A and B being the fitting coefficients. These functions are straight lines in the log C_N vs. log E_o plot, which describes correctly the calculated data (see similar plots in [11], Fig. 4).

Experimental data were taken from the literature. In general, all the experimental data are subject to random and systematic errors. From the statistical point of view, it is possible to minimize these by using a relatively large set of experimental results (as in [11]). Thus, experimental values of the backscattering coefficient, η, were taken from several papers [8, 13–16]; in order to recalculate "experimental" values of C_N.

Then agreement between calculated and experimental values was assessed. For each model and for a

Table 1. The calculated fit of all the experimental and calculated values of parameter C_N by means of function $C_N = A\,E_o^{-B}$ and the average sum of logarithmic residuals R_s, giving agreement between the experimental and calculated values of C_N; *a*) elastic models, *b*) inelastic models

Table 1a.

Model For Calculation	Element Al			Cu			Au		
	A	B	R_s	A	B	R_s	A	B	R_s
Experiment	12913	1.922		2035	1.392		2949	1.549	
Rutherford	13535	2.027	0.89	11096	1.913	0.46	11780	1.947	0.45
Mott-Reimer	9938	1.917	0.64	6704	1.760	0.35	1802	1.476	0.41
Hartree-Fock	10215	1.849	0.34	6403	1.678	0.31	1819	1.354	0.41

Table 1b.

Model For Calculation	Element Al			Cu			Au		
	A	B	R_s	A	B	R_s	A	B	R_s
Experiment	12913	1.922		2035	1.392		2949	1.549	
EXPONENT	9463	1.815	0.37	6607	1.683	0.32	1801	1.350	0.42
HYPERB	10215	1.849	0.34	6403	1.678	0.31	1819	1.354	0.41
MODHYP	8874	1.837	0.58	6210	1.683	0.23	1720	1.339	0.42
MODCSD	9096	1.847	0.37	6117	1.677	0.29	1752	1.345	0.40

given element, the sum of squares of the residuals R_s is calculated by taken from [25]

$$R_s = \sum_{i=1}^{n} (\log(C_i) - \log(AE_i^{-B}))^2, \qquad (22)$$

where C_i are the experimental values of C_N at primary energy E_i, A and B are the 'best fit' coefficients to the C_N values, and n is the number of measurements. R_s characterizes the agreement between the calculated and experimental values of C_N for the whole energy range; the best model is that having the lowest value of R_s. Due to the power dependence of C_N on E_o, we used logarithmic terms in formula (22). The results are summarized in Table 1a, b for Al, Cu, Au, where A, B and R_s values are presented. Data for elastic models were added for comparison (Table 1a); they were recalculated with logarithmic terms, without averaging, unlike [11]. The models of elastic scattering as given in [11] are presented: the Rutherford formula with screened atomic potential (RUTHERFORD); scattering according to the Mott theory, by Reimer (MOTT-REIMER); tables calculated using Hartree-Fock atomic model (HARTREE-FOCK), and very similar results were obtained. The HYPERB model of inelastic scattering was combined with them.

For inelastic models (Table 1b), it is seen the agreement of experimental and MC values of C_N is worst for Al and best for Cu particularly when using the MODHYP model. For all the elements and all models except these two cases, the results are very similar. The low or high value of R_s for MODHYP shows the importance of taking into account deflections at inelastic collision for the elements in question. However in general, because of the small differences between models, it is difficult to decide which is the best for all the elements.

Therefore, we tried to find a quantity more dependent upon the element and model type. We calculated the energy distributions of the transmitted electrons and determined the most probable energy loss W_p in the thickness range 150–450 µg/cm^2 for Al and Cu and 150–650 µg/cm^2 for Au. These values were compared with data from [15, 24, 26]. Calculation was performed at the primary electron energy 20 keV; experimental data are available for various energies from 6 keV [15] or 15 keV [24], but for energies 20, 30, and 40 keV the largest set is in [26], although for 18 and 20 keV there are also some data in [15] and [24], respectively.

Unfortunately, the values of W_p were measured in [26] and [15] by an electron spectrometer with relatively low solid angle (0.75 10^{-2} and 2.5 10^{-2} rad, i.e., about 0.5 and 1.5 degree, respectively). Using these angles in the MC calculation, gives an energy distribution with very poor statistics. To overcome this problem we calculated the dependence of W_p on decreasing the output angle of the electrons. Practically no dependence was found which is in reasonable agreement with [15], where only very weak dependence on output angle (up to 55 degrees) was observed. Thus, in our calculations we took into account electrons scattered up to an output angle of 90 degrees, i.e., one half of the full solid angle, a similar solid angle is subtended by the detector in [24].

Comparison of results for all the inelastic models are shown in Figs. 1–3, for Al, Cu, and Au, respectively. The coefficient R_s' was calculated in a slightly different way, by direct comparison of experimental and calculated values of W_p at all thicknesses, i.e.

$$R_s' = \frac{1}{n}\sum_{i=1}^{n}(W_{i,\mathrm{exp}} - W_{i,\mathrm{calc}})^2, \tag{23}$$

where $W_{i,\mathrm{calc}}$ and $W_{i,\mathrm{exp}}$ are the calculated and measured (in this case from [26] only) values of the most probable electron energy loss for i-th film thickness. The results for inelastic models are in Table 2a. The MODCSD and model EXPONENT give the worst agreement; giving predictions which are far from reality. The energy distribution of transmitted electrons, calculated using MODCSD, shows a very steep increase in W_p caused by energy losses occuring even without collision of electrons inside the film. It is seen that for Al and Au, the best agreement between experimental and calculated values of W_p was reached using the HYPERB model whereas for Cu, the best results are obtained using MODHYP. In the cases of Al and Cu, the differences between MODHYP and HYPERB models are not large.

For all elements studied, the cross-sections calculated using different models of elastic scattering were very similar at 20 keV [11] and thus we concluded they would produce little influence on the energy distribution. Nevertheless, we calculated R_s' for the elastic models (Table 3a) and it is seen that the dependence of R_s' on the element and model type is lower than for inelastic models.

For completeness, the most probable output angle, θ_p, was calculated by MC using various models of elastic and inelastic scattering and compared with experimental data [15] (primary electron energy 20 keV, thickness approximately 20–500 μg/cm^2). R_s' was evaluated by Eq. (23), where W_p was simply replaced by θ_p. The results are shown in Tables 2b and 3b where the dependence of R_s' on the model type is much stronger for elastic models than was supposed.

In the numerical comparisons of W_p, we used only experimental data from [26]. In Figs. 1–3, it is seen

Table 2a. Values of the average sum of residuals R_s' giving agreement between the experimental and calculated values *a)* of the most probable energy loss W_p and *b)* of the most probable output angle θ_p of electrons transmitted through a thin films of Al, Cu, and Au, at an electron energy of 20 keV – inelastic models

Model for calculation	Element		
	Al	Cu	Au
EXPONENT	3.59	2.35	2.73
HYPERB	0.04	0.22	0.64
MODHYP	0.06	0.19	1.02
MODCSD	2.21	2.35	3.05

Table 2b.

Model for calculation	Element		
	Al	Cu	Au
EXPONENT	11.66	16.99	14.00
HYPERB	5.52	6.35	8.79
MODHYP	1.84	6.03	14.81
MODCSD	9.06	6.63	12.63

Table 3a. Values of the average sum of residuals R_s' giving agreement between the experimental and calculated values *a)* of the most probable energy loss W_p and *b)* of the most probable output angle of electrons transmitted through a thin films of Al, Cu, and Au, at an electron energy of 20 keV – elastic models

Model for calculation	Element		
	Al	Cu	Au
Rutherford	0.08	0.04	1.87
Mott-Reimer	0.02	0.06	0.45
Hartree-Fock	0.04	0.22	0.64

Table 3b.

Model for calculation	Element		
	Al	Cu	Au
Rutherford	10.13	16.37	47.82
Mott-Reimer	6.56	8.27	38.07
Hartree-Fock	5.52	6.35	8.79

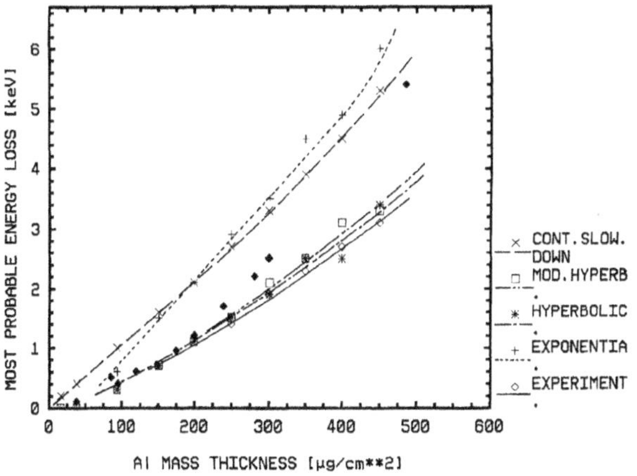

Fig. 1. The dependence of the most probable electron energy loss W_p of electrons transmitted through thin Al films, at an electron energy of 20 keV, thin film thickness in the range 150–450 μg/cm^2. Experiment – data points ◇, full line [26], data points ◆ [15, 24]; MC simulations – dashed lines

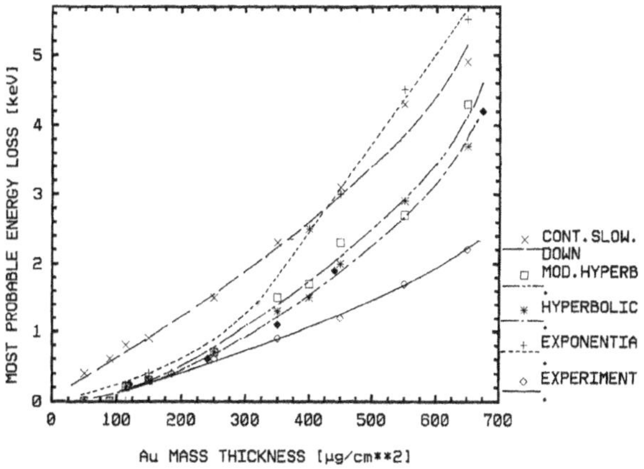

Fig. 3. The dependence of the most probable electron energy loss W_p of electrons transmitted through thin Au films, at an electron energy of 20 keV, thin film thickness in the range 150–650 μg/cm^2. Experiment – data points ◇, full line [26], data points ◆ [15, 24]; MC simulations – dashed lines

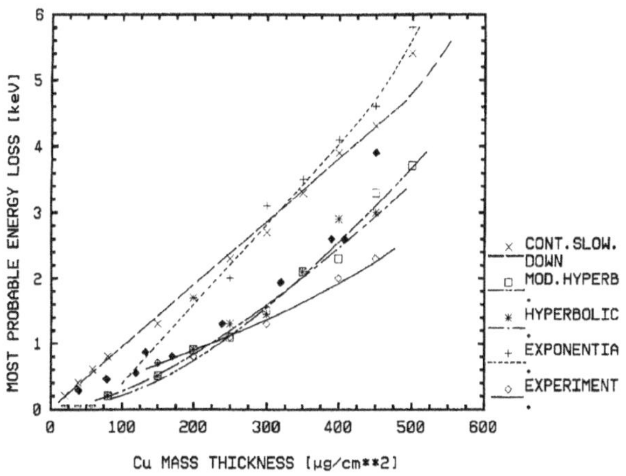

Fig. 2. The dependence of the most probable electron energy loss W_p of electrons transmitted through thin Cu films, at an electron energy of 20 keV, thin film thickness in the range 150–450 μg/cm^2. Experiment – data points ◇, full line [26], data points ◆ [15, 24]; MC simulations – dashed lines

that the problem of comparison is more complicated – we have two sets of experimental data [15] + [24] and [26], which do not agree with one another. Thus, it is necessary to compare critically these sets, which can be done only with more measurements. So, we can only remark on the agreement of our results with each set of measurements.

Very simple models of inelastic scattering were used. The values of λ_{in} calculated during MC modelling differ from the values of λ_{in} reported in the literature [27, 28]. In EXPONENT, the values of λ_{in} were calculated by Eq. (15) with $W_m \cong J_B$ and were systematically overestimated, especially for heavier elements (Cu, Au) and higher energy. In HYPERB and MODHYP, the value of W_m depends on actual electron energy. But even at the highest value of electron energy, this value does not reach that of the mean ionization potential J_B (or J_J), thus λ_{in} is much lower than in EXPONENT, but mostly larger than that referred to in [27, 28]. Only for Au are the values of λ_{in} underestimated.

Because the scattering angle of inelastic collisions calculated by Eq. (16) increases the path length in the film, it may be supposed that W_p for the HYPERB model should be larger than that for MODHYP. Nevertheless, from the figures it is seen that the values of W_p are practically the same for both models.

If there were a longer inelastic MFP in the model than in reality then a lower number of inelastic collisions would take place but the number of elastic collisions should stay nearly the same. Thus for all models (except for Au with the HYPERB and MODHYP models), the calculated average energy loss would be lower than the measured one. For this reason also, the calculated W_p should be lower than experimental values. This is true for Al and Cu to some extent data from [15, 24]. For Au, the calculated and experimental [15] values of W_p are in relatively good agreement with HYPERB and MODHYP as currently defined.

In fact, only for Al are our results very near to the data from [26]. For Cu and Au, the agreement is poor and bad, respectively, and the calculated values of W_p

are always higher than those measured in [26]. As mentioned, data in [26] (but also in [15]) were measured by an electron spectrometer with a relatively low input angle. Thus, the reason for disagreement may be associated with different angular distributions of transmitted electrons at higher thicknesses in the Al, Cu, and Au films. We assume that electrons deflected the most have undergone more elastic scattering events, but the chance to lose energy is the same whether the electron is deflected a large or a small amount; thus energy loss and deflection angle should be practically independent. However when comparing Au with Al, the elastic scattering is much greater than inelastic scattering. It causes the saturated value of θ_p to be reached sooner for Au than for Al, Cu being an intermediate case. By this reasoning, our supposition may not be valid and a substantial number of electrons which have large output angles may have lost rather more energy due to their longer path in the film. Thus different conditions of measurement and calculation can cause the observed discrepancy. Another reason could be the shape of the energy loss distribution. In the energy loss spectra calculated by HYPERB or MODHYP, higher energy losses ($W > 8-10$ keV) are more frequent then in measured spectra. So, it seems that a change of degree in the hyperbolic shape of the energy loss function to a higher one (i.e., from quadratic to cubic or more) could decrease the calculated values of W_p and give better agreement with the experimental results of [26].

Moreover, in HYPERB an MODHYP, the same dependence $F(W)$ and the same value of $W_{\min}$ was assumed for all elements, which is not true in reality. In these models, the dependence of λ_{in} on the element is given by the stopping power according to Eq. (3). Values of λ_{in} calculated by both models agree better with values from [27, 28] if $W_{\min}$ is adjusted to fit the proper value of λ_{in}; the fitting is supported by a simple physical reason – the different minimal bonding energies of valence electrons. Unfortunately, it will not be possible to reach agreement with both sets of experimental results simultaneously.

Conclusions

We compared the experimental values of Niedrig's coefficient (calculated directly from the backscattering coefficient) and the most probable electron energy loss with those calculated by Monte-Carlo methods using several models of inelastic scattering. Only simple models of inelastic scattering were assessed. The comparison was carried out by a defined statistical method.

The results of the comparison between calculated and measured values of Niedrig's coefficient, for the systems studied were not usually very different. Even though comparison among the models by means of the sum of residuals can be taken as being more exact, it can not easily be decided which model is best for all elements.

The energy distribution of transmitted electrons was also generated by MC and the most probable electron energy loss W_p was compared with experimental values. By the method of residuals the best agreement for Al and Au occured when using the model with the hyperbolic shape of electron energy loss (HYPERB); for Cu it was the model that produced no deflection at inelastic scattering events (MODHYP). For heavy elements (Au), the deflection at inelastic events is more important whilst for medium and light elements (Cu, Al), it is less so and the differences observed between models were small. Thus, the best of the models for all the elements studied was that with a hyperbolic shape of electron energy loss and with deflection at inelastic events. This was selected for future calculations.

Acknowledgements. For a critical reading of this manuscript and for helpful discussions, I would like to thank Dr. E. Tomková; for advice in the statistical evaluation I should like to thank Dr. K. Zvára, and for the English review R. A. Healey, M. A. is gratefully acknowledged. The work was supported by GA-CR grant No. 106/97/S008.

References

[1] L. Reichelt, A. Engel, *Ultramicroscopy* **1994**, *13*, 279.
[2] D. C. Joy, *J. Microscopy* **1987**, *147*, 51.
[3] D. F. Kyser, in: *Introduction to Analytical Electron Microscopy* (J. J. Hren, J. I. Goldstein, D. C. Joy, eds.), Plenum, New York 1987, p. 199.
[4] P. Karduck, N. Ammann, W. P. Rehback, *Microbeam Analysis 1990* (J. R. Michael, P. Ingram, eds.), San Francisco Press, San Francisco, 1990, p. 21.
[5] R. Shimizu, Z.-J. Ding, *Rep. Progr. Phys.* **1992**, *55*, 487.
[6] L. Reimer, *Microchimica Acta [suppl.]* **1996**, *13*, 1.
[7] L. Reimer, E. R. Krefting, in: *Use of Monte Carlo Calculations in Electron Probe Microanalysis and Scanning Electron Microscopy* (K. F. J. Heinrich et al., eds.), NBS Special Publication 460, U.S. Dept. of Comm., Washington, DC, 1976, p. 45.
[8] H. Seidel in: Rasterelektronen-mikroskopie (L. Reimer, G. Pfefferkorn, eds.), Springer, Berlin Heidelberg New York, 1983, p. 38.
[9] H. Kulenkampff, W. Spyra, *Z. Phys.* **1954**, *137*, 416.

[10] H. Kanter, *Ann. Physik* **1957**, *20*, 144.
[11] V. Starý, *Microchimica Acta [Suppl.]* **1996**, *13*, 559.
[12] H. Niedrig, *J. Appl. Phys.* **1982**, *53*, R15.
[13] H. Niedrig, P. Sieber, *Z. Angew. Phys.* **1971**, *31*, 27.
[14] G. Neubert, S. Rogaschewski, *J. Phys. D: Appl. Phys.* **1984**, *17*, 2439.
[15] V. E. Cosslett, R. N. Thomas, *Br. J. Appl. Phys.* **1964**, *15*, 883, 1283.
V. E. Cosslett, R. N. Thomas, *Br. J. Appl. Phys.* **1965**, *16*, 779.
[16] T. Doležel, *Diploma Thesis*, Charles University, Prague, 1982.
[17] M. E. Riley, C. J. MacCallum, F. Biggs, *At. Data Nucl. Data Tables* **1975**, *15*, 443.
[18] D. Liljequist, *J. Phys. D: Appl. Phys.* **1983**, *16*, 1567.
[19] M. J. Berger, S. M. Seltzer, *Nuclear Science Series Report No. 39*, NAS-NRC Publication No. 1133, National Academy of Science, Washington, DC, 1964, p. 205.
[20] T. S. Rao-Sahib, D. B. Wittry, *J. Appl. Phys* **1974**, *45*, 5060.
[21] D. C. Joy, S. Luo, *Scanning* **1989**, *11*, 176.
[22] D. Liljequist, *J. Phys. D: Appl. Phys.* **1978**, *11*, 839.
[23] A. Špalek, *Nucl. Instr. Meth.* **1982**, *198*, 399.
[24] R. Shimizu, Y. Kataoka, T. Matsukawa, T. Ikuta, K. Murata, H. Hashimoto, *J. Phys. D: Appl. Phys.* **1975**, *8*, 820.
R. Shimizu, Y. Kataoka, T. Ikuta, T. Koshikawa, H. Hashimoto, *J. Phys. D: Appl. Phys.* **1976**, *9*, 101.
[25] N. R. Draper, H. Smith, Applied Regression Analysis, Willey, New York, 1966.
[26] L. Reimer, K. Brockmann, U. Rhein, *J. Phys. D: Appl. Phys.* **1978**, *11*, 2151.
[27] D. L. Misell, *Adv. Electr. Phys.* **1973**, *32*, 63.
[28] C. J. Tung, J. C. Ashley, R. H. Ritchie, *Surf. Sci.* **1979**, *81*, 427.

Mikrochim. Acta [Suppl.] 15, 351–363 (1998)

A Rapid Comparison of Matrix Corrections in AES and XPS by Means of Computer Programs

Alexander G. Fitzgerald

Carnegie Laboratory of Physics, Department of Applied Physics and Electronic and Mechanical Engineering, University of Dundee, Dundee DD1 4HN, Scotland, U.K.

Abstract. The principles of qualitative and quantitative surface analysis by Auger electron spectroscopy and x-ray photoelectron spectroscopy are discussed and the various matrix-dependent parameters which influence the emission of Auger electrons and x-ray photoelectrons are described. Programs which have been developed to permit rapid surface analysis by Auger electron spectroscopy and x-ray photoelectron spectroscopy from raw spectral intensities are described. These programs are shown to be important in comparing different calculations for a particular matrix dependent factor. The surface composition of a number of materials have been determined using the programs from Auger electron spectra and x-ray photoelectron spectra.

Key words: Auger electron spectroscopy, x-ray photoelectron spectroscopy, matrix factor, attenuation length, backscatter factor.

Until recently the development of software to determine matrix factor corrections in microbeam analysis has been confined mainly to electron probe microbeam analysis. In both Auger electron spectroscopy (AES) and x-ray photoelectron spectroscopy (XPS) it is clear that the number of electrons emitted is dependent on several physical effects; for example, in addition to the generation of Auger electrons by the primary electron beam, backscattered electrons can also give rise to Auger electrons. The magnitude of the electron backscatter factor is matrix dependent. The attenuation length (AL), which is the average distance (in nanometres) that an electron travels between successive inelastic collisions also affects the number of Auger electrons or x-ray photoelectrons which will be able to escape from the surface. The energies of emitted x-ray photoelectrons and Auger electrons are also matrix dependent being related to the atomic species in the material. Other matrix dependent parameters are the atomic volume, related to the density of a material (important in AES and XPS), and the photoionisation cross-section (important in XPS). It is clear that any accurate determination of surface composition by XPS or AES must take account of these complex matrix dependent parameters [1, 2, 3] and the most efficient method of doing this is to develop programs with libraries of them. Walker et al. [1] and Watton et al. [3] have developed computer programs which deal with matrix effects and also considered the problem of quantification of a thin film on a substrate by AES and XPS respectively. It should be noted however, that other factors can have a strong influence on the measured surface composition determined by AES and XPS, for example preferential sputtering surface roughness, electron/ion surface composition modification and modification due to surface cleaning.

Qualitative Determination of Surface Composition by AES and XPS

To suppress the large background in Auger electron spectra the electron energy distribution can be electronically differentiated to obtain a plot of $dN(E)/dE$ against kinetic energy. The peak-to-peak magnitude of an Auger peak in a differentiated spectrum generally is directly related to the surface concentration of the element which produces the Auger electrons. With improvements in electron detection instrumentation, Auger electron spectra can now be conveniently displayed in the *N(E)* mode.

A qualitative determination of surface concentration, C_X of an element X in a sample can be obtained by measuring the Auger electron intensity from an element X in the sample and in a standard under the same conditions. The qualitative concentration is given by

$$C_X = \frac{I_X}{I_X^{std}} \tag{1a}$$

where I_X and I_X^{std} are the peak-to-peak Auger amplitudes from the sample and standard respectively.

If pure standards are available the concentration C_X can be obtained from the relation

$$C_X = \frac{I_X/I_X^{std}}{\sum_i I_i/I_i^{std}}. \tag{1b}$$

Another qualitative method which avoids the need for a large number of pure elemental standards, is to compare the signal from the sample with that from a pure silver target. The relative sensitivity, S_X, between any element, X, and silver can be obtained from standard Auger spectra recorded previously or collected in a handbook [4–6]. The atomic concentration of an element X is given by

$$C_X = \frac{I_X}{I_{Ag} S_X D_X} \tag{2}$$

where I_X is the peak-to-peak amplitude from the element X in the sample, I_{Ag} is the peak-to-peak amplitude from the silver standard and D_X is a relative scale factor between the spectra for the sample and silver which takes account of differences in beam current and lock-in amplifier modulation between the two measurements

$$D_X = \frac{L_X E_{m,X} I_{p,X}}{L_{Ag} E_{m,Ag} I_{p,Ag}} \tag{3}$$

where L_X is the lock-in amplifier sensitivity, $E_{m,X}$ is the modulation energy and $I_{p,X}$ is the primary beam current used to obtain the sample spectrum and L_{Ag}, $E_{m,Ag}$ and $I_{p,Ag}$ are the equivalent quantities for the silver standard.

In XPS a qualitative composition can also be obtained by substituting core photoelectron intensities in the sample and in standards for the equivalent Auger intensities in equation (1b).

Quantitative Determination of Surface Composition by XPS using Standards

To obtain a quantitative determination of atomic concentration which takes account of matrix dependent factors requires the measurement of x-ray photoelectron peak intensities from the sample and from a suitable standard [7]. The general expression for the signal strength observed from a measured peak from an element A in a simple matrix m can be written as

$$I_A = \sigma_A(hv) \int_\gamma \gamma \int_x \int_y \int_\Omega J_o(xy) T(x, y, \Omega, E_A) D(E_A \Omega) \times \int_z N_A(xyz) \exp\left\{\frac{-z}{\lambda_m(E_A)\cos\theta}\right\} dz d\Omega dx dy d\gamma \tag{4}$$

for XPS [8]. In general m represents the "unknown" matrix, A, B and C are elements in the unknown matrix, x and y are co-ordinates in the plane of the sample surface, z is the depth from the sample surface, $N_A(xyz)$ is the number density of atoms of element A, $X_A(xyz)$ is the atomic fraction of element A, a_m^3 is the average atomic volume in matrix m, ρ is the density in $g\,cm^{-3}$, Ω is the solid angle of detection of the spectrometer entrance aperture, θ is the angle of electron emission with respect to the sample normal, E is the electron energy in eV, $T(xy\Omega E)$ is the spectrometer transmission function and $D(E\Omega)$ is the efficiency of the electron detector.

In XPS, I_A is the photoelectron current from element A, I_A^{std} is the photoelectron current measured from a standard material, σ_A is the cross-section for photoemission from the particular inner shell of atom A at an incident photon energy hv, E_A is the kinetic energy of the photoelectrons from element A, γ is the photon/photoelectron angle, $L_A(\gamma)$ is the angle distribution of emitted photoelectrons relative to incident x-ray beam, $J_o(xy)$ is the incident x-ray flux which is assumed not to vary with depth due to the high (~100 X) inelastic mean free path of photons with respect to photoelectrons and $\lambda_m(E_A)$ is the escape depth of photoelectrons of energy E_A in matrix m.

In XPS if the specimen is of uniform chemical composition in the region sampled and if it is assumed that the x-ray flux is uniform and that the electron spectrometer primarily accepts electrons along the angle θ then Eq. (4) becomes

$$I_A = I_o \sigma_A(hv) L_A(\gamma) N_A \lambda_m(E_A) \cos\theta T(E_A) D(E_A) \tag{5}$$

the $\exp(-z/\lambda_m(E)\cos\theta)$ term integrating to $\lambda_m(E)\cos\theta$ as $z \to \infty$

The assumption regarding spectrometer acceptance angle is of no consequence if use is made of a standard spectrum recorded under identical conditions on the same instrument, as these terms will cancel. In a similar way, the intensity measured from a pure standard can be written as

$$I_A^{std} = I_o\sigma_A(h\nu)L_A(\gamma)N_A^{std}\lambda_m^{std}(E_A)\cos\theta T(E_A)D(E_A) \quad (6)$$

By taking the ratio I_A/I_A^{std} and thereby cancelling out many of the terms this equation can be simplified giving

$$\frac{I_A}{I_A^{std}} = \frac{N_A\lambda_m(E_A)}{N_A^{std}\lambda_m^{std}(E_A)} \quad (7)$$

The term N_A is actually related to the atomic concentration of element A in the matrix by the relation $N_A = X_A/a_m^3$, where a_m^3 is an average atomic volume for the matrix. This leads to

$$\frac{I_A}{I_A^{std}} = \frac{X_A(a_m)^{-3}\lambda_m(E_A)}{X_A^{std}(a_A^{std})^{-3}\lambda_m^{std}(E_A)} \quad (8)$$

where X_A^{std} represents the fractional content of element A in the standard used for that element. This acknowledges the fact that the intensity I_A^{std} may not originate from a pure standard sample. In order to simplify the equations that follow it has been assumed that the I_A^{std} terms have been normalised by multiplying by the concentration factor. If two elements A and B in the sample are considered, then their intensities are related by

$$\frac{I_A/I_A^{std}}{I_B/I_B^{std}} = \frac{X_A\lambda_m(E_A)}{X_B\lambda_m(E_B)}\frac{(a_A^{std})^3\lambda_B^{std}(E_B)}{(a_B^{std})^3\lambda_A^{std}(E_A)} \quad (9)$$

This equation can be rearranged to give the ratio of compositions X_A/X_B. This then allows a "matrix factor" F_{AB} to be defined as

$$\frac{X_A}{X_B} = F_{AB}\frac{I_A/I_A^{std}}{I_B/I_B^{std}} \quad (10)$$

where

$$F_{AB} = \frac{\lambda_m(E_B)}{\lambda_m(E_A)}\frac{\lambda_A^{std}(E_A)}{\lambda_B^{std}(E_B)}\left[\frac{(a_B^{std})}{(a_A^{std})}\right]^3 \quad (11)$$

The general expression for a system of n elements, giving atomic concentrations directly, rather than ratios, can be derived because $\sum_{i=1}^{n} X_i = 100\%$, thus

$$X_A = \frac{I_A/I_A^{std}}{\sum_i F_{AB}I_i/I_i^{std}} \quad (12)$$

Provided the functions $T(E)$ and $D(E)$ are known, which is possible for the constant analyser energy mode used in XPS, then standardless analysis is possible

$$\frac{I_A}{I_B} = \frac{\sigma_A(h\nu)L_A(\gamma)\lambda_m(E_A)T(E_A)D(E_A)}{\sigma_B(h\nu)L_B(\gamma)\lambda_m(E_B)T(E_B)D(E_B)}\frac{X_A}{X_B} = S_{AB}\frac{X_A}{X_B} \quad (13)$$

Quantitative Determination of Surface Composition by AES using Standards

The general expression for the signal strength observed from a measured peak in an Auger spectrum from an element A in a simple matrix m can be written as [9],

$$\begin{aligned} I_A = &I_o\sigma_{AXYZ}(E_p)\sec\alpha[1 + r_m(E_A, E_P, \alpha)] \\ &\times \int_y\int_x\int_\Omega T(x, y, \Omega, E_A)D(E_A)D(E_A\Omega) \\ &\times \int_z N_A(xyz)\exp\left\{\frac{-z}{\lambda_m(E_A)\cos\theta}\right\}dzd\Omega dy \quad (14) \end{aligned}$$

where the parameters in this equation have the same meanings as in the equivalent XPS Eq. (4) above.

In AES, I_A is the Auger electron current, I_A^{std} is the Auger electron current measured from the standard material, I_o is the incident electron beam current, α is the angle of incidence of the electron beam relative to the sample normal, E_p is the primary electron beam energy, E_A is the kinetic energy of the Auger electrons, E_A^b is the ionisation energy of the initial level involved in Auger transition, $\sigma_{AXYZ}(E_p)$ is the electron impact cross-section for emission of an Auger electron from level XYZ of element A, incident energy E_p, $[1 + r_m(E_A^b, E_p, \alpha)]$ is the backscatter factor relevant to the ionisation energy E_A^b at angle of incidence α and $\lambda_m(E_A)$ is the escape depth of Auger electrons of energy E_A in matrix m.

The equivalent expression for quantitative Auger electron spectroscopy can be derived by making similar assumptions regarding the homogeneity of the specimen under analysis. The equivalent Auger equation to the XPS Eq. (6) is

$$\begin{aligned} I_A = &I_o\sigma_{AXYZ}(E_p)[1 + r_m(E_A^bE_p\alpha)]N_A\lambda_m(E_A) \\ &\times \cos\theta\, T(E_A)D(E_A) \quad (15) \end{aligned}$$

The degree of homogeneity with depth is more important in AES due to the electron backscattering factor r_m. Electron backscattering is usually regarded as a "bulk" rather than a surface property and to regard it as constant as in the above equation, presupposes a constant composition to a depth of about 1000 Å [10].

By means of an analysis parallel to that for XPS, the equivalent expression for the Auger matrix factor F_{AB} which can be used in the AES equivalent of equation (11) is

$$F_{AB} = \frac{[1 + r_A^{std}(E_A^b)]\,[1 + r_m(E_B^b)]}{[1 + r_B^{std}(E_B^b)]\,[1 + r_m(E_A^b)]} \times \frac{\lambda_m(E_B)\,\lambda_A^{std}(E_A)}{\lambda_m(E_A)\,\lambda_B^{std}(E_B)} \left[\frac{(a_B^{std})}{(a_A^{std})}\right]^{3/2} \quad (16)$$

Matrix Correction Factors

Atomic Size Correction

The term $\left[a_B^{std}/a_A^{std}\right]$ in Eqs. (11) and (16) can be regarded as a correction for the different atomic densities of the standards used. If one standard has a higher atomic density (smaller atomic volume) then the AES or XPS intensity will be higher merely by virtue of the fact that more atoms are available in the sampling volume (neglecting any other effects such as change in escape depth with atomic volume – this is corrected separately). The effect of the correction is to reduce the intensity from the standard with the lower atomic volume. There is no correction of this type corresponding to the atomic size of the sample under analysis as, in this case, the intensities come from the same matrix and the atomic volume term cancels.

The average volume (i.e. neglecting structure), a^3, occupied by each atom in a matrix can be determined from the relationship between atomic weight, A, and bulk density, ρ, thus

$$a^3 = \frac{A}{\rho N_A} \times 10^{21}\,(\text{nm}^3) \quad (17)$$

where the density is expressed in units of g cm^{-3}. For compound standards an average atomic volume can be calculated by taking the atomic weight to be used in equation (17) as $A = \sum_i X_i A_i$ the "weighted sum" of the constituent elements.

Since the materials used as compound standards should be well characterised the density should be well known in most cases. Where the bulk density for a compound is not known the atomic volume has been calculated by summing the individual atomic volumes, thus

$$a^3 = \sum_i X_i a_i^3.$$

Escape Depth Correction

The term $[\lambda_m(E_B)\lambda_A^{std}(E_A)/\lambda_m(E_A)\lambda_B^{std}(E_B)]$ in Eq. (16) is designed to correct the measured intensities, both from standards and unknown, for differences between the surface sensitivity of electrons of different energy. If an Auger or photoelectron line is measured from a material in which this electron has a narrow escape depth then its intensity will be reduced as the deeper atoms cannot contribute. An ability to predict both the material and energy dependence of the electron escape depth is required.

The correct expression to be used in matrix factor calculation for estimates of escape depth is the attenuation length (AL) of electrons in the solid. The attenuation length is defined as the average distance (in nanometres) that an electron with a given energy travels between successive inelastic collisions as derived from a particular model in which the elastic electron scattering is assumed to be insignificant [11]. The reason that this is the correct expression to use is that the attenuation length is derived from overlayer-film measurements [12] in which the assumptions made are identical to those made in the electron spectroscopy experiment itself. In both cases the electrons are assumed to travel from their point of generation to the surface along straight line paths. The effects of elastic scattering, however, will modify the electron trajectories. It is common to assume [13] that the AL and the inelastic mean free path (IMFP) have similar dependencies on material and energy although the difference between these quantities will also depend on material and energy. Werner [14] and Jablonski and Tougaard [15] have simulated the overlayer experiments by means of a Monte Carlo model. They found that the AL also depends on the overlayer thickness and the geometry of the measurement. Cumpson [16] and more recently Jackson et al. [17] have used Monte Carlo simulations to calculate attenuation lengths and also simulate the emission of Auger electrons from bilayers and trilayer structures

The inelastic scattering of electrons is described by the IMFP. This is defined simply as the average distance that an electron travels between successive inelastic collisions. Several expressions exist for calculation of the IMFP in solids and these are often used in quantitative surface analysis work. Their use, however, is not strictly valid as the elastic scattering effects, inherently accounted for in AL measurements, are not considered. It is estimated that the IMFP may be greater than a corresponding AL by an amount in the range 15–30% [18]. The actual difference depends on the differential cross-section for elastic scattering, which in turn depends on electron energy and atomic number. No analytical expression exists for the conversion of IMFP to AL as yet [19].

One of the most commonly cited works in quantitative surface analysis papers is the attenuation length compilation of Seah and Dench [20] in which the available attenuation length data (some 350 values in total) are divided into three main groups: elements, inorganic compounds and organic compounds. In the two major groups of interest (elements and inorganic compounds) the best fit to a "universal curve" when AL is plotted against electron energy was found to be when the AL were expressed in terms of an "average monolayer thickness", a. The parameter a is defined exactly as in the atomic volume parameter (Eq. 11). The curves fitted to the data were as follows

$$\lambda(E) = A\left[\frac{a}{E^2}\right] + BE^{1/2} \quad (nm) \qquad (18)$$

where A is a constant and B is a material parameter equal to $ka^{3/2}$ where k is a constant which differs for elements and inorganic compounds. The electron energy E in each case has been expressed in units of eV. In most cases the first term in the equation can be considered negligible as electron energies are likely to be greater than 150 eV.

If the high electron energy form of Eq. (18) is substituted into the matrix factor equation for XPS (Eq. (11)) then F_{AB} for XPS is greatly simplified to

$$F_{AB} = \left[\frac{(a_B^{std})}{(a_A^{std})}\right]^{3/2} \qquad (19)$$

as all the energy-dependent terms now cancel. A similarly simplified expression exists for the escape depth and atomic size correction contribution to the AES matrix factor.

The inelastic mean free path (IMFP) is obtained from theoretical calculations. These calculations involve the total inelastic scattering obtained by summing contributions from valence electrons and those from the less important core scattering events [21]. A form of the Bethe equation for inelastic scattering was proposed for the IMFP by Penn [22] and has the form

$$\lambda(E) = \frac{E}{a(\ln E + b)} \quad (\text{Å}). \qquad (20)$$

Values of a and b have been tabulated for core and valence contributions and for the total IMFP over the energy range $200 \leq E \leq 2400$ eV for the elements. The model used is relevant to free-electron-like solids. The error in $\lambda(E)$ for non-free-electron-like materials may be as high as 40%, compared to 5% for materials of the free-electron type.

More recently a number of IMFP calculations have been made. For example, Tanuma et al. [23] have carried out IMFP calculations for 27 elements and 4 compounds, considering electrons in the energy range 200 to 2000 eV. The values obtained were then used to obtain a fit to a modified form of the Bethe equation:

$$\lambda(E) = \frac{E}{E_p^2 \beta \ln(\gamma E)} \quad (\text{Å}) \qquad (21)$$

where $E_p = 28.8[N_V \rho/A]^{1/2}$ is the free electron plasmon energy in (eV), N_V is the number of valence electrons per atom or molecule, ρ is the density in g/cm^3, and A is the atomic or molecular weight.

The two parameters β and γ in Eq. (21) were determined by fits to the calculated IMFP values. The calculations were later extended to lower energies (down to 50 eV) and the formulae modified to improve the fit over the entire 50 eV to 2000 eV range. The Bethe equation was modified to

$$\lambda(E) = \frac{E}{E_p^2[\beta \ln(\gamma E) - C/E + D/E^2]} \quad (\text{Å}) \qquad (22)$$

[19] where the free electron plasmon energy is defined as before, the remaining material-dependent parameters being defined as

$$\beta = -0.0216 + 0.944/(E_p^2 + E_g^2)^{1/2} + 7.39 \times 10^{-4}\rho \qquad (23)$$

$$\gamma = 0.191\rho^{-0.5} \qquad (24)$$

$$C = 0.0650/U^2 - 0.130/U + 1.11 \qquad (25)$$

$$D = 1.91/U^2 - 5.12/U + 35.3 \qquad (26)$$

where E_g is the bandgap energy for a non-conductor and $U = N_V \rho/A$.

Equation (22) is a general formula for IMFPs over the 50 to 2000 eV range.

Backscatter Factor

In AES a proportion of the incident primary electrons entering the sample will undergo inelastic and elastic collisions which result in them passing through the surface region again and emerging as backscattered electrons. Providing the energy of the backscattered electron is high enough Auger electrons may be produced and give rise to an Auger current which adds to that produced by the incident primaries. As the degree of backscattering depends on the material it will vary between sample and standard. The backscattering correction term $[(1 + r_A^{std}(E_A^b))(1 + r_m(E_B^b))/(1 + r_B^{std}(E_B^b))(1 + r_m(E_A^b)]$ attempts to correct for this.

Assumptions have already been made in the equations quoted here regarding the nature of ionisation of core levels by incident and rediffused primary electrons. The $[1 + r(E^b)]$ term is actually a simplification of the electron distribution function $\Phi(z, E)$ [24]. One of the assumptions made is that the depth over which Auger generation is important (i.e. the escape depth), is negligible when compared to the depth from which backscattered electrons can originate. This is equivalent to stating that the number of primaries backscattered from the "surface region" is negligible. It is also necessary to assume that the angular distribution of backscattered electrons does not vary with their energy.

Some differences exist in the way various terms are defined depending on whether x-ray microanalysis or AES is being considered. It has become customary in surface science to refer to $[1 + r_m(E_A^b, E_p, \alpha)]$ as the "backscatter factor", with $r_m(E_A^b, E_p, \alpha)$ defined as the ratio of ionisations produced by backscattered electrons to those produced by incident electrons:

$$r_m = \frac{\int_{E_A^b}^{E_p} I_o \left[\frac{d\eta}{dE}\right] \sigma_A(E) dE}{I_o \sigma_A(E_p) \sec\alpha} \tag{27}$$

where $I_o[d\eta/dE]$ is the energy distribution of backscattered electrons and α is the incident angle of the primary electron beam (relative to the sample normal). The integration in Eq. (27) corrects for the fact that the backscattered electrons will have a distribution of energies below the primary E_p, and consequently a different ionisation cross-section. The definition of the backscatter factor in this way allows $[1 + r_m]$ to be inserted in equations such as Eq. (15) as an "effective primary beam current".

Hall and Moribito [25] in calculating the AES matrix factors for dilute binary alloys used the empirically-derived backscatter factor equation of Reuter [26]

$$1 + r_m(E_A^b, E_p) = 1 + 2.8(1 - 0.9E_A^b/E_p)\eta_m \tag{28}$$

where the matrix dependent η_m term is the backscatter coefficient, or fraction, defined as the fraction of electrons backscattered and given by Reuter [26] as

$$\eta_m = -0.254 + 0.016Z - 0.000186Z^2 + 8.3 \times 10^{-7} Z^3 \tag{29}$$

for an element of atomic number Z. This equation was obtained as the result of a polynomial fit to data of Heinrich [27] at a primary beam energy of 20 keV and normal incidence.

More recently, Ichimura and Shimizu [28] have made use of a Monte Carlo simulation model [29] to estimate the electron backscattering effect in over twenty-five materials including pure elements, compounds and alloys of practical importance. The results obtained are more applicable to AES than Reuter's equation (Eq. (29)) because the primary beam energies used were 3, 5, 7.5 and 10 keV, as opposed to the 20 keV of Reuter. Furthermore the calculations were carried out at three angles of incidence: 0°, 30° and 45°. A functional representation of the backscatter factor obtained in these Monte Carlo calculations was given by Ichimura et al. [30]. This enabled the backscattering effects predicted in this way to be of practical use. As with Reuter's equation it was assumed that the backscatter factor could be described by the overvoltage ratio ($U = E_p/E^b$) and the atomic number Z. The equations proposed by Ichimura et al. were later given slightly modified coefficients by Shimizu [24]. The final form being

$$1 + r_m(E_A^b E_p) = 1 + (a - bZ^x)U^y + (cZ^w - d) \tag{30}$$

where a, b, c, d and x, y, w depend on the angle of incidence.

Quantification Programs

Construction of AES and XPS Libraries

The basic aim of the set of programs developed for quantification, and described here, was to provide a means whereby the matrix correction factors necessary for improved accuracy in quantification could be applied to raw spectral data. The programs were

Table 1. Breakdown of the various terms in the simplified equations for XPS and AES detected current that can be calculated from basic elemental parameters

Technique	Term	Parameters required
XPS/AES	Atomic size	Density, atomic weight
XPS/AES	Attenuation length	Density, atomic weight Kinetic energy Penn's tabulated parameters
XPS	Photoelectron ionisation cross-section	Tabulated cross-section for individual lines
XPS	Angular asymmetry of photoelectron production	Tabulated asymmetry parameters
AES	Electron backscattering	Atomic number, binding energy of initial core level
AES	Electron impact ionisation cross-section	Binding energy of initial core level

developed on a stand alone basis to work alongside any commercially supplied spectrometer control software.

From the above discussion of quantitative surface analysis it is clear that various parameters from reference tables are required for each of the elements of interest in a sample. For example, in estimating matrix factors for quantitative AES, element densities, atomic numbers, atomic weights and ionisation energies are just some of the parameters that may be required. When used in conjunction with computer programs to perform matrix calculations automatically, a database of elemental information can help to cut down on tedious amounts of user input. Table 1 shows some of the parameters required for quantification in XPS and AES.

As a number of the parameters which appear are common to both XPS and AES two separate databases were created but no information was duplicated. The two library files *QUAX.LIB* and *AQUA.LIB* for XPS and AES respectively were constructed. These names correspond to the names given to the actual quantification programs – *QUAX* for Quantitative XPS and *AQUA* for Auger quantification. The arrangement of these library files as separate entities from the main programs allows for the creation of additional files containing, say, values of cross-section determined by a different author. The likely lines for analysis for any particular element were determined from "The Handbook of Auger Electron Spectroscopy" [4] and "The Handbook of x-ray Photoelectron Spectroscopy" [31].

Many of the binding energies entered into the XPS library are also relevant to Auger production, for example the binding energy of an initial level. A simple program was therefore written which can read a line from a record in the Auger library and recognise the correct binding energy and spectral notation in the XPS library.

System Set-Up

As well as the information about each element held in the library files, there are certain parameters which are constants for a particular calculation which may occasionally be altered to suit a slightly different analysis configuration. Parameters such as incident beam energy in AES and the incident and take-off angles for both AES and XPS fall into this category. There are also a number of choices which must be made regarding methods of calculation and the use of library files other than the default set-up. A simple method, whereby the so-called "current system set-up" can be altered, has therefore been made available as part of the system. The investigation described here will show that this is especially useful for comparisons of the various routes to quantification. For example, altering the AL, a practice which is generally accepted as being desirable in order to obtain a feel for the errors concerned in calculating the matrix factors.

When the system set-up option has been selected from either *QUAX* or *AQUA* the main menu gives the following simple options: (1) Set General Parameters; (2) Set Backscatter Formula; (3) Set Attenuation Length Formula; (4) Set Depth Profile Parameters; (5) Quit and Save Parameters; (6) Quit Without Saving.

The "Set General Parameters" option above presents the user with an editing screen. The system has been designed for universal use on all types of surface analysis systems. When the system of programs is used in applying matrix corrections, variables such as the electron beam incident energy in AES and the photon/photoelectron angle in XPS can be altered simply by accessing this screen editor. When the system of programs is used in calculations where the prediction of instrument energy response is required (i.e. "standardless calculations") it is assumed that this function can be described as proportional to E^m, where E is the electron energy and m is a constant for the analyser at a particular constant analyser energy (CAE). For the CAE mode commonly used in XPS this simple power law is used for standardless calculations at present.

Setting up Data Processing Parameters

For some parameters used in quantifying the raw spectral data, a range of possible calculated values exist, for example a choice of photoionisation cross-section. For the backscatter correction either the backscatter formula of Ichimura et al. [29] or the backscatter formula of Reuter [26] can be used in the quantification of surface composition by AES. At present the *AQUA* programs have been set up to allow the user to choose between these two formulae from within the "Set Backscatter Equation" option on the "Main Setup Menu". Since the results of Ichimura et al. [29] produce equations for three angles of incidence a choice is available within the "Set Backscatter Equation" menu option for each of these three equations.

In the menu for the choice of formula to be used for calculating the AL, apart from the "No Correction" option there is a choice of three basic methods. The Seah and Dench [20] formula for AL and the IMFP formulae of Penn [22] and Tanuma et al. [32]. As mentioned above the AL rather than the IMFP is the preferred parameter for use in matrix factor calculations.

Experimental

XPS and AES spectra were acquired using a VG HB100 Multilab SEM-equipped with a CLAM 100 concentric hemispherical analyser. The spectrometer control unit was operated by a PC running VG Scientific VGS2000 software. XPS spectra were recorded with the spectrometer operating in the CAE mode at a pass energy of 20 eV with an energy window of 0.4 eV. XPS peak intensities were obtained from peak areas using both straight line background subtraction and the Shirley background subtraction method [33]. For Auger analysis, the CRR mode was used, with the retard ratio normally set at R = 2. AES spectra could be collected in either the direct N(E) form or in the differentiated dN(E)/dE mode, in which case a phase-sensitive detector was used with an analyser modulation of 10.6 V at 50 eV kinetic energy increasing linearly to 15.6 V at 1050 eV kinetic energy.

In the past in AES spectra have been measured in the differential mode but recently the direct mode has become more popular. The peak-to-peak value of the signal obtained in the differential mode is perfectly valid and with three main restrictions may be used in quantification. These restrictions are that the peak shapes are the same in the sample and the standard spectrum, that the analysers used for both measurements have the same resolution and that, in both cases, the same modulation is used. These conditions have been met in spectra recorded in the work described here. Peak shape changes can occur as a result of changes in chemical environment. In the direct mode the Auger electron signal is related to the area under the appropriate peak in the direct N(E) spectrum. In this mode the sloping background is acquired along with the Auger peak and it is often difficult to record the peak. With the increasing use of computers to record and process spectra, the problem of background removal and the display of the Auger electron peaks in the direct mode is relatively straightforward.

A rastered argon ion beam from a differentially-pumped VG Ionex AG60 source with a maximum spot current density of $\approx 500\,\mu A\,cm^{-2}$ and an ion energy of 4 keV was used for specimen cleaning. Ion beam etching was carried out until the surface impurity peaks in spectra were reduced to <1% of the main peaks. This was usually attained after a period of two minutes.

Results and Discussion

Qualitative and Quantitative AES and XPS Analysis of Silicon-Germanium Films

Table 2 shows the raw Auger peak-to-peak intensity data obtained from silicon-germanium films prepared by r.f. sputter deposition with a range of compositions. The results of qualitative surface analysis and quantitative surface analysis obtained using the *AQUA* program are also included in this table. The composition of the four silicon-germanium films has also been measured by x-ray photoelectron spectroscopy. The results are given in Table 3. Using pure standards of silicon and germanium the difference in calculated composition obtained from Auger measurements compared to XPS measurements ranged from approximately 2 at.% to 13 at.% when the *QUAX* and *AQUA* software are used. The large difference between surface compositions obtained by AES and XPS for the SiGe56 sample may result from local variations in surface composition which are detected in AES. Because of the larger sampling area in XPS only an average surface composition is measured. The differ-

Table 2. Qualitative and quantitative surface compositions of a number of SiGe films obtained by Auger electron spectroscopy using the LMM Auger lines with the Reuter backscatter correction and Penn (elements) inelastic mean free path

Sample	Si LMM intensity	Ge LMM intensity	Qualitative silicon composition atomic %	Quantitative silicon composition atomic %
Si standard	19712	–	–	–
Ge standard	–	16896	–	–
SiGe51	15360	7680	63.2	67.0
SiGe56	50176	46592	48.0	52.3
SiGe58	18256	12416	55.8	59.9
SiGe63	7040	14848	28.9	32.7

Table 3. Qualitative and quantitative surface compositions of a number of SiGe films obtained by x-ray photoelectron spectroscopy. IMFP correction, Penn (elements)

Sample	Si 2 s	Ge 3 d	Qualitative surface composition silicon at.%	Surface composition (standardless) silicon at.%	Quantitative surface composition silicon at.%
Si standard	5.916	–	–	–	–
Ge standard	–	6.02	–	–	–
SiGe51	3.71	1.90	66.5	76.5	74.5
SiGe56	2.97	2.38	55.9	67.5	65.2
SiGe58	2.15	2.93	42.8	55.0	52.4
SiGe63	1.14	3.98	22.6	32.3	30.0

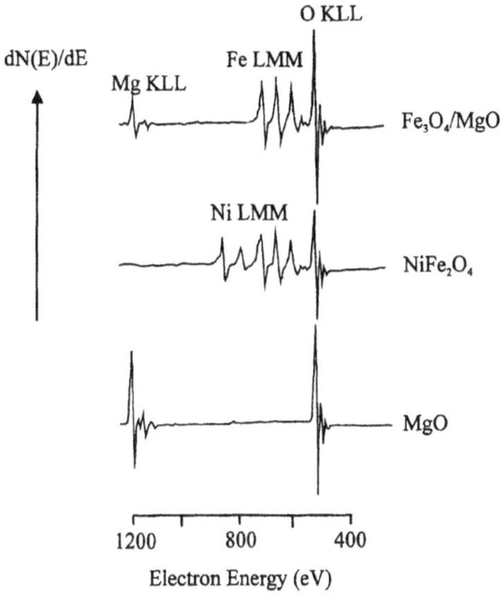

Fig. 1. AES spectra from CVD-deposited Fe_3O_4/MgO, polycrystalline $NiFe_2O_4$ and MgO. The $NiFe_2O_4$ and MgO were used as standards for quantitative calculations. All the spectra were taken after the surface had been cleaned by a short (120 s) Ar ion etch

ence between qualitative and quantitative composition is larger for XPS. This probably results from the smaller difference between the kinetic energies of Si 2*s* and Ge 3*d* photoelectrons by comparison with the kinetic energy difference between Auger electrons from these elements.

Ferrite Films on Magnesium Oxide

Figure 1 shows the Auger spectra from a CVD-deposited Fe_3O_4 film on an MgO substrate, polycrystalline $NiFe_2O_4$ and MgO. The $NiFe_2O_4$ and a separate MgO single crystal were used as standards for iron, oxygen and magnesium in quantitative calculations. The measured peak-to-peak intensities for the samples and standards are given in Table 4. As a check the composition of the $NiFe_2O_4$ standard was confirmed by EDX to within 0.1%.

It is clear from the results given in Table 4 that the surface region of the film contains magnesium. This magnesium originates from the MgO single crystal substrate, which is held at 1200 °C and probably interacts with the depositing species during the CVD reaction. It can also be seen that for this particular analysis, where the standards are similar in structure and composition to the sample being analysed, the difference between the qualitative and the quantitative composition obtained using *AQUA* is of the order of one atomic percent.

Surface Analysis of Alloys

The Auger electron intensities obtained from PdAg and PdNi alloys by Sekine et al. [34] are reproduced in Table 5 together with a comparison of the results they obtained for the surface composition of these alloys and the results obtained using the *AQUA* program, first of all using the attenuation length and backscattering equations they used, and then substituting the Seah and Dench attenuation length equation (for elements) for the purpose of a comparison. The slightly different compositions obtained by the *AQUA* system when using the same equations as Sekine et al. can be attributed to slightly different values of the atomic density used for pure palladium. The *AQUA* program calculates the atomic density in pure Pd from a density $\rho = 12.02\ \mathrm{g\,cm^{-3}}$ and atomic weight $A = 106.4$. This gives $N_{Pd} = 68.0$ atoms nm^{-3} compared to a values of $N_{Pd} = 68.6$ atoms nm^{-3} quoted by Sekine et al. [34].

Table 4. Raw AES *dN(E)/dE* peak-to-peak intensities from Fe_3O_4/MgO and $NiFe_2O_4$ and MgO standards and the surface composition obtained by AES from a film of Fe_3O_4 on MgO (a) from raw peak-to-peak intensities and (b) after applying *AQUA* with the Penn IMFP [22] equation and the Ichimura et al. [30] 45° backscatter factor

	Auger peak-to-peak intensity (arb. units)				Atomic % (qualitative) (a)			Atomic % (matrix) (b)		
	Fe	Mg	O	Ni	Fe	Mg	O	Fe	Mg	O
Fe_3O_4 on MgO	30	27	118	–	24.3	12	63.7	23.8	13.8	62.4
MgO standard	–	80	117	–	–	–	–	–	–	–
$NiFe_2O_4$ standard	25	–	75	30	–	–	–	–	–	–

Table 5. Comparison of the quantitative AES calculations of Sekine et al. [24] for fractured Pd/Ag and scribed Pd/Ni surfaces with those obtained with the *AQUA* AES program by two methods (a) using the Penn elemental for IMFP [22] and (b) "Seah and Dench Elemental" for Al [20]. Both (a) and (b) use the equation of Ichimura et al. (30°) [30] for backscattering. Primary beam energy 5 keV

	I/I_{std}	At.% (no correction)	At.% (Sekine et al.)	*AQUA* at.% (a)	*AQUA* at.% (b)	Chemical analysis (Sekine et al.)
Pd	0.576	64.9	67.3	67.1	66.4	70.3
Ag	0.312	35.1	32.7	32.9	33.6	29.7
Pd	1.356	57.6	54.5	54.3	56.2	50.0
Ni	1	42.4	45.5	45.7	43.8	50.0

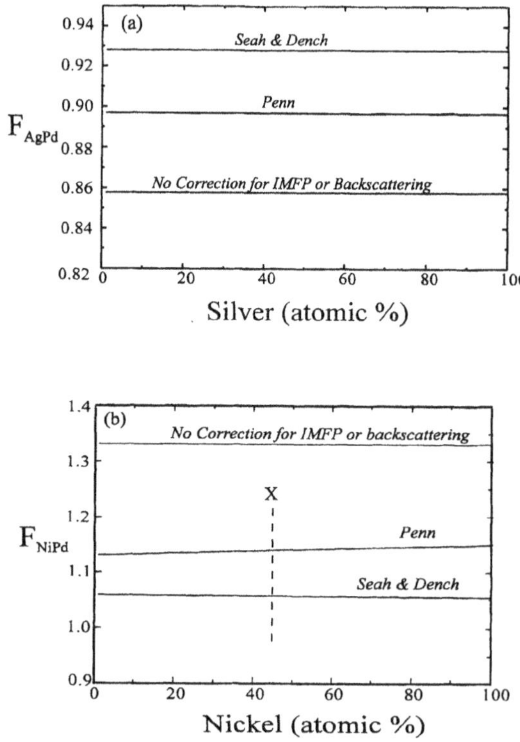

Fig. 2. Variation of AES matrix factor with (**a**) silver concentration for PdAg alloys and (**b**) nickel concentration for PdNi alloys, backscattering correction Ichimura et al. (30°)

Although the data shows that by application of the correction factors the concentrations move closer to those obtained by a chemical analysis, the PdAg alloy cannot be considered to be a test of the effectiveness of the convergence of the iterative system. By use of the AES "Plot Matrix Factors" option in *AQUA* we can obtain the graph shown in Fig. 2a which immediately indicates that F_{AgPd} is calculated as approximately constant over the entire range. The fact that Pd and Ag differ by only 1 in atomic number means that the backscatter estimated from a weighted average will be approximately equal to that of the pure standards. When combining this with the fact that the parameters for use in Penn's equation are as follows: $a_{Pd} = 20.6$; $a_{Ag} = 19.5$; $b_{Pd} = -2.98$; $b_{Ag} = -2.95$; $E_{Pd} = 330$ eV; $E_{Ag} = 367$ eV, we see that the IMFP will not show appreciable variation over the composition range either. Figure 2a shows the result obtained using Penn's equation for the inelastic mean free path, and for comparison that obtained using the Seah and Dench (Elements) equation for the attenuation length. As in Table 5 the backscatter equation of Ichimura et al. at 30° is used in both cases. A larger variation is found with the PdNi system and Fig. 2b plots the matrix factor F_{NiPd} against Ni atomic concentration. On this graph, X represents approximately the composition of the matrix in question indicating a difference between the two methods of 0.084. The actual composition difference of 1.9 at.% obtained is shown in Table 5.

Surface Composition of MnBi Films

To show how simply analysis can be carried out by all of the possible routes, and to give an example of how

Table 6. Surface composition of an MnBi film obtained by AES and XPS using various IMFP correction methods. For AES the back-scattering correction is that of Ichimura et al. in all cases

IMFP/AL method	AES		XPS		XPS (standardless)	
	Mn At.%	F_{MnBi}	Mn At.%	F_{MnBi}	Mn At.%	F_{MnBi}
Seah and Dench	89.3	1.4060	90.5	1.6983	84.6	0.6897
Penn	90.9	1.6819	87.1	1.1592	86.4	0.5988
Tanuma, Powell and Penn [BiN$_V$ = 5]	90.8	1.6646	87.1	1.1415	86.4	0.5952
Tanuma, Powell and Penn [Bi N$_V$ = 15]	93.2	2.3142	90.4	1.6093	86.4	0.5952

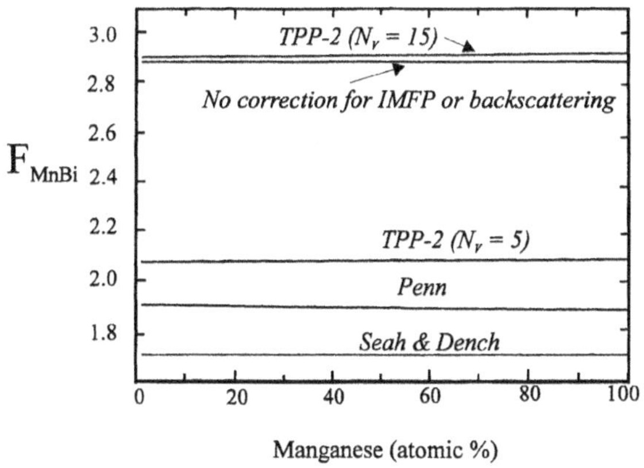

Fig. 3. Variation of XPS matrix factors with manganese concentration for MnBi.

the spread of results can be used to give an estimate of quantification error, Table 6 shows the results of a quantification of a single manganese/bismuth film prepared by vacuum evaporation and annealing. This example is chosen as it has the added problem of having an element present where one is unsure of the number of valence electrons (N_V) to use in the TPP-2 formula. In the table the compositions are calculated using both $N_V = 5$ and $N_V = 15$ for comparison. Figure 3 is a plot of the variation in XPS matrix factor with manganese concentration. A considerable difference in the matrix factor and in the eventual estimate of composition is obtained for the TPP-2 formula for $N_V = 5$ by comparison with $N_V = 15$. Most of the total inelastic scattering cross-section for electron energies of interest in AES and XPS is due to valence electron excitations typically between 5 and 50 eV. Tanuma et al [32] suggest from their calculations that for bismuth a value of $N_V = 5$ should be assumed.

Comparison of Surface Compositions Obtained with Different Matrix Factors

It is not possible to completely determine the importance of the differences between matrix factor corrections in terms of concentration difference from plots of the type given in Fig. 3. It is necessary to convert a variation in matrix factor to a variation in composition. From data of the type plotted in Figs. 2 and 3 it is possible, by means of a spreadsheet, to compare the composition obtained using a matrix factor with that obtained when the attenuation length correction or backscatter correction is omitted.

The composition changes obtained when different inelastic mean free path calculations are used are shown graphically in Fig. 4a. The approximately six atomic percent increase in silicon concentration, by comparison with no IMFP correction, at compositions close to sixty atomic percent silicon when the Penn inelastic mean free path calculation is used, would be difficult to deduce from graphs such as Figs. 2 and 3. The approximately five atomic percent decrease in silicon concentration observed for silicon concentrations close to fifty atomic percent silicon (Fig. 4b) when the Reuter backscatter is used by comparison with no backscatter correction would also be difficult to deduce from a straightforward plot of matrix correction against concentration. The plots, shown in Fig. 4, underline the importance of having software of this type to enable an estimation of the range of uncertainty in surface composition arising from the use of different calculations of matrix-dependent parameters.

Composition of a-SiC$_x$ Films Obtained by XPS

Apart from the analysis of Fe_3O_4 films on MgO substrates, the surface composition analyses have involved using elemental standards. Figure 5 compares surface compositions obtained from raw x-ray photoelectron intensities from films of a-SiC$_x$ which have been obtained under different silane and methane gas flow ratios. The compositions were obtained using the *QUAX* program in standardless mode, a silicon carbide compound standard and also with elemental standards. The data points for the elemental standards case represent the mean surface composition obtained by using all possible IMFPs. The analyses with elemental standards gave consistently lower silicon concentrations than the two other methods.

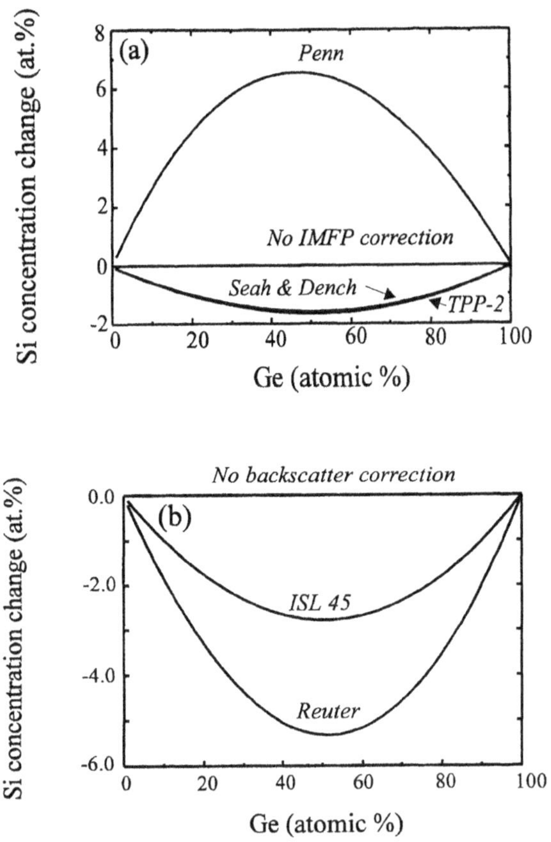

Fig. 4. Effects on calculation of the composition of a-SiGe alloys on (**a**) using various attenuation lengths (Ichimura et al. (45°) backscatter correction (ISL 45)) compared with the composition obtained with the IMFP/AL correction omitted and on (**b**) using various electron backscatter corrections (Seah and Dench attenuation length)

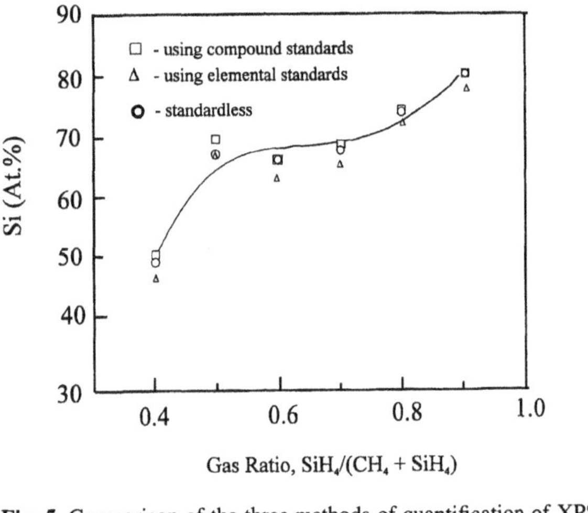

Fig. 5. Comparison of the three methods of quantification of XPS data for a-SiC_x films. The line through the points is drawn as an aid to the eye

In conclusion, the matrix factor corrections contained in the *QUAX* and *AQUA* programs lead to markedly different surface compositions for materials by comparison with those obtained from qualitative calculations. Compositions obtained by Auger electron spectroscopy have been compared with those obtained by x-ray photoelectron spectroscopy using these programs. In the MnBi system good agreement was obtained, however, for the $SiGe_x$ films larger differences in composition (8 at.% on average) were obtained from AES and XPS measurements with the largest difference of 13 at.% occurring for one sample. For these materials it is believed that this difference may arise from local composition variations detected by the small probe size in AES.

The system of programs has also been used to investigate the importance of matrix-dependent parameters. The effects of using different calculations of the inelastic mean free path or attenuation length on the matrix correction factor have been determined for a number of materials. These different matrix factors have been translated into changes in surface composition for $SiGe_x$ films. At particular compositions for films of this material, significant differences in the calculated composition are obtained when different calculations for matrix-dependent parameters are used.

The *QUAX* and *AQUA* programs have been shown to give a rapid means of determining quantitative surface compositions. They have also been shown to give a rapid means of investigating the effects on composition of different calculations of matrix-dependent parameters. These calculations give useful information about the range of error in surface composition which might be expected from x-ray photoelectron spectroscopy and Auger electron spectroscopy.

References

[1] C. G. H. Walker, D. C. Peacock, M. Prutton, M. M. El Gomati, *Surf. Interface Anal.* **1988**, *11*, 266.
[2] P. A. Moir, A. G. Fitzgerald, B. E. Storey, *Surf. Interface Anal.* **1989**, *14*, 295.
[3] H. L. L. Watton, A. G. Fitzgerald, P. A. Moir, *Surf. Interface Anal.* **1991**, *17*, 320.
[4] L. E. Davis, N. C. MacDonald, P. W. Palmberg, G. E. Riach, R. E. Weber, *Handbook of Auger Electron Spectroscopy, 2nd Ed.*, Physical Electronics, Minnesota, 1976.
[5] *Handbook of Auger Electron Spectroscopy*, JEOL, Tokyo, 1982.
[6] G. E. McGuire, *Auger Electron Spectroscopy Reference Manual*, Plenum New York, 1979.

[7] D. Briggs, M. P. Seah (eds.), *Practical Surface Analysis, 2nd Ed., Vol. 1, Auger and X-ray Photoelectron Spectroscopy*, Wiley, Chichester, 1996.

[8] M. P. Seah, Surf. Interface Anal. **1980**, *1*, 86.

[9] M. P. Seah, *Scanning Electron Microsc.* **1983**, *II*, 521.

[10] H. E. Bishop, *J. Appl. Phys.* **1967**, *18*, 703.

[11] C. J. Powell, *J. Elect. Spectrosc. Relat. Phenom.* **1988**, *47*, 197.

[12] K. Siegbahn, C. Nordling, A. Fahlman, R. Nordberg, K. Hanrin, J. Hedman, G. Johansson, T. Bergmark, S. E. Karlsson, I. Lindgren, B. Lindberg, *ESCA: Atomic Molecular and Solid State Structure Studied by Means of Electron Spectroscopy*, Almquist and Wiksells, Upsalla, 1967.

[13] C. J. Powell, *Scanning Electron Microsc.* **1984**, *IV*, 1649.

[14] W. S. M. Werner, *Surf. Sci.* **1991**, *257*, 319.

[15] A. Jablonski, S. Tougaard, *J. Vac. Sci. Technol.* **1990**, *A8*, 106.

[16] P. J. Cumpson, Surf. Interface Anal. **1993**, *20*, 727.

[17] A. R. Jackson, M. M. El Gomati, J. A. D. Matthew, P. J. Cumpson, *Surface Interface Anal.* **1997**, *25*, 341.

[18] A. Jablonski, *Surf. Sci.* **1987**, *188*, 164.

[19] S. Tanuma, C. J. Powell, D. R. Penn, *J. Electron Spectrosc. Relat. Phenom.* **1990**, *52*, 285.

[20] M. P. Seah, W. A. Dench, *Surf. Interface Anal.* **1979**, *1*, 2.

[21] D. R. Penn, *Phys. Rev.* **1976**, *B13*, 5248.

[22] D. R. Penn, *J. Elect. Spectrosc. Relat. Phenom.* **1976**, *9*, 29.

[23] S. Tanuma, C. J. Powell, D. R. Penn, *Surf. Interface Anal.* **1988**, *11*, 577.

[24] R. Shimizu, *Jpn. J. Appl. Phys.*, **1983**, *22*, 1631.

[25] P. M. Hall, J. M. Moribito, *Surf. Sci.* **1979**, *83*, 391.

[26] W. Reuter, *Proc. 6th Intern. Conf. on X-ray Optics and Microanalysis* (G. Shinoda, K. Kohara, I. Ichinokawa, eds.), University of Tokyo Press, 1971 p. 121.

[27] K. J. F. Heinrich, *X-ray Optics and Microanalysis*, (R. Castaing, P. Deschamps J. Philibert, eds.), Herman, Paris, 1966, p. 159.

[28] S, Ishimura, R. Shimizu, *Surf. Sci.* **1981**, *112*, 386.

[29] S, Ichimura, M. Aratama, R. Shimizu, *J. Appl. Phys.* **1980**, *51*, 2853.

[30] S Ichimura, R. Shimizu, J. P. Langeron, *Surf. Sci.* **1983**, *124*, L49.

[31] C. D. Wagner, W. M. Riggs, L. E. Davis, J. E. Moulder, G. E. Mullenberg, *Handbook of X-ray Photoelectron Spectroscopy*, Perkin Elmer Corporation, Minnesota, 1979.

[32] S. Tanuma, C. J. Powell, D. R. Penn, *Surf. Interface Anal.* **1991**, *17*, 911.

[33] D. A. Shirley, *Phys. Rev.* **1972**, *B5*, 4709.

[34] T. Sekine, K. Hirata, A. Mogami, *Surf. Sci.* **1983**, *125*, 565.

Mikrochim. Acta [Suppl.] 15, 365–369 (1998)

Fractals and $BaTiO_3$-Ceramic Microstructure Analysis

Vojislav V. Mitić[1,*], **Ljubisa M. Kocić**[1], **Miroslav Miljković**[2], and **Ivana Petković**[1]

[1] Faculty of Electronic Engineering, P.O.B. 73, 18000 Niš, Yugoslavia
[2] Laboratory of Electronic Microscopy, University of Niš, 18000 Niš, Yugoslavia

Abstract. The complex grainy structure of $BaTiO_3$ is difficult to describe by using traditional analytical or geometrical methods. Here, an attempt is made to establish new mathematical models based on both Euclidean and fractal geometrical methods. The first uses ellipsoidal approximations for barium-titanate grains describing the contact area as surface patches with an ellipsoidal boundary. The fractal method offers a better approach in describing the irregular and wavy surface of such contact zones and therefore, it defines aspects that are complementary to those defined by Euclidean methods. Electric, ferroelectric and opto-electric properties of $BaTiO_3$ are influenced strongly by the fractal dimension of the intergranular contact surfaces. This parameter is measured in a laboratory environment using SEM micrographs with different magnifications. The $BaTiO_3$ samples examined have been sintered using different sintering pressures and temperatures. Then, the fractal dimension is calculated from a log-log diagram, and the grain contours are constructed using Iterated Function Systems connected with fractal interpolation.

Key words: Ceramic grains, Boundary interpolation, Fractal method

The fractal method has been systematically employed for describing and studying hierarchical rules and patterns in different fields of science, by many authors, such as K. Weierstrass, H. Poincaré, G. Julia, P. Fatou, B. van der Pol, V. Arnold, M. Figenbaum, B. Mandelbrot, R. Barnsley etc. The basics of fractal theory have been created mainly by mathematicians but accepted and developed primarily by physicists during the first half of the century. However, the giant task of systematizing a huge quantity of results has been performed by Benoit Mandelbrot and his collaborators. Putting the theory into a more formal framework was the next painstaking process, one which was excellently accomplished by the expert group supervised by Richard Barnsley.

In addition to the Euclidean concept of symmetry, a fractal object has a special kind of inner, deeper symmetry, characterized by the Hutchinson's operator: the union of contractions of a metric space. Being a fixed point of such an operator (a contraction in itself), a fractal set possesses a fairly complicated structure and the unique feature of self-similarity. The next characteristic of fractal sets (fractals) is its fractal or Hausdorf-Besicovitch dimension (a generalization of the classical concept of dimension), which is, as a rule, a fractional number. This number determines the complexity of the particular fractal.

These features make fractal theory an essential tool in the study of physical reality. In this paper we shall focus our attention on applying fractal theory and in making it an efficient and useful model for stereological structure of $BaTiO_3$-ceramics, These ceramics have been obtained by a sintering process using different sintering parameters (pressing pressure, temperature and sintering time). The sintering conditions have a strong influence on the distribution of grains, pores and internal stresses as well as the chemical and electrical properties in the grain boundary areas [1–4].

This theory helps us to establish the fractal dimension of $BaTiO_3$-ceramic grains, to use it in a

* To whom correspondence should be addressed

mathematical model of grain distribution, an approximate area of contact between the grains, and to determine its influence on the electrical properties of ceramics [5–8].

Experimental

Different stereological constituents, such as grains and pores, are responsible for the distribution of areas in grain boundary intergranular phases. The intergranular and integral surfaces are the basis for the different electric properties, especially for capacitance. The geometry of grains and grain boundary surfaces are very important for space charge and processes on the electric level near to grain boundaries. The contact surface patches are small in size, but highly complicated in form. So, in order to be as close as possible to the real structure of $BaTiO_3$-ceramics, the grains are approximated by ellipsoids of different sizes and orientations. Then, intersections between two ellipsoids are determined and used as support for the fractal patch which is created in accordance with the fractal dimension estimated by using SEM (JOEL-JSM-T20) micrographs taken at different magnifications.

The smaples were prepared from "Murata" barium-titanate powder with the composition 65.24% BaO and 34.7% TiO_2. To investigate the influence of the grain growth on the electrical characteristics the following pressures have been used: 86–150 MPa. The specimens were sintered in a tunnel furnace type "CT-10 MURATA" from 1240–1370 °C for 2 hours. The capacitance was measured on the specially prepared components, based on the sintered samples using a "Hewlett-Packard" device.

Grains of $BaTiO_3$-ceramic samples can be approximated by ellipsoids scattered throughout the material's volume. Let us consider the ellipsoid E_1. It is determined by the centre coordinates (x, y, z) and semi-axes vectors **a**, **b**, **c**. By affine transformation E_1 can be put in the canonical forma $x^2/a^2 + y^2/b^2 + z^2/c^2 = 1$, where $a \geq b \geq c > 0$ are semi-axes' lengths. Suppose another ellipsoid E_2 with semi-axes p, q, r has the point (0, d, 0) (d>0) as the centre and the axes vectors **p**, **q**, **r** are pair-wise collinear with **a**, **b**, **c** in the same direction. These two ellipsoids can be seen as a model of two grains in contact. Actually, due to sintering pressure and the sintering process, one grain partly penetrates into another, forming a small contact area that can be fairly accurately approximated by the intersection of ellipsoids E_1 and E_2. Our aim is to determine the value of this area as a function of the distance apart, d, of the grain centres (Fig. 1).

In fact, ellipsoids E_1 and E_2 have a non-empty intersection if, and only if, $|b - q| \leq d \leq b + q$. Determining the area S of intersection is a hard task, even for our simplified arrangement; the power of analytical methods is limited indeed. For example, if we take $p = ka$, $r = kc$, which is a very special case, an analytical approach yields

$$S = \pi ac\left[1 - (\alpha d - \beta\sqrt{y + d^2})^2\right],$$

where

$$\alpha = \frac{k^2 b}{k^2 b^2 - q^2},$$
$$\beta = \frac{kq}{k^2 b^2 - q^2},$$
$$\gamma = \frac{(1 - k^2)(q^2 - k^2 b^2)}{k^2}.$$

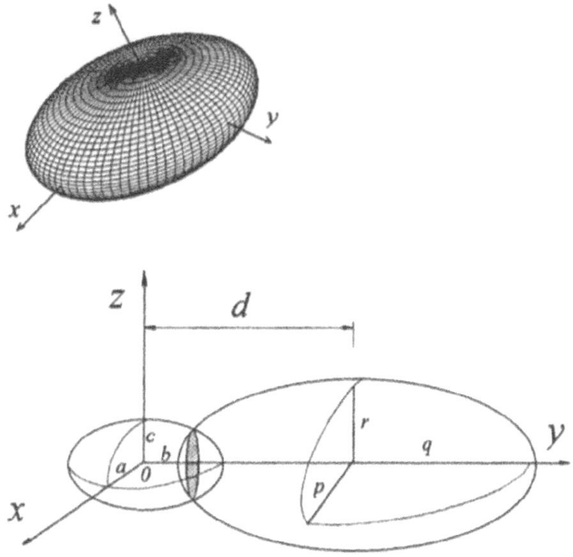

Fig. 1. Ellipsoid as the $BaTiO_3$-ceramic grain approximation

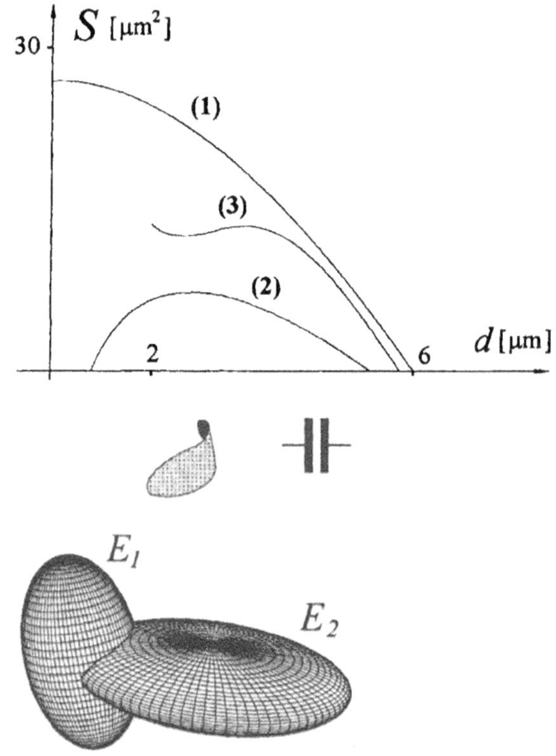

Fig. 2. Value of the area of the contact zone and the microcapacitor formed in this zone

This example reveals the complicated functional dependence of S in ellipsoids geometric parameters and their separation. Note that the general situation is even more complicated so that the analytical tool is useless. However we can, of course, use symbolic program packages such as "Mathematica" for this area determination.

A typical dependence of S as a function of d is shown in Fig. 2 (left). The curve labelled 1 is the graph of this function for coaxial ellipsoids with proportional semi-axes lengths, curve 2 is for non-proportional semi-axes lengths while curve 3 expresses this relation for an arbitrary position of the ellipsoids. Two grains partly penetrating each into another, and the corresponding microcapacitor model, formed by the contact area is shown in Fig. 2 (right). This area has a fractal structure, which means that the effective size of S is greater than if it was smooth. A striking example of Le Méhauté et al. concerning lithium batteries is given in Mandelbrot [10, p. 462]. For this reason, we will focus in the next section on methods that can help in identifying the type of fractal which is characteristic of $BaTIO_3$-ceramics.

Fractal Method

Irregularity of the surface of ceramic grains can be expressed using a term called the *fractal dimension* (see [10, 11]). Being attached to some fractal set G, this number $D = \dim(G)$ strictly exceeds its topological dimension. The fractal dimension can be used in two ways: to improve the numerical value of the contact area size, S, and to make a geometrical model of the grain that faithfully represents it. So, the initial problem is to extract the fractal dimension from the measured data. For this purpose we use Richardson's law ([10]) which gives the relationship between the measurement scale ε and the length $L(\varepsilon)$ of some irregular contour $L(\varepsilon) = K e^{1-D}$, where K is some positive constant and D is just the fractal dimension of this contour. Using the micrographs, we measure $L(\varepsilon)$ for different magnification, ε. The data points $(Ln(L(\varepsilon_j)), Ln(\varepsilon_j))$ are approximated by a linear function using the least squares method. The slope of this line is approximately $1\text{-}D$. By this technique we obtained the following dimensions for six different grains: $D_1 \approx 1.071890$, $D_2 \approx 1.071993$, $D_3 \approx 1.061486$, $D_4 \approx 1.0732408$, $D_5 \approx 1.116055$ and $D_6 \approx 1.0119546$. The average value $D \approx 1.0677699$ can be accepted as an approximation of the fractal dimension for the contour line of $BaTiO_3$-ceramic grains. Fig. 3 (right) shows the log-log diagram with the approximating line, the slope of which is used to estimate the dimension from the five measurements. For similar methods in powder metallurgy see [12]. Also, as an example for comparison purposes, the results of Mandelbrot, Passoja and Paullay [10, p. 459–461] concerning the fractal dimension of level-line contours in the surface of tensile steel specimens should be mentioned. The resulting fractal dimension is estimated to range from 1.118 to 1.164. The contour of the grain obtained by fractal interpolation is given in the same figure (left). This method is discussed next.

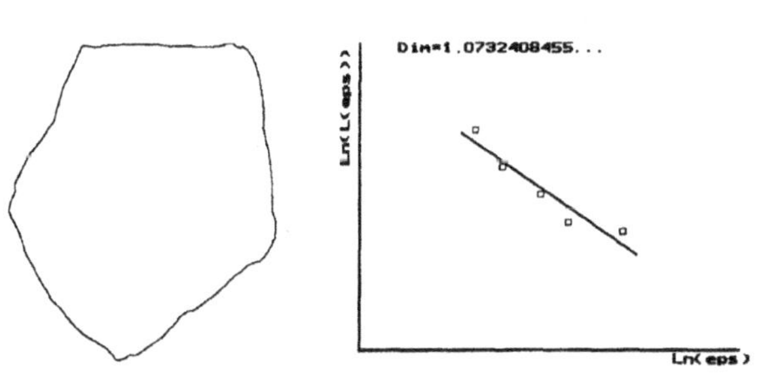

Fig. 3. The grain boundary and its fractal dimension

Let $\{x_i, y_i\}$, $i = 0, 1, \ldots, N$ be the set of interpolation points in $\mathbf{R}^2$ (taken from the boundary of a ceramics grain). We need to find a fractal interpolation curve, being the graph of parametric mapping F: $[0, 1] \to \mathbf{R}^2$, i.e. the function F such that $F(t_i) = [x_i y_i]^T$, where $0 = t_0 < t_1 < \cdots < t_N = 1$, are interpolation knots. For simplicity, it is sufficient to take $t_i = i/N$. For this purpose, we construct the IFS $H = \{\mathbf{X}; w_1, \ldots, w_N\}$ where $\mathbf{X} = [0, 1] \times \mathbf{R}^2$, and w_i: $\mathbf{X} \to \mathbf{X}$ is the affine mapping, given by

$$\omega_i : \begin{bmatrix} t \\ x \\ y \end{bmatrix} \to \begin{bmatrix} 1/N & 0 & 0 \\ x_i - x_{i-1} & d_i & 0 \\ y_i - y_{i-1} & 0 & e_i \end{bmatrix} \begin{bmatrix} t \\ x \\ y \end{bmatrix} + \begin{bmatrix} (i-1)/N \\ x_{i-1} - d_i x_0 \\ y_{i-1} - e_i y_0 \end{bmatrix},$$

where $\{d_i\}$ and $\{e_i\}$ are free constants that satisfy $0 \leq d_i < 1$, and $0 \leq e_i < 1$.

For fixed values d_i and e_i $(i = 1, \ldots, N)$, the attractor of the IFS H is a C^0 curve G that interpolates points $\{x_i, y_i\}$. Such interpolation, is known as *hidden variable fractal interpolation* and is introduced in [9]. The question is how to choose the free constants d_i and $e_i (i = 1, \ldots, N)$? It is known that the fractal dimension of G is given by

$$D = 1 + \ln \mu / \ln N \qquad (1)$$

where $\mu = \sum_{i=1}^{N} \max(|d_i|, |e_i|)$. The fractal dimension is known, so if we suppose that all d_i and e_i are equal, from Eq. (1) it follows that $d_i = e_i = \mu/N$. When the fractal curve G is constructed (the Random Iteration Algorithm [9] is used), it may look different than G_0 – the contour that we have obtained from the micrographs of the $BaTiO_3$-ceramic samples. By changing the values of d_i and e_i around the uniform distribution μ/N, the curve G can be interactively adjusted to be closer to G_0. According to our experience, it is useful to locate the interpolation nodes at the points where G_0 bends sharply. In this case, a slight modification of the d_i and e_i values is sufficient to correct the graph properly. Note that D, given by Eq. (1), is an

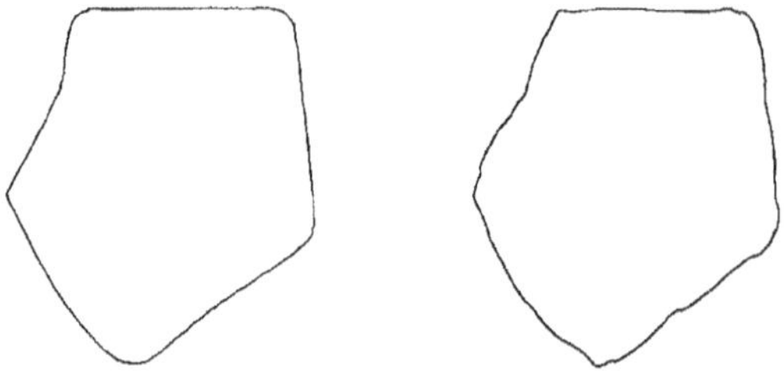

Fig. 4. Akima vs. fractal interpolation method

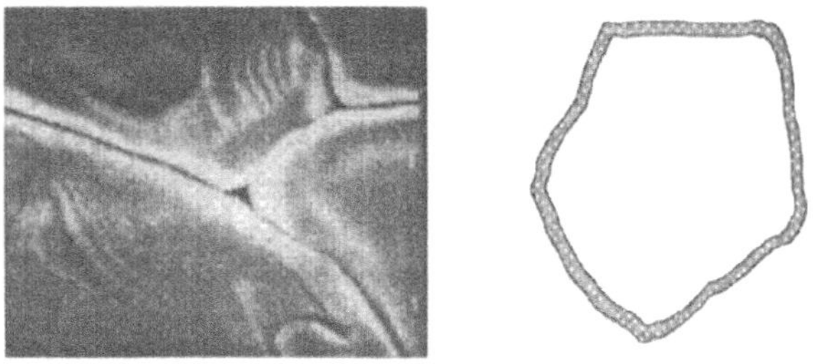

Fig. 5. Sintered $BaTiO_3$; $T_{sint} = 1370\,°C$, $\tau = 120'$, $P = 105$ *MPa* ($\times 5000$) and its fractal model

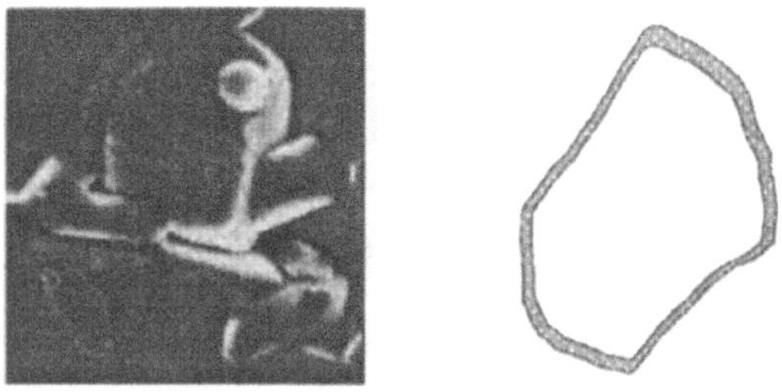

Fig. 6. Sintered $BaTiO_3$; $T_{sint} = 1240\,°C$, $\tau = 120'$, $P = 150$ *MPa* ($\times 5000$) and its fractal model

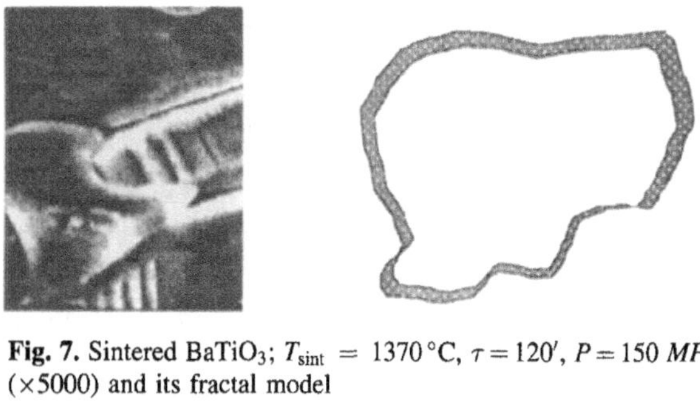

Fig. 7. Sintered $BaTiO_3$; $T_{sint} = 1370\,°C$, $\tau = 120'$, $P = 150$ *MPa* ($\times 5000$) and its fractal model

increasing function of the variables d_i and $e_i (i = 1, \ldots, N)$. So, increasing d_i and e_i will increase oscillations of G at the j-th segment of interpolation.

Besides the fractal method, some conventional methods for grain boundaries are also available. For example, the Akima method [13] gives the piece-wise

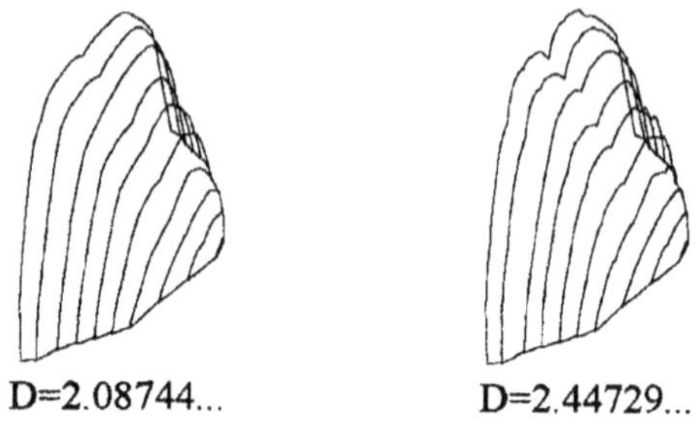

Fig. 8. Grains' surfaces with different fractal dimension

cubic C^1 interpolant, which means that each segment of the boundary between two data points (x_{i+1}, y_{i+1}) and (x_i, y_i) is given by

$$x(u) = a_0 u^3 + a_1 u^2 + a_2 u + a_3,$$
$$y(u) = b_0 u^3 + b_1 u^2 + b_2 u + b_3,$$

where $u = Nt - i$ is a local variable, $t \in [0, 1]$. The coefficients a_k and b_k are determined using a local procedure described in [13]. Figure 4 shows juxtaposed contours obtained by Akima interpolation (left) and with fractal interpolation (right). It is obvious that the fractal method gives a much more realistic picture.

The shapes of the grains used in the experiment are illustrated in Figs. 5, 6, and 7. On the right side of each figure, the contact grain area, obtained by the fractal method is shown. This area is responsible for the electric properties of $BatiO_3$-ceramics. Our task is to obtain models of grains' surfaces. A simple model can be constructed by making use of level-lines. Each level-line is produced by the fractal interpolation technique explained above, and based upon separate sets of interpolated data. Figure 8 shows fragments of two grains with different fractal dimension. It is bigger than 2 which is the value of an ordinary (topological) dimension of a smooth surface.

Finally, let us discuss the influence of fractal dimension on increasing the size of contact surfaces. Suppose the fractal dimension of the surface S is D_s. If $D_s = 2$, the surface does not have fractal structure and the size of the area of S is $|S| < +\infty$. For fractal-surfaces, $D_s > 2$ and $|S| < +\infty$. The size of the real surface of some physical material cannot be unbounded. So, it can be seen as a stage in the iterative process

$$S_j = \bigcup_{i=1}^{N} \omega_i(S_{j-1}), \quad j = 1, 2, \ldots, \; S_0 = S, \qquad (2)$$

where ω_i: $R^3 \rightarrow R^3$ is a similarity contractive mapping with contractive factor s_i. So, after the first iteration the resulting surface will be $S_1 = \bigcup_{i=1}^{N} \omega_i(S_0)$ with a size $|S_1| = |S_0| \sum s_i$. Consequently, the size of k-th iteration is $|S_k| = |S_0|(\sum s_i)^k$. It is known that $\sum s_i > 1$ ([9]), so the value of surface area increases at each iteration. This poses the problem of determining an iteration number k such that the surface S_k fulfils the criterion $d_h(S_0, S_k) \rightarrow \min$. A study of this problem is currently being undertaken.

Conclusion

This paper deals with various mathematical models that can be helpful in understanding the complex structure of $BaTiO_3$-ceramics and its influence on the electric properties of this material. The geometry of real grains can be replaced by ellipsoidal approximations. These ellipsoids are distributed throughout the volume, making contacts that have fractal character. The fractal dimension – a parameter that reflects irregularity of contact surfaces – can be estimated by using SEM micrographs taken at different magnifications. Then, the fractal dimension is calculated (approximately) from Richardson's law. Examples from some $BaTiO_3$-ceramic specimens support the theory.

Acknowledgement. This research has been supported by the projects: "Prognosis of material characteristics" financed by the Ministry of Science and Technology of the Republic of Serbia and "Physical chemistry processes of technology of modern materials" financed by Serbian Academy of Science and Arts.

References

[1] M. Kuwabara, T. Matsunaga, S. Takahashi, H. Shimooka, *97th Annual Meeting*, Cincinnati, 1995.
[2] V. V. Mitić, *Ph. D. Thesis*, University of Niš, 1995.
[3] V. V. Mitić, Z. S. Nikolić, A. M. Gokhale, M. M. Ristić, *97th Annual Meeting of the American Ceramic Society*, Cincinnati, 1995.
[4] V. V. Mitić, Z. S. Nikolić, M. M. Ristić, *97th Annual Meeting of the American Ceramic Society*, Cincinnati, 1995.
[5] V. V. Mitić, Lj. Kocić, B. Jordović, *98th Annual Meeting of the American Ceramic Society*, Advanced Ceramic Materials, Electronic Division, Indianapolis, USA, 1996.
[6] V. V. Mitić, Lj. Kocić, Lj. Živković, M. M. Ristić, *98th Annual Meeting of the American Ceramic Society*, Advanced Ceramic Materials, Electronic Division, Indianapolis, USA, 1996.
[7] V. V. Mitić, Lj. Kocić, B. Jordović, *2nd International Meeting of Pacific Rim Ceramic Societies – PacRim2*, Cairns, Australia, 15–17 July, 1996.
[8] V. V. Mitić, Lj. Kocić, M. M. Ristić, *2nd International Meeting of Pacific Rim Ceramic Societies – PacRim2*, Cairns, Australia, 15–17 July, 1996.
[9] M. F. Barnsley, *Fractals Everywhere*, Academic Press, 1988.
[10] B. Mandelbrot, *The Fractal Geometry of Nature*, Freeman, San Francisco, 1983.
[11] K. Falconer, *Fractal Geometry, Mathematical Foundation and Applications*, Wiley, Chichester, 1993.
[12] N. Jumate, M. F. Danca, *Fractalia* **1996**, *19*, 28–29.
[13] H. Akima *J. ACM* **1970**, *17*, 589.

Mikrochim. Acta [Suppl.] 15, 371–377 (1998)

Fragmentation of Sputtered Cluster Ions of Transition Metals: Distributions of Lifetimes and Internal Energies

A. D. Bekkerman[1], **N. Kh. Dzhemilev**[1], **S. V. Verkhoturov**[1], **I. V. Veryovkin**[1,2,*], and **A. Adriaens**[2]

[1] Arifov Institute of Electronics, Uzbek Academy of Sciences, 700143 Tashkent, Uzbekistan
[2] Department of Chemistry, University of Antwerp, B-2610 Antwerpen, Belgium

Abstract. Dissociation energies, lifetime distributions and internal energies have been determined experimentally for Fe_n^+ ($4 \leq n \leq 10$) clusters produced by sputtering under 8.5 keV Xe^+ ion bombardment. The average lifetimes and the average internal energies were calculated from the experimental distributions. A comparison between the results of the present work on Ta_n^+ and Nb_n^+ clusters and those obtained earlier has been made. For all sputtered clusters compared, the linear dependence of the average internal energy $\bar{E}_{int}$ on the number of atoms, n, in the cluster has been identified. Good qualitative agreement has been found between experimental results and theoretical values determined from MD-simulations of cluster sputtering processes.

Key words: cluster sputtering, kinetic energy spectra, unimolecular fragmentation, fragmentation rate constants, lifetimes, kinetic energy release distributions, internal energy distributions.

The formation of clusters under ion bombardment of solids is one of the most interesting phenomena in ion-surface interaction physics. During the last decade, much new qualitative experimental data has been gathered on this subject [1–17]. Recent reviews of published results can be found in Refs. [18, 19]. One of the greatest difficulties in developing an adequate theoretical mechanism [20–22] for cluster formation is to compare directly such basic properties as yields, kinetic and internal energy distributions calculated for "nascent" sputtered clusters with values determined experimentally from detectable "final" clusters. This is because the high vibrational excitation of nascent clusters results in their evolution over time, the causes being the competitive processes of unimolecular fragmentation [1–8, 15] and ionisation [9–14, 16, 17].

One effective approach is to consider this problem using a combination of Molecular Dynamics (MD) and Monte Carlo (MC) computer simulations of cluster sputtering [23, 24]. Recently, very promising and realistic results for Ag_n clusters have been obtained [25] by using a so-called MD/MC-CEM interaction potential (where CEM stands for "Corrected Effective Medium") [26]. It has been shown that the nascent Ag_n clusters sputtered by sub-keV Ar^+ ions had an average internal energy of about 1 eV per constituent atom. The decomposition pathway of unstable sputtered clusters was the branching or chain decay which resulted in a total time of decomposition ranging from 10 ps to 50 ps. Such a short lifetime of evolution of Ag_n clusters was determined by their low dissociation energy and high internal energy.

Nevertheless, accurate experimental studies of the cluster emission process are the only reliable criterion for testing the theoretical model. Thus, the importance of experimental studies of a stochastic process of nascent cluster fragmentation is obvious. The fragmentation of sputtered clusters as a stochastic process has been previously investigated [27–30] and the decay rate constants and lifetime distributions of excited sputtered Ta_n^+ [27–29] and Nb_n^+ [30] clusters have been measured. It was established that within the time range of $10^{-9}-10^{-4}$ s the dependence was a

* To whom correspondence should be addressed

power function rather than an exponential one. In addition, the internal energy distributions and corresponding average internal energies were calculated by the use of Rice-Ramsperger-Kassel (RRK) theory. In contrast to the above MD/MC simulations, longer average lifetimes for Ta_n^+ and Nb_n^+ sputtered clusters were determined in the experiments and explained by their higher dissociation energy of several eV [29, 30].

In the present work the dissociation energies, the lifetime distributions and the internal energy distributions together with average values of lifetimes and internal energies were determined for Fe_n^+ ($n \leq 10$) clusters sputtered by 8.5 keV Xe^+ ions. The common regularities observed for fragmentation of the Fe_n^+, Nb_n^+ and Ta_n^+ sputtered clusters within the time range of $10^{-9}-10^{-4}$ s are considered and discussed.

Experimental

The double focusing SIMS instrument with reverse geometry employed in the present work has been described in detail previously [31]. This instrument (Fig. 1) was designed and produced as a prototype of the IMT3 ion microprobe, which was in serial production in SELMI (Sumy, Ukraine). Briefly, Xe^+ primary ions generated by a duoplasmatron ion source {1} are accelerated and mass separated in a Wien filter {3}. The focused ion beam (100 μm) bombards the sample at an angle of 45° with respect to the surface normal. The secondary ion flux is accelerated to energies, eU_0, of 2.5 keV and 5 keV and is focused by the immersion lens {8} onto the entrance slit of the magnetic mass analyser (the stigmatic magnetic prism). The energy analyser {12} is located behind the magnetic prism and consists of a three-electrode electrostatic mirror. In the sample chamber and along the entire ion flight path from the sample to the detecting system (the electron multiplier {14}), the pressure of the residual gases ($2 \cdot 10^{-8}$ Torr) was maintained by using a differential pumping system. The kinetic energy distributions of the secondary ions were determined by varying the secondary ion accelerating voltage [6].

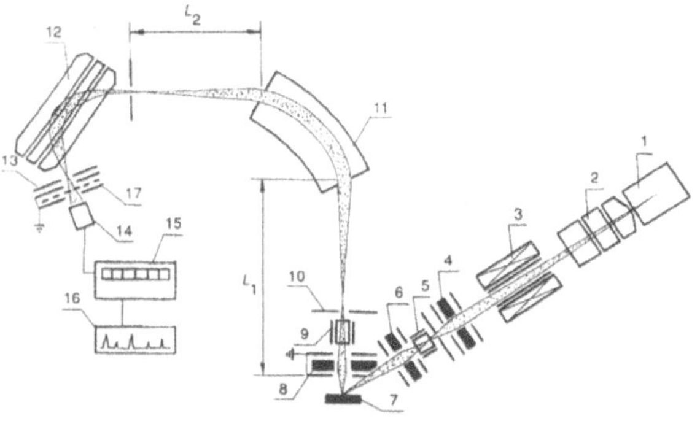

Fig. 1. Scheme of the secondary ion mass spectrometer: *1* primary ion source (duoplasmatron); *2* condenser lens; *3* Wien filter; *4* intermediate lens; *5* deflector; *6* objective lens; *7* sample; *8* immersion lens; *9* dynamic transfer system; *10* entrance slit; *11* magnetic mass analyzer; *12* electrostatic energy analyzer; *13* exit slit; *14* electron multiplier; *15* pulse counting system; *16* plotter register; *17* grid system. L_1 and L_2 are the field-free zones of the mass spectrometer

Results and Discussion

For the time range of $10^{-9}-10^{-4}$ s the fragmentation reaction identified for sputtered Fe_n^+ cluster ions ($3 \leq n \leq 10$) as well as for Nb_n^+ ($5 \leq n \leq 10$) [30] and Ta_n^+ ($4 \leq n \leq 8$) [27–29] cluster ions can be expressed as the following "quasi chemical" formula:

$$Me_n^+ \rightarrow Me_{n-1}^+ + Me \tag{1}$$

where Me_n^+ stands for the corresponding metal clusters.

The cluster fragmentation which takes place within the accelerating electric field between the sample {7} and the accelerating electrode {8} was studied by measuring the kinetic energy distributions of *sputtered* cluster ions. For the fragmentation reactions within a time range of $\cong 10^{-8}-10^{-6}$ s, the kinetic energy of clusters gained in the accelerating field is distributed between the fragments proportionally to their masses. In this case, the cluster ion must have a kinetic energy deficit at the exit from the zone. This deficit can be determined by kinetic energy distribution measurements [6, 15]. Such distributions of Fe_n^+ secondary cluster ions are shown in Fig. 2. The extended "tails" of these distributions over the range of energies $<eU_0$ correspond to larger cluster ions (Fe_{n+1}^+) fragmented within the accelerating zone. Moreover, the cluster fragmentation times can be determined from the "tails" because different points correspond to different times [27–30].

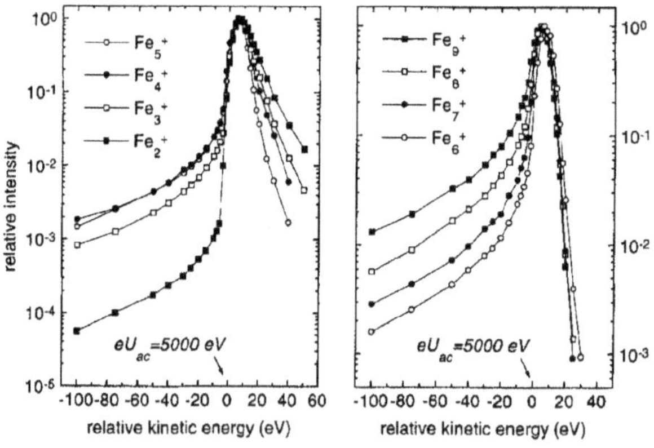

Fig. 2. Kinetic energy distributions of Fe_n^+ cluster ions ($2 \leq n \leq 9$) sputtered by 8.5 keV Xe^+ primary ions

Fragmentation within the L_2 field-free drift zone of the SIMS instrument between the mass analyser {11} and energy analyser {12} was studied by measuring the kinetic energy distributions of *fragment* ions formed by fragmentation of the sputtered clusters within the time range $\cong 5 \cdot 10^{-6} - 10^{-4}$ s. These measurements were carried out by realigning the energy analyser for the transmission of ions with energies of $eU_0 \pm 250$ eV. As seen below, such methods provide information also on the dissociation energies of the sputtered clusters.

Considered as a whole, these two ways of studying fragmentation allowed a total experimental time range of $\cong 10^{-8} - 10^{-4}$ s. In order to widen each time range, measurements were carried out for two values of the secondary accelerating voltage: $U_0 = 2.5$ kV and $U_0 = 5$ kV.

The dependence of fragmentation rate on the time immediately following cluster emission was obtained by transposing the energy scale to a time scale: $dN/dt = dN/dE \cdot dE/dt$. In this case the magnitude of *dN/dE* was determined from the kinetic energy distributions of the sputtered cluster ions using the extended "tails", as well as from the kinetic energy distributions of fragment clusters. The *dE/dt* derivative was determined from a solution of the ion movement equation for the accelerating electric field configuration. For the plane target, this electric field can be described by an approximation of a plane capacitor [15]. As seen from Fig. 3, over the entire experimental time range the dependence $F(t) = dN/dt$ can be best fitted by a power function (not by an exponential one, as might be expected):

$$F(t) = b \cdot (t + c)^{-a} \tag{2}$$

where the b parameter is given by:

$$b = N_0 \cdot (a - 1) \cdot c^{a-1} \tag{3}$$

and where N_0 is the number of fragmented clusters determined by the magnitude of the area under the curve of the *dN/dE* distribution. As usual, the a and c parameters were calculated by the least squares method (see, for instance, [27–30]). For this purpose we used the following normalised function:

$$f(t) = F(t)/N_0 \tag{4}$$

The non-exponential nature of the dependence, *F(t)*, clearly indicates the existence of clusters having different k rate constants for the fragmentation reaction. This difference is considered to be a result of the internal energy distribution of the sputtered clusters.

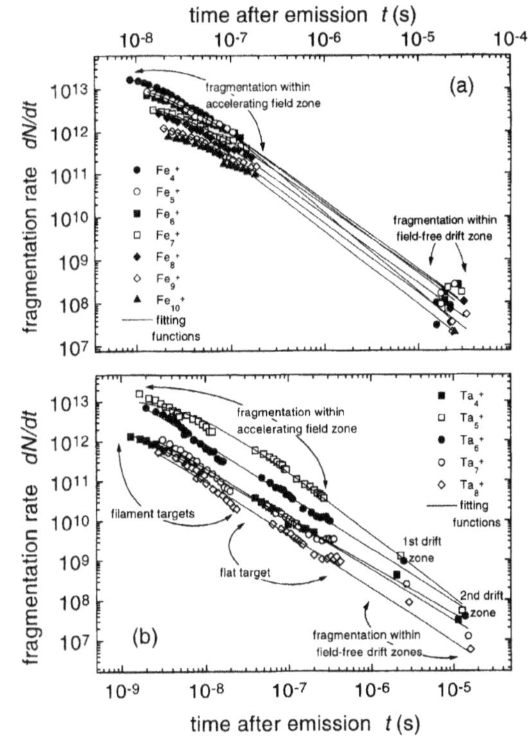

Fig. 3. Fragmentation rate dependency on the time following cluster emission for cluster ions sputtered by 8.5 keV Xe^+ ions: **a** Fe_n^+ cluster ions (the present work); **b** Ta_n^+ cluster ions (Ref. [29])

By introducing the $\varphi(k)$ distribution for the rate constants and taking into account the statistical character of the decay process of the clusters, one can consider the $f(t)$ function as a superposition of the exponential functions [28]:

$$\int_0^{k_{max}} k \cdot \varphi(k) \exp(-tk) dk \tag{5}$$

where $k_{max} = 10^{13}\,s^{-1}$ is the maximum rate of the decay reaction (this value is proportional to the characteristic frequency of atomic oscillations). It should be noted that, within the time range of the experiment, the $t \cdot k$ product is $\cong 10^4 - 10^9$ and the function under the integral (5) approaches zero at $k = k_{max}$. In this case the upper limit in integral (5) can be substituted by infinity and equation (5) can be solved analytically.

The procedure described has allowed determination of the $\varphi(k)$ rate constants distributions:

$$\varphi(k) = c^{a-1} \cdot \exp(-ck) \cdot k^{a-2} / \Gamma(a - 1) \tag{6}$$

Table 1. τ average lifetimes, E_d dissociation energies (activation energies of $Me_n^+ \to Me_{n-1}^+ + Me$ fragmentation reaction) and $\bar{E}_{int}/n$ average internal energy per constituent atom determined for Ta_n^+, Nb_n^+ and Fe_n^+ clusters sputtered by Xe^+ ions

n	Ta_n^+ [29]			Nb_n^+ [30]			Fe_n^+		
	$\bar{\tau}$, s	E_d, eV	$\bar{E}_{int}/n$, eV/at	$\bar{\tau}$, s	E_d, eV	$\bar{E}_{int}/n$, eV/at	$\bar{\tau}$, s	E_d, eV	$\bar{E}_{int}/n$, eV/at
3					3.96				
4	$1.93 \cdot 10^{-9}$	6.60	1.8		2.83[a]		$1.75 \cdot 10^{-8}$	1.09	0.568
5	$1.07 \cdot 10^{-8}$	7.02	1.8	$8.68 \cdot 10^{-8}$	5.69	1.104	$2.34 \cdot 10^{-8}$	2.43	0.673
6	$0.31 \cdot 10^{-9}$	5.52	1.6	$6.18 \cdot 10^{-8}$	6.06	1.122	$3.04 \cdot 10^{-8}$	2.23	0.902
7	$0.36 \cdot 10^{-9}$	8.34	2.3	$1.71 \cdot 10^{-8}$	7.5	1.32	$4.08 \cdot 10^{-8}$	2.45	0.831
8	$0.91 \cdot 10^{-9}$			$1.29 \cdot 10^{-8}$	5.84	1.512	$3.06 \cdot 10^{-8}$	2.29	0.711
9				$2.04 \cdot 10^{-8}$	5.87	1.573	$4.65 \cdot 10^{-8}$	2.45	0.73
10				$3.69 \cdot 10^{-8}$	6.88	1.722	$5.23 \cdot 10^{-8}$	2.82	0.784
11								2.96	
12								3.2	

[a] Activation energy for $Nb_4^+ \to Nb_2^+ + Nb_2$ fragmentation.

as well as the $\Phi(\tau)$ lifetime distribution:

$$\Phi(\tau) = c^{a-1} \cdot \exp(-c/\tau) \cdot \tau^{-a}/\Gamma(a-1) \quad (7)$$

where $\tau = 1/k$ is the lifetime of the sputtered cluster. The average magnitude of the decay rate constant and the lifetime of secondary cluster ions can be determined from the distributions (6) and (7), respectively. Thus, experimental studies of the cluster fragmentation processes over a wide time range allow $\varphi(k)$ and $\Phi(\tau)$ distributions to be determined without the need for any *a priori* theoretical consideration of the process. The average lifetimes $\bar{\tau}$ of Fe_n^+ clusters are shown in Table 1 together with the corresponding values for Nb_n^+ [30] and Ta_n^+ [27–29] sputtered clusters.

The $\varphi(k)$ distribution provides information about the internal energy distribution of sputtered clusters. As shown in Ref. [28], exact knowledge of the k rate constant dependence on the $\varepsilon = E_{int}/E_d$ relative internal energy (where E_{int} is internal energy and E_d is dissociation energy) can allow the relative internal energy distribution $\Psi(\varepsilon)$ to be established. In order to estimate $\Psi(\varepsilon)$ we used the quantum version of the Rice-Ramsperger-Kassel (RRK) theory [32]. The quantum approximation of the RRK theory provides the dependence of the k rate constant on the energy of the $E_{int} = i \cdot h\nu$ vibrational excitation, taking into account the $E_d = j \cdot h\nu$ dissociation energy for the given reaction channel:

$$k = k_0 \cdot \frac{i! \cdot (i-j+s-1)!}{(i-j)! \cdot (i+s-1)!} \quad (8)$$

where $s = 3n-6$ is the number of vibrational degrees of freedom of the cluster, n is the number of atoms in it, $k_0 = 10^{13}\,s^{-1}$, $h\nu$ is a quantum of energy, ν is the average vibration frequency of the excited cluster ($\nu \cong 10^{13}\,s^{-1}$). Using Eqs. (6) and (8), one can obtain analytically the distribution of the clusters over the relative internal energies $\varepsilon = E_{int}/E_d = i/j$:

$$\Psi(\varepsilon) = \frac{(ck_0)^{a-1} j^2 (s-1)}{\Gamma(a-1)(i-j+1)(i+s)} \times \left[\frac{i!(i-j+s-1)!}{(i-j)!(i+s-1)!}\right]^{a-1} \times \exp\left[-ck_0 \frac{i!(i-j+s-1)!}{(i-j)!(i+s-1)!}\right] \quad (9)$$

For calculating the $\Psi(E_{int})$ internal energy distribution, values of the dissociation energies must be determined. The information about E_d can be obtained from the *kinetic energy release* (KER) distribution of the fragmenting clusters [33]. In the fragmentation process, this distribution characterises the part of the cluster internal energy that is expended in translational movement of fragments. The measurements of kinetic energy distributions of *fragment* cluster ions formed in the field-free zone of a mass-spectrometer allow determination of KER-distributions for the fragmenting *sputtered* clusters. In order to transform kinetic energy distributions of fragments corresponding to the laboratory reference system into KER-distributions of fragmenting clusters corresponding to the cluster centre-of-mass system, we used a data processing procedure similar to that used earlier by

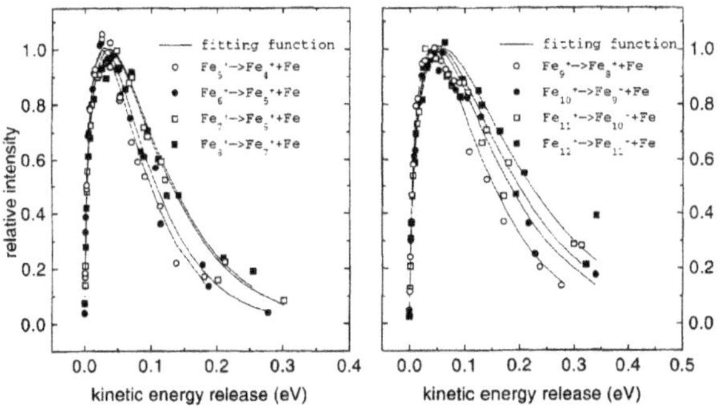

Fig. 4. Kinetic energy release distributions for the unimolecular fragmentation of Fe_n^+ cluster ions ($5 \leq n \leq 12$) within field-free zone (L_2) of the mass spectrometer

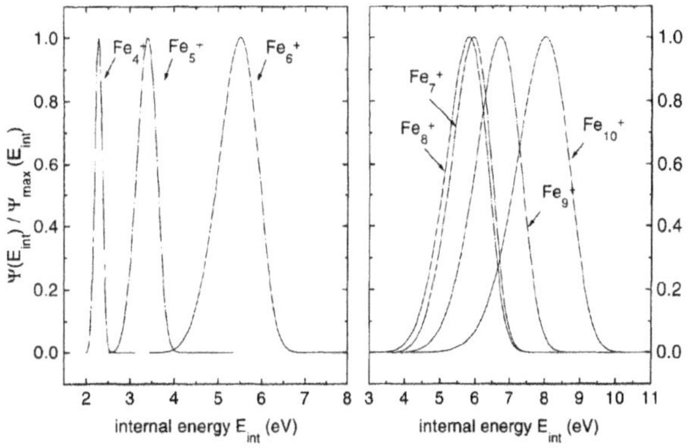

Fig. 5. Internal energy distributions of Fe_n^+ cluster ions ($4 \leq n \leq 10$) sputtered by 8.5 keV Xe^+ primary ions

various authors [34, 35] (for details, see Ref. [33]). KER-distributions of Fe_n^+ clusters are shown in Fig. 4. The dissociation energy (the activation energy for a given cluster decay channel) can be determined within the framework of a so-called "evaporation ensemble" conception [36–38]. Then the KER-distribution of fragment ions is described by the following expression.

$$dN/dE \propto E^l \exp(-E/k_B T^{\#}) \tag{10}$$

where E is the kinetic energy of the fragment, k_B is Boltzman's constant, $T^{\#}$ is the temperature of the excited cluster (this temperature corresponds to the excitation energy per degree of freedom). As before, we obtained l and $T^{\#}$ parameters using the least squares method for approximating the experimental KER-distributions by expression (10). The relationship between the dissociation energy E_d and the temperature $T^{\#}$ is described in terms of the "evaporative ensemble" conception by the following expression:

$$E_d = \frac{\gamma(n) \cdot k_B T^{\#}}{1 - \gamma(n)/2C(n) + [\gamma(n)/C(n)]^2/12 + \ldots} \tag{11}$$

where $\gamma(n)$ is the Gspann's parameter dependent on the cluster size n and $C(n)$ is the heat capacity expressed in units of k_B. In our case $\gamma(n) \approx 22.6 \pm 0.4$.

The E_d dissociation energies calculated for Fe_n^+ as well as for Nb_n^+ [30] and Ta_n^+ [29] sputtered clusters are shown in Table 1. The reliability of the procedure was first tested on sputtered Nb_n^+ clusters ($3 \leq n \leq 10$) [30]. Generally, values of the dissociation energy of the sputtered Nb_n^+ clusters were in agreement

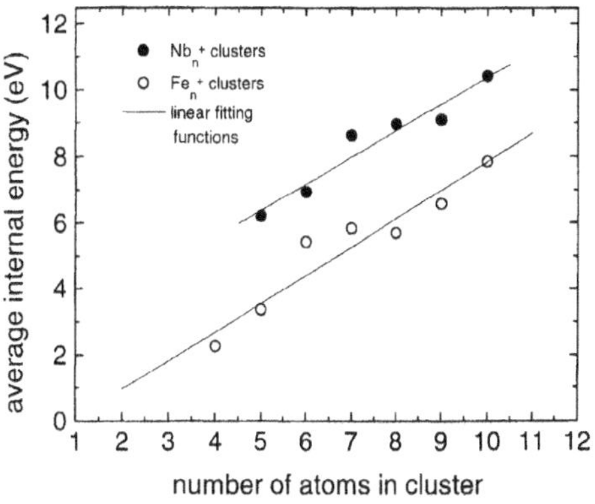

Fig. 6. Dependency of the average internal energy of Fe_n^+ and Nb_n^+ [30] sputtered clusters on the number of atoms in the cluster

($\cong 10\%$) with the data obtained in the experiments on collision-induced cluster dissociation [39, 40].

Internal energy distributions of Fe_n^+ clusters ($4 \leq n \leq 10$) are shown in Fig. 5. The $\bar{E}_{int}/n$ average internal energies (per constituent atom) calculated from the internal energy distribution are shown in Table 1 together with the corresponding values for Nb_n^+ [30] and Ta_n^+ [29] clusters. In Fig. 6, the dependency of $\bar{E}_{int}$ on the number of atoms in the cluster is shown for Nb_n^+ [30] and Fe_n^+ sputtered clusters. As seen from the plots and Table 1, the dependency is a linear one and the $\bar{E}_{int}/n$ values depend weakly on cluster size, being approximately equal to 1.88 eV/at, 1.39 eV/at and 0.75 eV/at for Ta, Nb and Fe, respectively. A similar result was obtained by the MD simulation of the sputtering processes [23, 25]. The data in Ref. [23, 25] were used on nascent clusters of Ag_n ($n \leq 31$) sputtered by keV and sub-keV

Ar^+ ions. In the simulations, these neutral clusters were considered immediately after the moment of emission. The $\bar{E}_{int}/n$ values for Ag_n clusters were found to be 1.5 eV/at, approximately. In contrast to the MD/MC simulation which demonstrates the branching or chain decay characterised by a decomposition time of 10^{-11}–$5 \cdot 10^{-11}$ s, it has been shown [29, 30] that the possibility of rapid chain decay is very questionable for Ta_n^+ and Nb_n^+ charged sputtered clusters because their dissociation energies are approximately twice as large as those of Ag clusters [25]. In this case, the internal energy of small Ta_n^+ and Nb_n^+ nascent clusters may be insufficient for a further chain of decay if both the ionisation act and the first decay event have taken place already. Such a time evolution of Ta_n and Nb_n clusters is considerably different from that described earlier [23, 25] but the dependence of $\bar{E}_{int}$ on cluster size is in qualitative agreement with the data of Ref. [23, 25]. The higher $\bar{E}_{int}/n$ average internal energies of Ta_n^+ cluster ions compared with Nb_n^+ clusters may be explained by the differences in time range of the fragmentation studies [15, 28, 29] and the influence of the time range on the experimental results of the sputtered cluster fragmentation study will be considered in detail in a following paper.

In contrast to the results obtained for Ta_n^+ and Nb_n^+ cluster ions, those for Fe_n^+ clusters demonstrate a very good agreement with the results of the MD/MC simulation in Ref. [25]. This is because the dissociation energies of Fe_n^+ cluster ions (see Table 1) are of the same order as those for Ag_n clusters. The similarity allows us to suppose that the time evolution of the nascent Fe_n clusters can be accurately simulated using the MD/MC-CEM interaction potential. In this connection, the agreement between experimental values of $\bar{E}_{int}/n$ for Fe_n^+ cluster ions and those predicted theoretically in Ref. [25] for "final" Ag_n neutral clusters should be noted. For the Fe_n^+ cluster ions, comparison between our experimental results and the MD/MC simulations allow us to suggest the possibility of chain decay steps in cluster time evolution.

In spite of the experimental limitations of $5 \cdot 10^{-10}$–10^{-9} s for the achievable time range of direct observations of the sputtered cluster fragmentation, the methods to study this stochastic process developed in Refs. [29, 30] and used in the present work can be extrapolated to shorter fragmentation times. The procedure is to consider the sputtered cluster time evolution as well as the cluster internal energy filtering taking place in the instruments. This approach will be presented in our next paper referred to above.

Conclusion

At the moment, it seems possible that there is some agreement regarding the process of sputtered cluster time evolution. This includes the emission of vibrationally excited nascent clusters that are transformed into experimentally detectable final cluster ions in the processes of ionisation and fragmentation. This is why modern theoretical and experimental approaches to study this process demonstrate a good agreement when results are compared. In summary, future developments in the study of cluster emission during sputtering must include a close interaction between experimental work on cluster fragmentation and theoretical MD/MC simulation of cluster emission and time evolution. One can expect that this will result in a realistic physical explanation of the phenomena.

Acknowledgements. The authors would like to acknowledge the support from INTAS (Project 96-0470). Two of the authors would also like to thank Belgian organisations IUAP and DWTC (I. V. Veryovkin) and FWO (A. Adriaens) for providing an opportunity to conduct this study.

References

[1] W. Ens, R. Beavis, K. G. Standing, *Phys. Rev. Lett.* **1983**, *50*, 27.
[2] N. Kh. Dzhemilev, S. V. Verkhoturov, *Izv. Akad. Nauk SSSR Ser. Fiz.* **1985**, *49*, 1831.
[3] I. Katakuze, T. Ichihara, Y. Fujita, T. Matsuo, T. Sakurai, H. Matsuda, *Int. J. Mass Spectrom. Ion Proc.* **1985**, *67*, 229.
[4] W. Begemann, K. H. Meiwes-Broer, H. O. Lutz, *Phys. Rev. Lett.* **1985**, *56*, 2248.
[5] N. Kh. Dzhemilev, S. V. Verkhoturov, U. Kh. Rasulev, *Nucl. Instr. Meth.* **1987**, *B29*, 531.
[6] N. Kh. Dzhemilev, S. V. Verkhoturov, I. V. Veriovkin, *Nucl. Instr. Meth.* **1990**, *B51*, 219.
[7] I. Katakuze, H. Ito, T. Ichihara, *Int. J. Mass Spectrum. Ion Proc.* **1990**, *97*, 47.
[8] W. Begemann, K. H. Meiwes-Broer, H. O. Lutz, *J. Phys. (Paris)* **1989**, *50*, 133.
[9] H. Gnaser, W. O. Hofer, *Appl. Phys.* **1989**, *A48*, 261.
[10] H. Oechner, *Int. J. Mass Spectrom. Ion Proc.* **1990**, *103*, 31.
[11] B. N. Makarenko, A. B. Popov, A. A. Shaporenko, A. P. Shergin, *Rad. Eff. Def. Solids* **1991**, *116*, 15.
[12] S. R. Coon, W. F. Calaway, J. W. Burnett, M. J. Pellin, D. M. Gruen, D. R. Spiegel, J. M. White, *Surf. Sci.* **1991**, *259*, 275.
[13] S. R. Coon, W. F. Calaway, M. J. Pellin, G. A. Curlec, J. M. White, *Nucl. Instr. Meth.* **1993**, *B82*, 329.

[14] A. Wucher, M. Wahl, H. Oechner, *Nucl. Instr. Meth.* **1993**, *B82*, 337.

[15] N. Kh. Dzhemilev, A. M. Goldenberg, I. V. Veriovkin, S. V. Verkhoturov, *Int. J. Mass Spectrom. Ion Proc.* **1991**, *107*, R19.

[16] A. Wucher, M. Wahl, in: *Secondary Ion Mass Spectrometry SIMS XI* (A. Bennighoven, H. W. Werner, B. Hegenhoff, eds.) Wiley, Chichester, 1997, p. 65.

[17] A. Wucher, M. Wahl, *Nucl. Instr. Meth.* **1996**, *B115*, 581.

[18] H. M. Urbassek, W. O. Hofer, *K. Danske Vidensk Selsk. Mat. Fys. Medd.* **1993**, *43*, 97.

[19] W. O. Hofer, in: *Sputtering by Particle Bombardment, Vol. 3* (R. Behrish, K. Wittmaack, eds.) Springer, Berlin Heidelberg New York Tokyo, 1991.

[20] H. M. Urbassek, *Rad. Eff. Def. Solids* **1989**, *109*, 293.

[21] I. S. Bitensky, E. S. Parilis, *Nucl. Instr. Meth.* **1987**, *B21*, 26.

[22] B. V. King, I. S. T. Tsong, S. H. Lin, *Int. J. Mass Spectrom. Ion Proc.* **1987**, *78*, 341.

[23] A. Wucher, B. J. Garrison, *Phys. Rev.* **1992**, *B46*, 4885.

[24] A. Wucher, *Nucl. Instr. Meth.* **1993**, *B83*, 79.

[25] A. Wucher, B. J. Garrison, *J. Chem. Phys.* **1996**, *105*(14), 5999.

[26] C. L. Kelchner, D. M. Halstead, L. S. Perkins, N. M. Wallace, A. E. DePristo, *Surf. Sci.* **1994**, *310*, 425.

[27] N. Kh. Dzhemilev, I. V. Veriovkin, S. V. Verkhoturov, A. M. Goldenberg, *Izv. Akad. Nauk SSSR Ser. Fiz.* **1991**, *55*, 2418.

[28] N. Kh. Dzhemilev, A. M. Goldenberg, I. V. Veriovkin, S. V. Verkhoturov, *Int. J. Mass Spectrom. Ion Proc.* **1995**, *141*, 209.

[29] N. Kh. Dzhemilev, A. M. Goldenberg, I. V. Veriovkin, S. V. Verkhoturov. *Nucl. Instr. Meth.* **1996**, *B114*, 245.

[30] I. V. Veryovkin, S. V. Verkhoturov, A. M. Goldenberg, N. Kh. Dzhemilev, *Izv. Akad, Nauk Ser. Fiz.*, **1994**, *58(10)*, 63 and *Bull. Russ. Acad. Sci.–Phys.* **1994**, *58(10)*, 1635.

[31] A. D. Bekkerman, N. Kh. Dzhemilev, V. M. Rotstein, *Surf. Interf. Anal.* **1990**, *15*, 587.

[32] P. J. Robinson, K. A. Holbrook, *Unimolecular Reactions*, Wiley, London, 1972.

[33] I. V. Veryovkin, S. V. Verkhoturov, A. M. Goldeberg and N. Kh. Dzhemilev, *Izv. Akad. Nauk Ser. Fiz.* **1994**, *58(3)*, 56 and *Bull. Russ. Acad. Sci.–Phys.* **1994**, *58(3)*, 416.

[34] M. F. Jarrold, A. J. Illies, N. J. Kirchner, W. Wagner-Redeker, M. T. Bowers, M. L. Mandich, J. L. Beauchamp, *J. Phys. Chem.* **1983**, *87*, 2213.

[35] P. P. Radi, M. E. Rincon, M. T. Hsu, J. Brodbelt-Lustig, P. Kemper, M. T. Bowers, *J. Phys. Chem.* **1989**, *93*, 6187.

[36] C. E. Klots, *Int. J. Mass Spectrom. Ion Proc.* **1990**, *100*, 457.

[37] C. E. Klots, *Z. Phys.* **1991**, *D20*, 105.

[38] C. E. Klots, *Z. Phys.* **1991**, *D21*, 335.

[39] D. A. Hales, L. Lian, P. B. Armentrout, *Int. J. Mass Spectrom. Ion Proc.* **1990**, *102*, 269.

[40] S. K. Loh. D. A. Hales, Li Lian, P. B. Atmentrout, *J. Chem. Phys.* **1989**, *90(10)*, 5466.

Mikrochim. Acta [Suppl.] 15, 379–385 (1998)

Sputtering of Tantalum by Atomic and Molecular Gold Ions: Comparative Study of Yields and Kinetic Energy Distributions of Atomic and Cluster Ions

S. F. Belykh, U. Kh. Rasulev, A. V. Samartsev, S. V. Verkhoturov, and **I. V. Veryovkin***

Arifov Institute of Electronic, Academy of Science of Uzbekistan, 700143 Tashkent, Uzbekistan

Abstract. Relative secondary ion yields and kinetic energy spectra of Ta_n^+ ions ($n \leq 12$) sputtered from a tantalum target by Au^- atomic ions with energies of 6 keV and 12 keV as well as by 12 keV Au_2^- molecular ions have been measured under identical experimental conditions. Under molecular ion bombardment, an anomalous high non-additive sputtering yield was observed for large metal clusters. In this case, for instance, the yield of Ta_{10}^+ clusters was determined to be approximately 320 times higher than that under corresponding atomic ion bombardment (6 keV per constituent atom of the projectile). Comparison between mass spectra and kinetic energy spectra determined under molecular ion bombardment with those for atomic ion bombardment showed both qualitative and quantitative differences. Taking into account the experimentally observed regularities, a hypothesis regarding the existence of two different cluster ion emission mechanisms for small clusters ($n < 5$) and for larger ones ($n > 5$) has been proposed.

Key words: Cluster bombardment, cluster sputtering, cluster mass distributions, cluster kinetic energy spectra, unimolecular fragmentation, non-additive effects in sputtering.

It is well known that the bombardment of solids with energetic ions leads to emission of a variety of secondary particles from the surface. The flux of sputtered particles may contain an abundant fraction of agglomerates of several atoms. Since their detection about four decades ago [1] these sputtered clusters have attracted much attention. Recent reviews of published works can be found in Refs. [2, 3]. During the last decade many new qualitative experimental results on the emission of cluster ions [4–11] and cluster neutrals [12–19] under atomic ion bombardment has been obtained. Nevertheless, the fundamental processes leading to the formation of these sputtered clusters are still not completely understood. A promising experimental approach to the problem is a comparative study of cluster emission under bombardment by different projectiles, especially molecular ones. The data gathered in such experiments allow the effective examination of existing theoretical models as well as encouraging the generation of new ones.

Bombardment by molecular ions results in non-additive effects that manifest themselves as a non-linear dependence of the sputtering and ion emission yields on the number of atoms in the impinging projectile. Molecular ion bombardment produces extreme sputtering conditions because of the high degree of collision cascade superposition. Simple estimations demonstrate that such conditions are possible only at a current density of primary atomic ions of $j_0 \cong 10^{12}$ A/cm^2, a value that is practically unachievable.

Sputtering of various samples under bombardment by molecular ions with energies in the range 10–100 keV has been mainly involved with studies of atomic secondary particles [20]. Recently, non-linear enhancement of the emission yield from different targets (organic, CsI and Au) induced by keV cluster ion impact has been observed [21]. The yield of

* To whom correspondence should be addressed

complex ions like AuI^- and AuI_2^- was found to increase more rapidly with the number of constituents in the projectile than with the yield of atomic ions. The conclusion drawn was that cluster projectiles produce a larger emission volume which itself enhances the probability of cluster emission compared to that of single atomic projectiles. Strong non-linear effects in emission of large size clusters from inorganic and organic solids bombarded by MeV polyatomic ions were reported in a recent paper [22]. The emission yields of $(CsI)_nI^-$ for $n > 10$ were observed to differ by several orders of magnitudes under bombardment of the CsI target by 0.5 MeV/atom Au_4^+ clusters in comparison with Au^+ single ion impact at the same energy per atom.

In our previous paper [23] the emission of Ta_n^+ ($n = 1 - 5$) ions from a Ta surface bombarded by Au^- and Au_2^- ions with initial energy 7 keV/atom has been investigated and non-linear effects in yields of charged clusters have been observed. These effects were found to be more pronounced for the larger emitted clusters, testifying that collective processes responsible for cluster emission were closely connected with the density of energy deposited near the solid surface. For the first time, the rapid increase of cluster ion yield ($n > 2$) with the number of atoms in the cluster was observed for Au_2^- molecular ion bombardment. It was shown that when sputtering with molecular projectiles the fragmentation of large clusters occurring far from the solid surface may contribute significantly to the enhancement, not only of particular cluster yields but the total yield of secondary ions.

In the present work, relative yields and kinetic energy spectra of Ta_n^+ ions ($n \leq 12$) sputtered from a tantalum target by Au^- atomic ions with energies of 6 keV and 12 keV as well as by 12 keV Au_2^- ions have been measured under identical experimental conditions. The features observed in mass spectra and in kinetic energy spectra are discussed in terms of non-additive effects in cluster ion sputtering under molecular ion bombardment.

Experimental

The secondary ion mass spectrometer used in the experiments is a modified commercial MI-1201 mass spectrometer produced in SELMI (Sumy, Ukraine). To study the cluster ion emission under molecular ion bombardment, this instrument was equipped with a Au_m^- negative ion source ($m = 1$–4) [24]. A diagram of the experimental apparatus is shown in Fig. 1 and its detailed

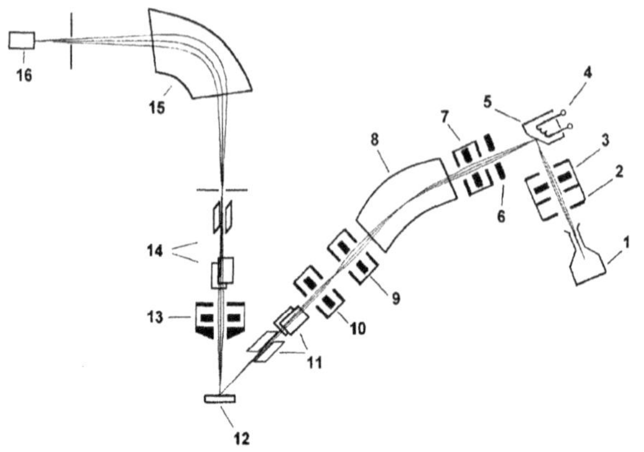

Fig. 1. Scheme of the experimental apparatus: *1* the surface ionisation Cs^+ ion source; *2* the accelerating electrode; *3* Einzel electrostatic lens; *4* the target heater; *5* the Au target (gold foil); *6* accelerating electrode; *7* Einzel lens; *8* the stigmatic magnetic mass separator; *9, 10* Einzel electostatic lenses; *11* primary ion deflection system; *12* the Ta target (tantalum foil); *13* secondary ion lens; *14* secondary ion deflecting system; *15* magnetic mass spectrometer; *16* electron multiplier

description can be found in Ref. [23]. Briefly, the apparatus consists of the negative ion source {1–5}, a mass separator for the primary ion beam {8}, primary ion optical system {6, 7, 9–11} for acceleration, formation and deflection of the primary ion beam, target assembly {12}, secondary ion optical system {13, 14} and the MI-1201 mass spectrometer {15, 16}.

The negative ion source operated as follows. The Cs^+ ions extracted from the surface ionisation ion source {1} were accelerated {2} and then focused {3} by the lens onto the target, which consisted of a good foil {5} and a heater {4} mounted on a drive mechanism. The Cs atoms cover the gold foil due to implantation and adsorption under steady state conditions of ion bombardment. Because of this, a relatively high yield of Au_m^- secondary ions was obtained. The Au_m^- ions were extracted and accelerated {6} and then were focused {7} by the lens and selected in masses by the stigmatic magnetic mass separator {8}. The ion beam impinged on the Ta target after passing two einzel electrostatic lenses {9, 10} and the deflection system {11}. The incident energy E_0 of atomic Au^- and dimer Au_2^- projectiles varied in the range 4–12.5 keV/atom.

Heating the Ta target up a temperature of 2300 °C for a long time was found to cause a significant decrease in the concentration of contaminants not only on the surface but also in the bulk of the solid. Then the surface was cleaned by sputtering using Au^-ion bombardment. In addition, the sample surface was cleaned by heating for a short time (5–10 min) to the above temperature prior to every acquisition of data. The Ta surface prepared in such a way was believed to be a "clean surface". The yield of TaO^+ molecules has been measured before and after the cleaning of the Ta surface. The yield was observed to drop by three orders of magnitude after the cleaning procedure. Since the probability of chemical reactions increases on a hot solid surface we believe that the drop in the TaO^+ yield indicates a corresponding decrease in the oxygen concentration on the target surface. The temperature of the Ta sample during the experiment was maintained at 2000 °C and under working conditions, the residual gases pressure did not exceed 8.10^{-8} Torr.

Kinetic energy spectra of secondary ions were measured by the variation of the U_{ac} accelerating voltage over the range ± 300V relative to $U_{ac} = 2000$ V. In order to determine the energy resolution of the SIMS instrument, kinetic energy spectra of alkaline metal ions (Na^+, K^+ and Cs^+) evaporated from the heated Ta surface were measured in the absence of the primary ion beam. The full width at half maximum (FWHM) determined for energy spectra of the thermal alkaline metal ions was of $\cong 5.5$ eV.

As mentioned above, the mass distributions of Ta_n^+ secondary ions were determined by the calculation of areas under the kinetic energy spectra curves. The advantage of such a procedure over a conventional mass spectra measurement is the ability to take into account not only the difference in energy spectra widths for the different secondary ions but also the cluster fragmentation processes taking place within the accelerating zone of the mass spectrometer between the target {12} and the lens {13}. In order to make a comparison between secondary cluster ion yields, the cluster ion intensities measured under bombardment by 12 keV Au^- and 12 keV Au_2^- ions were normalized according to the ratio of corresponding primary ion currents (1.14 nA and 6.81 nA, respectively) to 6 keV Au^- ion current (58.74 nA).

Results and Discussion

In the present work, kinetic energy spectra of Ta_n^+ secondary ions ($n \leq 10$) sputtered by 6 keV Au^- ions and Ta_n^+ ions ($n \leq 12$) sputtered by 12 keV Au^- ions and 12 keV Au_2^- ions have been measured. Typical energy spectra for $n = 1, 4, 7$ are shown in Fig. 2. In general, the energy spectra measured under atomic ion bombardment are in a good agreement with results of Ref. [9] for Ta_n^+ cluster ions sputtered by 8.5 keV Xe^+ atomic ions. For Ta_n^+ cluster ions with $n \geq 3$, the extended energy spectra "tails" in the energy range of $-300 - 0$ eV (relative to $eU_{ac} = 2000$ eV) correspond to $Ta_{n+1}^+ \rightarrow Ta_n^+ + Ta$ unimolecular fragmentation within the acceleration zone of the mass spectrometer (between the target {12} and the lens {13}). In addition extended symmetric peaks, corresponding to Ta_n^+ ions formed as a result of the fragmentation within the field-free zone of the mass spectrometer (between the lens {13} and the mass analyser {15}) have been observed for larger cluster ($n \geq 7$) (Fig. 2c) in the range of the "tails". These peaks are formed by the *kinetic energy release* in the fragmentation process [25]. The apparent absence of such fragment ion peaks in spectra "tails" of smaller clusters ($3 \leq n < 7$) is explained by the limited range of energy measurement (for instance, simple estimations predict for Ta_6^+ clusters the appearance of such a peak at -333 eV; which is outside the experimental energy range). It has been observed that the relative part of the fragment ions in the energy spectra increased with n as well as during a change of bombardment conditions from 6 keV $Au^- \rightarrow$ 12 keV $Au^- \rightarrow$ 12 keV Au_2^-. The higher intensity of the fragment ions was observed under molecular Au_2^- ion bombardment.

In order to make a detailed comparison between the kinetic energy spectra, their full widths at half maximum (FWHM) have been determined and the high energy ranges (> 10 eV) have been fitted by a $f(E) = \beta \cdot E^{\alpha}$ power function. The parameters of the kinetic energy spectra (FWHM and α exponent) are shown in Table 1 and their dependencies on the number of atoms in the sputtered cluster are plotted

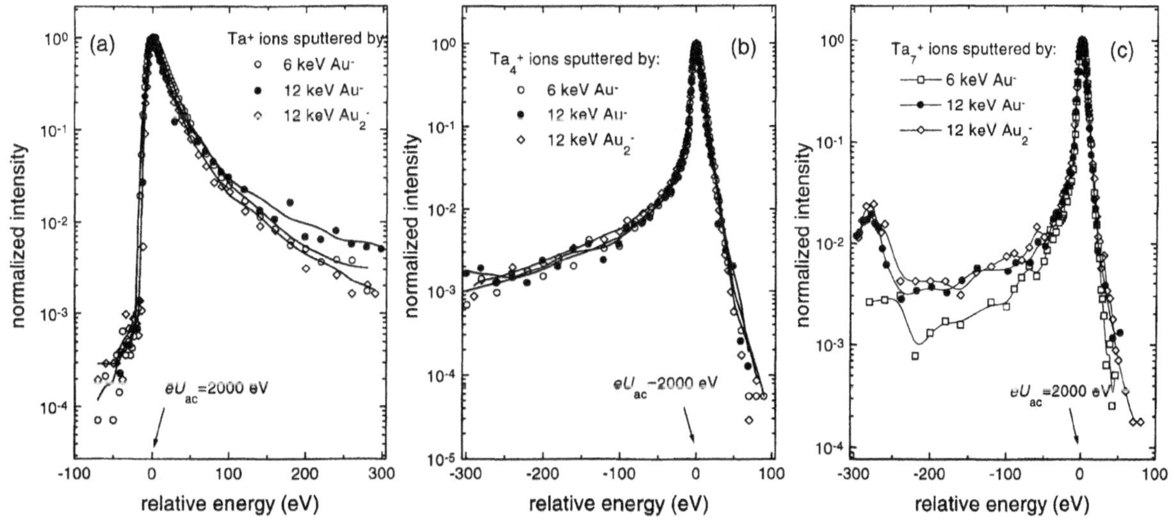

Fig. 2. Kinetic energy spectra of Ta^+ (**a**), Ta_4^+ (**b**), and Ta_7^+ (**c**) secondary ions sputtered by 6 keV Au^-, 12 keV Au^- and 12 keV Au_2^- primary ions

Table 1. Kinetic energy spectra parameters for Ta_n^+ cluster ions sputtered by the different primary ions: 6 keV Au^-, 12 keV Au^- and 12 keV Au_2^-

	6 keV $Au^- \rightarrow Ta$		12 keV $Au_2^- \rightarrow Ta$		12 keV $Au^- \rightarrow Ta$	
n number of atoms in Ta_n^+	α exponent of power function	FWHM (eV)	α exponent of power function	FWHM (eV)	α exponent of power function	FWHM (eV)
1	-2.38 ± 0.05	24.85 ± 1.28	-2.33 ± 0.13	19.88 ± 0.98	-1.94 ± 0.06	20.12 ± 1.11
2	-4.20 ± 0.26	14.78 ± 1.1	-3.89 ± 0.16	13.23 ± 0.84	-3.56 ± 0.33	12.88 ± 1.04
3	-4.29 ± 0.07	12.16 ± 0.6	-3.87 ± 0.11	11.40 ± 0.49	-4.05 ± 0.41	10.55 ± 0.6
4	-5.17 ± 0.42	11.75 ± 0.35	-5.99 ± 0.17	11.40 ± 0.32	-4.83 ± 0.28	10.81 ± 0.36
5	-4.21 ± 0.15	10.95 ± 0.25	-5.63 ± 0.42	10.87 ± 0.26	-4.84 ± 0.34	10.38 ± 0.33
6	-4.58 ± 0.31	10.84 ± 0.2	-4.79 ± 0.31	11.11 ± 0.48	-3.37 ± 0.09	9.61 ± 0.47
7	-4.30 ± 0.67	10.77 ± 0.11	-3.69 ± 0.21	11.33 ± 0.4	-3.36 ± 0.14	10.53 ± 0.4
8	-3.08 ± 0.18	11.51 ± 0.46	-3.67 ± 0.24	11.39 ± 0.37	-2.69 ± 0.19	10.59 ± 0.4
9	-3.35 ± 0.52	10.65 ± 0.44	-3.09 ± 0.10	11.64 ± 0.32	-2.45 ± 0.17	10.93 ± 0.49
10	-3.44 ± 0.34	11.14 ± 0.45	-2.87 ± 0.31	11.66 ± 0.38	-2.41 ± 0.11	10.19 ± 0.56
11			-3.71 ± 0.44	11.48 ± 0.41	-2.57 ± 0.04	10.62 ± 0.64
12			-4.17 ± 0.14	10.97 ± 0.52	-2.53 ± 0.07	11.54 ± 1.13

in Fig. 3. As seen Fig. 3, in the range of $n \geq 5$ the corresponding $\alpha(n)$ and FWHM (n) dependencies are very similar for the bombardment by 6 keV Au^- and 12 keV Au_2^- ions but those for 12 keV Au^- ion bombardment differ. In contrast, in the range of $n<5$ the above dependency trends are similar for bombardment by Au^- and Au_2^- ions with an energy of 12 keV and different from those at 6 keV Au^- ion bombardment. In our opinion, special emphasis should be placed on the range around $n = 5$ as it corresponds to the minimum in the $\alpha(n)$ dependency as well as determining a minimal cluster size since the FWHM values becomes practically constant and independent on the number of atoms when $n>5$. Similar behaviour of the α exponents was observed in Ref. [18] for kinetic energy spectra of neutral Ag_n clusters: $\alpha(2) = -2.9, \alpha(3) = -4.4, \alpha(4) = -4.4, \alpha(n > 4) \approx -4$ and the authors considered this fact confirming the inaccurate description of the large cluster formation in the framework of the *atomic combination model* [26] that predicted a monotonic decrease of the exponent as $\alpha(n) = -2.5n + 0.5$.

The kinetic energy spectra regularities observed in the present work for Ta_n^+ clusters allow us to distinguish mechanisms of the small cluster formation($n < 5$) from those for larger cluster. It seems probable that the emission of small clusters depends mainly on the energy characteristics of the impinging

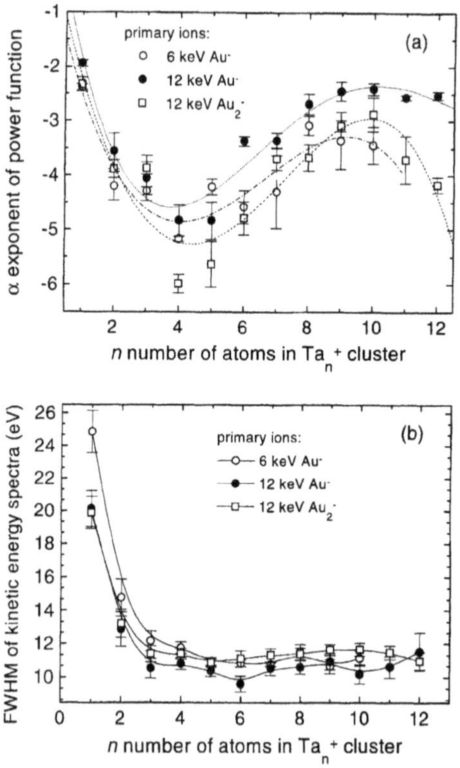

Fig. 3. Dependencies of kinetic energy spectra parameters on the n number of atoms in Ta_n^+ secondary ions sputtered by 6 keV Au^-, 12 keV Au^- and 12 keV Au_2^- primary ions: **a** α exponents of power functions describing the asymptotic behaviour of kinetic energy spectra in the high energy range (>10 eV); **b** full widths at half maximum (FWHM) of kinetic energy spectra

projectile (i.e. on the absolute value of the energy dissipated in the collision cascade area) and does not depend (or depends weakly) on the trajectories in solids (i.e. on the geometric from and on the location of the area of high energy density). In contrast to small clusters, the emission of the larger ones seems to be dependent on the latter factor mainly. The influence of the form and the location of the "hot" area in the parameters of kinetic energy spectra of large sputtered clusters could be demonstrated by comparing Fig. 3a and Fig. 3b where large clusters sputtered by 12 keV Au^- primary ions demonstrate lower FWHM values but higher values of the α exponent (i.e. intensities of high energy clusters decrease more slowly). For the 12 keV Au^- ion bombardment, simple estimations in the framework of the collision cascade model predict a deeper location of the "hot" area centre compared to that for the 6 keV/atom projectiles. This means that clusters should be ejected from this area into a sharper solid angle which, in turn, results in a larger weighting of high energy clusters in the energy spectra as well as in lower FWHM values.

To make an all-round comparison between the results obtained, dependencies of the Ta_n^+ cluster ion yield on the number of atoms in the cluster have been determined from the kinetic energy spectra. As mentioned above, this method of calculating the mass distribution allowed us to determine these dependencies for two registration times: of $\cong 10^{-6}$ s (without considering short fragmentation processes within the accelerating zone of the mass spectrometer) and of $\cong 10^{-8}$ s (taking them into account). In the latter case, the mass distributions were determined by an addition of the area under the fragmentation "tail" of the corresponding energy spectra of the $(n+1)$-atomic cluster and the area under the main energy spectra peak of a corresponding n-atomic cluster. Mass distributions of Ta_n^+ clusters determined are shown in Fig. 4. Two ranges of mass spectra (cluster sizes) can be considered separately: the $n \leq 5$ range and the $n \geq 5$ range. The latter one can be fitted very well by the $Y(n) \propto n^{-\delta}$ power function. The values of the δ exponent for the fitting function are given in Table 2. It is seen from Fig. 4 and Table 2 that for the larger cluster range ($n \geq 5$) the fragmentation processes change the observed mass spectra significantly. A comparison of the δ parameter obtained for 6 keV Au^- projectiles with that obtained in Ref. [19] for neutral Ta_n sputtered clusters under similar bombardment

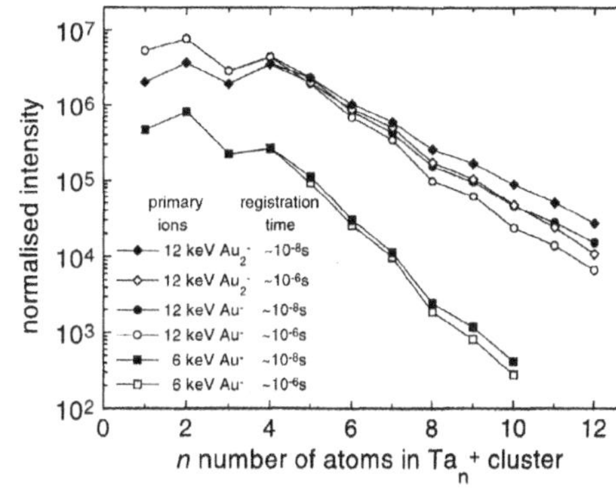

Fig. 4. Mass spectra of Ta_n^+ secondary ions sputtered by 6 keV Au^-, 12 keV Au^- and 12 keV Au_2^- primary ions. Mass spectra with the registration times of $\cong 10^{-8}$ s were determined from kinetic energy spectra taking into account the cluster ion fragmentation within the accelerating zone of the mass spectrometer. Mass spectra with the registration time of $\cong 10^{-6}$ s correspond to the areas of the main peaks of kinetic energy spectra

conditions (5 keV Ar^+ ions) shows a very good agreement between the parameter: -8.46 for cluster ions and -8.5 for cluster neutrals. It should be noted that the mass spectra obtained in Ref. [19] corresponded also to registration times of $\cong 10^{-6}$ s and the faster cluster fragmentation processes have not been taken into account. (As known to us from private communications with authors of Ref. [19], the experimental procedure will be developed and be applied to mass distributions of cluster neutrals soon to make possible a more detailed comparison between data for ions and neutrals.). The almost equal value of δ exponents for secondary neutrals and ions can be explained by the fact that within this range the sputtered cluster ionisation cross-section becomes practically independent on the cluster size [19]. In our opinion the rather good reproducibility of the results obtained using different experimental instruments confirms the applicability of the mass spectra claculation procedure described above.

Our results demonstrate the highest normalised yield of larger clusters under diatomic ion bombardment. The bombardment by atomic and diatomic primary ions produced both qualitative and quantitative differences in the corresponding mass distribution. In spite of the above similarity between kinetic energy spectra, mass distribution variations were observed to be different. In contrast to the results for the neutral Ta_n clusters, the cluster ion yield dependence is non-monotonic for $n \leq 5$. This can be explained by the complex influence of the size-

Table 2. The δ exponent of the power fitting function for the $n \geq 5$ range of mass spectra and the γ factor of the exponential function characterising the increase of normalised secondary ion yields under bombardment by 12 keV Au^- and 12 keV Au_2^- primary ions; data cross compared and compared also to 6 keV Au^- ions

	δ exponent of the power function $y(n) \propto n^{-\delta}$			γ factor of the exponential function "ratio" $\exp(\gamma \cdot n)$		
Mass spectra registration time (s)	6 keV $Au^- \rightarrow Ta$	12 keV $Au_2^- \rightarrow Ta$	12 keV $Au^- \rightarrow Ta$	$\frac{Y(Ta_n^+\|12\,keV Au_2^-)}{Y(Ta_n^+\|6\,keV Au^-)}$	$\frac{Y(Ta_n^+\|12\,keV Au^-)}{Y(Ta_n^+\|6\,keV Au^-)}$	$\frac{Y(Ta_n^+\|12\,keV Au_2^-)}{Y(Ta_n^+\|12\,keV Au^-)}$
$\cong 10^{-6}$	-8.46 ± 0.43	-5.93 ± 0.3	-6.5 ± 0.22	0.439 ± 0.014	0.254 ± 0.016	0.185 ± 0.014
$\cong 10^{-8}$	-8.10 ± 0.33	-5.02 ± 0.18	-5.72 ± 0.12	0.462 ± 0.014	0.278 ± 0.020	0.184 ± 0.012

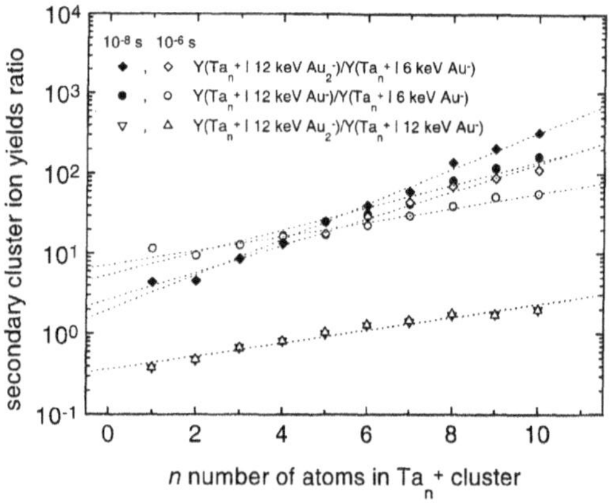

Fig. 5. Ratios characterised the enhancement of Ta_n^+ secondary ion emission under bombardment by 12 keV Au^- and 12 keV Au_2^- primary ions in comparison to bombardment by 6 keV Au^- ions

dependent cluster ionisation probability [19] as well as the cluster intensity redistribution because of fast fragmentation processes taking place outside of the experimental range (limited by times of $\cong 10^{-8}$ s – in this connection see Ref. [25]). In the range of $n \leq 5$ the following common relationships between ion yields were observed: (1) for *atomic ion bombardment:* $Y(Ta^+) > Y(Ta_3^+) > Y(Ta_5^+)$; $Y(Ta_2^+) > Y(Ta_4^+)$ and the highest yield corresponded to Ta_2^+ cluster ions; (2) for *diatomic ion bombardment:* $Y(Ta^+) < Y(Ta_3^+) < Y(Ta_5^+)$; $Y(Ta_2^+) \approx Y(Ta_4^+)$ and the highest yield corresponded to Ta_4^+ cluster ions. This means that the molecular ion bombardment influences significantly the enhancement of sputtered cluster yields, resulting in qualitative changes in mass distribution.

To consider the above enhancement phenomena quantitatively, the following ion yield ratios have been determined from the mass spectra obtained for two registration times:

$$Y(Ta_n^+|12\,keV\,Au_2^-)/Y(Ta_n^+|6\,keV\,Au^-),$$
$$Y(Ta_n^+|12\,keV\,Au^-)/Y(Ta_n^+|6\,keV\,Au^-) \text{ and}$$
$$Y(Ta_n^+|12\,keV\,Au_2^-)/Y(Ta_n^+|12\,keV\,Au^-).$$

Figure 5 represents the dependency of the above secondary ion yield ratios on n, the number of atoms in the cluster. As seen, the ratios increase proportionally to the $\exp(\gamma n)$ function (dotted lines on the plot correspond to the exponential fitting functions) as n values increase from $n = 1$ to $n = 10$. The γ exponent factors determined for the ratios listed above are given in Table 2. It is interesting that the $Y(Ta_n^+|12\,keV\,Au_2^-)/Y(Ta_n^+|6\,keV\,Au^-)$ ratio is lower than the $Y(Ta_n^+|12\,keV\,Au^-)/Y(Ta_n^+|6\,keV\,Au^-)$ one for $n < 5$ but becomes higher when $n > 5$. In our opinion, a possible explanation of this experimental result is the competition between two different mechanisms of cluster formation during sputtering.

It should be noted that measurements for the 12 keV atomic ion bombardment (in addition to those for bombardment by 12 keV diatomic ions) have been conducted to make a difference between the influence of the projectile energy (6 keV or 12 keV, for instance) and the projectile properties (atomic or molecular one) in the emission of secondary cluster ions. As seen from the results, non-additive effects in cluster ion emission under molecular ion bombardment are qualitatively and quantitatively different from (additive) effects taking place under atomic ion bombardment of solids. In our opinion, the present work should be considered as one of the initial attempts to study the sputtered cluster formation comparatively for atomic and molecular primary ions. In order to understand the physics of the complex processes of cluster sputtering, it is necessary to continue comparative studies using larger molecular projectiles and higher energy atomic projectiles. Future studies of the process should also emphasise detailed consideration of the fragmentation "tails" of kinetic energy spectra because of the possibility of determining the internal energies of sputtered clusters [25].

Conclusion

Consideration of the results of the present work allows us to state the following:

- non-additive effects in cluster ion emission under molecular ion bombardment are qualitatively and quantitatively different from (additive) effects taking place under atomic ion bombardment; these effects manifest themselves in mass spectra as a dramatic increase of the relative yield of larger clusters ($n > 5$) as well as in the qualitatively different spectra behaviour in the smaller clusters range ($n < 5$);
- for identical bombardment energies (6 keV) per constituent atom of the projectile, the influence of the molecular ion bombardment on the *parameters* of kinetic energy spectra of Ta_n^+ sputtered clusters is not noticeable in the $n \geq 5$ range; in contrast, for the range of $n \leq 5$, kinetic energy spectra parameters are found to be similar for identical absolute energies (12 keV) of primary ions (atomic and molecular ones); the higher relative part of fragmented particles was observed in the kinetic energy spectra of sputtered clusters under the molecular ion bombardment;
- the hypothesis regarding the existence of two different cluster ion emission mechanisms for $n < 5$ and for $n > 5$ seems to be probable.

Acknowledgements. This study was partially supported by the following organisations: Fundamental Research Support Foundation of the Academy of Sciences (Grant 29-96, Uzbekistan), INTAS (Project 96-0470), and NATO (Science for Peace Program, Project SfP 97.1929). I. V. Veryovkin would like to thank IUAP (Belgium) and DWTC (Belgium) for their support of this work. The authors would like to thank I. S. Bitensky for helpful discussions.

References

[1] R. E. Honig, *J. Appl. Phys.* **1958**, *29*, 549.
[2] H. M. Urbassek, W. O. Hofer, K *Danske Vidensk. Selsk. Mat. Fys. Medd.* **1993**, *43*, 97.
[3] W. O. Hofer, in: *Sputtering by Particle Bombardment, Vol. 3* (R. Behrish, K. Wittmaack, eds.), Springer, Berlin Heidelberg New York Tokyo 1991.
[4] W. Ens, R. Beavis, K. G. Standing, *Phys. Rev. Lett.* **1983**, *50*, 27.
[5] N. Kh. Dzehemilev, S. V. Verkhoturov, U. Kh. Rasulev, *Poverkhnost. Fizika, Chimija, Mekhanika.* **1986**, **2**, 86.
[6] I. Katakuze, T. Ichihara, Y. Fujita, T. Matsuo, T. Sakurai, H. Matsuda, *Int. J. Mass Spectrom. Ion Proc.* **1985**, *67*, 229.
[7] W. Begemann, K. H. Meiwes-Broer, H. O. Lutz, *Phys. Rev. Lett.***1985**, *56*, 2248.
[8] N. Kh. Dzhemilev, U. Kh. Rasulev, S. V. Verkhoturov, *Nucl. Instr. Meth.* **1987**, *B29*, 531.
[9] N. Kh. Dzhemilev, S. V. Verkhoturov, I. V. Veriovkin, *Nucl. Instr. Meth.* **1990**, *B51*, 219
[10] I. Katakuze, H. Ito, T. Ichihara, *Int. J. Mass Spectrom. Ion Proc.* **1990**, *97*, 47.
[11] W. Begemann, K. H. Meiwes-Broer, H. O. Lutz. *J. Phys. (Paris)* **1989**, *50*, 133.
[12] H. Gnaser, W. O. Hofer, *Appl. Phys.* **1989**, *A48*, 261.
[13] H. Oechner, *Int. J. Mass Spctrom. Ion Proc.* **1990**, *103*, 31.
[14] B. N. Makarenko, A. B. Popov, A. A. Shaporenko, A. P. Shergin, *Rad. Eff. Def. Solids* **1991**, *116*, 15.
[15] S. R. Coon, W. F. Calaway, J. W. Burnett, M. J. Pellin, D. M. Gruen, D. R. Spiegel, J. M. White, *Surf. Sci.* **1991**, *259*, 275.
[16] S. R. Coon, W. F. Calaway, M. J. Pellin, G. A. Curlec, J. M. White, *Nucl. Instr. Meth.* **1993**, *B82*, 329.
[17] A. Wucher, M. Wahl, H. Oechner, *Nucl. Instr. Meth.* **1993**, *B82*, 337.
[18] A. Wucher, M. Wahl, in: *Secondary Ion Mass Spectrometry SIMS X* (A. Bennighoven, H. W. Werner, B. Hegenhoff, eds.), Wiley, Chichester, 1997, p. 65.
[19] A. Wucher, M. Wahl, *Nucl. Instr. Meth.* **1996**, *B115*, 581.
[20] D. A. Thompson, *Rad. Eff* **1981**, *56*, 105.
[21] M. Benguerba, A. Brunelle, M. G. Blain, E. A. Schweikert, G. Ben Assayag, P. Sudraud, *Nucl. Instr. Meth.* **1991**, *B62*, 8.
[22] K. Baudin, A. Brunelle, S. Della-Negra, D. Jacquet, P. Hakansson, Y. Le Beyec, M. Pautrat, R. R. Pinho, Ch. Schoppmann, *Nucl. Instr. Meth.* **1996**, *B112*, 59.
[23] S. F. Belykh, I. S. Bitensky, D. Mullajanov, U. Kh. Rasulev, *Nucl. Instr. Meth.* **1997**, *B129*, 451.
[24] S. F. Belykh, R. N. Evtukhov, J. N. Lysenko, U. Kh. Rasulev, *Rev. Sci. Instr.* **1992**, *63*, 2458.
[25] N. Kh. Dzhemilev, A. M. Goldenberg, I. V. Veriovkin, S. V. Verkhoturov, *Nucl. Instr. Meth.* **1996**, *B114*, 245.
[26] G. P. Können, A. Tip, A. E. de Vries, *Rad. Eff.* **1974**, *21*, 269 and **1975**, *26*, 23.

Mikrochim. Acta [Suppl.] 15, 387–392 (1998)

The Standards, Measurements and Testing Programme (SMT), the European Support to Standardisation, Measurements and Testing Projects

Achim Boenke

European Commission, Standards, Measurements and Testing Programme (SMT), 200, Rue de la Loi, B-1049 Brussels, Belgium

Abstract. Chemical, biological and physical types of measurement and testing procedures need to be reliable, accurate, precise and comparable in order to allow industry and trade to function successfully. In addition, Community legislation ranging from the common agricultural policy, the Single Market up to worker protection cannot be successfully implemented without these. The EC, SMT-Programme is therefore fully in line with the principles of subsidiarity and cohesion, and supports European industrial competitiveness as well as employment.

The objectives of the EC, SMT-Programme are summarised as follows: (a) to improve the competitive position of all sectors of European industry (including in particular SMEs) and the quality of European products by including written standards for industry via RTD activities related to the development of generic measurement and testing methods, instrumentation, more efficient in-process measurements, reference materials, etc. (b) to promote research and other technical support necessary for the development and implementation of other Community policies (Single Market, environment, agriculture, food, fisheries, health, transport and protection of the Community's external frontiers). (c) to promote research in support of the activities of CEN, CENELEC, ETSI and other European bodies which seek to maintain or establish quality standards via either new and exacting written standards or codes of practice. (d) to support the further development of the European measurement infrastructure via the maintenance and development of the meterological system and to support laboratory quality assurance and accreditation. (e) to promote the dissemination and application of good measurement practice throughout Europe, particularly in the less favoured regions, for example, by the organisation of training courses and by the establishment of networks.

When attempting to meet the above objectives, the SMT-Programme will carry out its activities in close collaboration with other specific research programme of the European Commission.

Key words: research, industry, standardisation, intercomparison, certified reference materials (CRMs), measurements, testing, quality control/assurance, analytical chemistry.

The EC, SMT-Programme is carrying out its activities in close collaboration with other specific research programme of the European Commission.

The article will briefly inform about the overall and different specific objectives of the EC, SMT-Programme. More detailed and specific information about the individual THEMES, Technology Stimulation for Small and Medium Enterprises (SMEs), Thematic Networks, and Preparatory, Accompanying and Support Measures can be found in the SMT-Work Programme and the SMT-Information Package. Both are obtainable on request by sending a fax to the EC, SMT-Programme (FAX nos.:+32/2/2958072 or +32/2/2966757).

Discussion

Basic, Scientific and Technical Objectives of the EC, SMT-Programme (1994–1998)

The promotion of the development of a harmonised system of accurate, reproducible and repeatable

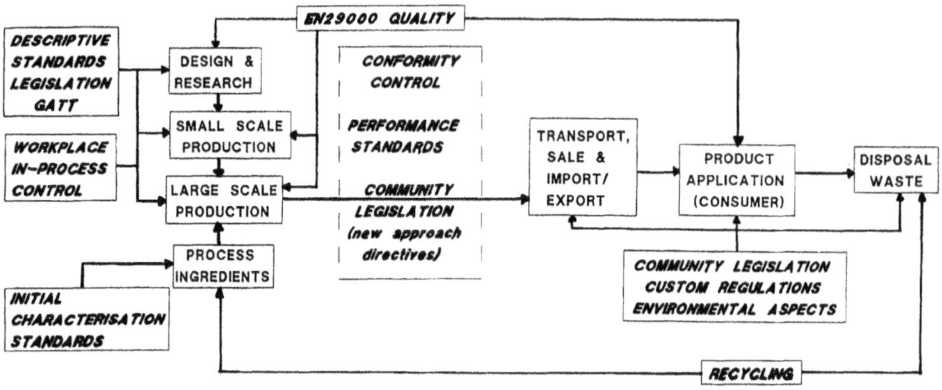

Fig. 1. Overview on the product life cycle versus quality control related to [1, 2]

measurements, chemical analyses and tests necessary to increase the competitive position of European industry and to develop and implement EU-Policies, form the basic objectives of the EC, SMT-Programme (see Fig. 1) [1].

Consequently, the scientific and technical objectives follow these two basic objectives. These are summarised in the following three THEMES of research:

THEME I : Measurements for quality European products including written standards for industry;

THEME II : Research related to written standards and technical support to trade:

THEME III : Measurements related to the needs of society.

In addition to these three THEMES, the SMT-Programme consists of another vital element which is entitled: "*Preparatory, Accompanying and Support Measures*". The basic aims of these measures are to increase the accessibility of the SMT-Programme to potential participants and to disseminate as well as to exploit project results in other areas of industry/legislation, and to allow promotion of the application of good measurement practices throughout Europe including, in particular, its less favoured regions.

The objectives described are implemented by means of shared-cost, concerted actions, and accompanying measures. The research, technological, and development (RTD) activities are carried out by third parties such as industry, universities, and research institutes/organisations/laboratories or by the EC, Joint Research Centres (JRCs) in association with at least two third parties. The Programme could be addressed via the conventional time limited call for proposals, while important objectives in support of Community Policy and European Standardisation could be addressed via dedicated calls for proposals. These dedicated calls were separately identified and included more detailed specified topics. Each of these specific topics was accompanied by a background document prepared by the body that had requested it (e.g. the different EC Directorate Generals or the European Standardisation Bodies (CEN, CENELEC, etc.)).

Measurements for Quality European Products Including Written Standards for Industry (THEME I)

This THEME of the SMT-Programme took the individual items of the intra-and inter-industrial aspects of research needs related to standardisation, measurements and testing into account (e.g. product planning, its design, its preparation via small and large scale production, purchasing of quality process ingredients, product manufacturing including its inspection, its sales as well as distribution, and quality control/assurance). Such standardisation, measurement and testing types of research were divided in the following sub-themes:

- measurements in support of the research phase;
- measurements and testing in the industrial development phase;
- measurements and testing for the control of production;
- technical support for the achievement of total quality in measurements related to industrial products.

The SMT-Programme includes also specific technology stimulation for Small and Medium Enterprises

(SMEs) related to THEME I. These activities consist of the following two types of measures which are envisaged for encouraging and facilitating participation of SMEs in RTD activities:

Exploratory awards: to enable SMEs to carry out the exploratory phase of an RTD activity (i.e. an industrial research project or a co-operative research project), including the search for partners during a period of 12 months;

Co-operative research projects: whereby SMEs having similar technical problems in the field of measurements and testing but not having adequate research facilities of their own, engage other legal entities to carry out RTD on their behalf.

This specific technology stimulation is *open for all SMEs*, not only for High-Tech-SMEs, and is accessible via a continuously open call until the 8 April 1998.

Research Related to Written Standards and Technical Support to Trade (THEME II)

This THEME took research needs related to standardisation, measurements and testing problems into account which arose during the implementation of EU-regulations and European written standards. It was also directed towards the improvement of comparability of measurements to allow European legislation to be applied uniformly and to facilitate trade via mutual recognition of measurement and testing results. It consisted of the following sub-themes:

- research in support to European trade;
- European measurement and testing infrastructure including the maintenance and development of the meterological system and the support for laboratory quality assurance as well as accreditation;
- support for the needs of custom laboratories.

The last call for proposals on THEME II/3/ covered only the following sub-themes:

- European measurement and testing infrastructure;
- support for the needs of custom laboratories.

Measurements Related to the Needs of Society (THEME III)

As measurements and tests form a vital part in guaranteeing the quality of life, this THEME is related to the following sub-themes:

- health and safety, including an increase in measurement quality for public and animal health, consumer protection and workplace monitoring;
- methods and reference materials (RMs) for environmental monitoring which is related to the implementation of the European Environmental Policy (5th Environment Programme);
- protection of the cultural heritage, including measurement and testing research needs involving the adequate protection, proper conservation, and restoration of this heritage;
- justice system, which includes measurement and testing needs to allow a free circulation of persons, goods, and capital due to the need for co-operation in the area of actions against crime such as fraud, drug traffic and terrorism.

The last call for proposals on THEME III/3/ covered the following sub-themes:

- health and safety, which is restricted to novel and innovative calibration tools (including certified reference materials) and the development of novel measurement methods;
- methods and reference materials (RMs) for environmental monitoring, which is restricted to novel and innovative calibration tools (including certified reference materials) and the development of novel measurement methods;
- protection of the cultural heritage;
- justice system.

Thermatic Networks

The specific aim of Thematic Networks was to bring together research carried out in industry, universities and research centres/organisations/laboratories in a specific area of measurement, analysis and testing in order to facilitate the transfer of technology and/or knowledge, training, and the mobility of researchers.

This part of the SMT-Programme, like the specific technology stimulation for SMEs, was accessible via a continuously open call until the end of 1997 in the following two step approach:

Step 1: It included the proposal of the concept and justification – to be carried out by at least two non-affiliated organisations. This step facilitated the participation of organisations concerned and the definition of activities to be performed.

Step 2: If the Step 1 proposal was accepted by the Commission, the proposers were invited to develop a full proposal for its implementation.

Related to the critical examination of the state of the art and to different end-user's research needs in the area of standardisation, measurements and testing, such a Thematic Network could help to carry out complementary research supported by the SMT-Programme.

Preparatory, Accompanying and Support Measures

Such measures are either an essential tool operated by the Commission which only offer opportunities for specific contracts or financial support, or they are closely co-ordinated with on-going SMT-project activities. These different type of measures are summarised below:

- support to conferences, which includes support for exchanges of information, conferences, seminars, workshops, or several different kinds of meeting;
- specific training, which can be divided into: (a) training through participation in research projects; (b) training for exploitation and technology transfer;
- study contracts in support of this programme and in preparation for future activities;
- programme evaluation according to Article 4 of the Council decision [1];
- "*Targeted Socio-economic Research*" analysis of the SMT-Programme;
- activities for the dissemination, promotion and exploitation of results;
- SME-networks to increase the awareness and provide decentralised assistance to SMEs, in co-ordination with the complementary actions carried out by VALUE Relay Centre-activities, the Euro-management auditing activity RTD, etc.;
- dissemination and exploitation of project results by means of: (a) evaluation of completed SMT-Projects; (b) financial support to market research studies (licences, patents, etc); (c) demonstration awards after the evaluation of the results following the termination of the SMT-Projects.

Financial Provisions and Calls for Proposals of the EC, SMT-Programme (1994–1988)

The following Table 1 will provide an overview on the financial provisions, the timing and duration of the different calls for proposals in the SMT-Programme.

Conclusions

The possibilities for different end-users to collaborate in research related to the development, improvement and standardisation of analytical methods as well as in the preparation and certification of Reference Materials (RMs) was laid down in the Council Decision of 1994 in which the EC, Standards, Measurements and Testing Programme (SMT) is described [1]. This research programme offers the possibility, in particular, to different industries, to SMEs, but also to public and private research institutes/laboratories, official control laboratories and universities of submitting any research proposal under the areas of standards, measurements and testing described above in response to the published call for proposals.

In particular, the participation of third parties in the call for proposals for THEMES I, II and III, technology stimulation for SMEs, Thematic Networks, the dedicated calls for important needs of different EC-General-Directorates as well as European Standardisation Bodies (e.g. CEN, CENELEC, etc.), and for accompanying measures in the areas explained above is possible. It is therefore stressed that the currently running call for proposals for technology stimulation with respect to SMEs (deadline 8th April 1998, 12:00 hrs local time), and accompanying measures (deadline 30th July 1998, 12:00 hrs local time) are continuously open. The last dedicated call also opened on the 17th June with its deadline also on the 27th November 1997. The call for proposals for THEMES II & III had been limited as follows to avoid over subscription of proposals:

(a) THEME II: European measurement and testing infrastructure; Support for the needs of customs laboratories.

(b) THEME III: Health and safety, restricted to novel and innovative calibration tools (including Certified Reference Materials (CRMs)) and the development of novel measurement methods; Methods and Reference Materials (RMs) for environmental monitoring, also restricted to the same areas as explained in the 1st indent under (b); Protection of cultural heritage; Justice system.

Table 1. Financial provisions and calls for proposals of the EC, SMT-programme [1]

	Indicative contracts budget (mecu)				Call opens	Deadline	Proposal review and Selection	Likely start of contracts
	Th. I	Th. II	Th. III	TOTAL				
Shared cost actions Open Call in Support of Community Policy	1	19	14	34	December 1994	Continuously open until July 1998	Continuous	March 1995
Theme I: Measurements for Quality European Products & Prenormative Research for Industry – first call	22			22	December 1994	April 1995	May/June 1995	November 1995
Theme II and III: European Measurement Infrastructure & Measurements Related to the Needs of Society – first call		15	9	24	June 1995	November 1995	January/ February 1996	August 1996
Theme I: Measurements for Quality European Products &Written Standards for Industry – second call	20			20	June 1996	November 1996	January/ February 1997	August 1997
Themes II and III: European Measurement Infrastructure & Measurements Related to the Needs of Society – second call		11	10	21	June 1997	November 1997	January/ February 1998	August 1998
Technology stimulation: Exploratory awards and Cooperative research	15			15	December 1994	Continuously open until April 1998	From April 1995, 3 times per year	Cooperative research – from October 1995 Exploratory Awards – from July 1995
Coordinated activities Concerted actions Networks	 1 	 2 3	 2 2	 4 6	December 1994	Continuously open until the end of 1997	From April 1995, once/ twice per year	From September 1995
Accompanying measures				8.5	June 1995	Continuously open until the 30 July 1998 (12:00 hrs)	In combination with others of the above	In combination with others of the above
Total mecu	59	50	37	154.5				
Programme administration				18.5				

The open call for proposals on accompanying measures, in particular, still welcomes proposals on the such topics as:

- support for exchange of information, conferences, seminars, workshops or other scientific or technical meetings;
- scientific publications and activities for the dissemination, promotion and exploitation of SMT-project results,

The submission and participation in possible SMT-research projects within the area of standardisation, measurements and testing therefore contributed to:

- an increase in the competitive position of European Industries;
- pre-normative research;
- a better implementation of EU-Policies and European written standards;
- an improved facilitation of trade within the European Single Market in particular;
- a better quality of life in the EU;
- a high quality of analytical results with respect to accuracy, precision, comparability, and traceability;
- a high quality of trained and motivated personal;
- a high quality of European products and increased satisfaction of end-users and consumers;
- an improvement of the productivity in European Industry.

References

[1] Council Decision 94/803/EC 1994. Adopting a Specific Programme for Research and Technological Development, Including Demonstration, in the Field of Standards, Measurements and Testing (1994 to 1998). Official Journal of the European Communities, No.L334/47:47.

[2] O'Conner R. M. 1992. Monitor-Sast Activity – Standards, Technical Regulations and Quality Assurance – What will change ? What Implications for Community S&T Policy? Overall Review. European Commission, Directorate General for Information Technologies and Industries, and Telecommunications. Luxembourg. EUR-14703-EN. Catalogue Number: CD-NA-14703-EN-C.

[3] European Commission 1997. Call for proposals for RTD actions for the specific programme for research and technological development, including demonstration, in the field of standards, measurements and testing (1994 to 1998). Official Journal of European Communities No. C183/16, 17/06/1997.

GPSR Compliance

The European Union's (EU) General Product Safety Regulation (GPSR) is a set of rules that requires consumer products to be safe and our obligations to ensure this.

If you have any concerns about our products, you can contact us on ProductSafety@springernature.com

In case Publisher is established outside the EU, the EU authorized representative is:

Springer Nature Customer Service Center GmbH
Europaplatz 3
69115 Heidelberg, Germany

Batch number: 10732714

Printed by Printforce, the Netherlands